Accession no.
36138908

WITHDRAWN

KU-282-060

80.

The Analysis
of Biological Data

The Analysis of Biological Data

Second Edition

LIS - LIBRARY

Date	Fund
19.9.14	b-che

Order No.

2346693

University of Chester

Michael C. Whitlock and Dolph Schluter

ROBERTS AND COMPANY PUBLISHERS
Greenwood Village, Colorado

The Analysis of Biological Data, **Second Edition**

Roberts and Company Publishers, Inc.
4950 South Yosemite Street, F2 #197
Greenwood Village, CO 80111 USA
Tel: (303) 221-3325
Fax: (303) 221-3326
Web: www.roberts-publishers.com
Email: info@roberts-publishers.com

Publisher: Ben Roberts
Proofreader: Kathi Townes
Art Studio: Lineworks, Inc.; Lori Heckelman
Cover Photographer: Christopher Marley/Form and Pheromone
Cover Designer: Emiko Paul
Permissions Coordinators: Terri Wright and Austin MacRae, www.terriwright.com
Compositor: Kristina Elliott at TECHarts

©2015 by Roberts and Company Publishers

Reproduction or translation of any part of this work beyond that permitted by Section 107 or 108 of the 1976 United States Copyright Act without permission of the copyright owner is unlawful. Requests for permission or further information should be addressed to the Permissions Department at Roberts and Company Publishers.

ISBN: 978-1-936221-48-6

Library of Congress Cataloging-in-Publication Data
Whitlock, Michael, author.
 The analysis of biological data / Michael C. Whitlock and Dolph Schluter. -- Second edition.
   pages cm
 Includes bibliographical references and index.
 ISBN 978-1-936221-48-6
 1. Biometry--Textbooks.   I. Schluter, Dolph, author.   II. Title.
 QH323.5.W48 2015
 570.1'5195--dc23
                                       2014010300

10 9 8 7 6 5 4 3 2 1

To Sally and Wilson, Andrea and Maggie

Contents in brief

Contents

PART 3:   COMPARING NUMERICAL VALUES

PART **5**   MODERN STATISTICAL METHODS

Preface

Modern biologists need the powerful tools of data analysis. As a result, an increasing number of universities offer, or even require, a basic data analysis course for all their biology and premedical students. We have been teaching such a course at the University of British Columbia for the last two decades. Over this period, we have sought a textbook that covered the material we needed in an introductory course at just the right level. We found that most texts were too technical and encyclopedic, or else they didn't go far enough, missing methods that were crucial to the practice of modern biology and medicine. We wanted a book that had a strong emphasis on intuitive understanding to convey meaning, rather than an overreliance on formulas. We wanted to teach by example, and the examples needed to be interesting. Most importantly, we needed a biology book, addressing topics important to biologists and health care providers handling real data.

We couldn't find the book that we needed, so we decided to write this one to fill the gap. We include several unusual features that we have discovered to be helpful for effectively reaching our audience:

Interesting biology examples. Our teaching has shown us that biology students learn data analysis best in the context of interesting examples drawn from the medical and biological literature. Statistics is a means to an end, a tool to learn about the human and natural world. By emphasizing what we can learn about the science, the power and value of statistics becomes plain. Plus, it's just more fun for everyone concerned.

Every chapter has several biological or medical examples of key concepts, and each example is prefaced by a substantial description of the biological setting. The examples are illustrated with photos of the real organisms, so that students can look at what they're learning about. The emphasis on real and interesting examples carries into the problem sets; for each chapter, there are dozens of questions based on real data about biological and medical issues. In this Second Edition, we have added approximately 200 new problems. These include a new type of problem, called Calculation Practice, that takes the student step-by-step through the important procedures described in the chapter. The corresponding answers are provided at the back of the book so that the students can check their success at each step. Two or three such problems are at the start of most of the chapter problem sets. We've also added three new sections of review problems (after Chapters 9, 13, and 17) to allow cumulative review of the important concepts up to that point in the book.

Intuitive explanations of key concepts. Statistical reasoning requires a lot of new ways of thinking. Students can get lost in the barrage of new jargon and multitudinous tests. We have found that starting from an intuitive foundation, away from all the details, is extremely valuable. We take an intuitive approach to basic questions: What's a good sample? What's a confidence interval? Why do an experiment? The first several chapters establish this basic knowledge, and the rest of the book builds on it.

Practical data analysis. As its title suggests, this book focuses on data rather than the mathematical foundations of statistics. We teach how to make good graphical displays, and we emphasize that a good graph is the beginning point of any good data analysis. We give equal time to estimation and hypothesis testing, and we avoid treating the *P*-value as an end in itself. The book does not demand a knowledge of mathematics beyond simple algebra. We focus on practicality over nuance, on biological usefulness over theoretical hand wringing. We not only teach the "right" way of doing something but also highlight some of the pitfalls that might be encountered.

We demonstrate how to carry out the calculations for most methods so that the steps are not mysterious to students. At the same time, we know that a computer will be available for most calculations. Hence, we focus on the concepts of biological data analysis and how statistics can help extract scientific insight from data. With the power of modern computers at hand, the challenge in analyzing data becomes knowing what method to use and why.[1] We imagine and hope that every course using this book will have a component encouraging students to use computer statistical packages. We are also aware that the diversity of such packages is immense, and so we have not tied the book to any particular program. We have heard most often from instructors who use the R package with this book, and we now provide R scripts for all the examples in this book at the book's web site (http://whitlockschluter.zoology.ubc.ca).

Practical experimental design. A biologist cannot do good statistics—or good science—without a practical understanding of experimental design. Unlike most books, our book covers basic topics in experimental design, such as controls, randomization, pseudoreplication, and blocking, and we do it in a practical, intuitive way.

Up to date on the basics. Believe it or not, the best confidence interval for the proportion is not the one you probably learned as an undergraduate. Nonparametric statistics do not effectively test for differences in means (or medians, for that matter) without some fairly strong assumptions that we normally hear little about. For these and many other topics, we have updated the coverage of basic, everyday topics in statistics.

Coverage of modern topics. Modern biology uses a larger toolkit than the one available a generation ago. In this book, we go beyond most introductory books by

---

1. "A computer lets you make more mistakes faster than any invention in human history—with the possible exceptions of handguns and tequila." (Mitch Ratcliffe, in *Technology Review,* 1992).

establishing the conceptual principles of important topics, such as likelihood, non-linear regression, permutation tests, meta-analysis, and the bootstrap.

Useful summaries. Near the end of each chapter is a short, clear summary of the key concepts, and most chapters end with Quick Formula Summaries that put most equations in one easy-to-find place.

Interleaves. Between chapters are short essays that we call interleaves. These interleaves cover a variety of conceptual topics in a nontechnical way—ideas that are nevertheless crucial for the interpretation of statistical results in scientific research. Several of them focus on common mistakes in the analysis and interpretation of biological data and how to avoid them. Although the interleaves are pressed between the chapters, they complement the material in the core chapters in important ways. We believe that the interleaves discuss some of the most important topics of the book, such as the meaning of statistical significance (Interleaf 3), the difference between correlation and causation (Interleaf 4), why control treatments are necessary (Interleaf 6), and the distortions caused by publication bias (Interleaf 10). Interleaf 7 summarizes what statistical test should be used and when, and it is a good place to start when reviewing for exams.

After five years of writing—and a couple more years of work on this new edition—the result is now in your hands. We think *The Analysis of Biological Data* provides a good background in data analysis for biologists and medical practitioners, covering a broad range of topics in a practical and intuitive way. It works for our classes; we hope that it works for yours, too.

Organization of the book

The Analysis of Biological Data is divided into five parts, each with a handful of chapters as indicated in the table of contents. We recommend starting with the first part, because it introduces many basic concepts that are used throughout the book, such as sampling, drawing a good graph, describing data, estimating, hypothesis testing, and concepts in probability. These early chapters are meant to be read in their entirety.

After the first block, most chapters progress from the most general topics at the start to more specialized topics by the end. Each chapter is structured so that a basic understanding of the topic may be obtained from the earliest sections. For example, in the chapter on analysis of variance (Chapter 15), the basics are taught in the first two sections; reading Sections 15.1 and 15.2 gives roughly the same material that most introductory statistics texts provide about this method. Sections 15.3–15.6 explain additional twists and other interesting applications.

The last block of chapters (Chapters 18–21) is mainly for the adventurous and the curious. These chapters introduce topics, such as likelihood, bootstrapping, and meta-analysis, that are commonly encountered in the biological and medical literature but not often mentioned in an introductory course. These chapters introduce the basic principles of each topic, discuss how the methods work, and point to where you might look to find out more.

A basic course could be taught by using only Chapters 1–17 and, within this subset of chapters, by stopping after Sections 5.8, 7.3, 8.4, 9.4, 12.6, 13.6, 15.2, 16.4, and 17.6 in their respective chapters. We suggest that all courses highlight specific topics covered only in the interleaves.

Each chapter ends with a series of problems that are designed to test students' understanding of the concepts and the practical application of statistics. The problems are divided into Practice Problems and Assignment Problems. Three cumulative review problem sets have been added to the Second Edition, placed approximately at locations convenient for midterm and final exam review. Short answers to all Practice Problems and Review Problems are provided in the back of the book; answers to the Assignment Problems are available to instructors only from the publisher. For a copy, contact Ben Roberts at info@roberts-publishers.com or (303) 594-2221. Other teaching resources for the book are available online at http://whitlockschluter.zoology.ubc.ca.

A word about the data

The data used in this book are real, with a few well-marked exceptions. For the most part, these data were obtained directly from published papers. In some cases, we contacted the authors of articles who generously provided the raw data for our use. Often, when raw data were not provided in the original paper, we resorted to obtaining data points by digitizing graphical depictions, such as scatter plots and histograms. Inevitably, the numbers we extracted differ slightly from the original numbers because of measurement error. In rare cases, we generated data by computer that matched the statistical summaries in the paper. In all cases, the results we present are consistent with the conclusions of the original papers. Most of the data sets are available online at http://whitlockschluter.zoology.ubc.ca.

Acknowledgments

This book would not have happened had the two of us had been left to do it by ourselves. Many people contributed in substantial ways, and we are forever grateful. The clarity and accuracy of its contents were improved by the careful attention of a lot of generous readers, including Arianne Albert, Brad Anholt, Cécile Ané, Eric Baack, Arthur Berg, Chad Brassil, James Bryant, Martin Buntinas, Mark Butler, Carrie Case, C. Ray Chandler, Mark Clemens, Bradley Cosentino, Perry de Valpine, Flo Débarre, Christiana Drake, Jonathan Dushoff, Steven George, Aleeza Gerstein, George Gilchrist, Brett Goodwin, Steven Green, Tim Hanson, Mike Hickerson, Lisa Hines, Darren Irwin, Nusrat Jahan, Philip Johns, Roger Johnson, Istvan Karsai, Robert Keen, John Kelly, Rex Kenner, Ben Kerr, Laura Kubatko, Joseph G. Kunkel, Bret Larget, Theo Light, Todd Livdahl, Heather Masonjones, Brian C. McCarthy, Kevin Middleton, Eli Minkoff, Robert Montgomerie, Spencer Muse, Courtney Murren, Claudia Neuhauser, Liam O'Brien, Maria Orive, Patrick C. Phillips, Jay Pitocchelli, Danielle Presgraves, James Robinson, Simon Robson, Michael Rosenberg, Noah Rosenberg, Nathan Rank, Bruce Rannala, Mark Rizzardi, Colin Rundel, Michael Russell, Ronald W. Russell, Andrew Schaffner, Andrea Schluter, James Scott, Joel Shore, John Soghigian, William Thomas, Michael Travisano, Thomas Valone, Bruce Walsh, Claus Wilke, Michael Wunder, Grace A. Wyngaard, and Sam Yeaman. Many of these good people read multiple chapters, and we thank all for their invaluable aid and considerate forebearance. Sally Otto, Allan Stewart-Oaten, and Maria Orive earned our undying gratitude by reading and commenting on the entire book. Of course, any errors that remain are our own fault; we didn't always take everyone's advice, even perhaps when we should have. If we have forgotten anyone, you have our thanks even if our memories are poor.

We owe a debt to the students of BIOL 300 at the University of British Columbia, who class-tested this book over the last several years. The book also benefited by class testing at several colleges and universities in courses by Helen Alexander, Brad Anholt, Eric Baack, Carol Baskauf, Peter Dunn, Matthias Foellmer, Marie-Josée Fortin, George Gilchrist, Michael Grant, Joe Hardin, Karen Harper, Scott Harrison, Mike Hickerson , Stephen Hudman, Nusrat Jahan, Laura Kapitula, Randy Larsen, Terri Lacourse, Susan Lehman, Todd Livdahl, Kelly McCoy, Jean Richardson, Simon Robson, Chris Sanford, Tom Short, Andrew Tierman, Steve Vamosi, Liette Vasseur, Jason Wilson, and Grace A. Wyngaard. George Gilchrist and his students gave us a very detailed and extremely helpful set of comments at a crucial stage of the book. The following students from UBC and other institutions uncovered errors in earlier versions of the book: Jessica Beaubier, Chad Brassil, Edward Cheung, Lorena Cheung, Stephanie Cheung, Denise Choi, Peter Dunn, Maryam Garabedian,

Samrad Ghavimi, Inderjit Grewal, Sarah Hamanishi, Gurpreet Khaira, Jung Min Kim, Arleigh Lambert, Alexander Leung, Mira Li, Flora Liu, Dianna Louie, Johnston Mak, Giovanni Marchetti, Sarah Neumann, Jarad Niemi, Tyler Ng, Ruth Ogbamichael, Jasmine Ono, Marion Pearson, Trevor Schofield, Meredith Soon, Erin Stacey, Michelle Uzelac, John Wakeley, Hillary Ward, Chris Wong, Irene Yu, Anush Zakaryan, Paul Zhou, and Jon-Paul Zacharias. We send special thanks to Nick Cox, who graciously read a previous printing of this book with extraordinary care.

A number of researchers freely sent us their original data, including Matt Arnegard, Angela Attwood, Audrey Barker-Plotkin, Cynthia Beall, Butch Brodie, Pamela Colosimo, Kimmo Eriksson, Kevin Fowler, Suzanne Gray, Chris Harley, Luke Harmon, Andrew Hendry, Peter Keightley, Fredrik Liljeros, Jean Thierry-Mieg, Jeffrey S. Mogil, Patrik Nosil, Margarita Ramos, Rick Relyea, Locke Rowe, Natarajan Singaravelan, Jake Socha, Jan Soumanan, Brian Starzomski, Richard Svanback, David Tilman, Andrew Trites, Neils van de Ven, Christoph von Beeren, Yitong Wang, Jason Weir, Jack Werren, and Martin Wikelski.

The book has benefited enormously from the efforts of a large team of able people: copyediting by John Murdzek (1st ed.), Gunder Hefta (1st ed.), and Kathi Townes (2nd ed.); photo research by Laura Roberts (1st ed.), Terri Wright and Austin MacRae (2nd ed.); art by Tom Webster (1st ed.) and Lori Heckelman (2nd ed.); and design and composition by Mark Ong (1st ed.), Kathi Townes (2nd ed.), and Kristina Elliott (2nd ed.). Eric Baack has our special appreciation for slaving over the problem sets to create the answer keys, as does Holly Kindsvater, who carefully checked all the answer keys for the 2nd edition. Steven Green pointed out several improvements to the answer key and reviewed the answers to the review problem sets. Aleeza Gerstein corrected numerous errors with her careful proofreading. Finally, Ben Roberts deserves our greatest thanks, for all of his support and vision in making this book happen, and especially for Clause 24.

The book was started while MCW was supported by the Peter Wall Institute for Advanced Studies at UBC as a Distinguished-Scholar-in-Residence, and the majority of the final stages of the book were written while he was a Sabbatical Scholar at the National Evolutionary Synthesis Center in North Carolina (NSF #EF-0423641). DS began working on the book while a visiting professor in Developmental Biology at Stanford University. The second edition was aided by a sabbatical leave at the University of Texas (MCW) and a Canada Council Senior Killam Fellowship (DS), which included a wonderful stay at La Selva Biological Station. The scholarly support and environment provided by each of these institutions was exceptional—and greatly appreciated.

Finally, we would like to give great thanks to all of the people that have taught us the most over the years. MCW would like to thank Dave McCauley, Mike Wade, Nick Barton, Ben Pierce, Kevin Fowler, Patrick Phillips, Sally Otto, and Betty Whitlock.

About the authors

Michael Whitlock is an evolutionary biologist and population geneticist known for his work on evolution in spatially structured populations. He is a Professor of Zoology at the University of British Columbia, where he has taught statistics to biology students since 1995. He is a fellow of the American Association for the Advancement of Science and of the American Academy of Arts and Science.

Dolph Schluter is Professor and Canada Research Chair in the Zoology Department and Biodiversity Research Center at the University of British Columbia. He is known for his research on the ecology and evolution of Galápagos finches and threespine stickleback. He is a fellow of the Royal Societies of Canada and London and a foreign member of the American Academy of Arts and Sciences.

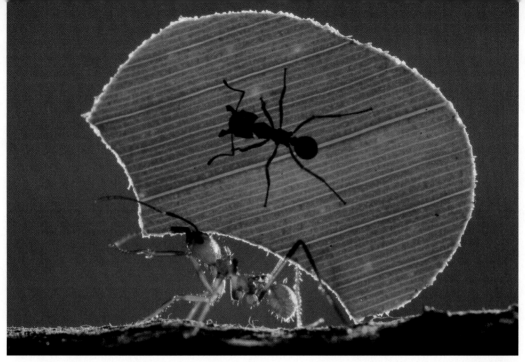

Leafcutter ant

1 Statistics and samples

1.1 What is statistics?

Biologists study the properties of living things. Measuring these properties is a challenge, though, because no two individuals from the same biological population are ever exactly alike. We can't measure everyone in the population, either, so we are constrained by time and funding to limit our measurements to a *sample* of individuals drawn from the population. Sampling brings uncertainty to the project because, by chance, properties of the sample are *not* the same as the true values in the population. Thus, measurements made from a sample are affected by who happened to get sampled and who did not.

Statistics is the study of methods to describe and measure aspects of nature from samples. Crucially, statistics gives us tools to *quantify the uncertainty* of these measures—that is, statistics makes it possible to determine the likely magnitude of their departure from the truth.

Statistics is about **estimation**, the process of inferring an unknown quantity of a target population using sample data. Properly applied, the tools for estimation allow us to approximate almost everything about populations using only samples. Examples range from the average flying speed of bumblebees, to the risks of exposure to cell phones, to the variation in beak size of finches on a remote Galápagos island. We can estimate the proportion of people with a particular disease who die per year and the fraction who recover when treated.

Most importantly, we can assess differences between groups and relationships between variables. For example, we can estimate the effects of different drugs on the possibility of recovery, we can measure the association between the lengths of horns on male antelopes and their success at attracting mates, and we can determine by how much the survival of women and children during shipwrecks differs from that of men.

> *Estimation* is the process of inferring an unknown quantity of a population using sample data.

All of these quantities describing populations—namely, averages, proportions, measures of variation, and measures of relationship—are called **parameters**. Statistical methods tell us how best to estimate these parameters using our measurements of a sample. The parameter is the truth, and the estimate (also known as the **statistic**) is an approximation of the truth, subject to error. If we were able to measure every possible member of the population, we could know the parameter without error, but this is rarely possible. Instead, we use estimates based on incomplete data to approximate this true value. With the right statistical tools, we can determine just how good our approximations are.

> A *parameter* is a quantity describing a population, whereas an *estimate* or *statistic* is a related quantity calculated from a sample.

Statistics is also about **hypothesis testing**. A **statistical hypothesis** is a specific claim regarding a population parameter. Hypothesis testing uses data to evaluate evidence for or against statistical hypotheses. Examples are "The mean effect of this new drug is not different from that of its predecessor," and "Inhibition of the *Tbx4* gene changes the rate of limb development in chick embryos." Biological data usually get more interesting and informative if they can resolve competing claims about a population parameter.

Statistical methods have become essential in almost every area of biology—as indispensable as the PCR machine, calipers, binoculars, and the microscope. This book presents the ideas and methods needed to use statistics effectively, so that we can improve our understanding of nature.

Chapter 1 begins with an overview of samples—how they should be gathered and the conclusions that can be drawn from them. We also discuss the types of variables that can be measured from samples, introducing terms that will be used throughout the book.

1.2 Sampling populations

Our ability to obtain reliable measures of population characteristics—and to assess the uncertainty of these measures—depends critically on how we sample populations. It is often at this early step in an investigation that the fate of a study is sealed, for better or worse, as Example 1.2 demonstrates.

EXAMPLE Raining cats

1.2 In an article published in the *Journal of the American Veterinary Medical Association,* Whitney and Mehlhaff (1987) presented results on the injury rates of cats that had plummeted from buildings in New York City, according to the number of floors they had fallen. Fear not: no experimental scientist tossed cats from different altitudes to obtain the data for this study. Rather, the cats had fallen (or jumped) of their own accord. The researchers were merely recording the fates of the sample of cats that ended up at a veterinary hospital for repair. The damage caused by such falls was dubbed Feline High-Rise Syndrome, or FHRS.[1]

Not surprisingly, cats that fell five floors fared worse than those dropping only two, and those falling seven or eight floors tended to suffer even more (see Figure 1.2-1). But the astonishing result was that things got better after that. On average, the number of injuries was reduced in cats that fell more than nine floors. This was true in every injury category. Their injury rates approached that of cats that had fallen only two floors! One cat fell 32 floors and walked away with only a chipped tooth.

FIGURE 1.2-1
A graph plotting the average number of injuries sustained per cat according to the number of stories fallen. Numbers in parentheses indicate number of cats. Modified from Diamond (1988).

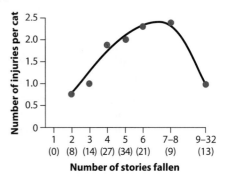

---

1. "The diagnosis of high-rise syndrome is not difficult. Typically, the cat is found outdoors, several stories below, and a nearby window or patio door is open." (Ruben 2006).

This effect cannot be attributed to the ability of cats to right themselves so as to land on their feet—a cat needs less than one story to do that. The authors of the article put forth a more surprising explanation. They proposed that after a cat attains terminal velocity, which happens after it has dropped six or seven floors, the falling cat relaxes, and this change to its muscles cushions the impact when the cat finally meets the pavement.

Remarkable as these results seem, aspects of the sampling procedure raise questions. A clue to the problem is provided by the number of cats that fell a particular number of floors, indicated along the horizontal axis of Figure 1.2-1. No cats fell just one floor, and the number of cats falling increases with each floor from the second floor to the fifth. Yet, surely, every building in New York that has a fifth floor has a fourth floor, too, with open windows no less inviting. What can explain this curious trend?

To answer this, keep in mind that the data are a *sample* of cats. The study was not carried out on the whole population of cats that fell from New York buildings. Our strong suspicion is that the sample is biased. Not all fallen cats were taken to the vet, and the chance of a cat making it to the vet might have been affected by the number of stories it had fallen. Perhaps most cats that tumble out of a first- or second-floor window suffer only indignity, which is untreatable. Any cat appearing to suffer no physical damage from a fall of even a few stories may likewise skip a trip to the vet. At the other extreme, a cat fatally plunging 20 stories might also avoid a trip to the vet, heading to the nearest pet cemetery instead.

This example illustrates the kinds of questions of interpretation that arise if samples are biased. If the sample of cats delivered to the vet clinic is, as we suspect, a distorted subset of all the cats that fell, then the measures of injury rate and injury severity will also be distorted. We cannot say whether this bias is enough to cause the surprising downturn in injuries at a high number of stories fallen. At the very least, though, we can say that, if the chances of a cat making it to the vet depends on the number of stories fallen, the relationship between injury rate and number of floors fallen will be distorted.

Good samples are a foundation of good science. In the rest of this section we give an overview of the concept of sampling, what we are trying to accomplish when we take a sample, and the inferences that are possible when researchers get it right.

Populations and samples

The first step in collecting any biological data is to decide on the target population. A **population** is the entire collection of individual units that a researcher is interested in. Ordinarily, a population is composed of a large number of individuals—so many that it is not possible to measure them all. Examples of populations include

- all cats that have fallen from buildings in New York City,
- all the genes in the human genome,
- all individuals of voting age in Australia,

- all paradise flying snakes in Borneo, and
- all children in Vancouver, Canada, suffering from asthma.

A **sample** is a much smaller set of individuals selected from the population.[2] The researcher uses this sample to draw conclusions that, hopefully, apply to the whole population. Examples include

- the fallen cats brought to one veterinary clinic in New York City,
- a selection of 20 human genes,
- all voters in an Australian pub,
- eight paradise flying snakes caught by researchers in Borneo, and
- a selection of 50 children in Vancouver, Canada, suffering from asthma.

> A *population* is all the individual units of interest, whereas a *sample* is a subset of units taken from the population.

In most of the above examples, the basic unit of sampling is literally a single individual. However, in one example, the sampling unit was a single gene. Sometimes the basic unit of sampling is a *group* of individuals, in which case a sample consists of a set of such groups. Examples of units that are groups of individuals include a single family, a colony of microbes, a plot of ground in a field, an aquarium of fish, and a cage of mice. Scientists use several terms to indicate the sampling unit, such as *unit, individual, subject,* or *replicate*.

Properties of good samples

Estimates based on samples are doomed to depart somewhat from the true population characteristics simply by chance. This chance difference from the truth is called **sampling error**. The spread of estimates resulting from sampling error indicates the **precision** of an estimate. The lower the sampling error, the higher the precision. Larger samples are less affected by chance and so, all else being equal, larger samples will have lower sampling error and higher precision than smaller samples.

> *Sampling error* is the difference between an estimate and the population parameter being estimated caused by chance.

Ideally, our estimate is **accurate** (or **unbiased**), meaning that the average of estimates that we might obtain is centered on the true population value. If a sample is not properly taken, measurements made on it might systematically underestimate (or overestimate) the population parameter. This is a second kind of error called **bias**.

---

2. In biology, a "blood sample" or a "tissue sample" might refer to a substance taken from a single individual. In statistics, we reserve the word *sample* to refer to a subset of individuals drawn from a population.

Bias is a systematic discrepancy between the estimates we would obtain, if we could sample a population again and again, and the true population characteristic.

The major goal of sampling is to minimize sampling error and bias in estimates. Figure 1.2-2 illustrates these goals by analogy with shooting at a target. Each point represents an estimate of the population bull's-eye (i.e., of the true characteristic). Multiple points represent different estimates that we might obtain if we could sample the population repeatedly. Ideally, all the estimates we might obtain are tightly grouped, indicating low sampling error, and they are centered on the bull's-eye, indicating low bias. Estimates are *precise* if the values we might obtain are tightly grouped and highly repeatable, with different samples giving similar answers. Estimates are *accurate* if they are centered on the bull's-eye. Estimates are imprecise, on the other hand, if they are spread out, and they are biased (inaccurate) if they are displaced systematically to one side of the bull's-eye. The shots (estimates) on the upper right-hand target in Figure 1.2-2 are widely spread out but centered on the bull's-eye, so we say that the estimates are accurate but imprecise. The shots on the lower left-hand target are tightly grouped but not near the bull's-eye, so we say that they are precise but inaccurate. Both precision and accuracy are important, because a lack of either means that an estimate is likely to differ greatly from the truth.

FIGURE 1.2-2
Analogy between estimation and target shooting. An accurate estimate is centered around the bull's-eye, whereas a precise estimate has low spread.

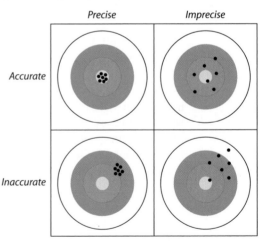

With sampling, we also want to be able to quantify the precision of an estimate. There are several quantities available to measure precision, which we discuss in Chapter 4.

The sample of cats in Example 1.2 falls short in achieving some of these goals. If uninjured and dead cats do not make it to the pet hospital, then estimates of injury rate are biased. Injury rates for cats falling only two or three floors are likely to be overestimated, whereas injury rates for cats falling many stories might be underestimated.

Random sampling

The common requirement of the methods presented in this book is that the data come from a **random sample**. A random sample is a sample from a population that fulfills two criteria.

First, every unit in the population must have an **equal chance** of being included in the sample. This is not as easy as it sounds. A botanist estimating plant growth might be more likely to find the taller individual plants or to collect those closer to the road. Some members of animal or human populations may be difficult to collect because they are shy of traps, never answer the phone, ignore questionnaires, or live at greater depths or distances than other members. These hard-to-sample individuals might differ in their characteristics from those of the rest of the population, so underrepresenting them in samples would lead to bias.

Second, the selection of units must be **independent**. In other words, the selection of any one member of the population must in no way influence the selection of any other member.[3] This, too, is not easy to ensure. Imagine, for example, that a sample of adults is chosen for a survey of consumer preferences. Because of the effort required to contact and visit each household to conduct an interview, the lazy researcher is tempted to record the preferences of multiple adults in each household and add their responses to those of other adults in the sample. This approach violates the criterion of independence, because the selection of one individual has increased the probability that another individual from the same household will also be selected. This will distort the sampling error in the data if individuals from the same household have preferences more similar to one another than would individuals randomly chosen from the population at large. With non-independent sampling, our sample size is effectively smaller than we think. This, in turn, will cause us to miscalculate the precision of the estimates.

In a *random sample,* each member of a population has an equal and independent chance of being selected.

In general, the surest way to minimize bias and allow sampling error to be quantified is to obtain a random sample.[4]

Random sampling minimizes bias and makes it possible to measure the amount of sampling error.

---

3. Other than by the removal from the population of those individuals already selected, which prevents them from being sampled again.

4. Methods are available for more complicated sampling designs incorporating non-random sampling, but we don't discuss them in this book.

How to take a random sample

Obtaining a random sample is easy in principle but can be challenging in practice. A random sample can be obtained by using the following procedure:

1. Create a list of every unit in the population of interest, and give each unit a number between one and the total population size.
2. Decide on the number of units to be sampled (call this number n).
3. Using a random-number generator,[5] generate n random integers between one and the total number of units in the population.
4. Sample the units whose numbers match those produced by the random-number generator.

An example of this process is shown in Figure 1.2-3. In both panels of the figure, we've drawn the locations of all 5699 trees present in 2001 in a carefully mapped tract of Harvard Forest in Massachussets, USA (Barker-Plotkin et al. 2006). Every tree in this population has a unique number between 1 and 5699 to identify it. We used a computerized random-number generator to pick $n = 20$ random integers between 1 and 5699, where 20 is the desired sample size. The 20 random integers, after sorting, are as follows:

156, 167, 232, 246, 826, 1106, 1476, 1968, 2084, 2222, 2223, 2284, 2790, 2898, 3103, 3739, 4315, 4978, 5258, 5500

These 20 randomly chosen trees are identified by red dots in the left panel of Figure 1.2-3.

How realistic are the requirements of a random sample? Creating a numbered list of every individual member of a population might be feasible for patients recorded in a hospital database, for children registered in an elementary-school system, or for

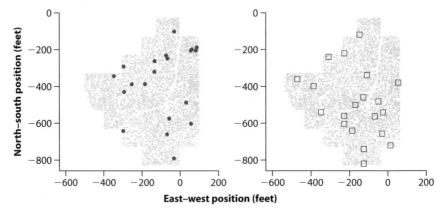

FIGURE 1.2-3 The locations of all 5699 trees present in the Prospect Hill Tract of Harvard Forest in 2001 (green circles). The red dots in the left panel are a random sample of 20 trees. The squares in the right panel are a random sample of 20 quadrats (each 20 feet on a side).

5. For example, one is available at www.random.org.

some other populations for which a registry has been built. The feat is impractical for most plant populations, however, and unimaginable for most populations of animals or microbes. What can be done in such cases?

One answer is that the basic unit of sampling doesn't have to be a single individual—it can be a group, instead. For example, it is easier to use a map to divide a forest tract into many equal-sized blocks or plots and then to create a numbered list of these plots than it is to produce a numbered list of every tree in the forest. To illustrate this second approach, we divided the Harvard Forest tract into 836 plots of 400 square feet each. With the aid of a random-number generator, we then identified a random sample of 20 plots, which are identified by the squares in the right panel of Figure 1.2-3.

The trees contained within a random sample of plots do *not* constitute a random sample of trees, for the same reason that all of the adults inhabiting a random sample of households do not constitute a random sample of adults. Trees in the same plot are not sampled independently; this can cause problems if trees growing next to one another in the same plot are more similar (or more different) in their traits than trees chosen randomly from the forest. The data in this case must be handled carefully. A simple technique is to take the average of the measurements of all of the individuals within a unit as the single independent observation for that unit.

Random numbers should always be generated with the aid of a computer. Haphazard numbers made up by the researcher are not likely to be random (see Example 19.1). Most spreadsheet programs and statistical software packages on the computer include random-number generators.

The sample of convenience

One undesirable alternative to the random sample is the **sample of convenience**, a sample based on individuals that are easily available to the researcher. The researchers must assume (i.e., dream) that a sample of convenience is unbiased and independent like a random sample, but there is no way to guarantee it.

> A *sample of convenience* is a collection of individuals that are easily available to the researcher.

The main problem with the sample of convenience is bias, as the following examples illustrate:

- If the only cats measured are those brought to a veterinary clinic, then the injury rate of cats that have fallen from high-rise buildings is likely to be underestimated. Uninjured and fatally injured cats are less likely to make it to the vet and into the sample.
- The spectacular collapse of the North Atlantic cod fishery in the last century was caused in part by overestimating cod densities in the sea, which led to excessive allowable catches by fishing boats (Walters and Maguire 1996).

Density estimates were too high because they relied heavily on the rates at which the fishing boats were observed to capture cod. However, the fishing boats tended to concentrate in the few remaining areas where cod were still numerous, and they did not randomly sample the entire fishing area (Rose and Kulka 1999).

A sample of convenience might also violate the assumption of independence if individuals in the sample are more similar to one another in their characteristics than individuals chosen randomly from the whole population. This is likely if, for example, the sample includes a disproportionate number of individuals who are friends or who are related to one another.

Volunteer bias

Human studies in particular must deal with the possibility of **volunteer bias**, which is a bias resulting from a systematic difference between the pool of volunteers (the **volunteer sample**) and the population to which they belong. The problem arises when the behavior of the subjects affects their chance of being sampled.

In a large experiment to test the benefits of a polio vaccine, for example, participating schoolchildren were randomly chosen to receive either the vaccine or a saline solution (serving as the control). The vaccine proved effective, but the rate at which children in the saline group contracted polio was found to be higher than in the general population. Perhaps parents of children who had not been exposed to polio prior to the study, and therefore had no immunity, were more likely to volunteer their children for the study than parents of kids who had been exposed (Brownlee 1955, Bland 2000).

Compared with the rest of the population, volunteers might be

- more health conscious and more proactive;
- low-income (if volunteers are paid);
- more ill, particularly if the therapy involves risk, because individuals who are dying anyway might try anything;
- more likely to have time on their hands (e.g., retirees and the unemployed are more likely to answer telephone surveys);
- more angry, because people who are upset are sometimes more likely to speak up; or
- less prudish, because people with liberal opinions about sex are more likely to speak to surveyors about sex.

Such differences can cause substantial bias in the results of studies. Bias can be minimized, however, by careful handling of the volunteer sample, but the resulting sample is still inferior to a random sample.

Data in the real world

In this book we use real data, hard-won from observational or experimental studies in the lab and field and published in the literature. Do the samples on which the studies are based conform to the ideals outlined above? Alas, the answer is often no. Random samples are desired but often not achieved by researchers working in the trenches. Sometimes, the only data available are a sample of convenience or a volunteer sample, as the falling cats in Example 1.2 demonstrate.

Scientists deal with this problem by taking every possible step to obtain random samples. If random sampling is impossible, then it is important to acknowledge that the problem exists and to point out where biases might arise in their studies.[6] Ultimately, further studies should be carried out that attempt to control for any sampling problems evident in earlier work.

1.3 Types of data and variables

With a sample in hand, we can begin to measure variables. A **variable** is any characteristic or measurement that differs from individual to individual. Examples include running speed, reproductive rate, and genotype. Estimates (e.g., average running speed of a random sample of 10 lizards) are also variables, because they differ by chance from sample to sample. **Data** are the measurements of one or more variables made on a sample of individuals.

> *Variables* are characteristics that differ among individuals.

Categorical and numerical variables

Variables can be categorical or numerical. **Categorical variables** describe membership in a category or group. They describe qualitative characteristics of individuals that do not correspond to a degree of difference on a numerical scale. Categorical variables are also called attribute or qualitative variables. Examples of categorical variables include

- survival (alive or dead),
- sex chromosome genotype (e.g., XX, XY, XO, XXY, or XYY),
- method of disease transmission (e.g., water, air, animal vector, or direct contact),
- predominant language spoken (e.g., English, Mandarin, Spanish, Indonesian, etc.),

---

6. We biologists are generally happier to find such flaws in other researchers' data than in our own.

- life stage (e.g., egg, larva, juvenile, subadult, or adult),
- snakebite severity score (e.g., minimal severity, moderate severity, or very severe), and
- size class (e.g., small, medium, or large).

A categorical variable is *nominal* if the different categories have no inherent order. Nominal means "name." Sex chromosome genotype, method of disease transmission, and predominant language spoken are nominal variables. In contrast, the values of an *ordinal* categorical variable can be ordered. Unlike numerical data, the magnitude of the difference between consecutive values is not known. Ordinal means "having an order." Life stage, snakebite severity score, and size class are ordinal categorical variables.

> *Categorical data* are qualitative characteristics of individuals that do not have magnitude on a numerical scale.

A variable is **numerical** when measurements of individuals are quantitative and have magnitude. These variables are numbers. Measurements that are counts, dimensions, angles, rates, and percentages are numerical. Examples of numerical variables include

- core body temperature (e.g., degrees Celsius, °C),
- territory size (e.g., hectares),
- cigarette consumption rate (e.g., average number per day),
- age at death (e.g., years),
- number of mates, and
- number of amino acids in a protein.

> *Numerical data* are quantitative measurements that have magnitude on a numerical scale.

Numerical data are either continuous or discrete. *Continuous* numerical data can take on any real-number value within some range. Between any two values of a continuous variable, an infinite number of other values are possible. In practice, continuous data are rounded to a predetermined number of digits, set for convenience or by the limitations of the instrument used to take the measurements. Core body temperature, territory size, and cigarette consumption rate are continuous variables.

In contrast, *discrete* numerical data come in indivisible units. Number of amino acids in a protein and numerical rating of a statistics professor in a student evaluation are discrete numerical measurements. Number of cigarettes consumed on a specific day is a discrete variable, but the rate of cigarette consumption is a continuous variable when calculated as an average number per day over a large number of days. In practice, discrete numerical variables are often analyzed as though they were continuous, if there is a large number of possible values.

Just because a variable is indexed by a number does not mean it is a numerical variable. Numbers might also be used to name categories (e.g., family 1, family 2, etc.). Numerical data can be reduced to categorical data by grouping, though the result contains less information (e.g., "above average" and "below average").

Explanatory and response variables

One major use of statistics is to relate one variable to another, by examining associations between variables and differences between groups. Measuring an association is equivalent to measuring a difference, statistically speaking. Showing that "the proportion of survivors *differs* between treatment categories" is the same as showing that the variables "survival" and "treatment" are *associated*.

Often when association between two variables is investigated, a goal is to assess how well one of the variables, deemed the **explanatory variable**, predicts or affects the other variable, called the **response variable**. When conducting an experiment, the treatment variable (the one manipulated by the researcher) is the explanatory variable, and the measured effect of the treatment is the response variable. For example, the administered dose of a toxin in a toxicology experiment would be the explanatory variable, and organism survival would be the response variable. When neither variable is manipulated by the researcher, their association might nevertheless be described by the "effect" of one of the variables (the explanatory) on the other (the response), even though the association itself is not direct evidence for causation. For example, when exploring the possibility that high blood pressure affects the risk of stroke in a sample of people, blood pressure is the explanatory variable and incidence of strokes is the response variable. When natural groups of organisms, such as populations or species, are compared in some measurement, such as body mass, the group variable (population or species) is typically the explanatory variable and the measurement is the response variable. In more complicated studies involving more than two variables, there may be more than one explanatory or response variable.

Sometimes you will hear variables referred to as "independent" and "dependent." These are the same as explanatory and response variables, respectively. Strictly speaking, if one variable depends on the other, then neither is independent, so we prefer to use *explanatory* and *response* throughout this book.

1.4 Frequency distributions and probability distributions

Different individuals in a sample will have different measurements. We can see this variability with a frequency distribution. The **frequency** of a specific measurement in a sample is the number of observations having a particular value of the measurement.[7] The **frequency distribution** shows how often each value of the variable occurs in the sample.

> The *frequency distribution* describes the number of times each value of a variable occurs in a sample.

Figure 1.4-1 shows the frequency distribution for the measured beak depths of a sample of 100 finches from a Galápagos island population.[8]

The large-beaked ground finch on the Galápagos Islands.

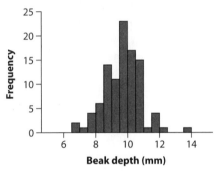

FIGURE 1.4-1   The frequency distribution of beak depths in a sample of 100 finches from a Galápagos island population (Boag and Grant 1984). The vertical axis indicates the frequency, the number of observations in each 0.5-mm interval of beak depth.

We use the frequency distribution of a sample to inform us about the distribution of the variable (beak depth) in the population from which it came. Looking at a frequency distribution gives us an intuitive understanding of the variable. For example, we can see which values of beak depth are common and which are rare, we can get an idea of the average beak depth, and we can start to understand how variable beak depth is among the finches living on the island.

The distribution of a variable in the whole population is called its **probability distribution**. The real probability distribution of a population in nature is almost never known. Researchers typically use theoretical probability distributions to approximate the real probability distribution. For a continuous variable like beak depth, the distribution in the population is often approximated by a theoretical probability distribution known as the **normal distribution**. The normal distribution drawn

---

7. The **absolute frequency** is the number of times that a value is observed. The **relative frequency** is the proportion of individuals which have that value.

8. Beak depth is the height of the beak near its base.

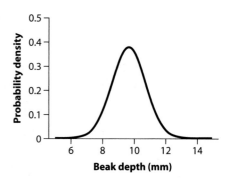

FIGURE 1.4-2
A normal distribution. This probability distribution is often used to approximate the distribution of a variable in the population from which a sample has been drawn.

in Figure 1.4-2, for example, approximates the probability distribution of beak depths in the finch population from which the sample of 100 birds was drawn.

The normal distribution is the familiar "bell curve." It is the most important probability distribution in all of statistics. You'll learn a lot more about it in Chapter 10. Most of the methods presented in this book for analyzing data depend on the normal distribution in some way.

1.5   Types of studies

Data in biology are obtained from either an experimental study or an observational study. In an **experimental study**, the researcher assigns different treatment groups or values of an explanatory variable randomly to the individual units of study. A classic example is the clinical trial, where different treatments are assigned randomly to patients in order to compare responses. In an **observational study**, on the other hand, the researcher has no control over which units fall into which groups.

Studies of the health consequences of cigarette smoking in humans are all observational studies, because it is ethically impossible to assign smoking and nonsmoking treatments to human beings to assess the effects of smoking. The individuals in the sample have made the decision themselves about whether to smoke. The only experimental studies of the health consequences of smoking have been carried out on nonhuman subjects, such as mice, where researchers can assign smoking and nonsmoking treatments randomly to individuals.

> A study is *experimental* if the researcher assigns treatments randomly to individuals, whereas a study is *observational* if the assignment of treatments is *not* made by the researcher.

The distinction between experimental studies and observational studies is that experimental studies can determine cause-and-effect relationships between variables, whereas observational studies can only point to associations. An association between smoking and lung cancer might be due to the effects of smoking per se, or perhaps to an underlying predisposition to lung cancer in those individuals prone to smoking. It is difficult to distinguish these alternatives with observational studies alone. For this reason, experimental studies of the health hazards of smoking in nonhuman animals have helped make the case that cigarette smoking is dangerous to human health. But experimental studies are not always possible, even on animals. Smoking in humans, for example, is also associated with a higher suicide rate (Hemmingsson and Kriebel 2003). Is this association caused by the effects of smoking, or is it caused by the effects of some other variable?

Just because a study was carried out in the laboratory does not mean that the study is an experimental study in the sense described here. A complex laboratory apparatus and careful conditions may be necessary to obtain measurements of interest, but such a study is still observational if the assignment of treatments is out of the control of the researcher.

1.6 Summary

- Statistics is the study of methods for measuring aspects of populations from samples and for quantifying the uncertainty of the measurements.
- Much of statistics is about estimation, which infers an unknown quantity of a population using sample data.
- Statistics also allows hypothesis testing, a method to determine how well hypotheses about a population parameter fit the sample data.
- Sampling error is the chance difference between an estimate describing a sample and the corresponding parameter of the whole population. Bias is a systematic discrepancy between an estimate and the population quantity.
- The goals of sampling are to increase the accuracy and precision of estimates and to ensure that it is possible to quantify precision.
- In a random sample, every individual in a population has the same chance of being selected, and the selection of individuals is independent.
- A sample of convenience is a collection of individuals easily available to a researcher, but it is not usually a random sample.
- Volunteer bias is a systematic discrepancy in a quantity between the pool of volunteers and the population.
- Variables are measurements that differ among individuals.
- Variables are either categorical or numerical. A categorical variable describes which category an individual belongs to, whereas a numerical variable is expressed as a number.
- The frequency distribution describes the number of times each value of a variable occurs in a sample. A probability distribution describes the number of times each value occurs in a population. Probability distributions in populations can often be approximated by a normal distribution.

- In studies of association between two variables, one variable is typically used to predict the value of another variable and is designated as the explanatory variable. The other variable is designated as the response variable.
- In experimental studies, the researcher is able to assign subjects randomly to different treatments or groups. In observational studies, the assignment of individuals to treatments is not controlled by the researcher.

PRACTICE PROBLEMS

Answers to the practice problems are provided at the end of the book, starting on page 747.

1. Which of the following numerical variables are continuous? Which are discrete?
   a. Number of injuries sustained in a fall
   b. Fraction of birds in a large sample infected with avian flu virus
   c. Number of crimes committed by a randomly sampled individual
   d. Logarithm of body mass

2. The peppered moth (*Biston betularia*) occurs in two types: peppered (speckled black and white) and melanic (black). A researcher wished to measure the proportion of melanic individuals in the peppered moth population in England, to examine how this proportion changed from year to year in the past. To accomplish this, she photographed all the peppered moth specimens available in museums and large private collections and grouped them by the year in which they had been collected. Based on this sample, she calculated the proportion of melanic individuals in every year. The people who collected the specimens, she knew, would prefer to collect whichever type was rarest in any given year, since those would be the most valuable.
   a. Can the specimens from any given year be considered a random sample from the moth population?
   b. If not a random sample, what type of sample is it?
   c. What type of error might be introduced by the sampling method when estimating the proportion of melanic moths?

3. What feature of an estimate—precision or accuracy—is most strongly affected when individuals differing in the variable of interest do not have an equal chance of being selected?

4. In a study of stress levels in U.S. army recruits stationed in Iraq, researchers obtained a complete list of the names of recruits in Iraq at the time of the study. They listed the recruits alphabetically and then numbered them consecutively. One hundred random numbers between one and the total number of recruits were then generated using a random-number generator on a computer. The 100 recruits whose numbers corresponded to those generated by the computer were interviewed for the study.
   a. What is the population of interest in this study?
   b. The 100 recruits interviewed were randomly sampled as described. Is the sample affected by sampling error? Explain.
   c. What are the main advantages of random sampling in this example?
   d. What effect would a larger sample size have had on sampling error?

5. An important quantity in conservation biology is the number of plant and animal species inhabiting a given area. To survey the community of small mammals inhabiting Kruger National Park in South Africa, a large series of live traps were placed randomly throughout the park for one week in the main dry season of 2004. Traps were set each evening and checked the following morning. Individuals caught were identified, tagged (so that new captures could be distinguished from recaptures), and released. At the end of the survey, the total number of small mammal species in the park was estimated by the total number of species captured in the survey.
   a. What is the parameter being estimated in the survey?
   b. Is the sample of individuals captured in the traps likely to be a random sample? Why or why not? In your answer, address the two criteria that define a sample as random.
   c. Is the number of species in the sample likely to be an unbiased estimate of the total number of small mammal species in the park? Why or why not?

6. In a recent study, researchers took electrophysiological measurements from the brains of two rhesus macaques (monkeys). Forty neurons were tested in each monkey, yielding a total of 80 measurements.
   a. Do the 80 neurons constitute a random sample? Why or why not?
   b. If the 80 measurements were analyzed as though they constituted a random sample, what consequences would this have for the estimate of the measurement in the monkey population?

7. Identify which of the following variables are discrete and which are continuous:
   a. Number of warts on a toad
   b. Survival time after poisoning
   c. Temperature of porridge
   d. Number of bread crumbs in 10 meters of trail
   e. Length of wolves' canines

8. A study was carried out in women to determine whether the psychological consequences of having an abortion differ from those experienced by women who have lost their fetuses by other causes at the same stage of pregnancy.
   a. Which is the explanatory variable in this study, and which is the response variable?
   b. Was this an observational study or an experimental study? Explain.

9. For each of the following studies, say which is the explanatory variable and which is the response variable. Also, say whether the study is observational or experimental.
   a. Forestry researchers wanted to compare the growth rates of trees growing at high altitude to that of trees growing at low altitude. They measured growth rates using the space between tree rings in a set of trees harvested from a natural forest.
   b. Researchers randomly assign diabetes patients to two groups. In the first group, the patients receive a new drug tasploglutide, whereas patients in the other group receive standard treatment without the new drug. The researchers compared the rate of insulin release in the two groups.
   c. Psychologists tested whether the frequency of illegal drug use differs between people suffering from schizophrenia and those not having the disease. They measured drug use in a group of schizophrenia patients and compared it with that in a similar sized group of randomly chosen people.
   d. Spinner Hansen et al. (2008) studied a species of spider whose females often eat males that are trying to mate with them. The researchers removed a leg from each male spider in one group (to make them weaker and more vulnerable to being eaten) and left the males in another group undamaged. They studied whether survival of males in the two groups differed during courtship.
   e. Bowen et al. (2012) studied the effects of advanced communication therapy for patients whose communication skills had been affected by previous strokes. They randomly assigned two therapies to stroke patients. One group received advanced communication therapy and the other received only social visits without formal therapy. Both groups otherwise received normal,

best-practice care. After six months, the communication ability (as measured by a standardized quantitative test score) was measured on all patients.

10. Each of the examples (a–e) in problem 9 involves estimating or testing an association between two variables. For each of the examples, list the two variables and state whether each is categorical or numerical.

11. A random sample of 500 households was identified in a major North American city using the municipal voter registration list. Five hundred questionnaires went out, directed at one adult in each household, and surveyed respondents about attitudes regarding the municipal recycling program. Eighty of the 500 surveys were filled out and returned to the researchers.

   a. Can the 80 households that returned questionnaires be regarded as a random sample of households? Explain.

   b. What type of bias might affect the survey outcome?

12. State whether the following represent cases of estimation or hypothesis testing.

   a. A random sample of quadrats in Olympic National Forest is taken to determine the average density of *Ensatina* salamanders.

   b. A study is carried out to determine whether there is extrasensory perception (ESP), by counting the number of cards guessed correctly by a subject isolated from a second person who is drawing cards randomly from a normal card deck. The number of correct guesses is compared with the number we would expect by chance if there were no such thing as ESP.

   c. A trapping study measures the rate of fruit fall in forest clear-cuts.

   d. An experiment is conducted to calculate the optimal number of calories per day to feed captive sugar gliders (*Petaurus breviceps*) to maintain normal body mass and good health.

   e. A clinical trial is carried out to determine whether taking large doses of vitamin C benefits health of advanced cancer patients.

   f. An observational study is carried out to determine whether hospital emergency room admissions increase during nights with a full moon compared with other nights.

13. A researcher dissected the retinas of 20 randomly sampled fish belonging to each of two subspecies of guppy in Venezuela. Using a sophisticated laboratory apparatus, he measured the two groups of fish to find the wavelengths of visible light to which the cones of the retina were most sensitive. The goal was to explore whether fish from the two subspecies differed in the wavelength of light that they were most sensitive to.

   a. What are the two variables being associated in this study?

   b. Which is the explanatory variable and which is the response variable?

   c. Is this an experimental study or an observational study? Why?

ASSIGNMENT PROBLEMS

14. Identify whether the following variables are numerical or categorical. If numerical, state whether the variable is discrete or continuous. If categorical, state whether the categories have a natural order (ordinal) or not (nominal).

   a. Number of sexual partners in a year

   b. Petal area of rose flowers

   c. Heartbeats per minute of a Tour de France cyclist, averaged over the duration of the race

   d. Birth weight

   e. Stage of fruit ripeness (e.g., underripe, ripe, or overripe)

   f. Angle of flower orientation relative to position of the sun

   g. Tree species

   h. Year of birth

   i. Gender

15. In the vermilion flycatcher, the males are brightly colored and sing frequently and prominently. Females are more dull-colored and make less sound. In a field study of this bird, a researcher attempted to estimate the fraction of individuals of each sex in the population. She based her estimate on the number of individuals of each sex detected while walking through suitable habitat. Is her sample of birds detected likely to be a random sample? Why or why not?

16. Not all telephone polls carried out to estimate voter or consumer preferences make calls to cell phones. One reason is that in the USA, automated calls ("robocalls") to cell phones are not permitted and interviews conducted by humans are more costly.
  a. How might the strategy of leaving out cell phones affect the goal of obtaining a random sample of voters or consumers?
  b. Which criterion of random sampling is most likely to be violated by the problems you identified in part (a): equal chance of being selected, or the independence of the selection of individuals?
  c. Which attribute of estimated consumer preference is most affected by the problem you identified in (a): accuracy or precision?

17. The average age of piñon pine trees in the coast ranges of California was investigated by placing 500 ten-hectare plots randomly on a distribution map of the species in California using a computer. Researchers then found the location of each random plot in the field, and then they measured the age of every piñon pine tree within each of the 10-hectare plots. The average age within the plot was used as the unit measurement. These unit measurements were then used to estimate the average age of California piñon pines.

a. What is the population of interest in this study?
b. Why did the researchers take an average of the ages of trees within each plot as their unit measurement, rather than combine into a single sample the ages of all the trees from all the plots?

18. Refer to problem 17.
  a. Is the estimate of age based on 500 plots influenced by sampling error? Why?
  b. How would the sampling error of the estimate of mean age change if the investigators had used a sample of only 100 plots?

19. In each of the following examples, indicate which variable is the explanatory variable and which is the response variable.
  a. The anticoagulant warfarin is often used as a pesticide against house mice, *Mus musculus*. Some populations of the house mouse have acquired a mutation in the *vkorc1* gene from hybridizing with the Algerian mouse, *M. spretus* (Song et al. 2011). In the Algerian mice, this gene confers resistance to warfarin. In a hypothetical follow-up study, researchers collected a sample of house mice to determine whether presence of the *vkorc1* mutation is associated with warfarin resistance in house mice as well. They fed warfarin to all the mice in a sample and compared survival between the individuals possessing the mutation and those not possessing the mutation.
  b. Cooley et al. (2009) randomly assigned either of two treatments, naturopathic care (diet counseling, breathing techniques, vitamins, and a herbal medicine) or standardized psychotherapy (psychotherapy with breathing techniques and a placebo added), to 81 individuals having moderate to severe anxiety. Anxiety scores decreased an average of 57% in the naturopathic group and 31% in the psychotherapy group.
  c. Individuals highly sensitive to rewards tend to experience more food cravings and are more likely to be overweight or develop eating disorders than other people. Beaver et al. (2006) recruited 14 healthy volunteers

and scored their reward sensitivity using a questionnaire (they were asked to answer yes or no to questions like: "I'm always willing to try something new if I think it will be fun"). The subjects were then presented with images of appetizing foods (e.g., chocolate cake, pizza) while activity of their fronto–striatal–amygdala–midbrain was measured using functional MRI. Reward sensitivity of subjects was found to correlate with brain activity in response to the images.

d. Endostatin is a naturally occurring protein in humans and mice that inhibits the growth of blood vessels. O'Reilly et al. (1997) investigated its effects on growth of cancer tumors, whose growth and spread requires blood vessel proliferation. Mice having lung carcinoma tumors were randomly divided into groups that were treated with doses of 0, 2.5, 10, or 20 mg/kg of endostatin injected once daily. They found that higher doses of endostatin led to inhibition of tumor growth.

20. For each of the studies presented in problem 19, indicate whether the study is an experimental or observational study.

21. In a study of heart rate in ocean-diving birds, researchers harnessed 10 randomly sampled, wild-caught cormorants to a laboratory contraption that monitored vital signs. Each cormorant was subjected to six artificial "dives" over the following week (one dive per day). A dive consisted of rapidly immersing the head of the bird in water by tipping the harness. In this way, a sample of 60 measurements of heart rate in diving birds was obtained. Do these 60 measurements represent a random sample? Why or why not?

22. Researchers sent out a survey to U.S. war veterans that asked a series of questions, including whether individuals surveyed were smokers or nonsmokers (Seltzer et al. 1974). They found that nonsmokers were 27% more likely than smokers to respond to a survey within 30 days (based on the much larger number of smokers and nonsmokers who eventually responded). Hypothetically, if the study had ended after 30 days, what effect would this have on the estimate of the proportion of veterans who smoke? (Use terminology you learned in this chapter to describe the effect.)

23. During World War II, the British Royal Air Force estimated the density of bullet holes on different sections of planes returning to base from aerial sorties. Their goal was to use this information to determine which plane sections most needed additional protective shields. (It was not possible to reinforce the whole plane, because it would weigh too much.) They found that the density of holes was highest on the wings and lowest on the engines and near the cockpit, where the pilot sits (their initial conclusion, that therefore the wings should be reinforced, was later shown to be mistaken). What is the main problem with the sample: bias or large sampling error? What part of the plane should have been reinforced?

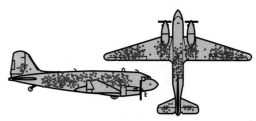

24. In a study of diet preferences in leafcutter ants, a researcher presented 20 randomly chosen ant colonies with leaves from the two most common tree species in the surrounding forest. The leaves were placed in piles of 100, one pile for each tree species, close to colony entrances. Leaves were cut so that each was small enough to be carried by a single ant. After 24 hours, the researcher returned and counted the number of leaves remaining of the original 100 of each species. Some of the results are shown in the following table.

Tree species	Number of leaves removed
Spondius mombin	1561
Sapium thelocarpum	851
Total	2412

Using these results, the researcher estimated the proportion of *Spondius* leaves taken as 0.65 and concluded that the ants have a preference for leaves of this species.

a. Identify the two variables whose association is displayed in the table. Which is the explanatory variable and which is the response variable? Are they numeric or categorical?

b. Why do the 2412 leaves used in the calculation of the proportion not represent a random sample?

c. Would treating the 2412 leaves as a random sample most likely affect the accuracy of the estimate of diet preference or the precision of the estimate?

d. If not the leaves, what units were randomly sampled in the study?

25. Garaci et al. (2012) examined a sample of people with and without multiple sclerosis (MS) to test the controversial idea that the disease is caused by blood flow restriction resulting from a vein condition known as chronic cerebrospinal venous insufficiency (CCSVI). Of 39 randomly sampled patients with MS, 25 were found to have CCSVI and 14 were not. Of 26 healthy control subjects, 14 were found to have CCSVI and 12 were not. The researchers found an association between CCSVI and MS.

a. What is the explanatory variable and what is the response variable?

b. Is this an experimental study or an observational study?

c. Where might hypothesis testing have been used in the study?

Biology and the history of statistics

The formal study of probability began in the mid-17th century when Blaise Pascal and Pierre de Fermat started to describe the mathematical rules for determining the best gambling strategies. Gambling and insurance continued to motivate the development of probability for the next couple of centuries.

The application of probability to data did not happen until much later. The importance of variation in the natural world, and by extension to samples from that world, was made obvious by Charles Darwin's theory of natural selection. Darwin highlighted the importance of variation in biological systems as the raw material of evolution. Early followers of Darwin saw the need for quantitative descriptions of variation and the need to incorporate the effects of sampling error in biology. This led to the development of modern statistics. In many ways, therefore, modern statistics was an offshoot of evolutionary biology.

One of the first pioneers in statistical data analysis was Francis Galton, who began to apply probability thinking to the study of all sorts of data. Galton was a real polymath, thinking about nearly everything and collecting and analyzing data at every chance. He said, "Whenever you can, count."[1] He invented the study of fingerprints, he tested whether prayer increased the life span of preachers compared with others of the middle class, and he quantified the heritable nature of many important traits. He once recorded his idea of the attractiveness of women seen from the window of a train headed from London to Glasgow, finding that "attractiveness" (at least according to Galton) declined as a function of distance from London. Galton is best known, though, for his twin interests in data analysis and evolution (he was the first cousin of Darwin). He invented the idea of regression, which we will learn more about in Chapter 17. Galton was also responsible for establishing a lab that brought more researchers into the study of both statistics and biology, including Karl Pearson and Ronald Fisher.

Karl Pearson, like Galton, was interested in many spheres of knowledge. Pearson was motivated by biological data—in particular, by heredity and evolution. Pearson was

Sir Ronald Fisher

1. Quoted J. R. Newman, *The World of Mathematics* (New York: Simon & Schuster, 1956).

23

responsible for our most often-used measure of the correlation between numerical variables. In fact, the correlation coefficient that we will learn about in Chapter 16 is often referred to as Pearson's correlation coefficient. He also made many contributions to the study of regression analysis and invented the χ^2 contingency test (Chapter 9). He also invented the term *standard deviation* (Chapter 3).

Last, but definitely not least, Ronald Fisher was one of the great geniuses of the 20th century. Fisher is well known in evolutionary biology as one of the three founders of the field of theoretical population genetics. Among his many contributions are the demonstration that Mendelian inheritance is compatible with the continuous variation we see in many traits, the accepted theory for why most animals conceive equal numbers of male and female offspring, and a great deal of the mathematical machinery we use to describe the process of evolution. But his contributions did not end there. He is probably also the most important figure in the history of statistics. He developed the analysis of variance (Chapter 15), likelihood (Chapter 20), the *P*-value (Chapter 6), randomized experiments (Chapter 14), multiple regression, and many other tools used in data analysis. His mathematical knowledge was made practical by a lifelong association with biologists, particularly agricultural scientists at the Rothamsted Experimental Station in England. Fisher solved problems associated with the analysis of real data as he encountered them, and he generalized them for application to many other related questions. Moreover, Fisher developed experimental designs that would give more information than might otherwise be possible.

What this short hagiography is intended to demonstrate is that the early history of statistics is tightly bound up with biology. Biological questions motivated the development of most of the basic statistical tools, and biologists helped to develop those tools. Biology and statistics naturally go hand in hand.

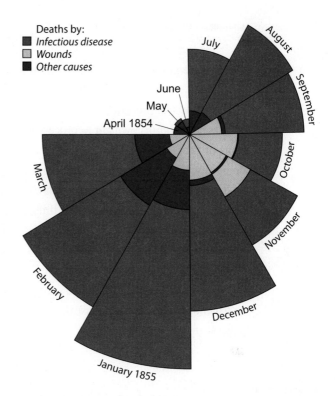

Deaths by:
■ *Infectious disease*
□ *Wounds*
■ *Other causes*

2 Displaying data

The human eye is a natural pattern detector, adept at spotting trends and exceptions in visual displays. For this reason, biologists spend hours creating and examining visual summaries of their data—graphs and, to a lesser extent, tables. Effective graphs enable visual comparisons of measurements between groups, and they expose relationships between different variables. They are also the principal means of communicating results to a wider audience.

Florence Nightingale (1858) was one of the first persons to put graphs to good use. In her famous wedge diagrams, redrawn in the figure above, she visualized the causes of death of British troops during the Crimean War. The number of cases is indicated by the area of a wedge, and the cause of death by color. The diagrams showed convincingly that disease was the main cause of soldier deaths during the wars, not wounds or other causes. With these vivid graphs, she successfully campaigned for military and public health measures that saved many lives.

Effective graphs are a prerequisite for good data analysis, revealing general patterns in the data that bare numbers cannot. Therefore, the first step in any data analysis or statistical procedure is to graph the data and look at it. Humans are a visual species, with brains evolved to process visual information. Take advantage of millions of years of evolution, and look at visual representations of your data before doing anything else. We'll follow this prescription throughout the book.

In this chapter, we explain how to produce effective graphical displays of data and how to avoid common pitfalls. We'll then review which types of graphs best show the data. The choice will depend on the type of data, numerical or categorical, and whether the goal is to show measurements of one variable or the association between two variables. There is often more than one way to show the same pattern in data, and we will compare and evaluate successful and unsuccessful approaches. We will also mention a few tips for constructing tables, which should also be laid out to show patterns in data.

2.1 Guidelines for effective graphs

Graphs are vital tools for analyzing data. They are also used to communicate patterns in data to a wider audience in the form of reports, slide shows, and web content. The two purposes, analysis and presentation, are largely coincident because the most revealing displays will be the best both for identifying patterns in the data and for communicating these patterns to others. Both purposes require displays that are clear, honest, and efficient.

To motivate principles of effective graphs, let's highlight some common ways in which researchers might get it wrong.

How to draw a bad graph

Figure 2.1-1 shows the results of an experiment in which maize plants were grown in pots under three nitrogen regimes and two soil water contents. Height of bars represents the average maize yield (dry weight per plant) at the end of the experiment under the six combinations of water and nitrogen. The data are real (Quaye et al. 2009), but we made the bad graph intentionally to highlight four common defects. Examine the graph before reading further and try to recognize some of them. Many graphics packages on the computer make it easy to produce flawed graphs like this one, which is probably why we still encounter them so often.

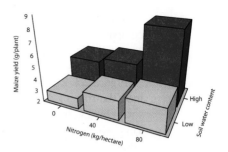

FIGURE 2.1-1
An example of a defective graph showing mean plant height of maize grown in pots under different nitrogen and water treatments.

Mistake #1: Where are the data? Each bar in Figure 2.1-1 represents average yield of four plant pots assigned to that nitrogen and water treatment. The data points—yields of all the experimental units (pots)—are nowhere to be seen. This means we are unable to see the variation in yield between pots and compare it with the magnitude of differences between treatments. It means that any unusual observations that might distort the calculation of average yield remain hidden. It would be a challenge to add the data points to this particular graph because the bars are in the way. We'll say more later about when bars are appropriate and when they are not.

Mistake #2: Patterns in the data are difficult to see. The three dimensions and angled perspective make it difficult to judge bar height by eye, which means that average plant growth is difficult to compare between treatments. In his classic book on information graphics, Tufte (1983) referred to 3-D and other visual embellishments as "chartjunk". Chartjunk adds clutter that dilutes information and interferes with the ability of the eye and brain to "see" patterns in data.

Mistake #3: Magnitudes are distorted. The vertical axis on the graph, plant yield, ranges from 2 to 9 g/plant, rather than 0 to 9, which means that bar height is out of proportion to actual magnitudes.

Mistake #4: Graphical elements are unclear. Text and other figure elements are too small to read easily.

How to draw a good graph

A few straightforward principles will help to make sure that your graphs do not end up with the kind of problems illustrated in Figure 2.1-1. We follow these four rules ourselves in the remainder of the book.

- Show the data.
- Make patterns in the data easy to see.
- Represent magnitudes honestly.
- Draw graphical elements clearly.

Show the data, first and foremost (Tufte 1983). A good graph allows you to visualize the measurements and helps the eye detect patterns in the data. Showing the data makes it possible to evaluate the shape of the distribution of data points and to compare measurements between groups. It helps you to spot potential problems, such as extreme observations, which will be useful as you decide the next step of your data analysis.

Figure 2.1-2 gives an example of what it means to show data. The study examined the role of the neurotransmitter serotonin[1] in bringing about a transition in social behavior, from solitary to gregarious, in a desert locust (Anstey et al. 2009). This behavior change is a critical point in the production of huge locust swarms that blacken skies and ravage crops in many parts of the world. Each data point is the serotonin level of one of 30 locusts experimentally caged at high density for 0, 1, or 2 hours, with 0 representing the control. The panel on the left of Figure 2.1-2 *shows* the data (this type of graph is called a strip chart or dot plot). The panel on the right of Figure 2.1-2 *hides* the data, using bars to show only treatment averages.

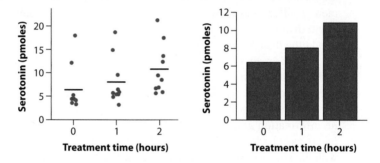

FIGURE 2.1-2 A graph that shows the data (*left*) and a graph that hides the data (*right*). Data points are serotonin levels in the central nervous system of desert locusts, *Schistocerca gregaria,* that were experimentally crowded for 0 (the control group), 1, and 2 hours. The data points in the left panel were perturbed a small amount to the left or right to minimize overlap and make each point easier to see. The horizontal bars in the left panel indicate the mean (average) serotonin level in each group. The graph on the right shows only the mean serotonin level in each treatment (indicated by bar height). Note that the vertical axis does not have the same scale in the two graphs.

In the left panel of Figure 2.1-2, we can see lots of scatter in the data in each treatment group and plenty of overlap between groups. We see that most points fall below the treatment average, and that each group has a few extreme observations. Nevertheless, we can see a clear shift in serotonin levels of locusts between treatments. All this information is missing from the right panel of Figure 2.1-2, which uses more ink yet shows only the averages of each treatment group.

Make patterns easy to see. Try displaying your data in different ways, possibly with different types of graphs, to find the best way to communicate the findings. Is the main pattern in the data recognizable right away? If not, try again with a different method. Stay away from 3-D effects and elaborate chartjunk that obscures the

---

1. Serotonin is a neurotransmitter in most animals, including humans. Some antidepressant drugs improve feelings of well-being by manipulating serotonin levels.

patterns in the data. In the rest of the chapter we'll compare alternative ways of graphing the same data sets and discuss their effectiveness.

Avoid putting too much information into one graph. Remember the purpose of a graph: to communicate essential patterns to eyes and brains. The purpose is not to cram as much data as possible into each graph. Think about getting the main point across with one or two key graphs in the main body of your presentation. Put the remainder into an appendix or online supplement if it is important to show them to a subset of your audience.

Represent magnitudes honestly. This sounds easy, but misleading graphics are common in the scientific literature. One of the most important decisions concerns the smallest value on the vertical axis of a graph (the "baseline"). A bar graph must always have a baseline at zero, because the eye instinctively reads bar height and area as proportional to magnitude. The upper bar graph in Figure 2.1-3 shows an example, depicting government spending on education each year since 1998 in British Columbia. The area of each bar is not proportional to the magnitude of the value displayed. As a result, the graph exaggerates the differences. The figure falsely suggests that spending increased twenty-fold over time, but the real increase is less than 20%. It is more honest to plot the bars with a baseline of zero, as in the lower graph in

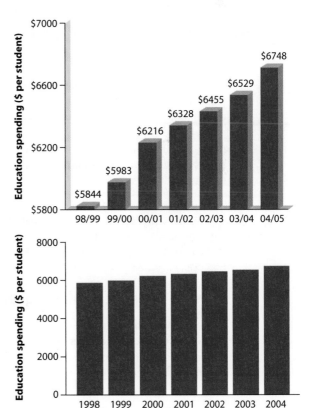

FIGURE 2.1-3
Upper graph: A bar graph, taken from a British Columbia government brochure, indicating education spending per student in different years. *Lower graph:* A revised presentation of the same data, in which the magnitude of the spending is proportional to the height and area of bars. This revision also removed the 3-D effects and the numbers above bars to make the pattern easier to see. The upper graph is modified from British Columbia Ministry of Education (2004).

Figure 2.1-3 (the revised graph also removed the 3-D effects and the numbers above bars to make the pattern easier to see).

Other types of graphs, such as strip charts, don't always need a zero baseline if the main goal is to show differences between treatments rather than proportional magnitudes.

Draw graphical elements clearly. Clearly label the axes and choose unadorned, simple typefaces and colors. Text should be legible even after the graph is shrunk to fit the final document. Always provide the units of measurement in the axis label. Use clearly distinguishable graphical symbols if you plot with more than one kind. Don't always accept the default output of statistical or spreadsheet programs.

Up to a tenth of your male audience is red-green color-blind, so choose colors that differ in intensity and apply redundant coding to distinguish groups (for example, use distinctive shapes or patterns as well as different colors).[2]

A good graph is like a good paragraph. It conveys information clearly, concisely, and without distortion. A good graph requires careful editing. Just as in writing, the first draft is rarely as good as the final product.

2.2 Showing data for one variable

To examine data for single variable, we show its **frequency distribution**. Recall from Chapter 1 that the frequency of occurrence of a specific measurement in a sample is the number of observations having that particular measurement. The frequency distribution of a variable is the number of occurrences of all values in the data.

Relative frequency is the proportion of observations having a given measurement, calculated as the frequency divided by the total number of observations. The **relative frequency distribution** is the proportion of occurrences of each value in the data set.

> The *relative frequency distribution* describes the fraction of occurrences of each value of a variable.

Showing categorical data: frequency table and bar graph

Let's start with displays for a categorical variable. A **frequency table** is a text display of the number of occurrences of each category in the data set. A **bar graph** uses the height of rectangular bars to visualize the frequency (or relative frequency) of occurrence of each category.

---

2. If you are in doubt, load your graphic file into a colorblindness simulator such as Vischeck (vischeck.com).

A *bar graph* uses the height of rectangular bars to display the frequency distribution (or relative frequency distribution) of a categorical variable.

Example 2.2A illustrates both kinds of displays.

EXAMPLE 2.2A Crouching tiger

Conflict between humans and tigers threatens tiger populations, kills people, and reduces public support for conservation. Gurung et al. (2008) investigated causes of human deaths by tigers near the protected area of Chitwan National Park, Nepal. Eighty-eight people were killed by 36 individual tigers between 1979 and 2006, mainly within 1 km of the park edge. Table 2.2-1 lists the main activities of people at the time they were killed. Such information may be helpful to identify activities that increase vulnerability to attack.

Table 2.1-1 is a frequency table showing the number of deaths associated with each activity. Here, alternative values of the variable "activity" are listed in a single column, and frequencies of occurrence are listed next to them in a second column. The categories have no intrinsic order, but comparing the frequencies of each activity is made easier by *arranging the categories in order of their importance,* from the most frequent at the top to the least frequent at the bottom.

TABLE 2.2-1 Frequency table showing the activities of 88 people at the time they were attacked and killed by tigers near Chitwan National Park, Nepal, from 1979 to 2006.

Activity	Frequency (number of people)
Collecting grass or fodder for livestock	44
Collecting non-timber forest products	11
Fishing	8
Herding livestock	7
Disturbing tiger at its kill	5
Collecting fuel wood or timber	5
Sleeping in a house	5
Walking in forest	3
Using an outside toilet	2
Total	88

The table shows that more people were killed while collecting grass and fodder for their livestock than when doing any other activity. The number of deaths under this activity was four times that of the next category of activity (collecting non-timber forest products) and is related to the amount of time people spend carrying out these activities.

The differences in frequency stand out even more vividly in the bar graph shown in Figure 2.2-1. In a bar graph, frequency is depicted by the height of rectangular bars. Unlike a frequency table, a bar graph does not usually present the actual numbers. Instead, the graph gives a clear picture of how steeply the numbers drop between categories. Some activities are much more common than others, and we don't need the actual numbers to see this.

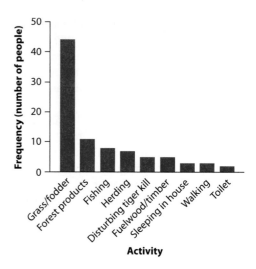

FIGURE 2.2-1
Bar graph showing the activities of people at the time they were attacked and killed by tigers near Chitwan National Park, Nepal, between 1979 and 2006. Total number of deaths: $n = 88$. The frequencies are taken from Table 2.2-1, which also gives more detailed labels of activities.

Making a good bar graph

The top edge of each bar conveys all the information about frequency, but the eye also compares the areas of the bars, which must therefore be of equal width. It is crucial that the baseline of the y-axis is at zero—otherwise, the area and height of bars are out of proportion with actual magnitudes and so are misleading.

When the categorical variable is nominal, as in Figure 2.2-1 and Table 2.2-1, the best way to arrange categories is by frequency of occurrence. The most frequent category goes first, the next most frequent category goes second, and so on. This aids in the visual presentation of the information. For an ordinal categorical variable, such as snakebite severity score, the values should be in the natural order (e.g., minimally severe, moderately severe, and very severe). Bars should stand apart, not be fused together. It is a good habit to provide the total number of observations (n) in the figure legend.

A bar graph is usually better than a pie chart

The pie chart is another type of graph often used to display frequencies of a categorical variable. This method uses colored wedges around the circumference of a circle to represent frequency or relative frequency. Figure 2.2-2 shows the tiger data again, this time in a pie chart. This graphical method is reminiscent of Florence Nightingale's wedge diagram shown at the beginning of this chapter.

FIGURE 2.2-2
Pie chart of the activities of people at the time they were attacked and killed by tigers near Chitwan National Park, Nepal. The frequencies are taken from Table 2.2-1.Total number of deaths: $n = 88$.

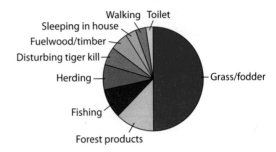

The pie chart has received a lot of criticism from experts in information graphics. One reason is that while it is straightforward to visualize the frequency of deaths in the first and most frequent category (Collecting grass/fodder), it is more difficult to compare frequencies in the remaining categories by eye. This problem worsens as the number of categories increases. Another reason is that it is very difficult to compare frequencies between two or more pie charts side by side, especially when there are many categories. To compensate, pie charts are often drawn with the frequencies added as text around the circle perimeter. The result is not better than a table. The shape of a frequency distribution is more readily perceived in a bar graph than a pie chart, and it is easier to compare frequencies between two or more bar graphs than between pie charts. Use the bar graph instead of the pie chart for showing frequencies in categorical data.

Showing numerical data: frequency table and histogram

A frequency distribution for a numerical variable can be displayed either in a frequency table or in a **histogram**. A histogram uses area of rectangular bars to display frequency. The data values are split into consecutive intervals, or "bins," usually of equal width, and the frequency of observations falling into each bin is displayed.

> A *histogram* uses the area of rectangular bars to display the frequency distribution (or relative frequency distribution) of a numerical variable.

We discuss how histograms are made in greater detail using the data in Example 2.2B.

EXAMPLE   Abundance of desert bird species

2.2B

How many species are common in nature and how many are rare? One way to address this question is to construct a frequency distribution of species abundance. The data in Table 2.2-2 are from a survey of the breeding birds of Organ Pipe Cactus National Monument in southern Arizona, USA. The measurements were extracted from the North American Breeding Bird Survey, a continent-wide data set of estimated bird numbers (Sauer et al. 2003).

TABLE 2.2-2 Data on the abundance of each species of bird encountered during four surveys in Organ Pipe Cactus National Monument.

Species	Abundance	Species	Abundance
Greater roadrunner	1	Turkey vulture	23
Black-chinned hummingbird	1	Violet-green swallow	23
Western kingbird	1	Lesser nighthawk	25
Great-tailed grackle	1	Scott's oriole	28
Bronzed cowbird	1	Purple martin	33
Great horned owl	2	Black-throated sparrow	33
Costa's hummingbird	2	Brown-headed cowbird	59
Canyon wren	2	Black vulture	64
Canyon towhee	2	Lucy's warbler	67
Harris's hawk	3	Gilded flicker	77
Loggerhead shrike	3	Brown-crested flycatcher	128
Hooded oriole	4	Mourning dove	135
Northern mockingbird	5	Gambel's quail	148
American kestrel	7	Black-tailed gnatcatcher	152
Rock dove	7	Ash-throated flycatcher	173
Bell's vireo	10	Curve-billed thrasher	173
Common raven	12	Cactus wren	230
Northern cardinal	13	Verdin	282
House sparrow	14	House finch	297
Ladder-backed woodpecker	15	Gila woodpecker	300
Red-tailed hawk	16	White-winged dove	625
Phainopepla	18		

We treated each bird species in the survey as the unit of interest and the abundance of a species in the survey as its measurement. The range of abundance values was divided into 13 intervals of equal width (0–50, 50–100, and so on), and the number of species falling into each abundance interval was counted and presented in a frequency table to help see patterns (Table 2.2-3).

TABLE 2.2-3 Frequency distribution of bird species abundance at Organ Pipe Cactus National Monument.

Abundance	Frequency (Number of species)
0–50	28
50–100	4
100–150	3
150–200	3
200–250	1
250–300	2
300–350	1
350–400	0
400–450	0
450–500	0
500–550	0
550–600	0
600–650	1
Total	43

Source: Data are from Table 2.2-2.

Although the table shows the numbers, the shape of the frequency distribution is more obvious in a histogram of these same data (Figure 2.2-3). Here, frequency (number of species) in each abundance interval is perceived as bar area.

The frequency table and histogram of the bird abundance data reveal that the majority of bird species have low abundance. Frequency falls steeply with increasing abundance.[3] The white-winged dove (pictured in Example 2.2B) is exceptionally

FIGURE 2.2-3
Histogram illustrating the frequency distribution of bird species abundance at Organ Pipe Cactus National Monument. Total number of bird species: $n = 43$.

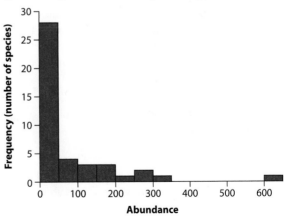

3. This pattern is a remarkably general one in nature, found in many types of organisms. Typically, only a few species are common, whereas most species are rare.

abundant at Organ Pipe Cactus National Monument, accounting for a large fraction of all individual birds encountered in the survey.

Describing the shape of a histogram

The histogram reveals the shape of a frequency distribution. Some of the most common shapes are displayed in Figure 2.2-4. Any interval of the frequency distribution that is noticeably more frequent than surrounding intervals is called a peak. The **mode** is the interval corresponding to the highest peak. For example, a bell-shaped frequency distribution has a single peak (the mode) in the center of the range of observations. A frequency distribution having two distinct peaks is said to be **bimodal**.

> The *mode* is the interval corresponding to the highest peak in the frequency distribution.

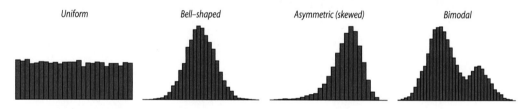

FIGURE 2.2-4 Some possible shapes of frequency distributions.

A frequency distribution is **symmetric** if the pattern of frequencies on the left half of the histogram is the mirror image of the pattern on the right half. The uniform distribution and the bell-shaped distribution in Figure 2.2-4 are symmetric. If a frequency distribution is not symmetric, we say that it is **skewed**. The distribution in Figure 2.2-4 labeled "Asymmetric" has left or negative skew: it has a long tail extending to the left. The distribution in Figure 2.2-4 labeled "Bimodal" is also asymmetric but is positively skewed: its long tail is to the right.[4] The abundance data for desert bird species also have positive skew (Figure 2.2-3), which means they have a long tail extending to the right.

> *Skew* refers to asymmetry in the shape of a frequency distribution for a numerical variable.

Extreme data points lying well away from the rest of the data are called **outliers**. The histogram of desert bird abundance (Figure 2.2-3) includes one extreme observa-

---

4. The nomenclature of skew seems backward to many people. Focus on the sharp tail of the distribution extending to the left in the third distribution in Figure 2.2-4. We say it is skewed left (or has a negative skew) because it seems to have a "skewer" sticking out to the left toward negative numbers, like the skewer through a shish kebab.

tion (the white-winged dove) that falls well outside the range of abundance of other bird species. The white-winged dove, therefore, is an outlier. Outliers are common in biological data. They can result from mistakes in recording the data or, as in the case of the white-winged dove, they may represent real features of nature. Outliers should always be investigated. They should be removed from the data only if they are found to be errors.

An *outlier* is an observation well outside the range of values of other observations in the data set.

How to draw a good histogram

When drawing a histogram, the choice of interval width must be made carefully because it can affect the conclusions. For example, Figure 2.2-5 shows three different histograms that depict the body mass of 228 female sockeye salmon (*Oncorhynchus nerka*) from Pick Creek, Alaska, in 1996 (Hendry et al. 1999). The leftmost histogram of Figure 2.2-5 was drawn using a narrow interval width. The result is a somewhat bumpy frequency distribution that suggests the existence of two or even more peaks. The rightmost histogram uses a wide interval. The result is a smoother frequency distribution that masks the second of the two dominant peaks. The middle histogram uses an intermediate interval that shows two distinct body-size groups. The fluctuations from interval to interval within size groups are less noticeable.

To choose the ideal interval width we must decide whether the two distinct body-size groups are likely to be "real," in which case the histogram should show both, or whether a bimodal shape is an artifact produced by too few observations.[5]

When you draw a histogram, each bar must rise from a baseline of zero, so that the area of each bar is proportional to frequency. Unlike bar graphs, adjacent histogram bars are contiguous, with no spaces between them. This reinforces the perception of

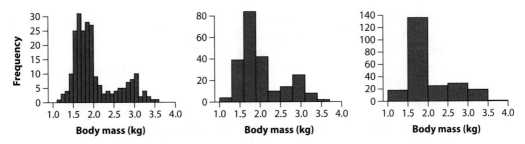

FIGURE 2.2-5 Body mass of 228 female sockeye salmon sampled from Pick Creek in Alaska (Hendry et al. 1999). The same data are shown in each case, but the interval widths are different: 0.1 kg (*left*), 0.3 kg (*middle*), and 0.5 kg (*right*).

5. The two distinct body-size classes in this salmon population correspond to two age groups.

a numerical scale, with bars grading from one into the next. In this book we follow convention by placing an observation whose value is exactly at the boundary of two successive intervals into the higher interval. For example, the Gila woodpecker, with a total of 300 individuals observed (Table 2.2-2), is recorded in the interval 300–350, not in the interval 250–300.

There are no strict rules about the number of intervals that should be used in frequency tables and histograms. Some computer programs use Sturges's rule of thumb, in which the number of intervals is $1 + \ln(n)/\ln(2)$, where n is the number of observations and ln is the natural logarithm. The resulting number is then rounded up to the higher integer (Venables and Ripley 2002). Many regard this rule as overly conservative, and in this book we tend to use a few more intervals than Sturges. The number of intervals should be chosen to best show patterns and exceptions in the data, and this requires good judgment rather than strict rules. Computers allow you to try several alternatives to help you determine the best option.

When breaking the data into intervals for the histogram, use readable numbers for breakpoints—for example, break at 0.5 rather than 0.483. Finally, it is a good idea to provide the total number of individuals in the accompanying legend.

Other graphs for numerical data

The histogram is recommended for showing the frequency distribution of a single numerical variable. The *box plot* and the *strip chart* are alternatives, but most often these are used to show differences when there are data from two or more groups. We describe these graphs in the next section. Another type of graph, the *cumulative frequency distribution,* is explained in Chapter 3.

2.3 Showing association between two variables

Here we illustrate how to show data for two variables simultaneously, rather than one at a time. The goal is to create an image that visualizes association or correlation between two variables and differences between groups. The most suitable type of graph depends on whether both variables are categorical, both are numerical, or one is of each data type.

Showing association between categorical variables

If two categorical variables are associated, the relative frequencies for one variable will differ among categories of the other variable. To reveal such association, show the frequencies using a contingency table, a mosaic plot, or a grouped bar graph. Here's an example.

EXAMPLE   **Reproductive effort and avian malaria**

2.3A   Is reproduction hazardous to health? If not, then it is difficult to
explain why adults in many organisms seem to hold back on the
number of offspring they raise. Oppliger et al. (1996) investigated the impact of reproductive effort on the susceptibility to
malaria[6] in wild great tits (*Parus major*) breeding in nest boxes.
They divided 65 nesting females into two treatment groups. In
one group of 30 females, each bird had two eggs stolen from her

nest, causing the female to lay an additional egg. The extra effort required might increase
stress on these females. The remaining 35 females were left alone, establishing the control group. A blood sample was taken from each female 14 days after her eggs hatched to
test for infection by avian malaria.

The association between experimental treatment and the incidence of malaria is
displayed in Table 2.3-1. This table is known as a **contingency table**, a frequency
table for two (or more) categorical variables. It is called a contingency table because
it shows how the frequencies of the categories in a response variable (the incidence of
malaria, in this case) are contingent upon the value of an explanatory variable (the
experimental treatment group).

TABLE 2.3-1 Contingency table showing the incidence
of malaria in female great tits in relation to experimental
treatment.

| | Experimental treatment group | | |
	Control group	Egg-removal group	Row total
Malaria	7	15	22
No malaria	28	15	43
Column total	35	30	65

Each experimental unit (bird) is counted exactly once in the four "cells" of Table
2.3-1, and so the total count (65) is the number of birds in the study. A cell is one combination of categories of the row and column variables in the table. The explanatory
variable (experimental treatment) is displayed in the columns, whereas the response
variable, the variable being predicted (incidence of malaria), is displayed in the rows.
The frequency of subjects in each treatment group is given in the column totals, and
the frequency of subjects with and without malaria is given in the row totals.

According to Table 2.3-1, malaria was detected in 15 of the 30 birds subjected to
egg removal, but in only seven of the 35 control birds. This difference between treat-

---

6. Malaria is a common cause of death in humans, but avian forms of the disease are even more prevalent
   in many bird species. For example, many native bird species in Hawaii are threatened with extinction
   after the inadvertent introduction of mosquitoes and avian malaria by humans in the 19th century.

ments suggests that the stress of egg removal, or the effort involved in producing one extra egg, increases female susceptibility to avian malaria.

> A *contingency table* gives the frequency of occurrence of all combinations of two (or more) categorical variables.

Table 2.3-1 is an example of a 2×2 ("two-by-two") contingency table, because it displays the frequency of occurrence of all combinations of two variables, each having exactly two categories. Larger contingency tables are possible if the variables have more than two categories.

Two types of graph work best for displaying the relationship between a pair of categorical variables. The **grouped bar graph** uses heights of rectangles to graph the frequency of occurrence of all combinations of two (or more) categorical variables. Figure 2.3-1 shows the grouped bar graph for the avian malaria experiments. Grouped bar graphs are like bar graphs for single variables, except that different categories of the response variable (e.g., malaria and no malaria) are indicated by different colors or shades. Bars are grouped by the categories of the explanatory variable treatment (control and egg removal), so make sure that the spaces between bars from different groups are wider than the spaces between bars separating categories of the response variable. We can see from the grouped bar graph in Figure 2.3-1 that incidence of malaria is associated with treatment, because the relative heights of the bars for malaria and no malaria differ between treatments. Most birds in the control group had no malaria (the gold bar is much taller than the red bar), whereas in the experimental group, the frequency of subjects with and without malaria was equal.

FIGURE 2.3-1
Grouped bar graph for reproductive effort and avian malaria in great tits. The data are from Table 2.3-1, where $n = 65$ birds.

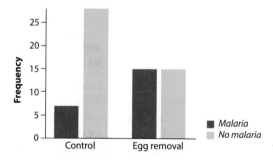

> A *grouped bar graph* uses the height of rectangular bars to display the frequency distributions (or relative frequency distributions) of two or more categorical variables.

A **mosaic plot** is similar to a grouped bar plot except that bars within treatment groups are stacked on top of one another (Figure 2.3-2). Within a stack, bar area and height indicate the relative frequencies (i.e., the proportion) of the responses. This makes it easy to see the association between treatment and response variables:

if an association is present in the data, then the vertical position at which the colors meet will differ between stacks. If no association is present, then the meeting point between the colors will be at the same vertical position between stacks. In Figure 2.3-2, for example, few individuals in the control group were infected with malaria, so the red bar (malaria) meets the gold bar (no malaria) at a higher vertical position than in the egg removal stack, where the incidence of malaria was greater.

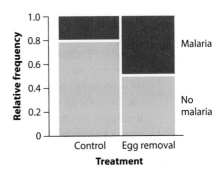

FIGURE 2.3-2
Mosaic plot for reproductive effort and avian malaria in great tits. Red indicates birds with malaria, whereas gold indicates birds free of malaria. The data are from Table 2.3-1, where $n = 65$ birds.

Another feature of the mosaic plot is that the width of each vertical stack is proportional to the number of observations in that group. In Figure 2.3-2, the wider stack for the control group reflects the greater total number of individuals in this treatment (35) compared with the number in the egg-removal treatment (30). As a result, the total area of each box is proportional to the relative frequency of that combination of variables in the whole data set.

A mosaic plot provides only relative frequencies, not the absolute frequency of occurrence in each combination of variables. This might be considered a drawback, but keep in mind that the most important goal of graphs is to depict the *pattern* in the data rather than exact figures. Here, the pattern is the association between treatment and response variables: the difference in the relative frequencies of diseased birds in the two treatments.

> The *mosaic plot* uses the area of rectangles to display the relative frequency of occurrence of all combinations of two categorical variables.

Of the three methods for presenting the same data—the contingency table, the mosaic plot, and the grouped bar graph—which is best? The answer depends on the circumstances, and it is a good idea to try all three to evaluate their effectiveness in any data set. It is usually easier to see differences in relative frequency between groups when the data are visualized in a grouped bar plot or mosaic plot than in a contingency table. On the other hand, a contingency table might work best if one of the response categories is vastly more frequent than the other, making it difficult to see the bars corresponding to rare categories in a graph, or if the explanatory and response variables have many categories, thus increasing the complexity of the graph.

We find that association, or lack of association, is easier to see in a mosaic plot than in a grouped bar graph, but this will not always be the case. Deciding which type of display is most effective for a given circumstance is best done by trying several methods and choosing among them on the basis of information, clarity, and simplicity.

Showing association between numerical variables: scatter plot

Use a scatter plot to show the association between two numerical variables. Position along the horizontal axis (the *x*-axis) indicates the measurement of the explanatory variable. The position along the vertical axis (the *y*-axis) indicates the measurement of the response variable. The pattern in the resulting cloud of points indicates whether an association between the two variables is positive (in which case the points tend to run from the lower left to the upper right of the graph), negative (the points run from the upper left to the lower right), or absent (no discernible pattern). Example 2.3B shows an example.

EXAMPLE 2.3B   **Sins of the father**

The bright colors and elaborate courtship displays of the males of many species posed a problem for Charles Darwin: how can such elaborate traits evolve? His answer was that they evolved because females are attracted to them when choosing a mate. But why would females choose those kinds of males? One possible answer: females that choose fancy males have attractive sons as well. A recent laboratory study examined how attractive traits in guppies are inherited from father to son (Brooks 2000). The attractiveness of sons (a score representing the rate of visits by females to corralled males, relative to a standard) was compared with their fathers' ornamentation (a composite index of several aspects of male color and brightness). The father's ornamentation is the explanatory variable in the resulting scatter plot of these data (Figure 2.3-3).

FIGURE 2.3-3
Scatter plot showing the relationship between the ornamentation of male guppies and the average attractiveness of their sons. Total number of families: $n = 36$.

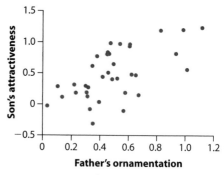

Each dot in the scatter plot is a father-son pair. The father's ornamentation is the explanatory variable and the son's attractiveness is the response variable. The plot shows a *positive* association between these variables (note how the points tend to run from the lower left to the upper right of the graph). Thus, the sexiest sons come from the most gloriously ornamented fathers, whereas unadorned fathers produce less attractive sons on average.

> A *scatter plot* is a graphical display of two numerical variables in which each observation is represented as a point on a graph with two axes.

Showing association between a numerical and a categorical variable

There are several good methods to show an association between a numerical variable and a categorical variable. Three that we recommend are the *strip chart* (which we first saw in Figure 2.1-2), the *box plot,* and the *multiple histograms* method. Here we compare these methods with an example. We recommend against the common practice of using a bar graph because the bars make it difficult to show the data (bar graphs are ideal for frequency data). Showing an association between a numerical and a categorical variable is the same as showing a difference in the numerical variable between groups.

EXAMPLE   **Blood responses to high elevation**

2.3C   The amount of oxygen obtained in each breath at high altitude can be as low as one-third that obtained at sea level. Studies have begun to examine whether indigenous people living at high elevations have physiological attributes that compensate for the reduced availability of oxygen. A reasonable expectation is that they should have more hemoglobin, the molecule that binds and transports oxygen in the blood. To test this, researchers sampled blood from males in three high-altitude human populations: the high Andes, high-elevation Ethiopia, and Tibet, along with a sea-level population from the USA (Beall et al. 2002). Results are shown in Figures 2.3-4 and 2.3-5.

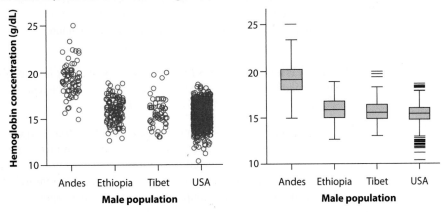

FIGURE 2.3-4 Strip chart (*left*) and box plot (*right*) showing hemoglobin concentration in males living at high altitude in three different parts of the world: the Andes (71), Ethiopia (128), and Tibet (59). A fourth population of 1704 males living at sea level (USA) is included as a control.

The left panel of Figure 2.3-4 shows the hemoglobin data with a **strip chart** (sometimes also called a dot plot). In a strip chart, each observation is represented as a dot on a graph showing its numerical measurement on one axis (here, the vertical or *y*-axis) and the category (group) to which it belongs on the other (here, horizontal or *x*-axis). A strip chart is like a scatter plot except the explanatory variable is categorical rather than numerical. It is usually necessary to spread, or "jitter," the points along the horizontal axis to reduce overlap of points, so that they can be more easily seen. The strip chart method worked well in Figure 2.1-2, where there were few data points in each group. However, several of the male populations in Example 2.3C have so many observations that the points overlap too much in the strip chart, making it difficult to see the individual dots and their distribution (left panel of Figure 2.3-4).

> The *strip chart* is a graphical display of a numerical variable and a categorical variable in which each observation is represented as a dot.

An alternative method to "show the data" is the **box plot**, which uses lines and rectangles to display a compact summary of the frequency distribution of the data (right panel of Figure 2.3-4). A box plot doesn't show most of the data points, but instead uses lines and boxes to indicate where the bulk of the observations lie. The scale of the vertical axis is the same in both panels of Figure 2.3-4 so that you can see the correspondence between boxes and data points.

The line inside each box (right panel of Figure 2.3-4) is the *median,* the middle measurement of the group of observations. Half the observations lie above the median and half lie below. The lower and upper edge of each box are first and third quartiles. One-fourth of the observations lie below the *first quartile* (three-fourths lie above). Conversely, three-fourths of the observations lie below the *third quartile* (one-quarter lie above). Two lines, called whiskers, extend outward from a box at each end. The whiskers stop at the smallest and largest "non-extreme" values in the data. Extreme values are plotted as isolated dots past the ends of the whiskers. We explain these quantities in more detail, and how to calculate them, in Chapter 3.

> A *box plot* is a graph that uses lines and a rectangular box to display the median, quartiles, range, and extreme measurements of the data.

The box plot in Figure 2.3-4 shows key features of the four frequency distributions using just a few graphical elements. We can clearly see in this graph how only men from the high Andes had elevated hemoglobin concentrations, whereas men from high-elevation Ethiopia and Tibet were not noticeably different in hemoglobin concentration from the sea-level group.[7] We can see that the shapes of distributions are

---

7. Mysteriously, oxygen levels in the blood of highland Ethiopian men are just as high as those in men living at sea level (data not shown), despite their similar concentrations of hemoglobin. The physiological mechanism behind this feat is not yet known.

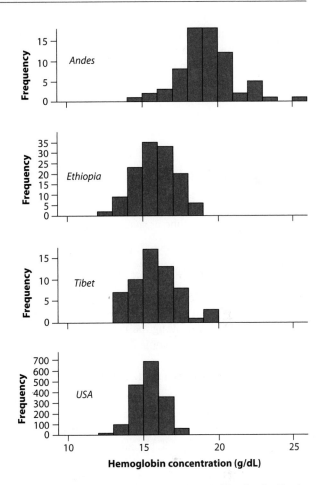

FIGURE 2.3-5
Multiple histograms showing the hemoglobin concentration in males of the four populations. The number of measurements in each group is given in Figure 2.3-4.

relatively similar in the four groups of males—the span of the boxes is similar in the four groups, although greatest in the Andean males and least in the USA males. The lengths of the whiskers are also fairly similar. USA males have the most extreme observations, but the shape of the distribution, as indicated by the box and whiskers, is similar to that in the other groups.

The third method uses multiple histograms, one for each category, to show the data, as shown in Figure 2.3-5. It is important that the histograms be stacked above one another as shown so that the position and spread of the data are most easily compared. Side-by-side histograms lose most of the advantages of the multiple histogram method for visualizing association, because differences in the position of bars between groups are difficult to see. Use the same scale along the horizontal axis to allow comparison.

Of the three methods for showing association between a numeric and a categorical variable (difference between groups), which is the best? The strip chart shows all the data points, which is ideal when there are only a few observations in each category. The box plot picks out a few of the most important features of the frequency distri-

LIBRARY, UNIVERSITY O CHESIⅬⅭ

bution and is more suitable when the number of observations is large. The multiple histogram plot shows more features of the frequency distribution but takes up more space than the other two options. It works best when there are only a few categories. As usual, the best strategy is to try all three methods on your data and judge for that situation which method shows the association most clearly.

2.4 Showing trends in time and space

Often a variable of interest represents a summary measurement taken at consecutive points in time or space. In this case, *line graphs* and *maps* are excellent visual tools.

Line graph

A **line graph** is a powerful tool for displaying trends in time or other ordered series. Typically, one *y*-measurement is displayed for each point in time or space, which is displayed along the *x*-axis. Adjacent points along the *x*-axis are connected by a line segment. Example 2.4A illustrates the line graph.

EXAMPLE 2.4A

Bad science can be deadly

Since the introduction of a measles vaccine, the number of cases in the U.K. dropped dramatically. A disease that once killed hundreds of people per year in the U.K. became a negligible risk as most of the population was immunized. However, recent declines in the fraction of people vaccinated, in part from unfounded fears concerning the safety of the vaccine,[8] has caused a resurgence in the number of cases of measles (Jansen et al. 2003). The number of cases quarterly between 1995 and 2011 is shown in a line plot in Figure 2.4-1 (data from Health Protection Agency 2012).

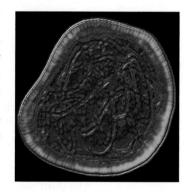

---

8. Vaccination rates dropped below 85% in the years after a 1998 publication that appeared to link the measles, mumps, and rubella vaccine to an increased risk of autism (e.g., see Gilmour et al. 2011). This controversial conclusion was refuted in subsequent reviews (e.g., Canadian Paediatric Society 2007). The original article has since been retracted by most of its authors and by the medical journal that published it. In 2010, the General Medical Council of the U.K. found the senior author of the research guilty of ethical breaches, dishonesty, and conflict of interest and banned him from practicing medicine.

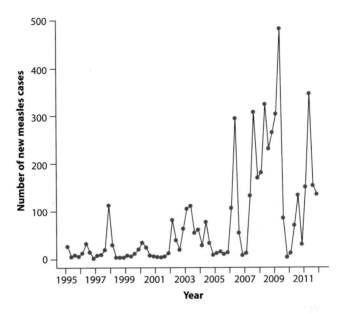

FIGURE 2.4-1 Confirmed cases of measles in England and Wales from 1995 to 2011. The four numbers in each year refer to new cases in each quarter.

The trends in the number of measles cases over time are made more visible by the lines connecting the points in Figure 2.4-1. The steepness of the line segments reflects the speed of change in the number of cases from one quarter-year to the next. Notice, for example, how steeply the number of cases rises when an outbreak begins, and then how cases decline just as quickly afterward, as immunity spreads. When the baseline for the vertical axis is zero, as in the present example, the area under the curve between two time points is proportional to the total number of new cases in that period.

> A *line graph* uses dots connected by line segments to display trends in a measurement over time or other ordered series.

Maps

A **map** is the spatial equivalent of the line graph, using a color gradient to display a numerical response variable at multiple locations on a surface. The explanatory variable is location in space. One measurement is displayed for each point or interval of the surface, as shown in Example 2.4B.

EXAMPLE **Biodiversity hotspots**

2.4B South America is renowned for its extraordinary numbers of species. We tend to think of the vast expanse of lowland Amazon rainforest as the seat of this abundance. The data shown in Figure 2.4-2 are the numbers of plant species recorded at many points on a fine grid covering the northern part of South America. Points are colored such that "hotter" colors represent more plant species at each point. The image shows that peak diversities actually occur at the northwest coast, the nearby inland where the Andes mountains meet the lowland rainforest, and along the southeast coast of Brazil.

FIGURE 2.4-2
Map displaying numbers of plant species in northern South America. Colors reflect the numbers of species estimated at many points in a fine grid, with each point consisting of an area 100 km x 100 km. The color scale is on the right, with hotter colors reflecting more species. The horizontal gray line is the equator. Modified from Bass et al. (2010).

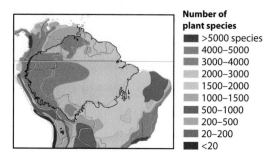

Number of plant species
- >5000 species
- 4000–5000
- 3000–4000
- 2000–3000
- 1500–2000
- 1000–1500
- 500–1000
- 200–500
- 20–200
- <20

The map in Figure 2.4-2 contains an enormous amount of data, yet the pattern is easy to see. The regions of peak diversity, and those of relatively low diversity, are clearly evident.

Maps can be used for measurements at points on any two-dimensional surface, including a spatial grid (such as in the plant species richness example) or at political or geological boundaries on a map. They can also be used to indicate measurements at locations on the surface of two- or three-dimensional objects, such as the brain or the body. For example, a visual representation of an MRI scan is also a map.

2.5 How to make good tables

Tables have two functions: to communicate patterns and to record quantitative data summaries for further use. When the main function of a table is to display the patterns in the data—a "display table"—numerical detail is less important than the effective communication of results. This is the kind of table that would appear in the main body of a report or publication. Compact frequency tables are examples of display tables; for example, look at Table 2.2-3, which shows the number of bird species in a sequence of abundance categories. In this section we summarize strategies for making good display tables.

The purpose of the second kind of table is to store raw data and numerical summaries for reference purposes. Such "data tables" are often large and are not ideal for recognizing patterns in data. They are inappropriate for communicating general findings to a wider audience. They are nevertheless often invaluable. Table 2.2-2, which lists the abundances of every bird species in a survey, is an example. Data tables aren't usually included in the main body of a report. When published, they usually appear as appendices or online supplements, so specialized readers interested in more details can find them.

Follow similar principles for display tables

Producing clear, honest, and efficient display tables should follow many of the same principles discussed already for graphs. In particular,

- Make patterns in the data easy to see.
- Represent magnitudes honestly.
- Draw table elements clearly.

Make patterns easy to see. Make the table compact and present as few significant digits as are necessary to communicate the pattern. Avoid putting too much data into one table. Arrange the rows and columns of numbers to facilitate pattern detection. For example, a series of numbers listed above one another in a single column are easier to compare with one another than the same numbers listed side by side in different columns. Our earlier recommendations for frequency tables apply here (Section 2.2). For example, list unordered categorical (nominal) data in order of importance (frequency) rather than alphabetically or otherwise. If the categorical variables have a natural order (such as life stages: zygote, fetus, newborn, adolescent, adult), they should be listed in that order.

Represent magnitudes honestly. For example, when combining numbers into bins in frequency tables, use intervals of equal width so that the numbers can be more accurately compared.

Draw table elements clearly. Clearly label row and column headers, and always provide units of measurement. Choose unadorned, simple fonts.

Let's look at an example of a display table and then consider how it might be improved. The data in Table 2.5-1 were put together by Alvarez et al. (2009) to investigate the idea that a strong preference for consanguineous marriages (inbreeding) within the line of Spanish Habsburg kings, which ruled Spain from 1516 to 1700, contributed to its downfall. The quantity F is a measure of inbreeding in the offspring. F is zero if king and queen were unrelated, and F is 0.25 if they were brother and sister whose own parents were unrelated. Values may be lower or higher if there was inbreeding further generations back.

There is a tendency for less related kings and queens to produce a higher proportion of surviving offspring, but it is not so easy to see this in Table 2.5-1. Before reading any further, examine the table and make a note of any deficiencies. How might these deficiencies be overcome by modifying the table?

TABLE 2.5-1 Inbreeding coefficient (*F*) of Spanish Habsburg kings and queens and survival of their progeny.

King/Queen	F	Pregnan-cies	Miscarriages & stillbirths	Neonatal deaths	Later deaths	Survivors to age 10	Survival (total)	Survival (postnatal)
Ferdinand of Aragon								
Elizabeth of Castile	0.039	7	2	0	0	5	0.714	1.000
Philip I								
Joanna I	0.037	6	0	0	0	6	1.000	1.000
Charles I								
Isabella of Portugal	0.123	7	1	1	2	3	0.429	0.600
Philip II								
Elizabeth of Valois	0.008	4	1	1	0	2	0.500	1.000
Anna of Austria	0.218	6	1	0	4	1	0.167	0.200
Philip III								
Margaret of Austria	0.115	8	0	0	3	5	0.625	0.625
Philip IV								
Elizabeth of Bourbon	0.050	7	0	3	2	2	0.286	0.500
Mariana of Austria	0.254	6	0	1	3	2	0.333	0.400

Source: Data are from Alvarez et al. (2009).

Let's apply the principles of effective display to improve this table. Consider that the main goal of producing the table should be to show a pattern, in this case a possible association between *F* and offspring survival. Here is a list of features of Table 2.5-1 that we felt made it difficult to see this pattern.

- King/queen pairs are not ordered in such a way as to make it easy for the eye to see any association.
- The main variables of interest, *F* and survival, are separated by intervening columns.
- Blank lines are inserted for every new king listed, fragmenting any pattern.
- The number of decimal places is overly large, making it difficult to read the numbers.

To overcome these problems, we have extracted the most crucial columns and reorganized them in Table 2.5-2. In this revised table, king and queen pairs are ordered by *F* value of the offspring, and survival has been placed in the adjacent column. Blank lines have been eliminated along with unnecessary columns, and decimals have been rounded to two places.

The revised Table 2.5-2 suggests that survival of more inbred progeny tends to be lower, at least when measured as postnatal survival. The trend appears weaker for total survival, which includes prenatal and neonatal survival.

Just as for a graph, a good table must convey information clearly, concisely, and without distortion. A good table requires careful editing. See Ehrenberg (1977) for further insights into how to draw tables.

TABLE 2.5-2 Inbreeding coefficient (*F*) of Spanish kings and queens and survival of their progeny. These data are extracted and reorganized from Table 2.5-1.

King/Queen	*F*	Survival (postnatal)	Survival (total)	Number of pregnancies
Philip II/Elizabeth of Valois	0.01	1.00	0.50	4
Philip I/Joanna I	0.04	1.00	1.00	6
Ferdinand/Elizabeth of Castile	0.04	1.00	0.71	7
Philip IV/Elizabeth of Bourbon	0.05	0.50	0.29	7
Philip III/Margaret of Austria	0.12	0.63	0.63	8
Charles I/Isabella of Portugal	0.12	0.60	0.43	7
Philip II/Anna of Austria	0.22	0.20	0.17	6
Philip IV/Mariana of Austria	0.25	0.40	0.33	6

2.6 Summary

- Graphical displays must be clear, honest, and efficient.
- Strive to show the data, to make patterns in the data easy to see, to represent magnitudes honestly, and to draw graphical elements clearly.
- Follow the same rules when constructing tables to reveal patterns in the data.
- A frequency table is used to display a frequency distribution for categorical or numerical data.
- Bar graphs and histograms are recommended graphical methods for displaying frequency distributions of categorical and numerical variables:

Type of data	Graphical method
Categorical data	Bar graph
Numerical data	Histogram

- Contingency tables describe the association between two (or more) categorical variables by displaying frequencies of all combinations of categories.
- Recommended graphical methods for displaying associations between variables and differences between groups include the following:

Types of data	Graphical method
Two numerical variables	Scatter plot
	Line plot (space or time)
	Map (space)
Two categorical variables	Grouped bar graph
	Mosaic plot
One numerical variable and one categorical variable	Strip chart
	Box plot
	Multiple histograms
	Cumulative frequency distributions (Chapter 3)

PRACTICE PROBLEMS

1. Estimate by eye the relative frequency of the shaded areas in each of the following histograms.

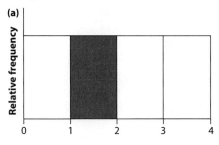

(a)

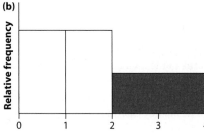

(b)

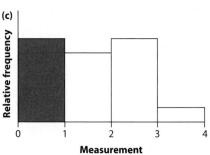

(c)

2. Using a graphical method from this chapter, draw three frequency distributions: one that is symmetric, one that is skewed, and one that is bimodal.
   a. Identify the mode in each of your frequency distributions.
   b. Does your skewed distribution have negative or positive skew?
   c. Is your bimodal distribution skewed or symmetric?

3. In the southern elephant seal, males defend harems that may contain hundreds of reproductively active females. Modig (1996) recorded the numbers of females in harems in a population on South Georgia Island. The histograms of the data (below, drawn from data in Modig 1996) are unusual because the rarer, larger harems have been divided into wider intervals. In the upper histogram, bar *height* indicates the relative frequency of harems in the interval. In the lower histogram, bar height is adjusted such that bar *area* indicates relative frequency. Which histogram is correct? Why?

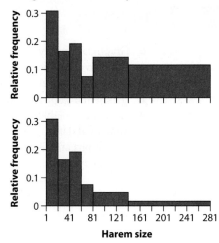

4. Draw scatter plots for invented data that illustrate the following patterns:
   a. Two numerical variables that are positively associated
   b. Two numerical variables that are negatively associated
   c. Two numerical variables whose relationship is nonlinear

5. A study by Miller et al. (2004) compared the survival of two kinds of Lake Superior rainbow trout fry (babies). Four thousand fry were from a government hatchery on the lake, whereas 4000 more fry came from wild trout. All 8000 fry were released into a stream flowing into the

lake, where they remained for one year. After one year, the researchers found 78 survivors. Of these, 27 were hatchery fish and 51 were wild. Display these results in the most appropriate table. Identify the type of table you used.

6. The following data are the occurrences in 2012 of the different taxa in the list of endangered and threatened species under the U.S. Endangered Species Act (U.S. Fish and Wildlife Service 2012). The taxa are listed in no particular order in the table.

Taxon	Number of species
Birds	93
Clams	83
Reptiles	36
Fish	152
Crustaceans	22
Mammals	85
Snails	40
Flowering plants	782
Amphibians	26
Insects	66
Arachnids	12

a. Rewrite the table, but list the taxa in a more revealing order. Explain your reasons behind the ordering you choose.
b. What kind of table did you construct in part (a)?
c. Choosing the most appropriate graphical method, display the number of species in each taxon. What kind of graph did you choose? Why?
d. Should the baseline for the number of species in your graph in part (c) be 0 or 12, the smallest number in the data set? Why?
e. Create a version of this table that shows the relative frequency of endangered species by taxon.

7. Can environmental factors influence the incidence of schizophrenia? A recent project measured the incidence of the disease among children born in a region of eastern China:

192 of 13,748 babies born in the midst of a severe famine in the region in 1960 later developed schizophrenia. This compared with 483 schizophrenics out of 59,088 births in 1956, before the famine, and 695 out of 83,536 births in 1965, after the famine (St. Clair et al. 2005).
a. What two variables are compared in this example?
b. Are the variables numerical or categorical? If numerical, are they continuous or discrete; if categorical, are they nominal or ordinal?
c. Effectively display the findings in a table. What kind of table did you use?
d. In each of the three years, calculate the relative frequency (proportion) of children born who later developed schizophrenia. Plot these proportions in a line graph. What pattern is revealed?

8. Human diseases differ in their virulence, which is defined as their ability to cause harm. Scientists are interested in determining what features of different diseases make some more dangerous to their hosts than others. The graph below depicts the frequency distribution of virulence measurements, on a log base 10 scale, of a sample of human diseases (modified from Ewald 1993). Diseases that spread from one victim to

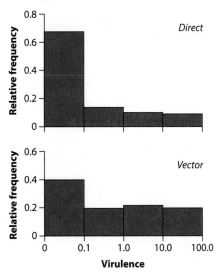

another by direct contact between people are shown in the upper graph. Those transmitted from person to person by insect vectors are shown in the lower graph.

a. Identify the type of graph displayed.

b. What are the two groups being compared in this graph?

c. What variable is being compared between the two groups? Is it numerical or categorical?

d. Explain the units on the vertical (*y*) axis.

e. What is the main result depicted by this graph?

9. Examine the figure below, which indicates the date of first occurrence of rabies in raccoons in the townships of Connecticut, measured by the number of months following March 1, 1991 (modified from Smith et al. 2002).

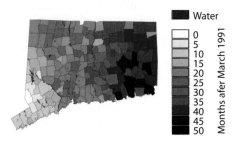

a. Identify the type of graph shown.

b. What is the response variable?

c. What is the explanatory variable?

d. What was the direction of spread of the disease (from where, to where, approximately)?

10. The following graph is taken from a study of married women who had been raised by adoptive parents (Bereczkei et al. 2004). It shows the facial resemblance between the women and their husbands (first bar), between their husbands and the women's adoptive fathers (second bar), and between their husbands and the women's adoptive mothers (third bar). Facial resemblance of a given woman to her husband was scored by 242 "judges," each of whom was given a photograph of the woman, photos of three other women, and a photo of her husband. The judges were asked to decide from the photos which of the four women was the wife of the husband based on

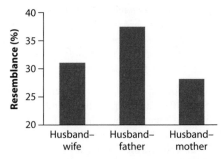

facial similarity. Her resemblance was scored as the percentage of judges who chose correctly. If there was no resemblance between a given woman and her husband, then by chance only one in four judges (25%) should have chosen correctly. Resemblance of the woman's husband to the wife's adoptive father and to the wife's adoptive mother was measured in the same way.

a. Describe the essential findings displayed in the figure.

b. Which two principles of good graph design are violated in this figure?

11. Each of the following graphs illustrates an association between two variables. For each graph identify (1) the type of graph, (2) the explanatory and response variables, and (3) the type of data (whether numerical or categorical) for each variable.

a. Observed fruiting of individual plants in a population of *Campanula americana* according to the number of fruits produced previously (Richardson and Stephenson 1991):

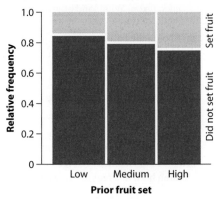

b. The maximum density of wood produced at the end of the growing season in white spruce trees in Alaska in different years (modified from Barber et al. 2000):

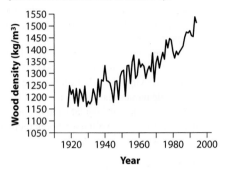

c. Relative expression levels of *Neuropeptide Y* (*NPY*), a gene whose activity correlates with anxiety and is induced by stress, in the brains of people differing in their genotypes at the locus (Zhou et al. 2008).

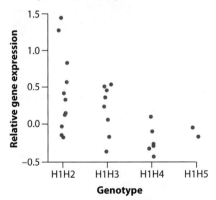

12. The following data are from the Cambridge Study in Delinquent Development (see Problem 22). They examine the relationship between the occurrence of convictions by the end of the study and the family income of each boy when growing up. Three categories described income level: inadequate, adequate, and comfortable.

	Income level		
	Inadequate	Adequate	Comfortable
No convictions	47	128	90
Convicted	43	57	30

a. What type of table is this?
b. Display these same data in a mosaic plot.
c. What type of variable is "income level"? How should this affect the arrangement of groups in your mosaic plot in part (b)?
d. By viewing the table above and the graph in part (b), describe any apparent association between family income and later convictions.
e. In answering part (d), which method (the table or the graph) better revealed the association between conviction status and income level? Explain.

13. Each of the following graphs illustrates an association between two variables. For each graph, identify (1) the type of graph, (2) the explanatory and response variables, and (3) the type of data (whether numerical or categorical) for each variable.

a. Taste sensitivity to phenylthiocarbamide (PTC) in a sample of human subjects grouped according to their genotype at the PTC gene—namely, *AA*, *Aa*, or *aa* (Kim et al. 2003):

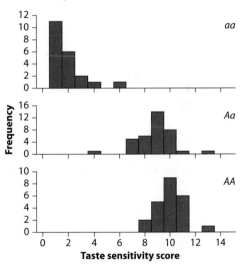

b. Migratory activity (hours of nighttime restlessness) of young captive blackcaps (*Sylvia atricapilla*) compared with the migratory activity of their parents (Berthold and Pulido 1994):

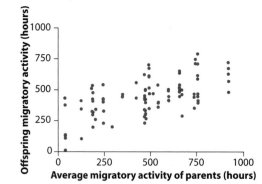

c. Sizes of the second appendage (middle leg) of embryos of water striders, in 10 control embryos and 10 embryos dosed with RNAi for the developmental gene *Ultrabithorax* (Khila et al. 2009).

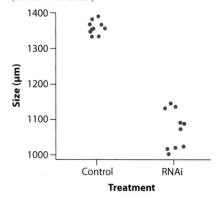

d. The frequency of injection-heroin users that share or do not share needles according to their known HIV infection status (Wood et al. 2001):

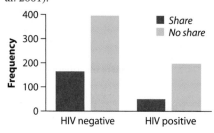

14. *Spot the flaw.* In an experimental study of gender and wages, Moss-Racusin et al. (2012) presented professors from research-intensive universities each with a job application for a laboratory manager position. The application was randomly assigned a male or female name, and the professors were asked to state the starting salary they would offer the candidate if hired. The average starting salary reported is compared in the following figure between applications with male names and female names. (The vertical lines at the top edge of each bar are "standard error bars"—we'll learn about them in Chapter 4).

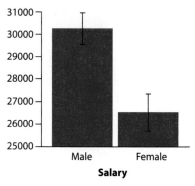

a. Identify at least two of the four principles of good graph design that are violated.
b. What alternative graph type is ideal for these data?
c. Identify the main pattern in the data (interestingly, this pattern was similar when male professors and female professors were examined separately).

15. How do the insects that pollinate flowers distinguish individual flowers with nectar from empty flowers? One possibility is that they can detect the slightly higher humidity of the air—produced by evaporation—in flowers that contain nectar. von Arx et al. (2012) recently tested this idea by manipulating the humidity of air emitted from artificial flowers that were otherwise identical. The following graph summarizes the number of visits to the two types of flowers by hawk moths (*Hyles lineata*).

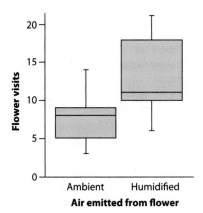

a. What type of graph is this?

b. What does the horizontal line in the center of each rectangle represent?

c. What do the top and bottom edges of each rectangle represent?

d. What are the vertical lines extending above and below each rectangle?

e. Is an association apparent between the variables plotted? Explain.

16. For each of the graphs shown below, based on hypothetical data, identify the type of graph and say whether or not the two variables exhibit an association. Explain your answer in each case.

FIGURE FOR PROBLEM 16

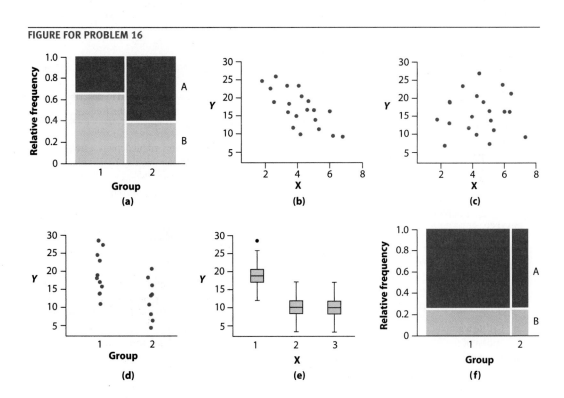

17. "Animal personality" has been defined as the presence of consistent differences between individuals in behaviors that persist over time. Do sea anemones have it? To investigate, Briffa and Greenaway (2011) measured the consistency of the startle response of individuals of wild beadlet anemones, *Actinia equina,* in tide pools in the U.K. When disturbed, such as with a mild jet of water (the method used in this study), the anemones retract their feeding tentacles to cover the oral disc, opening them again some time later. The accompanying table records the duration of the startle response (time to reopen,

in seconds) of 12 individual anemones. Each anemone was measured twice, 14 days apart.

a. Choose the best method, and make a graph to show the association between the first and second measurements of startle response.

b. Is a strong association present? In other words, does the beadlet anemone have animal personality?

18. Refer to the previous question.

a. Draw a frequency distribution of startle durations measured on the first occasion.

b. Describe the shape of the frequency distribution: is it skewed or symmetric? If skewed, say whether skew is positive or negative.

TABLE FOR PROBLEM 17

Anemone:	1	2	3	4	5	6	7	8	9	10	11	12
Occasion one	1065	248	436	350	378	410	232	201	267	687	688	980
Occasion two	939	268	460	261	368	467	303	188	401	690	711	571

ASSIGNMENT PROBLEMS

19. Male fireflies of the species *Photinus ignitus* attract females with pulses of light. Flashes of longer duration seem to attract the most females. During mating, the male transfers a spermatophore to the female. Besides containing sperm, the spermatophore is rich in protein that is distributed by the female to her fertilized eggs. The data below are measurements of spermatophore mass (in mg) of 35 males (Cratsley and Lewis 2003).

0.047, 0.037, 0.041, 0.045, 0.039, 0.064, 0.064, 0.065, 0.079, 0.070, 0.066, 0.059, 0.075, 0.079, 0.090, 0.069, 0.066, 0.078, 0.066, 0.066, 0.055, 0.046, 0.056, 0.067, 0.075, 0.048, 0.077, 0.081, 0.066, 0.172, 0.080, 0.078, 0.048, 0.096, 0.097

a. Create a graph depicting the frequency distribution of the 35 mass measurements.

b. What type of graph did you choose in part (a)? Why?

c. Describe the shape of the frequency distribution. What are its main features?

d. What term would be used to describe the largest measurement in the frequency distribution?

20. The accompanying graph depicts a frequency distribution of beak widths of 1017 black-bellied seedcrackers, *Pyrenestes ostrinus,* a finch from West Africa (Smith 1993).

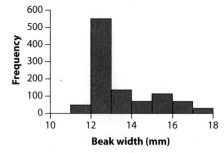

a. What is the mode of the frequency distribution?

b. Estimate by eye the fraction of birds whose measurements are in the interval representing the mode.

c. There is a hint of a second peak in the frequency distribution between 15 and 16 mm. What strategy would you recommend be used to explore more fully the possibility of a second peak?

d. What name is given to a frequency distribution having two distinct peaks?

21. When its numbers increase following favorable environmental conditions, the desert locust, *Schistocerca gregaria,* undergoes a dramatic transformation from a solitary, cryptic form into a gregarious form that swarms by the billions. The transition is triggered by mechanical stimulation—locusts bumping into one another. The accompanying figure shows the results of a laboratory study investigating the degree of gregariousness resulting from mechanical stimulation of different parts of the body (modified from Simpson et al. 2001).

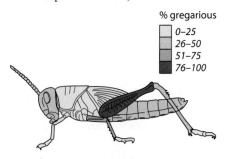

% gregarious
- 0–25
- 26–50
- 51–75
- 76–100

a. Identify the type of graph displayed.
b. Identify the explanatory and response variables.

22. The *Cambridge Study in Delinquent Development* was undertaken in north London (U.K.) to investigate the links between criminal behavior in young men and the socioeconomic factors of their upbringing (Farrington 1994). A cohort of 395 boys was followed for about 20 years, starting at the age of 8 or 9. All of the boys attended six schools located near the research office. The following table shows the total number of criminal convictions by the boys between the start and end of the study.

Number of convictions	Frequency
0	265
1	49
2	21
3	19
4	10
5	10
6	2
7	2
8	4
9	2
10	1
11	4
12	3
13	1
14	2
	Total: 395

a. What type of table is this?
b. How many variables are presented in this table?
c. How many boys had exactly two convictions by the end of the study?
d. What fraction of boys had no convictions?
e. Display the frequency distribution in a graph. Which type of graph is most appropriate? Why?
f. Describe the shape of the frequency distribution. Is it skewed or is it symmetric? Is it unimodal or bimodal? Where is the mode in number of criminal convictions? Are there outliers in the number of convictions?
g. Does the sample of boys used in this study represent a random sample of British boys? Why or why not?

23. Swordfish have a unique "heater organ" that maintains elevated eye and brain temperatures when hunting in deep, cold water. The following graph illustrates the results of a study by Fritsches et al. (2005) that measured how the ability of swordfish retinas to detect rapid motion, measured by the flicker fusion frequency, changes with eye temperature.

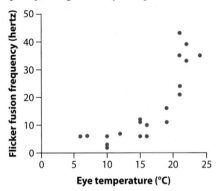

a. What types of variables are displayed?
b. What type of graph is this?
c. Describe the association between the two variables. Is the relationship between flicker fusion frequency and temperature positive or negative? Is the relationship linear or nonlinear?
d. The 20 points in the graph were obtained from measurements of six swordfish. Can we treat the 20 measurements as a random sample? Why or why not?

24. The following graph displays the net number of species listed under the U.S. Endangered Species Act between 1980 and 2002 (U.S. Fish and Wildlife Service 2001):

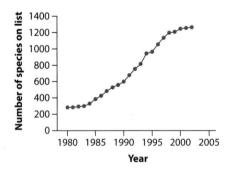

a. What type of graph is this?

b. What does the steepness of each line segment indicate?
c. Explain what the graph tells us about the relationship between the number of species listed and time.

25. *Spot the flaw.* Examine the following figure, which displays the frequency distribution of similarity values (the percentage of amino acids that are the same) between equivalent (homologous) proteins in humans and pufferfish of the genus *Fugu* (modified from Aparicio et al. 2002).

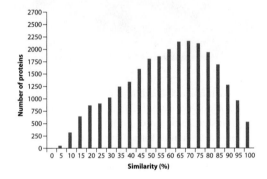

a. What type of graph is this?
b. Identify the main flaw in the construction of this figure.
c. What are the main results displayed in the figure?
d. Describe the shape of the frequency distribution shown.
e. What is the mode in the frequency distribution?

26. The following data give the photosynthetic capacity of nine individual females of the neotropical tree *Ocotea tenera*, according to the number of fruits produced in the previous reproductive season (Wheelwright and Logan 2004). The goal of the study was to investigate how reproductive effort in females of these trees impacts subsequent growth and photosynthesis.

Number of fruits produced previously	Photosynthetic capacity ($\mu mol\ O_2/m^2/s$)
10	13.0
14	11.9
5	11.5
24	10.6
50	11.1
37	9.4
89	9.3
162	9.1
149	7.3

a. Graph the association between these two variables using the most appropriate method. Identify the type of graph you used.

b. Which variable is the explanatory variable in your graph? Why?

c. Describe the association between the two variables in words, as revealed by your graph.

27. Examine the accompanying figure, which displays the percentage of adults over 18 with a "body mass index" greater than 25 in different

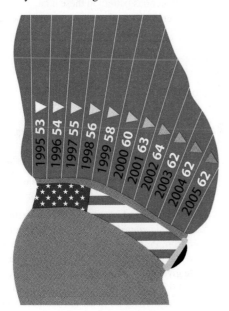

years (modified from *The Economist* 2005, with permission). Body mass index is a measure of weight relative to height.

a. What is the main result displayed in this figure?

b. Which of the four principles for drawing good graphs are violated here? How are they violated?

c. Redraw the figure using the most appropriate method discussed in this chapter. What type of graph did you use?

28. When a courting male of the small Indonesian fish *Telmatherina sarasinorum* spawns with a female, other males sometimes sneak in and release sperm, too. The result is that not all of the female's eggs are fertilized by the courting male. Gray et al. (2007) noticed that courting males occasionally cannibalize fertilized eggs immediately after spawning. Egg eating took place by 61 of 450 courting males who fathered the entire batch; the remaining 389 males did not cannibalize eggs. In contrast, 18 of 35 courting males ate eggs when a single sneaking male also participated in the spawning event. Finally, 16 of 20 males ate eggs when two or more sneaking males were present.

a. Display these results in a table that best shows the association between cannibalism and the number of sneaking males. Identify the type of table you used.

b. Illustrate the same results using a graphical technique instead. Identify the type of graph you used.

29. The graph at the top of page 62, in red, shows the number of new cases of influenza in New York, Pennsylvania, and New Jersey, according to data from the Centers for Disease Control (Ginsberg et al. 2009). The black line shows predictions based on the number of Google searches of words like "flu" or "influenza."

a. What type of graph is this?

b. Describe some of the scientific conclusions you might draw from looking at this graph.

FIGURE FOR PROBLEM 29

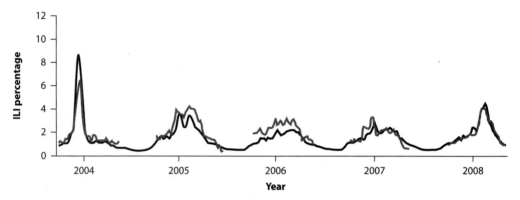

Source: Jeremy Ginsberg et al., "Detecting Influenza Epidemics Using Search Engine Query Data," *Nature* 457 (2009): 1012–1014.

c. Can you suggest an improvement to the axes labels?

30. The following graph was drawn using a very popular spreadsheet program in an attempt to show the frequencies of observations in four hypothetical groups. Before reading further, estimate by eye the frequencies in each of the four groups.
   a. Identify two features of this graph that cause it to violate the principle, "Make patterns in the data easy to see."
   b. Identify at least two other features of the graph that make it difficult to interpret.
   c. The actual frequencies are 10, 20, 30, and 40. Draw a graph that overcomes the problems identified above.

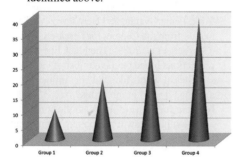

31. In Poland, students are required to achieve a score of 21 or higher on the high-school Polish language "maturity exam" to be eligible for university. The following graph shows the frequency distribution of scores (Freakonomics 2011).
   a. Examine the graph and identify the most conspicuous pattern in these data.
   b. Generate a hypothesis to explain the pattern.

32. More than 10% of people carry the parasite *Toxoplasma gondii*. The following table gives data from Prague on 15- to 29-year-old drivers who had been involved in an accident. The table gives the number of drivers who were infected with *Toxoplasma gondii* and who were uninfected. These numbers are compared with a control sample of 249 drivers of the same age living in the same area who had not been in an accident.

	Infected	Uninfected
Drivers with accidents	21	38
Controls	38	211

a. What type of table is this?

b. What are the two variables being compared? Which is the explanatory variable and which is the response?

c. Depict the data in a graph. Use the results to answer the question: are the two variables associated in this data set?

33. The cutoff birth date for school entry in British Columbia, Canada, is December 31. As a result, children born in December tend to be the youngest in their grade, whereas those born in January tend to be the oldest. Morrow et al. (2012) examined how this relative age difference influenced diagnosis and treatment of attention deficit/hyperactivity disorder (ADHD). A total of 39,136 boys aged 6 to 12 years and registered in school in 1997–1998 had January birth dates. Of these, 2219 were diagnosed with ADHD in that year. A total of 38,977 boys had December birth dates, of which 2870 were diagnosed with ADHD in that year. Display the association between birth month and ADHD diagnosis using a table or graphical method from this chapter. Is there an association?

34. Examine the following figure, which displays hypothetical measurements of a sample of individuals from several groups.

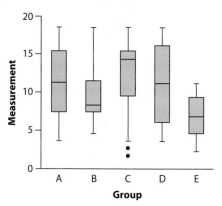

a. What type of graph is this?

b. In which of the groups is the frequency distribution of measurements approximately symmetric?

c. Which of the frequency distributions show positive skew?

d. Which of the frequency distributions show negative skew?

e. Which group has the largest value for the upper quartile?

f. Which group has the smallest value for the median?

g. Which group has the most extreme observation?

35. The following data are from Mattison et al. (2012), who carried out an experiment with rhesus monkeys to test whether a reduction in food intake extends life span (as measured in years). The data are the life spans of 19 male and 15 female monkeys who were randomly assigned a normal nutritious diet or a similar diet reduced in amount by 30%. All monkeys were adults at the start of the study.

Females—reduced: 16.5, 18.9, 22.6, 27.8, 30.2, 30.7, 35.9

Females—control: 23.7, 24.5, 24.7, 26.1, 28.1, 33.4, 33.7, 35.2

Males—reduced: 23.7, 28.1, 29.8, 31.1, 36.3, 37.7, 39.9, 39.9, 40.2, 40.2

Males—control: 24.9, 25.2, 29.6, 33.2, 34.1, 35.4, 38.1, 38.8, 40.7

a. Graph the results, using the most appropriate method and following the four principles of good graph design.

b. According to your graph, which difference in life span is greater: that between the sexes, or that between diet groups?

36. The accompanying graph indicates the amount of time (latency) that female subjects were willing to leave their hand in icy water while they were swearing ("words you might use after hitting yourself on the thumb with a hammer") or while not swearing, using other words instead ("words to describe a table"). The data are from Stephens et al. (2009).

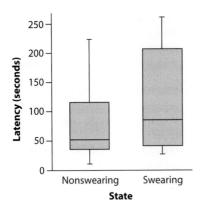

a. Identify the type of graph.

b. Is any association apparent between the variables? Explain.

c. What do the "whiskers" indicate in this graph?

d. List two other types of graphs that would also be appropriate for showing these results.

37. Following is a list of all the named hurricanes in the Atlantic between 2001 and 2010, along with their category on the Saffir-Simpson Hurricane Scale,[9] which categorizes each hurricane by a label from 1 to 5 depending on its power.

2001: Erin, 3; Feliz, 3; Gabrielle, 1; Humberto, 2; Iris, 4; Karen, 1; Michelle, 4; Noel, 1; Olga, 1.

2002: Gustav, 2; Isidore, 3; Kyle, 1; Lili, 4;

2003: Claudette, 1; Danny, 1; Erika, 1; Fabian, 4; Isabel, 5; Juan, 2; Kate, 3.

2004: Alex, 3; Charley, 4; Danielle, 2; Frances, 4; Gaston, 1; Ivan, 5; Jeanne, 3; Karl, 4; Lisa, 1.

2005: Cindy, 1; Dennis, 4; Emily, 5; Irene, 2; Katrina, 5; Maria, 3; Nate, 1; Ophelia, 1; Philippe 1; Rita, 5; Stan, 1; Vince, 1; Wilma, 5; Beta, 3; Epsilon, 1.

2006: Ernesto, 1; Florence, 1; Gordon, 3; Helene, 3; Isaac, 1.

2007: Dean, 5; Felix, 5; Humberto, 1; Karen, 1; Lorenzo, 1; Noel, 1.

2008: Bertha, 3; Dolly, 2; Gustav, 4; Hanna, 1; Ike, 4; Kyle, 1; Omar, 4; Paloma, 4.

2009: Bill, 4; Fred, 3; Ida, 2.

2010: Alex, 2; Danielle, 4; Earl, 4; Igor, 4, Julia, 4; Karl, 3; Lisa, 1; Otto, 1; Paula, 2; Richard, 2; Shary, 1; Toas, 2.

a. Make a frequency table showing the frequency of hurricanes in each severity categories during the decade.

b. Make a frequency table that shows the frequency of hurricanes in each year.

c. Explain how you chose to order the categories in your tables.

9. Data taken from http://en.wikipedia.org/wiki/2010_Atlantic_hurricane_season and similar articles for other years.

Saiga

3 Describing data

Descriptive statistics, or summary statistics, are quantities that capture important features of frequency distributions. Whereas graphs reveal shapes and patterns
in the data, descriptive statistics provide hard numbers. The most important descriptive statistics for numerical data are those measuring the **location** of a frequency
distribution and its **spread**. The location tells us something about the average or
typical individual—where the observations are centered. The spread tells us how
variable the measurements are from individual to individual—how widely scattered
the observations are around the center. The **proportion** is the most important
descriptive statistic for a categorical variable, measuring the fraction of observations
in a given category.

The importance of calculating the location of a distribution seems obvious.
How else do we address questions like "Which species is larger?" or "Which drug
yielded the greatest response?" The importance of describing distribution spread is

less obvious but no less crucial, at least in biology. In some fields of science, variability around a central value is instrument noise or measurement error, but in biology much of the variability signifies real differences among individuals. Different individuals respond differently to treatments, and this variability begs measurement. Measuring variability also gives us perspective. We can ask, "How large are the differences between groups compared with variations within groups?" Biologists also appreciate variation as the stuff of evolution—we wouldn't be here without variation.

In this chapter, we review the most common statistics to measure the location and spread of a frequency distribution and to calculate a proportion. We introduce the use of mathematical symbols to represent values of a variable, and we show formulas to calculate each summary statistic.

3.1 **Arithmetic mean and standard deviation**

The arithmetic mean is the most common metric to describe the location of a frequency distribution. It is the average of a set of measurements. The standard deviation is the most commonly used measure of distribution spread. Example 3.1 illustrates the basic calculations for means and standard deviations.

EXAMPLE Gliding snakes

3.1 When a paradise tree snake (*Chrysopelea paradisi*) flings itself from a treetop, it flattens its body everywhere except for the region around the heart. As it gains downward speed, the snake forms a tight horizontal S shape and then begins to undulate widely from side to side. This generates lift, causing the snake to glide away from the source tree. By orienting the head and anterior part of the body, the snake can change direction during a glide to avoid trees, reach a preferred landing site, and even chase aerial

prey. To better understand how lift is generated, Socha (2002) videotaped the glides of eight snakes leaping from a 10-m tower.[1] Among the measurements taken was the rate of side-to-side undulation on each snake. Undulation rates of the eight snakes, measured in hertz (cycles per second), were as follows:

0.9, 1.4, 1.2, 1.2, 1.3, 2.0, 1.4, 1.6

---

1. See films of these snakes flying at http://www.flyingsnake.org/video/video.html.

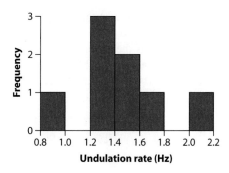

FIGURE 3.1-1
A histogram of the undulation rate of gliding paradise tree snakes. $n = 8$ snakes.

A histogram of these data is shown in Figure 3.1-1. The frequency distribution has a single peak between 1.2 and 1.4 Hz.

The sample mean

The **sample mean** is the average of the measurements in the sample, the sum of all the observations divided by the number of observations. To show its calculation, we use the symbol Y to refer to the variable and Y_i to represent the measurement of individual i. For the gliding snake data, i takes on values between 1 and 8, because there are eight snakes. Thus, $Y_1 = 0.9$, $Y_2 = 1.4$, $Y_3 = 1.2$, $Y_4 = 1.2$, and so on.[2]
The sample mean, symbolized as $\overline{Y}$ (and pronounced "Y-bar"), is calculated as

$$\overline{Y} = \frac{\sum_{i=1}^{n} Y_i}{n},$$

where n is the number of observations. The symbol Σ (uppercase Greek letter sigma) indicates a sum. The "$i = 1$" under the Σ and the "n" over it indicate that we are summing over all values of i between 1 and n, inclusive:

$$\sum_{i=1}^{n} Y_i = Y_1 + Y_2 + Y_3 + \cdots + Y_n.$$

When it is clear that i refers to individuals $1, 2, 3, \ldots, n$, the formula is often written more succinctly as

$$\overline{Y} = \frac{\sum Y_i}{n}.$$

Applying this formula to the snake data yields the mean undulation rate:

$$\overline{Y} = \frac{0.9 + 1.2 + 1.2 + 2.0 + 1.6 + 1.3 + 1.4 + 1.4}{8} = 1.375 \text{ Hz}.$$

---

2. We have adopted the simple convention of using uppercase letters (e.g., Y) when referring to both variable names and data, and prefer to distinguish the two by context. This is a departure from mathematical convention, which reserves uppercase exclusively for random variables.

Based on the histogram in Figure 3.1-1, we see that the value of the sample mean is close to the middle of the distribution. Note that the sample mean has the same units as the observations used to calculate it. In Section 3.6, we review how the sample mean is affected when the units of the observations are changed, such as by adding a constant or multiplying by a constant.

> The *sample mean* is the sum of all the observations in a sample divided by n, the number of observations.

Variance and standard deviation

The **standard deviation** is a commonly used measure of the spread of a distribution. It measures how far from the mean the observations typically are. The standard deviation is large if most observations are far from the mean, and it is small if most measurements lie close to the mean.

The standard deviation is calculated from the **variance**, another measure of spread. The standard deviation is simply the square root of the variance. The standard deviation is a more intuitive measure of the spread of a distribution (in part because it has the same units as the variable itself), but the variance has mathematical properties that make it useful sometimes as well. The standard deviation from a sample is usually represented by the symbol s, and the sample variance is written as s^2.

To calculate the variance from a sample of data, we must first compute the deviations. A deviation from the mean is the difference between a measurement and the mean $(Y_i - \overline{Y})$. Deviations for the measurements of snake undulation rate are listed in Table 3.1-1.

The best measure of the spread of this distribution isn't just the average of the deviations $(Y_i - \overline{Y})$, because this average is always zero (the negative deviations

TABLE 3.1-1   Quantities needed to calculate the standard deviation and variance of snake undulation rate ($\overline{Y} = 1.375$ Hz).

Observations (Y_i)	Deviations ($Y_i - \overline{Y}$)	Squared deviations ($Y_i - \overline{Y})^2$
0.9	−0.475	0.225625
1.2	−0.175	0.030625
1.2	−0.175	0.030625
1.3	−0.075	0.005625
1.4	0.025	0.000625
1.4	0.025	0.000625
1.6	0.225	0.050625
2.0	0.625	0.390625
Sum	0.000	0.735

cancel the positive deviations). Instead, we need to average the *squared* deviations (the third column in Table 3.1-1) to find the variance:

$$s^2 = \frac{\sum_{i=1}^{n}(Y_i - \overline{Y})^2}{n - 1}.$$

By squaring each number, deviations above and below the mean contribute equally[3] to the variance. The summation in the numerator (top part) of the formula, $\sum(Y_i - \overline{Y})^2$, is called the **sum of squares** of Y. Note that the denominator (bottom part) is $n - 1$ instead of n, the total number of observations. Dividing by $n - 1$ gives a more accurate estimate of the population variance.[4] We provide a shortcut formula for the variance in the Quick Formula Summary (Section 3.7).

For the snake undulation data, the variance (rounded to hundredths) is

$$s^2 = \frac{0.735}{7} = 0.11 \ \text{Hz}^2.$$

The variance has units equal to the square of the units of the original data. To obtain the standard deviation, we take the square root of the variance:

$$s = \sqrt{\frac{\sum(Y_i - \overline{Y})^2}{n - 1}}.$$

For the snake undulation data,

$$s = \sqrt{\frac{0.735}{7}} = 0.324037 \ \text{Hz}.$$

The standard deviation is never negative and has the same units as the observations from which it was calculated.

The *standard deviation* is a common measure of the spread of a distribution. It indicates just how different measurements typically are from the mean.

The standard deviation has a straightforward connection to the frequency distribution. If the frequency distribution is bell shaped, like the example in Figure 2.2-4, then about two-thirds of the observations will lie within one standard deviation of the mean, and about 95% will lie within two standard deviations. In other words, about 67% of the data will fall between $\overline{Y} - s$ and $\overline{Y} + s$, and about 95% will fall between $\overline{Y} - 2s$ and $\overline{Y} + 2s$. For an in-depth discussion of standard deviation, see Chapter 10.

---

3. We could have averaged the absolute values of the deviations instead, to yield the mean absolute deviation. Averaging the square of the deviations is more common because the result, the variance, has many more useful mathematical properties.

4. The reason is that the sample mean is itself calculated using each data point. Therefore, the measurements in the sample are slightly closer on average to the sample mean than they are to the true population mean. This causes a bias that is corrected by dividing by $n-1$ instead of by n.

This straightforward connection between the standard deviation and the frequency distribution diminishes when the frequency distribution deviates from the bell-shaped (normal) distribution. In such cases, the standard deviation is less informative about where the data lie in relation to the mean. This point is explored in greater detail in Section 3.3.

Rounding means, standard deviations, and other quantities

To avoid rounding errors when carrying out calculations of means, standard deviations, and other descriptive statistics, always retain as many significant digits as your calculator or computer can provide. Intermediate results written down on a page should also retain as many digits as feasible. Final results, however, should be rounded before being presented.

There are no strict rules on the number of significant digits that should be retained when rounding. A common strategy, which we adopt here, is to round descriptive statistics to one decimal place more than the measurements themselves. For example, the undulation rates in snakes were measured to a single decimal place (tenths). We therefore present descriptive statistics with two decimals (hundredths). The mean rate of undulation for the eight snakes, calculated as 1.375 Hz, would be communicated as

$$\overline{Y} = 1.38 \text{ Hz.}$$

Similarly, the standard deviation, calculated as 0.324037 Hz, would be reported as

$$s = 0.32 \text{ Hz.}$$

Note that even though we report the rounded value of the mean as $\overline{Y} = 1.38$, we used the more exact value, $\overline{Y} = 1.375$, in the calculation of s to avoid rounding errors.

Coefficient of variation

For many traits, standard deviation and mean change together when organisms of different sizes are compared. Elephants have greater mass than mice and also more variability in mass. For many purposes, we care more about the relative variation among individuals. A gain of 10 g for an elephant is inconsequential, but it would double the mass of a mouse. On the other hand, an elephant that is 10% larger than the elephant mean may have something in common with a mouse that is 10% larger than the mouse mean. For these reasons, it is sometimes useful to express the standard deviation relative to the mean. The **coefficient of variation** (CV) calculates the standard deviation as a percentage of the mean:

$$\text{CV} = \frac{s}{\overline{Y}} \times 100\%.$$

A higher CV means that there is more variability, whereas a lower CV means that individuals are more consistently the same. For the snake undulation data, the coefficient of variation is

$$CV = \frac{0.324}{1.375} \ 100\% = 24\%.$$

The coefficient of variation makes sense only when all of the measurements are greater than or equal to zero.

> The *coefficient of variation* is the standard deviation expressed as a percentage of the mean.

The coefficient of variation can also be used to compare the variability of traits that do not have the same units. If we wanted to ask, "What is more variable in elephants, body mass or life span?" then the standard deviation is not very informative, because mass is measured in kilograms and life span is measured in years. The coefficient of variation would allow us to make this comparison.

Calculating mean and standard deviation from a frequency table

Sometimes the data include many tied observations and are given in a frequency table. The frequency table in Table 3.1-2, for example, lists the number of criminal

TABLE 3.1-2 Number of criminal convictions of a cohort of 395 boys.

Number of convictions	Frequency
0	265
1	49
2	21
3	19
4	10
5	10
6	2
7	2
8	4
9	2
10	1
11	4
12	3
13	1
14	2
Total	395

convictions of a cohort of 395 boys (Farrington 1994; see Assignment Problem 22 in Chapter 2).

To calculate the mean and standard deviation of the number of convictions, notice first that the sample size is *not* 15, the number of rows in Table 3.1-2, but 395, the frequency total:

$$n = 265 + 49 + 21 + 19 + \cdots + 2 = 395.$$

Calculating the mean thus requires that the measurement of "0" be represented 265 times, the number "1" be represented 49 times, and so on. The sum of the measurements is thus

$$\sum Y_i = (265 \times 0) + (49 \times 1) + (21 \times 2) + (19 \times 3)$$
$$+ \cdots + (2 \times 14) = 445.$$

The mean of these data is then

$$\overline{Y} = \frac{445}{395} = 1.126582,$$

which we round to $\overline{Y} = 1.1$ when presenting the results.

The calculation of standard deviation must also take into account the number of individuals with each value. The sum of the squared deviations is

$$\sum (Y_i - \overline{Y})^2 = 265(0 - \overline{Y})^2 + 49(1 - \overline{Y})^2 + 21(2 - \overline{Y})^2 + \cdots$$

$$+ 2(14 - \overline{Y})^2 = 2377.671.$$

The standard deviation for these data is therefore

$$s = \sqrt{\frac{2377.671}{395 - 1}} = 2.4566,$$

which we present as $s = 2.5$.

These calculations assume that all the data are presented in the table. This approach would not work, however, for frequency tables in which the data are grouped into intervals, such as Table 2.2-3.

Effect of changing measurement scale

Results may need to be converted to a different scale than the one in which they were originally measured. For example, if temperature measurements were made in °F, it may be necessary to convert results to °C. The snake data were measured in hertz (cycles per second), but in some cases hertz must be converted to angular velocity (radians per second) instead. The good news is that we don't need to start over by converting the raw data. Instead, we can convert the descriptive statistics directly, as follows.

Briefly, here are the rules (we summarize them in the Quick Formula Summary at the end of this chapter). If converting data to a new scale, Y', involves multiplying the data, Y, by a constant, c,

$$Y' = cY,$$

then multiply the original mean $\overline{Y}$ by the same constant to obtain the new mean, and multiply the original standard deviation s by the absolute value of c to get the new standard deviation:

$$\overline{Y}' = c\overline{Y}$$
$$s' = |c|s.$$

However, the variance s^2 is converted by multiplying by c^2:

$$s'^2 = c^2 s^2.$$

If converting data to a new scale, Y', involves *adding* a constant, c, then the mean is converted by adding the same constant,

$$\overline{Y}' = \overline{Y} + c,$$

whereas the standard deviation and variance are unchanged:

$$s' = s$$
$$s'^2 = s^2.$$

This makes sense. Adding a constant to the data changes the location of the frequency distribution by the same amount but does not alter its spread.

For example, converting degrees Fahrenheit to degrees Celsius uses the transformation

$$°C = (5/9)°F - 17.8.$$

Therefore, if the mean temperature in a data set is $\overline{Y} = 80°F$, with a standard deviation of $s = 3°F$, then the new mean temperature is

$$\overline{Y}' = (5/9)80 - 17.8 = 26.6°C$$

and the new standard deviation is

$$s' = (5/9)3 = 1.7°C.$$

The new variance is

$$s'^2 = (5/9)^2(3)^2 = 2.8°C^2.$$

3.2 Median and interquartile range

After the sample mean, the *median* is the next most common metric used to describe the location of a frequency distribution. As we showed in Chapter 2, the median is often displayed in a box plot alongside the span between the first and third quartiles, or *interquartile range,* another measure of the spread of the distribution. We define and demonstrate these concepts with the help of Example 3.2.

EXAMPLE
3.2

I'd give my right arm for a female

Male spiders in the genus *Tidarren* are tiny, weighing only about 1% as much as females. They also have disproportionately large pedipalps, copulatory organs that make up about 10% of a male's mass. (See the adjacent photo; the pedipalps are indicated by arrows.) Males load the pedipalps with sperm and then search for females to inseminate. Astonishingly, male *Tidarren* spiders voluntarily amputate one of their two organs, right

or left, just before sexual maturity. Why do they do this? Perhaps speed is important to males searching for females, and amputation increases running performance. To test this hypothesis, Ramos et al. (2004) used video to measure the running speed of males on strands of spider silk. The data are presented in Table 3.2-1.

TABLE 3.2-1 Running speed (cm/s) of male *Tidarren* spiders before and after voluntary amputation of a pedipalp.

Spider	Speed before	Speed after	Spider	Speed before	Speed after
1	1.25	2.40	9	2.98	3.70
2	2.94	3.50	10	3.55	4.70
3	2.38	4.49	11	2.84	4.94
4	3.09	3.17	12	1.64	5.06
5	3.41	5.26	13	3.22	3.22
6	3.00	3.22	14	2.87	3.52
7	2.31	2.32	15	2.37	5.45
8	2.93	3.31	16	1.91	3.40

The median

The **median** is the middle observation in a set of data, the measurement that partitions the ordered measurements into two halves. To calculate the median, first sort the sample observations from smallest to largest. The sorted measurements of running speed of male spiders before amputation (Table 3.2-1) are 1.25, 1.64, 1.91, 2.31, 2.37, 2.38, 2.84, 2.87, 2.93, 2.94, 2.98, 3.00, 3.09, 3.22, 3.41, 3.55 in cm/s. Let $Y_{(i)}$ refer to the *i*th sorted observation, so $Y_{(1)}$ is 1.25, $Y_{(2)}$ is 1.64, $Y_{(3)}$ is 1.91, and so on. If the number of observations (*n*) is odd, then the median is the middle observation:

$$\text{Median} = Y_{([n+1]/2)}.$$

If the number of observations is even, as in the spider data, then the median is the average of the middle pair:

$$\text{Median} = [Y_{(n/2)} + Y_{(n/2+1)}]/2.$$

Thus, $n/2 = 8$, $Y_{(8)} = 2.87$, and $Y_{(9)} = 2.93$ for the spider data (before amputation). The median is the average of these two numbers:

$$\text{Median} = (2.87 + 2.93)/2 = 2.90 \text{ cm/s.}$$

> The *median* is the middle measurement of a set of observations.

The interquartile range

Quartiles are values that partition the data into quarters. The first quartile is the middle value of the measurements lying below the median. The second quartile is the median. The third quartile is the middle value of the measurements larger than the median. The **interquartile range** (*IQR*) is the span of the middle half of the data, from the first quartile to the third quartile:

$$\text{Interquartile range} = \text{third quartile} - \text{first quartile.}$$

Figure 3.2-1 shows the meaning of the median, first quartile, third quartile, and interquartile range for the spider data set (before amputation).

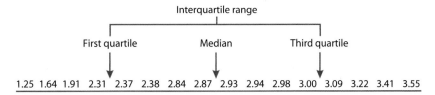

FIGURE 3.2-1 The first quartile, median, and third quartile break the data set into four equal portions. The median is the middle value, and the first and third quartiles are the middles of the first and second halves of the data. The interquartile range is the span of the middle half of the data.

The first step in calculating the interquartile range is to compute the first and third quartiles, as follows.[5]
For the first quartile, calculate

$$j = 0.25n,$$

---

5. Don't be surprised if your computer program gives slightly different values from ours for the quartiles and the interquartile range. The method given here is simple to calculate, but it does not give the most accurate estimates of the population quantities. Several improved methods are available (Hyndman and Fan 1996).

where n is the number of observations. If j is an integer then the first quartile is the average of $Y_{(j)}$ and $Y_{(j+1)}$:

$$\text{First quartile} = (Y_{(j)} + Y_{(j+1)})/2,$$

where $Y_{(j)}$ is the jth sorted observation. For the sorted spider data,

$$j = (0.25)(16) = 4,$$

which is an integer. Therefore, the first quartile is the average of $Y_{(4)}$ and $Y_{(5)}$:

$$\text{First quartile} = (2.31 + 2.37)/2 = 2.34.$$

If j is not an integer, then convert j to an integer by replacing it with the next integer that exceeds it (i.e., round j up to the nearest integer). The first quartile is then

$$\text{First quartile} = Y_{(j)},$$

where j is now the integer you rounded to.

The third quartile is computed similarly. Calculate

$$k = 0.75n.$$

If k is an integer, then the third quartile is the average of $Y_{(k)}$ and $Y_{(k+1)}$:

$$\text{Third quartile} = (Y_{(k)} + Y_{(k+1)})/2,$$

where $Y_{(k)}$ is the kth sorted observation. For the sorted spider data,

$$j = (0.75)(16) = 12,$$

which is an integer. Therefore, the third quartile is the average of $Y_{(12)}$ and $Y_{(13)}$:

$$\text{Third quartile} = (3.00 + 3.09)/2 = 3.045.$$

If k is not an integer, then convert k to an integer by replacing it with the next integer that exceeds it (i.e., round k up to the nearest integer). The third quartile is then

$$\text{Third quartile} = Y_{(k)},$$

where k is the integer you rounded to.

The interquartile range is then

$$\text{Interquartile range} = 3.045 - 2.34 = 0.705 \text{ cm/s}.$$

The *interquartile range* is the difference between the third and first quartiles of the data. It is the span of the middle 50% of the data.

The box plot

A **box plot** displays the median and interquartile range, along with other quantities of the frequency distribution. We introduced the box plot in Chapter 2. Figure 3.2-2 shows a box plot for the spider running speeds, with data before and after amputation plotted separately. The lower and upper edges of the box are the first and third quartiles. Thus, the interquartile range is visualized by the span of the box. The horizontal line dividing each box is the median. The whiskers extend outward from the box at each end, stopping at the smallest and largest "non-extreme" values in the data. "Extreme" values are defined as those lying farther from the box edge than 1.5 times the interquartile range. Extreme values are plotted as isolated dots past the ends of the whiskers.[6] There is one extreme value in the box plots shown in Figure 3.2-2, the smallest measurement for running speed before amputation.

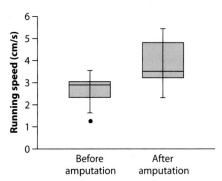

FIGURE 3.2-2
Box plot of the running speeds of 16 male spiders before and after self-amputation of a pedipalp.

3.3 How measures of location and spread compare

Which measure of location, the sample mean or the median, is most revealing about the center of a distribution of measurements? And which measure of spread, the standard deviation or the interquartile range, best describes how widely the observations are scattered about the center? The answer depends on the shape of the frequency distribution. These alternative measures of location and of spread yield similar information when the frequency distribution is symmetric and unimodal. The mean and standard deviation become less informative than the median and interquartile range when the data are strongly skewed or include extreme observations. We compare these measures using Example 3.3.

EXAMPLE Disarming fish

3.3 The marine threespine stickleback is a small coastal fish named for its defensive armor. It has three sharp spines down its back, two pelvic spines under the belly, and a series

---

6. Some computer programs extend whiskers all the way to the most extreme values on each end and do not indicate extreme values with isolated dots. There is no universally agreed-upon method for drawing whiskers.

of lateral bony plates down both sides. The armor seems to reduce mortality from predatory fish and diving birds. In contrast, in lakes and streams, where predators are fewer, stickleback populations have reduced armor. (See the photo at the right for examples of different types. Bony tissue has been stained red to make it more visible.) Colosimo et al. (2004) measured the grandchildren of a cross made between a marine and a freshwater stickleback. The study found that much of the difference in number of plates is caused by a single gene, *Ectodysplasin*.[7] Fish inheriting two copies of the gene from the marine grandparent, called *MM* fish, had many plates (the top histogram in Figure 3.3-1). Fish inheriting both copies of the gene from the freshwater grandparent (*mm*) had few plates (the bottom histogram in Figure 3.3-1). Fish having one copy from each grandparent (*Mm*) had any of a wide range of plate numbers (the middle histogram in Figure 3.3-1).

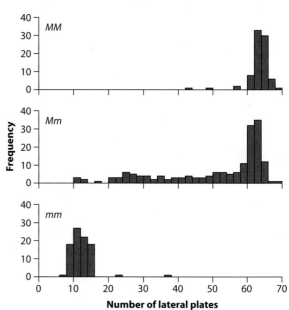

FIGURE 3.3-1
Frequency distributions of lateral plate number in three genotypes of stickleback, *MM, Mm,* and *mm*, descended from a cross between marine and freshwater grandparents. Plates are counted as the total number down the left and right sides of the fish. The total number of fish: 82 (*MM*), 174 (*Mm*), and 88 (*mm*).

Mean versus median

The mean and median of the three distributions in Figure 3.3-1 are compared in Table 3.3-1. The two measures of location give similar values in the case of the *MM* and *mm* genotypes, whose distributions are fairly symmetric, although one or two outliers are present. The mean is smaller than the median in the case of the *Mm* fish, whose distribution is strongly asymmetric.

---

7. Mutations in the same gene in humans cause the loss of hair, teeth, and sweat glands.

TABLE 3.3-1 Descriptive statistics for the number of lateral plates of the three genotypes of threespine sticklebacks[8] discussed in Example 3.3.

Genotype	n	Mean	Median	Standard deviation	Interquartile range
MM	82	62.8	63	3.4	2
Mm	174	50.4	59	15.1	21
mm	88	11.7	11	3.6	3

Why are the median and mean different from one another when the distribution is asymmetric? The answer, shown in Figure 3.3-2, is that the median is the middle measurement of a distribution, whereas the mean is the "center of gravity."

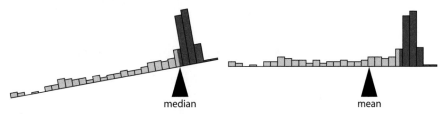

FIGURE 3.3-2 Comparison between the median and the mean using the frequency distribution for the *Mm* genotype (middle panel of Figure 3.3-1). The median is the middle measurement of the distribution (different colors represent the two halves of the distribution). The mean is the center of gravity, the point at which the frequency distribution would be balanced (if observations had weight).

The balancing act illustrated in Figure 3.3-2 suggests that the mean is sensitive to extreme observations. To demonstrate, imagine taking the four smallest observations of the *MM* genotype (top panel in Figure 3.3-1) and moving them far to the left. The median would be completely unaffected, but the mean would shift leftward to a point near the edge of the range of most observations (Figure 3.3-3).

FIGURE 3.3-3
Sensitivity of the mean to extreme observations using the frequency distribution of the *MM* genotypes (see the upper panel in Figure 3.3-1). The two different colors represent the two halves of the distribution. When the four smallest observations of the *MM* genotype are shifted far to the left (*lower panel*), the mean is displaced downward, to the edge of the range of the bulk of the observations. The median, on the other hand, which is located where the two colors meet, is unaffected by the shift.

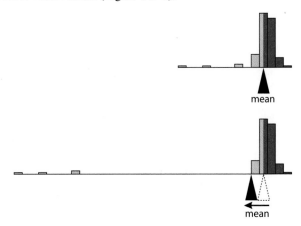

8. When listing descriptive statistics in tables, put the same quantity calculated on different groups into one column. Numbers stacked in a single column are easier to compare than numbers placed side by side in a row. Exchange the columns and rows in Table 3.3-1, and you'll see what we mean.

> Median and mean measure different aspects of the location of a distribution. The *median* is the middle value of the data, whereas the *mean* is its center of gravity.

Thus, the mean is displaced from the location of the "typical" measurement when the frequency distribution is strongly skewed, particularly when there are extreme observations. The mean is still useful as a description of the data as a whole, but it no longer indicates where most of the observations are located. The median is less sensitive to extreme observations, and hence the median is the more informative descriptor of the typical observation in such instances. However, the mean has better mathematical properties, and it is easier to calculate measures of the reliability of estimates of the mean.

Standard deviation versus interquartile range

Because it is calculated from the square of the deviations, the standard deviation is even more sensitive to extreme observations than is the mean. When the four smallest observations of the *MM* genotype are shifted far to the left, such that the smallest is set to zero (Figure 3.3-3), the standard deviation jumps from 3.4 to 12.0, whereas the interquartile range is not affected. For this reason, the interquartile range is a better indicator of the spread of the main part of a distribution than the standard deviation when the data are strongly skewed to one side or the other, especially when there are extreme observations. On the other hand, the standard deviation reflects the variation among all of the data points.

◼◼◼ 3.4 Cumulative frequency distribution

The median and quartiles are examples of percentiles, or quantiles, of the frequency distribution for a numerical variable. Plotting all the quantiles using the cumulative frequency distribution is another way to compare the shapes and positions of two or more frequency distributions.

Percentiles and quantiles

The Xth **percentile** of a sample is the value below which X percent of the individuals lie. For example, the median, the measurement that splits a frequency distribution into equal halves, is the 50th percentile. Ten percent of the observations lie below the 10th percentile, and the other 90% of observations exceed it. The first and third quartiles are the 25th and 75th percentiles, respectively.

The same information in a percentile is sometimes represented as a **quantile**. This only means that the proportion less than or equal to the given value is represented as a decimal rather than as a percentage. For example, the 10th percentile is the 0.10

quantile, and the median is the 0.50 quantile. Be careful not to mix up the words *quantile* and *quartile* (note the difference in the fourth letters). The first and third quartiles are the 0.25 and 0.75 quantiles.

> The *percentile* of a measurement specifies the percentage of observations less than or equal to it; the remaining observations exceed it. The *quantile* of a measurement specifies the fraction of observations less than or equal to it.

Displaying cumulative relative frequencies

All the quantiles of a numerical variable can be displayed by graphing the **cumulative frequency distribution**.

Figure 3.4-1 shows the cumulative frequency distribution of the running speeds of male spiders before amputation. The raw data are from Table 3.2-1. To make this graph, all the measurements of running speed (before amputation) were sorted from smallest to largest. Next, the fraction of observations less than or equal to each data value was calculated. This fraction, which is called the cumulative relative frequency, is indicated by the height of the curve in Figure 3.4-1 at the corresponding data value. Finally, these points were connected with straight lines to form an ascending curve. The result is an irregular sequence of "steps" from the smallest data value to the largest data value. Each step is flat, but the curve jumps up by $1/n$ at every observed measurement, where n is the total number of observations (here, 16 spiders), to a maximum of 1. There may be multiple jumps at one measurement if multiple data points have the same measurement.

> *Cumulative relative frequency* at a given measurement is the fraction of observations less than or equal to that measurement.

FIGURE 3.4-1
The cumulative frequency distribution of male spiders before amputation (solid curve). Horizontal dotted lines indicate the cumulative relative frequencies 0.25 (*lower*) and 0.75 (*upper*); vertical lines indicate corresponding 0.25 and 0.75 quantiles of running speed (2.34 and 3.045). The data are from Table 3.2-1. $n = 16$ spiders.

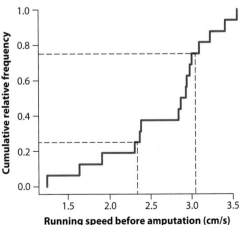

The curve in Figure 3.4-1 shows a lot of information because all the data points are represented. We can see that one-fourth of the observations (corresponding to a cumulative relative frequency of 0.25) had running speeds below 2.34, which is the value of the first quartile calculated earlier. Three-fourths of all observations lie below 3.045, which is the value of the third quartile calculated earlier. Both these values are indicated in Figure 3.4-1 with the dashed lines.

Because of their simplicity and ease of interpretation, the histogram and box plot are usually superior to the cumulative frequency distribution for showing the data. However, with practice, the cumulative frequency distribution can be very useful, especially to compare frequency distributions of multiple groups.

3.5 Proportions

The proportion is the most important descriptive statistic for a categorical variable.

Calculating a proportion

The **proportion** of observations in a given category, symbolized $\hat{p}$, is calculated as

$$\hat{p} = \frac{\text{Number in category}}{n},$$

where the numerator is the number of observations in the category of interest, and n is the total number of observations in all categories combined.[9]

For example, of the 344 individual sticklebacks in Example 3.3, 82 had genotype *MM*, 88 were *mm*, and 174 were *Mm* (Table 3.3-1). The proportion of *MM* fish is

$$\hat{p} = \frac{82}{344} = 0.238.$$

The other proportions are calculated similarly, and all three proportions are listed in Table 3.5-1.

TABLE 3.5-1 The number of fish of each genotype from a cross between a marine stickleback and a freshwater stickleback (Example 3.3). As written, the sum of the proportions does not add precisely to one because of rounding.

Genotype	Frequency	Proportion
MM	82	0.24
Mm	174	0.51
mm	88	0.26
Total	344	1.00

9. The "hat" in $\hat{p}$ is used to indicate an estimate of the true proportion p.

The proportion is like a sample mean

The proportion $\hat{p}$ has properties in common with the arithmetic mean. To see this, let's create a new numerical variable Y for the stickleback study. Give individual fish i the value $Y_i = 1$ if it has the *MM* genotype, and give it the value $Y_i = 0$ otherwise. The sum of all the ones and zeroes, ΣY_i, is the frequency of fish having genotype *MM*.

The mean of the ones and zeroes is

$$\overline{Y} = \frac{\Sigma Y_i}{n} = \frac{82}{344} = 0.238,$$

which is just $\hat{p}$, the proportion of observations in the first category. If we imagine the Y-measurements to have weight, then the proportion is their center of gravity (Figure 3.5-1).

FIGURE 3.5-1
The distribution of *Y*, where *Y* = 1 if a stickleback is genotype *MM* and 0 otherwise. The mean of *Y* is the proportion of *MM* individuals in the sample (0.238).

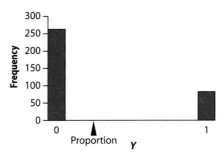

3.6 Summary

- The location of a distribution for a numerical variable can be measured by its mean or by its median. The mean gives the center of gravity of the distribution and is calculated as the sum of all measurements divided by the number of measurements. The median gives the middle value.
- The standard deviation measures the spread of a distribution for a numerical variable. It is a measure of the typical distance between observations and the mean. The variance is the square of the standard deviation.
- The quartiles break the ordered observations into four equal parts. The interquartile range, the difference between the first and third quartiles, is another measure of the spread of a frequency distribution.
- The mean and median yield similar information when the frequency distribution of the measurements is symmetric and unimodal. The mean and standard deviation become less informative about the location and spread of typical observations than the median and interquartile range when the data include extreme observations.

- The percentile of a measurement specifies the percentage of observations less than or equal to it. The quantile of a measurement specifies the fraction of observations less than or equal to it.
- All the quantiles of a sample of data can be shown using a graph of the cumulative frequency distribution.
- The proportion is the most important descriptive statistic for a categorical variable. It is calculated by dividing the number of observations in the category of interest by n, the total number of observations in all categories combined.

3.7 Quick Formula Summary

Table of formulas for descriptive statistics

Quantity	Formula
Sample size	n
Mean	$\bar{Y} = \dfrac{\Sigma Y}{n}$
Variance	$s^2 = \dfrac{\Sigma(Y_i - \bar{Y})^2}{n - 1}$
shortcut formula:	$s^2 = \dfrac{\Sigma(Y_i^2) - n\bar{Y}^2}{n - 1}$
Standard deviation	$s = \sqrt{\dfrac{\Sigma(Y_i - \bar{Y})^2}{n - 1}}$
shortcut formula:	$s = \sqrt{\dfrac{\Sigma(Y_i^2) - n\bar{Y}^2}{n - 1}}$
Sum of squares	$\Sigma(Y_i - \bar{Y})^2 = \Sigma(Y_i^2) - n\bar{Y}^2$
Coefficient of variation	$\text{CV} = \dfrac{s}{\bar{Y}} \times 100\%$
Median	$Y_{([n+1]/2)}$ (if n is odd)
	$[Y_{(n/2)} + Y_{(n/2+1)}]/2$ (if n is even)
	where $Y_{(1)}, Y_{(2)}, \ldots, Y_{(n)}$ are the ordered observations
Proportion	$\hat{p} = \dfrac{\text{Number in category}}{n}$

Effect of arithmetic operations on descriptive statistics

The table below lists the effect on the descriptive statistics of adding or multiplying all the measurements by a constant. The rules listed in the table are useful when converting measurements from one system of units to another, such as English to metric or degrees Fahrenheit to degrees Celsius.

Statistic	Value	Adding a constant c to all the measurements, $Y' = Y + c$	Multiplying all the measurements by a constant c, $Y' = cY$
Mean	$\bar{Y}$	$\bar{Y}' = \bar{Y} + c$	$\bar{Y}' = c\bar{Y}$
Standard deviation	s	$s' = s$	$s' = \|c\|s$
Variance	s^2	$s'^2 = s^2$	$s'^2 = c^2s^2$
Median	M	$M' = M + c$	$M' = cM$
Interquartile range	IQR	$IQR' = IQR$	$IQR' = \|c\|IQR$

PRACTICE PROBLEMS

1. **Calculation practice: Basic descriptive stats.** Systolic blood pressure was measured (in units of mm Hg) during preventative health examinations on people in Dallas, Texas. Here are the measurements for a subset of these patients.

   112, 128, 108, 129, 125, 153, 155, 132, 137

   a. How many individuals are in the sample (i.e., what is the sample size, n)?
   b. What is the sum of all of the observations?
   c. What is the mean of this sample? *Here and forever after, provide units with your answer.*
   d. What is the sum of the squares of the measurements?
   e. What is the variance of this sample?
   f. What is the standard deviation of this sample?
   g. What is the coefficient of variation for this sample?

2. **Calculation practice: Box plots.** Here is another sample of systolic blood pressure (in units of mm Hg), this time with all 101 data points. The mean is 122.73 and the standard deviation is 13.83.

   88, 88, 92, 96, 96, 100, 102, 102, 104, 104, 105, 105, 105, 107, 107, 108, 110, 110, 110, 111, 111, 112, 113, 114, 114, 115, 115, 116, 116, 117, 117, 117, 117, 117, 117, 119, 119, 120,

   121, 121, 121, 121, 121, 121, 122, 122, 123, 123, 123, 123, 123, 124, 124, 124, 124, 125, 125, 125, 126, 126, 126, 126, 126, 127, 127, 128, 128, 128, 128, 129, 129, 130, 131, 131, 131, 131, 131, 131, 133, 133, 133, 134, 135, 136, 136, 136, 138, 138, 139, 139, 141, 142, 142, 142, 143, 144, 146, 146, 147, 155, 156

   a. What is the median of this sample?
   b. What is the upper (third) quartile (or 75th percentile)?
   c. What is the lower (first) quartile (or 25th percentile)?
   d. What is the interquartile range (IQR)?
   e. Calculate the upper quartile plus 1.5 times the IQR. Is this greater than the largest value in the data set?
   f. Calculate the lower quartile minus 1.5 times the IQR. Is this less than the smallest value in the data set?
   g. Plot the data in a box plot. (A rough sketch by hand is appropriate, as long as the correct values are shown for each critical point.)

3. A review of the performance of hospital gynecologists in two regions of England measured the outcomes of patient admissions under each doctor's care (Harley et al. 2005). One measurement taken was the percentage of patient admis-

sions made up of women under 25 years old who were sterilized. We are interested in describing what constitutes a typical rate of sterilization, so that the behavior of atypical doctors can be better scrutinized. The frequency distribution of this measurement for all doctors is plotted in the following graph.

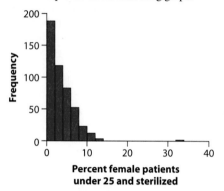

Percent female patients under 25 and sterilized

a. Explain what the vertical axis measures.

b. What would be the best choice to describe the location of this frequency distribution, the mean or the median, if our goal was to describe the typical individual? Why?

c. Do you see any evidence that might lead to further investigation of any of the doctors?

4. The data displayed in the plot below are from a nearly complete record of body masses of the world's native mammals (in grams, then converted to log base 10; Smith et al. 2003). The data were divided into three groups: those surviving from the last ice age to the present day ($n = 4061$), those who went extinct around the end of the last ice age ($n = 227$), and those driven extinct within the last 300 years (recent; $n = 44$).

a. What type of graph is this?

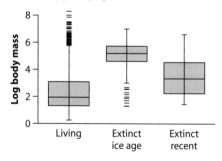

b. What does the horizontal line in the center of each rectangle represent?

c. What are the horizontal lines at the top and bottom edges of each rectangle supposed to represent?

d. What are the data points (indicated by "—") lying outside the rectangle?

e. What are the vertical lines extending above and below each rectangle?

f. Compare the locations of the three body-size distributions. How do they differ?

g. Compare the shapes of the three frequency distributions. Which are relatively symmetric and which are asymmetric? Explain your reasoning.

h. Which group's frequency distribution has the lowest spread? Explain your reasoning.

i. What has been the likely effect of ice-age and recent extinctions on the median body size of mammals?

5. Mehl et al. (2007) wired 396 volunteers with electronically activated recorders that allowed the researchers to calculate the number of words each individual spoke, on average, per 17-hour waking day. They found that the mean number of words spoken was only slightly higher for the 210 women (16,215 words) than for the 186 men (15,669) in the sample. The frequency distribution of the number of words spoken by all

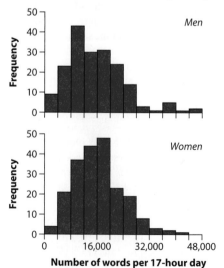

Number of words per 17-hour day

individuals of each sex is shown in the accompanying graphs (modified from Mehl et al. 2007).

a. What type of graph is shown?

b. What are the explanatory and response variables in the figure?

c. What is the mode of the frequency distribution of each sex?

d. Which sex likely has the higher median number of words spoken per day?

e. Which sex had the highest variance in number of words spoken per day?

6. The following data are measurements of body mass, in grams, of finches captured in mist nets during a survey of various habitats in Kenya, East Africa (Schluter 1988).

Crimson-rumped waxbill	8, 8, 8, 8, 8, 8, 8, 6, 7, 7, 7, 8, 8, 8, 7, 7, 7
Cutthroat finch	16, 16, 16, 12, 16, 15, 15, 17, 15, 16, 15, 16
White-browed sparrow weaver	40, 43, 37, 38, 43, 33, 35, 37, 36, 42, 36, 36, 39, 37, 34, 41

a. Calculate the mean body mass of each of these three finch species. Which species is largest, and which is smallest?

b. Which species has the greatest standard deviation in body mass? Which has the least?

c. Calculate the coefficient of variation (CV) in mass for each finch species. How different are the coefficients between the species? Compare the difference in CVs with the differences in standard deviation calculated in part (b).

d. The following measurements are of another trait, beak length, in mm, of the 16 white-browed sparrow weavers. Which measurement is more variable in this species (relative to the mean), body mass or beak length?

10.6, 10.8, 10.9, 11.3, 10.9, 10.1, 10.7, 10.7, 10.9, 11.4, 10.8, 11.2, 10.7, 10.0, 10.1, 10.7

7. The spider data in Example 3.2 consist of *pairs* of measurements made on the same subjects. One measurement is running speed before amputation and the second is running speed after amputation. Calculate a new variable called "change in speed," defined as the speed of each spider after amputation minus its speed before amputation.

a. What are the units of the new variable?

b. Draw a box plot for the change in running speed. Use the method outlined in Section 3.2 to calculate the quartiles.

c. Based on your drawing in part (b), is the frequency distribution of the change in running speed symmetric or asymmetric? Explain how you decided this.

d. What is the quantity measured by the span of the box in part (b)?

e. Calculate the mean change in running speed. Is it the same as the median? Why or why not?

f. Calculate the variance of the change in running speed.

g. What fraction of observations fall within one standard deviation above and below the mean?

8. Refer to the previous problem. If you were to convert all of the observations of change in running speed from cm/s into mm/s, how would this change

a. the mean?

b. the standard deviation?

c. the median?

d. the interquartile range?

e. the coefficient of variation?

f. the variance?

9. Niderkorn's (1872; from Pounder 1995) measurements on 114 human corpses provided the first quantitative study on the development of

rigor mortis.[10] The data in the following table give the number of bodies achieving rigor mortis in each hour after death, recorded in one-hour intervals.

Hours	Number of bodies
1	0
2	2
3	14
4	31
5	14
6	20
7	11
8	7
9	4
10	7
11	1
12	1
13	2
Total	114

a. Calculate the mean number of hours after death that it took for rigor mortis to set in.
b. Calculate the standard deviation in the number of hours until rigor mortis.
c. What fraction of observations lie within one standard deviation of the mean (i.e., between the value $\overline{Y} - s$ and the value $\overline{Y} + s$)?
d. Calculate the median number of hours until rigor mortis sets in. What is the likely explanation for the difference between the median and the mean?

10. The following graph shows the population growth rates of the 204 countries recognized by the United Nations. Growth rate is measured as the average annual percent change in the total human population between 2000 and 2004 (United Nations Statistics Division 2004).

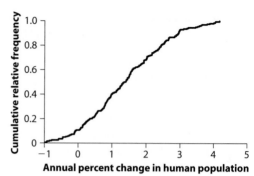

a. Identify the type of graph depicted.
b. Explain the quantity along the y-axis.
c. Approximately what percentage of countries had a negative change in population?
d. Identify by eye the 0.10, 0.50, and 0.90 quantiles of change in population size.
e. Identify by eye the 60th percentile of change in population size.

11. Refer to the previous problem.
a. Draw a box plot using the information provided in the graph in that problem.
b. Label three features of this box plot.

12. *Spot the flaw.* The accompanying table shows means and standard deviations for the length of migration on a microgel of 20 lymphocyte cells exposed to X-irradiation. The length of migration is an indication of DNA damage suffered by the cells. The data are from Singh et al. (1988).
a. Identify the main flaw in the construction of this table.
b. Redraw the table following the principles recommended in this chapter and Chapter 2.

13. The following graph illustrates an association between two variables. It shows percent changes in the range sizes of different species of native

TABLE FOR PROBLEM 12

X-ray dose	Control	25 rads	50 rads	100 rads	200 rads
Mean	3.70	5.27	12.37	23.30	29.80
Standard deviation	1.10	1.19	4.69	3.27	2.99

---

10. Rigor mortis is the muscular stiffening that occurs after death. It is caused by linkages forming between actin and myosin in muscle when muscle glycogen is depleted, pH drops, and the level of ATP falls below a critical level.

butterflies (red), birds (blue), and plants (black) of Britain over the past two to four decades (modified from Thomas et al. 2004). Identify (a) the type of graph, (b) the explanatory and response variables, and (c) the type of data (whether numerical or categorical) for each variable.

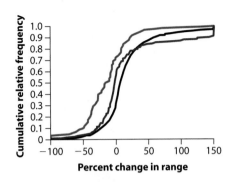

ASSIGNMENT PROBLEMS

14. The gene for the vasopressin receptor *V1a* is expressed at higher levels in the forebrain of monogamous vole species than in promiscuous vole species.[11] Can expression of this gene influence monogamy? To test this, Lim et al. (2004) experimentally enhanced *V1a* expression in the forebrain of 11 males of the meadow vole, a solitary promiscuous species. The percentage of time each male spent huddling with the female provided to him (an index of monogamy) was recorded. The same measurements were taken in 20 control males left untreated.

Control males: 98, 96, 94, 88, 86, 82, 77, 74, 70, 60, 59, 52, 50, 47, 40, 35, 29, 13, 6, 5

V1a-enhanced males: 100, 97, 96, 97, 93, 89, 88, 84, 77, 67, 61

  a. Display these data in a graph. Explain your choice of graph.

  b. Which group has the higher mean percentage of time spent huddling with females?

  c. Which group has the higher standard deviation in percentage of time spent huddling with females?

15. The data in the accompanying table are from an ecological study of the entire rainforest community at El Verde in Puerto Rico (Waide and Reagan 1996). Diet breadth is the number of types of food eaten by an animal species. The number of animal species having each diet breadth is shown in the second column. The total number of species listed is $n = 127$.

Diet breadth (number of prey types eaten)	Frequency (number of species)
1	21
2	8
3	9
4	10
5	8
6	3
7	4
8	8
9	4
10	4
11	4
12	2
13	5
14	2
15	1
16	1
17	2
18	1
19	3
20	2
>20	25
Total	127

  a. Calculate the median number of prey types consumed by animal species in the community.

---

11. In monogamous vole species, single males and females form stable mating pairs. In promiscuous voles, no stable pairs form and voles might mate with multiple partners.

b. What is the interquartile range in the number of prey types? Use the method outlined in Section 3.2 to calculate the quartiles.

c. Can you calculate the mean number of prey types in the diet? Explain.

16. Francis Galton (1894) presented the following data on the flight speeds of 3207 "old" homing pigeons traveling at least 90 miles.

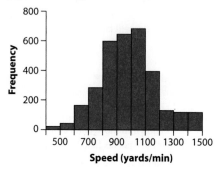

a. What type of graph is this?

b. Examine the graph and visually determine the approximate value of the mean (to the nearest 100 yards per minute). Explain how you obtained your estimate.

c. Examine the graph and visually determine the approximate value of the median (to the nearest 100 yards per minute). Explain how you obtained your estimate.

d. Examine the graph and visually determine the approximate value of the mode (to the nearest 100 yards per minute). Explain how you obtained your estimate.

e. Examine the graph and visually determine the approximate value of the standard deviation (to the nearest 50 yards per minute). Explain how you obtained your estimate.

17. A study of the endangered saiga antelope (pictured at the beginning of the chapter) recorded the fraction of females in the population that were fecund in each year between 1993 and 2001 (Milner-Gulland et al. 2003). A graph of the data is as follows:

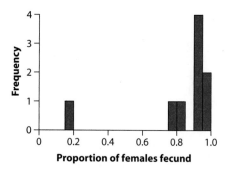

a. Assume that you want to describe the "typical" fraction of females that are fecund in a typical year, based on these data. What would be the better choice to describe this typical fraction, the mean or the median of the measurements? Why?

b. With the same goal in mind, what would be the better choice to describe the spread of measurements around their center, the standard deviation or the interquartile range? Why?

18. Accurate prediction of the timing of death in patients with a terminal illness is important for their care. The following graph compares the survival times of terminally ill cancer patients with the clinical prediction of their survival times (modified from Glare et al. 2003).

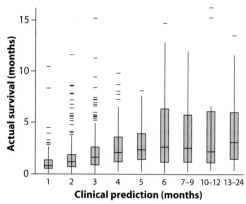

a. Describe in words what features most of the frequency distributions of actual survival times have in common, based on the box plots for each group.

b. Describe the differences in shape of actual survival time distributions between those for one to five months predicted survival times and those for six to 24 months.

c. Describe the trend in median actual survival time with increasing predicted number.

d. The predicted survival times of terminally ill cancer patients tend to overestimate the medians of actual survival times. Are the *means* of actual survival times likely to be closer to, further from, or no different from the predicted times than the medians? Explain.

19. Measurements of lifetime reproductive success (LRS) of individual wild animals reveal the disparate contributions they make to the next generation. Jensen et al. (2004) estimated LRS of male and female house sparrows in an island population in Norway. They measured LRS of an individual as the total number of "recruits" produced in its lifetime, where a recruit is an offspring that survives to breed one year after birth. Parentage of recruits was determined from blood samples using DNA techniques. Their results are tabulated as follows:

Lifetime reproductive success	Frequency	
	Females	Males
0	30	38
1	25	17
2	3	7
3	6	6
4	8	4
5	4	10
6	0	2
7	4	0
8	1	0
>8	0	0
Total	81	84

a. Which sex has the higher mean lifetime reproductive success?

b. Every recruit must have both a father and a mother, so it is not easy to see why male and female LRS should differ. Can you think of a biological explanation?

c. Which sex has the higher variance in reproductive success?

20. If all the measurements in a sample of data are equal, what is the variance of the measurements in the sample?

21. Researchers have created every possible "knockout" line in yeast. Each line has exactly one gene deleted and all the other genes present (Steinmetz et al. 2002). The growth rate—how fast the number of cells increases per hour—of each of these yeast lines has also been measured, expressed as a multiple of the growth rate of the wild type that has all the genes present. In other words, a growth rate greater than 1 means that a given knockout line grows faster than the wild type, whereas a growth rate less than 1 means it grows more slowly. Below is the growth rate of a random sample of knockout lines:

0.86, 1.02, 1.02, 1.01, 1.02, 1, 0.99, 1.01, 0.91, 0.83, 1.01

a. What is the mean growth rate of this sample of yeast lines?

b. What is the median growth rate of this sample?

c. What is the variance of growth rate of the sample?

d. What is the standard deviation of growth rate of the sample?

22. As in other vertebrates, individual zebrafish differ from one another along the shy–bold behavioral spectrum. In addition to other differences, bolder individuals tend to be more aggressive, whereas shy individuals tend to be less aggressive. Norton et al. (2011) compared several behaviors associated with this syndrome between zebrafish that had the *spiegeldanio* (*spd*) mutant at the *Fgfr1a* gene (reduced fibroblast growth factor receptor 1a) and the "wild type" lacking the mutation. The data below are measurements of the amount of time, in seconds, that individual zebrafish with and without this mutation spent in aggressive activity over 5 minutes when presented with a mirror image.

Wild type: 0, 21, 22, 28, 60, 80, 99, 101, 106, 129, 168

Spd **mutant:** 96, 97, 100, 127, 128, 156, 162, 170, 190, 195

a. Draw a boxplot to compare the frequency distributions of aggression score in the two groups of zebrafish. According to the box plot, which genotype has the higher aggression scores?

b. According to the box plot, which sample spans the higher range of values for aggression scores?

c. Which sample has the larger interquartile range?

d. What are the vertical lines projecting outward above and below each box?

23. A eunuch is a castrated human male. Eunuchs were often used as servants and guards in harems in Asia and the Middle East. In males of some mammal species, castration increases life span. Do male eunuchs also have long lives compared to other men? The accompanying graph shows data on life spans of 81 male eunuchs from the Korean Chosun Dynasty between about 1400 and 1900, according to historical records. These data are compared with life spans of non-eunuch males who lived at the same time, and who belonged to families of sim-

ilar social status ($n = 1126, 1414,$ and 49 for the three families shown). Modified from Min et al. (2012), with permission.

a. What type of graph is this?

b. What do the upper and lower margins of the boxes indicate?

c. Which male group had the highest median longevity?

d. Although the mean is not indicated on the graph, which sample of men probably had the highest mean longevity? Explain your reasoning.

24. As the Arctic warms and winters become shorter, hibernation patterns of arctic mammals are expected to change. Sheriff et al. (2011) investigated emergence dates from hibernation of Arctic ground squirrels at sites in the Brooks Range of northern Alaska. The measurements shown in the following figure are emergence dates in a sample of male and female ground squirrels at one of their study sites.

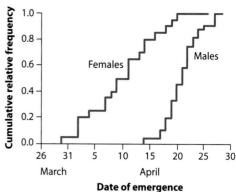

a. What type of graph is this?

b. Which sex, males or females, has the earliest median emergence date? Explain how you obtained your answer.

c. Which sex, male or female, has the greater interquartile range in emergence date? Explain how you obtained your answer.

25. Convert the following statistics, calculated on samples in English units, to the metric equivalents. (Conversion factors are given as well.)

a. Mean: 100 miles (1 km = 0.62 miles)

b. Standard deviation: 17 miles (1 km = 0.62 miles)

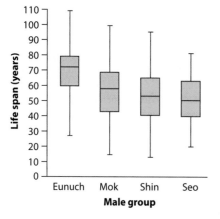

c. Variance: 289 miles^2 (1 km $=0.62$ miles)

d. Coefficient of variation: 17% (1 km $=0.62$ miles)

e. Mean: 23 pounds (1 kg $=2.2$ pounds)

f. Standard deviation: 1.2 ounces (1 g $=0.032$ ounces)

g. Variance: 550 gallons^2 (1 liter $=0.227$ gallons)

26. The snake undulation data of Example 3.1 were measured in Hz, which has units of $1/s$ (cycles per second). Often frequency measurements are expressed instead as angular velocity, which is measured in radians per second. To convert measurements from Hz to angular velocity (rad/s), multiply by 2π, where $\pi = 3.14159$.

a. The sample mean undulation rate in the snake sample was 1.375 Hz. Calculate the sample mean in units of angular velocity.

b. The sample variance of undulation rate in the snake sample was 0.105 Hz^2. Calculate the sample variance if the data were in units of angular velocity.

c. The sample standard deviation of undulation rate in the snake sample was 0.324 Hz. Calculate the sample standard deviation in units of angular velocity. Provide the appropriate units with your answer.

27. *Spot the flaw.* Crohn's disease is an autoimmune inflammatory disorder. The accompanying table shows medians and interquartile ranges for three response variables in 62 Crohn's disease patients randomly assigned either the immuno-suppressant drug azathioprine ($n = 32$) or a placebo ($n = 30$) in a clinical trial. Response variables are measured as change from baseline. *IQR* is interquartile range. The data are from Candy et al. 1995.

a. Identify the main flaw in the construction of this table.

b. Redraw the table following the principles recommended in this book.

28. Reproduction in sea urchins involves the release of sperm and eggs in the open ocean. Fertilization begins when a sperm bumps into an egg and the sperm protein bindin attaches to recognition sites on the egg surface. Gene sequences of bindin and egg-surface proteins vary greatly between closely related urchin species, and eggs can identify and discriminate between different sperm. In the burrowing sea urchin, *Echinometra mathaei,* the protein sequence for bindin varies even between populations within the same species. Do these differences affect fertilization? To test this, Palumbi (1999) carried out trials in which a mixture of sperm from AA and BB males, referring to two populations differing in bindin gene sequence, were added to dishes containing eggs from a female from either the AA or the BB population. The results below indicate the fraction of fertilizations of eggs of each of the two types by AA sperm (remaining eggs were fertilized by BB sperm).

AA females: 0.58, 0.59, 0.69, 0.72, 0.78, 0.78, 0.81, 0.85, 0.85, 0.92, 0.93, 0.95

BB females: 0.15, 0.22, 0.30, 0.37, 0.38, 0.50, 0.95

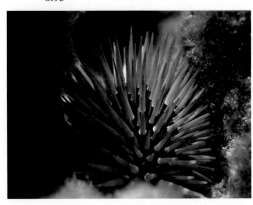

TABLE FOR PROBLEM 27

Response Variable	Azathioprine		Placebo	
	Median	*IQR*	Median	*IQR*
Crohn's Disease Activity Index	191.5	211	50.0	230
Erythrocyte sedimentation rate (mm/hr)	15.5	30	−6.5	26
Serum C reactive protein (%)	30.0	53	0.0	27

a. Plot the data using a method other than the box plot. Is there an association in these data between female type and fertilizations by AA sperm?

b. Inspect the plot. On this basis, which method from this chapter (mean or median) would be best to compare the locations of the frequency distributions for the two groups? Explain your reasoning. Calculate and compare locations using this method.

c. Which method would be best to compare the spread of the frequency distributions for the two groups? Explain your reasoning. Calculate and compare spread using this method.

29. The following graph illustrates an association between two variables. The graph shows density of fine roots in Monterey pines (*Pinus radiata*) planted in three different years of study (redrawn from Moir and Bachelard 1969, with permission). Identify (a) the type of graph, (b) the explanatory and response variables, and (c) the type of data (whether numerical or categorical) for each variable.

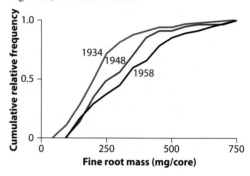

DNA crystal

4 Estimating with uncertainty

When biologists carry out a study, their goals are usually more ambitious than mere description of the resulting data. Rather, data are gathered so that something may be discovered about the larger population from which the sample came. The descriptive statistics measured on a sample are used to estimate parameters of the population. Such estimation is possible when the sample is a random sample. Recall from Chapter 1 that in a random sample, all individuals in the population have an equal chance of being selected and individuals are sampled independently.

For example, the sample mean $\overline{Y}$ is used to estimate the true mean of the population, symbolized using the Greek letter μ (mu; pronounced "mew"). The sample standard deviation s is used to estimate the population standard deviation, symbolized by σ (lowercase Greek letter sigma). Likewise, the sample proportion $\hat{p}$ (pronounced "p-hat") is an estimate[1] of the population proportion p.

For an estimate of a population parameter to be useful, we also need to quantify its precision. We need a measure of how far the estimate is likely to be from the target parameter being estimated. If precision is high, then our uncertainty is low. We can be reasonably confident that our estimate is close to the truth. If instead precision is low, then our uncertainty is high and we'll need more data to reduce it.

In Chapter 4, we explain the basics of estimation and the uncertainties involved when making generalizations about a population from a random sample. We introduce the standard error, a key measure of the precision of an estimate, and we demonstrate it in the case of the sample mean. We add a brief, conceptual introduction to the confidence interval, another important means of describing the precision of an estimate.

4.1 The sampling distribution of an estimate

Estimation is the process of inferring a population parameter from sample data. The value of an estimate calculated from data is almost never exactly the same as the value of the population parameter being estimated, because sampling is influenced by chance. The crucial question is, "In the face of chance, how much can we trust an estimate?" In other words, what is its *precision*? To answer this question, we need to know something about how the sampling process might affect the estimates we get. We use the **sampling distribution** of the estimate, which is the probability distribution of all the values for an estimate that we *might* have obtained when we sampled the population. We illustrate the concept of a sampling distribution using samples from a known population, the genes of the human genome.

EXAMPLE The length of human genes

4.1 The international Human Genome Project was the largest coordinated research effort in the history of biology. It yielded the DNA sequence of all 23 human chromosomes, each consisting of millions of nucleotides chained end to end.[2] These encode the genes whose products—RNA and proteins—shape the growth and development of each individual.

We obtained the lengths of all 20,290 known and predicted genes of the published genome sequence (Hubbard et al. 2005).[3] The length of a gene refers to the total number

---

1. We will often use Greek letters (like μ and σ) to describe parameters and roman letters (e.g., $\overline{Y}$ and s) for their estimates. In addition, it is common to put a circumflex, or "hat" (^), over a variable name to show it is an estimate (e.g., $\hat{p}$ is an estimate for p).

2. The photo at the beginning of this chapter is a crystal of pure DNA.

3. We used the largest known transcript of each gene in release 35 of the human genome (http://www. ensembl.org/Homo_sapiens). Accessed Dec 2, 2005.

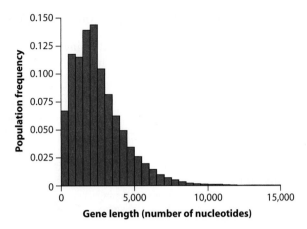

FIGURE 4.1-1
Distribution of gene lengths in the known human genome. The graph is truncated at 15,000 nucleotides; 26 larger genes are too rare to be visible in this histogram.

of nucleotides comprising the coding regions. The frequency distribution of gene lengths in the population of genes is shown in Figure 4.1-1. The figure includes only genes up to 15,000 nucleotides long; in addition, there are 26 longer genes. [4]

The histogram in Figure 4.1-1 is like those we have seen before, except that it shows the distribution of lengths in the *population* of genes, not simply those in a *sample* of genes. Because it is the population distribution, the relative frequency of genes of a given length interval in Figure 4.1-1 represents the *probability* of obtaining a gene of that length when sampling a single gene at random. The probability distribution of gene lengths is positively skewed, having a long tail extending to the right.

The population mean and standard deviation of gene length in the human genome are listed in Table 4.1-1. These quantities are referred to as *parameters* because they are quantities that describe the population.

TABLE 4.1-1 Population mean and standard deviation of gene length in the known human genome.

Name	Parameter	Value (nucleotides)
Mean	μ	2622.0
Standard deviation	σ	2036.9

In real life, we would not usually know the parameter values of the study population, but in this case we do. We'll take advantage of this to illustrate the process of sampling.

---

4. The longest human gene, with nearly 100,000 nucleotides, encodes the gigantic protein titin, which is expressed in heart and skeletal muscle. The protein was named for the titans of Greek mythology, giants who ruled the earth until overthrown by the Olympians. Some mutations in the *titin* gene cause heart muscle disease and muscular dystrophy.

Estimating mean gene length with a random sample

To begin, we collected a single random sample of $n = 100$ genes from the known human genome.[5] A histogram of the lengths of the resulting sample of genes is shown in Figure 4.1-2.

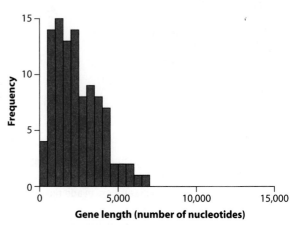

FIGURE 4.1-2
Frequency distribution of gene lengths in a unique random sample of $n = 100$ genes from the human genome.

The frequency distribution of the random sample (Figure 4.1-2) is not an exact replica of the population distribution (Figure 4.1-1), because of chance. The two distributions nevertheless share important features, including approximate location, spread, and shape. For example, the sample frequency distribution is skewed to the right like the true population distribution.

The sample mean and standard deviation of gene length from the sample of 100 genes are listed in Table 4.1-2. How close are these estimates to the population mean and standard deviation listed in Table 4.1-1? The sample mean is $\overline{Y} = 2411.8$, which is about 200 nucleotides shorter than the true value, the population mean of $\mu = 2622.0$. The sample standard deviation ($s = 1463.5$) is also different from the standard deviation of gene length in the population ($\sigma = 2036.9$). We shouldn't be surprised that the sample estimates differ from the parameter (population) values. Such differences are virtually inevitable because of chance in the random sampling process.

TABLE 4.1-2 Mean and standard deviation of gene length Y in our unique random sample of $n = 100$ genes from the human genome.

Name	Statistic	Sample value (number of nucleotides)
Mean	$\overline{Y}$	2411.8
Standard deviation	s	1463.5

5. All genes were listed in a file, one gene per line, with lines numbered from 1 to 20,290. We then used a computer program to generate 100 random integers between the values 1 and 20,290, without allowing duplicates. Each random number was then used to draw a gene according to its line number in the list of genes.

The sampling distribution of $\overline{Y}$

We obtained $\overline{Y} = 2411.8$ nucleotides in our single sample, but by chance we might have obtained a different value. When we took a second random sample of 100 genes, we found $\overline{Y} = 2643.5$. Each new sample will usually generate a different estimate of the same parameter. If we were able to repeat this sampling an infinite number of times, we could create the probability distribution of our estimate. The probability distribution of values we might obtain for an estimate make up the estimate's **sampling distribution.**

> The *sampling distribution* is the probability distribution of all values for an estimate that we might obtain when we sample a population.

The sampling distribution represents the "population" of values for an estimate. It is not a real population, like the squirrels in Muir Woods or all the retirees basking in the Florida sunshine. Rather, the sampling distribution is an imaginary population of values for an estimate. Taking a random sample of n observations from a population and calculating $\overline{Y}$ is equivalent to randomly sampling a *single* value of $\overline{Y}$ from its sampling distribution.

To visualize the sampling distribution for mean gene length, we used the computer to take a vast number of random samples of $n = 100$ genes from the human genome. We calculated the sample mean $\overline{Y}$ each time. The resulting histogram in Figure 4.1-3 shows the values of $\overline{Y}$ that might be obtained when randomly sampling 100 genes, together with their probabilities.

FIGURE 4.1-3
The sampling distribution of mean gene length, $\overline{Y}$, when $n = 100$. Note the change in scale from Figure 4.1-2.

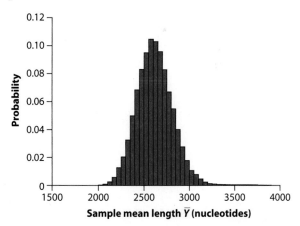

Figure 4.1-3 makes plain that, although the population mean μ is a constant (2622.0), its estimate $\overline{Y}$ is a variable. Each new sample yields a different $\overline{Y}$ value from the one before. We don't ever see the sampling distribution of $\overline{Y}$ because ordinarily we have only one sample, and therefore only one $\overline{Y}$. Notice that the sampling

distribution for $\overline{Y}$ is centered exactly on the true mean, μ. This means that $\overline{Y}$ is an unbiased estimate of μ.[6]

The spread of the sampling distribution of an estimate depends on the sample size. The sampling distribution of $\overline{Y}$ based on $n = 100$ is narrower than that based on $n = 20$, and that based on $n = 500$ is narrower still (Figure 4.1-4). The larger the sample size, the narrower the sampling distribution. And the narrower the sampling distribution, the more precise the estimate. Thus, larger samples are desirable whenever possible because they yield more precise estimates. The same is true for the sampling distributions of estimates of other population quantities, not just $\overline{Y}$.

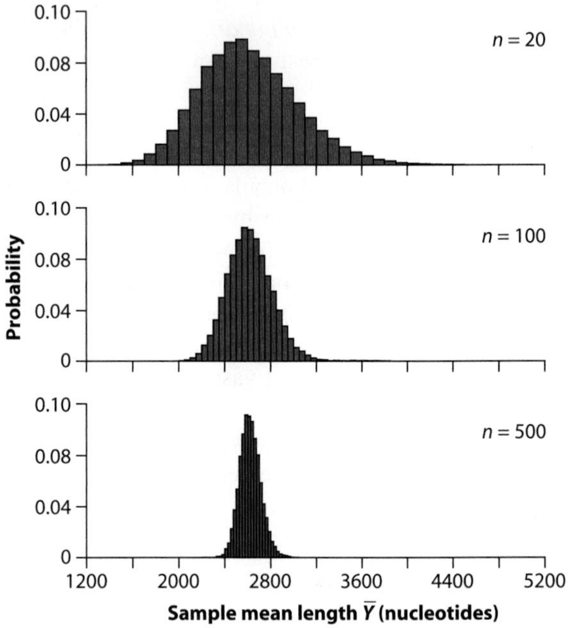

FIGURE 4.1-4
Comparison of the sampling distributions of mean gene length, $\overline{Y}$, when $n = 20$, 100, and 500.

Increasing sample size reduces the spread of the sampling distribution of an estimate, increasing precision.

4.2 Measuring the uncertainty of an estimate

In this section we show how the sampling distribution is used to measure the uncertainty of an estimate.

---

6. Notice how the shape of the sampling distribution resembles that of a normal distribution, or "bell curve," more so than the distribution of gene lengths themselves. Most of the methods presented in this book rely on the normal distribution to approximate the sampling distribution of an estimate.

Standard error

The standard deviation of the sampling distribution of an estimate is called the **standard error**. Because it reflects the differences between an estimate and the target parameter, the standard error reflects the precision of an estimate. Estimates with smaller standard errors are more precise than those with larger standard errors. The smaller the standard error, the less uncertainty there is about the target parameter in the population.

> The *standard error* of an estimate is the standard deviation of the estimate's sampling distribution.

The standard error of $\overline{Y}$

The standard error of the sample mean is particularly simple to calculate, so we show it here. We can represent the standard error of the mean with the symbol $\sigma_{\overline{Y}}$. It has a remarkably straightforward relationship with σ, the population standard deviation of the variable Y:

$$\sigma_{\overline{Y}} = \frac{\sigma}{\sqrt{n}}.$$

The standard error decreases with increasing sample size. Table 4.2-1 lists the standard error of the sample mean based on random samples of $n = 20$, 100, and 500 from the known human genome.

TABLE 4.2-1 Standard error of the sampling distributions of mean gene length $\overline{Y}$ according to sample size. These measure the spread of the three sampling distributions in Figure 4.1-4.

Sample size, n	Standard error, $\sigma_{\overline{Y}}$ (nucleotides)
20	455.5
100	203.7
500	91.1

The standard error of $\overline{Y}$ from data

The trouble with the formula for the standard error of the mean ($\sigma_{\overline{Y}}$) is that we almost never know the value of the population standard deviation (σ), and so we cannot calculate $\sigma_{\overline{Y}}$. The next best thing is to approximate the standard error of the mean by using the sample standard deviation (s) as an estimate of σ. To show that it is approximate, we will use the symbol $\mathrm{SE}_{\overline{Y}}$. The approximate standard error of the mean is

$$\mathrm{SE}_{\overline{Y}} = \frac{s}{\sqrt{n}}.$$

According to this simple relationship, all we need is one random sample to approximate the spread of the entire sampling distribution for $\overline{Y}$. The quantity $SE_{\overline{Y}}$ is usually called the "standard error of the mean."

> The *standard error of the mean* is estimated from data as the sample standard deviation (s) divided by the square root of the sample size (n).

Calculating $SE_{\overline{Y}}$ is so routine in biology that a sample mean should never be reported without it. For example, if we were submitting the results of our unique random sample of 100 genes in Figure 4.1-2 for publication, we would calculate $SE_{\overline{Y}}$ from the results in Table 4.1-2 as follows:

$$SE_{\overline{Y}} = \frac{s}{\sqrt{n}} = \frac{1463.5}{\sqrt{100}} = 146.3.$$

We would then report the sample mean in the text of the paper as $2411.8 \pm 146.3(SE)$.

Every estimate, not just the mean, has a sampling distribution with a standard error, including the proportion, median, correlation, difference between means, and so on. In the rest of this book, we will give formulas to calculate standard errors of many kinds of estimates. The standard error is the usual way to indicate uncertainty of an estimate.

4.3   Confidence intervals

The **confidence interval** is another common way to quantify uncertainty about the value of a parameter. It is a range of numbers, calculated from the data, that is likely to contain within its span the unknown value of the target parameter. In this section, we introduce the concept without showing exact calculations. Confidence intervals can be calculated for means, proportions, correlations, differences between means, and other population parameters, as later chapters will demonstrate.

> A *confidence interval* is a range of values surrounding the sample estimate that is likely to contain the population parameter.

An example is the **95% confidence interval for the mean**. This confidence interval is a range likely to contain the value of the true population mean μ. It is calculated from the data and extends above and below the sample estimate $\overline{Y}$. You will encounter confidence intervals frequently in the biological literature. We'll show you in Chapter 11 how to calculate an exact confidence interval for the mean, but for now we give you the result and its interpretation. The 95% confidence interval for the mean calculated from the unique sample of 100 genes (Table 4.1-2) is

$$2121.4 < \mu < 2702.2.$$

For this example, 2121.4 is the *lower limit* of the confidence interval, whereas 2702.2 is the *upper limit*. This calculation allows us to say, "We are 95% confident that the true mean lies between 2121.4 and 2702.2 nucleotides." We do *not* say that "there is a 95% probability that the population mean falls between 2121.4 and 2702.2 nucleotides," which is a common misinterpretation of the confidence interval (2121.4 and 2702.2 are both constants, and the true mean either is or is not between them, so there's no probability involved).

To better understand the correct interpretation of "95% confidence," imagine that 20 researchers independently take unique random samples of $n = 100$ genes from the human genome. Each researcher calculates an estimate $\overline{Y}$ and then a 95% confidence interval for the parameter (the population mean, μ). Each researcher ends up with a different estimate and a different 95% confidence interval, because by chance their samples are not the same (Figure 4.3-1). On average, however, 19 out of 20 (95%) of the researchers' intervals will contain the value of the population parameter. On average, therefore, one out of 20 intervals (5%) will *not* contain the parameter value. None of the researchers will know for sure whether his or her own confidence interval contains the value of the unknown parameter, but each can be "95% confident" that it does.

FIGURE 4.3-1
The 95% confidence intervals for the mean calculated from 20 separate random samples of $n = 100$ genes from the known human genome. Dots indicate the sample means $\overline{Y}$. The vertical line (gold) represents the population mean, $\mu = 2622.0$. In this example, 19 of 20 intervals included the population mean, whereas one interval did not.

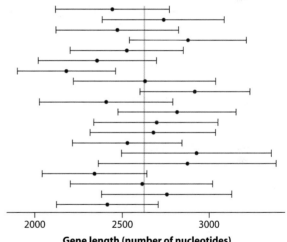

Gene length (number of nucleotides)

All numbers falling between the lower and upper bounds of a confidence interval can be regarded as the most plausible values for the parameter, given the data sampled. Values falling outside the confidence interval are less plausible. For example, on the basis of our random sample of 100 genes we can say that a mean gene length of 2000 nucleotides in the whole genome is implausible, because it falls outside the 95% confidence interval, $2121.4 < \mu < 2702.2$. However, a mean gene length of 2500 nucleotides falls within the 95% confidence interval, and so remains plausible.

In general, the width of the 95% confidence interval is a good measure of our uncertainty about the true value of the parameter. If the confidence interval is broad, then uncertainty is high and the data are not very informative about the location of the population parameter. If the confidence interval is narrow, on the other hand, then we can be confident that the parameter is close to the estimated value.

> The *95% confidence interval* provides a most-plausible range for a parameter. Values lying within the interval are most plausible, whereas those outside are less plausible, based on the data.

The 2SE rule of thumb

A good "quick-and-dirty" approximation to the 95% confidence interval for the population mean is obtained by adding and subtracting two standard errors from the sample mean (the so-called 2SE rule of thumb). This calculation assumes that the sample is a random sample.

> A rough approximation to the 95% confidence interval for a mean can be calculated as the sample mean plus and minus two standard errors.

For our unique random sample of 100 genes (Figure 4.1-2), for example, the sample mean of gene length was $\overline{Y} = 2411.8$ nucleotides, and its standard error was $SE_{\overline{Y}} = 146.3$ nucleotides. Two standard errors below the mean provides the lower limit of the approximate confidence interval:

$$\overline{Y} - 2SE_{\overline{Y}} = 2411.8 - (2 \times 146.3) = 2119.2$$

and two standard errors above the mean provides the upper limit:

$$\overline{Y} + 2SE_{\overline{Y}} = 2411.8 + (2 \times 146.3) = 2704.4.$$

According to the 2SE rule of thumb, then, the 95% confidence interval for the mean gene length in the population can be approximated as

$$2119.2 < \mu < 2704.4.$$

This is not too far off from the more exact confidence interval (i.e., between 2121.4 and 2702.2 nucleotides) that we calculated previously. Although approximate, the 2SE rule is simple and works reasonably well.

4.4 Error bars

Standard errors or confidence intervals for the mean (and other parameters) are often illustrated graphically with "error bars." Error bars are lines on a graph that extend outward from the sample estimate to illustrate the precision of estimates, reflecting uncertainty about the value of the parameter being estimated. For example, Figure 4.4-1 reproduces the strip chart of locust serotonin data shown previously in Chapter 2 (Figure 2.1-2) but adds error bars to illustrate the standard error of the sample mean serotonin level in each of the three experimental treatments. The lines projecting outward from the sample mean indicate one standard error above the mean and one standard error below the mean. Remember that standard error bars, unlike the whiskers on a box plot, are not intended to span a specified fraction of the data. Error bars indicate uncertainty about the population parameter, not variability in the data (even though variability in the data contributes to uncertainty).

FIGURE 4.4-1
Strip chart of locust serotonin data (from Figure 2.1-2) with error bars added to illustrate the standard error (SE) of the mean for each treatment. Each filled black dot indicates the sample mean. Lines projecting outward indicate one SE above the mean and one SE below the mean.

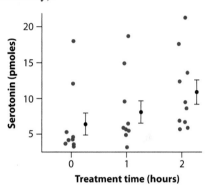

> *Error bars* are lines on a graph extending outward from the sample estimate to illustrate uncertainty about the value of the parameter being estimated.

Error bars are used for multiple purposes and so don't always show the same measure of precision. Often they are used to illustrate standard errors, but sometimes error bars show confidence intervals instead. They may even indicate *two* standard errors rather than one. For example, Figure 4.4.2 draws an error bar for each of these three measures of precision for the mean of the same random sample. Notice that because they are measuring different quantities, the three error bars do not have the same span. Therefore it is crucial to read carefully the caption of any figure that has error bars to determine which measure of uncertainty is being shown. (And when you draw graphs yourself, it is important to give this information clearly in the figure legend.) Most commonly, error bars are used either for 95% confidence intervals or for standard errors. Because these two quantities differ approximately by a factor of two, you can see how knowing the meaning of the error bars is important.

Finally, error bars are sometimes used to indicate the standard deviation of the data, but we recommend against this practice to minimize confusion. Error bars are a poor method for illustrating variability in the data, and they are redundant if you show the data. We added an error bar for the standard deviation to Figure 4.4-2 only to show how different—and potentially misleading—it can be. Use error bars only to illustrate the precision of estimates, not variability in the data.

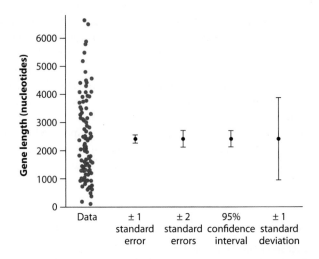

FIGURE 4.4-2 Comparison of alternative error bars calculated from gene lengths in the random sample of *n* = 100 genes (Example 4.1). The data are plotted as a strip chart on the left. The filled black circles indicate the sample mean of gene length, 2411.8 nucleotides. The leftmost error bar visualizes one standard error of the mean (SE) above and below the sample mean. The line extending above the black dot indicates one SE above the mean; the line extending below indicates one SE below the mean. The adjacent error bar indicates two standard errors above and below the mean. The third error bar indicates the 95% confidence interval for the mean. The rightmost error bar indicates one standard deviation above and below the sample mean.

▮4.5▮ Summary

- Estimation is the process of inferring a population parameter from sample data.
- All estimates have a sampling distribution, which is the probability distribution of all the possible values of the estimate that might be obtained under random sampling with a given sample size.
- The standard error of an estimate is the standard deviation of its sampling distribution. The standard error measures precision. The smaller the standard error, the more precise is the estimate.
- The usual formulas for standard errors and confidence intervals assume that sampling is random.
- The standard error of an estimate declines with increasing sample size.
- The confidence interval is a range of values calculated from sample data that is likely to contain within its span the value of the target parameter. On average, 95% confidence intervals calculated from independent random samples will include the value of the parameter 19 times out of 20.

- The 2SE rule of thumb (i.e., the sample mean plus or minus two standard errors) provides a rough approximation to the 95% confidence interval for a mean.
- Add error bars to graphs to illustrate standard errors or confidence intervals. Make sure to clarify which is being illustrated in the figure legend.

4.6 Quick Formula Summary

Standard error of the mean

What is it for? Measuring the precision of the sample estimate $\overline{Y}$ of the population mean μ.

What does it assume? The sample is a random sample.

Estimate: $\mathrm{SE}_{\overline{Y}}$

Parameter: $\sigma_{\overline{Y}}$

Formula: $\mathrm{SE}_{\overline{Y}} = \dfrac{s}{\sqrt{n}}$

where s is the sample standard deviation and n is the sample size. $\mathrm{SE}_{\overline{Y}}$ estimates the quantity $\sigma_{\overline{Y}} = \dfrac{\sigma}{\sqrt{n}}$, where $\sigma_{\overline{Y}}$ is the standard error of the sample mean, and σ is the standard deviation of Y in the population.

PRACTICE PROBLEMS

1. **Calculation practice: Standard error of the mean and approximate confidence intervals for the mean.** We will use the same data for systolic blood pressure collected for Calculation Practice Problem 1 in Chapter 3. Here again are the data points:

   112, 128, 108, 129, 125, 153, 155, 132, 137

   The mean is 131.0 mm Hg and the variance is 254.5.

   a. What is s, the standard deviation of these data?

   b. What is n, the sample size?

   c. Calculate the standard error of the mean.

   d. Using the 2SE rule of thumb, calculate an approximate 95% confidence interval for the mean. Provide the lower and upper limits.

2. Examine the times to rigor mortis of the 114 human corpses tabulated in Practice Problem 9 of Chapter 3.

   a. What is the standard error of the mean time to rigor mortis?

   b. The standard error calculated in part (a) measures the spread of what frequency distribution?

   c. What assumption does your calculation in part (a) require?

3. Examine the frequency distribution of gene lengths in the human genome displayed in Figure 4.1-1. Is the population median gene length in the human genome likely to be larger, smaller, or equal to the population mean? Explain.

4. As a general rule, is the spread of the sampling distribution for the sample mean mainly determined by the magnitude of the mean or by the sample size?

5. Seven of the 100 human genes we sampled randomly from the human genome (in Example 4.1) were found to occur on the X chromosome. The sample fraction of genes on the X was thus $\hat{p} = 7/100 = 0.07$. For each of the following statements, specify whether it is true or false:
- **a.** $\hat{p} = 0.07$ is the fraction of all human genes on the X chromosome.
- **b.** $\hat{p} = 0.07$ estimates p, the fraction of all human genes on the X chromosome.
- **c.** $\hat{p}$ has a sampling distribution representing the frequency distribution of values of $\hat{p}$ that we might obtain when we randomly sample 100 genes from the human genome.
- **d.** The fraction of all human genes on the X chromosome has a sampling distribution.
- **e.** The standard deviation of the sampling distribution of $\hat{p}$ is the standard error of $\hat{p}$.

6. In a poll of 1641 people carried out in Canada in November 2005, 73% of people surveyed agreed with the statement that "you don't really expect that politicians will keep their election promises once they are in power" (CBC News 2005).
- **a.** What is the parameter being estimated?
- **b.** What is the value of the sample estimate?
- **c.** What is the sample size?
- **d.** The poll also reported that "the results are considered accurate within 2.5 percentage points, 19 times out of 20." Explain what this statement likely refers to.

7. The following data are flash durations, in milliseconds, of a sample of 35 male fireflies of the species *Photinus ignitus* (Cratsley and Lewis 2003; see Assignment Problem 19 in Chapter 2):

79, 80, 82, 83, 86, 85, 86, 86, 88, 87, 89, 89, 90, 92, 94, 92, 94, 96, 95, 95, 95, 96, 98, 98, 98, 101, 103, 106, 108, 109, 112, 113, 118, 116, 119

- **a.** Estimate the sample mean flash duration. What does this quantity estimate?
- **b.** Is the estimate in part (a) likely to equal the population parameter? Why or why not?
- **c.** Calculate a standard error for your sample estimate.
- **d.** What does the quantity in part (c) measure?
- **e.** Using an approximate method, calculate a rough 95% confidence interval for the population mean.
- **f.** Provide an interpretation for the interval you calculated in part (e).

8. Imagine that the results of a study calculated a sample mean of zero, with a narrow 95% confidence interval for the population mean (modified from Borenstein 1997). The most appropriate conclusion is that (choose one):
- **a.** The population mean is likely to be zero or close to zero.
- **b.** The population mean is probably zero, but there is some chance that it is either slightly less than zero or slightly greater than zero.
- **c.** We can be reasonably certain the mean differs from zero.

9. One of the great discoveries of biology is that organisms have a class of genes called "regulatory genes," whose only job is to regulate the activity of other genes. How many genes does the typical regulatory gene regulate? A study of interaction networks in yeast (*S. cerevesiae*) came up with the following data for 109 regulatory genes (Guelzim et al. 2002).

Number of genes regulated	Frequency
1	20
2	10
3	7
4	7
5	8
6	8
7	5
8	2
9	4
10	4
11	3
12	4
13	5
14	1
15	2
16	1
17	3
18	2
19	2
20	3
22	3
25	1
26	1
28	1
29	1
37	1
Total	109

a. What type of graph should be used to display these data?

b. What is the estimated mean number of genes regulated by a regulatory gene in the yeast genome?

c. What is the standard error of the mean?

d. Explain what this standard error measures.

e. What assumption are you making in part (c)?

10. Refer to the previous problem (Practice Problem 9).

a. Using an approximate method, provide a rough 95% confidence interval for the population mean.

b. Provide an interpretation of the interval you calculated in part (a).

11. Goldman et al. (1988) analyzed data on 405 patients with white blood cell cancer (chronic myelogenous leukemia) who received bone marrow transplants. They estimated the probability of relapse within 4 years of treatment to be 0.19, with a 95% confidence of 0.12 to 0.28. Which of the following statements are true?

a. The population proportion is 0.19.

b. The population proportion is likely to be between 0.12 and 0.28.

c. There is a 95% chance that the population proportion is between 0.12 and 0.28.

d. A population proportion of 0.30 is plausible.

12. An absentminded (and not too clever) scientist friend of yours has just analyzed his data, and he has two numbers—25.4 and 2.54—written on a scrap of paper. He says: "I remember that one of these is the standard deviation of my data and the other is the standard error of the mean, but I can't remember which is which. Can you help?"

a. Which number is the standard deviation and which is the standard error of the mean?

b. What was your friend's sample size?

13. The following is a list of sample means for human adult height, in cm. Each was calculated in the same way from samples taken from the same hypothetical population.

160.5, 162.5, 161.7, 160.2, 163.7, 159.8, 160.6, 161.1

The true mean of the population is 158.7 cm. Use the jargon of estimation to describe the likely type of problem in the sampling process.

14. When planning to obtain a sample from a population of interest, what can you do to make the standard error of the mean smaller?

ASSIGNMENT PROBLEMS

15. A massive survey of sexual attitudes and behavior in Britain between 1999 and 2001 contacted 16,998 households and interviewed 11,161 respondents aged 16–44 years (one per responding household). The frequency distributions of ages of men and women respondents were the same. The following results were reported on the number of heterosexual partners individuals had had over the previous five-year period (Johnson et al. 2001).

	Sample size, n	Mean	Standard deviation
Men	4620	3.8	6.7
Women	6228	2.4	4.6

a. What is the standard error of the mean in men? What is it in women? Assume that the sampling was random.

b. Which is a better descriptor of the variation among men in the number of sexual partners, the standard deviation or the standard error? Why?

c. Which is a better descriptor of uncertainty in the estimated mean number of partners in women, the standard deviation or the standard error? Why?

d. A mysterious result of the study is the discrepancy between the mean number of partners of heterosexual men and women. If each sex obtains its partners from the other sex, then the true mean number of heterosexual partners should be identical. Considering aspects of the study design, suggest an explanation for the discrepancy.

16. Our unique random sample of 100 human genes from the human genome (Example 4.1) was found to have a median length of 2150 nucleo-

tides. Specify whether each of the following statements is true or false.

a. The median gene length of all human genes is 2150 nucleotides.

b. The median gene length of all human genes is estimated to be 2150 nucleotides.

c. The sample median has a sampling distribution with a standard error.

d. A random sample of 1000 genes would likely yield an estimate of the median closer to the population median than a random sample of 100 genes.

17. The following figure is from the website of a U.S. national environmental laboratory.[7] It displays sample mean concentrations, with 95% confidence intervals, of three radioactive substances. The text accompanying the figure explained that *"the first plotted mean is 2.0 ± 1.1, so there is a 95% chance that the actual result is between 0.9 and 3.1, a 2.5% chance it is less than 0.9, and a 2.5% chance it is greater than 3.1."* Is this a correct interpretation of a confidence interval? Explain.

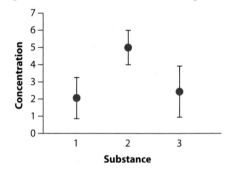

18. *Amorphophallus johnsonii* is a plant growing in West Africa, and it is better known as a "corpse-flower." Its common name comes from the fact

---

7. Redrawn from a figure at http://www.pnl.gov/env/Helpful2.html. Accessed January 15, 2006.

that when it flowers, it gives off a "powerful aroma of rotting fish and faeces" (Baeth 1996). The flowers smell this way because their principal pollinators are carrion beetles, who are attracted to such a smell. Baeth (1996) observed the number of carrion beetles (*Phaeochrous amplus*) that arrive per night to flowers of this species. The data are as follows:

51, 45, 61, 76, 11, 117, 7, 132, 52, 149

a. What is the mean and standard deviation of beetles per flower?
b. What is the standard error of this estimate of the mean?
c. Give an approximate 95% confidence interval of the mean. Provide lower and upper limits.
d. If you had been given 25 data points instead of 10, would you expect the mean to be greater, less than, or about the same as the mean of this sample?
e. If you had been given 25 data points instead of 10, would you have expected the standard deviation to be greater, less than, or about the same as this sample?
f. If you had been given 25 data points instead of 10, would you have expected the standard error of the mean to be greater, less than, or about the same as this sample?

19. The following three histograms (A, B, and C) plot information about the number of hours of sleep adult Europeans get per night (Roenneberg 2012). One of them shows the frequency distribution of individual values in a random sample. Another shows the distribution of sample means for samples of size 10 taken from the same population. Another shows the distribution of sample means for samples of size 100.

A.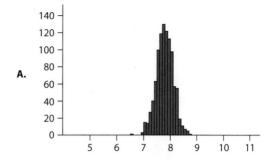

B.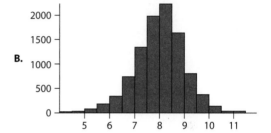

C.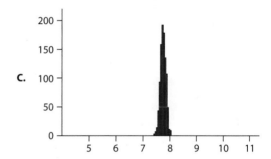

a. Identify which graph goes with which distribution.
b. What features of these distributions allowed you to distinguish which was which?
c. Estimate by eye the approximate population mean of the number of hours of sleep using the distribution for the data.
d. Estimate by eye the approximate mean of the distributions of sample means.

20. The following figure shows two alternative ways of presenting means with standard error bars in a graph. The data are from Daborn et al. (2002), who showed that elevated expression of the gene *Cyp6g1* in *Drosophila* causes resistance to DDT. Expression levels of the gene (relative to a standard) were measured in 12 resistant strains of *Drosophila* and 6 susceptible strains. Which graphical method is superior? Explain.

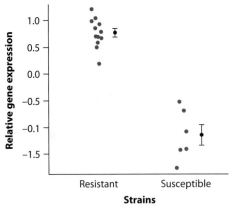

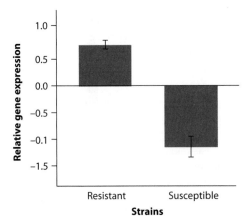

21. Is sleep necessary? To investigate, Lesku et al. (2012) measured the activity patterns of breeding pectoral sandpipers (*Calidris melanotos*) in the high Arctic in summer, when the sun never sets. The accompanying figure shows the observed percent time that individual males were awake and active in a 2008 field study. The data are on the left. To right of the data are the sample mean (filled circle) and error bars for the standard deviation, the standard error of the mean, and a 95% confidence interval for the mean (in no particular order).

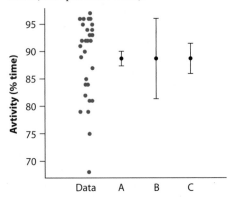

a. Which of the error bars indicates the standard deviation?
b. Which error bar indicates the standard error of the mean?
c. Which error bar indicates a 95% confidence interval for the mean?
d. Estimate by eye the smallest plausible value for the mean activity (% time) of male pectoral sandpipers. Using this smallest plausible value, calculate approximately the maximum plausible number of hours (out of 24 hours) that males spend inactive or asleep.

22. How long do you hug somebody? Nagy (2011) measured the duration of spontaneous embraces at the 2008 Summer Olympic Games in Beijing, China. The data are the durations of hugs, in seconds, of athletes immediately after competing in the finals of an event. Hugs were either with their coach, a supporter (e.g., a team member), or a competitor. Descriptive statistics calculated from the data are in the following table. *n* refers to the sample size.

Relationship	Mean	Standard deviation	*n*
Coach	3.77	3.96	77
Supporter	3.16	2.76	75
Competitor	1.81	1.13	33

a. According to the values in the table, which relationship group gets the longest hugs, on average, and which gets the briefest hugs?

Do the values shown represent parameters or sample estimates? Explain.

b. Using the numbers in the table, calculate the standard error of the mean hug duration for each relationship group. What do these values measure?

c. What assumption(s) about the samples are you making in (b)?

d. Using the numbers in the table, calculate an approximate 95% confidence interval for the mean hug duration when athletes embrace competitors. Provide the lower and upper limits of the confidence interval.

e. In light of your results in (d), consider the most-plausible values for the mean duration of hugs with competitors in the population of athletes. Is 2 seconds among the most plausible values for the population mean hug duration?

f. For which of the relationship groups is the possibility of a 3-second mean hug duration in the population plausible?

23. Pitcher plants of the genus *Nepenthes* are typically carnivorous, obtaining a great deal of their nutrition from insects that become trapped in the pitcher, die, and decay. *N. lowii*, a pitcher plant from Borneo, produces a second type of pitcher that attracts tree shrews (*Tupaia montana*), which provide nutrients by defecating into the pitcher [8] while they feed on a substance secreted by the plant. Based on measurements of 20 plants, Clarke et al. (2009) calculated a 95% confidence interval for the mean fraction of total leaf nitrogen in the plant species derived from tree shrews: $0.57 < \mu < 1.0$.

a. Does this result imply that individual plants receive between 57% and 100% of their leaf nitrogen from tree shrews? Explain.

b. Is the confidence interval meant to bracket the sample mean or the population mean?

c. Identify two values for the mean shrew fraction of total leaf nitrogen that the analysis suggests are among the most plausible.

d. Identify two values for the mean shrew fraction of total leaf nitrogen that the analysis suggests are less plausible.

24. Hagen et al. (2011) estimated the home range sizes of 4 bumblebees (*Bombus*) by fitting them with tiny radio transmitters and tracking their

8. See video at http://rsbl.royalsocietypublishing.org/content/suppl/2009/06/03/rsbl.2009.0311.DC1/rsbl20090311supp3.mpg. Accessed February 3, 2014.

positions by plane and ground surveys. They estimated the mean home range size to be 20.7 ± 11.6 ha, where the number after the $\pm$ sign refers to standard error of the mean.

a. Provide a description for the standard error. What does it measure?

b. What assumption are we making when calculating the standard error of the mean?

c. What would you recommend the researchers do next to reduce the standard error of their estimate of the mean home range size?

25. The following definition of a confidence interval was found on a web page at the National Institute of Standards and Technology. *"Confidence intervals are constructed at a confidence level, such as 95%, selected by the user. ... It means that if the same population is sampled on numerous occasions and interval estimates are made on each occasion, the resulting intervals would bracket the true population parameter in approximately 95% of the cases."* [9] Is this a correct interpretation of the confidence interval? Explain.

26. When a female jewel wasp (*Ampulex compressa*) encounters a cockroach (*Periplaneta americana*), she stings and injects neurotoxins into its head that render the insect unable to initiate self-movement but not paralyzed. The wasp then holds the compliant (zombie) cockroach by the antenna and leads it to her nest, where it will become live food for her larval offspring. The following graph (modified from Gal and Libersat 2010) compares the mean self-initiated walking duration of stung and control cockroaches during the first 30 minutes after treatment. The error bars indicate approximate 95% confidence intervals. $n = 5$ in each group.

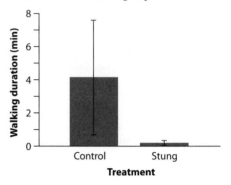

a. Estimate the lower and upper limits of the confidence intervals for the control group.

b. Approximate the value of the standard error for the control group.

c. Identify two values of the population mean duration for the control group that are among the most plausible.

d. Identify two values of the population mean duration for the stung group that are less plausible.

e. Identify the main weakness in the construction of the graph. How would you improve it?

9. http://www.itl.nist.gov/div898/handbook/prc/section1/prc14.htm. Accessed February 3, 2014.

Pseudoreplication

Most statistical techniques assume that the data are a random sample from the population, in which individuals are sampled independently and with equal probability. Unfortunately, many experiments are conducted and analyzed in a way that violates the assumption of independence.

For example, consider the problem of how to measure differences between male birdsongs in their attractiveness to females. In some species male birds vary in their song complexity. Some males have songs with multiple notes and phrases and other males have simpler songs with fewer notes. A classic way to measure song attractiveness is to play a tape recording of a song to a female and watch what she does. (Females sometimes have recognizable courtship behaviors that can be used to indicate their interest.) In one study[1] researchers recorded the complex song of one male and the relatively simple song of another male, and they played these same two songs to each of 40 different females. The goal of the study was to determine by how much on average females preferred complex songs over simple songs.

A confidence interval for the mean attractiveness of the two male songs was calculated

> **Pseudoreplication is probably the single most common fault in the design and analysis of ecological field experiments. It is at least equally common in many other areas of research.**
>
> —Stuart Hurlbert (1984)

based on the responses of the 40 females. The result was a very narrow range of plausible values. But something has gone wrong. The 40 measurements of attractiveness of simple and complex songs were not independent. All females listened to the songs of the same two males. When females listened to the "complex" song, they were listening to one male, and when they listened to the "simple" song, females were listening to another single male. In reality, the sample size of complex songs was just $n = 1$, and the sample size for simple songs was $n = 1$ also. As a result, the study only showed that females liked the song of one of the two males better than that of the other, not that complex songs are more attractive in general. To ensure independence, the study should have recorded the songs of 40 males with complex songs, and 40 males with simple songs. Each female should have listened to a unique pair of songs, one simple and one complex.

The birdsong study is an example of pseudoreplication. **Pseudoreplication** occurs whenever individual measurements that are not independent are analyzed as if they are independent of one another. In the birdsong study, the 40 measurements of females were treated as 40 independent data points, when they were not independent. The problem with pseudoreplication is that measurements

---

1. We'll keep the specific example fictional, to avoid singling out one set of authors unfairly, but this is based on examples in the literature.

115

obtained from individuals not sampled independently might be more similar to one another than measurements made on individuals sampled independently from the population. "Replication" by itself is good—it refers to the sampling of multiple independent units from a population, which makes it possible to estimate population characteristics and the precision of those estimates. In general, the larger the level of replication, the greater our confidence in our results. But if we analyze non-independent data points as if they were independent, then we are making a false claim about the amount of replication, hence the "pseudo-" in *pseudoreplication*.[2]

Most statistical techniques, including almost everything in this book, assume that each data point is independent of the others. Independence is, after all, built into the definition of a random sample. When we assume that data points are independent of each other, we give each data point equal credence and weigh its information as heavily as every other point. If two data points are not independent, though, then treating them as independent makes it seem as if we have more information than we really do. We would be treating the data set as if it were larger than it really is, and as a result we would calculate confidence intervals that were too narrow and P-values (see Section 6.2) that were too small.

Imagine that we wanted to know the blood sugar levels of diabetes patients. An overzealous phlebotomist takes 15 samples from each of 10 patients, yielding a total of 150 measurements. How can we treat these 150 data points? If we threw them all together and analyzed them as if we had 150 independent data points, we would commit the

sin of pseudoreplication. If the patients were randomly chosen, then we would have only 10 independent data points. We needn't throw out any of the 15 samples per patient. Rather, for each patient, we could take the average of the 15 measurements and use this as the independent observation. Doing so gives us a more reliable estimate of the blood sugar level of each patient, which can reduce the amount of sampling variation in our estimate compared with an estimate based on only a single measurement per patient. But no matter how many measurements are taken from a patient, the total sample size of this study is still only $n = 10$. By recognizing that the patients, not the measurements, represent our unit of replication, we can analyze the data appropriately.

Pseudoreplication can often be avoided by summarizing the information on each independently sampled individual and using those summaries as the data for the analysis.

Pseudoreplication is often subtle, and it remains a major source of mistakes in the analysis of experiments.

The rate of pseudoreplication has been estimated to be one in every eight field studies in ecology (Hurlbert and White 1993, Heffner et al. 1996). When reading the scientific literature, keep in mind the possibility of pseudoreplication. Be on the lookout for features that group individual data points during the sampling process. Watch out for multiple measurements taken on the same individuals or the same experimental unit. If the number of measurements, and not the number of independent individuals, is counted as the sample size in a statistical analysis, there could be a problem.

Suggested reading

Hurlbert, S. H. 1984. Pseudoreplication and the design of ecological field experiments. *Ecological Monographs* 54: 187–211.

---

2. The term *pseudoreplication* was coined by Hurlbert (1984) in an extremely readable exposé of the shocking ubiquity of pseudoreplication in biology.

White-breasted nuthatch

5 Probability

The concept of probability is important in almost every field of science, including biology. Probability is the backbone of data analysis. In Chapter 4, we made statements about probability to quantify the uncertainty of parameter estimates, and we will see even more uses of probability throughout the book.

Probability is essential to biology because we almost always look at the natural world by way of a sample, and, as we have seen, chance can play a major role in the properties of samples. In this chapter, we will discuss the basic principles of probability and basic probability calculations. In Chapter 6, we will begin to apply these concepts to data analysis.

5.1 The probability of an event

Imagine that you have 1000 songs on your phone, each of them recorded exactly once. When you push the "shuffle" button, the phone plays a song at random from the list of 1000. The probability that the first song played is your single favorite song is 1/1000, or 0.001. The probability that the first song played is not your favorite song is 999/1000, or 0.999. What exactly do these numbers mean?

The concept of probability rests on the notion of a **random trial**. A random trial is a process or experiment that has two or more possible outcomes whose occurrence cannot be predicted with certainty. Only one outcome is observed from each repetition of a random trial. In the phone songs example, a random trial consists of pushing the shuffle button once. The specified outcome is "your favorite song is played," which is one of 1000 possible outcomes. Other examples of random trials include

- Flipping a coin to see if heads or tails comes up,
- Rolling a pair of dice to see what the sum of their numbers is,
- Rolling a die 10 times to measure the proportion of times a "6" comes up.

What is probability? To answer this, we need to define the **event** of interest and the list of all possible **outcomes** of a random trial. An event is any potential subset of all the possible outcomes. For example, there are six possible outcomes if we roll a six-sided die—the numbers one through six. The event of interest could be "the result is an even number," "the result is a number greater than three," or even the simple event "the result is four." As the last example shows, an outcome is also an event. However, events can be more complicated subsets of outcomes, so we will define principles of probability mainly in terms of events.

The **probability** of an event is the *proportion* of all random trials in which the specified event occurs when the same random process is repeated over and over again independently and under the same conditions.[1] In an infinite number of random trials carried out in exactly the same way, the probability of an event is the fraction of the trials in which the event occurs.

> The *probability* of an event is the proportion of times the event would occur if we repeated a random trial over and over again under the same conditions. Probability ranges between zero and one.

A useful shorthand is the following:

Pr[*A*] means "the probability of event *A*."

Thus, if we want to state the probability of "rolling a four" with a six-sided die, then we can write

$$\Pr[rolling\ a\ four] = 1/6.$$

Because probabilities are proportions, they must always fall between zero and one, inclusive. An event has probability zero if it *never* happens, and an event has probability one if it *always* happens.

---

1. Believe it or not, the word "probability" has other definitions, even within statistics. In one alternative, "probability" refers to a subjective state of belief by the researcher about the truth. For example, "There's a 95% probability that I turned the gas off before we left for vacation." In this book, though, we define probability only as a proportion.

Flipping coins and rolling dice are not biological processes, but they are simple and familiar and the probabilities are known. Their relevance to biology is nevertheless high because they mimic the process of sampling. Randomly sampling 10 new babies and counting the number that are female is mimicked by flipping a coin 10 times and counting the number of heads (assuming both have probability 1/2). Randomly sampling 100 individuals from a human population and counting the proportion that are left-handed is mimicked by rolling a six-sided die 100 times and counting the proportion of sixes (assuming that a "six" and a "left-handed person" both have probability 1/6).

In other words, randomly sampling a population represents a random trial just like rolling a die or flipping a coin. The value of a variable measured on a randomly sampled individual is an outcome of a random trial. The following are therefore also random trials:

- Randomly sampling an individual from a population of sockeye salmon to see what its weight is, and
- Randomly sampling 1000 fetuses from clinics in a large North American city to measure the proportion that have Down syndrome.

▮5.2▮ Venn diagrams

One useful way to think about the probabilities of events is with a graphical tool called a **Venn diagram**. The area of the diagram represents all possible outcomes of a random trial, and we can represent various events as areas within the diagram. The probability of an event is proportional to the area it occupies in the diagram.

Figure 5.2-1 shows a Venn diagram for one roll of a fair six-sided die. The six possible outcomes fill the diagram, indicating that these are all possible results. The area of the box for each outcome is the same, showing that these outcomes are equally probable in this particular example. They each contain 1/6 of the area of the Venn diagram.

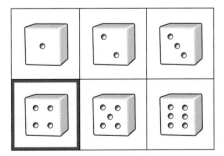

FIGURE 5.2-1
A Venn diagram for the possible outcomes of a roll of a six-sided die. The area corresponding to the event "the result is a four" is highlighted in red.

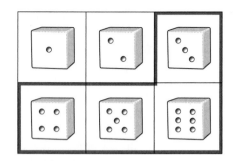

FIGURE 5.2-2
A Venn diagram showing the event "the result is greater than two" highlighted in red. The probability of this event is $4/6 = 2/3$, equal to the area of the red region.

We can use Venn diagrams to show more complicated events as well. In Figure 5.2-2, for example, the event "the result is greater than two" is shown.

5.3 Mutually exclusive events

When two events cannot both occur at the same time, we say that they are **mutually exclusive**. For example, a single die rolled once cannot yield both a one and a six. The events "one" and "six" are mutually exclusive events for this random trial.

> Two events are *mutually exclusive* if they cannot both occur at the same time.

Sometimes physical constraints explain why certain events are mutually exclusive. It is impossible, for example, for more than one number to result from a single roll of a die. Sometimes events are mutually exclusive because they never occur simultaneously in nature. For example, "has teeth" and "has feathers" are mutually exclusive events when we randomly sample a single living animal species from a list of all existing animal species, because no living animals have both teeth and feathers. If we sample a living animal species at random, the probability that it has both teeth and feathers is zero, although plenty of animals have teeth *or* feathers.

In mathematical terms, two events A and B are mutually exclusive if

$$\Pr[A \text{ and } B] = 0.$$

Here, $\Pr[A \text{ and } B]$ means the probability that both A and B occur.

5.4 Probability distributions

A probability distribution describes the probabilities of each of the possible outcomes of a random trial. Some probability distributions can be described mathematically, while others are just a list of the possible outcomes and their probabilities. The precise meaning of a probability distribution depends on whether the variable is discrete or continuous.

> A *probability distribution* is a list of the probabilities of all mutually exclusive outcomes of a random trial.

Discrete probability distributions

A discrete numerical variable is measured in indivisible units. Categorical variables are discrete, as are many numerical variables. A discrete probability distribution gives the probability of each possible value of a discrete variable. Categorical and discrete numerical variables have discrete probability distributions. For example, the probability distribution of outcomes for the single roll of a fair die is given in Figure 5.4-1. In this case, all integers between one and six are equally probable outcomes (probability $= 1/6 = 0.167$). The histogram in Figure 5.4-2 shows the probability distribution for the sum of the two numbers resulting from a roll of two dice. Here the different outcomes are *not* equally probable.

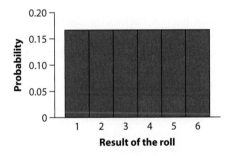

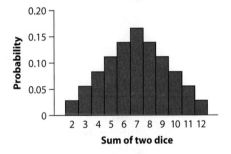

FIGURE 5.4-1 The probability distribution of outcomes resulting from the roll of a single six-sided fair die. The probability of each possible outcome is $1/6 = 0.167$.

FIGURE 5.4-2 The probability distribution for the sum of the numbers resulting from rolling two 6-sided fair dice.

Because all possible outcomes are taken into account, the sum of all probabilities in a probability distribution must add to one. This is because a probability distribution has to describe all possible outcomes, and the probability that *some* outcome occurs from a random trial is one.

Continuous probability distributions

Unlike discrete variables, continuous numerical variables can take on any real number value within some range. Between any two values of a continuous variable (call the variable Y), an infinite number of other values are possible. We describe a continuous probability distribution with a curve whose height is **probability density**. A probability density allows us to describe the probability of any range of values for Y.

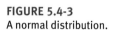

FIGURE 5.4-3
A normal distribution.

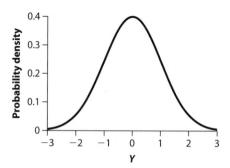

The normal distribution, first introduced in Section 1.4, is a continuous probability distribution. It is bell-shaped like the curve shown in Figure 5.4-3. We'll see much more of this distribution in Chapter 10 and later.

Imagine that we sample a random number from this distribution—let's call the number Y. Unlike discrete probability distributions, the height of a continuous probability curve at the value of $Y = 2.4$ does not give the probability of obtaining $Y = 2.4$. Because a continuous probability distribution covers an infinite number of possible outcomes, the probability of obtaining any specific outcome is infinitesimally small and therefore zero.

With continuous probability distributions, such as the normal curve, it makes more sense to talk about the probability of obtaining a value of Y within some range. The probability of obtaining a value of Y within some range is indicated by the area under the curve. For example, the probability that a single randomly chosen individual has a measurement lying between the two numbers a and b equals the area under the curve between a and b (Figure 5.4-4).

The area under the curve between a and b is calculated by integrating[2] the probability density function between the values a and b. Integration is the continuous analog of summation, so integrating the probability density function from a to b is analogous to adding together the areas of very many narrow rectangles under the curve between a and b (see the right panel in Figure 5.4-4).

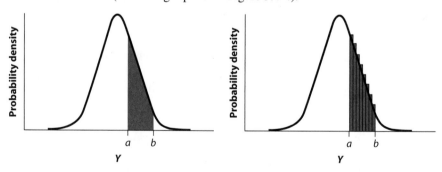

FIGURE 5.4-4 The probability that a randomly chosen Y-measurement lies between a and b is the area under the probability density curve between a and b (*left panel*). In the right panel we approximate the same area using discrete bars.

---

2. Don't panic! We won't ask you to carry out this integration in this book.

For any probability distribution, the area under the entire curve of a continuous probability density function is always equal to one. Finally, because the probability of any individual Y-value is infinitesimally small under a continuous probability density distribution, $\Pr[a \leq Y \leq b]$ is the same as $\Pr[a < Y < b]$.

5.5 Either this or that: adding probabilities

Very often, we want to know the probability that we get *either* one event *or* another. For example, the probability that a randomly chosen North American has a particular ABO blood type and Rh factor (+ or –) is shown in Table 5.5-1 (Stanford Blood Center 2012). What is the probability that an American has blood type O? A person is blood type O if she is either O+ or O–. We can use the addition rule to calculate this probability.

TABLE 5.5-1
Probability that a randomly chosen American will have a given blood type. A, B, and O refer to ABO blood type, and "+" and "–" refer to Rh factor.

Blood type	Probability
O+	0.374
O–	0.066
A+	0.357
A–	0.063
B+	0.085
B–	0.015
AB+	0.034
AB–	0.006

The addition rule

If two events are mutually exclusive, then calculating the probability of one or the other event occurring is both intuitive and easy. The probability of getting either of two mutually exclusive events is simply the sum of the probabilities of each of those events separately. Having blood type O– and having blood type O+ are mutually exclusive events. Therefore, the chance of a person being either O– or O+ is the chance of being O– plus the chance of being O+:

$$\Pr[O- \text{ or } O+] = \Pr[O-] + \Pr[O+]$$
$$= 0.374 + 0.066 = 0.440.$$

Figure 5.5-1 illustrates the probability of O– or O+ with a Venn diagram.

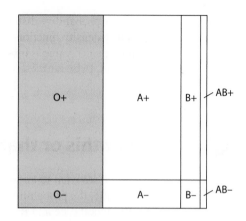

FIGURE 5.5-1
The probability of O– or O+ is equal to the probability of O– plus the probability of O+, because the two events are mutually exclusive.

This additive property of the probabilities of mutually exclusive events is called the **addition rule**.

> The *addition rule*: If two events *A* and *B* are mutually exclusive, then
>
> $$\Pr[A \text{ or } B] = \Pr[A] + \Pr[B].$$

The addition rule extends to more than two events as long as they are all mutually exclusive. Let's say that your blood type is B–, and we want to know the probability that you could safely donate blood to a randomly sampled American in the event of emergency. B– blood can be safely donated to anyone who is B+, B–, AB+, or AB–. These four possibilities are mutually exclusive, because a randomly sampled American cannot have more than one of these blood types at the same time. Thus, the probability that an American is able to receive your B– blood safely can be calculated as follows using the addition rule:

$$\Pr[\text{B+ or B– or AB+ or AB–}] = \Pr[\text{B+}] + \Pr[\text{B–}] + \Pr[\text{AB+}] + \Pr[\text{AB–}]$$
$$= 0.085 + 0.015 + 0.034 + 0.006$$
$$= 0.140.$$

The addition rule is about "or" statements. If two events are mutually exclusive and we want to know the probability of being *either* one *or* the other, we can use the addition rule. This property is vital to analyzing data because it allows us to calculate the probabilities of different outcomes of random sampling when they are mutually exclusive.

The probabilities of all possible mutually exclusive outcomes add to one

The probabilities of all possible mutually exclusive outcomes of a random trial must add to one. With blood type, for example, there are eight possible outcomes. Therefore, the sum of the probabilities of all outcomes is

$$\Pr[\text{O+ or O– or A+ or A– or B+ or B– or AB+ or AB–}]$$
$$= \Pr[\text{O+}] + \Pr[\text{O–}] + \Pr[\text{A+}] + \Pr[\text{A–}] + \Pr[\text{B+}] + \Pr[\text{B–}] + \Pr[\text{AB+}] + \Pr[\text{AB–}]$$
$$= 0.374 + 0.066 + 0.357 + 0.063 + 0.085 + 0.015 + 0.034 + 0.006$$
$$= 1.$$

This means that the probability of an outcome or event *not* occurring is simply one minus the probability that it occurs. For example, the probability that you do *not* get O+ when you type the blood of a randomly sampled American is

$$\Pr[\text{not O+}] = 1 - \Pr[\text{O+}] = 1 - 0.374 = 0.626.$$

This calculation is much easier than summing the probabilities of all outcomes other than O+.

> The probability of an event not occurring is one minus the probability that it occurs.
> $$\Pr[not\ A] = 1 - \Pr[A].$$

The general addition rule

Not all events, though, are mutually exclusive. It is possible, for example, for the ABO blood type of a randomly sampled American to be O and his or her Rh factor to be positive (+). If the two events are not mutually exclusive, how do we calculate the probability of either one or the other event occurring?

In mathematical notation, a **general addition rule** can be written as

$$\Pr[A \text{ or } B] = \Pr[A] + \Pr[B] - \Pr[A \text{ and } B].$$

This calculates the probability that either A or B (or both) occur.

When events A and B are mutually exclusive, $\Pr[A \text{ and } B] = 0$, so the generalized addition rule reduces to the addition rule for mutually exclusive events introduced previously. The reason we have to subtract the probability of both A and B occuring is illustrated in Figure 5.5-2. If we do *not* subtract the probability of both A and B occurring, then we will double-count those outcomes where both A and B occur.

So, for example, the probability that a randomly chosen American has either the most common ABO type (O) or the most common Rh factor (+) is

$$\Pr[\text{O}] + \Pr[\text{+}] - \Pr[\text{O and +}] = 0.440 + 0.850 - 0.374 = 0.916.$$

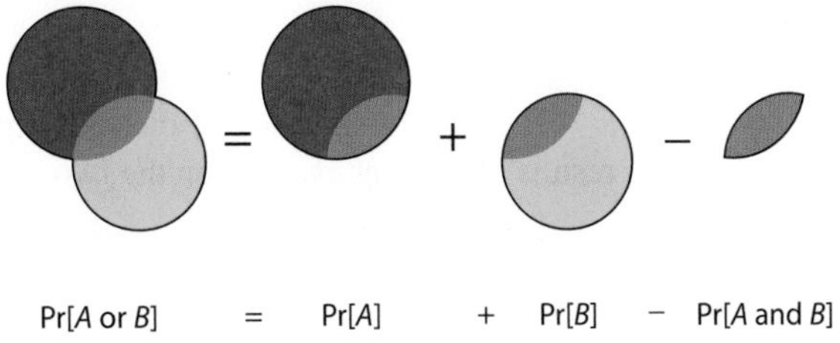

FIGURE 5.5-2
The general addition rule. Pr[A and B] is subtracted from Pr[A] + Pr[B] so that the outcomes where both A and B occur (the tan shaded areas) are not counted twice.

| Pr[A or B] | = | Pr[A] | + | Pr[B] | – | Pr[A and B] |

5.6 Independence and the multiplication rule

Science is the study of patterns, and patterns are generated by relationships between events. Men are more likely to be tall, more likely to have a beard, more likely to die young, and more likely to go to prison than women. In other words, height, beardedness, age at death, and criminal conduct are not independent of sex in the human population.

Sometimes, though, the chance of one event occurring does *not* depend on another event. If we roll two dice, for example, the number on one die does not affect the number on the other die. If knowing one event gives us no information about another event, then these two events are called independent.

Two events are **independent** if the occurrence of one does not in any way inform us about the probability that the other will also occur. When rolling the same fair die twice in a row, for example, the probability that the first roll gives a three is 1/6, as we saw previously:

$$\Pr[\textit{first roll is three}] = 1/6.$$

What is the probability that the next roll will also be a three? The probability of rolling a three on the second roll is still 1/6, regardless of whether the first roll was a three or not. Because the outcome of the first roll does not give any information about the probability of rolling a three on the second roll, we can say that the two events are independent (Figure 5.6-1).

FIGURE 5.6-1
A Venn diagram for all the possible outcomes of rolling two 6-sided dice. The first digit of each pair shows the result of the roll of the first die, and the second number shows the result of the roll of the second die. Rolling a three on the first roll is shown in the blue row. The probability of rolling a three on the second roll is shown in the green column and is the same (1/6) regardless of the result of the first roll.

1,1	1,2	1,3	1,4	1,5	1,6
2,1	2,2	2,3	2,4	2,5	2,6
3,1	3,2	3,3	3,4	3,5	3,6
4,1	4,2	4,3	4,4	4,5	4,6
5,1	5,2	5,3	5,4	5,5	5,6
6,1	6,2	6,3	6,4	6,5	6,6

Probability of rolling a 3 on the first roll is 1/6.

Probability of rolling a 3 on the second roll is 1/6.

Two events are *independent* if the occurrence of one does not inform us about the probability that the second will occur.

When the occurrence of one event provides at least some information about the results of another event, then the two events are **dependent**.

Multiplication rule

When two events are independent, then the probability that they both occur is the probability of the first event multiplied by the probability of the second event. This is called the **multiplication rule**. When we analyze data, we use this multiplication rule to determine what to expect when two variables are independent.

> The *multiplication rule*: If two events A and B are independent, then
>
> $$\Pr[A \text{ and } B] = \Pr[A] \times \Pr[B].$$

We can see the basis of the multiplication rule in Figure 5.6-1. The area of the Venn diagram that corresponds to "rolling a three on the first die" and "rolling a three on the second die" is the region of overlap between the blue and green areas. Because the two events are independent, the area of this overlap zone is just the probability of being in the blue times the probability of being in the green:

$$\Pr[(\textit{first roll is a three}) \text{ and } (\textit{second roll is a three})]$$
$$= \Pr[\textit{first roll is a three}] \times \Pr[\textit{second roll is a three}]$$
$$= 1/6 \times 1/6$$
$$= 1/36.$$

The multiplication rule pertains to combinations with "and"—that is, that *both* events occur. If we want to know the probability of *this* and *that* occurring, and if the two events are independent, we can multiply the probabilities of each to get the probability of both occurring. Example 5.6A applies the multiplication rule to a study about smoking and high blood pressure.

EXAMPLE   Smoking and high blood pressure

5.6A   Both smoking and high blood pressure are risk factors for strokes and other vascular diseases. In the United States, approximately 17% of adults smoke and about 22% have high blood pressure. Research has shown that high blood pressure is not associated with smoking; that is, they seem to be independent of each other (Liang et al. 2001). What is the probability that a randomly chosen American adult has both of these risk factors?

Because these two events are independent, the probability of a randomly sampled individual both "smoking" and "having high blood pressure" is the probability of smoking times the probability of high blood pressure:

$$\Pr[\textit{smoking and high blood pressure}] = \Pr[\textit{smoking}] \times \Pr[\textit{high blood pressure}]$$
$$= 0.17 \times 0.22$$
$$= 0.037.$$

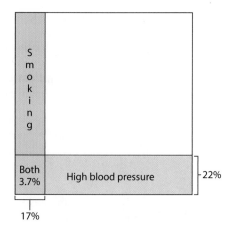

FIGURE 5.6-2
Venn diagram for the two independent factors smoking and high blood pressure. The probability of having both risk factors is proportional to the area of the rectangle in the bottom left corner.

Therefore, 3.7% of adult Americans will have both of these risk factors for strokes. This calculation is shown geometrically in the Venn diagram in Figure 5.6-2.

"And" versus "or"

Probability statements involving "and" or "or" statements are common enough, and confusing enough, that it is worth summarizing them together:

- The probability of A or B involves addition. That is,
  $\Pr[A \text{ or } B] = \Pr[A] + \Pr[B]$ if the two events A and B are mutually exclusive.
- The probability of A and B involves multiplication. That is,
  $\Pr[A \text{ and } B] = \Pr[A] \times \Pr[B]$ if A and B are independent.

What may be confusing is that the statement involving "and" requires multiplication, not addition.

Independence of more than two events

The multiplication rule also applies to more than two events, as Example 5.6B demonstrates. If several events are all independent, then the probability of all events occurring is the product of the probabilities that each one occurs.

EXAMPLE Mendel's peas

5.6B Like blue eyes in humans, yellow pods in peas is a recessive trait. That is, pea pods are yellow only if both copies of the gene code for yellow. A plant having only one yellow copy and one green copy (a "heterozygote") has green pods just like the pods of plants having two green copies of the gene (a green "homozygote"). Gregor Mendel devised a method to determine whether a green plant was a heterozygote or a homozygote. He crossed the test plant to itself and assessed the pod color of 10 randomly chosen offspring. If all 10 were green, he inferred the plant was a homozygote, but if even one offspring was yellow, the test plant was classified as a heterozygote. However, he might have missed some heterozygotes, if by chance not a single yellow offspring was chosen. What is the chance of missing a heterozygote by Mendel's method? If the test plant is a homozygote, every offspring is green. If the test plant is a heterozygote, on the other hand, the chance of an offspring being green is 3/4 and the chance of it being yellow is only 1/4. What is the chance that all 10 offspring from a heterozygote test plant are green?

Mendel didn't carry out these calculations, but we can use our rules of probability to figure out the reliability of his approach. The chance that any one of a heterozygote's offspring is green is 3/4. Because the genotype of each offspring is independent of the genotypes of other offspring, the probability that all 10 are green can be calculated using the multiplication rule.

$$\Pr[\textit{all 10 green}] = \Pr[\textit{first is green}] \times \Pr[\textit{second is green}] \times \Pr[\textit{third is green}] \times \cdots$$
$$= 3/4 \times 3/4 \times 3/4 \times \cdots = (3/4)^{10} = 0.056.$$

Thus, Mendel likely misidentified about 5.6% of heterozygous individuals. On the other hand, his method correctly identified heterozygotes with probability $(1 - 0.056) = 0.944$.

5.7   Probability trees

A **probability tree** is a diagram that can be used to calculate the probabilities of combinations of events resulting from multiple random trials. We show how to use probability trees with Example 5.7.

EXAMPLE

5.7

Sex and birth order

Some couples planning a new family would prefer to have at least one child of each sex. The probability that a couple's first child is a boy[3] is 0.512. In the absence of technological intervention, the probability that their second child is a boy is independent of the sex of their first child, and so remains 0.512. Imagine that you are helping a new couple with their planning. If the couple plans to have only two children, what is the probability of getting one child of each sex?

This question requires that we know the probabilities of all mutually exclusive values of two separate variables. The first variable is "the sex of the first child." The second variable is "the sex of the second child." We can start building a probability

---

3. This excess of boys is a highly repeatable pattern, measured over tens of millions of births. More boys than girls are born. The fraction of males is even higher at conception than at birth, but this fraction declines during pregnancy because male fetuses die at a higher rate than female fetuses. It is thought that sperm bearing a Y chromosome might swim faster—and thus reach the egg sooner—than X-bearing sperm, which would account for the excess of boys.

tree by considering the two variables in sequence. Let's start with the sex of the first child. Two mutually exclusive outcomes are possible—namely, "boy" and "girl"—which we list vertically, one below the other (see figure at right). We then draw arrows from a single point on the left to both possible outcomes. Along each arrow we write the probability of occurrence of each outcome (0.512 for "boy" and 0.488 for "girl").

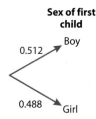

Now, we list all possible values for the second variable, but we do so separately for each possible value of the first variable. For example, for the value "boy" for the first child, we list both possible values (i.e., "boy" and "girl") for the sex of the second child. Next, we draw arrows originating from the value "boy" for the first variable to both possible values for the second variable. Then we write the probability of each value for the second variable along each arrow. We repeat this process for the case when "girl" is the value of the first variable. The resulting probability tree is shown in Figure 5.7-1.

FIGURE 5.7-1
A probability tree for all possible values of a two-child family.

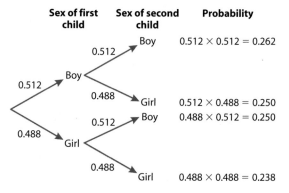

At this point, we should check that our probabilities are written down correctly. For instance, the probabilities along all arrows originating from a single point must sum to one (within rounding error) because they represent all the mutually exclusive possibilities. If they don't sum to one, we've forgotten to include some possibilities or we've written down the probabilities incorrectly.

With a probability tree, we can calculate the probability of every possible sequence of values of the two variables. A sequence of values is represented by a path along the arrows of the tree that begins at the root at the far left and ends at one of the branch tips on the right. The probability of a given sequence is calculated by multiplying all of the probabilities along the path taken from the root to the tip. For example, the sequence "boy then girl" in Figure 5.7-1 has a probability of 0.512 × 0.488 = 0.250. On our probability tree, we usually list the probabilities of each sequence of values in a column to the right of the tree tips, as shown in Figure 5.7-1.

Each tree tip defines a unique and mutually exclusive sequence of events. Check Figure 5.7-1 (or any probability tree) to make sure that the probabilities of all possible sequences add to one. If they don't add to one (within rounding error), then something has gone wrong in the construction of the tree.

What is the probability of having one child of each sex in a family of two children? According to the probability tree, two of the four possible sequences result in the birth of one boy and one girl. In the first sequence, the boy is born first, followed by the girl, whereas in the second sequence, the girl is born first and the boy is born second. These two different sequences are mutually exclusive, and we are looking for the probability of either the first *or* the second sequence. By the addition rule, therefore, the probability of getting exactly one boy and one girl when having two children is the sum of the probabilities of the two alternative sequences leading to this event: $0.250 + 0.250 = 0.500$.

We could also use the probability tree in Figure 5.7-1 to calculate probabilities of the following events:

- The probability that at least one girl is born,
- The probability that at least one boy is born, and
- The probability that both children are the same sex.

Calculate these probabilities yourself to test your understanding.[4]

It is not essential to use probability trees when calculating the probabilities of sequences of events, but they are a helpful tool for making sure that you have accounted for all of the possibilities.

5.8 Dependent events

Independent events are mathematically convenient, but when the probability of one event depends on another, things get interesting. Much of science involves identifying variables that are associated.

Sex determination is more exotic in many insects than in humans. In many species, the mother can alter the relative numbers of male and female offspring depending on the local environment. In this case, "sex of offspring" and "environment" are dependent events, as Example 5.8 demonstrates.

EXAMPLE
5.8

Is this meat taken?

The jewel wasp, *Nasonia vitripennis*, is a parasite, laying its eggs on its host, the pupae of flies. The larval *Nasonia* hatch inside the pupal case, feed on the live host, and grow until they emerge as adults from the now dead, emaciated host. Emerging males and females, possibly brother and sister, mate on the spot. *Nasonia* females have a remarkable ability to manipulate the sex of the eggs that they lay.[5] When a

---

4. Answers: Pr[*at least one girl*] = 0.738; Pr[*at least one boy*] = 0.762; Pr[*both same sex*] = 0.500.

5. Wasps, like ants and bees, have a very different mechanism than humans for determining the sex of their offspring. All a female has to do to determine the sex of an egg at the time of laying is to control whether or not she fertilizes it with the sperm she has stored. If she fertilizes it, it becomes a female. If not, it's a male.

female finds a fresh host that has not been previously parasitized, she lays mainly female eggs, producing only the few sons needed to fertilize all her daughters. But if the host has already been parasitized by a previous female, the next female responds by producing a higher proportion of sons.[6] Thus, the state of the host encountered by a female and the sex of an egg laid are dependent variables (Werren 1980).

Suppose that, when a given *Nasonia* female finds a host, there is a probability of 0.20 that the host already has eggs, laid by a previous female wasp. Presume that the female can detect previous infections without error. If the host is unparasitized, the female lays a male egg with probability 0.05 and a female egg with probability 0.95. If the host already has eggs, then the female lays a male egg with probability 0.90 and a female egg with probability 0.10. Figure 5.8-1 shows a mosaic plot of these probabilities.

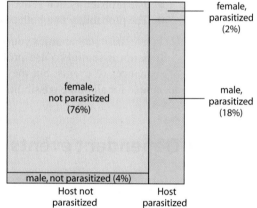

FIGURE 5.8-1

A mosaic plot showing that the sex of eggs laid by *Nasonia* females depends on the state of the host.

Based on Figure 5.8-1, the events "host is previously parasitized" and "producing a male egg" are dependent. The probability of laying a male egg changes depending on whether the host has been previously parasitized. Suppose we want to know the probability that a new, randomly chosen egg is male. We can approach this question using a probability tree like the one shown in Figure 5.8-2.

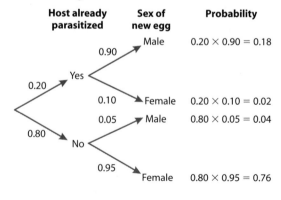

FIGURE 5.8-2

A probability tree for the sex of offspring laid by *Nasonia* according to whether the host has been previously parasitized.

---

6. The value to her of sons has risen in the second case because there are now plenty of unrelated females to mate with.

According to the probability tree, there are exactly two paths that yield a male egg. In the first, the host is already parasitized and the mother lays a male egg. This path has probability

Pr[*host already parasitized* and *sex of new egg is male*] = 0.20 × 0.90 = 0.18.

In the second path, the host is not previously parasitized and the female lays a male egg. This second path has probability

Pr[*host not already parasitized* and *sex of new egg is male*] = 0.80 × 0.05 = 0.04.

The probability of a new egg being male is the sum of the probabilities of these two mutually exclusive paths:

$$\Pr[male] = 0.18 + 0.04 = 0.22.$$

The probability of an egg being male in this population is 0.22. See Figure 5.8-3.

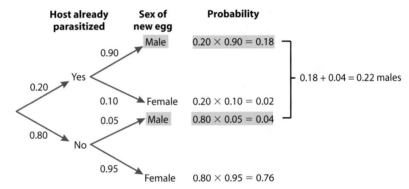

FIGURE 5.8-3
A probability tree
for the sex of an egg
laid by *Nasonia*.

The probability tree shows that the event "sex of new egg is male" depends on whether the host encountered by a mother has been previously parasitized. Does this mean that the events "host already parasitized" and "sex of new egg is male" are not independent? One way to confirm this is via the multiplication rule, which applies only to independent events. The probability that "the host already had been parasitized and the sex of the new egg is male" is 0.18. This is *not* what we would have expected assuming independence, though. If we multiply the probability that the new egg is a male (0.22, as we just calculated) and the probability that a host is already parasitized (0.20), we get 0.22 × 0.20 = 0.044, which is different from the actual probability of these two events (0.18). Based on the definition of "independence," then, these two events are *not* independent.

5.9 Conditional probability and Bayes' theorem

If we want to know the chance of an event, we need to take account of all existing information that might affect its outcome. If we want to know the probability that we will see an elephant on our afternoon stroll, for example, we would get a different answer depending on whether our walk was in the Serengeti or downtown Manhattan. The algebra of conditional probability lets us hone our statements about the chances of random events in the context of extra information.

Conditional probability

Conditional probability is the probability of an event given that another event occurs.

> The conditional probability of an event is the probability of that event occurring *given that* a condition is met.

In Example 5.8, the conditional probability that a jewel wasp will lay a male egg is 0.90 *given that* the host that she is laying on already has wasp eggs (i.e., has already been parasitized). Confirm this for yourself by looking at Figure 5.8-2 again. We write conditional probability in the following way:

$$\Pr[\textit{new egg is male} \mid \textit{host is previously parasitized}] = 0.90.$$

More generally, $\Pr[\textit{event} \mid \textit{condition}]$ represents the probability that the event will happen given that the condition is met. The vertical bar in the middle of this expression is a symbol that means "given that" or "when the following condition is met." (Be careful not to confuse it with a division sign.)

The Venn diagram in Figure 5.8-1 illustrates the meaning of this conditional probability. Ninety percent of the area corresponding to "host parasitized" represents the cases when the offspring is male, with the remaining 10% being females. The probability of a male is different under the condition "host parasitized" than under the condition "host not parasitized."

Conditional probability has many important applications. If we want to know the overall probability of a particular event, we sum its probability across every possible condition, weighted by the probability of that condition. This is known as the **law of total probability**.

> According to the *law of total probability*, the probability of an event, A, is
>
> $$\Pr[A] = \sum_{\textit{All values of B}} \Pr[B]\,\Pr[A \mid B],$$
>
> where B represents all possible mutually exclusive values of the conditions.

One way of thinking about this formula is that it gives the weighted average probability of A over all possible mutually exclusive conditions.

The Venn diagram in Figure 5.8-1 makes it possible to visualize this, too. The probability of being male is obtained by adding the two blue areas, one for the condition when the host is already parasitized and the other for when the host is not already parasitized. The width of these boxes is proportional to the probability of the condition; the height is proportional to Pr[*male* | *host condition*]. By multiplying the width by the height of each box we find its area (its probability), and by adding all such boxes together we find the total probability of males.

To calculate the probability that a new egg is a male, we must consider two possible conditions: (1) the host is already parasitized and (2) the host is not parasitized. Thus, we'll have two terms on the right side of our equation:

$$\Pr[\text{egg is male}] = \Pr[\text{host already parasitized}] \Pr[\text{egg is male} \mid \text{host already parasitized}]$$
$$+ \Pr[\text{host not parasitized}] \Pr[\text{egg is male} \mid \text{host not parasitized}]$$
$$= (0.20 \times 0.90) + (0.80 \times 0.05) = 0.22.$$

This is the same answer that we got from the probability tree, but now we can see how it can be derived from statements of conditional probability.

The general multiplication rule

With conditional probability statements, we can find the probability of a combination of two events even if they are not independent. When two events are not independent, the probability that both occur can be found by multiplying the probability of one event by the conditional probability of the second event, given that the first has occurred. This is the **general multiplication rule**.

> The *general multiplication rule* finds the probability that both of two events occur, even if the two are dependent:
>
> $$\Pr[A \text{ and } B] = \Pr[A] \Pr[B \mid A].$$

This rule makes sense, if we think it through. For two events (*A* and *B*) to occur, event *A* must occur. By definition, this happens with probability Pr[*A*]. Now that we know *A* has occurred, the probability that *B* also occurred is Pr[*B* | *A*]. Multiplying these together gives us the probability of both *A* and *B* occurring.

It doesn't matter which event we label *A* and which we label *B*. The reverse is also true; that is,

$$\Pr[A \text{ and } B] = \Pr[B] \Pr[A \mid B].$$

With the jewel wasps, for example, if we wanted to know the probability that a host had already been parasitized and that the mother wasp laid a male egg, we would multiply the probability that it had been parasitized (0.2) times the probability of a male egg *given that* the egg was already parasitized (0.9), to get 0.18. We can see the same probability by following the appropriate path (the top one) through the probability tree in Figure 5.8-2.

If A and B are independent, then having information about A gives no information about B, and therefore $\Pr[B \mid A] = \Pr[B]$. That is, the general multiplication rule reduces to the multiplication rule when the events are independent.

Sampling without replacement

One common use of conditional probability is **sampling without replacement**. This process occurs when the specific outcome of one random trial eliminates or depletes that outcome from the possibilities and so changes the probability distribution of values for subsequent random trials.

As a simple example, consider drawing cards randomly from a fair card deck in which the 52 ordinary cards have been shuffled and so are in random order. What is the probability of drawing three cards in the precise sequence "ace-2-3," ignoring card suit, without returning the cards to the deck? The probability that the first card drawn is an ace is $4/52$, because there are four aces out of the 52 cards. The key novelty is that the outcome of the first draw changes the probabilities of outcomes for later draws if the card is not returned to the deck. For example, if the first card is an ace, then the probability of a 2 in the next draw is changed because there are now only 51 cards in the deck. The chance that the second card is a 2 is now $4/51$. And if we have already taken an ace and a 2 from the deck, the probability that the third card is a 3 is $4/50$, because there are 50 cards left and four of them are 3's. So the probability that the first three draws are in the sequence ace-2-3 is $(4/52) \times (4/51) \times (4/50)$.

In contrast, when sampling with replacement, the sampled individual is not removed from the population after sampling. In this case, the frequencies of possible outcomes in the population are not changed by successive samples.

When sampling populations for biological study, we usually choose populations that are large enough that the sampling of each individual doesn't change the probability distribution of possible values in the individuals that remain. We assume that the effects of depletion are so slight that they don't matter. This will not always be the case, however.

Bayes' theorem

One powerful mathematical relationship about conditional probability is **Bayes' theorem.**[7]

According to Bayes' theorem, for two events A and B,

$$\Pr[A \mid B] = \frac{\Pr[B \mid A]\,\Pr[A]}{\Pr[B]}$$

---

7. This theorem is named after its discoverer, the Reverend Thomas Bayes, an 18th-century English Presbyterian minister.

Bayes' theorem may seem rather complicated, but it can be derived from the general multiplication rule. Because

$$\Pr[A \text{ and } B] = \Pr[B] \Pr[A \mid B]$$

and

$$\Pr[A \text{ and } B] = \Pr[A] \Pr[B \mid A],$$

it is also true that

$$\Pr[B] \Pr[A \mid B] = \Pr[A] \Pr[B \mid A].$$

Dividing both sides by $\Pr[B]$ gives Bayes' theorem. Example 5.9 applies Bayes' theorem to the detection of Down syndrome.

EXAMPLE   Detection of Down syndrome

5.9   Down syndrome (DS) is a chromosomal condition that occurs in about one in 1000 pregnancies. The most accurate test for DS in wide use requires amniocentesis, which unfortunately carries a risk of miscarriage (about one in 200). It would be better to have an accurate test of DS without the risks. One such test in common use is called the triple test, which screens for levels of three hormones in maternal blood at around 16 weeks of pregnancy.

The triple test is not perfect, however. It does not always correctly identify a fetus with DS (an error called a false negative), and sometimes it incorrectly identifies a fetus with a normal set of chromosomes as DS (an error called a false positive). Under normal conditions, the detection rate of the triple test (i.e., the probability that a fetus with DS will be correctly scored as having DS) is 0.60. The false-positive rate (i.e., the probability that a test would say incorrectly that a normal fetus had DS) is 0.05 (Newberger 2000).

Most people's intuition is that these numbers are acceptable. Based on the probabilities given, the triple test would seem to be right most of the time. But, if the test on a randomly chosen fetus gives a positive result (i.e., it indicates that the fetus has DS), what is the probability that this fetus actually has DS? Make a guess at the answer before we work it through.

To address this question, we need Bayes' theorem. We want to know a conditional probability—the probability that a fetus has DS given that its triple test showed a positive result. In other words, we want to know $\Pr[DS \mid positive\ result]$. Using Bayes' theorem,

$$\Pr[DS \mid positive\ result] = \frac{\Pr[positive\ result \mid DS] \; \Pr[DS]}{\Pr[positive\ result]}.$$

We've been given $\Pr[positive\ result \mid DS]$ and $\Pr[DS]$, the two factors in the numerator, but we haven't been given $\Pr[positive\ result]$, the term in the denominator. We can figure out the probability of a positive result, though, by using the law of total probability introduced earlier in this section. That is, we can sum over all the possibilities to find the probability of a positive result.

$$\begin{aligned}
\Pr[positive\ result] &= (\Pr[positive\ result \mid DS]\ \Pr[DS]) \\
&+ (\Pr[positive\ result \mid no\ DS]\ \Pr[no\ DS]) \\
&= (0.60 \times 0.001) + [0.05 \times (1 - 0.001)] = 0.05055.
\end{aligned}$$

The probability of something *not* occurring is equal to one minus the probability of it occurring, so the probability that a randomly chosen fetus does *not* have DS is one minus the probability that it has DS. According to Example 5.9, $\Pr[DS] = 0.001$, so $\Pr[no\ DS] = 1 - 0.001 = 0.999$ in the preceding equation.

Now, returning to Bayes' theorem, we can find the answer to our question.

$$\Pr[DS \mid positive\ result] = \frac{0.60 \times 0.001}{0.05055} = 0.012.$$

There is a very low probability (i.e., 1.2%) that a fetus with a positive score on the triple test actually has DS!

Many people find it more intuitive to think in terms of numbers rather than probabilities for these kinds of calculations. For every million fetuses tested, 1000 will have DS, and 999,000 will not. Of those 1000, 60% or 600 will test positive. Of the 999,000, 5% or 49,950 will test false-positive. Out of a million tests, therefore, there are 600 + 49,950 = 50,550 positive results, only 600 of which are true positives. The 600 true positives divided by the 50,550 total positives is 1.2%, the same answer as we got before. DS babies have a high probability of being detected, but they are a very small fraction of all babies. Thus, the true positive results get swamped by the false positives.

This high false-positive ratio is not unusual. Many diagnostic tools have high proportions of false positives among the positive cases. In this case, erring on the side of caution is appropriate because, when the triple test returns a positive result, it can be checked by amniocentesis.

Did you think that the probability of DS with a positive result would be higher? If so, you're not alone. A survey of practicing physicians found that their grasp of conditional probability with false positives was extremely poor (Elstein 1988). In a question about false-positive rates, where the correct answer was that 7.5% of patients with a positive test result had breast cancer, 95% of the doctors guessed that the answer was 75%! If these doctors had a better understanding of probability theory, they could avoid overstating the risks of serious disease to their patients, thus reducing unnecessary stress.

5.10 Summary

- Probability is an important concept in biology. One reason is that randomly sampling a population represents a random trial whose outcomes are governed by the rules of probability.
- A random trial is a process or experiment that has two or more possible outcomes whose occurrence cannot be predicted with certainty.

- The probability of an event is the proportion of times the event occurs if we repeat a random trial over and over again under the same conditions.
- A probability distribution describes the probabilities of all possible outcomes of a random trial.
- Two events (A and B) are mutually exclusive if they cannot both occur (i.e., $\Pr[A \text{ and } B] = 0$). If A and B are mutually exclusive, then the probability of A or B occurring is the sum of the probability of A occurring and the probability of B occurring (i.e., $\Pr[A \text{ or } B] = \Pr[A] + \Pr[B]$). This is the addition rule.
- The general addition rule gives the probability of either of two events occurring when the events are not mutually exclusive:

$$\Pr[A \text{ or } B] = \Pr[A] + \Pr[B] - \Pr[A \text{ and } B].$$

The general addition rule reduces to the addition rule when A and B are mutually exclusive, because then $\Pr[A \text{ and } B] = 0$.
- Two events are independent if knowing one outcome gives no information about the other outcome. More formally, A and B are independent if $\Pr[A \text{ and } B] = \Pr[A] \Pr[B]$. This is the multiplication rule.
- Probability trees are useful devices for calculating the probabilities of complicated series of events.
- If events are not independent, then they are said to be dependent. The probability of two dependent events both occurring is given by the general multiplication rule: $\Pr[A \text{ and } B] = \Pr[A] \Pr[B \mid A]$.
- The conditional probability of an event is the probability of that event occurring given some condition.
- Probability trees and Bayes' theorem are important tools for calculations involving conditional probabilities.
- The law of total probability, $\Pr[A] = \sum_{All\ values\ of\ B} \Pr[B] \Pr[A \mid B]$, makes it possible to calculate the probability of an event (A) from all of the conditional probabilities of that event. The law multiplies, for all possible conditions (B), the probability of that condition ($\Pr[B]$) times the conditional probability of the event assuming that condition ($\Pr[A \mid B]$).

PRACTICE PROBLEMS

1. **Calculation practice: Addition rule.** When women are asked how much they like Brussels sprouts, 30% say sprouts are "very repulsive," 20% say that they are "somewhat repulsive," 43% are "indifferent," 6% say sprouts are "somewhat delicious," and 1% claim they are "especially delicious." Only one answer per woman was allowed. The data are from Trinkaus and Dennis (1991).

   a. Are these five possible answers mutually exclusive? Explain.

   b. What is the probability that a woman would say that Brussels sprouts are either very repulsive or somewhat repulsive?

   c. What is the probability that a woman would say that Brussels sprouts are anything other than especially delicious?

2. **Calculation practice: Law of total probability.** The survey in the previous problem was conducted on men as well: 34% say Brussels sprouts are "very repulsive," 19%

say that they are "somewhat repulsive," 38% are "indifferent," 8% say they are "somewhat delicious" and 1% claim they are "especially delicious." Assume that in a given population, 52% of the adults are women. Use the following steps to build a probability tree and calculate the probability that a random adult says that Brussels sprouts are somewhat delicious or especially delicious using the law of total probability.

a. What is the probability that a randomly chosen adult is a man? What is the probability that a randomly chosen adult is a woman? Draw the first part of the probability tree for these two events.

b. What is the probability that a man says that Brussels sprouts are somewhat delicious or especially delicious?

c. Write (b) as a shorthand probability statement. Hint: (b) can also be stated as: What is the probability that a randomly chosen adult says that Brussels sprouts are somewhat delicious or especially delicious, given that he is a man?

d. What is the probability that a woman says that Brussels sprouts are somewhat delicious or especially delicious?

e. Complete the probability tree for the preceding events.

f. Apply the law of total probability to determine the probability that a random adult says that Brussels sprouts are somewhat delicious or especially delicious.

3. **Calculation practice: General addition rule.** Among women voluntarily tested for sexually transmitted diseases in one university, 24% tested positive for human papilloma virus (HPV) only, 2% tested positive for *Chlamydia* only, and 4% tested positive for both HPV and *Chlamydia* (Tábora et al. 2005). Use the following steps to calculate the probability that a woman from this population who gets tested would test positive for either HPV or *Chlamydia*.

a. Write the goal of the question as a probability statement.

b. Write the general addition rule with words specific to this example.

c. Calculate the probability that a randomly sampled woman would test positive for HPV or *Chlamydia*.

4. **Calculation practice: General multiplication rule.** In the 1980s in Canada, 52% of adult men smoked. It was estimated that male smokers had a lifetime probability of 17.2% of developing lung cancer, whereas a nonsmoker had a 1.3% chance of getting lung cancer during his life (Villeneuve and Mao 1994).[8]

a. What is the conditional probability of a Canadian man getting cancer, given that he smoked in the 1980s?

b. Draw a probability tree to show the probability of getting lung cancer conditional on smoking.

c. Using the tree, calculate the probability that a Canadian man in the 1980s both smoked and eventually contracted lung cancer.

d. Using the general multiplication rule, calculate the probability that a Canadian man in the 1980s both smoked and eventually contracted lung cancer. Did you get the same answer as in (c)?

e. Using the general multiplication rule, calculate the probability that a Canadian man in the 1980s both did not smoke and never contracted lung cancer.

5. **Calculation practice: Bayes' theorem.** Refer to Practice Problem 4. Use the following steps to calculate the probability that a Canadian man smoked, given that he had been diagnosed with lung cancer.

a. Write Bayes' theorem for the specific case described in this question.

b. Calculate the probability that a Canadian man in the late eighties would eventually develop lung cancer. (Use the law of total probability.)

c. Use Bayes' theorem to calculate the probability that a man from this population smoked, given that he eventually developed lung cancer.

---

8. Actually, later studies showed that smokers were about 23 times more likely than nonsmokers to get lung cancer, but for the purpose of this problem, we'll use the numbers given in this study.

6. The pizza below, ordered from the Venn Pizzeria on Bayes Street, is divided into eight slices:

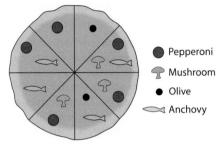

● Pepperoni

🍄 Mushroom

● Olive

🐟 Anchovy

The slices might have pepperoni, mushrooms, olives, and/or anchovies. Imagine that, late at night, you grab a slice of pizza totally at random (i.e., there is a 1/8 chance that you grabbed any one of the eight slices). Base your answers to the following questions on the drawing of the pizza.

a. What is the chance that your slice had pepperoni on it?

b. What is the chance that your slice had both pepperoni and anchovies on it?

c. What is the probability that your slice had either pepperoni or anchovies on it?

d. Are pepperoni and anchovies mutually exclusive on the slices from this pizza?

e. Are olives and mushrooms mutually exclusive on the slices from this pizza?

f. Are getting mushrooms and getting anchovies independent when choosing slices from this pizza?

g. If I pick a slice from this pizza and tell you that it has olives on it, what is the chance that it also has anchovies?

h. If I pick a slice from this pizza and tell you that it has anchovies on it, what is the chance that it also has olives?

i. Seven of your friends each choose a slice at random and eat them without telling you what toppings they had. What is the chance that the last slice left has olives on it?

j. You choose two slices at random from this pizza. What's the chance that they both have olives on them? (Be careful—after removing the first slice, the probability of choosing one of the remaining slices changes.)

k. What's the probability that a randomly chosen slice does *not* have pepperoni on it?

l. Draw a pizza for which mushrooms, olives, anchovies, and pepperoni are all mutually exclusive.

7. In the first hour of a hunting trip, the probability that a pride of Serengeti lions will encounter a Cape buffalo is 0.035. If it encounters a buffalo, the probability that the pride successfully captures it is 0.40 (numbers are from Scheel 1993). What is the probability that the next one-hour hunt for Cape buffalo by a pride of lions will end in a successful capture?

8. Cavities in trees are important nesting sites for a wide variety of wildlife, including the white-breasted nuthatch shown on the first page of this chapter. Cavities in trees are much more common in old-growth forests than in recently logged forests. A recent survey in Missouri found that 45 out of 273 trees in an old-growth area had cavities, while the rest did not (Fan et al. 2005). What is the probability that a randomly chosen tree in this area has a cavity?

9. The accompanying bar graph gives the relative frequency of letters in texts from the English language. Such charts are useful for deciphering simple codes.

a. If a letter were chosen at random from a book written in normal English, estimate by eye (and a bit of calculation) the probability that it is a vowel (i.e., A, E, I, O, or U).

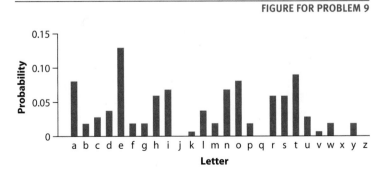

FIGURE FOR PROBLEM 9

b. Estimate by eye the probability that five letters chosen independently and at random from an English text would spell out (in order) "S-T-A-T-S".

c. Estimate by eye the probability that two letters chosen at random from an English text are both E's.

10. The gene *Prdm9* is thought to regulate hotspots of recombination (crossing over) in mammals, including humans. In the people of Han Chinese descent living in the Los Angeles area there are five alleles at the *Prdm9* gene, labeled A_1, A_2, A_3, A_4, and A_5. The relative frequencies with which these alleles occur in that population are 0.06, 0.03, 0.84, 0.03, and 0.04, respectively (Parvanov et al. 2010). Assume that in this population, the two alleles present in any individual are independently sampled from the population as a whole (this can happen if people in the community marry and produce children randomly with respect to *Prdm9* genotype).

a. What is the probability that a single allele chosen at random from this population is either A_1 or A_4?

b. What is the probability that an individual has two A_1 alleles (i.e., what is the probability that its first allele is A_1 and its second allele is A_1)?

c. What is the probability that an individual has one A_1 allele and one A_3 allele? (Note that this can happen if the first allele drawn is A_1 and the second is A_3, or if the first allele is A_3 and the second is A_1. A probability tree will help to keep track of all the possibilities.)

d. What is the probability that an individual is not A_1A_1 (i.e., does not have two A_1 alleles)?

e. What is the probability, if you drew two individuals at random from this population, that neither of them would have an A_1A_1 genotype?

f. What is the probability, if you drew two individuals at random from this population, that at least one of them would have an A_1A_1 genotype?

g. What is the probability that three randomly chosen individuals would have no A_2 or A_3 alleles? (Remember that each individual has two alleles.)

11. After graduating from your university with a biology degree, you are interviewed for a lucrative job as a snake handler in a circus sideshow. As part of your audition, you must pick up two rattlesnakes from a pit. The pit contains eight snakes, three of which have been defanged and are assumed to be harmless, but the other five are definitely still dangerous. Unfortunately, budget cuts have eliminated the herpetology course from the curriculum, so you have no way of telling in advance which snakes are dangerous and which are not. You pick up one snake with your left hand and another snake with your right.

a. What is the probability that you picked up *no* dangerous snakes?

b. Assume that any dangerous snake you pick up has a probability of biting you. This probability is the same for each snake: 0.8. The defanged snakes do not bite. What is the chance that, in picking up your two snakes, you are bitten at least once?

c. Still assume that the defanged snakes do not bite and the dangerous snakes have a probability of 0.8 of biting. If you picked up only one snake and it did not bite you, what is the probability that this snake is defanged?

12. Five different researchers independently take a random sample from the same population and calculate a 95% confidence interval for the same parameter.

   a. What is the probability that all five researchers have calculated an interval that includes the true value of the parameter?

   b. What is the probability that at least one does *not* include the true parameter value?

13. Schrödinger's cat lives under constant threat of death from the random release of a deadly poison. The probability of release of the poison is 0.01 per day, and the release is independent on successive days.

   a. What is the probability that the cat will survive one day?

   b. What is the probability that the cat will survive seven days?

   c. What is the probability that the cat will survive a year (365 days)?

   d. What is the probability that the cat will die by the end of a year?

14. Rapid HIV tests allow for quick diagnosis without expensive laboratory equipment. However, their efficacy has been called into question. In a population of 1517 tested individuals in Uganda, 4 had HIV but tested negative (false negatives), 166 had HIV and tested positive, 129 did not have HIV but tested positive (false positives), and 1218 did not have HIV and tested negative (Gray et al. 2007).

   a. What was the probability of a false-positive (also called the false-positive rate)?

   b. What was the false-negative rate?

   c. If a randomly sampled individual from this population tests positive on a rapid test, what is the probability that he or she has HIV?

15. Kalani et al. (2008) discovered cells responsive to *Wnt* proteins in the subventricular zone of developing brains of mouse embryos. These cells included a high fraction of self-renewing stem cells, which suggested that *Wnt* signaling occurs during brain cell self-renewal. In a particular cell preparation in vitro, 9% of subventricular brain cells were *Wnt*-responsive. If six cells are sampled randomly from the cell preparations, what is the probability of sampling *Wnt*-responsive (W) and nonresponsive (L) cells in the following orders, from a large population of cells?

   a. WWLWWW

   b. WWWWWL

   c. LWWWWW

   d. WLWLWL

   e. WWWLLL

   f. WWWWWW

   g. What is the probability of at least one nonresponsive brain cell when six cells are randomly sampled?

16. Studies have shown that the probability that a man washes his hands after using the restroom at an airport is 0.74, and the probability that a woman washes hers is 0.83 (American Society for Microbiology 2005). A waiting room in an airport contains 40 men and 60 women. Assume that individual men and women are equally likely to use the restroom. What is the probability that the next individual who goes to the restroom will wash his or her hands?

17. If you have ever tried to take a family photo, you know that it is very difficult to get a picture in which no one is blinking. It turns out that the probability of an individual blinking during a photo is about 0.04 (Svenson 2006).

   a. If you take a picture of one person, what is the probability that she will *not* be blinking?

   b. If you take a picture of 10 people, what is the probability that at least one person is blinking during the photo?

ASSIGNMENT PROBLEMS

18. Imagine that a collection of 1600 pea plants from one of Mendel's experiments had 900 that were tall plants with green pods, 300 that were tall with yellow pods, 300 that were short with green pods, and 100 that were short with yellow pods.

   a. Are "tall" and "green pods" mutually exclusive traits for this collection of plants?

   b. Are "tall" and "green pods" independent traits for this collection of plants?

19. A normal deck of cards has 52 cards, consisting of 13 each of four suits: spades, hearts, diamonds, and clubs. Hearts and diamonds are red, while spades and clubs are black. Each suit has an ace, nine cards numbered 2 through 10, and three face cards. The face cards are a jack, a queen, and a king. Answer the following questions for a single card drawn at random from a well-shuffled deck of cards.

   a. What is the probability of drawing a king of any suit?

   b. What is the probability of drawing a face card that is also a spade?

   c. What is the probability of drawing a card without a number on it?

   d. What is the probability of drawing a red card? What is the probability of drawing an ace? What is the probability of drawing a red ace? Are these events ("ace" and "red") mutually exclusive? Are they independent?

   e. List two events that are mutually exclusive for a single draw from a deck of cards.

   f. What is the probability of drawing a red king? What is the probability of drawing a face card in hearts? Are these two events mutually exclusive? Are they independent?

20. The human genome is composed of the four DNA nucleotides: A, T, G, and C. Some regions of the human genome are extremely G–C rich (i.e., a high proportion of the DNA nucleotides there are guanine and cytosine). Other regions are relatively A–T rich (i.e., a high proportion of the DNA nucleotides there are adenine and thymine). Imagine that you want to compare nucleotide sequences from two regions of the genome. Sixty percent of the nucleotides in the first region are G–C (30% each of guanine and cytosine) and 40% are A–T (20% each of adenine and thymine). The second region has 25% of each of the four nucleotides.

   a. If you choose a single nucleotide at random from each of the two regions, what is the probability that they are the same nucleotide?

   b. Assume that nucleotides over a single strand of DNA occur independently within regions and that you randomly sample a three-nucleotide sequence from each of the two regions. What is the chance that these two triplets are the same?

21. In Vancouver, British Columbia, the probability of rain during a winter day is 0.58, for a spring day is 0.38, for a summer day is 0.25, and for a fall day is 0.53. Each of these seasons lasts one quarter of the year.

   a. What is the probability of rain on a randomly chosen day in Vancouver?

   b. If you were told that on a particular day it was raining in Vancouver, what would be the probability that this day would be a winter day?

22. When asked an embarrassing question in a survey—such as whether the respondent has ever shoplifted—individuals may be reluctant to answer truthfully. However, answers might be more truthful if the survey incorporates a random component, such as a coin toss, that prevents the questioner from determining whether any given individual is guilty (Warner 1965). For example, consider a survey of a population in which 20% of individuals really have shoplifted at least once. The survey asks every participating individual to begin by flipping a fair coin twice. If the result of the first toss is heads, then the individual is instructed to answer honestly the question "did the second toss also yield heads?" If the first coin toss yields tails, however, the respondent is instructed to answer honestly the question "have you ever shoplifted?"

a. Draw a probability tree that describes all possible outcomes of such a survey and their probabilities.

b. What is the overall probability that a randomly sampled respondent answers yes?

23. Imagine that a long stretch of single-stranded DNA has 30% adenine, 25% thiamine, 15% cytosine, and 30% guanine. (These make up the nucleotides of the DNA.) What is the probability of randomly drawing 10 adenines in a row in a sample of 10 randomly chosen nucleotides?

24. The *Hox* genes are responsible for determining the anterior–posterior identity of body regions (segments) in the developing insect embryo. Different *Hox* genes are turned on (expressed) in different segments of the body, and in this way they determine which segments become head and which thorax, which develop legs and which antennas. One surprising thing about the *Hox* genes is that they usually occur in a row on the same chromosome and in the same order as the body regions that they control. For example, the fruit fly *Drosophila melanogaster* has eight *Hox* genes located on a chromosome in exactly the same order as the body regions in which they are expressed, from head to tail (see figure below; Lewis et al. 2003; Negre et al. 2005). If the eight genes were thrown randomly onto the same chromosome, what is the probability that they would line up in the same order as the segments in which they are expressed?

25. The flour beetle, *Tribolium castaneum*, has 10 chromosomes, roughly equal in size, and it

also has eight *Hox* genes (Brown et al. 2002; see Assignment Problem 24). If the eight *Hox* genes were randomly distributed throughout the genome of the beetle, what is the probability that all eight would land on the same chromosome?

26. A seed randomly blows around a complex habitat. It may land on any of three different soil types: a high-quality soil that gives a 0.8 chance of seed survival, a medium-quality soil that gives a 0.3 chance of survival, and a low-quality soil that gives only a 0.1 chance of survival. These three soil types (high, medium, and low) are present in the habitat in proportions of 30:20:50, respectively. The probability that a seed lands on a particular soil type is proportional to the frequency of that type in the habitat.

a. Draw a probability tree to determine the probabilities of survival under all possible circumstances.

b. What is the probability of survival of the seed, assuming that it lands?

c. Assume that the seed has a 0.2 chance of dying before it lands in a habitat. What is its overall probability of survival?

27. A flycatcher is trying to catch passing bugs. The probability that it catches a bug on any given try is 20%.

a. What is the probability that it catches its first bug on its fourth try?

b. What is the probability that it catches its first bug after at least four failures, assuming that it keeps trying until it is successful?

FIGURE FOR PROBLEM 24

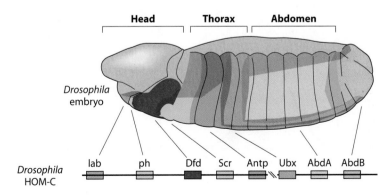

28. Blackjack is a game played with an ordinary deck of cards. (See Assignment Problem 19 for a description of such a deck.) "Blackjack" itself means that, of two cards dealt to a player, one is an ace and the other is either a 10, jack, queen, or king. If you are dealt two cards randomly from the same deck, what is the probability that you get blackjack? (Remember that, when a card is dealt, it is removed from the deck.)

29. Ignoring leap years, there are 365 days in a year.
  a. If people are born with equal probability on each of the 365 days, what is the probability that three randomly chosen people have different birthdates?
  b. If people are born with equal probability on each of the 365 days, what is the probability that 10 randomly chosen people all have different birthdates?
  c. If, as in fact turns out to be the case, birth rates are higher during some parts of the year than other times, would this increase or decrease the probability that 10 randomly chosen people have different birthdates, compared with your answer in part (b)?

30. During the Manhattan Project, the physicist Enrico Fermi asked Leslie R. Groves, the general in charge, "How do you define a 'great general'?" General Groves replied, "Any general who wins five battles in a row is great." He went on to say that only about 3% of generals are great. If battles are won entirely at random with a probability of 0.50 per side, what fraction of generals engaging in exactly five battles would be great by this definition? How does this compare to the percentage given by the general?

31. The figure at the bottom of the page shows the probability density of colony diameters (in mm) in a hypothetical population of *Paenibacillus* bacteria. The distribution is continuous, so the probability of sampling a colony within some range of diameter values is given by area under the curve. Numbers next to the curve indicate the area of the region indicated in red. Consider the case in which a single colony is randomly sampled from the population.
  a. Are the events "diameter is between 4 and 6" and "diameter is between 8 and 12" mutually exclusive? Explain.
  b. What is the probability that a randomly chosen colony diameter is between 4 and 6 or between 8 and 12?

FIGURE FOR PROBLEM 31

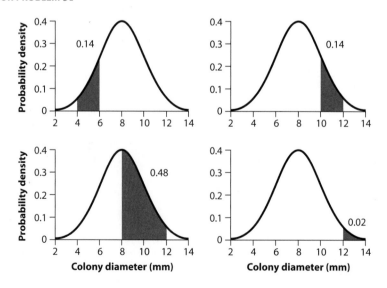

Colony diameter (mm)

c. What is the probability that a randomly chosen colony diameter is greater than or equal to 10?

d. What is the probability that a randomly chosen colony diameter is between 8 and 10?

e. What is the probability that a randomly chosen colony diameter is between 8 and 12 or greater than or equal to 10?

32. "After taking 10 mammograms, a patient has a 50% chance of having had at least one false alarm." (A false alarm is a false-positive result.) Given this information (from Elmore et al. 2005), and assuming that false alarms are independent of each other, what is the probability of a false alarm on a single mammogram?

33. A boy mentions that none of the 21 kids in his third-grade class has had a birthday since school started 56 days previously. Assume that kids in the class are drawn from a population whose birthdays have the same probability on all days of the year. What is the probability that 21 kids in such a class would not yet have a birthday in 56 days?

34. Refer to the figure accompanying Assignment Problem 31. Consider the case in which two colonies are randomly sampled from the probability distribution shown.

a. Are the events "the first diameter is between 4 and 6" and "the second diameter is between 8 and 12" mutually exclusive? Explain.

b. Are the events "the first diameter is between 4 and 6" and "the second diameter is between 8 and 12" independent? Explain.

c. What is the probability that the first diameter is between 4 and 6 and the second diameter is between 8 and 12?

d. What is the probability that the first diameter is between 8 and 12 or the second diameter is between 10 and 12?

35. Three variants of the gene encoding the β-globin component of hemoglobin occur in the human population of the Kassena-Nankana district of Ghana, West Africa. The most frequent allele, *A,* occurs at frequency 0.83. The two other variants, *S* ("sickle cell") and *C,* occur at fre-

quency 0.04 and 0.13, respectively (Ghansah et al. 2012). Each individual has two alleles, determining its genotype at the β-globin gene. Assume that knowing the identity of one of the alleles of any individual provides no information about the identity of the second allele (i.e., alleles occur independently in individuals).

a. *CC* individuals, having two copies of allele *C,* are slightly anemic. What is the probability that a randomly sampled individual from the population has two copies of the *C* allele (in other words, what is the probability that the individual's first allele is *C* and his or her second allele is also *C)?*

b. What is the probability that a randomly sampled individual is a homozygote (has two copies of the same allele)?

c. Compared with *AA* individuals, *AS* and *AC* individuals are largely resistant to malaria, which is endemic to the region. They also experience fewer deleterious side effects than *SS* and *CC* individuals. What is the probability that a randomly sampled individual is *AS?* (Remember that if an individual can be *AS* by getting *A* from mom and *S* from dad or by getting *S* from mom and *A* from dad.)

d. What is the probability that a randomly sampled individual is *AS* or *AC?*

36. Some people are hypersensitive to the smell of asparagus, and can even detect a strong odor in the urine of a person who has recently eaten asparagus. This trait turns out to have a simple genetic basis. An individual with one or two copies of the *A* allele at the gene

(*AA* or *Aa* genotypes) can smell asparagus in urine, whereas a person with two copies of the alternative "*a*" allele (*aa* genotypes) cannot (Online Mendelian Inheritance in Man, 2012). Assume that men and women in the population have the same allele frequencies at the asparagus-smelling gene and that marriage and child production are independent of the genotype at the gene. In the human population, 5% of alleles are *A* and 95% are *a*.

a. What is the probability that a randomly sampled individual from the population has two copies of the *a* allele (that is, that it has an *aa* genotype)?

b. What is the probability that both members of a randomly sampled married couple (man and woman) are *aa* at the asparagus-smelling gene?

c. What is the probability that both members of a randomly sampled married couple (man and woman) are heterozygotes at this locus (meaning that each person has one allele *A* and one allele *a*)?

d. Consider the type of couple described in (c). What is the probability that the first child of such a couple also has one *A* allele and one *a* allele (is a heterozygote)? Remember that the child must receive exactly one allele from each parent.

37. Refer to Assignment Problem 36. If a randomly sampled child has the *aa* genotype, what is the probability that both its parents were also *aa?*

38. Refer to Table 5.5-1. It turns out that blood type is controlled by two unlinked genes. One gene determines ABO blood type, and the other determines Rh factor (+ or −). Using the probabilities presented in Table 5.5-1, determine whether the events "individual is blood type O" and "Rh factor is −" are independent.

$P = \dfrac{\text{No of possibilities that meet my condition}}{\text{no of equally likely possibilities}}$

Sample space   all the possible outcomes

+ probability distribution is a list of the probabilities of all mutually exclusive outcomes of a random trial.

Bayes' theorem : the probability of a hypothesis · H conditional on a new piece of evidence. E

$P(H/E) = \dfrac{P(E/H) \times P(H)}{P(E)}$

probability of the evidence given the H — the prior probability of the hypothesis

prior prob. of the evidence

- tells you how to calc. conditional probabilities
- cond. prob. of a hypothesis is given new evidence, depends on 3 things
  1. conditional pr. of E given H
  2. the prior p. of the hypothesis
  3. the prior probability of the evidence.

Cyanella alba

6 Hypothesis testing

Hypothesis testing, like estimation, uses sample data to make inferences about the population from which the sample was taken. Unlike estimation, however, which puts bounds on the value of a population parameter, hypothesis testing asks only whether the parameter differs from a specific "null" expectation. Estimation asks, "How large is the effect?" Hypothesis testing asks, "Is there any effect at all?"

To better understand hypothesis testing, consider the polio vaccine developed by Jonas Salk. In 1954, Salk's vaccine was tested on elementary-school students across the United States and Canada. In the study, 401,974 students were divided randomly into two groups: kids in one group received the vaccine, whereas those in the other group (the control group) were injected with saline solution instead. The students were unaware of which group they were in. Of those who received the vaccine, 0.016% developed paralytic polio during the study, whereas 0.057% of the control group developed the disease (Brownlee 1955). The vaccine seemed to reduce the rate

149

of disease by two-thirds, but the difference between groups was quite small, only about four cases per 10,000. Did the vaccine work, or did such a small difference arise purely by chance?

Hypothesis testing uses probability to answer this question. The null hypothesis is that the vaccine didn't work, and that any observed difference between groups happened only by chance. Evaluating the null hypothesis involved calculating the probability, under the assumption that the vaccine has no effect, of getting a difference between groups as big or bigger than that observed. This probability turned out to be very small. Even though the rate of disease was not hugely different between the vaccine and control groups, the Salk vaccine trial was so large (over 400,000 participants) that it was able to demonstrate a real difference. Thus, the "null" hypothesis was rejected. The vaccine had an effect, sparing many kids from disease, which was borne out by the success of the vaccine in the ensuing decades.

Hypothesis testing quantifies how unusual the data are, assuming that the null hypothesis is true. If the data are too different from what is expected by the null hypothesis, then we reject the null hypothesis.

> *Hypothesis testing* compares data to what we would expect to see if a specific null hypothesis were true. If the data are too unusual, compared to what we would expect to see if the null hypothesis were true, then the null hypothesis is rejected.

In this chapter, we illustrate the basics of hypothesis testing in the simplest possible setting: a test about a proportion in a single population. Our goal is to present the main concepts with a minimum of calculation. The rest of this book will present many specific methods of hypothesis testing.

6.1   Making and using statistical hypotheses

Formal hypothesis testing begins with clear statements of two hypotheses—the null and alternative hypotheses—about a population. The null hypothesis is the default, whereas the alternative hypothesis usually includes every other possibility except the one stated in the null hypothesis. One of the two hypotheses is true, and the other must be false. We analyze the data to help determine which is which.

Both statistical hypotheses, the null and the alternative, are simple statements about a population. They are not to be confused with scientific hypotheses, which are statements about the existence and possible causes of natural phenomena. Scientists design experiments and observational studies to test predictions of scientific hypotheses. When applied to the resulting data, statistical hypotheses help to decide which predictions of these scientific hypotheses are met and which are not met.

Null hypothesis

The **null hypothesis** is a specific claim about the value of a population parameter. It is made for the purposes of argument and often embodies the skeptical point of view. Often, the null hypothesis is that the population parameter of interest is zero (i.e., no effect, no preference, no correlation, or no difference). In general, the null hypothesis is a statement that would be interesting to *reject*. For example, if we can reject the statement, "Medication X does not affect the average life span of patients suffering from illness Y," then we have learned something useful—that such patients do in fact live longer—or shorter—lives on average when taking medication X. Rejecting the null hypothesis would provide support for the scientific hypothesis that predicted a beneficial effect of medication X, whereas failing to reject the null hypothesis would not provide support.

The null hypothesis, which we can abbreviate as H_0 (pronounced "H-naught" or "H-zero"), is always *specific*; it identifies one particular value for the parameter being studied. In a study to investigate the impact of drift-net fishing on the density of dolphins, for example, a valid null hypothesis could be the following:

H_0: The density of dolphins *is the same* in areas with and without drift-net fishing.

A clinical trial designed to compare the effects of the antidepressant medication sertraline (Zoloft) and the older, tricyclic medication amitriptyline would state the null hypothesis as

H_0: The antidepressant effects of sertraline *do not differ* from those of amitriptyline.

In other cases, the null hypothesis might represent an expectation from theory or from prior knowledge. For example, the following are valid null hypotheses:

H_0: Brown-eyed parents, each of whom had one parent with blue eyes, have brown- and blue-eyed children in *a 3:1 ratio*.

H_0: The mean body temperature of healthy humans *is 98.6°F*.

> The *null hypothesis* is a specific statement about a population parameter made for the purposes of argument. A good null hypothesis is a statement that would be interesting to reject.

Alternative hypothesis

Every null hypothesis is paired with an **alternative hypothesis** (abbreviated H_A) that usually represents all other feasible parameter values except that stated in the null hypothesis. The alternative hypothesis typically includes possibilities that are biologically more interesting than that stated in the null hypothesis. The alternative hypothesis often includes parameter values predicted by a scientific hypothesis being evaluated. For this reason the alternative hypothesis is often, but not always, the statement that the researcher hopes is true.

The following are some alternative hypotheses that go with the null hypotheses stated previously:

H_A: The density of dolphins *differs* between areas with and without drift-net fishing.

H_A: The antidepressant effects of sertraline *differ* from those of amitriptyline.

H_A: Brown-eyed parents, each of whom had one parent with blue eyes, have brown- and blue-eyed children at *something other than a 3:1 ratio*.

H_A: The mean body temperature of healthy humans *is not 98.6°F.*

> The *alternative hypothesis* includes all other feasible values for the population parameter besides the value stated in the null hypothesis.

In contrast to the null hypothesis, the alternative hypothesis is nonspecific. Every possible value for a population characteristic or contrast is included, except that specified by the null hypothesis.

To reject or not to reject

Crucially, null and alternative hypotheses do not have equal standing. The null hypothesis is the only statement being tested with the data. If the data are consistent with the null hypothesis, then we say we have failed to reject it (we never "accept" the null hypothesis). If the data are inconsistent with the null hypothesis, we reject it and say the data support the alternative hypothesis.

Rejecting H_0 means that we have ruled out the null hypothesized value. It also tells us in which direction the true value likely lies, compared to the null hypothesized value. But rejecting a hypothesis by itself reveals nothing about the magnitude of the population parameter. We use estimation to provide magnitudes.

▆▆ 6.2 ▆ **Hypothesis testing: an example**

To show you the basic concepts and terminology of hypothesis testing, we'll take you through all the steps by using an example. Our goal is to illuminate the basic process without distraction from the details of the probability calculations. We'll get to plenty of the details in later chapters.

Four basic steps are involved in hypothesis testing:

1. State the hypotheses.
2. Compute the test statistic.
3. Determine the *P*-value.
4. Draw the appropriate conclusions.

We'll define the new terms we just used in this section.

Example 6.2 tests a hypothesis about a proportion, but hypothesis testing can address a wide variety of quantities, such as means, variances, differences in means, correlations, and so on. We'll try to emphasize the general over the specific here. Further details of how to test hypotheses about proportions are discussed in Chapter 7.

EXAMPLE   The right hand of toad

6.2   Humans are predominantly right-handed. Do other animals exhibit handedness as well? Bisazza et al. (1996) tested the possibility of handedness in European toads, *Bufo bufo*, by sampling and measuring 18 toads from the wild. We will assume that this was a random sample. The toads were brought to the lab and subjected one at a time to the same indignity: a balloon was wrapped around each individual's head. The researchers then recorded which forelimb each toad used to remove the

balloon. It was found that individual toads tended to use one forelimb more than the other. At this point the question became: do right-handed and left-handed toads occur with equal frequency in the toad population, or is one type more frequent than the other, as in the human population?

Of the 18 toads tested, 14 were right-handed and four were left-handed. Are these results evidence of a predominance of one type of handedness in toads?

Stating the hypotheses

The number of interest is the proportion of right-handed toads in the *population*. Let's call this proportion *p*. The default statement, the null hypothesis, is that the two types of handedness are equally frequent in the population, in which case $p = 0.5$.

H_0: Left- and right-handed toads are *equally frequent* in the population
(i.e., $p = 0.5$).

This is a specific statement about the state of the toad population, one that would be interesting to prove wrong. If this null hypothesis is wrong, then toads, like humans, on average favor one hand over the other. This statement establishes the alternative hypothesis:

H_A: Left- and right-handed toads are *not equally frequent* in the population
(i.e., $p \neq 0.5$).

The alternative hypothesis is **two-sided.** This just means that the alternative hypothesis allows for two possibilities: that p is greater than 0.5 (in which case right-handed toads outnumber left-handed toads in the population), or that p is less than 0.5 (i.e., left-handed toads predominate). Neither possibility can be ruled out before gathering the data, so both should be included in the alternative hypothesis.

> In a *two-sided* (or two-tailed) test, the alternative hypothesis includes parameter values on both sides of the parameter value specified by the null hypothesis.

"Two-tailed" has the same meaning as "two-sided." It refers to the tails of the sampling distribution, where a "tail" is the region at the upper or lower extreme of the distribution.

The test statistic

The **test statistic** is a number calculated from the data that is used to evaluate how compatible the results are with those expected under the null hypothesis.

For the toad study, we use the observed number of right-handed toads as our test statistic. On average, if the null hypothesis were correct, we would expect to observe nine right-handed toads out of the 18 sampled (and nine left-handed toads, too). Instead, we observed 14 right-handed toads out of the 18 sampled. Fourteen, then, is the value of our test statistic.

> The *test statistic* is a number calculated from the data that is used to evaluate how compatible the data are with the result expected under the null hypothesis.

The null distribution

Unfortunately, data do not always perfectly reflect the truth. Because of the effects of chance during sampling, we don't really expect to see exactly nine right-handed toads when we sample 18 from the population, even if the null hypothesis is true. There is usually a discrepancy, due to chance, between the observed result and that expected under H_0. The mismatch between the data and the expectation under H_0 can be quite large, even when H_0 is true, particularly if there are not many data. To decide whether the data are compatible with the null hypothesis, we must calculate the probability of a mismatch as extreme as or more extreme than that observed, assuming that the null hypothesis is true.

To obtain this probability, we need to determine the sampling distribution of the test statistic *assuming that the null hypothesis is true*. We need to determine what values of the test statistic are possible under H_0 and their associated probabilities. The probability distribution of values for the test statistic, assuming the null hypothesis is true, is called the "sampling distribution under H_0" or, more simply, the **null distribution.**

> The *null distribution* is the sampling distribution of outcomes for a test statistic under the assumption that the null hypothesis is true.

The tricky part is to figure out what the null distribution is for the test statistic. For the moment, let's use the power of a computer to do the calculations. (We'll learn a simpler and more elegant way to calculate this null distribution in Chapter 7.) Sampling 18 toads under H_0 is like tossing 18 coins into the air and counting the number of heads that turn up when they land (letting heads represent right-handed toads). Tossing coins mimics well the sampling process under this H_0 because the probability of obtaining heads in any one toss is 0.5, which matches the null hypothesis. When we tossed 18 coins a vast number of times with the aid of a computer and counted the number of heads (right-handed toads) each time, we obtained the sampling distribution of outcomes illustrated in Figure 6.2-1. The probabilities themselves are listed in Table 6.2-1.

FIGURE 6.2-1
The null distribution for the test statistic, the number of right-handed toads out of 18 sampled.

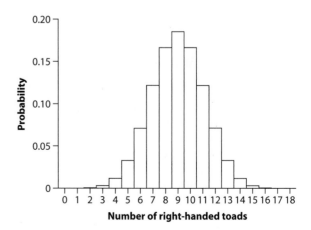

TABLE 6.2-1 All possible outcomes for the number of right-handed toads when 18 toads are sampled, and their probabilities under the null hypothesis.

Number of right-handed toads	Probability
0	0.000004
1	0.00007
2	0.0006
3	0.0031
4	0.0117
5	0.0327
6	0.0708
7	0.1214
8	0.1669
9	0.1855
10	0.1669
11	0.1214
12	0.0708
13	0.0327
14	0.0117
15	0.0031
16	0.0006
17	0.00007
18	0.000004
Total	1.0

Based on this null distribution, any number of right-handed toads between 0 and 18 is possible in a random sample of 18 individuals, but some numbers have a much higher probability of occurring than others.

Quantifying uncertainty: the *P*-value

Fourteen right-handed toads out of 18 total is not a perfect match to the expectation of the null hypothesis, but is the mismatch large enough to reject the possibility that chance alone is responsible? The usual way of describing the mismatch between data and a null hypothesis is to calculate the chance of getting those data, or data that are even more different from that expected, while assuming the null hypothesis. In other words, we want to know the probability of all results *as unusual as or more unusual than* that exhibited by the data. If this probability is small, then the null hypothesis is inconsistent with the data and we would reject the null hypothesis in favor of the alternative hypothesis. If the probability is not small, then we do not have enough evidence to doubt the null hypothesis, and we would not reject it.

The probability of obtaining the data (or data that are an even worse match to the null hypothesis), assuming the null hypothesis, is called the **P-value**. If the P-value is small, then the null hypothesis is inconsistent with the data and we reject it.[1] Otherwise, we do not reject the null hypothesis. In general, the smaller the P-value, the stronger is the evidence against the null hypothesis.

> The *P-value* is the probability of obtaining the data (or data showing as great or greater difference from the null hypothesis) if the null hypothesis were true.

The P-value is *not* the probability that the null hypothesis is true. (Hypotheses are not outcomes of random trials and so do not have probabilities.) The P-value refers to the probability of a specific event when sampling data under the null hypothesis: it is the probability of obtaining a result as extreme as or more extreme than that observed.

In practice, we calculate the P-value from the null distribution for the test statistic, shown for the toad data in Figure 6.2-2.

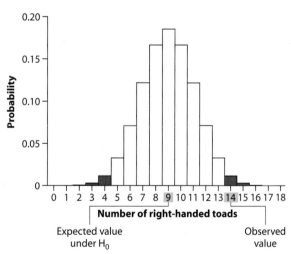

FIGURE 6.2-2
The null distribution for the number of right-handed toads out of the 18 sampled. Outcomes in red are values as different as, or more different from, the expectation under H_0 than 14, the number observed in the data.

According to Figure 6.2-2, a total of 14 or more right-handed toads out of 18 is fairly unusual, assuming the null hypothesis. These values lie at the right tail of the null distribution and have a low probability of occurring if H_0 is true. Equally unusual are 0, 1, 2, 3, or 4 right-handed toads, which are outcomes at the other tail of the null distribution. Remember that our alternative hypothesis H_A is two-sided: it allows for the possibility that right-handed toads outnumber left-handed toads in the population, and also the possibility that left-handed toads outnumber right-handed

---

1. In this book the P-value is denoted by an uppercase, italicized P, which stems from the word "probability." Don't confuse the P-value with the lowercase p, used here to indicate the proportion of right-handed toads in the population.

toads. Therefore, outcomes from both tails of the null distribution that are as unusual as the observed data, or even more unusual, must be accounted for in the calculation of the P-value.

Based on the data in Figure 6.2-2, the probability of 14 or more right-handed toads, assuming the null hypothesis is true, is

$$\Pr[\textit{14 or more right-handed toads}] = \Pr[14] + \Pr[15] + \Pr[16] + \Pr[17] + \Pr[18]$$
$$= 0.0155,$$

where $\Pr[14]$ is the probability of exactly 14 right-handed toads. We can add the probabilities of 14, 15, 16, 17, and 18 because each outcome is mutually exclusive. This sum is not the P-value, though, because it does not yet include the equally extreme results at the left tail of the null distribution—that is, those outcomes involving a predominance of left-handed toads. The quickest way to include the probabilities of the equally extreme results at the other tail is to take the above sum and multiply by two:

$$P = 2 \times (\Pr[14] + \Pr[15] + \Pr[16] + \Pr[17] + \Pr[18])$$
$$= 2 \times 0.0155$$
$$= 0.031.$$

This number is our P-value. In other words, the probability of an outcome as extreme as or more extreme than 14 right-handed toads out of 18 toads sampled is $P = 0.031$, assuming that the null hypothesis is true.

Draw the appropriate conclusion

Having calculated the P-value, what conclusion can we draw from it? On page 157, we said that if P is "small," we reject the null hypothesis; otherwise, we do not reject H_0. But what value of P is small enough? By convention in most areas of biological research, the boundary between small and not-small P-values is 0.05. That is, if P is less than or equal to 0.05, then we reject the null hypothesis; if $P > 0.05$, we do not reject it.

The P-value for the toad data, $P = 0.031$, is indeed less than 0.05, so we reject the null hypothesis that left-handed and right-handed toads are equally frequent in the toad population. We conclude from these data that most of the toads in the population are right-handed.

This decision threshold for P (i.e., $P = 0.05$) is called the **significance level**, which is signified by α (the lowercase Greek letter alpha). In biology, the most widely used significance level is $\alpha = 0.05$, but you will encounter some studies that use a different value for α. After $\alpha = 0.05$, the next most commonly used significance level is $\alpha = 0.01$. In Section 6.3, we explain the consequences of choosing a significance level and consider why $\alpha = 0.05$ is the most common choice.[2]

---

2. At the same time, let's not get carried away. A P-value of 0.051 is not much different from a P-value of 0.049. The boundary of 0.05 should be seen as a guide to interpretation, not as a clear boundary between truth and fiction.

> The *significance level,* α, is a probability used as a criterion for rejecting the null hypothesis. If the *P*-value is less than or equal to α, then the null hypothesis is rejected. If the *P*-value is greater than α, then the null hypothesis is *not* rejected.

Reporting the results

When writing up your results in a research paper or laboratory report, always include the following information in the summary of the results of a statistical test:

- the value of the test statistic
- the sample size
- the *P*-value

Leaving out any of these three values (e.g., presenting only the bare *P*-value), makes it difficult for the reader to determine how you obtained it. When writing up the results of the toad study, we would need to indicate that 14 out of 18 toads were right-handed (which in this case gives both the test statistic and the sample size) and that $P = 0.031$.

In addition, the best practice is to provide confidence intervals, or at least the standard errors, for the parameters of interest. This is because although the *P*-value indicates the weight of evidence against the null hypothesis (smaller *P* means stronger evidence), *P* does not measure the size of the effect. A very small *P*-value may result even when the size of the effect being measured is small. The confidence interval puts bounds on the estimated magnitude of effect. Using the methods we explain in Chapter 7, we calculated the following 95% confidence interval for the proportion p of right-handed toads in the study population:

$$0.54 < p < 0.91.$$

This calculation reveals that the range of most-plausible values for the true proportion of right-handed toads in the population is very broad. We would need a larger sample size to obtain a more precise estimate.

6.3 Errors in hypothesis testing

The most unsettling aspect of hypothesis testing is the possibility of errors. Rejecting H_0 does not necessarily mean that the null hypothesis is false. Similarly, failing to reject H_0 does not necessarily mean that the null hypothesis is true. This is because chance affects samples, sometimes with large impact. Some uncertainty can be quantified, though, if the data are a random sample, so making rational decisions is possible.

Type I and Type II errors

There are two kinds of errors in hypothesis testing, prosaically named Type I and Type II.

Rejecting a true null hypothesis is a **Type I error**. Failing to reject a false null hypothesis is a **Type II error**. Both types of error are summarized in Table 6.3-1.

TABLE 6.3-1 Types of error in hypothesis testing.

	Reality	
Conclusion	H_0 true	H_0 false
Reject H_0	Type I error	Correct
Do not reject H_0	Correct	Type II error

The significance level, α, gives us the probability of committing a Type I error. If we go along with convention and use a significance level of $\alpha = 0.05$, then we reject H_0 whenever P is less than or equal to 0.05. This means that, if the null hypothesis were true, we would reject it mistakenly one time in 20. Biologists typically regard this as an acceptable error rate.

> *Type I error* is rejecting a true null hypothesis. The significance level α sets the probability of committing a Type I error.

We could reduce our Type I error rate if we wanted to, simply by using a smaller significance level than 0.05. For example, a more cautious approach would be to use $\alpha = 0.01$ instead of 0.05—that is, reject the null hypothesis only if P is less than or equal to 0.01. This would have the beneficial effect of reducing the probability of committing a Type I error down to 0.01 (i.e., one time in 100). Unfortunately, this has the side effect of increasing the chance of committing a Type II error. Reducing α makes the null hypothesis more difficult to reject when true, but it also makes the null hypothesis more difficult to reject when *false*. For this reason the convention is to use a higher value such as $\alpha = 0.05$.

> *Type II error* is failing to reject a false null hypothesis.

If a null hypothesis is false, we need to reject it to get the right answer. Failure to reject a false null hypothesis is a Type II error. Because the Salk vaccine really did reduce the probability of catching polio, another study that by chance found the vaccine had no effect would have committed a Type II error.

A study that has a low probability of Type II error is said to have high **power**. Power is the probability that a random sample taken from a population will, when analyzed, lead to rejection of a false null hypothesis. All else being equal, a study is better if it has more power.

> The *power* of a test is the probability that a random sample will lead to rejection of a false null hypothesis.

Power is difficult to quantify, because the probability of rejecting a null hypothesis depends on how different the truth is from the null hypothesis. Detecting a small effect is more difficult than detecting a large effect. Because we never know how large the true value is, we usually cannot predict with any confidence how much power a study really has.

If we can guess the magnitude of the deviation from the null hypothesis, however, we can usually estimate the power of a study. A study has more power if the sample size is large, if the true discrepancy from the null hypothesis is large, or if the variability in the population is low. We discuss how to calculate power and how to design a study to optimize power when we study experimental design in Chapter 14.

6.4   When the null hypothesis is not rejected

Example 6.4 describes a study in which the null hypothesis is *not* rejected. We discuss how to interpret such a *nonsignificant* result.

EXAMPLE 6.4   The genetics of mirror-image flowers

Individuals of most plant species are hermaphrodites (with both male and female sexual organs) and are therefore prone to the worst kind of inbreeding: having sex with themselves. The mud plantain, *Heteranthera multiflora*, has a simple mechanism to avoid "selfing." The female sexual organ (the style) deflects to the left in some individuals and to the right in others (see the pair of flower images above). The male sexual organ (the anther) is on the opposite side. Bees visiting a left-handed plant are dusted with pollen on their right side, which then is deposited on the styles of only right-handed plants visited later. To investigate the genetics of this variation, Jesson and Barrett (2002) crossed pure strains of left- and right-handed flowers, yielding only right-handed plants in the next generation. These right-handed plants were then crossed to each other. The expectation under a simple model of inheritance would be that their offspring should consist of left- and right-handed individuals in a 1:3 ratio. Of 27 offspring measured from one such cross, six were left-handed and 21 were right-handed. Do these data support the simple genetic model?

Let's go through the four steps of hypothesis testing with this new example: state the hypotheses; compute the test statistic; determine the P-value; draw the appropriate conclusions.

The test

The null hypothesis states the expectation of the simple genetic model:

H_0: Left- and right-handed offspring occur at a 1:3 ratio (i.e., the proportion of left-handed individuals in the offspring population is $p = 1/4$).

The alternative hypothesis covers every other possibility:

H_A: Left- and right-handed offspring do not occur at a 1:3 ratio (i.e., $p \neq 1/4$).

As with the study of handedness in toads (Example 6.2), the test in this flower study is two-sided: under the alternative hypothesis, the proportion of left-handed offspring may be less than $1/4$ or it may be greater than $1/4$. Neither possibility can be ruled out before gathering the data, so both possibilities must be included.

The number of left-handed offspring (out of 27) is the test statistic. We also could have used the number of right-handed individuals as the test statistic; the choice between the two doesn't much matter as long as the null distribution reflects our choice. Under the null hypothesis, the expected number of left-handed offspring is $27 \times 1/4 = 6.75$. This expected frequency is a long-run average, because we don't really expect to find 0.75 of a left-handed flower.

To get the null distribution for the number of left-handed offspring, we again used the computer to take a vast number of random samples of 27 individuals from an imaginary population in which $1/4$ of the individuals were left-handed. The resulting null distribution is illustrated in Figure 6.4-1.

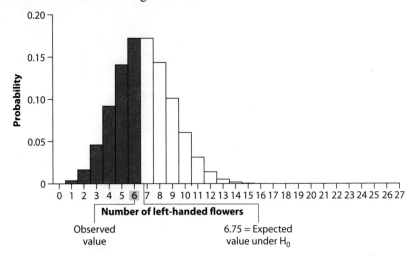

FIGURE 6.4-1 The null distribution for the number of left-handed individuals out of 27 sampled. Red bars on the left indicate outcomes less than or equal to six, the observed number of left-handed individuals. We've used red bars to indicate just one tail of the null distribution in this drawing. However, the test is two-sided and a equivalent portion of the right tail of the distribution must be incorporated when calculating the P-value for a two-sided test.

The observed number of left-handed flowers (6) is less than the expected number from the null hypothesis (6.75), so we begin the calculation of the *P*-value by determining the probability of obtaining *six or fewer* left-handed offspring, assuming that the null distribution is true:

$$\text{Pr}[\textit{number of left-handed flowers} \leqslant 6] = \text{Pr}[6] + \text{Pr}[5] + \cdots + \text{Pr}[0].$$

Summing these probabilities (given here by the height of the bars in Figure 6.4-1) yields

$$\text{Pr}[\textit{number of left-handed flowers} \leqslant 6] = 0.471.$$

This sum yields only the probability under the left tail of the null distribution (Figure 6.4-1). Because the test is two-sided, we also need to account for outcomes at the other tail of the distribution that are as unusual as or more unusual than the outcome observed. The most straightforward method to obtain *P* is to multiply the above sum by two (Yates 1984), yielding

$$P = 2\,\text{Pr}[\textit{number of left-handed flowers} \leqslant 6]$$
$$= 2 \times 0.471 = 0.942.$$

The *P*-value is quite high, and there is a high probability of getting data like these when the null hypothesis is true. The *P*-value is not less than or equal to the conventional significance level $\alpha = 0.05$, so our conclusion is that the null hypothesis is *not rejected*.

Interpreting a nonsignificant result

What does failure to reject H_0 mean? Can we conclude that the null hypothesis is true? Sadly, we can't, because it is always possible that the true value of the proportion *p* differs from 1/4 by a small or even moderate amount. You can tell this from the 95% confidence interval for *p*, the proportion of left-handed flowers in the population of possible offspring from the cross. Using methods we will detail in Chapter 7, we calculated this interval to be:

$$0.11 < p < 0.42.$$

This result indicates that 1/4 falls between the lower and upper limits for *p*, and so it is certainly among the most-plausible values for this proportion. However, many other possible values for the proportion also fall within this most-plausible range. In other words, it's possible that *p* is 1/4, but it is also possible that *p* differs from 1/4 even though H_0 was not rejected, because the power of the test was limited by the relatively small sample size of 27.

How, then, do we interpret the result? A valid interpretation is that the data are *compatible* or *consistent* with the null hypothesis; in other words, the data are compatible with the simple genetic model of inheritance of handedness in the mud plantain. There is no need or justification to build more complex genetic models of inheritance for flower handedness. A time may come when a new study—one with a larger sample size and more power—convincingly rejects the null hypothesis. If so, then researchers will need to develop a new genetic model. In the meantime, no evidence warrants a more complex scenario. This attitude treats the null hypothesis for what it is: the default until data show otherwise.

Keep in mind that an analysis should never be terminated when having only the results of a hypothesis test to show for it. Drawing conclusions about populations from data also requires that we estimate useful parameters and put bounds on these estimates (Chapter 4). Calculating 95% confidence intervals will help to identify the most plausible set of parameter values given the data. If a test fails to reject a null hypothesis, but the confidence interval is wide, then we know that we do not yet have enough information to draw a strong conclusion. But if the confidence interval is narrow and tightly bounded around the parameter value stated in the null hypothesis, then any real deviation from H_0 is likely to be either small or nonexistent.

6.5 One-sided tests

The studies of handedness in toads (Example 6.2) and flowers (Example 6.4) required two-sided tests, but **one-sided tests** are justified in some circumstances. In a one-sided test, the alternative hypothesis includes values for the population parameter exclusively to one side of the value stated in the null hypothesis.

> In a *one-sided* (or one-tailed) test, the alternative hypothesis includes parameter values on only one side of the value specified by the null hypothesis. H_0 is rejected only if the data depart from it in the direction stated by H_A.

For example, imagine a study designed to test whether daughters resemble their fathers. In each trial of the study a participant examines a photo of one girl and photos of two adult men, one of whom is the girl's father. The participant must guess which man is the father. If there is no daughter–father resemblance, then the probability that the participant guesses correctly is only $1/2$:

H_0: Participants pick the father correctly half the time ($p = 1/2$).

The only reasonable alternative hypothesis is that daughters indeed resemble their fathers, in which case the probability that the participant guesses correctly should *exceed* $1/2$:

H_A: Participants pick the father correctly *more frequently than* half the time ($p > 1/2$).

The test is "one-sided" because the alternative hypothesis includes values for the parameter on only one side of the value stated in the null hypothesis. This is justified if the values on the other side of the value stated in the null hypothesis are inconceivable for any reason other than chance. It is not really conceivable that daughters would on average resemble their fathers *less* than they resemble randomly chosen men.

The alternative hypothesis for a one-tailed test must be chosen before looking at the data. Data will always be on one side or another of the null hypothesis, so the data themselves should not be used to predict in which direction the deviation might be.

Now let's imagine that the study was carried out using 18 independent trials (which would require 18 different sets of photographs and participants) and that 13 out of 18 participants successfully guessed the father of the daughter. Under the null hypothesis, we would expect only $18 \times 1/2 = 9$ correct guesses on average. The null distribution is the same as that shown in Figure 6.2-1 and was obtained by taking a vast number of random samples of 18 participants from an imaginary population in which $1/2$ guess correctly. The actual probabilities are the same as those presented previously in Table 6.2-1.

Calculating a *P*-value for a one-sided test is different from the procedure used in a two-sided test. This is because the only outcomes that would cause us to reject H_0 and prefer H_A are those with an unusually large number of correct guesses (the direction stated in the alternative hypothesis). Therefore, we examine only the right tail of the null distribution (Figure 6.5-1).

FIGURE 6.5-1
The null distribution for the number of participants who correctly guessed the father of the daughter from photographs. Bars filled in red on the right tail correspond to values greater than or equal to 13, the observed number of correct guesses.

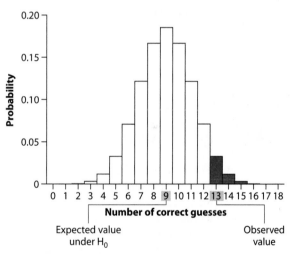

If 13 participants guessed the father correctly, then the *P*-value is

$$P = \Pr[\textit{number of correct guesses} \geqslant 13]$$
$$= \Pr[13] + \Pr[14] + \Pr[15] + \cdots + \Pr[18]$$
$$= 0.048,$$

assuming that H_0 is true. There is no need to multiply this number by two, as in the case of the two-sided tests presented earlier, because we are accounting for the probability in only one tail of the distribution.

To reiterate: the appropriate tail of the null distribution to use when calculating P is given by the alternative hypothesis. Had we observed only four correct guesses out of 18, the P-value would still be calculated using the probabilities in the right tail of the null distribution, even though four correct guesses is smaller than the null expectation:

$$P = \Pr[\textit{number of correct guesses} \geq 4]$$
$$= 0.996.$$

One-sided tests should be used sparingly because the decision whether to use a one-sided or a two-sided alternative hypothesis is usually less clear-cut than the daughter–father resemblance study, and it is therefore subjective. For example, what if we carried out a subsequent study to test whether daughters, when they marry, choose husbands who resemble their fathers? The null hypothesis is that there is no resemblance, but what is the alternative hypothesis? Should it be one-sided (husbands resemble fathers) or two-sided (husbands resemble fathers or husbands are very unlike fathers)? One researcher may have a clear theoretical basis for predicting a deviation from the null hypothesis in one direction, but a second researcher may have a different prediction. More importantly, even when we predict that a result may go in a particular direction, we may still be tempted to interpret a result in the tail of the opposite direction as a significant deviation from the null hypothesis. Two-tailed tests keep us honest.

For these reasons, we recommend using two-sided tests except in very special circumstances, and we adopt that policy in this book.

6.6 Hypothesis testing versus confidence intervals

The confidence interval puts bounds on the most-plausible values for a population parameter based on the data in a random sample (Chapter 4). Would confidence intervals and hypothesis tests on the same data, then, give the same answer? In other words, if the parameter value stated in the null hypothesis fell outside the 95% confidence interval estimated for the parameter, would H_0 be rejected by a hypothesis test at $\alpha = 0.05$? And, if the parameter value stated in the null hypothesis fell inside the 95% confidence interval, would a test at $\alpha = 0.05$ fail to reject H_0?

The answer is almost always "yes."[3] A 95% confidence interval usually contains all values of the parameter that would not be rejected if it were stated as a null hypothesis and tested with the same data at $\alpha = 0.05$.

---

3. In a few cases, the most powerful test method and the best method for calculating a confidence interval for the parameter use different approximations or slightly different sampling distributions, so they would not yield the identical answer in every instance.

Why, then, don't we just skip hypothesis testing altogether? The confidence interval does virtually everything that hypothesis testing can do, and it has the added benefit of being much more revealing about the actual magnitude of the parameter.

Hypothesis testing is used all the time in biology. And while it is fair to say that confidence intervals are not used enough in biological research, hypothesis testing is informative, too. Its main use is to decide whether sufficient evidence has been presented to support a scientific claim (Frick 1996). The kinds of claims addressed by hypothesis testing are largely qualitative, such as "this new drug is effective" or "this pollutant harms fish." Such claims are embodied by the alternative hypothesis, where "sufficient evidence" is defined as $P \leq 0.05$ (or some other significance level). For this reason, both hypothesis testing and confidence intervals are a big part of statistical analyses and of this book.

6.7 Summary

- The four steps of hypothesis testing are (1) state the hypotheses; (2) compute the test statistic; (3) determine the P-value; and (4) draw the appropriate conclusions.
- Hypothesis testing uses data to decide whether a parameter equals the value stated in a null hypothesis. If the data are too unusual, assuming the null hypothesis is true, then we reject the null hypothesis.
- The null hypothesis (H_0) is a specific claim about a parameter. The null hypothesis is the default hypothesis, the one assumed to be true unless the data lead us to reject it. A good null hypothesis would be interesting if rejected.
- The alternative hypothesis (H_A) usually includes all values for the parameter other than that stated in the null hypothesis.
- The test statistic is a quantity calculated from data, used to evaluate how compatible the data are with the null hypothesis.
- The null distribution is the sampling distribution of the test statistic under the assumption that the null hypothesis is true.
- The P-value is the probability of obtaining a difference from the null expectation as great as or greater than that observed in the data if the null hypothesis were true. If P is less than or equal to α, then H_0 is rejected.
- The threshold α is called the significance level of a test. Typically, α is set to 0.05.
- The P-value is not the probability that the null hypothesis is true or false.
- The P-value reflects the weight of evidence against the null hypothesis, but P does not measure the size of the effect. Use confidence intervals to put bounds on the magnitude of effect.
- A Type I error is rejecting a true null hypothesis. A Type II error is failing to reject a false null hypothesis:

	Reality	
Decision	**H_0 true**	**H_0 false**
Reject H_0	Type I error	(no error)
Do not reject H_0	(no error)	Type II error

- The probability of making a Type I error is set by the significance level, α. If $\alpha = 0.05$, then the probability of making a Type I error is 0.05.
- The power of a test is the probability that a random sample, when analyzed, leads to rejection of a false null hypothesis.
- Increasing sample size increases the power of a test.
- In a two-sided test, the alternative hypothesis includes parameter values on both sides of the parameter value stated by the null hypothesis. In a one-sided test, the alternative hypothesis includes parameter values on only one side of the parameter value stated by the null hypothesis.
- Most hypothesis tests are two-sided. One-sided tests should be restricted to rare instances in which a parameter value on one side of the null value is inconceivable.

PRACTICE PROBLEMS

1. A scientist tests the null hypothesis that the mean height of plants in his population is 0.75 meters, as it is in a nearby population observed with a complete census of all plants. Assume that this null hypothesis is true. However, the plants in his sample were chosen non-randomly, and the tallest plants were more likely to be chosen than expected by chance. Which of the following statements is true?
   a. A hypothesis test based on this biased sample would have an increased probability of making a Type I error.
   b. This sampling bias will not affect the probability of a Type I error.

2. Imagine that you are using a random sample of data to test a null hypothesis. Answer whether the following statement is true or false: A parameter estimate with high sampling error will result in a test with a higher Type I error rate compared to an estimate with low sampling error.

3. Answer the following questions:
   a. Define Type II error.

   b. Define significance level.
   c. If the significance level of a test is increased, will the probability of making a Type II error increase, decrease, or stay the same? Explain.

4. Assume that a null hypothesis for a statistical test is true. Say whether each of the following statements is true or false:
   a. Calculating the P-value assumes that the sample is a random sample.
   b. If you reject H_0 with a test, you will be making a Type II error.
   c. If you fail to reject H_0 with a test, you will be making a Type I error.

5. Do people have powers of extrasensory perception (ESP)? Some people claim to have such abilities or to know someone who has them. Some people claim that individuals who consistently get the wrong answer must have ESP too. Other people disbelieve such claims. Imagine that you are to set up an experiment to test the existence of ESP. Each trial in your experiment involves one person privately rolling a fair six-

sided die and holding the image of the face of the die that turned up firmly in her mind. In another room, a second person attempts to identify the result correctly without being shown the outcome.
   a. What would the null hypothesis be for your test?
   b. What would the alternative hypothesis be?

6. Identify whether each of the following statements is more appropriate as the null hypothesis or as the alternative hypothesis in a test:
   a. Hypothesis: The number of hours preschool children spend watching television affects how they behave with other children when at day care.
   b. Hypothesis: Most genetic mutations are deleterious to health.
   c. Hypothesis: A diet of fast foods has no effect on liver function.
   d. Hypothesis: Cigarette smoking influences risk of suicide.
   e. Hypothesis: Growth rates of forest trees are unaffected by increases in carbon dioxide levels in the atmosphere.

7. What effect does reducing the value of the significance level from 0.05 to 0.01 have on
   a. the probability of committing a Type I error?
   b. the probability of committing a Type II error?
   c. the power of a test?
   d. the sample size?

8. Assume a random sample. What effect does increasing the sample size have on
   a. the probability of committing a Type I error?
   b. the probability of committing a Type II error?
   c. the power of a test?
   d. the significance level?

9. In the toad experiment (Example 6.2), what would the P-value have been if
   a. 15 toads were right-handed and the rest were left-handed?
   b. 13 toads were right-handed and the rest were left-handed?
   c. 10 toads were right-handed and the rest were left-handed?

   d. 7 toads were right-handed and the rest were left-handed?

10. Why do we "fail to reject H_0" rather than "accept H_0" after a test in which the P-value is calculated to be greater than α?

11. An imaginary researcher examined the 18 largest mammal species in the Americas that occur on both the mainland and on islands. Sixteen of the mammal species were smaller on islands than on the mainland, such as the Channel Island pygmy mammoth shown next to a continental mammoth in the drawing. The remaining two species were larger on the islands than the mainland. With these data, test whether large mammals are likely to differ in size between islands and the mainland in a particular direction. Proceed through all four steps of the hypothesis testing procedure. Use the null sampling distribution in Table 6.2-1 to calculate your P-value. Apply the conventional significance level, $\alpha = 0.05$.

12. A clinical trial was carried out to test whether a new treatment affects the recovery rate of patients suffering from a debilitating disease. The null hypothesis "H_0: The treatment has no effect" was rejected with a P-value of 0.04. The researchers used a significance level of $\alpha = 0.05$. State whether each of the following conclusions is correct. If not, explain why.
   a. The treatment has only a small effect.
   b. The treatment has some effect.
   c. The probability of committing a Type I error is 0.04.
   d. The probability of committing a Type II error is 0.04.
   e. The null hypothesis would not have been rejected had the significance level been set to $\alpha = 0.01$ instead of 0.05.

13. As neuronal activity increases in the brain, blood flow to the brain also rises to meet the increasing demands for oxygen. Sheth et al. (2004) measured blood dynamics in the somatosensory cortex of rat brains to determine whether volume increased linearly with greater neuronal activity, or whether the relationship might be nonlinear, with blood volume to the brain increasing more steeply the greater the amount of neuronal activity. They estimated that the rate at which total hemoglobin to the tissue increased with increasing neuronal activity was 1.17. A rate of 1 is expected if the relationship between the two variables is linear, whereas a value different from one indicates a nonlinear relationship. The 95% confidence interval for the rate in the population was $0.74 \leq rate \leq 1.59$. The researchers also tested the null hypothesis that the relationship between the variables is linear (H_0: $rate = 1$). Did their test reject the null hypothesis or not? Explain.

14. Imagine that you have carried out a study to determine whether sons resemble their mothers. In each of 18 independent trials, you showed a participant a photo of one boy and photos of two adult women, one of whom is the boy's mother and the other of whom is randomly chosen. You ask the participant to guess which woman is the mother. If there is no son–mother resemblance, the probability that the participant guesses correctly is 1/2. If sons really do resemble their mothers, the probability that the participant guesses correctly is greater than 1/2. By answering the following questions, carry out the four steps of hypothesis testing.

   a. State the appropriate null and alternative hypotheses.
   b. Is this a one-sided or two-sided test? How can you tell?
   c. What would you use as the test statistic for this study?
   d. Suppose you found that 7 of the 18 participants guessed correctly and 11 guessed incorrectly. Calculate the P-value for this result. The null distribution is the same as that shown in Figure 6.2-1, and the probabilities of each outcome are given in Table 6.2-1.
   e. What is the conclusion from your test?
   f. What should be your next step, once you have completed the hypothesis test, to help interpret your findings and to determine the most-plausible range of values for the population parameter, p?

ASSIGNMENT PROBLEMS

15. For the following alternative hypotheses, give the appropriate null hypothesis.
   a. Pygmy mammoths and continental mammoths differ in their mean femur lengths.
   b. Patients who take phentermine and topiramate lose weight at a different rate than control patients without these drugs.
   c. Patients who take phentermine and topiramate have different proportions of their babies born with cleft palates than do patients not taking these drugs.
   d. Shoppers on average buy different amounts of candy when Christmas music is playing in the shop compared to when the usual type of music is playing.
   e. Male white-collared manakins (a tropical bird) dance more often when females are present than when they are absent.

16. Identify whether each of the following statements is more appropriate as a null hypothesis or an alternative hypothesis.
   a. Hypothesis: The number of hours that grade-school children spend doing homework predicts their future success on standardized tests.
   b. Hypothesis: King cheetahs on average run the same speed as standard spotted cheetahs.
   c. Hypothesis: The mean length of African elephant tusks has changed over the last 100 years.
   d. Hypothesis: The risk of facial clefts is equal for babies born to mothers who take folic acid supplements compared with those born to mothers who do not.
   e. Hypothesis: Caffeine intake during pregnancy affects mean birth weight.

17. State the most appropriate null and alternative hypothesis for each of the following experiments or observational studies:
   a. A test of whether cigarette smoking causes lung cancer
   b. An experiment to test whether mean herbivore damage to a genetically modified crop plant differs from that in the related unmodified crop
   c. A test of whether industrial effluents from a factory into the Mississippi River are affecting fish densities downstream
   d. A test of whether providing municipal safe-injection sites for drug addicts influences the rate of HIV transmission

18. Assume that a null hypothesis is *true*. Which one of the following statements is true?
   a. A study with a larger sample is more likely than a smaller study to get the result that $P < 0.05$.
   b. A study with a larger sample is less likely than a smaller study to get the result that $P < 0.05$.
   c. A study with a larger sample is equally likely compared to a smaller study to get the result that $P < 0.05$.

19. Assume that a null hypothesis is *false*. Which one of the following statements is true?
   a. A study with a larger sample is more likely than a smaller study to get the result that $P < 0.05$.

   b. A study with a larger sample is less likely than a smaller study to get the result that $P < 0.05$.
   c. A study with a larger sample is equally likely compared to a smaller study to get the result that $P < 0.05$.

20. Tikal National Park in Guatemala is heavily visited by tourists. Does the disturbance affect animal densities? To investigate, Hidinger (1996) compared the densities of various bird and mammal species in places immediately next to heavily visited ruins to places in the park that were rarely visited by tourists. The mean densities (in animals/km^2) are found in the accompanying table. The table also lists the P-value associated with a test of the null hypothesis that the two types of plots do not differ in mean density.

   a. Which species show a statistically significant reduction in mean density near heavily

TABLE FOR PROBLEM 20

Species	Mean density near ruins	Mean density far from ruins	P-value
Agouti	160.2	14.5	0.03
Coatimundi	99.4	1.0	0.01
Collared peccary	4.6	1.8	0.79
Deppes squirrel	32.3	2.2	0.54
Howler monkey	7.3	1.9	0.03
Spider monkey	170.8	15.0	0.88
Crested guan	0	49.4	0.001
Great curassow	10.8	72.0	0.048
Ocellated turkey	47.0	0	0.02
Tinamou	0	4.9	0.049

visited ruins? Use a significance level of $\alpha = 0.05$.

b. Which species show a significant increase in density near heavily visited ruins?

c. Which species provide no significant evidence of a difference in mean density between areas frequented by tourists and those rarely visited?

d. For which species is the evidence strongest for a change in density near tourist sites?

21. Imagine that two researchers independently carry out clinical trials to test the same null hypothesis, that COX-2 selective inhibitors (which are used to treat arthritis) have no effect on the risk of cardiac arrest. They use the same population for their study, but one experimenter uses a sample size of 60 participants, whereas the other uses a sample size of 100. Assume that all other aspects of the studies, including significance levels, are the same between the two studies.

a. Which study has the higher probability of a Type II error, the 60-participant study or the 100-participant study?

b. Which study has higher power?

c. Which study has the higher probability of a Type I error?

d. Should the tests be one-tailed or two-tailed? Explain.

22. A study showed that the sex ratio of children born to families in a native community of Ontario deviated significantly from the continental average ratio. A newspaper report about this study claimed that "there was only a 1-percent probability that the results were due to chance." [4] What do you think this statement refers to? Rewrite this statement so that it is correct.

23. A group of researchers tested whether snakes tend to choose a warm resting site when both a warm site and a cool site are presented to them. Their hypotheses were H_0: Snakes do not prefer the warmer site. H_A: Snakes prefer the warmer site. They carried out the experiment and with their data calculated a one-tailed P-value of

$P = 0.03$. They rejected their null hypothesis and concluded that snakes prefer the warmer sites.

a. Is a one-tailed test appropriate here? Explain.

b. What would their hypothesis statements have been had they used a two-tailed test instead?

c. What would their P-value have been had they used a two-tailed test instead?

24. Does being told how a suspenseful story will end ruin the experience for the reader? Movie and book critics are careful to avoid giving too much away in their reviews. But the impact of such "spoilers" has only recently been tested scientifically (Leavitt and Christenfeld 2011). Students were asked to read a variety of crime-mystery stories. Some were told the ending before starting to read whereas others were not told. At the end, readers were asked to rate their enjoyment of the stories on a scale from 1 to 10, with 10 indicating greatest enjoyment. The mean enjoyment score was 7.3 in the group of students told the endings beforehand, while average enjoyment score was 6.6 in the other group. A test of the statistical null hypothesis of no difference between the means of the two groups yielded $P = 0.001$.

a. State the appropriate conclusion of the test.

b. Do the data support the idea that knowing the ending reduces the mean enjoyment of the stories?

c. The authors used a two-sided test. Is this appropriate? Explain.

25. Can parents distinguish their own children by smell alone? To investigate, Porter and Moore (1981) gave new T-shirts to children of nine mothers. Each child wore his or her shirt to bed for three consecutive nights. During the day, from waking until bedtime, the shirts were kept in individually sealed plastic bags. No scented soaps or perfumes were used during the study. Each mother was then given the shirt of her child and that of another, randomly chosen child and asked to identify her own by smell. Eight of nine mothers identified their children correctly. Use this study to answer the following ques-

---

4. *Globe and Mail,* November 15, 2005.

tions, using a two-sided test and a significance level of $\alpha = 0.05$.

a. To carry out a statistical test based on these data, what is the appropriate null hypothesis?
b. What is the alternative hypothesis?
c. What test statistic should you use?
d. The following figure shows the null distribution for the number of mothers out of nine guessing correctly. The probability of each outcome is given above the bars. If the null hypothesis were true, what is the probability of exactly eight correct identifications?

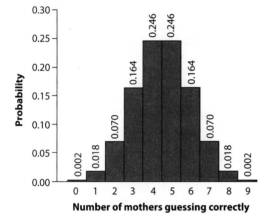

Number of mothers guessing correctly

e. If the null hypothesis were true, what is the probability of obtaining eight or more correct identifications?
f. What is the P-value for the test?
g. What is the appropriate conclusion?
h. As part of the analysis of these data, why would it be a good idea to calculate a 95% confidence interval for the true proportion of correct identifications?

26. Dondorp et al. (2010) carried out a large clinical trial that compared effectiveness of two drugs, artesunate and quinine, for treatment of African children having severe malaria. They used the results to test the null hypothesis that the proportion of children dying was the same in the two treatments against the alternative hypothesis that one treatment was better than the other. Which single answer below is correct? Explain your answer.

The null distribution used in the test of the null hypothesis …

a. describes the probability distribution of possible true benefits of each of the two drug treatments.
b. describes the possible probability distribution of true benefits and costs of one drug treatment or the other drug treatment.
c. describes the probability distribution of possible observed differences between the treatment groups if there were truly no difference between drug treatments.
d. describes the probability distribution of possible observed differences between treatments given that one of the drugs really is better than the other.

27. About 30% of people cannot detect any odor when they sniff the steroid androstenone, but they can become sensitive to its smell if exposed to the chemical repeatedly. Does this change in sensitivity happen in the nose or in the brain? Mainland et al. (2002) exposed one nostril of each of 12 non-detector participants to androstenone for short periods every day for 21 days. The other nostril was plugged and had humidified air flow to prevent androstenone from entering. After the 21 days, the researchers found that 10 of 12 participants had improved androstenone-detection accuracy in the plugged nostril, whereas two had reduced accuracy. This suggested that increases in sensitivity to androstenone happen in the brain rather than the nostril, since the epithelia of the nostrils are not connected. The authors conducted a statistical hypothesis test of whether accuracy in fact did change. Let p refer to the proportion of non-detectors in the population whose accuracy scores improve after 21 days. Under the null hypothesis, $p = 0.5$ (as many participants should improve as deteriorate in their accuracy after 21 days). The alternative hypothesis is that $p \neq 0.5$ (the proportion of participants increases or decreases after 21 days).

a. Did the authors carry out a one- or two-sided test? What justification might they provide?
b. The accompanying figure shows the null distribution for the number of participants out of 12 having an improved accuracy score. The probability of each outcome is given

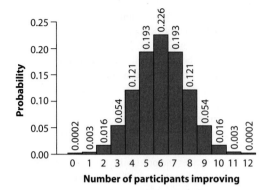

above the bars. To what do these probabilities refer?
c. What is the test statistic for the test?
d. What is the *P*-value for the test?
e. What is the appropriate conclusion? Use significance level $\alpha = 0.05$.

28. Refer to problem 27. The researchers also found that 11 of 12 participants showed improved accuracy in the *exposed* nostril after 21 days.
   a. Carry out the four steps of hypothesis testing, to test whether accuracy scores changed after 21 days in the exposed nostrils of participants. Use significance level $\alpha = 0.05$.
   b. Why it would be useful also to provide a confidence interval for the proportion of participants improving?

29. A team of researchers conducted 100 independent hypothesis tests using a significance level of $\alpha = 0.05$.
   a. If all 100 null hypotheses were true, what is the probability that the researchers would reject none of them?
   b. If all 100 null hypotheses were true, how many of these tests on average are expected to reject the null hypothesis?

30. In many animal species, individuals communicate using ultraviolet (UV) signals, such as bright patches of skin or feathers, that are visible to one another but invisible to humans. Secondi et al. (2012) investigated the role of UV colors as sexual signals in two closely related species of *Lissotriton* newt, by asking whether females find males of their own species more attractive

when UV radiation is present than when UV light is absent. Each trial consisted of confining a male newt to one end of an aquarium and measuring how much time a female of the same species chose to spend near the male (rather than at the opposite end of the aquarium). Each male was tested both under natural light conditions (UV light present) and when UV light was absent (by using a UV filter). The box plots in the following figure (modified from Secondi et al. 2012) show the difference between the two treatments in the time females spent near the males. A positive number indicates that a female spent more time near the male when UV light was present, and a negative number indicates that she spent less time near the male when UV light was present.

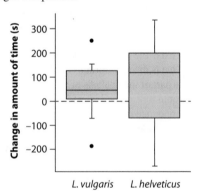

L. vulgaris   *L. helveticus*

For both species, the authors tested the null hypothesis that UV light treatment had no effect; that is, H_0: Mean change in time females spent close to males $= 0$. Test results for the two species are as follows, along with 95% confidence intervals for the mean change.

L. vulgaris: $\overline{X} = 50.7$, $s = 87.3$, $n = 23$, $P = 0.011$, $12.9 \leq \mu \leq 88.4$.

L. helveticus: $\overline{X} = 199.8$, $s = 587.0$, $n = 25$, $P = 0.102$, $-42.5 \leq \mu \leq 442.1$.

Using these results, indicate whether the following statements are true or false. Explain your answers.

a. The null hypothesis was rejected in the case of *L. vulgaris* but not in the case of *L. helveticus* at significance level 0.05.

b. UV light treatment affects the attractiveness of male *L. vulgaris* to females but has no effect on the attractiveness of male *L. helveticus*.

c. The weight of evidence against the null hypothesis was stronger in *L. vulgaris* than in *L. helveticus*.

d. The magnitude of the effect of UV treatment was greater in *L. vulgaris* than in *L. helveticus*.

Why statistical significance is not the same as biological importance

In the early 20th century, the word "significant" had one dominant meaning. If you said that something was "significant" you meant that it *signified* or showed something—that it had or conveyed a meaning. When R. A. Fisher said that a result was significant, he meant that the data showed some difference from the null hypothesis. In other words, we were able to learn something from those data. This sense of the word "significant" has persisted in the scientific literature. When discussing data, a "statistically significant" result means that a null hypothesis has been rejected.

But languages, like species, evolve. Over the 20th century, the word "significant" came to mean *important*. Today, outside of statistics, when we say that something is significant, we usually mean that it has value and import—that it ought to be paid attention to. This creates ambiguity when the term is applied now to scientific findings. When a newspaper article describes a new scientific result as "a significant new finding" or "a significant advance in our knowledge," for example, is that a statistical statement or a value judgment? Sometimes the meaning is blurred.

A statistically significant result is not the same as a biologically important result. An important result in biology refers to one whose effect is large enough to matter in some way. The mean lengths of the third molars may differ between two closely related species of mammals, but this doesn't by itself mean that the difference is vital. To address the importance of the difference, we need to know its magnitude and how it might matter.

The problem is that extremely small, biologically uninteresting effects can be statistically significant, as long as the sample size is sufficiently large. For example, automobile accidents increase during full moons, and this result is statistically significant (Templer et al. 1982). Such results attracted media attention because they bring to mind stories of werewolves and vampires wreaking havoc on the human population. But the size of the effect is a trivial 1% increase in the accident rate, far too small to make it worth cautioning you to check the phase of the moon before you go driving. In fact, almost any null hypothesis can be disproved with a large enough sample. There are likely few factors in biology whose effects on humans and other organisms are exactly zero. We should care about the effects only if they are large enough to matter.

We already have some sense that statistically significant does not mean important when we read of scientific studies that showed

> **Full of sound and fury, signifying nothing.**
>
> —Shakespeare, *Macbeth*

such earth-shattering facts as "teenagers like to listen to music sometimes," "driving fast makes the ride feel bumpier," and "spiders scare some people." Each of these results was backed by hard data and statistics, but each surprised no one.

On the other hand, a result can be important even if it is not statistically significant. Sometimes new data suggest a pattern that, if true, would be very important, provoking further study. For example, the first studies testing whether administering streptokinase prevents strokes did not reject the null hypothesis that this drug had no effect on mortality rate. But they were small studies that showed a suggestive pattern. As a result, further larger studies were conducted, and streptokinase was eventually shown to be effective in reducing mortality rates from strokes. This is why we don't "accept the null hypothesis"—a small study on a small effect will have a low probability of rejecting a false null hypothesis.

At the other extreme, sometimes it is important to show the lack of an effect. A large-scale study of the efficacy of hormone replacement therapy (HRT) showed no statistically significant evidence for a benefit

of HRT to post-menopausal women. Moreover, confidence intervals showed that any plausible effect was small. HRT had been in wide use with substantial money being invested and with some known deleterious side effects. Knowing that it had little benefit saved a great deal of resources and prevented many unnecessary side effects. This result was not "statistically significant," but it was very important medically.

When presenting data, we should always report the estimated magnitude of the effect with a confidence interval, not just the P-value. The confidence interval gives us a plausible range for the size of the effect, and if this interval includes values with greatly varying interpretations, we know that we have to revisit the question with further data. We should look at a graphical presentation of the data to gauge the magnitude of the effect. The importance of a result depends on the value of the question and the size of the effect. Statistical significance tells us merely how confidently we can reject a null hypothesis, but not how big or how important the effect is.

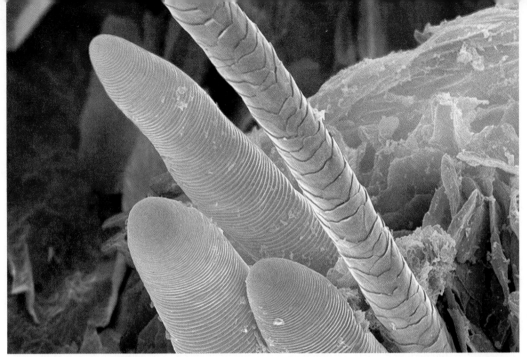

Follicle mites in human hair follicle

7 Analyzing proportions

What proportion of people with Lou Gehrig's disease will survive at least 10 years after diagnosis? What proportion of the North Carolina red wolf population is female? In what fraction of years does global temperature increase? Each of these questions is about a proportion, the fraction of the population that has a particular characteristic of interest. The proportion of individuals sharing some characteristic in a population is also the probability that an individual randomly sampled from that population will have that attribute. A proportion can range from zero to one.

In this chapter, we'll describe how best to estimate a population proportion using a random sample, including how to calculate its confidence interval. We continue the development, begun in Chapter 4, of one of the major themes of data analysis: estimation with confidence intervals.

We also outline the best way to test hypotheses about a population proportion. In Chapter 6, we used a computer to take a vast number of random samples to obtain the null distribution for a proportion. Here in Chapter 7 we show a much quicker method to test hypotheses about a population proportion. This method, called the *binomial test*, provides exact *P*-values.

The key to estimation and hypothesis testing is an understanding of the sampling distribution for a proportion. Therefore, we begin this chapter by exploring the properties of random samples from a population when each individual can be categorized into one of two types.

7.1 The binomial distribution

Consider a measurement made on individuals that divides them into two mutually exclusive groups, such as success or failure, alive or dead, left-handed or right-handed, or diabetic or nondiabetic. In the population, a fixed proportion *p* of individuals fall into one of the two groups (call it "success") and the remaining individuals fall into the other group (call it "failure"). Calling one of the categories "success" and the other "failure" is a convenience, not a value judgment.[1]

If we take a random sample of *n* individuals from this population, the sampling distribution for the number of individuals falling into the success category is described by the **binomial distribution**. The term "binomial" reveals its meaning: there are only two (*bi-*) possible outcomes, and both are named (-*nomial*) categories.

> The *binomial distribution* provides the probability distribution for the number of "successes" in a fixed number of independent trials, when the probability of success is the same in each trial.

Formula for the binomial distribution

The binomial formula gives the probability of *X* successes in *n* trials, where the outcome of any single trial is either success or failure. The binomial distribution assumes that

- the number of trials (*n*) is fixed,
- separate trials are independent, and
- the probability of success (*p*) is the same in every trial.

---

1. For example, if we are measuring the fraction of hurricanes that hit Florida in a given year, then we might call a hurricane hit a success and a miss a failure. We are clearly not cheering for the hurricane; the categories are for convenience only. The terms have their origins in gambling, where success and failure correspond more appropriately to winning and losing, respectively.

Under these conditions, the probability of getting X successes in n trials is

$$\Pr[X\ successes] = \binom{n}{X}p^X(1 - p)^{n-X}.$$

The left side of this equation, $\Pr[X\ successes]$, means the "probability of X successes," where X is an integer between 0 and n. On the right-hand side, the quantity $\binom{n}{X}$ is read "n choose X." This represents the number of unique ordered sequences of successes and failures that yield exactly X successes in n trials.[2] The term is shorthand for

$$\binom{n}{X} = \frac{n!}{X!(n - X)!},$$

where $n!$ is called "n factorial" and refers to the product

$$n! = n \times (n - 1) \times (n - 2) \times (n - 3) \times \cdots \times 2 \times 1.$$

Similarly, $X!$ is "X factorial" and $(n - X)!$ is "$(n - X)$ factorial." By definition, $0! = 1$, so $\binom{n}{0}$ and $\binom{n}{n}$ are both equal to 1. Factorials get very large very fast. For example, $5! = 120$, but $20! = 2{,}432{,}902{,}008{,}176{,}640{,}000$. Even with a reasonably small number of trials, calculating the binomial coefficient can require a good calculator or computer.[3]

Suppose, for example, that you randomly sample $n = 5$ wasps from a population (representing five independent trials), where each wasp has probability $p = 0.2$ of being male. The probability, then, that exactly three of the wasps in your sample are male is

$$\Pr[3\ males] = \binom{5}{3}(0.2)^3(1 - 0.2)^{5-3}$$

$$= \frac{(5 \times 4 \times 3 \times 2 \times 1)}{(3 \times 2 \times 1)(2 \times 1)}(0.2)^3(0.8)^2$$

$$= 0.0512.$$

The binomial distribution describes the sampling distribution for the *number* of successes in a random sample of n trials, but it also describes the *proportion* of successes. Because the number of trials is fixed at n, the probability that the sample has the proportion X/n successes is the same as the probability that the sample has X successes. For example, the probability that a proportion 0.6 of the five wasps are male is the same as the probability that exactly three are male—namely, 0.0512.

---

2. This term is also called the *binomial coefficient*.

3. Some simple multiplication tricks will keep your calculator from maxing out. For example, $\frac{20!}{18!2!}$ can be rewritten as $\frac{20 \times 19 \times 18!}{18!2!}$, which simplifies to give $\frac{20 \times 19}{2!} = 190$, because 18! cancels out.

Number of successes in a random sample

Let's look at a complete binomial distribution. Imagine, for example, that we are randomly sampling $n = 27$ individuals from a population in which $p = 0.25$ of the individuals are successes and $1 - p = 0.75$ of the individuals are failures. What is the probability that the sample contains exactly X successes? This scenario is identical to the one stated by the null hypothesis in Example 6.4 (the study of mirror-image flowers). Recall that we sampled a population of mud plantains in which $p = 0.25$ had left-handed flowers and $1 - p = 0.75$ had right-handed flowers. The process of taking a random sample from this population exactly matches the assumptions of the binomial distribution: n independent random trials, where the probability of success in each trial p is equal in every trial. Hence, the binomial distribution can be used to determine the probability of any given number of successes.

For example, the probability of getting exactly six left-handed flowers with $n = 27$ and $p = 0.25$ is

$$Pr[6\ \textit{left-handed flowers}] = \binom{27}{6}(0.25)^6(1 - 0.25)^{27-6}.$$

The binomial coefficient, "27 choose 6," is

$$\binom{27}{6} = \frac{27!}{6!(27 - 6)!} = \frac{27!}{6!\ 21!}$$

$$= \frac{27 \times 26 \times 25 \times \cdots \times 3 \times 2 \times 1}{(6 \times 5 \times 4 \times 3 \times 2 \times 1)(21 \times 20 \times 19 \times \cdots \times 3 \times 2 \times 1)}$$

$$= 296{,}010.$$

Therefore,

$$Pr[6\ \textit{left-handed flowers}] = 296{,}010\ (0.25)^6\ (0.75)^{21} = 0.1719$$

In other words, there is about a 17% chance that exactly six of 27 randomly chosen flowers are left-handed, if the proportion of left-handed flowers in the population is 0.25.

We continue in this way to calculate all the probabilities of the sampling distribution (Table 7.1-1). Check some of the probabilities in this table for practice.

The probability distribution given in Table 7.1-1 is plotted in Figure 7.1-1. The distribution in Figure 7.1-1 looks the same as the distribution in Figure 6.4-1, which was obtained by using a computer to take a vast number of random samples of $n = 27$ from the population. In other words, the mathematical formula for the binomial distribution predicts the distribution generated using a computer simulation of the sampling process, but it does so much more rapidly and easily. The binomial distribution also gives *exact* probabilities. As a result, we can use the binomial distribution in place of repeated sampling on the computer to test null hypotheses.

TABLE 7.1-1 The probability of obtaining X left-handed flowers out of $n = 27$ randomly sampled, if the proportion of left-handed plants in the population is 0.25.

Number of left-handed flowers (X)	Pr[X]	Number of left-handed flowers (X)	Pr[X]
0	4.2×10^{-4}	14	0.0018
1	0.0038	15	5.1×10^{-4}
2	0.0165	16	1.3×10^{-4}
3	0.0459	17	2.8×10^{-5}
4	0.0917	18	5.1×10^{-6}
5	0.1406	19	8.1×10^{-7}
6	0.1719	20	1.1×10^{-7}
7	0.1719	21	1.2×10^{-8}
8	0.1432	22	1.1×10^{-9}
9	0.1008	23	7.9×10^{-11}
10	0.0605	24	4.4×10^{-12}
11	0.0312	25	1.8×10^{-13}
12	0.0138	26	4.5×10^{-15}
13	0.0053	27	5.5×10^{-17}

FIGURE 7.1-1
Plot of the probabilities given in Table 7.1-1 that were calculated from the binomial distribution. The plot shows the probability of obtaining X left-handed flowers out of $n = 27$ randomly sampled, if the proportion of left-handed plants in the population is 0.25.

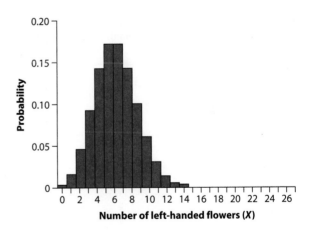

Sampling distribution of the proportion

If there are X successes out of n trials in a random sample, then the estimated proportion of successes is $\hat{p}$:

$$\hat{p} = \frac{X}{n}.$$

(We pronounce $\hat{p}$ as "p-hat." Recall that p refers to the proportion in the population, whereas $\hat{p}$ refers to the *sample* proportion.)

We can use the same hypothetical population of flowers, having a true proportion of $p = 0.25$ successes, to illustrate the sampling distribution for the sample proportion $\hat{p}$. The panel on the top in Figure 7.1-2 is the sampling distribution when $n = 10$ (a relatively small sample size), whereas the panel on the bottom is the distribution for a larger sample size, $n = 100$. Both are based on binomial distributions, but rather than showing the number of successes X, we have divided X by n to yield $\hat{p}$.

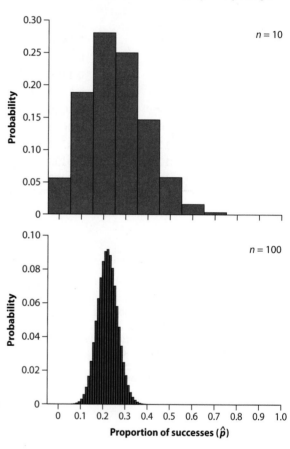

FIGURE 7.1-2
The sampling distribution for the proportion of successes $\hat{p}$ for sample size $n = 10$ (*top*) and $n = 100$ (*bottom*). In both of these graphs, the population proportion is $p = 0.25$. The distribution is narrower (smaller standard deviation) when n is larger.

The mean of the sampling distribution of $\hat{p}$, the proportion of successes in a random sample of size n, is p. In other words, the proportion of successes in random samples is the same *on average* as the proportion of successes in the population. Therefore, $\hat{p}$ is an unbiased estimate of the population proportion—on average, it gives the right answer.

Notice in Figure 7.1-2 how the sample size affects the width of the sampling distribution for $\hat{p}$. When n is large (bottom panel), the sampling distribution is narrow. This effect is quantified in the formula for the standard error of $\hat{p}$. (Remember from

Section 4.2 that the standard error of an estimate is the standard deviation of its sampling distribution.) The standard error of $\hat{p}$ (designated by $\sigma_{\hat{p}}$) is

$$\sigma_{\hat{p}} = \sqrt{\frac{p(1-p)}{n}}.$$

The sample size (n) is in the denominator of the standard error equation, so the standard error decreases as the sample size increases. That is why the estimates from samples of size 10 in Figure 7.1-2 (top panel) are more spread out than the estimates based on 100 individuals (bottom panel). Larger samples yield more precise estimates. The improvement in precision as sample size increases is called the **law of large numbers**.

7.2   Testing a proportion: the binomial test

The **binomial test** applies the binomial sampling distribution to hypothesis testing for a proportion. The types of questions it is suitable for have already been encountered in Chapter 6. The binomial test is used when a variable in a population has two possible states (i.e., "success" and "failure"), and we wish to test whether the relative frequency of successes in the population (p) matches a null expectation (p_0). The hypothesis statements look like this:

H_0: The relative frequency of successes in the population is p_0.

H_A: The relative frequency of successes in the population is not p_0.

The null expectation (p_0) can be any specific proportion between zero and one, inclusive.

> The *binomial test* uses data to test whether a population proportion (p) matches a null expectation (p_0) for the proportion.

Example 7.2 shows how to apply the binomial test to real data.

EXAMPLE   Sex and the X

7.2   A study of 25 genes involved in spermatogenesis (sperm formation) found their locations in the mouse genome. The study was carried out to test a prediction of evolutionary theory that such genes should occur disproportionately often on the X chromosome.[4] As it turned out, 10 of the 25 spermatogenesis genes (40%) were on the X chromosome (Wang et al.

---

4. New, recessive mutations are expressed only in males if they occur on the X chromosome, because they are not masked by the old allele at a companion chromosome. As a result, recessive mutations that benefit males are more likely to be seen by natural selection if they are on the X chromosome than if they are on other chromosomes.

2001; see Figure 7.2-1). If genes for spermatogenesis occurred "randomly" throughout the genome, then we would expect only 6.1% of them to fall on the X chromosome, because the X chromosome contains 6.1% of the genes in the genome. Do the results support the hypothesis that spermatogenesis genes occur preferentially on the X chromosome?

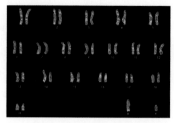

FIGURE 7.2-1
Cartoon of the mouse genome. Each vertical line represents one of the mouse chromosomes and indicates its length relative to the others. Each mark on a line indicates a single gene involved in spermatogenesis. Note the abundance of these genes on the X chromosome.

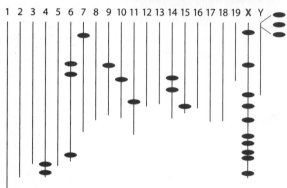

The null hypothesis is that the spermatogenesis genes would be on the X chromosome about 6.1% of the time, if they were randomly spread around the genome. To express this in terms of the binomial distribution, let's call the placement of each gene in the sample a "trial," and if the gene is on the X chromosome we'll call it a success. The null hypothesis is that the probability of success (p) is 0.061. The more interesting alternative hypothesis is that the probability of success (p) is *not* 0.061—that is, either spermatogenesis genes occur *more* frequently than 0.061 or they occur *less* frequently on the X chromosome than expected by chance.

We can write these hypotheses more formally as follows:

H_0: The probability that a spermatogenesis gene falls on the X chromosome is $p = 0.061$.

H_A: The probability that a spermatogenesis gene falls on the X chromosome is something other than 0.061 ($p \neq 0.061$).

Note once again the asymmetry of these two hypotheses. The null hypothesis is very specific, while the alternative hypothesis is not specific, referring to every other possibility. Also note that there are two ways to reject the null hypothesis: there can be an excess of spermatogenesis genes on the X chromosome (i.e., $p > 0.061$) or there can be too few (i.e., $p < 0.061$). Too few is not inconceivable, so it should also be included in the alternative hypothesis. Therefore, the test is two-sided.

The next step is to identify the test statistic that will be used to compare the observed result with the null expectation. In the case of the binomial test, the test statistic is the observed number of successes. For the data in Example 7.2, that would be 10 spermatogenesis genes on the X chromosome. The null expectation is $0.061 \times 25 = 1.525$. On average, we expect the fraction 0.061 of the 25 spermatogenesis genes sampled—namely, 1.525—to be located on the X chromosome if H_0 is true. Therefore, we know that in the *sample* more genes were found on the X chromosome than were expected by the null hypothesis.

The question now is whether we are likely to get such an excess by chance alone if the null hypothesis were true. To decide this we need the null distribution, the sampling distribution for the test statistic assuming that the null hypothesis is true. As mentioned previously, the sampling distribution for the number of successes X in a random sample of n individuals from a population having the proportion p of successes is described by the binomial distribution. Under the null hypothesis, the proportion is $p = 0.061$, so, for the above data (where $n = 25$ genes), the null distribution is given by

$$\Pr\left[X \; successes\right] = \binom{25}{X}(0.061)^X(1 - 0.061)^{25-X}.$$

This null distribution allows us to calculate the *P*-value, the probability of getting a result as extreme as, or more extreme than, 10 spermatogenesis genes on the X chromosome when the null expectation is 1.525. Because the test is two-tailed, *P* is the probability of getting 10 or more genes on the X chromosome plus the probability of similarly extreme results at the other tail of the null distribution, corresponding to too *few* genes on the X chromosome. We account for all the extreme outcomes by doubling the probability of getting 10 or more:

$$P = 2\Pr\left[number \; of \; successes \; \geq 10\right].$$

The probability of getting exactly 10 out of 25 on the X chromosome, when the probability of being on the X chromosome is 0.061, is

$$\Pr\left[10 \; successes\right] = \binom{25}{10}(0.061)^{10}(1 - 0.061)^{15} = 9.07 \times 10^{-7}.$$

This calculation is listed in Table 7.2-1, along with the probability of more than 10 genes on the X chromosome, if the null hypothesis were true.

TABLE 7.2-1 Probabilities in the right-hand tail of the binomial distribution with $n = 25$ and $p = 0.061$.

Number of genes on X	Probability under the null hypothesis	Number of genes on X	Probability under the null hypothesis
10	9.1×10^{-7}	18	4.2×10^{-17}
11	8.0×10^{-8}	19	1.0×10^{-18}
12	6.1×10^{-9}	20	2.0×10^{-20}
13	4.0×10^{-10}	21	3.1×10^{-22}
14	2.2×10^{-11}	22	3.6×10^{-24}
15	1.0×10^{-12}	23	3.1×10^{-26}
16	4.3×10^{-14}	24	1.7×10^{-28}
17	1.5×10^{-15}	25	4.3×10^{-31}

The probability of getting 10 *or more* spermatogenesis genes on the X chromosome, assuming the null hypothesis is true, is the sum over all of these mutually exclusive possibilities:

$$\Pr[\textit{number of successes} \geq 10] = \Pr[10] + \Pr[11] + \Pr[12] + \cdots + \Pr[25]$$

$$= 9.9 \times 10^{-7}.$$

The final *P*-value is

$$P = 2\Pr[\textit{number of successes} \geq 10] = 2\,(9.9 \times 10^{-7}) = 1.98 \times 10^{-6}.$$

This *P*-value[5] is well below the conventional significance level of $\alpha = 0.05$. The probability of getting a result as extreme as, or more extreme, than the observed result is very low if the null hypothesis were true. Therefore, we reject the null hypothesis and conclude that there *is* a disproportionate number of spermatogenesis genes on the X chromosome. Our best estimate of the proportion of spermatogenesis genes that are located on the mouse X chromosome is

$$\hat{p} = \frac{10}{25} = 0.40,$$

which is much greater than 0.061, the proportion stated in the null hypothesis. These results might be stated in a scientific report: "A disproportionately large proportion of spermatogenesis genes occur on the X chromosome (0.40, SE = 0.10; binomial test, $n = 25$, $P < 0.001$)." This statement includes the standard error of the proportion, which we show you how to calculate in the next section.

Approximations for the binomial test

The binomial test gives us an exact *P*-value and can be applied in principle to any data that fits into two categories. But calculating *P*-values for the binomial test can be tedious without a computer, especially when *n* is large. Alternatives that are faster to calculate can be used under certain situations. They yield only approximate *P*-values, but they can save a lot of time when appropriate. We discuss two of them in subsequent chapters, but here we just want to let you know that they exist. The first is the χ^2 goodness-of-fit test (Chapter 8), and the second is the normal approximation to the binomial test (Chapter 10).

7.3 Estimating proportions

Here we show you how to measure the precision of an estimate of a population proportion. We explain how to estimate a standard error for a sample proportion and how to calculate a confidence interval for a population proportion. We'll motivate this endeavor with the data from Example 7.3 throughout.

---

5. Some computer programs for the binomial test give a slightly different value for *P* for these same data, because they calculate the probability of extreme results at both tails using a different method.

EXAMPLE   Radiologists' missing sons

7.3   Male radiologists have long suspected that they tend to have fewer sons than daughters. What is the proportion of males among the offspring of radiologists? In a sample of 87 offspring of "highly irradiated" male radiologists, 30 were male (Hama et al. 2001). Assume that this was a random sample.

The best estimate of the proportion of male offspring in this population is the sample proportion

$$\hat{p} = \frac{X}{n} = \frac{30}{87} = 0.345.$$

Estimating the standard error of a proportion

As we learned in Section 4.2, the standard deviation of the sampling distribution for an estimate is known as the standard error of that estimate. We have seen in Section 7.1 that the standard deviation of $\hat{p}$ (and therefore its standard error) is

$$\sigma_{\hat{p}} = \sqrt{\frac{p(1-p)}{n}}.$$

We cannot usually calculate $\sigma_{\hat{p}}$, because we don't know the population parameter p. However, we can approximate this standard error with $SE_{\hat{p}}$, which uses the estimate of the proportion:[6]

$$SE_{\hat{p}} = \sqrt{\frac{\hat{p}(1-\hat{p})}{n}}.$$

In the sample of offspring of radiologists, the standard error of the estimate of the proportion who are male is approximated by

$$SE_{\hat{p}} = \sqrt{\frac{\hat{p}(1-\hat{p})}{n}} = \sqrt{\frac{0.345(1-0.345)}{87}} = 0.051.$$

This value tells us how close, on average, our sample estimate ($\hat{p}$) is likely to be to the population proportion (p).

Confidence intervals for proportions— the Agresti–Coull method

Recall from Section 4.3 that a confidence interval is the range of most-plausible values of the parameter we are trying to estimate, based on the data. The 95% confidence interval of a proportion will enclose the true value of the proportion 95% of the time that it is calculated from new data.

---

6. A few textbooks use $(n-1)$ rather than n in the denominator of the formula for $SE_{\hat{p}}$. Using $(n-1)$ creates an estimate of the standard error that is less biased, but using n gives an estimate that is closer to $\sigma_{\hat{p}}$, on average.

There are many methods in the statistical literature for calculating an approximate confidence interval for a proportion. We recommend using the **Agresti–Coull method** (Agresti and Coull 1998). For the 95% confidence interval, we first calculate a number called p':

$$p' = \frac{X + 2}{n + 4}.$$

This p' is just an intermediate calculation needed in the next equation, not an estimate of the proportion. The 95% confidence interval for a proportion is

$$p' - 1.96\sqrt{\frac{p'(1 - p')}{n + 4}} < p < p' + 1.96\sqrt{\frac{p'(1 - p')}{n + 4}}.$$

Getting back to the male radiologists and their many daughters, recall that our best estimate of the true proportion of sons among the offspring of highly irradiated radiologists is 0.345. We can calculate the 95% confidence interval for the population proportion using $X = 30$ and $n = 87$ in the formula for the Agresti–Coull method:

$$p' = \frac{X + 2}{n + 4} = \frac{30 + 2}{87 + 4} = 0.352.$$

The 95% confidence interval is

$$p' - 1.96\sqrt{\frac{p'(1 - p')}{n + 4}} < p < p' + 1.96\sqrt{\frac{p'(1 - p')}{n + 4}}$$

$$0.352 - 1.96\sqrt{\frac{0.352(1 - 0.352)}{87 + 4}} < p < 0.352 + 1.96\sqrt{\frac{0.352(1 - 0.352)}{87 + 4}}$$

$$0.254 < p < 0.450.$$

This interval does *not* include the value 0.512, which is the proportion of sons typically found in the human population. In other words, 0.512 is not one of the most-plausible values for the proportion of sons of radiologists. We can be reasonably confident, therefore, that the proportion of sons of radiologists is lower than the human average, assuming that the data are indeed a random sample. The reason for so few sons is not known.

Confidence intervals for proportions—the Wald method

The most commonly used method to determine a confidence interval for a proportion is called the **Wald method**. In fact, most statistics packages for the computer still use the Wald method. We show it here only because it is used so often, but we do not recommend using it, because it is not accurate in some commonly encountered situations.

The Wald method brackets the population estimate $\hat{p}$ by a multiple of its standard error. By the Wald method, the 95% confidence interval of the proportion is

$$\hat{p} - 1.96\,\mathrm{SE}_{\hat{p}} < p < \hat{p} + 1.96\,\mathrm{SE}_{\hat{p}}.$$

You can see that this formula is approximately the same as the 95% confidence interval calculated using the 2SE rule (Chapter 4). For the radiologist data in Example 7.3, for instance, the 95% confidence interval calculated using the Wald method is $0.244 < p < 0.445$.

Unfortunately, the method is accurate only when n is large and the population p is not close to 0 or 1. A 95% confidence interval should bracket the true population parameter in 95% of samples. Unfortunately, when n is small, or when p is close to 0 or 1, the Wald confidence interval for the proportion contains the true proportion less than 95% of the time. We recommend using the Agresti–Coull method instead.

7.4 Deriving the binomial distribution

We eventually use several probability distributions in this book. Each of these has been mathematically derived from first principles, and often this derivation is quite challenging. The binomial distribution, on the other hand, is relatively straightforward to derive. Here in this section, we sketch out how it is done.

Imagine that we randomly sample n individuals from a population and that we take them in order, one at a time, from 1 to n. We want to calculate the probability of getting X successes. The first step in calculating this probability is to determine the number of sequences of successes and failures that lead to X successes in total.

For most values of X, many different sequences of successes and failures will yield a total of exactly X successes. For example, imagine that we sample five children and we want to know the probability of getting exactly three boys (B) and two girls (G). For a sample this small, it is relatively easy to list all of the possible sequences of $n = 5$ trials that yield three boys:

BBBGG BBGBG BBGGB BGBBG BGBGB
BGGBB GBBBG GBBGB GBGBB GGBBB

There are exactly 10 such sequences.

In general, the number of different sequences of n events yielding exactly X successes is given by the binomial coefficient, $\binom{n}{X}$. In the above example of $X = 3$ boys out of $n = 5$ children, the binomial coefficient is

$$\binom{5}{3} = \frac{5!}{3!(5-3)!} = \frac{5 \times 4 \times 3 \times 2 \times 1}{(3 \times 2 \times 1)(2 \times 1)} = 10.$$

The next step in finding the probability of X successes in n random trials relies on the assumption that separate trials are independent. In this case, each success happens with the same probability (p), and each failure happens with probability $1 - p$. Thus, from the multiplication rule, the probability of any string of successes and failures is the product of these probabilities for each event. Thus, a single sequence that has X successes and $n - X$ failures has probability

$$p^X(1 - p)^{n-X}.$$

Under independence, each sequence of trials yielding exactly X successes has this same probability of occurring.

The last step is to add up the probabilities of all sequences yielding exactly X successes (i.e., we apply the addition rule). Because each of the alternative sequences is mutually exclusive and each has the same probability, we find the overall probability of X successes in n trials by multiplying the probability of any given sequence by the number of sequences:

$$\Pr\left[X \; successes\right] = \binom{n}{X}p^X(1 - p)^{n-X}.$$

This is the formula for the binomial distribution that we first introduced in Section 7.1 and used thereafter.

7.5 Summary

- The binomial distribution expresses the probability of getting X successes out of n trials, assuming that each trial is independent and has the same probability (p) of a success.
- The best estimate of a population proportion is the sample proportion.
- According to the law of large numbers, very large samples will have a proportion of successes that is close to the true proportion in the population.
- The binomial test compares the observed number of successes in a data set to that expected under a null hypothesis. The null distribution of the number of successes under H_0 is the binomial distribution, and so the binomial formula can be used to calculate the P-value for the test.
- A confidence interval for a proportion can be found using the Agresti–Coull method.

7.6 Quick Formula Summary

Binomial distribution

Formula: $\Pr\left[X \; successes\right] = \binom{n}{X}p^X(1 - p)^{n-X}$, where p is the probability of success in each trial, X is the number of successes, and n is the number of trials.

Proportion

Estimate: $\hat{p} = \dfrac{X}{n}$

Standard error: The standard error of a proportion is estimated by
$\text{SE}_{\hat{p}} = \sqrt{\hat{p}\,(1 - \hat{p})/n}$.

Agresti–Coull 95% confidence interval for a proportion

What does it assume? A random sample.

Formula: $\left(p' - 1.96\sqrt{\dfrac{p'(1 - p')}{n + 4}}\right) < p < \left(p' + 1.96\sqrt{\dfrac{p'(1 - p')}{n + 4}}\right),$

where

$p' = \dfrac{X + 2}{n + 4}$, X is the number of successes in the sample, and n is the sample size.

Binomial test

What is it for? Tests whether a population proportion (p) matches a null expectation (p_0) for the proportion.

What does it assume? A random sample.

Test statistic: The observed number of successes, X.

Formula: $P = 2\left(\displaystyle\sum_{i=X}^{n} \Pr[i\ successes]\right)$ for $X/n > p_0$ or $P = 2\left(\displaystyle\sum_{i=0}^{X} \Pr[i\ successes]\right)$

for $X/n < p_0$, where X is the observed number of successes, n is the sample size, and $\Pr[i\ successes]$ is the probability of obtaining i successes from n trials given by the binomial distribution.

PRACTICE PROBLEMS

1. **Calculation practice: Binomial probabilities.** *Enterococcus* bacteria are part of the normal intestinal flora of humans, but some strains can cause disease. In U.S. hospitals, 30% of pathogenic isolates are resistant to the antibiotic vancomycin (Wenzel 2004). Assume that seven independent pathogenic isolates have been extracted from patients and tested for resistance. Using the following steps, calculate the probability that five or more of the isolates are resistant to vancomycin:

a. What are the assumptions of the binomial distribution? Does this example match those assumptions?

b. Using the binomial distribution, what is the probability of success for this example? What is n?

c. Calculate the probability of exactly five resistant isolates using the binomial distribution.

d. Calculate the probability of exactly six resistant isolates, and calculate the probability of exactly seven resistant isolates.

e. Using the addition principle, combine the information from the previous answers to calculate the probability of five or more resistant isolates out of seven.

2. **Calculation practice: Binomial test.** Do people typically use a particular ear preferentially when listening to strangers? Marzoli and Tomassi (2009) had a researcher approach and speak to strangers in a noisy nightclub. An observer scored whether the person approached turned either the left or right ear toward the questioner. Of 25 participants, 19 turned the right ear toward the questioner and 6 offered the left ear. Is this evidence of population difference from 50% for each ear? Use the following steps to help answer this question with a binomial test. Assume that the assumptions of the binomial test are met in this study.

a. State the null and alternative hypotheses for the binomial test.

b. What is the observed value of the test statistic?

c. Under the null hypothesis, calculate the probability of getting exactly 19 right ears and six left ears.

d. List all possible outcomes in which the number of right ears is greater than the 19 observed.

e. Calculate the probability under the null hypothesis of each of the extreme outcomes listed in (d).

f. Use the addition rule to calculate the probability of 19 or more right-eared turns under the null hypothesis.

g. Give the two-tailed P-value based on your answer to (f).

h. Interpret this P-value. What does it indicate?

i. State the conclusion from your test.

3. **Calculation practice: Confidence interval for a population proportion.** In a study in Scotland (as reported by Devlin 2009), researchers left a total of 240 wallets around Edinburgh, as though the wallets were lost. Each contained contact information including an address. Of the wallets, 101 were returned by the people who found them. With the following steps, use the data to estimate the proportion of lost wallets that are returned, and give a 95% confidence interval for this estimate.

a. What is the observed proportion of wallets that were returned?

b. Calculate p' to use in the Agresti–Coull method of calculating a 95% confidence interval for the population proportion.

c. Calculate the lower bound of the 95% confidence interval.

d. Calculate the upper bound of the 95% confidence interval.

e. Provide two values for p that lie within the most-plausible range, according to these data, and two that lie outside.

f. If the authors had tested the null hypothesis that p was $1/2$ at a significance level 0.05, is it likely that they would have rejected the null hypothesis (base your answer only on your results above)?

4. In 1955, John Wayne played Genghis Khan in a movie called *The Conqueror*. It was not an artistic success. More unfortunately, the movie was filmed downwind of the site of 11 previous aboveground nuclear bomb tests. Of the 220 people who worked on this movie, 91 had been diagnosed with cancer by the early 1980s, including Wayne, his costars, and the director. According to large-scale epidemiological data, only about 14% of people of this age group, on average, should have been stricken with cancer within this time frame.[7] We want to know whether there is evidence for an increased cancer risk of people associated with this film.

---

7. See http://www.straightdope.com/columns/read/374/did-john-wayne-die-of-cancer-caused-by-a-radioactive-movie-set.

Assume that this probability is the same for each member of the cast.

a. What is the best estimate of the probability of a member of the cast or crew getting cancer within the study interval?

b. What is the standard error of your estimate? What does this quantity measure?

c. What is the 95% confidence interval for this probability estimate? Does this interval bracket the typical cancer rate of 14% for people of the same age group? Interpret this result.

5. In the United States, paper currency often comes into contact with cocaine either directly, during drug deals or usage, or in counting machines where it wears off from one bill to another. A forensic survey collected fifty $1 bills and measured the cocaine content of the bills. Forty-six of the bills had measurable amounts of cocaine on them (Jenkins 2001). Assume that the sample of bills was a random sample.

a. From these data, what is the best estimate of the proportion of U.S. $1 bills that have a measurable cocaine content?

b. What is a 95% confidence interval for the estimate in part (a)?

c. What is the correct interpretation of your 95% confidence interval?

6. For the following scenarios, state whether the binomial distribution would describe the probability distribution of possible outcomes. If not, explain why not.

a. The number of red cards out of the first five cards drawn from the top of a regular, randomly shuffled deck of cards.

b. The number of red balls out of 10 drawn one by one from a vat of 50 red and blue balls, if the balls are replaced and mixed after each draw.

c. The number of red balls out of 10 drawn one by one from a vat of 50 red and blue balls, if the balls are *not* replaced after each draw.

d. The number of red-eyed flies among 200 *Drosophila* individuals drawn at random from a large population having both red- and black-eyed flies.

e. The total number of red-eyed flies in five *Drosophila* families, each of 40 individuals,

with the families chosen at random from a large population.

7. The mite *Demodex folliculorum* lives in the hair follicles of humans, including the follicles of the eyelashes. (The bluish creatures in the photo on the first page of this chapter are these mites. The yellowish shaft in the photo is a human hair.) Having heard that "most people" have these mites living in their skin and eyelashes, we wanted to know what "most" really meant. The only data we could find was in a paper that compared 16 North American women with a skin condition called rosacea to 16 women that did not have this skin condition (call this the "control" group). Of the 16 women in the control group, 15 of the 16 had these mites. All 16 of the women in the rosacea group had mites (Al Am et al. 1997).

a. From these data, a researcher estimated the proportion of people in North America who have these mites as 31/32. What is wrong with this estimate?

b. What is your best estimate for the proportion of North Americans without rosacea who have these mites? Assume random sampling.

c. What is the 95% confidence interval for this estimate?

d. What is the best estimate (with a 95% confidence interval) for the proportion of women with rosacea who have these mites?

8. In the garden spider *Araneus diadematus*, the female often attempts to eat the male before or after mating, making sex a daunting prospect for the males. (This seemingly bizarre behavior, called sexual cannibalism, is not uncommon in the animal world.) In a series of mating observations, the courting male was captured and eaten by the female in 21 of 52 independent mating trials (Elgar and Nash 1988).

a. Based on the sample, estimate the proportion of males that are eaten by females, and give a 95% confidence interval for this estimate.

b. Is this estimate and confidence interval consistent with a true proportion of 50% capture of males? Is it consistent with 10% capture?

c. If the sample were much larger and the data showed instead that 210 out of 520 matings involved sexual cannibalism, would this change the estimated proportion? Would it change the confidence interval? How? (You don't need to do the calculations—just answer qualitatively.)

9. As our planet warms, as it has done for the last century or so, the change in temperature will have major effects on life. There are basically three possibilities for what might happen to a species: it can evolve to be better adapted to the new temperature, it can move closer to the poles so that it experiences temperatures closer to what it has experienced in the past, or it can go extinct. There have been a large number of studies of the second possibility. A recent study of the range limits of European butterflies found that, of 24 species that had changed their ranges in the last 100 years, 22 of them had moved further north and only two had moved further south (Parmesan et al. 1999). Assume that these 24 are a random sample of butterfly species with altered ranges. Test the hypothesis that the fraction of butterfly species moving north is different from the fraction moving south.

10. Imagine that there were two studies of the prevalence of melanism (solid black coat color). One estimated that the proportion of black leopards in this population was 52%, with a 95% confidence interval that ranged from 46% to 58%. The other study estimated the proportion to be 64% with a 95% confidence interval that ranged from 35% to 85%.
    a. Which study most likely had a larger sample size?
    b. Which estimate is more believable?

11. Mice have litters of several pups at once. The pups are arranged in a line within the mother's uterus, so many fetuses lie between two of their siblings. It has been shown that female fetuses located between two male fetuses (2M) experience higher testosterone levels than those adjacent to no male fetuses (0M), because the hormone is produced by the males and diffuses across the fetal membranes and through the amniotic fluid to adjacent females. The higher fetal testosterone levels are known to have several effects on the females later in life, including increased aggression levels, growth rates, and

even which eye opens first. One study wanted to measure whether fetal testosterone levels affected how attractive these female mice were to male mice as adults (vom Saal and Bronson 1980). Twenty-four male mice were given a choice between a female that was 0M and another unrelated female that was 2M. (Both the males and females were randomly chosen from their populations.) Each male was placed on a platform, and he could jump into the cage of whichever female he preferred. Nineteen of the 24 males chose the 0M female.

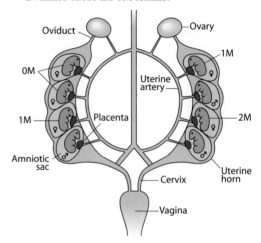

a. Is this evidence that the males prefer one type of female over the other?
b. If the two females presented to each male in a given trial had been sisters, would this have been a better or worse experimental design? Why?

12. A giant vat contains large numbers of two types of bacteria, called strain A and strain B. Assume that 30% of the bacteria in the vat are strain A, and the other 70% are strain B. The vat is well mixed. Twenty technicians each collect a random sample of 15 cells from the vat and then determine which strain each cell belongs to.
    a. What will the proportion of strain A cells be in these 20 samples, on average?
    b. Each technician counts the number of strain A bacteria cells in his or her sample. To what distribution should the number of strain A bacteria in samples conform?

c. Each technician counts the *proportion* of strain A cells in his or her sample. What should the standard deviation be among samples in this proportion?

d. Each technician correctly calculates a 95% confidence interval for the proportion of A cells in the vat. On average, what fraction of technicians will construct an interval that includes the value 0.3?

13. Twelve six-sided dice are rolled. Assume all of the dice are fair.

a. How many threes do you expect, on average, out of these 12 rolls?

b. What is the probability of rolling no threes out of the 12 rolls?

c. What is the probability of rolling exactly 3 threes out of the 12 rolls?

d. On average, what is the sum of the number of spots showing on top of the 12 dice?

e. What is the probability that all the dice show only ones and sixes (in any proportion)?

14. A common perception is that people suffering from chronic illness may be able to delay their deaths until after a special upcoming event, like Christmas. Out of 12,028 deaths from cancer in either the week before or after Christmas, 6052 happened in the week before (Young and Hade 2004).

a. What is the best estimate of the proportion of deaths out of this time interval that occurred in the week before Christmas?

b. What is the 95% confidence interval for this estimate?

c. Use this confidence interval to ask, "Are these data consistent with a true value of 50% for the percentage of deaths in the week before Christmas?" Do these data support the common perception?

15. In 1964, Ehrlich and Raven proposed that plant chemical defenses against attack by herbivores would spur plant diversification. In a test of this idea (Farrell et al. 1991), the number of species in 16 pairs of sister clades[8] whose species differed in their level of chemical protection were counted. In each pair of sister clades, the plants of one clade were defended by a gooey latex or resin that was exuded when the leaf was damaged, whereas the plants of the other clade lacked this defense. In 13 of the 16 pairs, the clade with latex/resin was found to be the more diverse (had more species), whereas in the other three pairs the sister clade lacking latex/resin was found to be the more diverse.

a. With these data, test whether the clade having latex/resin and its sister clade lacking it are equally likely to be more diverse.

b. Is this a controlled experiment or an observational study? Why?

16. In the same survey of the cocaine content of currency discussed in Practice Problem 5, heroin was detected on seven of the 50 bills.

a. What is the best estimate of the proportion of U.S. \$1 bills that have heroin on them?

b. What is the standard error of the estimate? What does this quantity measure?

c. What is the 95% confidence interval for this estimate?

d. If the proportion that you estimated from these data were in fact the true proportion in the population, what would be the probability of getting exactly seven bills with heroin when 50 bills are randomly sampled?

17. The Global Amphibian Assessment in 2005 declared that 1856 out of 5743 of the known species of amphibians worldwide are at risk of extinction. (Approximately 122 species have gone extinct already since 1980.) Amphibians are vulnerable to environmental change, so they are thought to be a bellwether of coming changes for other species.

a. What is the proportion of amphibian species that are at risk of extinction?

b. Would it make sense to calculate a confidence interval for this proportion in the usual way? Why or why not?

---

8. A clade is a group of species all descended from the same ancestor. For example, all the rodents form a clade. Sister clades are two clades that are each other's closest relatives. For example, rodents and rabbits are each other's closest relatives, so they are sister clades.

ASSIGNMENT PROBLEMS

18. We all believe that we see most of what goes on around us, at least the most obvious things. Recently, however, psychologists have identified a phenomenon called "selective looking" which means that, if our attention is drawn to one aspect of what we see, we can miss even seemingly obvious features presented at the same time. In a striking demonstration of this phenomenon, a series of randomly chosen students was shown a video of six people throwing a basketball around, and they were asked to count how many times the people in white shirts threw the ball (Simons and Chabris 1999). In the middle of this video,[9] a woman dressed as a gorilla walked through the shot, pausing in the center to thump her chest, and then walked out of the shot. Look at the photo, and you will realize that nothing could be more obvious. Or was it? Of the 12 students watching the video, only five noticed the gorilla.

a. What is the best estimate from these data of the proportion of students in the population who notice the woman in the gorilla suit?

b. What is the 95% confidence interval for the proportion of students in the population who notice the woman in the gorilla suit?

Figure provided by Daniel Simons. Simons, D. J., & Chabris, C. F. (1999). "Gorillas in our midst: Sustained inattentional blindness for dynamic events." *Perception* 28: 1059–1074.

c. What is the best estimate from these data of the proportion of students who *fail to notice* the woman in the gorilla suit?

19. A survey in the U.K. interviewed shoppers encountered in grocery stores about whether they had ever received injuries as a result of food or drink packaging, such as cuts sustained while cleaning up broken glass containers (Winder et al. 2002). Of the 200 who agreed to participate, 109 had received injuries "over the last few years" (27% of those injuries were significant enough to be treated by a doctor or emergency room).

a. What is the best estimate, and 95% confidence interval, for the proportion of shoppers in the population who have injured themselves with their food or drinks?

b. Do you think this was a random sample of all U.K. consumers? What factors might have rendered it a non-random sample?

20. A Royal Society for the Prevention of Cruelty to Animals (RSPCA) survey of 200 randomly chosen Australian pet owners found that 10 said that they had met their partner through owning the pet (RSPCA 2005). Find the 95% confidence interval for the proportion of Australian pet owners who find love through their pets.

21. One classical experiment on ESP (extrasensory perception) tests for the ability of an individual to show telepathy—to read the mind of another individual. This test uses five cards with different designs, all known to both participants. In a trial, the "sender" sees a randomly chosen card and concentrates on its design. The "receiver" attempts to guess the identity of the card. Each of the five cards is equally likely to be chosen, and only one card is the correct answer at any point.

a. Out of 10 trials, a receiver got four cards correct. What is her success rate? What is her expected rate of success, assuming she is only guessing?

---

9. See the video at http://viscog.beckman.uiuc.edu/grafs/demos/15.html.

b. Is her higher actual success rate reliable evidence that the receiver has telepathic abilities? Carry out the appropriate hypothesis test.

c. Assume another (extremely hypothetical) individual tried to guess the ESP cards 1000 times and was correct 350 of those times. This is very significantly different from the chance rate, yet the proportion of her successes is lower than the individual in part (a). Explain this apparent contradiction.

22. In a test of Murphy's law, pieces of toast were buttered on one side and then dropped. Murphy's law predicts that they will land butter-side down. Out of 9821 slices of toast dropped, 6101 landed butter-side down. (Believe it or not, these are real data![10])

a. What is a 95% confidence interval for the probability of a piece of toast landing butter-side down?

b. Using the results of part (a), is it plausible that there is a 50:50 chance of the toast landing butter-side down or butter-side up?

23. Out of 67,410 surgeries tracked in a study in the U.K., 2832 were followed by surgical site infections (Coello et al. 2005).

a. What is a 95% confidence interval for the proportion of surgeries followed by surgical site infection in the U.K.? Assume that the data are a random sample.

b. If the study had followed a total of only 674 surgeries from the same population, would the confidence interval have been wider or narrower?

24. Each member of a large genetics class grows 12 pea plants from an independent pea family. Each family is expected to have $3/4$ plants with smooth peas and $1/4$ plants with wrinkled peas.

a. On average, how many wrinkled pea plants will a student see in her 12 plants?

b. What is the standard deviation of the *proportion* of wrinkled pea plants per student?

c. What is the variance of the proportion of wrinkled pea plants per student?

d. Predict what proportion of the students saw exactly two wrinkled pea plants in their sample.

25. Juvenile long-tailed tits (*Aegithalos caudatus*), a European relative of the chickadee, "help" adult birds raise offspring, such as by feeding their nestlings. What is the evolutionary advantage of helping behavior: practice for parenthood; increased changes of inheriting the adults' territory in future; or indirect genetic benefits via increased success of kin? To investigate, Russell and Hatchwell (2001) monitored the behavior of 17 juveniles, each of which lived equidistant from two nests of adult birds. In each case, one nest was parented by a relative of the helper, and the other was parented by non-kin adults. Sixteen of the juveniles helped at the nest of their kin, whereas one helped at the non-kin nest. Do these results provide evidence for preferential helping at the nests of kin? Conduct the appropriate test.

26. In a blind taste test, do people prefer pâté or dog food? To investigate, Bohannon et al. (2010) presented 18 college-educated adults with unlabeled samples of dog food (Newman's Own Organics Canned Turkey & Chicken) and four meat products meant for humans (duck liver mousse, pork liver pâté, liverwurst, and Spam). Participants were asked to rank their preferences. Two of 18 participants ranked the dog food first, whereas the other 16 participants chose one of the other items. Based on these results, can we conclude that people are less

---

10. http://www.counton.org/thesum/issue-07/issue-07-page-05.htm. The cause is probably not the butter, since writing a "B" on the top of a piece of toast also yields an excess of outcomes with the B-side down.

likely to prefer dog food over all human food than would be expected by chance?

27. Mood variation is related to photoperiod in some people, and the likelihood of depression increases in the winter months. As a result, people often assume that suicide rates increase in winter. A study in Finland (Valtonen et al. 2006) divided the year 1997 into equal halves and compared the number of suicides in "winter" (24 September to 19 March) and "summer" (remainder of year). Out of a total of 1636 suicides, 766 were in winter and 870 were in the summer. Based on these data, estimate the proportion of suicides that occurred in winter, assuming that the suicides were independent. Are the data compatible with a greater suicide rate in winter than summer, based on a 95% confidence interval?

28. Biff and Dilara were having an argument over what fraction of people would likely go out of their way to drive over a live organism if it were standing innocently by the side of the road. Dilara, whose heart is pure, guessed that fewer than 2% of people would behave that badly—roughly the proportion of people who score as psychopaths in standard testing. Biff, who isn't revealing what he knows, guessed that the fraction would be higher, perhaps 5%. To settle the debate they analyzed data from an experiment in which a rubber facsimile of a turtle, a tarantula spider, a snake, or a leaf were placed on the paved shoulder of a two-way road (Rober 2012).[11] Of 1000 vehicles observed to drive by, 60 swerved onto the shoulder in an effort to drive over the rubber organism. Let's assume (perhaps unrealistically) that each vehicle represents an independent trial and that the proba-

bility of someone attempting to flatten the rubber organism is the same for each organism. Are these data consistent with a fraction of 2%? Are they consistent with a fraction of 5%?

29. Refer to Practice Problem 1.
   a. Produce a graph illustrating the probability distribution of the number of vancomycin-resistant isolates when seven isolates are randomly sampled from the U.S. hospital *Enterococcus* population.
   b. Does the distribution in (a) represent an ordinary sampling distribution or, more specifically, a null sampling distribution? Explain your answer.

30. Biff and Dilara were arguing over a news feed about a population of catfish (*Silurus glanis*) in France that had figured out how to hunt feral pigeons. A submerged catfish lying in wait would rush into shallow water (like a killer whale beaching itself when hunting seals), grab a pigeon that had been drinking or bathing at the water's edge, drag it into deeper water, and swallow it. Biff was skeptical and suggested that if the story had merit, the chance of a successful capture would surely be low: less than 10%. Dilara suspected that for the strategy to be profitable, the success rate would need to be fairly high, perhaps as high as 40%. To investigate, they tracked down the original article (Cucherousset et al. 2012), which reported that 15 out of 54 attempts were successful. Are these data consistent with Dilara's conjecture, Biff's argument, both, or neither?

31. Refer to Practice Problem 15.
   a. Plot the null probability distribution for this case.
   b. What is the meaning of this distribution?

---

11. http://www.youtube.com/watch?v=k-Fp7flAWMA&list=SP45865A763BAB32CA&index=5.

Correlation does not require causation

cience is aimed at understanding how the world works. Identifying the causes of events is a key part of the process of science. A first step in that process is the discovery of patterns, which usually involves noticing associations between events. When two events are associated, the possibility is raised that one is the cause of the other.

For example, our ancestors noticed that chewing on willow bark made sufferers of headaches and fevers feel better. We now know that the association between bark-chewing and pain relief is explained by the presence of salicylic acid in the bark. The acid blocks the release of prostaglandins in the body, which mediate pain and inflammation. This association led to the invention of aspirin.

An association (or correlation) between variables is a powerful clue that there may be a causal relationship between those variables. Smoking is associated with lung cancer; drinking is associated with fatal automobile accidents; taking streptomycin is associated with reduced bacterial infection. These associations exist because one thing causes the other, and the causal relationship was discovered because someone noticed their correlation.

The problem is that two variables can be correlated without one being the cause of the other. A correlation between two variables can result instead from a common cause. For example, the number of violent crimes tends to increase when ice cream sales increase. Does this mean that violence instills a deep need for ice cream? Or does it mean that squabbles over who ate the last of the Chunky Monkey® escalate to violence? Perhaps, but the more likely explanation for this association is that they share a common cause: hot weather encourages both ice cream consumption and bad behavior. Ice cream sales and violence are correlated, but this doesn't mean that one causes the other.

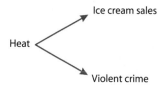

If we plot the mean life expectancy of the people of a country against the number of televisions per person in that country, we see quite a strong relationship (Rossman 1994):

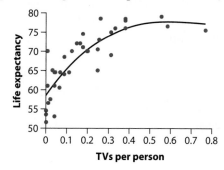

But the magical healing powers of the TV have yet to be demonstrated. Instead, it is likely that both televisions per capita and life expectancy have a common cause in the overall wealth of the citizens of the country.

These examples demonstrate the problems posed by **confounding variables**. A confounding variable is an unmeasured variable that changes in tandem with one or more of the measured variables; this gives the false appearance of a causal relationship between the measured variables. The apparent relationship between violence and ice cream sales is probably the result of the confounding variable *temperature. Overall wealth*, or something related to it, is probably the confounding variable in the correlation between the number of televisions and life expectancy.

Even more confusingly, a correlation between two variables may actually result from reverse causation. That is, the variable identified as the effect by the researcher may actually be the cause. For example, studies have repeatedly shown that babies who are breast-fed grow slightly more slowly than babies fed with formula. This has been interpreted as evidence that breast-feeding causes slow growth, but the truth turns out to be the reverse. Babies who grow rapidly are more likely to feed more, to be more demanding, and to be moved off the breast onto formula to give the poor mothers a break. Experimental studies confirm that exclusive breast-feeding has no measurable effect on infant growth (Kramer and Kakuma 2002).

These examples demonstrate the limitations of **observational studies,** which illuminate patterns but are unable to fully disentangle the effects of measured explanatory variables and unmeasured confounding variables. The main purpose of **experimental studies** is to disentangle such effects. In an experiment, the researcher is able to assign participants randomly to different treatment groups. Random assignment breaks the association between the confounding variable and the explanatory variable, allowing the causal relationship between treatment and response variables to be assessed.

Sir Ronald Fisher, our hero in many other respects, never believed that smoking caused lung cancer; instead, he thought that smoking may be caused by a genetic predisposition and that this genetic effect might also predispose one to cancer.[1] In other words, he thought that the genotype of an individual was a confounding factor. Fisher himself invented experimental design, which would make it possible to test his claim. In theory, one could assign participants randomly to smoking and nonsmoking treatments and any underlying correlation with genetics would be broken, because on average, equally as many participants genetically predisposed to cancer would be assigned to both treatments. If Fisher's hypothesis were correct, the smokers would not have an increased cancer rate. Such an experiment is not ethically possible with humans, but it has been done with other species.

Finding correlations and associations between variables is the first step in developing a scientific view of the world. The next step is determining whether these relationships are causal or coincidental. This requires careful experimentation.

1. Fisher was a lifelong smoker who consulted with the Tobacco Manufacturers' Standing Committee at the time he made this claim.

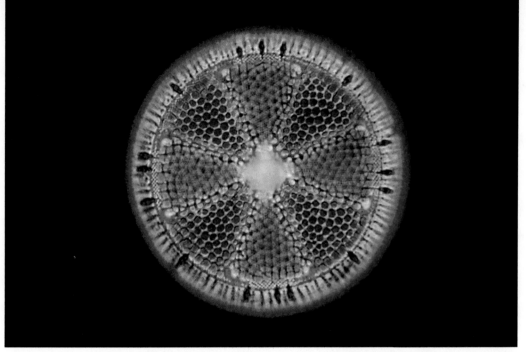

Fossil marine diatom, *Actinoptychus heliopelta*

8 Fitting probability models to frequency data

The binomial test, introduced in the preceding chapter, is an example of a "goodness-of-fit test." A **goodness-of-fit test** is a method for comparing an observed frequency distribution with the frequency distribution that would be expected under a simple probability model governing the occurrence of different outcomes. In Chapter 6, for example, we tested the simplest probability model imaginable: whether left- and right-handed toads occur with equal probability in a population. In Chapter 7, we tested whether the frequency of sperm genes on the X chromosome is proportional to the size of that chromosome. Rejecting the null hypothesis in both cases confirmed real patterns in nature.

The binomial test, however, is limited to categorical variables with only two possible outcomes. In this chapter, we introduce a more general goodness-of-fit test that allows us to handle categorical and discrete numerical variables having more than two outcomes. It also allows us to assess the fit of more complex probability models. We also show how to test the fit between observed frequency data and the frequencies predicted by simple probability models and how to interpret the results if the null hypothesis is rejected.

8.1 Example of a probability model: the proportional model

The **proportional model** is a simple probability model in which the frequency of occurrence of events is proportional to the number of opportunities. We encountered the proportional model in Example 7.2 when we tested whether the frequency of sperm genes on the mouse X chromosome was proportional to the size of that chromosome. Here we will explore the χ^2 goodness-of-fit test on a proportional model by using the data from Example 8.1.

EXAMPLE No weekend getaway

8.1 The U.S. National Center for Health Statistics records information on each new baby born, such as time and date of birth, weight, and sex (Ventura et al. 2001). One bit of information available from these data is the day of the week on which each baby was born. Under the proportional model, we would expect that babies should be born at the same frequency on all seven days of the week. But is this true? Table 8.1-1 lists the number of babies born on each day of the week in a random sample of 350 births from 1999. Figure 8.1-1 displays these data in a bar graph.

FIGURE 8.1-1
Bar graph of the day of the week for 350 births in the U.S. in 1999.

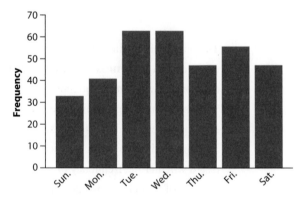

The data show a lot of variation in the number of births from one day to the next during the week, from a low of 33 on Sundays to a high of 63 on Tuesdays and Wednesdays. Under the proportional model, which will be the null hypothesis, the number

TABLE 8.1-1 Day of the week for 350 births in the U.S. in 1999.

Day	Number of births
Sunday	33
Monday	41
Tuesday	63
Wednesday	63
Thursday	47
Friday	56
Saturday	47
Total	350

of births on Monday should be proportional to the numbers of Mondays in 1999, except for chance differences. The same should be true for the other days of the week. Does the variation among days evident in Table 8.1-1 represent only chance variation? We can test the fit of the proportional model to the data with a χ^2 goodness-of-fit test.

8.2 χ^2 goodness-of-fit test

The χ^2 **goodness-of-fit test** uses a test statistic called χ^2 to measure the discrepancy between an observed frequency distribution and the frequencies expected under a simple probability model serving as the null hypothesis. (χ is the Greek letter "chi" and is usually pronounced "kye" in English.) The simple model is rejected if the discrepancy, χ^2, is too large.

> The χ^2 *goodness-of-fit test* compares frequency data to a probability model stated by the null hypothesis.

Null and alternative hypotheses

Under the proportional model, each day of the week should have the same probability of a birth, that is, $1/7$ (see Example 8.1). This is the simplest possible model, so it's our null hypothesis:

H_0: The probability of birth is the same on every day of the week.

H_A: The probability of birth is *not* the same on every day of the week.

Once again, the null (H_0) and alternative (H_A) hypotheses are statements about the population from which the data are a random sample. The null hypothesis is very specific, describing the expectation under the simple probability model. The alternative hypothesis is not specific, because it includes every other possibility.

Observed and expected frequencies

Because the proportional model is the null hypothesis, we use it to generate the expected frequency of births on each day of the week. We expect the accumulated number of births on each day of the week to reflect the number of times each day of the week occurred in 1999. It turns out that in 1999 every day of the week occurred 52 times—except Friday, which occurred 53 times. In Table 8.2-1, we divide these numbers by 365, the total number of days in 1999, yielding proportions.

TABLE 8.2-1   Expected frequency of births on each day of the week in 1999 under the proportional model.

Day	Number of days in 1999	Proportion of days in 1999	Expected frequency of births
Sunday	52	52/365	49.863
Monday	52	52/365	49.863
Tuesday	52	52/365	49.863
Wednesday	52	52/365	49.863
Thursday	52	52/365	49.863
Friday	53	53/365	50.822
Saturday	52	52/365	49.863
Sum	365	1	350

We can now use these proportions to calculate the expected frequencies of births for each day of the week under the proportional model. For example, there were 350 total births in the data set, and under H_0 the fraction $52/365$ of them should have occurred on Sundays. The expected frequency of births for Sunday is therefore

$$Expected = 350 \times \frac{52}{365}$$

$$= 49.863.$$

Note that expected frequencies can have fractional components, even if, in any given case, the number of individuals per category will be an integer. The expected frequencies are the average values expected with the null model.

The sum of the expected values should be the same as the sum of the observed values (i.e., 350, except for rounding error). If this isn't the case, you need to check your calculations for errors.

The χ^2 test statistic

The χ^2 statistic measures the discrepancy between the observed and expected frequencies. It is calculated by the following sum:

$$\chi^2 = \sum_i \frac{(Observed_i - Expected_i)^2}{Expected_i}.$$

$Observed_i$ is the frequency of individuals observed in the ith category, and $Expected_i$ is the frequency expected in that category under the null hypothesis. The numerator of this quantity is a difference between the data and what was expected, which is squared so that positive and negative deviations are treated equally. When we divide this squared deviation by the expected value, the deviation of the observed and expected is scaled to the expected value.

> The χ^2 statistic measures the discrepancy between observed frequencies from the data and expected frequencies from the null hypothesis.

It's important to notice that the χ^2 calculations use the *absolute* frequencies (i.e., counts) for the observed and expected frequencies, not proportions or *relative* frequencies. Using proportions in the calculation of χ^2 will give the wrong answer.

In the Example 8.1 data set, i can take on the values 1 through 7, where Sunday $= 1$, Monday $= 2$, and so on. If the observed frequencies in all categories exactly matched the expected frequencies under the null hypothesis, χ^2 would be zero. The larger χ^2 is, the greater is the discrepancy between the data and the frequencies expected under the null hypothesis.

To determine χ^2, we must calculate $(Observed_i - Expected_i)^2 / Expected_i$ for each day of the week. For Sundays, for example, $i = 1$ and

$$\frac{(Observed_1 - Expected_1)^2}{Expected_1} = \frac{(33 - 49.863)^2}{49.863} = 5.70.$$

Repeating this calculation for the rest of the days, we obtain the values shown in the last column of Table 8.2-2. (Make sure you can obtain these same values for yourself.) Note that this discrepancy is largest for Sundays, which has the largest difference between the number of observed births and the number of expected births.

Adding these up, we get

$$\chi^2 = 5.70 + 1.58 + 3.46 + 3.46 + 0.16 + 0.53 + 0.16 = 15.05.$$

The χ^2 statistic is the test statistic for the χ^2 goodness-of-fit test, the quantity measuring the level of agreement between the data and the null hypothesis. All we need now is the sampling distribution of the χ^2 test statistic under H_0. This will allow us to decide whether $\chi^2 = 15.05$ is large enough to warrant rejection of the null hypothesis.

TABLE 8.2-2 Observed and expected numbers of births on each day of the week under the proportional model.

Day	Observed number of births	Expected number of births	$\dfrac{(Observed - Expected)^2}{Expected}$
Sunday	33	49.863	5.70
Monday	41	49.863	1.58
Tuesday	63	49.863	3.46
Wednesday	63	49.863	3.46
Thursday	47	49.863	0.16
Friday	56	50.822	0.53
Saturday	47	49.863	0.16
Sum	350	350	15.05

The sampling distribution of χ^2 under the null hypothesis

Recall from Chapter 6 that we can determine the sampling distribution for a test statistic under the null hypothesis by computer simulation.[1] This approach is tedious and not recommended, but we ran the simulation for these data just to show you what the approximate null distribution for the χ^2 statistic looks like (see the histogram in Figure 8.2-1).

FIGURE 8.2-1
A histogram showing the approximate sampling distribution of χ^2 values under the null hypothesis that births in 1999 occurred with the same probability on each day of the week (Example 8.1). The solid black curve shows the theoretical χ^2 probability distribution with six degrees of freedom. The curve provides an excellent approximation of the sampling distribution of the χ^2 test statistic under the null hypothesis.

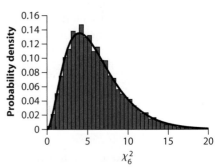

Fortunately, there's an easier way to obtain the null sampling distribution for the χ^2 statistic. The null distribution is well approximated by the theoretical χ^2 distribution, which has a known mathematical form. (See the solid curve superimposed on the histogram in Figure 8.2-1.) Happily, the key features of this theoretical χ^2 distribution (hereafter referred to as "the χ^2 distribution") have been compiled in tables that are easy

---

1. In each step of the simulation, assign each of 350 births with equal probability to the 365 days of 1999, count the number of births that fall on each of the seven days of the week, and then calculate χ^2 as the discrepancy between the simulated frequencies and the frequencies expected under the proportional model. Repeat this procedure a vast number of times, each time recording the χ^2 value. This procedure yields a close approximation to the sampling distribution of the χ^2 statistic under H_0, provided that the number of repetitions is very large.

and quick to use. We demonstrate the use of the tables in the next two subsections of Section 8.2.

The χ^2 distribution is a mathematical function, and to use it we need to specify a number called the **degrees of freedom** (*df*, for short).

> The number of *degrees of freedom* of a χ^2 statistic specifies which χ^2 distribution to use as the null distribution.

The degrees of freedom for a χ^2 goodness-of-fit test is calculated using the following formula.[2]

$df =$ (Number of categories) $- 1 -$ (Number of parameters estimated from the data).

Ignore the last term of this formula (i.e., "Number of parameters estimated from the data") for now, because it is zero for the birth data. We explain what this term means in Sections 8.5 and 8.6, where we first use it.

The birth data have seven categories (one for each day of the week), so the number of degrees of freedom is

$$df = 7 - 1 = 6.$$

This tells us that we need to compare our χ^2 value calculated from the birth data ($\chi^2 = 15.05$) to the χ_6^2 distribution with six degrees of freedom. (The subscript 6 on χ_6^2 indicates the number of degrees of freedom.)

The χ^2 distribution with six degrees of freedom is shown as the black curve in Figure 8.2-1. Note how similar the χ^2 distribution (the solid line) is to the simulated distribution (the histogram), only smoother. The smallest possible value for χ^2 is zero. On the right, the χ^2 distribution extends to positive infinity. This distribution is the one we will use to calculate the *P*-value for our test.

Calculating the *P*-value

The *P*-value for a test is the probability of getting a result as extreme as, or more extreme than, the result observed if the null hypothesis were true. For the χ^2 goodness-of-fit test, the *P*-value is the probability of getting a χ^2 value *greater* than the observed χ^2 value calculated from the data ($\chi^2 = 15.05$ for the birth data). Remember that if the data exactly matched the expectation of the null hypothesis, χ^2 would be zero. A deviation in either direction between an observed frequency and the expected frequency causes χ^2 to be greater than zero. Greater deviations from the null

---

2. The number of degrees of freedom for the χ^2 goodness-of-fit test is calculated as the number of categories minus the number of constraints imposed on the expected frequencies. In a goodness-of-fit test, the sum of the expected frequencies is constrained to be the same as the total number of observations; we always lose a degree of freedom because of this constraint. Additional constraints are imposed if we use the data to generate other numbers needed to calculate the expected frequencies, as Examples 8.5 and 8.6 will show. The logic behind this calculation is that every constraint causes the expected frequencies to be that much more similar to the observed frequencies, and we must adjust the degrees of freedom to compensate.

expectation result in a higher value of χ^2. As a result, we use only the right tail of the χ^2 distribution to calculate P.

The χ^2 distribution is a continuous probability distribution, so probability is measured by the area under the curve, not the height of the curve (Chapter 5). The probability of getting a χ^2 value greater than or equal to a single specified value, which is what we need to calculate a P-value, is equal to the area under the curve to the right of that value extending to positive infinity.

The data from Example 8.1 yielded $\chi^2 = 15.05$. The probability of getting a χ^2 value of 15.05 or greater is equal to the area under the χ^2_6 curve beyond 15.05, as shown by the region highlighted in red in Figure 8.2-2.

FIGURE 8.2-2
The χ^2 distribution with six degrees of freedom. The red area shows the probability of getting a χ^2 value greater than or equal to 15.05.

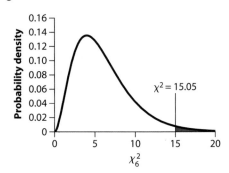

How do we go about finding this area beyond the measured χ^2 value? Two options are available. First, most computer statistical packages will provide the P-value directly: $P = 0.0199$. At the standard significance level of $\alpha = 0.05$, such a small P-value leads us to reject our null hypothesis. That is, these data provide evidence that births are not "randomly" distributed over the days of the week. The variation among days of the week in number of births listed in Table 8.1-1 is simply too large to be explained by chance.

The second option for assessing the P-value uses critical values, as we discuss in the next subsection.

Critical values for the χ^2 distribution

The second way to calculate the P-value for a χ^2 statistic does not require a computer. This method uses tables of critical values to set bounds on the P-value. A **critical value** is the value of a test statistic that marks the boundary of a specified area in the tail (or tails) of the sampling distribution under H_0. If we want a significance level of $\alpha = 0.05$, for example, we would need to know the value of χ^2 for which the area under the curve to its right is 0.05. This value of χ^2 is called the "critical value corresponding to $\alpha = 0.05$."

A *critical value* is the value of a test statistic that marks the boundary of a specified area in the tail (or tails) of the sampling distribution under H_0.

TABLE 8.2-3 An excerpt from the table of χ^2 critical values (Statistical Table A). Numbers down the left side are the number of degrees of freedom (*df*). Numbers across the top are significance levels (α). The critical value for a χ^2 distribution with *df* = 6 and α = 0.05 is 12.59 (indicated in red).

					Significance level α					
df	0.999	0.995	0.99	0.975	0.95	0.05	0.025	0.01	0.005	0.001
1	0.000002	0.00004	0.00016	0.00098	0.00393	3.84	5.02	6.63	7.88	10.83
2	0.002	0.01	0.02	0.05	0.10	5.99	7.38	9.21	10.6	13.82
3	0.02	0.07	0.11	0.22	0.35	7.81	9.35	11.34	12.84	16.27
4	0.09	0.21	0.30	0.48	0.71	9.49	11.14	13.28	14.86	18.47
5	0.21	0.41	0.55	0.83	1.15	11.07	12.83	15.09	16.75	20.52
6	0.38	0.68	0.87	1.24	1.64	12.59	14.45	16.81	18.55	22.46
7	0.60	0.99	1.24	1.69	2.17	14.07	16.01	18.48	20.28	24.32
8	0.86	1.34	1.65	2.18	2.73	15.51	17.53	20.09	21.95	26.12

Statistical Table A at the back of this book (p. 703) gives critical values for the χ^2 distribution. An excerpt from this table is shown in Table 8.2-3. To read the table, first find the column corresponding to the significance level of interest (we will use the standard α = 0.05). Then find the row corresponding to the number of degrees of freedom for the test statistic (*df* = 6 for Example 8.1). The number in the corresponding cell of the table is the critical value $\left[\chi^2_{0.05,\,6} = 12.59\right]$. Under the null hypothesis, the probability of obtaining a χ^2 value as extreme as, or more extreme than, 12.59 is 0.05:

$$\Pr\left[\chi^2_6 \geq 12.59\right] = 0.05.$$

Figure 8.2-3 illustrates the area under the curve to the right of 12.59.

FIGURE 8.2-3
The χ^2 distribution with six degrees of freedom. The area under the right tail of the curve in red is 5% of the total area under the curve. This is the region to the right of χ^2 = 12.59. Under the null hypothesis, χ^2_6 will be greater than 12.59 with probability 0.05.

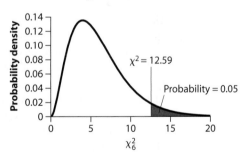

Because our observed χ^2 value (15.05) is greater than 12.59 (i.e., further out in the right tail of the distribution), χ^2 values of 15.05 or greater occur more rarely under the null hypothesis than 5% of the time. Therefore, our *P*-value must be less than 0.05,

$$P = \Pr\left[\chi^2_6 \geq 15.05\right] < 0.05,$$

so we reject the null hypothesis.

We can use Statistical Table A to get closer to the true *P*-value. Note that Statistical Table A also includes columns of critical values for other values of α.

Our observed χ^2 test statistic (15.05) falls between the critical values for $\alpha = 0.025$ $[\chi^2_{0.025, 6} = 14.45]$ and that for $\alpha = 0.01$ $[\chi^2_{0.01, 6} = 16.81]$. Our observed test statistic is greater than 14.45 but less than 16.81. Thus, Statistical Table A makes it possible to bound the P-value as

$$0.01 < P < 0.025.$$

This would be reported as just $P < 0.025$ in a scientific paper. Note that this conclusion is consistent with the P-value of 0.0199 calculated by a computer statistics package. Based on this analysis, we can conclude that births are not equitably distributed over the days of the week.[3]

8.3   Assumptions of the χ^2 goodness-of-fit test

The χ^2 goodness-of-fit test assumes that the individuals in the data set are a random sample from the whole population. This means that each individual was chosen independently of all of the others and that each member of the population was equally likely to find its way into the sample. This is an assumption of every test described in this book.

The sampling distribution of the χ^2 statistic follows a χ^2 distribution only approximately. The approximation is excellent, as long as the following rules[4] are obeyed:

- None of the categories should have an expected frequency less than one.
- No more than 20% of the categories should have expected frequencies less than five.

Notice that these restrictions refer to the *expected* frequencies, not to the *observed* frequencies. If these conditions are not met, then the test becomes unreliable.

If one of these conditions is not met, then we have two options. One option, if possible, is to combine some of the categories having small expected frequencies to yield fewer categories having larger expected frequencies (remember to change the degrees of freedom accordingly). We can do this only if the new combined categories make biological sense. We'll see examples of this approach in Section 8.6. A second option is to find an alternative to the χ^2 goodness-of-fit test, perhaps making use of computer simulation (Chapter 19).

---

3. This discrepancy is largely due to scheduled C-sections and induced labor, but these do not explain the effect completely (Ventura et al. 2001).

4. These restrictions are conservative, because the χ^2 goodness-of-fit test has been shown to work well even with smaller expected values. An alternative rule of thumb is that the average expected value should be at least five (Roscoe and Byars 1971).

8.4 Goodness-of-fit tests when there are only two categories

The χ^2 goodness-of-fit test works even when there are only two categories, so it's a quick substitute for the binomial test (Chapter 7), provided that the assumptions of the χ^2 test are met. The calculations are much quicker than those required for the binomial test, although they are less exact. We demonstrate these calculations using Example 8.4.

EXAMPLE   Gene content of the human X chromosome

8.4   The sex chromosomes are inherited in a very different pattern from that of the other chromosomes, which is known to affect their evolution in many ways. Are they unusual in other ways? For example, are there as many genes on the human X chromosome as we would expect from its size? The Human Genome Project has found 781 genes on the human X chromosome, out of 20,290 genes found so far in the entire genome.[5] The X chromosome represents 5.2% of the DNA content of the whole human genome, so under the proportional model we would expect 5.2% of the genes to be on the X chromosome. Is this what we observe?

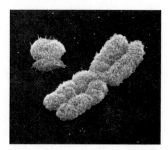

The null and alternative hypotheses are

H_0: The percentage of human genes on the X chromosome is 5.2%.
H_A: The percentage of human genes on the X chromosome is *not* 5.2%.

Observed frequencies and the frequencies expected under H_0 are listed in Table 8.4-1. The expected number of genes on the X chromosome, under the null hypothesis, is $20{,}290 \times 0.052 = 1055$. We observed only 781, which is substantially fewer. What is the probability of a result as extreme as, or more extreme than, the result observed assuming the null hypothesis?

TABLE 8.4-1 Numbers of genes on the human X chromosome and on the rest of the genome.

Chromosome	Observed	Expected
X	781	1,055
Not X	19,509	19,235
Total	20,290	20,290

5. We used release 35 of the human genome, available on the ENSEMBL website in December 2005 (http://www.ensembl.org/Homo_sapiens).

It would be a challenge to calculate the P-value using the binomial test. We would need to calculate

$$P = 2 \times \Pr[X \le 781].$$

When the number of trials (genes) is $n = 20{,}290$ and the probability of a gene being on the X chromosome is $p = 0.052$, this number P would be calculated as

$$P = 2 \times (\Pr[X = 0] + \Pr[X = 1] + \Pr[X = 2] + \cdots + \Pr[X = 781]).$$

The tedium of this sum causes the mind to boggle.[6]

It would be much faster to calculate the P-value using the χ^2 goodness-of-fit test. This procedure yields

$$\chi^2 = \frac{(781 - 1055)^2}{1055} + \frac{(19{,}509 - 19{,}235)^2}{19{,}235} = 75.1.$$

This test statistic has two categories and, therefore, only one degree of freedom:

$$df = 2 - 1 = 1.$$

From Statistical Table A, we see that the critical value of χ_1^2 for a significance level $\alpha = 0.05$ is 3.84. Because our observed $\chi^2 = 75.1$ is greater than 3.84, we know that $P < 0.05$, and we reject the null hypothesis. In fact, we can use Statistical Table A to be even more precise. Because our calculated χ^2 is greater than the largest critical value given for one df ($\chi_{0.001,\,1}^2 = 10.83$), we can say that $P < 0.001$. Thus, there are *significantly* fewer genes on the X chromosome in humans than would be expected from its size.

Remember, the P-value reflects the weight of evidence against the null hypothesis, not how big the difference is between the true proportion and the null expectation of 0.052. We can estimate the true proportion of genes on the X chromosome as $\hat{p} = 781/20{,}290 = 0.038$, yielding a 95% confidence interval of $0.036 < p < 0.041$. This reveals that the proportion of genes on the X chromosome is modestly smaller than the expectation of 0.052.

When there are only two categories, the binomial test is the best option when n is small and when the expected frequencies are too low to meet the assumptions of the χ^2 goodness-of-fit test. Even when n is large, though, the binomial test is preferred when a computer is available, because it yields an exact P-value.[7]

8.5 Fitting the binomial distribution

The proportional model is not the only probability model that can be tested using a goodness-of-fit approach. Biologists often fit their data to other probability distributions that also represent simple models for how nature behaves. By "model" we mean

---

6. We couldn't resist: the answer is $P = 2.64 \times 10^{-84}$.

7. This is not always true. Many statistical packages for computers cheat and use an approximation.

a mathematical description that mimics how we think a natural process works, or at least how it would work in the absence of complications. For this reason, probability distributions are used as null hypotheses in many branches of biology.

EXAMPLE
8.5

Designer two-child families?

In Chapter 5, we claimed that the sex of consecutive children is independent in humans. For example, having had one boy already does not change the probability that the next child will also be a boy. In the absence of complications, then, we expect the numbers of sons and daughters in families containing two children to match a binomial distribution with $n = 2$ and p equal to the probability of having a son in any single trial. Is this what we see? Rodgers and Doughty (2001) tested this hypothesis using data from the National Longitudinal Survey of Youth, which compiles data on the sex of children in a random sample of families of different sizes. Table 8.5-1 lists the number of sons in 2444 two-child families.

TABLE 8.5-1 The frequency distribution of the number of boys in families with two children.

Number of boys	Observed number of families
0	530
1	1332
2	582
Total	2444

There are three possible outcomes for families containing exactly two children: zero, one, or two boys.

We can test the fit of the binomial distribution to the data in Table 8.5-1 using the χ^2 goodness-of-fit test. The null and alternative hypotheses are as follows:

H_0: The number of boys in two-child families has a binomial distribution.
H_A: The number of boys in two-child families does not have a binomial distribution.

Here we are testing the fit of a distribution to the data on multiple families. We are not testing a hypothesis about the true proportion of boys. When testing the fit to a binomial distribution, we are fitting the results of multiple *sets* of trials and comparing the frequencies of sets having different numbers of successes to the expectation of the binomial distribution. This is different from using the binomial distribution to test a null hypothesis about a proportion. In a binomial test, we have only one set of trials.

Notice that in this case our null hypothesis does not specify p, the probability that an individual offspring is a boy. This complicates our task slightly, because we must first estimate p from the data before we can calculate the expected frequencies.

Here's how we estimate p from the data. There are 4888 children in the study, a value obtained by multiplying the number of families (2444) by the family size (2).

The total number of sons is $(2 \times 582) + 1332 = 2496$. Thus, we can estimate the probability of a child being a boy as

$$\hat{p} = 2496/4888 = 0.5106.$$

Next, we use this estimate of p and the binomial distribution with $n = 2$ to calculate the expected frequencies under the null hypothesis. For example, the expected fraction of two-child families having exactly one boy is

$$\Pr[1 \ boy] = \binom{2}{1}(0.5106)^1(1 - 0.5106)^1 = 0.49977.$$

Thus, the expected frequency of 2444 two-child families having exactly one boy is

$$Expected \ [1 \ boy] = 2444 \times 0.49977 = 1221.4.$$

Table 8.5-2 lists the expected frequencies for all possible outcomes, and Figure 8.5-1 shows expected values alongside the data. Surprisingly, the observed frequencies don't seem to match the frequencies expected under the binomial distribution. There is a shortage of two-child families having either no boys or two boys compared with the expectation, and an excess of families having exactly one boy. The differences between observed and expected frequencies are not huge; but can they be explained by chance, or must the null hypothesis be rejected?

TABLE 8.5-2 Observed and expected number of boys in two-child families.

Number of boys	Observed number of families	Expected number of families
0	530	585.3
1	1332	1221.4
2	582	637.3
Total	2444	2444.0

FIGURE 8.5-1
The observed number of two-child families with a given number of boys (*red*) compared with the frequency expected from a binomial distribution (*gold*). Compared with expected frequencies, there is an excess of two-child families with exactly one boy and a shortage of families with no boys or two boys.

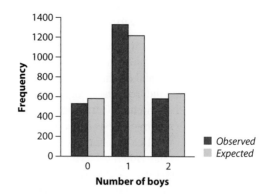

The formula to calculate χ^2, first introduced in Section 8.2, gives us

$$\chi^2 = \frac{(530 - 585.3)^2}{585.3} + \frac{(1332 - 1221.4)^2}{1221.4} + \frac{(582 - 637.3)^2}{637.3} = 20.04.$$

Next, we need to calculate the number of degrees of freedom for our test. There are three categories, which would ordinarily leave us with two degrees of freedom. However, we needed to estimate one parameter from the data to generate the expected frequencies (the probability of boys, p). Using this estimate costs us an extra degree of freedom.[8] As a result, the number of degrees of freedom is

$$df = 3 - 1 - 1 = 1.$$

The critical value for the χ^2_1 distribution having one degree of freedom and a significance level $\alpha = 0.05$ is 3.84 (see Statistical Table A). Because 20.04 is further into the tail of the distribution than 3.84, we know that $P < 0.05$; therefore, we reject the null hypothesis. If we probe Statistical Table A further, we find that 20.04 is greater even than the critical value corresponding to $\alpha = 0.001$, so $P < 0.001$. A statistics package on the computer gave us a more exact value, $P = 1.2 \times 10^{-7}$.

These data show that the frequency distribution of the number of boys and girls in two-child families does *not* match the binomial distribution. This means that one of the assumptions of the binomial distribution is not met in these data. Either the probability of having a son varies from one family to the next, or the individuals within a family are not independent of each other, or both.

What is the reason for the poor fit of the binomial distribution to the number of boys in two-child families? Is the sex of the second child not independent of that of the first, as we've assumed? Are parents manipulating the sex of their children? One likely explanation is that many parents of two-child families having either no boys or two boys are unsatisfied with not having at least one child of each sex and decide to have a third child, thus "removing" their family from the set of two-child families.

8.6 Random in space or time: the Poisson distribution

When the dust settled after the 1980 explosion of Mount St. Helens, spiders were among the first organisms to recolonize the moonscape-like terrain. They dropped out of the airstream and grew fat on insects that arrived in the same way. Let's imagine the frequency distribution of spider landings across the landscape. What would it

---

8. Why do we lose another degree of freedom? Using the data to estimate a parameter of the probability model of the null hypothesis (here, the binomial distribution) unfairly improves the fit of the model to the data. We compensate for this by removing one degree of freedom for the test statistic for every parameter we estimate. If we estimate too many parameters, then the number of degrees of freedom would drop to zero and we couldn't do the test.

look like if spider landings were completely "random" in space? The assumptions we need are as follows:

- The probability that a spider lands at a given point on the continuous landscape is the same everywhere (i.e., they aren't more likely to land some places than others).
- Whether a spider lands at a given point on the landscape is independent of landings everywhere else (i.e., spiders don't clump together or repel one another).

To count spiders, let's place a large grid across the landscape, breaking it up into equal-sized blocks. (The block size doesn't matter as long as they're large enough to accommodate many potential landing sites.) If both assumptions listed previously are met, then the frequency distribution of the number of spiders landing in blocks will follow a **Poisson distribution**.

> The *Poisson distribution* describes the number of successes in blocks of time or space, when successes happen independently of each other and occur with equal probability at every instant in time or point in space.

The Poisson distribution is a useful tool for asking whether events or objects occur randomly in continuous time and space. A Poisson distribution is a reasonable expectation for certain biological counts, such as the number of mutations carried by each individual in a population, the number of salmon caught on a given day by each sport fisher, or the number of seeds successfully germinated by each mother plant. For the biologist, the Poisson distribution is just a *model* for how successes may be distributed in time and space in nature. Life gets interesting when the model doesn't fit the data, because then we learn that one or more of the main assumptions is false, hinting at the existence of interesting biological processes. (For example, some individuals may actually be more prone to mutations than others, some fishers may be better catchers than others, or some plants may produce better-quality seeds.)

The alternative to the Poisson distribution is that successes are distributed in some nonrandom way in time or space. Successes can be **clumped**, for example, in which case they occur closer together than expected by chance (see the left panel in Figure 8.6-1), or successes can be **dispersed**, meaning they are spread out more evenly than expected by chance (see the right panel in Figure 8.6-1). A clumped distribution may arise when the presence of one success increases the probability of other successes occurring nearby. Outbreaks of contagious disease, for example, often lead to a clumped spatial distribution of cases, because individuals catch the disease from their neighbors. A dispersed distribution happens when the presence of one success decreases the probability of another success occurring nearby. Territorial animals are often more dispersed in space than would be expected by chance, for example, because individuals chase each other away. Deviations from the random pattern can therefore help us identify interesting biological processes that create the patterns.

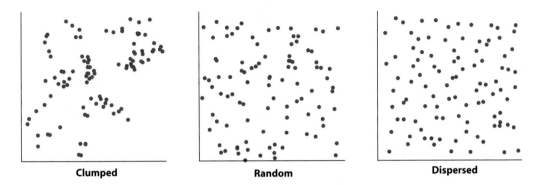

Clumped   **Random**   **Dispersed**

FIGURE 8.6-1 Distributions of points in space that follow a clumped distribution (*left*), a random distribution (*center*), or a dispersed distribution (*right*). In the "random" distribution, each point is independent and has an equal probability of appearing anywhere in the space. If the random graph were divided into a grid of equal-sized squares, the number of points per square would follow a Poisson probability distribution.

Formula for the Poisson distribution

The Poisson distribution was derived by Siméon Denis Poisson,[9] a French mathematical physicist. He showed that the probability of X successes occurring in any given block of time or space is

$$\Pr[X \ successes] = \frac{e^{-\mu}\mu^X}{X!},$$

where μ is the mean number of independent successes in time or space (expressed as a count per unit time or a count per unit space). Here e, the base of the natural log, is a constant approximately equal to 2.718, and X! is X factorial.

Testing randomness with the Poisson distribution

The main use of the Poisson distribution in biology is to provide a null hypothesis to test whether successes occur "randomly" in time or space. In practice, we usually don't know the exact rate at which successes may occur. So, to make predictions about the probability of different outcomes from a Poisson distribution, we must first estimate the rate from the data.

EXAMPLE   Mass extinctions

8.6   Do extinctions occur randomly through the long fossil record of Earth's history, or are there periods in which extinction rates are unusually high ("mass extinctions") compared with background rates? The best record of extinctions through Earth's history comes from fossil marine invertebrates,

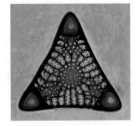

---

9. Poisson is famous for saying, "Life is good for only two things, doing mathematics and teaching mathematics," an opinion no doubt shared by most readers of this book.

because they have hard shells and therefore tend to preserve well. Table 8.6-1 lists the number of recorded extinctions of marine invertebrate families in 76 blocks of time of similar duration through the fossil record (Raup and Sepkoski 1982).

TABLE 8.6-1 The frequency of time blocks in the fossil record in which an observed number of marine invertebrate families went extinct.

Number of extinctions (X)	Frequency
0	0
1	13
2	15
3	16
4	7
5	10
6	4
7	2
8	1
9	2
10	1
11	1
12	0
13	0
14	1
15	0
16	2
17	0
18	0
19	0
20	1
> 20	0
Total	**76**

If the occurrence of family extinctions is "random" in time through the fossil record, then the number of extinctions per block of time should follow a Poisson distribution. Departures from the Poisson distribution could indicate that extinctions tend to be clumped in time and occur in bursts (mass extinctions). Another possibility is that extinctions are more evenly spread over time than we would expect if they occurred randomly.

The easiest way to test the randomness of family extinctions is to compare the frequency distribution of extinctions to that expected from a Poisson distribution using a χ^2 goodness-of-fit test. Our hypotheses are as follows:

H_0: The number of extinctions per time interval has a Poisson distribution.

H_A: The number of extinctions per time interval does *not* have a Poisson distribution.

To begin the test, we need to estimate μ, the mean number of extinctions per time interval. This is obtained using the sample mean,

$$\overline{X} = \frac{(0 \times 0) + (13 \times 1) + (15 \times 2) + \cdots}{76} = 4.21.$$

(See Section 3.1 to review how to calculate a mean from a frequency table. Remember that there are $n = 76$ separate data points here, not the smaller number indicated by the number of rows in Table 8.6-1.) This sample mean ($\overline{X}$) is used in place of μ in the formula for the Poisson distribution to generate the expected frequencies. We show the calculations next, but first look at the result graphically in Figure 8.6-2.

The histogram in Figure 8.6-2 gives the observed frequency distribution of extinctions per time interval, whereas the line connects the expected frequencies under the null hypothesis (the Poisson distribution). If you look closely, there is a discrepancy. Compared with the Poisson distribution, the fossil record shows too many time intervals with large numbers of extinctions and too many intervals having very few extinctions, compared with the Poisson distribution. But is the discrepancy between the observed and expected distributions greater than expected by chance? We will use the χ^2 goodness-of-fit test to find out.

FIGURE 8.6-2
The frequency distribution of the number of extinctions (*histogram*) compared with the frequencies expected from the Poisson distribution having the same mean (*curve*).

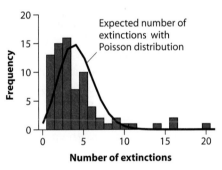

In Table 8.6-2, we have tabulated the observed and expected frequencies. The expected frequency for all but the last category of extinctions is computed by applying the formula for the Poisson distribution to get the expected probability and then multiplying this probability by the total number of intervals (76) to yield the expected frequency. For example, the expected probability of three extinctions in a time interval is

$$\Pr\left[3 \ extinctions\right] = \frac{e^{-\overline{X}}\overline{X}^3}{3!} = \frac{e^{-4.21}(4.21)^3}{3!} = 0.1846.$$

Multiplying this result times the total number of intervals in the data set (76), we find the expected number of intervals with three extinctions:

$$Expected\left[3 \ extinctions\right] = 76 \times 0.1846 = 14.03.$$

You may want to try to calculate some of the other expected values in Table 8.6-2 for practice.

We have grouped all $X \geq 10$ extinctions into the last category because the expected frequency of larger numbers is getting very small. This grouping makes sense,

TABLE 8.6-2 The observed frequency distribution of extinctions of marine invertebrate families compared with the number expected under the Poisson distribution.

Number of extinctions (X)	Observed frequency of time intervals	Expected frequency of time intervals
0	0	1.13
1	13	4.75
2	15	10.00
3	16	14.03
4	7	14.77
5	10	12.44
6	4	8.72
7	2	5.24
8	1	2.76
9	2	1.29
≥ 10	6	0.86
Total	76	76

because these are all large values with a similar biological meaning. The expected frequency for the final category is computed by subtracting the sum of all the previous expected values from 76, the total number of time intervals.

Unfortunately, the expected frequencies fail to meet the assumptions of the χ^2 test: one of them is less than one, and more than 20% are less than five. When this happens, we can group categories and try again. For example, combining $X = 0$ and $X = 1$ into a single class and combining all classes with $X \geq 8$ make sense, because the classes grouped are similar. The resulting data, listed in Table 8.6-3, have eight categories.

TABLE 8.6-3 The observed and expected frequencies of time intervals with a given number of extinctions of marine invertebrate families.

Number of extinctions (X)	Observed frequency of time intervals	Expected frequency of time intervals
0 or 1	13	5.88
2	15	10.00
3	16	14.03
4	7	14.77
5	10	12.44
6	4	8.72
7	2	5.24
≥ 8	9	4.91
Total	76	76

Using the standard formula for the χ^2 statistic, we compute

$$\chi^2 = \frac{(13 - 5.88)^2}{5.88} + \frac{(15 - 10.00)^2}{10.00} + \frac{(16 - 14.03)^2}{14.03} + \cdots = 23.93.$$

We have six degrees of freedom for this test, accounting for the one parameter, μ, that we had to estimate from the data:

df = (Number of categories) $-$ 1 $-$ (Number of parameters estimated from the data)
  = 8 $-$ 1 $-$ 1 = 6.

The critical value for $\chi^2_{0.05, 6}$ is 12.59 (see Statistical Table A). Our χ^2 statistic of 23.93 is further into the tail of the distribution than this critical value is, so our P-value is less than 0.05. More specifically, we can see that $P < 0.001$, because 23.93 is also greater than 22.46, the critical value corresponding to $\alpha = 0.001$. We reject the null hypothesis, therefore, and conclude that extinctions in this fossil record do *not* fit a Poisson distribution.

Comparing the variance to the mean

How can we describe the way that a pattern deviates from the Poisson distribution? One unusual property of the Poisson distribution is that the variance in the number of successes per block of time (the square of the standard deviation) is equal to the mean (μ). In an observed frequency distribution, if the variance is greater than the mean, then the distribution is clumped. There are more blocks with many successes, and more with few successes, than expected from the Poisson distribution. If the variance is less than the mean, then the distribution is dispersed. The ratio of the variance to the mean number of successes is therefore a measure of clumping or dispersion.

For the extinction data, the sample mean number of extinctions is 4.21. The sample variance in the number of extinctions is

$$s^2 = \frac{(0 - 4.21)^2(0) + (1 - 4.21)^2(13) + (2 - 4.21)^2(15) + \cdots}{76 - 1} = 13.72.$$

Because the sample variance (13.72) greatly exceeds the sample mean (4.21), the ratio of the variance to the mean (3.56) is greater than 1, and we conclude that the distribution of extinction events in time is highly clumped. That is, extinctions tend to occur in bursts (mass extinctions) rather than randomly or evenly in time.

8.7   Summary

- The χ^2 goodness-of-fit test compares the frequency distribution of a discrete or categorical variable with the frequencies expected from a probability model.

- The χ^2 goodness-of-fit test is more general than the binomial test because it can handle more than two categories. It is also easier to compute, even when there are only two categories.
- Goodness of fit is measured with the χ^2 test statistic.
- The χ^2 test statistic has a null distribution that is approximated by the theoretical χ^2 distribution. The approximation is excellent as long as no expected frequencies are less than one and no more than 20% of the expected frequencies are less than five. It may be necessary to combine categories to meet these criteria.
- The theoretical χ^2 distribution is a continuous distribution. Probability is measured by the area under the curve.
- The null hypothesis is rejected at significance level α if the observed χ^2 statistic exceeds the critical value of the χ^2 distribution corresponding to α.
- Under the proportional probability model, events fall in different categories in proportion to the number of opportunities. Rejecting H_0 implies that the probabilities are not proportional.
- If trials are independent, and the probability p of a success is the same for each trial, then the frequency distribution of the number of successes should follow a binomial distribution. Rejecting the null hypothesis that the number of successes follows a binomial distribution implies that trials are not independent or that the probability of success is not the same for all trials.
- The Poisson distribution describes the frequency distribution of successes in blocks of time or space when successes happen independently and with equal probability over time or space. Rejecting a null hypothesis of a Poisson distribution of successes implies that successes are not independent or that the probability of a success occurring is not constant over time or space.
- Comparing the variance of the number of successes per block of time or space to the mean number of successes measures the direction of departure from randomness in time or space. If the variance is *greater* than the mean, the successes are clumped; if the variance is *less* than the mean, successes are more evenly distributed than expected by the Poisson distribution.

8.8 Quick Formula Summary

χ^2 Goodness-of-fit test

What is it for? Compares observed frequencies in categories of a single variable to the expected frequencies under a random model.

What does it assume? Random samples. Also that the expected count of each cell is greater than one and that no more than 20% of the cells have expected counts less than five.

Test statistic: χ^2

Distribution under the null hypothesis: χ^2 distributed with df = (Number of categories) − 1 − (Number of parameters estimated from the data).

Formula: $\chi^2 = \sum_i \dfrac{(Observed_i - Expected_i)^2}{Expected_i}.$

Poisson distribution

What is it for? Describes the number of independent events that occur per unit of time or space.

Formula: $\Pr\left[X\ events\right] = \dfrac{e^{-\mu}\mu^X}{X!},$

where X is the number of events and μ is the mean number of events per unit time or space.

PRACTICE PROBLEMS

1. **Calculation problem: χ^2 goodness-of-fit test to a Poisson distribution.** Your friend is writing a computer program to place individuals randomly on a spatial landscape, where every individual is placed independently of all the others and probability is equal everywhere. He finds that many of the individuals land near each other and many other areas are empty, and he becomes concerned that the program is not behaving as intended. You offer to check his results against the Poisson distribution, which is the expected distribution for the number of individuals in equal-area blocks placed over the landscape according to his assumptions. The following data show the number of individuals placed by the program into 200 such blocks. We've ordered the numbers for your convenience.

0, 0, 0, 0, 0, 0, 0, 0, 0, 0, 0, 0, 0, 0, 0, 0, 0, 0, 0, 0,
0, 0, 0, 0, 0, 0, 0, 0, 0, 0, 0, 0, 0, 0, 0, 0, 0, 0, 0, 0,
0, 0, 0, 0, 0, 0, 0, 0, 0, 0, 0, 0, 0, 0, 0, 0, 0, 0, 0, 0,
0, 0, 0, 0, 0, 0, 0, 0, 0, 0, 0, 0, 0, 0, 0, 0, 0, 0, 0, 0,
0, 0, 0, 0, 0, 0, 0, 0, 0, 0, 0, 0, 0, 0, 0, 0, 0, 0, 0, 0,
0, 0, 0, 0, 0, 0, 0, 0, 0, 0, 0, 0, 0, 0, 0, 0, 0, 0, 1, 1,
1, 1, 1, 1, 1, 1, 1, 1, 1, 1, 1, 1, 1, 1, 1, 1, 1, 1, 1, 1,
1, 1, 1, 1, 1, 1, 1, 1, 1, 1, 1, 1, 1, 1, 1, 1, 1, 1, 1, 1,
1, 1, 1, 1, 1, 1, 1, 1, 1, 1, 1, 1, 1, 1, 1, 1, 1, 1, 1, 1,
1, 1, 2, 2, 2, 2, 2, 2, 2, 2, 2, 2, 2, 2, 2, 2, 2, 2, 3, 3

a. Explain why the Poisson distribution is the appropriate distribution to compare these results to.
b. Write the null and alternate hypotheses for this test.
c. Make a frequency table for the data.
d. Calculate the mean number of individuals per area in the data.
e. Using this mean, calculate the probability of 0, 1, 2, and 3 individuals per block assuming a Poisson distribution.
f. Calculate the expected numbers of blocks with zero to three individuals.
g. Are the expected values from part (f) suitable for the χ^2 goodness-of-fit test? Consider the requirements of the χ^2 test for the minimum expected values.
h. Combine categories (if necessary) to meet the minimum expected value requirements for the χ^2 test.
i. How many degrees of freedom will this test have?
j. Calculate the χ^2 test statistic with the observed and expected values.
k. Find the critical value and determine an approximate P-value for this test. (Here and

always, provide an exact P if you are using a computer to answer this question.)

l. Interpret this result in terms of the original hypotheses and the question asked of the data.

2. The parasitic nematode *Camallanus oxycephalus* infects many freshwater fish, including shad. The following table gives the number of nematodes per fish (Shaw et al. 1998).

Number of nematodes	Number of fish
0	103
1	72
2	44
3	14
4	3
5	1
6	1

a. Produce a graph of the data. What type of graph is most appropriate?

b. Calculate the frequencies expected if nematodes infect fish "at random" (i.e., independently and with equal probability).

c. Overlay the expected frequencies onto your graph. What are the noticeable differences?

d. Is there evidence that nematodes do not worm their way into the fish at random? Here and always, show all four steps of hypothesis testing.

3. Luijckx et al. (2012) discovered that resistance to the bacterial parasite *Pasteuria ramosa* is genetically variable in the common freshwater crustacean, *Daphnia magna*. To investigate the genetic basis of this variation, they crossed a completely resistant lineage to a completely susceptible lineage. All the F_1 offspring were resistant. These offspring, when mature, were then crossed to each other to produce an F_2 generation. If resistance is the result of only a single gene with two forms (alleles), then resistant and susceptible F_2 offspring should occur in a 3:1 ratio. Of 71 F_2's tested, 57 were resistant and 14 were susceptible.

a. With these data, calculate the range of most-plausible values for the proportion of resistant offspring. Does the plausible range include the proportion predicted if resistance is determined by a single gene?

b. Give two other values for the proportion that are also consistent with the data.

c. Test the genetic hypothesis. Are the results compatible with your findings in part (a)?

d. On the basis of these results, is it correct to conclude that a single gene in *Daphnia magna* underlies resistance to the bacterium?

4. Soccer reaches its apex every four years at the World Cup, attracting worldwide attention and fanatic devotion. The World Cup is widely thought to be the event that decides the best soccer team in the world. But how much do skill differences determine the outcome? If the probability of a goal is the same for all teams and games, and if goals are independent, then we would expect the frequency distribution of goals per game to approximate a Poisson distribution. In contrast, if skill differences really matter, then we would expect more high scores and more low scores than predicted from the Poisson distribution. The following table tallies the number of goals scored by one of the two teams (randomly chosen) in every game of the knockout round of the World Cup from 1986 though 2010.

Number of goals	Frequency
0	37
1	44
2	21
3	10
4	4
5	1
>5	0
Total	112

a. Plot the frequency distribution of goals per team using the data in the table.

b. What is the mean number of goals per game?

c. Using the Poisson distribution, calculate the expected frequency of games and teams with 0, 1, 2, ..., 5 goals, assuming independence and equal probability of scoring.

d. Overlay the expected frequencies calculated in part (c) on the graph you created in part (a). Do they appear similar?

e. If skill differences do not matter, would you expect the variance in the number of goals per team and side to be less than, equal to, or greater than the mean number of goals?

Calculate the variance in the number of goals per team and side. How similar is it to the mean?

5. Each of the following examples could be addressed with a goodness-of-fit test. From the information given, how many categories and how many degrees of freedom would each test have? Explain your answers.
   a. A die is rolled 50 times to test whether it is fair—whether it has a $1/6$ chance of coming up on each of its six different sides.
   b. A set of 10 coins are flipped, and the number of heads is recorded. This experiment is repeated with the same coins 1000 times. The test compares the frequency of heads to a binomial distribution with $p = 0.5$.
   c. The scenario is the same as in part (b), except now the question is whether the frequency of heads follows a binomial distribution (p not specified).
   d. A food-protection agency counts the number of insect heads found per 100-gram batch of wheat flour. The researchers have 500 batches, and they want to know whether the frequency of insect heads in batches follows a Poisson distribution. The 500 batches included at least 5 batches having 0, 1, 2, 3, or 4 insect heads. No batches had more than four heads.

6. One thousand coins were each flipped eight times, and the number of heads was recorded for each coin. The results are as follows:

Number of heads	Number of coins
0	6
1	32
2	105
3	186
4	236
5	201
6	98
7	33
8	103

   a. Test whether the distribution of coin flips matches the expected frequencies from a binomial distribution assuming all fair coins. (A coin is fair if the probability of heads per flip is 0.5.)
   b. If the binomial distribution is a poor fit to the data, identify in what way the distribution does not match the expectation.
   c. Some two-headed coins (which always show heads on every flip) were mixed in with the fair coins. Can you say approximately how many two-headed coins there might have been out of this 1000?

7. Practice Problem 4 from Chapter 7 gave data about the death rates of people working on the movie *The Conqueror*. Test whether the cancer rates in this group were different from the expected rate of 14%.

8. Imagine that a small hospital's emergency room has an average of 20 admissions per Saturday night. If you were a doctor working overtime on such a Saturday night, you might want to know the probability of having a quiet night, one that would let you catch up on some much-needed sleep. Let's call a quiet night one in which five or fewer admissions take place. What is the chance that you get some sleep? Assume that admissions are independent of one another and are just as likely to land in one instant in time as another on a Saturday night.

9. The following list gives the number of degrees of freedom and the χ^2 test statistic for several goodness-of-fit tests. Find the P-value for each test as specifically as possible from Statistical Table A. If you can, find the P-values more exactly by using a computer program.

Degrees of freedom	χ^2
1	4.12
4	1.02
2	9.50
10	12.40
1	2.48

10. Practice Problem 14 from Chapter 7 gave data about death rates from cancer before and after Christmas. Use these data to test whether the holiday affects death rates.

ASSIGNMENT PROBLEMS

11. If each "success" happens independently of all other successes and with the same probability, what probability distribution is expected for each of the following?

  a. Number of flowers in square-meter blocks in an alpine field
  b. Number of heads out of 10 flips of a coin
  c. Number of bombs landing in city blocks in London in World War II
  d. Daily number of hits on a website
  e. Annual number of elephant attacks on humans in Serengeti National Park
  f. Number of red flowers in sets of 100 flowers in a field of multiple types of flowers

12. The white "Spirit" black bear (or Kermode), *Ursus americanus kermodei*, differs from the ordinary black bear by a single amino acid change in the *melanocortin 1 receptor* gene[10] (*MC1R*). In this population, the gene has two forms (or alleles): the "white " allele *b* and the "black" allele *B*. The trait is recessive: white bears have two copies of the white allele of this gene (*bb*), whereas a bear is black if it has one or two copies of the black allele (*Bb* or *BB*). Both color morphs and all three genotypes are found together in the bear population of the northwest coast of British Columbia. If possessing the white allele has no effect on growth, survival, reproductive success, or mating patterns of individual bears, then the frequency of individuals with 0, 1, or 2 copies of the white allele (*b*) in the population will follow a binomial distribution. To investigate, Hedrick and Ritland (2011) sampled and genotyped 87 bears from the northwest coast: 42 were *BB*, 24 were *Bb*, and 21 were *bb*. Assume that this is a random sample of bears.

  a. Calculate the fraction of *b* alleles in the population (remember, each bear has two copies of the gene).
  b. With your estimate of the fraction of *b* alleles, and assuming a binomial distribution,

calculate the expected frequency of bears with 0, 1, and 2 copies.
  c. Compare the observed and expected frequencies in a graph. Describe how they differ.

13. Refer to Assignment Problem 12. A formal hypothesis test was carried out to compare the observed and expected frequencies of genotypes. The procedure obtained $P = 0.0001$. Answer the following questions:

  a. What are the null and alternative hypotheses?
  b. What are the degrees of freedom for the test statistic?

  Say whether each of the following statements is true or false solely on the basis of these results.

  c. The observed frequencies are compatible with a binomial distribution.
  d. The difference between the observed and expected frequencies is statistically significant.
  e. The test statistic exceeds the critical value corresponding to $\alpha = 0.05$.
  f. The test statistic exceeds the critical value corresponding to $\alpha = 0.01$.
  g. The difference is large between the true genotype frequencies in the bear population and that expected under the binomial distribution.

14. In North America, between 100 million and 1 billion birds die each year by crashing into

---

10. Different mutations in the same gene cause red hair and freckles in humans, white fur in Florida beach mice, white skin in lizards of the White Sands of New Mexico, and coat color variation in dogs and horses. DNA sequences of woolly mammoths have even found variation in this gene.

windows on buildings, more than any other human-related cause. This figure represents up to 5% of all birds in the area. One possible solution is to construct windows angled downward slightly, so that they reflect the ground rather than an image of the sky to a flying bird. An experiment by Klem et al. (2004) compared the number of birds that died as a result of vertical windows, windows angled 20 degrees off vertical, and windows angled 40 degrees off vertical. The angles were randomly assigned with equal probability to six windows and changed daily; assume for this exercise that windows and window locations were identical in every respect except angle. Over the course of the experiment, 30 birds were killed by windows in the vertical orientation, 15 were killed by windows set at 20 degrees off vertical, and 8 were killed by windows set at 40 degrees off vertical.

a. Clearly state an appropriate null hypothesis and an alternative hypothesis.

b. What proportion of deaths occurred while the windows were set at a vertical orientation?

c. What statistical test would you use to test the null hypothesis?

d. Carry out the statistical test from part (c). Is there evidence that window angle affects the mortality rates of birds?

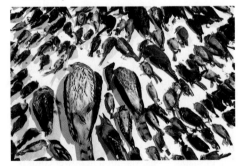

15. In the 19th century, cavalries were still an important part of the European military complex. While horses have many wonderful qualities, they can be dangerous beasts, especially if poorly treated. The Prussian army kept track of the number of fatalities caused by horse kicks to members of 10 of

their cavalry regiments over a 20-year time span. If these fatalities occurred independently and with equal probability for each regiment, then the number of deaths by horse kick per regiment per year should follow a Poisson distribution. On the other hand, if some regiments during some years consisted of particularly bad horsemen,[11] then the events would not occur with equal probability, in which case we would expect a frequency distribution different from the Poisson distribution. The following table shows the data, expressed as the number of fatalities per regiment-year (Bortkiewicz 1898).

Number of deaths (X)	Number of regiment-years
0	109
1	65
2	22
3	3
4	1
>4	0
Total	200

a. What is the mean number of deaths from horse kicks per regiment-year?

b. Test whether a Poisson distribution fits these data.

16. Are the outcomes of hospital care different on weekends than weekdays? In a random sample of 500 patients who experienced severe medical complications after admission to acute care wards in three U.S. states from 1999 and 2001, 119 had been admitted on a weekend and 381 had been admitted on a weekday. This compares with a large population of people at risk for such complications in which 14.8% are admitted on weekends and 85.2% are admitted on weekdays (Bendavid et al. 2007).

a. In the 500 sampled patients with severe complications, what fraction had been admitted on weekends? Is this higher or lower than the fraction of all at-risk patients admitted on weekends?

b. Name two statistical methods that could be used to test whether the probability of severe complications in at-risk patients admitted

---

11. Or if they were especially prone to standing behind their horses . . .

to hospitals differs between weekend and weekday. State the advantages and disadvantages of both.

c. State the null and alternative hypotheses for such a test.

d. Test the hypotheses. State your conclusions clearly.

17. Truffles are a great delicacy, sending thousands of mushroom hunters into the forest each fall to find them. A set of plots of equal size in an old-growth forest in Northern California was surveyed to count the number of truffles (Waters et al. 1997). The resulting distribution is presented in the following table. Are truffles randomly located around the forest? If not, are they clumped or dispersed? How can you tell? (The mean number of truffles per plot, calculated from these data, is 0.60.)

Number of truffles per plot	Frequency
0	203
1	39
2	18
3	13
> 3	15

18. The anemonefish *Amphiprion akallopisos* lives in small groups that defend territories consisting of a cluster of sea anemones, among the tentacles of which the anemonefish live (think *Nemo*). In a field study of the species at Aldabra Atoll in the Indian Ocean, Fricke (1979) noticed that each territory tended to have several males but just one female. Based on his counts, 20 territories of exactly four adult fish would have the following frequency distribution of female numbers.

Number of males	Number of females	Number of territories
0	4	0
1	3	20
> 2	< 3	0
	Total	20

a. Estimate the mean number of females per territory having four fish. Provide a standard error for this estimate.

b. Does the number of females in territories having four fish have a binomial distribution? Show all steps in carrying out your test.

c. If the number of females in territories does not have a binomial distribution, what is the likely statistical explanation (i.e., what assumption of the binomial distribution is likely violated)?

d. Optional: Can you suggest a biological explanation for a non-binomial pattern?

19. Hurricanes hit the United States often and hard, causing some loss of life and enormous economic costs. They are ranked in severity by the Saffir–Simpson scale, which ranges from Category 1 to Category 5, with 5 being the worst. In some years, as many as three hurricanes that rate a Category 3 or higher hit the U.S. coastline. In other years, no hurricane of this severity hits the United States. The following table lists the number of years that had 0, 1, 2, 3, or more hurricanes of at least Category 3 in severity, over the 100 years of the 20th century (Blake et al. 2005):

Number of hurricanes Category 3 or higher	Number of years
0	50
1	39
2	7
3	4
> 3	0

a. What is the mean number of severe hurricanes to hit the United States per year?

b. What model would describe the distribution of hurricanes per year, if they were to hit independently of each other and if the probability of a hurricane were the same in every year?

c. Test the fit of the model from part (b) to the data.

20. In snapdragons, variation in flower color is determined by a single gene (Hartl and Jones 2005). *RR* individuals are red, *Rr* (heterozygous) individuals are pink, and *rr* individuals are white. In a cross between heterozygous individuals, the expected ratio of red-flowered:pink-flowered:white-flowered offspring is 1:2:1.

a. The results of such a cross were 10 red-, 21 pink-, and 9 white-flowered offspring. Do these results differ significantly (at a 5% level) from the expected frequencies?

b. In another, larger experiment, you count 100 times as many flowers as in the experiment in part (a) and get 1000 red, 2100 pink, and 900 white. Do these results differ significantly from the expected 1:2:1 ratio?

c. Do the proportions observed in the two experiments [i.e., in parts (a) and (b)] differ? Did the results of the two hypothesis tests differ? Why or why not?

21. A more recent study of Feline High-Rise Syndrome (FHRS) (see Chapter 1, Example 1.2) included data on the month in which each of 119 cats fell (Vnuk et al. 2004).[12] The data are in the accompanying table. Can we infer that the rate of cat falling varies between months of the year?

Month	Number fallen
January	4
February	6
March	8
April	10
May	9
June	14
July	19
August	13
September	12
October	12
November	7
December	5

22. Consider an isolated population of humans in which some individuals are infected with a specific parasite species (e.g., malaria or a filarial worm). Think of two biological hypotheses for why the number of parasite individuals per person may not be well described by a Poisson distribution. Which assumptions of the Poisson distribution are likely violated by the process you propose, and how would the frequency distribution likely be affected?

23. *Spot the flaw.* Tabershaw and Lamm (1977) compared the observed and expected numbers

of different leukemia types in a study group of workers who had been exposed to benzene during their employment. They were testing a previous suggestion that exposure to benzene increases the probability of acute leukemia while not changing the occurrence of other leukemia types. Their expected numbers are based on the relative frequencies of these diagnoses in the population as a whole. The numbers are presented in the following table.

Type of leukemia	Observed	Expected
Chronic lymphocytic	0	2
Chronic myelogenous	1	1
Monocytic	2	1
Acute	4	3

The researchers applied a χ^2 goodness-of-fit test to the data and calculated a χ^2 value of 3.33, with $df = 3$. From the information given, what is the largest error made in this analysis?

24. Seedlings of the parasitic plant *Cuscuta pentagona* (dodder) hunt by directing growth preferentially toward nearby host plants. Lacking eyes, or even a nervous system, how do they detect their victims? To investigate the possibility that the parasite detects volatile chemicals produced by host plants, Runyon et al. (2006) placed individual dodder seedlings into a vial of water at the center of a circular paper disc. A chamber containing volatile extracts from tomato (a host plant) in solvent was placed at one edge of the disc, whereas a control chamber containing only solvent was placed at the opposite end. The researchers divided the disc onto equal-area quadrats to record in which direction the seedlings grew. Of 30 dodder plants tested, 17 seedlings grew toward the volatiles, 2 grew away (toward the solvent), 7 grew toward the left side, and 4 grew toward the right side.

a. Graph the *relative* frequency distribution for these results. What type of graph is ideal?

b. What are the relative frequencies expected if the parasite is unable to detect the plant volatiles or any other cues present? Add these

---

12. This newer study also found that the number of injuries sustained increased with increasing numbers of stories fallen, unlike the study reported in Example 1.2.

expected relative frequencies to your graph in part (a).

c. Using these data, calculate the fraction of seedlings that grow toward the volatiles. What does this fraction estimate?

d. Provide a standard error for your estimate. What does this standard error represent?

e. Calculate the range of most-plausible values for the fraction of dodder seedlings that grow toward the volatiles under these experimental conditions. Does it include or exclude the fraction expected if the parasite is unable to detect plant volatiles or other cues present?

25. Refer to Assignment Problem 24. The researchers used gas chromatographic analysis to extract and identify eight major volatile chemicals present in the host plants (tomato). They tested each of these chemicals separately using the same experimental design to determine whether dodder seedlings would preferentially orient their growth. Of 34 dodder plants tested with one of these chemicals, α-pinene (also present in pine resin, as its name suggests), 11 seedlings grew toward the volatiles, 6 grew away, 8 grew toward the left side, and 9 grew toward the right side. The authors compared these observed frequencies to those expected by the null hypothesis. Their test statistic was $\chi^2 = 1.53$. They obtained $P = 0.68$. Answer the following questions:

a. What are the null and alternate hypotheses?

b. Their χ^2 test statistic has how many degrees of freedom?

Based on these results, state whether each of the following statements is either true or false:

c. The observed frequencies are compatible with the proportional probability model.

d. The difference between the observed and expected frequencies is statistically significant.

e. The test statistic exceeds the critical value corresponding to $\alpha = 0.05$.

f. Dodder plants do not orient their growth toward the plant volatile α-pinene.

g. There is no evidence that dodder plants orient their growth toward α-pinene.

h. The difference between the proportion of individuals that grow toward α-pinene and that grow away from α-pinene in the dodder population is small.

Making a plan

Too often, experimenters do not carefully consider statistical issues until after the study is completed and the data are in hand. Sometimes a flaw in the experimental design becomes obvious only then, when they try to analyze the data. As Fisher once said, "To consult the statistician after an experiment is finished is often merely to ask him to conduct a post mortem examination. He can perhaps say what the experiment died of."[1] To ensure that your experiment is given a statistical clean bill of health and not a toe tag, it's important to plan the experiment carefully with statistics in mind and to follow that plan throughout the data collection.

Here are some guidelines to avoid a few common pitfalls. Chapter 14 delves into some of these issues in more detail than is possible here. For now, we list a few sensible procedures to help get you started.

1. *Develop a clear statement of the research question.* This needs to be as specific as possible. What is the scientific hypothesis? Is the question interesting? Has it already been addressed sufficiently in the literature?[2] Identify clear objectives for the experiment.

2. *List the possible outcomes of your experiment.* Once you have a preliminary plan for the treatments you *want* to compare, think of the outcomes you *might* obtain.

---

1. Indian Statistical Congress, Sankhyā, ca. 1938.
2. "A month in the laboratory can save an hour in the library." —Westheimer's Discovery

To consult the statistician after an experiment is finished is often merely to ask him to conduct a post mortem examination. He can perhaps say what the experiment died of.

—Fisher

Can you draw firm conclusions no matter what the outcome? Do these conclusions answer the questions? If not, then modify your design.

3. *Develop an experimental plan.* Write it down. Let it sit for a few days and then review it again.

4. *Keep the design of your experiment as simple as possible.* Do you really need 12 different treatments, or will two suffice? Simplifying the design will make it easier to keep track of your objectives, and it will avoid the need for complex statistical analyses.

5. *Check for common design problems.* Is there replication of treatments? Are these replicates truly independent? Will your sampling method yield random samples? Can you identify any confounding variables that will complicate the interpretation of the results?

6. *Is the sample size large enough?* Avoid getting to the end of an experiment before discovering that your sample size is only large enough to demonstrate an unrealistically large effect. Is the sample size

233

sufficient to produce a confidence interval narrow enough to permit conclusions, regardless of the size of the treatment effect? Chapter 14 discusses some methods to help in this planning.

7. *Discuss the design with other people.* Many brains think better than one, and others will often see a problem (and hopefully a solution) that wasn't obvious to you. It is better to get that feedback before doing all the work than to be told after the fact, when it's too late to do anything about it. Science is a social process, so take advantage of the brainpower you have around you.

By carefully considering these guidelines before starting the experiment, you can avoid a lot of wasted effort.

Gobiodon erythrospilus

9 Contingency analysis
associations between categorical variables

Biologists are keenly interested in associations between variables and differences between groups. Contingency tables (see Section 2.3) display how the frequencies of different values for one variable depend on the value of another variable when both are categorical. In this chapter, we analyze sample data for two categorical variables to infer associations between those variables in populations. We want to determine to what extent one variable is "contingent" on the other.

Analysis of contingency data can be used to answer questions such as the following:

- Do bright and drab butterflies differ in the probability of being eaten?
- How much more likely to drink are smokers than nonsmokers?
- Are heart attacks less likely among people who take aspirin daily?

In experimental studies, contingency data can help us establish whether the probability of living or dying differs between medical treatments. We can estimate the differences in these probabilities with odds ratios and with relative risk, which are explained in Sections 9.2 and 9.3. We can test hypotheses about differences in the probabilities using a contingency test. **Contingency analysis** allows us to determine whether, and to what degree, two (or more) categorical variables are associated. In other words, a contingency analysis helps us to decide whether the proportion of individuals falling into different categories of a response variable is the same for all groups.

> *Contingency analysis* estimates and tests for an association between two or more categorical variables.

At the heart of contingency analysis is the investigation of the *independence of variables*. If two variables are independent, then the state of one variable tells us nothing about the probability of the different values of the other variable.

9.1 Associating two categorical variables

An association between two categorical variables implies that the two variables are not independent. During the RMS *Titanic* disaster, for example, women had a lower probability of death than men. Sex and death were not independent; the sex of an individual predicts to some extent his or her probability of death. The mosaic plot on the left in Figure 9.1-1 shows the relationship between sex and death. If death had been independent of sex, then the probability of death would have been equal for both sexes, and the resulting mosaic plot would look like the one on the right in Figure 9.1-1.

9.2 Estimating association in 2 × 2 tables: odds ratio

The odds ratio measures the magnitude of association between two categorical variables when each variable has only two categories. One of the variables is the response variable—let's call its two categories "success" and "failure," where success just refers to the focal outcome of interest. The other variable is the explanatory variable, whose two categories identify the two groups whose probability of success is being compared. The odds ratio compares the proportion of successes and failures between the two groups.

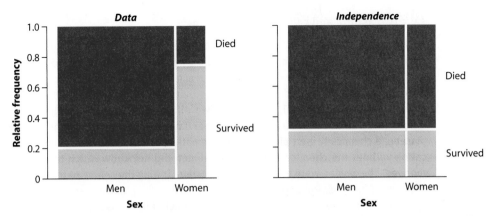

FIGURE 9.1-1 *Left:* Mosaic plot depicting the death of adult men and women passengers following the shipwreck of the *Titanic*. Survivors are represented by the gold and those that died by the red. The area of each box is proportional to the number of individuals in the sample with those attributes; $n = 2092$ individuals from data in Dawson (1995). *Right:* This is what the mosaic plot would have looked like if death and sex on the *Titanic* were perfectly independent. In reality, the probability of death differed between the sexes, so the two variables are not independent.

Odds

Consider a variable for which a single random trial yields one of two possible outcomes: success or failure. The probability of success is p and the probability of failure is $1 - p$. The **odds** of success (O) are the probability of success divided by the probability of failure:

$$O = \frac{p}{1 - p}.$$

If $O = 1$ (sometimes written as 1:1 or "the odds are one to one"), then one success occurs for every failure. If the odds are 10 (sometimes written as 10:1), then 10 trials result in success for every one that results in failure.

> The *odds* of success are the probability of success divided by the probability of failure.

The estimate of odds is calculated from a random sample of trials using the observed proportion of successes ($\hat{p}$) as follows:

$$\hat{O} = \frac{\hat{p}}{1 - \hat{p}}.$$

Example 9.2 shows how to estimate odds from sample data.

EXAMPLE   Take two aspirin and call me in the morning?

9.2   Aspirin, the medicine commonly used for headache and fever, has been shown to reduce the risk of stroke and heart attack in susceptible people. Observational studies have suggested that aspirin may also reduce the risk of cancer. A large, carefully designed experimental study was conducted to test this possibility (Cook et al. 2005). A total of 39,876 women were randomly assigned one of two different treatments. Of these, 19,934 women received 100 mg of aspirin every other day. The other 19,942 women received a placebo, a chemically inert pill that gives the patient the experience of taking the medication without the chemical effects. The women did not know which treatment they received. The women were monitored for 10 years. During the course of the study, 1438 of the women on aspirin and 1427 of those receiving the placebo were diagnosed with invasive cancer (Table 9.2-1). The results are depicted in a mosaic plot in Figure 9.2-1.

TABLE 9.2-1

2 × 2 contingency table for the aspirin and cancer experiment.

	Aspirin	Placebo
Cancer	1438	1427
No cancer	18,496	18,515

FIGURE 9.2-1

A mosaic plot showing the results of the study comparing cancer rates in women who took aspirin with women who did not take aspirin; $n = 39,876$.

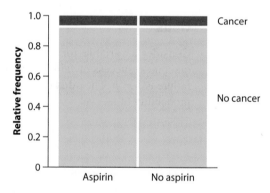

A glance at the mosaic plot suggests that cancer rates did not change much, if at all, as a result of taking aspirin. Let's focus on the outcome "getting cancer" and estimate the odds of this outcome in the two groups of women.[1] In the aspirin group (group 1), the estimated proportion that got cancer is

$$\hat{p}_1 = \frac{1438}{19{,}934} = 0.0721.$$

We have added the subscript "1" to identify the group. The estimated proportion of women who *did not* get cancer is

---

1. In medical studies such as this one, the convention is to calculate the odds of the outcome "diseased" or "died" rather than on the outcome "cured" or "survived." When calculating odds ratio, the convention is also to put the odds for the treatment group in the numerator and the odds for the control group in the denominator.

$$1 - \hat{p}_1 = 1 - 0.0721 = 0.9279.$$

So, the estimated odds of developing cancer while taking aspirin are

$$\hat{O}_1 = \frac{\hat{p}_1}{1 - \hat{p}_1} = \frac{0.0721}{0.9279} = 0.0777.$$

The odds of getting cancer while on aspirin are about 0.08:1, or about 1:13. In common speech, we would say that the odds are 13 to 1 that a woman who took aspirin would not get cancer in the next 10 years.

Similarly, the estimated probability that a woman on the placebo developed cancer is

$$\hat{p}_2 = 1427/19,942 = 0.0716.$$

So, the odds of a woman on the placebo getting cancer are

$$\hat{O}_2 = \frac{\hat{p}_2}{1 - \hat{p}_2} = \frac{0.0716}{0.9284} = 0.0771,$$

which is also about 1:13, similar to that in the aspirin group.

Odds ratio

We can use the odds ratio to quantify the difference between the odds of women developing cancer on aspirin and on the placebo. The **odds ratio** (OR) is just what it sounds like, the ratio of the odds of success between two groups. If O_1 is the odds of success in one group and O_2 is the odds in the other group, then the odds ratio is

$$OR = \frac{O_1}{O_2}.$$

> The *odds ratio* is the odds of success in one group divided by the odds of success in a second group.

If the odds ratio is equal to one, then the odds of success in the response variable are independent of treatment; the odds of success are the same for both groups. If the odds ratio is greater than one, then the event has higher odds in the first group than in the second group. Alternatively, if the odds ratio is less than one, then the odds are higher in the second group. The odds ratio is commonly calculated in medical research, where it is used to measure the change in the odds for a response variable resulting from medical intervention compared with a control treatment (the explanatory variable).

For the cancer/aspirin study described in Example 9.2, the estimated odds ratio is given by

$$\widehat{OR} = \frac{\hat{O}_1}{\hat{O}_2} = \frac{0.0777}{0.0771} = 1.008.$$

(The "hat" on $\widehat{OR}$ in the preceding equation indicates that it is an estimate of the population OR.) The odds of developing cancer while taking aspirin were about the same as the odds while taking the placebo. The estimated odds ratio is slightly greater than one, which means that in the data, the odds of getting cancer were slightly higher in the aspirin group than in the placebo group.

The following is a shortcut formula:

$$\widehat{OR} = \frac{a/c}{b/d} = \frac{ad}{bc},$$

where the symbols a, b, c, and d refer to the observed frequencies in the cells of the contingency table:

	Treatment	Control
Success (focal outcome)	a	b
Failure (alternative outcome)	c	d

When we apply this shortcut to the cancer data,

	Aspirin	Placebo
Cancer	$a = 1438$	$b = 1427$
No cancer	$c = 18{,}496$	$d = 18{,}515$

we get

$$\widehat{OR} = \frac{ad}{bc} = \frac{(1438)(18{,}515)}{(1427)(18{,}496)} = 1.009.$$

Our answer (1.009) here is slightly different from the one calculated earlier (1.008) because the shortcut reduces round-off error in the calculation of odds ratio.

Standard error and confidence interval for odds ratio

The sampling distribution for the odds ratio is highly skewed, and so we convert the odds ratio to its natural log, $\ln(\widehat{OR})$. We can calculate the standard error of the log-odds ratio as

$$SE\left[\ln(\widehat{OR})\right] = \sqrt{\frac{1}{a} + \frac{1}{b} + \frac{1}{c} + \frac{1}{d}}.$$

The symbols a, b, c, and d in this equation refer to the observed frequencies in the cells of the contingency table shown earlier.[2]

An approximate $100(1 - \alpha)\%$ confidence interval for the log-odds ratio is then given by

$$\ln(\widehat{OR}) - Z \times SE[\ln(\widehat{OR})] < \ln(OR) < \ln(\widehat{OR}) + Z \times SE[\ln(\widehat{OR})]$$

where $Z = 1.96$ for a 95% confidence interval and $Z = 2.58$ for a 99% confidence interval.[3] This formula for the confidence interval is an approximation that assumes the sample size is fairly large. To find the confidence interval for the odds ratio, rather than the log odds, we must take the antilog of the upper and lower limits of the interval for the log-odds ratio.

Let's calculate the 95% confidence interval for the aspirin data. We've already calculated $\widehat{OR} = 1.009$, so $\ln(\widehat{OR}) = 0.00896$. The standard error of this estimate of $\ln(\widehat{OR})$ is

$$SE[\ln(\widehat{OR})] = \sqrt{\frac{1}{a} + \frac{1}{b} + \frac{1}{c} + \frac{1}{d}}$$

$$= \sqrt{\frac{1}{1438} + \frac{1}{1427} + \frac{1}{18{,}496} + \frac{1}{18{,}515}}$$

$$= 0.03878.$$

With this standard error, we can calculate the 95% confidence interval. Using $Z = 1.96$ for a 95% confidence interval, we get

$$0.00896 - 1.96(0.03878) < \ln(OR) < 0.00896 + 1.96(0.03878).$$

The 95% confidence interval for the population log-odds ratio is therefore

$$-0.067 < \ln(OR) < 0.085.$$

To convert this to a confidence interval for the odds ratio, we must take the antilog of the limits of this interval by raising e to the power of each number:

$$e^{-0.067} < OR < e^{0.085}$$

or

$$0.93 < OR < 1.09.$$

The confidence interval for the odds ratio is tightly bounded around 1.00, so the data provide evidence that aspirin has little or no effect on the probability of developing cancer. The data are plausibly consistent with a small beneficial effect, a small deleterious effect, or no effect at all.

---

2. This formula will not work when a, b, c, or d is zero. One approximate solution in this case is to add $1/2$ to each of the four cells in the 2 × 2 table.

3. Z is the critical value for a standard normal distribution. The standard normal distribution is discussed in greater detail in Chapter 10.

9.3 Estimating association in 2 × 2 tables: relative risk

Relative risk is another commonly used measure of the association between two categorical variables when both have just two categories. It is especially appropriate when comparing the probability (risk) of a focal outcome between two treatments or groups. As the name implies, the focal outcome is usually the rarer and less desirable outcome. For example, we might use relative risk to estimate and compare the probabilities of sudden infant death syndrome between infants sleeping facedown and infants sleeping on their backs. The relative risk is the probability of the focal outcome in the treatment group divided by the probability in the control group.

If $\hat{p}_1$ and $\hat{p}_2$ are the estimates of the probability of the undesirable outcome in group 1 (treatment) and group 2 (control), respectively, then we calculate the relative risk, RR, as the ratio

$$\widehat{RR} = \frac{\hat{p}_1}{\hat{p}_2}.$$

> *Relative risk* is the probability of an undesired outcome in the treatment group divided by the probability of the same outcome in a control group.

Let's use the data in Example 9.2 to calculate the relative risk of cancer for women who take supplemental aspirin compared to those who do not. For women on the aspirin treatment, we calculated the estimated proportion of women with cancer as $\hat{p}_1 = 0.0721$. The proportion of women getting cancer in the placebo group was estimated as $\hat{p}_2 = 0.0716$. Putting the treatment group (with aspirin) in the numerator, we calculate the relative risk as

$$\widehat{RR} = \frac{0.0721}{0.0716} = 1.007.$$

The estimated probability of getting cancer is very similar in these two groups, hence the relative risk is close to one. The estimate is that the cancer rate is only slightly higher in the aspirin treatment group than in the control group; therefore, the relative risk is slightly greater than one. If aspirin had reduced the risk of cancer, we would have seen a relative risk less than one.

Calculations for standard errors and confidence intervals for relative risk are provided in the Quick Formula Summary on p. 256.

Odds ratio vs. relative risk

Which method is best for measuring association between two categorical variables: odds ratio or relative risk? Both methods provide a measure of the magnitude of the difference between two groups in the probability of a focal outcome, and both are used frequently in analyses of biological data. Relative risk, being simply the ratio of two proportions, is considered by many to be more intuitive than odds ratio. But you may have noticed that, when applied to the cancer and aspirin data, the values for $\widehat{RR}$ and $\widehat{OR}$ were almost identical (1.007 and 1.009, respectively). The values for odds ratio and relative risk will be similar whenever the focal outcome is rare, as is the probability of cancer in women in the aspirin study.

One advantage of the odds ratio is that it can be applied to data from case-control studies. A **case-control study** is a method of observational study in which a sample of individuals having a disease or other focal condition (the "cases") is compared to a second sample of individuals who do not have the condition (the "controls") but are otherwise similar in other characteristics that might also influence the results. The study allows investigators to examine whether the two samples differ in their exposure to one or more possible causal factors. In a case-control study, the total numbers of cases and controls in the samples are chosen by the experimenter rather than by sampling at random in the population, and thus the numbers of individuals with and without the disease or focal condition are not necessarily proportional to the frequency of the disease in the population. As a result, we cannot estimate risk. We can, however, ask whether the focal outcome is associated with another variable by using an odds ratio.

EXAMPLE   Your litter box and your brain

9.3   *Toxoplasma gondii* is a protozoan parasite that can infect the brains of many birds and mammals, including humans. *Toxoplasma*'s main hosts are cats, and humans may acquire the parasite via contact with cat feces. Roughly a quarter of all humans are infected. Because *Toxoplasma* tends to infect the brain of its victims, it seems likely that it affects the behavior of the

host as well. In humans, toxoplasmosis may be associated with some mental illnesses, and it may be associated with risky behavior.[4] Yereli et al. (2006) compared the prevalence of *Toxoplasma gondii* in a sample of 185 drivers between 21 and 40 years old who had been involved in a driving accident (cases) with a sample of 185 drivers of similar age and sex who had not had accidents (controls). The researchers were interested in whether *Toxoplasma* infection may cause a change in the probability of an accident. Their data are shown in Table 9.3-1 and Figure 9.3-1.

---

4. *Toxoplasma* is known to affect the behavior of rats and mice. Infected rats lose their fear of cats, and in fact may become attracted to cat smells (Vyas et al. 2007). In this way, the parasite has a higher probability of reaching its final host, the cat.

TABLE 9.3-1 The frequency of *Toxoplasma gondii* infection in a sample of drivers involved in driving accidents (cases) compared with a sample of drivers with no accidents (controls). From Yereli et al. (2006).

	Infected	Uninfected
Drivers with accidents	61	124
Drivers without accidents	16	169

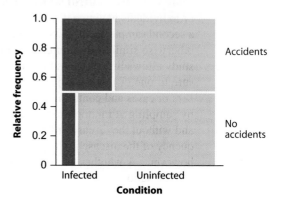

FIGURE 9.3-1

Mosaic plot illustrating relative frequency of *Toxoplasma gondii* infection in a sample of drivers involved in a driving accident (cases) compared with a sample of drivers with no accidents (controls).

In this example, *Toxoplasma* infection is the explanatory variable. Occurrence of an accident (cases vs. controls) is the response variable. The association between the two variables is illustrated with a mosaic plot in Figure 9.3-1. Notice that in Figure 9.3-1 we've illustrated separate rows for cases (accidents) and controls (no accidents), and divided each row according to the frequency of infected and uninfected individuals. This slightly different arrangement from previous mosaic plots takes into account the unusual sampling design of the case-control study, whereby individuals are sampled according to their value for the response variable (here, accidents vs. no accidents), and subsequently measured for their exposure to the explanatory variable (here, infected vs. uninfected). However, we have maintained the explanatory variable on the horizontal axis. We recommend this graphical convention for displaying case-control frequency data using mosaic plots, and we adopt it in the rest of this book.

Unfortunately, these data cannot be used to estimate the relative risk of an accident, comparing groups with and without infection. This is because the calculation of relative risk requires an unbiased estimate of p_1, the probability that an infected individual has an accident. It also requires an unbiased estimate of p_2, the probability that an uninfected individual has an accident. Such estimates are unavailable with this kind of data because of the case-control study design. Within each group, we do not have a random sample of drivers to use in estimating the probability of an accident. The data are enriched with drivers who have had accidents, compared to what we would see in a random sample of drivers from the population. As a result, the proportion of infected individuals in the study who were also in an accident is likely to be a severely biased estimate of the probability that an infected individual in the population has an accident. We are unable to calculate risk, and therefore we cannot calculate relative risk. (In contrast, if the researchers had first obtained two random samples of people with and without toxoplasmosis and then asked whether they had been in a car accident, we would be able to calculate the risk for both groups and then relative risk.)

> A *case-control study* is a type of observational study in which a sample of individuals with a focal condition (cases) is compared to a sample of subjects lacking the condition (controls).

Nevertheless, we can estimate the odds ratio of an accident, comparing infected and uninfected drivers. With odds ratios, the overall proportions of cases and controls cancel out in the ratio. As a result, the odds ratio is unaffected by having a sample of cases and controls that don't match the population proportions of cases and controls. Thus, a study with a case-control design can be analyzed even if the total numbers of cases and controls in the samples are chosen by the experimenter rather than by sampling at random from the population.

To finish this example, let's calculate the odds ratio with the usual formula:

$$\widehat{OR} = \frac{ad}{bc} = \frac{61 \times 169}{124 \times 16} = 5.20.$$

The odds of an accident are estimated to be about five times higher for drivers infected with *Toxoplasma* than for uninfected drivers. Recall that if the focal event is rare in the population, then relative risk and odds ratio are approximately the same magnitude. Hence, if driving accidents are rare in the population, the relative risk is also about fivefold. However, if accidents were common in this population, this interpretation would be inaccurate.

9.4 The χ^2 contingency test

Relative risk and odds ratio allow us to estimate the magnitude of association between two categorical variables. However, they do not directly test whether an association may be caused by chance alone. The χ^2 **contingency test** is the most commonly used test of association between two categorical variables. It tests the goodness of fit to the data of the null model of *independence* of variables.

> The χ^2 *contingency test* is the most commonly used test of association between two categorical variables.

Example 9.4 illustrates how the method works.

EXAMPLE   The gnarly worm gets the bird

9.4   Many parasites have more than one species of host, so the individual parasite must
get from one host to another to complete its
life cycle. Trematodes of the species *Euhap-
lorchis californiensis* use three hosts dur-

ing their life cycle. Worms mature in birds and lay eggs that pass out of the bird in its
feces. The horn snail *Cerithidea californica* eats these eggs, which hatch and grow
to another life stage in the snail, sterilizing the snail in the process. When an infected
snail is eaten by the California killifish *Fundulus parvipinnis,* the parasite develops to
the next life stage and encysts in the fish's braincase. Finally, when the killifish is eaten
by a bird, the worm becomes a mature adult and starts the cycle again.

Researchers have observed that infected fish spend excessive
time near the water surface, where they may be more vulnerable to
bird predation. This would certainly be to the worm's advantage, as
it would increase its chances of being ingested by a bird, its next host.
Lafferty and Morris (1996) tested the hypothesis that infection influences risk of predation by birds. A large outdoor tank was stocked with
three kinds of killifish: unparasitized, lightly infected, and heavily
infected. This tank was left open to foraging by birds, especially great
egrets, great blue herons (pictured), and snowy egrets. Table 9.4-1
lists the numbers of fish eaten according to their levels of parasitism.

TABLE 9.4-1 Observed frequencies of fish eaten or not eaten by birds
according to trematode infection level.

	Uninfected	Lightly infected	Highly infected	Row total
Eaten by birds	1	10	37	48
Not eaten by birds	49	35	9	93
Column total	50	45	46	141

We can use a mosaic plot to visualize the pattern in the data (Figure 9.4-1). Only
2% of the uninfected fish were eaten, while 22% and 80% of the lightly and heavily
infected fish, respectively, died from predation.

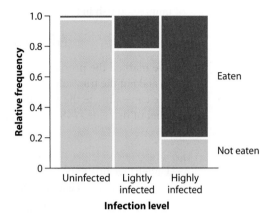

FIGURE 9.4-1

A mosaic plot for bird predation on killifish having different levels of trematode parasitism. The red areas represent the relative frequency of fish eaten by birds, and the gold areas are fish that escaped bird predation. A total of $n = 141$ fish are represented in these data.

Hypotheses

We want to test whether the probability of being eaten by birds differs according to infection status. In other words, we want to test whether the categorical variables "infection level" and "being eaten" are independent. The null and alternative hypotheses are as follows:

H_0: Parasite infection and being eaten are independent.

H_A: Parasite infection and being eaten are *not* independent.

To carry out the χ^2 contingency test, we need to calculate the expected frequencies for each of the cells in Table 9.4-1 under the null hypothesis of independence.

Expected frequencies assuming independence

Recall from Section 5.6 that, if two events are independent, then, by definition, the probability of both occurring is equal to the probability of one event occurring times the probability of the other event occurring (this is the multiplication rule). We use the multiplication rule to calculate the expected proportion of individual fish under each combination of events and then the expected frequencies under the null hypothesis. For example, if the infection status of a fish in our sample is independent of whether it's eaten, then

$$\Pr\left[\,uninfected \text{ and } eaten\,\right] = \Pr\left[uninfected\right] \times \Pr\left[eaten\right].$$

To calculate the expected fraction of fish both uninfected and eaten, though, we still need to estimate the probability that a fish in the sample is uninfected and the chance that a fish was eaten. We can estimate these probabilities from the data in

Table 9.4-1. The estimated probability that a fish was uninfected is the total number of uninfected fish in the sample (50) divided by the total number of fish (141):

$$\hat{\text{Pr}}[uninfected] = 50/141 = 0.3546.$$

We've marked the probability estimate with a "hat" (ˆ) to indicate that it is an estimate and not the true value.

We can estimate the probability of being eaten in the same way, by dividing the number of fish eaten (48) by the total number of fish (141):

$$\hat{\text{Pr}}[eaten] = 48/141 = 0.3404.$$

Under the null hypothesis of independence, therefore, the probability of a fish being uninfected *and* eaten is expected to be

$$\hat{\text{Pr}}[uninfected \text{ and } eaten] = 0.3546 \times 0.3404 = 0.1207.$$

This means that the *expected* frequency of fish both uninfected and eaten is this probability (0.1207) times the total number of individuals in the data set (141):

$$Expected[uninfected \text{ and } eaten] = 0.1207 \times 141 = 17.0.$$

Repeating the preceding procedure for the other cells in Table 9.4-1 gives the expected frequencies of all combinations of outcomes. These values are listed in Table 9.4-2, but you should make sure that you can calculate them on your own.

Note that the row and column totals in Table 9.4-2 match the totals in the actual data (Table 9.4-1). This must be true, because we used the proportions in the data themselves to generate our expected frequencies. If the row and column totals are *not* the same for the observed and expected frequencies (within rounding errors), a calculation error has been made.[5]

TABLE 9.4-2 Expected frequencies of fish eaten and not eaten by birds, according to trematode infection status.

	Uninfected	Lightly infected	Highly infected	Row totals
Eaten by birds	17.0	15.3	15.7	48
Not eaten by birds	33.0	29.7	30.3	93
Column totals	50	45	46	141

---

5. In this section, we calculated the expected frequencies for every combination of row and column by assuming that the data are a random sample from the study population. However, the same calculations work for study designs in which the researcher has fixed the number of individuals in each group for one of the two variables, such as the number of women in each of the two aspirin treatments in Example 9.2, provided that each group contains a random sample of subjects.

Moreover, the expected frequencies don't have to be integers. Remember that we use "expected" in the sense of "on average," so we don't necessarily expect an integer.

The χ^2 statistic

The observed frequencies in Table 9.4-1 are quite different from the expected frequencies in Table 9.4-2. From this point on, the χ^2 contingency analysis is just a special case of the χ^2 goodness-of-fit test. We calculate the χ^2 statistic to test whether these discrepancies are greater than expected by chance. Using c to represent the number of columns and r to represent the number of rows, we get

$$\chi^2 = \sum_{row=1}^{r} \sum_{column=1}^{c} \frac{\left[Observed(row, column) - Expected(row, column) \right]^2}{Expected(row, column)}.$$

This χ^2 calculation simply adds across all cells of the contingency table. When we apply it to the data, we get

$$\chi^2 = \frac{(1 - 17.0)^2}{17.0} + \frac{(49 - 33.0)^2}{33.0} + \frac{(10 - 15.3)^2}{15.3} + \frac{(35 - 29.7)^2}{29.7}$$

$$+ \frac{(37 - 15.7)^2}{15.7} + \frac{(9 - 30.3)^2}{30.3}$$

$$= 69.5.$$

Degrees of freedom

The sampling distribution of the χ^2 test statistic under the null hypothesis of independence is approximated by the theoretical χ^2 distribution. To calculate the degrees of freedom for the χ^2 distribution, we count the number of rows (r) and the number of columns (c) in the data table. The number of degrees of freedom, then, is given by

$$df = (r - 1)(c - 1).$$

For Example 9.4, Table 9.4-1 has two rows and three columns, so there are two degrees of freedom:

$$df = (2 - 1)(3 - 1) = 2.$$

P-value and conclusion

The critical value for the χ^2 distribution with $df = 2$ and significance level $\alpha = 0.05$ is 5.99 (see Statistical Table A). Our observed value ($\chi^2 = 69.5$) is further out in the tail of the distribution, much greater than the critical value of 5.99. We therefore reject the null hypothesis that infection level and probability of being eaten are independent in killifish. Instead, the probability of being eaten is contingent upon whether the fish was parasitized. Trematode parasitism in these killifish was associated with higher rates of predation by birds. We reach the same conclusion when we use a computer to calculate the P-value for a χ^2 value of 69.5 having two degrees of freedom: $P < 10^{-10}$.

How do we explain this result? If differences in infection are truly the cause of the differences in predation risk, then the most likely explanation is that the worms modify fish behavior or hinder their ability to escape, increasing their chances of being eaten by birds and thus completing the last transition in the worm's life cycle.[6]

A shortcut for calculating the expected frequencies

Here's a shortcut formula for calculating the expected frequencies in fewer keystrokes on your calculator. The expected cell value for a given row and column is

$$Expected\left[row\ i, column\ j\right] = \frac{(Row\ i\ total)(Column\ j\ total)}{Grand\ total}.$$

For the killifish parasite data, for example, the expected frequency for the top left cell in Table 9.4-2 (uninfected fish that were eaten) can be calculated by multiplying the total for its row (48) times the total for its column (50) and dividing by the overall total (141):

$$Expected\left[row\ 1, column\ 1\right] = (48 \times 50)/141 = 17.0.$$

The last cell in a row or column can also be computed by subtraction, because the sum of the expected frequencies for a given row or column is the same as the sum of the observed values. Thus, the expected frequency of the top right cell in Table 9.4-2 is $48 - 17.0 - 15.3 = 15.7$. (The number of cells that we *cannot* calculate by subtraction is the number of degrees of freedom for the test. The expected values of cells that are calculated by subtraction are fixed and not free to vary.)

This shortcut comes from the definition of independence and the way we estimate the probability of belonging to each row or column:

$$Expected = \Pr\left[row\right] \times \Pr\left[column\right] \times Grand\ total$$

$$= \left(\frac{Row\ total}{Grand\ total}\right) \times \left(\frac{Column\ total}{Grand\ total}\right) \times Grand\ total.$$

Canceling terms allows you to take the shortcut.

---

6. Many parasites modify the behavior of their hosts, with the result that their chances of making it to their next host are increased. Liver flukes make their ant hosts move to the top of grass blades, where they are more likely to be eaten by grazing cows, the host in which the flukes can reproduce. Wire worms make their cricket hosts find and jump into water, where the crickets drown but the worms complete their life cycle.

The χ^2 contingency test is a special case of the χ^2 goodness-of-fit test

You may have noticed that, once we specified the expected values, the χ^2 contingency test was very similar to the χ^2 goodness-of-fit test introduced in Chapter 8. This resemblance is no accident, because the χ^2 contingency test is a special application of the more general goodness-of-fit test for which the probability model being tested is the independence of variables. The number of degrees of freedom for the contingency test obeys the same rules as those for the goodness-of-fit test.[7]

Assumptions of the χ^2 contingency test

The χ^2 contingency test makes the same assumptions as the χ^2 goodness-of-fit test. No more than 20% of the cells can have an expected frequency less than five, and no cell can have an expected frequency less than one.

If these rules are not met, at least three options are available. First, if the table is bigger than 2×2, then two or more row categories (or two or more column categories) can be combined to produce larger expected frequencies. This should be done carefully, though, so that the resulting categories are still meaningful. (For example, the three categories of infection in a trematode predation experiment could have been collapsed into two categories—namely, "uninfected" and "infected," if necessary.) Second, if the table is 2×2, then Fisher's exact test should be used instead. Fisher's exact test is summarized in Section 9.5. Finally, a permutation test may be used instead of the χ^2 test, an approach that we discuss further in Chapter 13.

Correction for continuity

Some statisticians recommend a modified formula to calculate the χ^2 test statistic in the case of a 2×2 contingency table. The modification is known as the **Yates correction for continuity**:

$$\chi^2 = \sum_{row=1}^{2} \sum_{column=1}^{2} \frac{\left[\left| Observed(row, column) - Expected(row, column) \right| - \frac{1}{2} \right]^2}{Expected(row, column)}.$$

All other steps in the Yates corrected test are the same as in an ordinary χ^2 contingency test.

---

7. Recall from Section 8.2 that $df =$ (Number of categories) $- 1 -$ (Number of parameters estimated from the data). In a contingency table, the number of categories is $r \times c$. The number of parameters estimated from the data is $(r-1) + (c-1)$, reflecting the number of row and column totals needed to generate the expected cell frequencies. After some algebra, this calculation leads to $df = (r-1)(c-1)$.

We mention the Yates correction here because you will encounter it in the biological literature. However, we don't recommend that you use it. The correction makes the χ^2 contingency test too conservative (Maxwell 1976). That is, the Yates corrected test overestimates the correct P-value, with the result that the power of the test is reduced—it is less likely to reject a false null hypothesis.

9.5 Fisher's exact test

Fisher's exact test, named after Sir Ronald A. Fisher, provides an exact P-value for a test of association in a 2×2 contingency table. The test is an improvement over the χ^2 contingency test in cases where the expected cell frequencies are too low to meet the rules demanded by the χ^2 approximation.

> *Fisher's exact test* examines the independence of two categorical variables, even with small expected values.

The calculation of the P-value in Fisher's exact test is cumbersome and is best done with a computer statistical package. Therefore, we do not detail the calculations here, but we instead focus on what the test can do and when it is appropriate.

EXAMPLE 9.5   **The feeding habits of vampire bats**

In Costa Rica the common vampire bat, *Desmodus rotundus*, commonly feeds on the blood of domestic cattle. The bat prefers cows to bulls, which suggests that the bats might respond to a hormonal signal. To explore this behavior further, a researcher compared vampire bat attacks on cows in estrus ("in heat") with attacks on cows not in estrus[8] on a particular night (Turner

1975). The results are presented in Table 9.5-1. Do cows in estrus differ from cows not in estrus in their chance of being attacked?

---

8. His method for doing this was ingenious. It is difficult for humans to distinguish reliably whether cows are in heat, but bulls are very good at it. So, the researcher harnessed paint sponges onto the undersides of the bulls, which each night marked the cows that had been the object of the bull's affections.

TABLE 9.5-1 Numbers of cattle by estrus status and by vampire bat bite status.

	Cows in estrus	Cows not in estrus	Row totals
Bitten by vampire bat	15	6	21
Not bitten by vampire bat	7	322	329
Column totals	22	328	350

The null and alternative hypotheses are as follows:

H_0: State of estrus and vampire bat attack are independent.

H_A: State of estrus and vampire bat attack are *not* independent.

We are tempted to analyze these data with a χ^2 contingency test. However, if we calculate the expected values in the usual way, we find that the expected frequency for cows in estrus that were bitten by vampire bats is too low (Table 9.5-2).

TABLE 9.5-2 The expected frequency values for the vampire bat study.

	Cows in estrus	Cows not in estrus	Row totals
Bitten by vampire bat	1.3	19.7	21
Not bitten by vampire bat	20.7	308.3	329
Column totals	22	328	350

According to the null hypothesis, we expect to see only 1.3 cows that were both in heat and bitten by a vampire bat. This means that one out of the four cells (25%) has an expectation less than five, whereas the rule for a χ^2 test is that no more than 20% of cells should have expectations that low.

Because this is a 2×2 contingency table, we can turn to a Fisher's exact test. The null and alternative hypotheses remain the same as in the χ^2 test. Fisher's test proceeds by listing all 2×2 tables that are as extreme as or more extreme than the observed table of numbers under the null hypothesis of independence. For example, the following are the more extreme tables in one tail. Row and column totals remain the same in the alternative hypotheses, so we just change one of the values and adjust the others to match. Focus on the top right corner of each table:

16	5
6	323

17	4
5	324

18	3
4	325

19	2
3	326

20	1
2	327

21	0
1	328

The *P*-value for Fisher's exact test is the sum of the probabilities of all such extreme tables under the null hypothesis of independence, including equally or more extreme tables on the other tail of the distribution of tables. There are also confidence intervals for odds ratios based on small samples using the same logic as Fisher's exact test (see Agresti 2002).

We applied Fisher's exact test to the data in Table 9.5-1, using a statistical program on the computer, and we found that $P < 0.0001$. Thus, we can reject the null hypothesis of independence. Vampire bats evidently preferred the cows in estrus. The reasons for this are not clear.[9]

9.6 *G*-tests

The *G*-test is another contingency test often seen in the literature. The *G*-test is almost the same as the χ^2 test except that the following test statistic is used:

$$G = 2 \sum_{row=1}^{r} \sum_{column=1}^{c} Observed(row, column) \times \ln\left[\frac{Observed(row, column)}{Expected(row, column)}\right],$$

where "ln" refers to the natural logarithm. Under the null hypothesis of independence, the sampling distribution of the *G*-test statistic is approximately χ^2 with $(r - 1)(c - 1)$ degrees of freedom.

For the data in Example 9.4 on fish infection and bird predation, for example,

$$G = 2\left(1 \ln\left[\frac{1}{17.0}\right] + 49 \ln\left[\frac{49}{33.0}\right] + 10 \ln\left[\frac{10}{15.3}\right] + 35 \ln\left[\frac{35}{29.7}\right]\right.$$
$$\left. + 37 \ln\left[\frac{37}{15.7}\right] + 9 \ln\left[\frac{9}{30.3}\right]\right)$$
$$= 77.6.$$

This *G*-test statistic would be compared to the χ^2 distribution with

$$df = (r - 1)(c - 1) = 2.$$

Again, we would strongly reject the null hypothesis of independence.

The *G*-test is derived from principles of likelihood (see Chapter 20) and can be applied across a wider range of circumstances than the χ^2 contingency test. While the *G*-test is preferred by some statisticians, it has been shown to be less accurate for small sample sizes (Agresti 2002), and it is used somewhat less often than the χ^2 contingency test in the biological literature. The *G*-test has advantages, though, when analyzing complicated experimental designs involving multiple explanatory variables (see Sokal and Rohlf 1995 or Agresti 2002).

---

9. It has been speculated that, by drinking the blood of cows in heat, the bats minimize their intake of hormones, which are in higher concentration in the blood of non-estrous cows and may act as a birth control pill for bats, preventing reproduction.

9.7 Summary

- The odds of success are the probability of success divided by the probability of failure, where "success" refers to the outcome of interest.

- The odds ratio is the odds of success in one of two groups (the treatment group, if one is present) divided by the odds of success in the second group (the control group, if one is present). The odds ratio is used to quantify the magnitude of association between two categorical variables, each of which has two categories.

- Risk is the probability of an undesired event. Relative risk is risk in a treatment group divided by the risk in a control group. If relative risk is less than one, then the treatment is associated with reduced risk.

- The χ^2 contingency test makes it possible to test the null hypothesis that two categorical variables are independent.

- The sampling distribution of the χ^2 statistic under the null hypothesis is approximately χ^2 distributed with $(r-1)(c-1)$ degrees of freedom. The χ^2 approximation works well, provided that two rules are met: no more than 20% of the expected frequencies can be less than five, and none can be less than one.

- Fisher's exact test calculates an exact P-value for the test of independence of two variables in a 2×2 table. The test is especially useful when the rules for the χ^2 approximation are not met.

- The G-test is an alternative method for testing the null hypothesis of independence with contingency analysis.

9.8 Quick Formula Summary

Confidence interval for odds ratio

What does it assume? Random samples.

Formula: $\ln(\widehat{OR}) - Z\,\mathrm{SE}[\ln(\widehat{OR})] < \ln(OR) < \ln(\widehat{OR}) + Z\,\mathrm{SE}[\ln(\widehat{OR})]$,

where $\ln(\widehat{OR})$ is the natural logarithm of the estimate of odds ratio

$$\widehat{OR} = \frac{ad}{bc},$$

where a and b are the observed frequencies of the focal outcome ("success") in the two treatment groups, and c and d are the frequencies of the second category of the response variable (see table on p. 240). $\mathrm{SE}[\ln(\widehat{OR})]$ is the standard error of the log-odds ratio,

$$\mathrm{SE}[\ln(\widehat{OR})] = \sqrt{\frac{1}{a} + \frac{1}{b} + \frac{1}{c} + \frac{1}{d}},$$

and $Z = 1.96$ for a 95% confidence interval. The confidence interval for OR is found by taking the antilog of the limits of the confidence interval for $\ln(OR)$. When a, b, c, or d is zero, then add $1/2$ to all four values before calculating the estimate of the odds ratio and its confidence interval (Sweeting et al. 2004).

Confidence interval for relative risk

What does it assume? Random samples.

Formula: $\ln(\widehat{RR}) - Z\,\text{SE}[\ln(\widehat{RR})] < \ln(RR) < \ln(\widehat{RR}) + Z\,\text{SE}[\ln(\widehat{RR})]$,

where $\ln(\widehat{RR})$ is the natural logarithm of the estimate of relative risk,

$$\ln[\widehat{RR}] = \ln\left[\frac{\hat{p}_1}{\hat{p}_2}\right],$$

and $\hat{p}_1$ and $\hat{p}_2$ are the estimated proportions of the undesired outcome (risk) for the two groups, such that $\hat{p}_1 = \dfrac{a}{a+c}$ and $\hat{p}_2 = \dfrac{b}{b+d}$. When a, b, c, or d is zero, then add $1/2$ to all four values before calculating the estimate of the relative risk and its confidence interval.

$\text{SE}[\ln(\widehat{RR})]$ is the standard error of the log-relative risk,

$$\text{SE}[\ln(\widehat{RR})] = \sqrt{\frac{1}{a} + \frac{1}{b} - \frac{1}{a+c} - \frac{1}{b+d}},$$

and $Z = 1.96$ for 95% confidence interval. The confidence interval for RR is found by taking the antilog of the lower and upper limits of the confidence interval for $\ln(RR)$.

The χ^2 contingency test

What is it for? Testing the null hypothesis of no association between two or more categorical variables.

What does it assume? Random samples; the expected frequency of each cell is greater than one; no more than 20% of the cells have expected frequencies less than five.

Test statistic: χ^2

Sampling distribution under H_0: χ^2 distribution with $(r - 1)(c - 1)$ degrees of freedom, where r and c are the numbers of rows and columns, respectively.

Formula:

$$\chi^2 = \sum_{row=1}^{r} \sum_{column=1}^{c} \frac{\left[Observed(row, column) - Expected(row, column)\right]^2}{Expected(row, column)}.$$

Fisher's exact test

What is it for? Testing the null hypothesis of no association between two categorical variables, each having two categories. Appropriate with small expected values.

What does it assume? Random samples.

Formula: $P = 2 \sum\limits_{\substack{all\ equally\ or\ more \\ extreme\ tables}} \dfrac{R_1!R_2!C_1!C_2!}{a!b!c!d!n!}$,

where R_i and C_i are the row and column totals; a, b, c, and d are the cell values for each of the cells; and n is the total sample size. The summation is over all tables, including the observed table and any tables with the same row and column totals more different from H_0 than the observed table.

G-test

What is it for? Testing the null hypothesis of no association between two or more categorical variables.

What does it assume? Random samples; no more than 20% of cells have expected frequencies less than five.

Test statistic: G

Sampling distribution under H_0: χ^2 distribution with $(r-1)(c-1)$ degrees of freedom, where r and c are the numbers of rows and columns, respectively.

Formula:

$$G = 2 \sum_{row=1}^{r} \sum_{column=1}^{c} Observed(row, column) \times \ln\left[\frac{Observed(row, column)}{Expected(row, column)}\right]$$

PRACTICE PROBLEMS

1. **Calculation practice: Odds ratio and relative risk.** Wilson et al. (2011) followed a set of male health professionals for 20 years. Of all the men in the study, 7890 drank no coffee and 2492 drank on average more than 6 cups per day. In the "no coffee" group, 122 developed advanced prostate cancer during the course of the study, and 19 in the "high coffee" group did.
   a. Create a contingency table for these data. Follow the convention recommended in Chapter 2: the explanatory variable is in the

columns and the response variable is in the rows. What association is suggested?
   b. What is the estimated probability of advanced prostate cancer for the high-coffee group (i.e., what is the risk of advanced prostate cancer for men who drink more than six cups of coffee per day)?
   c. What is the estimated probability of advanced prostate cancer for the no-coffee group (i.e., what is the risk of advanced prostate cancer for non-coffee drinkers)?

d. What is the relative risk of advanced prostate cancer, comparing the treatment (high-coffee) and control (no-coffee) groups?

e. What are the odds of advanced prostate cancer for high-coffee consumers?

f. What are the odds of advanced prostate cancer for non-coffee drinkers?

g. What is the odds ratio of advanced prostate cancer, comparing these two groups?

h. What is the log odds ratio comparing these two groups?

i. What is the standard error of the log odds ratio in this case?

j. What is the 95% confidence interval for the log odds ratio?

k. What is the 95% confidence interval for the odds ratio?

l. Interpret this confidence interval for the odds ratio. Is it consistent with the possibility that coffee drinking and developing advanced prostate cancer are independent? Does coffee consumption tend to be associated with an increased or decreased probability of developing advanced prostate cancer?

2. Calculation practice: χ^2 contingency analysis. Married couples often split up after

one member is diagnosed with a catastrophic disease, such as terminal cancer or a brain tumor. Does the frequency of breakup depend on which member is diagnosed? Glantz et al. (2009) tallied divorces after such serious diagnoses in 515 patients in opposite-sex marriages. Of the 261 couples in which the man was ill, 7 divorced soon after diagnosis. Of 254 couples in which the woman was the patient, 53 divorced. Test for a difference between the two types of couples in the proportion divorcing after diagnosis.

a. Summarize the data in a contingency table and examine the frequencies. Do divorce frequencies appear similar between the two types of patients? If they differ, for which sex does diagnosis seem more often to lead to divorce?

b. State the null and alternate hypotheses for the test.

c. In what proportion of couples was the man the diagnosed patient?

d. What proportion of couples divorced?

e. Under the null hypothesis, what is the expected *proportion* of each of the four possible combinations of outcomes?

f. What is the total number of observations in the study?

g. Under the null hypothesis, what is the expected *number* (frequency) of observations in each combination?

h. Examine the expected numbers. Is it legitimate to use a χ^2 contingency test with these data? Why?

i. Calculate the test statistic, χ^2, for these data.

j. How many degrees of freedom does the χ^2 test statistic have?

k. What is the critical value for this test corresponding to significance level $\alpha = 0.05$?

l. Calculate the *P*-value for the test.

m. What is your conclusion? Are sex of the diagnosed patient and divorce independent?

3. The hypothetical plots at the top of the next page show the relative frequencies of subjects

FIGURE FOR PROBLEM 3

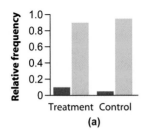

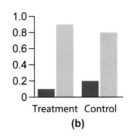

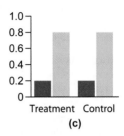

assigned to two experimental groups (treatment and control). The frequency of negative outcomes (red) and positive outcomes (gold) are illustrated. For each plot (a), (b), and (c), identify the correct value of relative risk from the following list of choices: 0.1, 0.5, 1, 2, or 9.

4. The common pigeon found in most American cities is derived from a domesticated European species released in North America. As a result, the pigeons in North America have variations in coloration caused by genes previously selected by pigeon fanciers. An example is the rump, whose feathers are white in wild European pigeons but blue in many pigeons in North America. It has been hypothesized that the white rump of pigeons serves to distract predators like peregrine falcons, and therefore it may be an adaptation to reduce predation. To test this, researchers followed the fates of 203 pigeons, 101 with white rumps and 102 with blue rumps. Nine of the white-rumped birds were captured and killed by falcons, while 92 of the blue-rumped birds were killed (Palleroni et al. 2005).

   a. Show the results in a frequency table. Follow the convention recommended in Chapter 2: put the explanatory variable in the columns and the response variable in the rows. What association is suggested?

   b. Do the two kinds of pigeons differ in their rate of capture by falcons? Carry out an appropriate test.

   c. What is the estimated odds ratio for capture of the two groups of pigeons? What is the 95% confidence interval for the odds ratio?

5. Malaria, which kills more than a million people each year worldwide, is caused by a *Plasmodium* that spreads between hosts by infected mosquitoes. The more people bitten by each infected mosquito, the higher the transmission rates of malaria. Does infection by plasmodium cause a mosquito to bite more people? To test this, researchers captured 262 mosquitoes that had human blood in their guts (Koella et al. 1998). They measured two attributes: whether mosquitoes were infected with malaria, and whether they had fed on the blood of more than one person (assessed by DNA fingerprinting of blood in mosquito guts). Of 173 uninfected mosquitoes, 16 had taken multiple blood meals. Of 89 infected mosquitoes, 20 had fed multiple times.

   a. Illustrate these results with a graph. What association is suggested?

   b. Do these data support the idea that infected mosquitoes behave differently than uninfected mosquitoes?

6. Female Australian redback spiders, *Latrodectus hasselti,* are about 50 times larger than males and often eat the males during mating. This might sound like a horrible accident, but males could gain some indirect advantage by being cannibalized in this way. Perhaps a female is more likely to accept the sperm of a male that she has eaten than that of a male that has escaped. Researchers watched the mating behavior of 32 virgin female redback spiders, recording whether each female ate her first mate and then whether she rejected advances from a second male later placed in her vicinity (Andrade 1996). The results were as follows:

	1st male eaten	1st male escapes
2nd male accepted	3	22
2nd male rejected	6	1

a. How does cannibalism affect the odds that the second male is accepted?

b. What method would you use to test the association between these two variables? Why?

7. Fires are a common and important part of many ecosystems. Many species have evolved mechanisms for dealing with fire. Reed frogs, a species living in West Africa, have been observed hopping away from grass fires long before the heat of the fire reached the area they were in. This finding led to the hypothesis that the frogs might hear the fire and respond well before the fire reaches them. To test this hypothesis, researchers played three types of sound to samples of reed frogs and recorded their response (Grafe et al. 2002). Twenty frogs were exposed to the sound of fire, 20 were exposed to the sound of fire played backward (to control for the range of sound frequencies present in the real sound), and 20 were exposed to equally loud white noise. Of these 60 frogs, 18 hopped away from the sound of fire, 6 hopped away from the sound of fire played backward, and 0 hopped away from the white noise.

a. Illustrate these data with a frequency table. What association is suggested?

b. Do the data provide evidence that reed frogs change their behavior in response to the sound of fire?

8. Some fish are famously able to develop into either sex, depending on social circumstances. One study of a coral reef fish, the goby *Gobiodon erythrospilus,*[10] placed juvenile fish with either an adult male or an adult female (Hobbs et al. 2004). Of the 12 juveniles placed with a male, 11 became female. Of the 10 juveniles placed with an adult female, six became male. What method can we use to test whether the social context of the juvenile fish affects which sex they become?

9. Between 20 and 25 violent acts are portrayed per hour in children's television programming. A study (Johnson et al. 2002) of the possible link between TV viewing and aggression followed the TV viewing habits of children between 1 and 10 years old. Of these children, 88 watched less than one hour of TV per day, 386 watched 1–3 hours per day, and 233 watched more than three hours per day. Eight years later, researchers evaluated the kids to see if they had a police record or had assaulted and injured another person. The number of aggressive individuals from the three TV-watching groups were 5, 87, and 67, respectively.

a. Estimate the proportion of kids in each TV-watching category who subsequently became violent. Give 95% confidence intervals for these estimates.

b. Is there evidence that childhood TV viewing is associated with future violence? Carry out an appropriate statistical test.

c. Does this prove that TV watching causes increased aggression in kids? Why or why not?

10. A study by Doll et al. (1994) examined the relationship between moderate intake of alcohol and the risk of heart disease. In all, 410 men (209 "abstainers" and 201 "moderate drinkers") were observed over a period of 10 years, and the number experiencing cardiac arrest over this period was recorded and compared with drinking habits. All men were 40 years of age at the start of the experiment. By the end of the experiment, 12 abstainers had experienced cardiac arrest and nine moderate drinkers had experienced cardiac arrest.

a. Test whether the relative frequency of cardiac arrest was different in the two groups of men.

b. Assume that you were unable to reject the null hypothesis in part (a). Would this imply that drinking has no effect on the risk of cardiac arrest? Why or why not?

11. Postnatal depression affects approximately 8–15% of new mothers. One theory about the onset of postnatal depression predicts that it may result from the stress of a complicated delivery. If so, then the rates of postnatal depression could be affected by the type of delivery. A study (Patel et al. 2005) of 10,934 women com-

---

10. This beautiful fish is shown on the opening page of this chapter.

pared the rates of postnatal depression in mothers who delivered vaginally to those who had voluntary cesarean sections (C-sections). Of the 10,545 women who delivered vaginally, 1025 suffered significant postnatal depression. Of the 389 who delivered by voluntary C-section, 48 developed postnatal depression.

a. Draw a graph of the association between postnatal depression and type of delivery. What is the pattern in this data?

b. How different are the odds of depression under the two procedures? Calculate the odds ratio of developing depression, comparing vaginal birth to C-section.

c. Calculate a 95% confidence interval for the odds ratio.

d. Based on your result in part (c), would the null hypothesis that postpartum depression is independent of the type of delivery likely be rejected if tested?

e. What is the relative risk of postpartum depression under the two procedures? Compare your estimate to the odds ratio calculated in part (b).

12. Migraine with aura is a potentially debilitating condition, yet little is known about its causes. A case-control study compared 93 people who suffer from chronic migraine with aura (cases) to a sample of 93 healthy patients (controls; Schwerzmann et al. 2005). The researchers used transesophageal echocardiography to look for cardiac shunts in all of these patients. (A cardiac shunt is a heart defect that causes blood to flow from the right to the left in the heart, causing poor oxygenation.) Forty-four of the migraine patients were found to have a cardiac shunt, while only 16 of the people without migraine symptoms had this heart defect.

a. Is this an observational or experimental study?

b. Show the association between migraines and cardiac shunts with a mosaic plot.

c. How strong is the association between migraines and cardiac shunts? Calculate the odds ratio for migraine, comparing the patients with and without cardiac shunts.

d. What is the 95% confidence interval for this odds ratio?

13. *Spot the flaw.* Since 1953, when Tenzing Norgay and Edmund Hillary reached the summit of Mount Everest, many climbers have attempted to scale the world's two highest mountains, Everest and K2. Norgay and Hillary aided their climb by bringing supplemental oxygen in tanks, and some later groups have attempted to outdo the originals by trying the ascent without supplemental oxygen. From 1978 to 1999, in fact, 159 teams comprising 1173 team members have attempted to climb either Everest or K2. The numbers of individuals who either survived or died[11] during those attempts is given in the following table (data from Huey and Eguskitza 2000):

	Supplemental O_2	No supplemental O_2	Row totals
Survived	1045	88	1133
Did not survive	32	8	40
Column totals:	1077	96	1173

A χ^2 contingency test on these data calculated $\chi^2 = 7.694$ with one degree of freedom, which corresponds to $P = 0.0055$. The null hypothesis is that oxygen use has *no* effect on survivorship during these expeditions. What's wrong with this analysis?[12]

14. A "Mediterranean diet" (high in fish, olive oil, red wine, etc.) has been touted as a key to a long life. A study by Trichopoulou et al. (2005) looked at the death rates of people according to whether their diet had a low component, a medium component, or a high component of foods that characterize a Mediterranean diet. In these kinds of studies, it is important to look for other confounding variables, such as smoking, that might be correlated with the main variable under study. For each person in the study, therefore, the team also recorded whether the person was a current smoker, a former smoker, or had

---

11. Almost all of the mortality occurred during the descents.

12. The original paper (Huey and Eguskitza 2000) did not make this mistake.

never smoked. We want to know if there is an association between diet and smoking. The data for men in the study are as follows, where the numbers represent the number of men in each category:

	Low Med. diet	Medium Med. diet	High Med. diet
Never smoked:	2516	2920	2417
Former smoker:	3657	4653	3449
Current smoker:	2012	1627	1294

a. Draw a mosaic plot of these data. What is the pattern?
b. Test whether there is an association between diet and smoking.
c. Comment on how the relationship you found in part (b) would affect the interpretation of a study that looked for health effects of switching to a Mediterranean diet.

ASSIGNMENT PROBLEMS

15. The hypothetical plots at the bottom of this page show the relative frequencies of subjects assigned to two experimental groups (treatment and control). The frequency of negative outcomes (red) and positive outcomes (gold) are illustrated. For each plot (a), (b), and (c), identify the correct value of the odds ratio: 0.1, 0.5, 1, 2, or 9.

16. In animals without paternal care, the number of offspring sired by a male increases as the number of females he mates with increases. This fact has driven the evolution of multiple matings in the males of many species. It is less obvious why females mate multiple times, because it would seem that the number of offspring that a female has would be limited by her resources

and not by the number of her mates, as long as she has at least one mate. To look for advantages of multiple mating, a study of the Gunnison's prairie dog followed females to find out how many times they mated (Hoogland 1998). They then followed the same females to discover whether they gave birth later. The results are compiled in the following table:

Number of times female mated:	1	2	3	4	5
Number who gave birth:	81	85	61	17	5
Number who didn't give birth:	6	8	0	0	0

FIGURE FOR PROBLEM 15

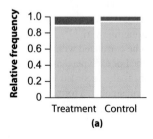
(a)

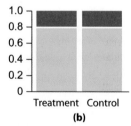
(b)

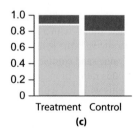
(c)

Did the number of times that a female mated affect her probability of giving birth?

a. Calculate expected frequencies for a contingency test.

b. Examine the expected frequencies. Do they meet the assumptions of the χ^2 contingency test? If not, what steps could you take to meet the assumptions and make a test?

c. An appropriate test shows that the number of mates of the female prairie dogs is associated with giving birth. Does this mean that mating with more males increases the probability of giving birth? Can you think of an alternative explanation?

17. Some people feel that they have good intuition about when others are lying, while others do not feel they have this ability. Are the more "intuitive" people better able to detect lies? Each of 100 people who thought they had good intuitive abilities was shown a video clip of a person stating the name of his or her favorite movie (Young 2002). The person was truthful in some of the clips shown, whereas in others the person was lying. Another 100 people who claimed not to have intuitive abilities were shown similar video clips. Fifty-nine of the 100 "intuitive" subjects correctly identified whether the person in the video was lying, whereas 69 of the 100 "nonintuitive" subjects correctly identified whether the person in the video was lying.

a. Draw a graph that best presents these data. What association is suggested?

b. Test whether the success rates of the two groups were different.

c. Are "intuitive" people better at detecting lies than "nonintuitive" people? Calculate an odds ratio and confidence interval for your answer. Interpret your result.

18. *Spot the flaw.* Scottish researchers compared rates of depression between 94 undergraduates who regularly kept diaries and 41 students who did not. They found that people who kept diaries were more likely to have depression than those who did not. The researchers said, "We expected diary-keepers to have some benefit, or be the same, but they were the worst off. You are probably much better off if you don't write anything

at all." Why is this an incorrect interpretation of these results?

19. It is well known, and scientifically documented, that yawning is contagious. When we see someone else yawn, or even think about someone yawning, we are very likely to yawn ourselves. (In fact, we predict that you are starting to want to yawn right now.) In a study of yawning contagion, researchers showed participants one of several pictures, including a picture of a man yawning, the same man smiling, a yawning man with his mouth covered, or a yawning man with his eyes obscured (Provine et al. 1989). Participants yawned much more often when shown the yawner than the smiler, but surprisingly an identical number also yawned when shown the picture with the mouth obscured. This suggests that something besides the mouth is an important trigger. What about the eyes? Seventeen of 30 participants yawned when confronted with a picture of a yawning man, while 11 of 30 independent participants yawned when shown a picture of a yawning man with his eyes covered. Is there evidence in these data that covering the yawning man's eyes in an image changes the occurrence of contagious yawns?

20. Day care centers expose children to a wider variety of germs than the children would be exposed to if they stayed at home more often. This has the obvious downside of more frequent colds and other illnesses, but it also serves to challenge the immune system of children at a critical stage in their development. A study by Gilham et al. (2005) tested whether social activity outside the house in young children affected their probability of later developing the white blood cell disease acute lymphoblastic leukemia (ALL), the most common cancer of children. They compared 1272 children with ALL to 6238 children without ALL. Of the ALL kids, 1020 had significant social activity outside the home (including day care) when younger. Of the kids without ALL, 5343 had significant social activity outside the home. The rest of the children in both groups lacked regular contact with children who were not in their immediate families.

a. Is this an experimental or observational study?

b. What are the proportions of children with significant social activity in children with and without ALL?

c. What is the odds ratio for ALL, comparing the groups with and without significant social activity?

d. What is the 95% confidence interval for this odds ratio?

e. Does this confidence interval indicate that amount of social activity is associated with ALL? If so, did the children with more social activity have a higher or lower occurrence of ALL?

f. The researchers interpreted their study results in terms of the differing immune system exposure of the children, but gave several alternative explanations for the pattern. Can you think of any possible confounding variables?

21. Aging workers of the Neotropical termite, *Neocapritermes taracua,* develop blue crystal-containing glands ("backpacks") on their backs. When they fight intruding termites and are hampered, these "blue" termites explode, and the glands spew a sticky liquid (Šobotník et al. 2012). The following data are from an experiment that measured the toxicity of the blue substance. A single drop of the liquid extracted from blue termites was placed on individuals of a second termite species, *Labiotermes labralis,* and the number that were immobilized (dead or paralyzed) within 60 minutes was recorded. The frequency of this outcome was compared with a control treatment in which liquid from glands of "white" termites lacking the blue crystals was dropped instead. Is the blue liquid toxic compared to liquid from white termites?

Liquid source	Unharmed	Immobilized
Blue workers	3	37
White workers	31	9

22. Keenan et al. (2001) used anesthesia to investigate which brain hemisphere is involved in self-recognition. Ten subjects were randomly assigned to two groups. The left hemisphere was anesthetized in one group, whereas the right hemisphere was anesthetized in the other group. Each subject was then shown a picture generated by averaging ("morphing") images of the face of a famous celebrity (e.g., Marilyn Monroe) and their own face, and told to remember the picture. After recovery from anesthesia, patients were presented with two pictures and asked to choose the one they had been shown earlier while under anesthesia. The two pictures were the original two images from which the morphed image had been generated (i.e., "self" and "celebrity," but separate this time). All five patients whose left hemisphere had been inactivated chose the picture of self. Four of the five patients whose right hemisphere had been anesthetized chose the celebrity picture, instead (the fifth chose self). State what test you would use to determine whether the treatment (left vs. right hemisphere anesthesia) influenced recognition of self versus celebrity. Explain why you would choose this test (don't carry out the test, just name it and justify your answer).

23. Eggebeen et al. (2010) found that men who at some point in their lives are fathers are more likely to have altruistic social relationships and be involved in community service organizations. This result was reported in the popular press (Jacobs 2009) as "Fatherhood… alters a man's neurochemistry, increasing his ability to cope with stress and generally making him a better mate. Just-published research suggests the benefits of this transformation extend far beyond one's immediate family and remain robust as the years go by." Are the conclusions drawn by the popular press article defendable? Why or why not?

24. Male *Drosophila* become sterile when exposed to moderately high temperatures, because sperm are damaged by heat at much lower temperatures than other cells. Rohmer et al. (2004) asked whether flies from warmer climates are adapted to higher temperatures. They collected flies from France (where it is relatively cool) and India (relatively warm) to test the effects of temperature on sterility. In one procedure, they raised male flies from both locations at a high temperature, 30.5°C, and recorded whether the flies were sterile or fertile. Thirty-two out of 50 flies from France were sterile at this temperature, whereas 20 of 50 flies from India were sterile.

a. Is this an observational or experimental study?

b. Draw a graph to illustrate the association between sterility and source location (India vs. France). What association is suggested?

c. Is there evidence that the populations of flies from these two locations differ in their probability of sterility at this temperature? Do the appropriate hypothesis test.

d. Estimate the relative risk of sterility at this temperature in the Indian population compared to the population in France (consider the Indian population to be the treatment group for this analysis). Include a 95% confidence interval.

25. Vampire bats, as their name implies, feed almost exclusively on blood. A bat must feed every day or it will starve to death, but bats are not always successful at finding a blood meal. Perhaps for this reason, they roost during the day in communal groups and sometimes share blood by regurgitative feeding.[13] Researchers measured whether hungry bats were more likely to receive regurgitated blood than were partially fed bats (Wilkinson 1984). Eight bats were captured in the evening before they had fed and were held without feeding until the next morning. As a control, six bats were captured after naturally feeding at night, and they were also held until the following morning. At that time, the bats were returned to their groups. Five of the eight hungry bats were given regurgitated blood meals by group-mates, but none of the six well-fed bats were given a blood meal by other bats. What statistical test would you use to address the question, "Does the probability of being fed by roost-mates vary according to hunger status?"

26. We are often happy to do favors for other people when they have a particular need. For example, we are more willing to let someone use a photocopier when they ask, "Can I go in front of you, because I am in a rush?" than when they give no reason: "Can I go in front if you?"

Some researchers believe that simply giving a reason—using the word "because"—may be enough to trigger this giving behavior, even when the reason is not a very good one. An experiment was done in which 60 people who were about to use a Xerox photocopy machine were approached (Langer et al. 1978). In 30 cases (randomly assigned; call these the "request only" group), an investigator asked, "May I use the Xerox machine?" Eighteen of these people allowed the investigator to go first. In the other group of 30 people (the "bad reason" group) an investigator said, "May I use the Xerox machine, because I have to make copies?" Of the 30, 28 allowed the investigator to go first. Test whether the bad-reason approach is better or worse than the request-only approach.

27. It is common wisdom that death of a spouse can lead to health deterioration of the partner left behind. Is common wisdom right or wrong in this case? To investigate, Maddison and Viola (1968) measured the degree of health deterioration of 132 widows in the Boston area, all of whose husbands had died at the age of 45–60 within a fixed six-month period before the study. A total of 28 of the 132 widows had experienced a marked deterioration in health, 47 had seen a moderate deterioration, and 57 had seen no deterioration in health. Of 98 control women with similar characteristics who had not lost their husbands, 7 saw a marked deterioration in health over the same time period, 31 experienced a moderate deterioration of health, and 60 saw no deterioration. Test whether the pattern of health deterioration was different between the two groups of women. Give the P-value as precisely as possible from the statistical tables, and interpret your result in words.

28. *Spot the flaw.* After golden monkeys fight with each other, opponents seem to reconcile. Ren et al. (1991) tested whether the behavior of golden monkeys after spontaneous conflicts differed from behavior at normal times. They recorded a large number of behaviors between individuals in two troops with a total of nine monkeys,

---

13. We're not making this up. See http://www.youtube.com/watch?v=q3gxpIJ-f2M, which would look very cute if you didn't know what was going on.

both after aggressive interactions and not after aggressive interactions ("control" periods). The observed frequencies of behaviors are given in the accompanying table. The team carried out a contingency test on these frequencies as given and rejected the null hypothesis that behaviors occurred at similar frequencies after conflicts compared to control periods. The expected values in this test are not large enough to justify a χ^2 test, but can you identify another problem with this analysis?

Behavior	Post conflict	Control periods
Open-mouth	46	1
Embrace	39	19
Groom	33	20
Contact sit	18	26
Hold-hand	8	3
Hold-lumbar	7	0
Crouch	21	0

29. Psychologists were interested in whether there is a "denomination effect"—are people more or less likely to spend money if that money is in large bills or in lots of small bills? Such research may help give advice to people with problems controlling their spending, for example. The researchers stopped 50 people at a U.S. convenience store and asked them three simple questions; as a reward, they gave each person $5, either as a $5 bill or as five $1 bills. Of the 25 people who were given the larger bill, four spent some money in the convenience store. Of the 25 people who were given five $1 bills, six spent some money.

a. Is there a significant difference in the probability of spending money depending on the denomination of the bills received?
b. What is the relative risk of spending money for small bills compared to large bills? Provide a 95% confidence interval for the relative risk. What step would you take next in this research if you wanted to produce a narrower confidence interval for the odds ratio?

30. Brent et al. (1993) carried out a study to investigate the possible association between firearms in the home and adolescent suicide. How big is the risk? They obtained information on 67 adolescent suicide victims, 70% of whom had died by firearms. The researchers compared this group to a control sample of 67 adolescents of similar race, age, gender, childhood behavior scores, and socioeconomic status and living in the same cities. They found that in 51 of the suicide cases, guns were kept in the home, whereas this was true of 16 of the controls.

a. Graph the association between suicide and the presence of guns in the home. What trend is suggested?
b. What is the estimated odds ratio of suicide with guns in the home, compared to homes without guns?
c. Provide a 95% confidence interval for the population odds ratio.
d. Under what circumstances can this estimate of odds ratio be used to approximate the relative risk of suicide with guns in the home, compared to homes without guns?

31. Darwin suggested that plants pollinated by long-tongued insects would benefit by having long flowers, because greater length would cause the long-tongued insects to press themselves further into the flower to reach the nectar, increasing the chance that pollen is removed and that pollen from other flowers is received. Some plants have evolved "nectar spurs"—a long projection off the back of the flower that has the nectar at the base. Several populations of the South African orchid, *Disa draconis,* evolved longer nectar spurs after switching pollinators from relatively short-tongued horseflies (tongue length about 30 mm) to long-tongued tanglewing flies

(57 mm). To measure the advantage of the long spurs, Johnson and Steiner (1997) randomly selected 59 of 118 long-spurred flowers and, using yarn, shortened the length of their spurs to that found in populations pollinated by horse-flies. The remaining 59 flowers retained their spurs at full length, but yarn was tied around the stigma as a control for any other effects of the yarn. One week later, the numbers of flowers that had or had not received pollen on their stigmas were recorded. Ten of the 59 flowers with shortened spurs had received pollen on their stigmas, whereas 27 of the 59 control flowers had received pollen.

a. Illustrate these results in a graph.

b. What is the estimated odds ratio of not receiving pollen after experimental shortening, as compared to control flowers? Provide a confidence interval for the population odds ratio.

32. Refer to Assignment Problem 31. If we carry out a χ^2 contingency test to compare the proportion of flowers in the two treatments that received pollen on the stigma, we obtain the following results: $\chi^2 = 11.38$, $df = 1$, $P = 0.0007$. But the researchers also tested whether there was an effect of treatment on the proportion of plants that had pollen *removed* by pollinating insects. The results are as follows: $\chi^2 = 6.38$, $df = 1$, $P = 0.012$. Is the following statement a true or false interpretation of these findings? "The effect of spur shortening on pollen receipt is stronger than the effect on pollinia removal." Explain your answer.

33. Kuru is a prion disease of the Fore people of highland New Guinea. Kuru is similar to Creutzfeldt–Jakob disease. It was once transmitted by the consumption of deceased relatives at mortuary feasts, a ritual that was ended by about 1960. Using archived tissue samples, Mead et al. (2009) investigated genetic variants that might confer resistance to kuru. The data in the accompanying table are genotypes at codon 129 of the prion protein gene of young and elderly individuals all having the disease. Since the elderly individuals have survived long exposure to kuru, unusually common genotypes in this group might indicate resistant genotypes.

	Genotypes at codon 129		
Age	MM	MV	VV
Elderly	13	77	14
Young	22	12	14

a. Illustrate these data with a grouped bar graph. Which genotype(s) are especially prevalent in the elderly compared with young individuals?

b. Test whether genotype frequencies differ between the two age groups.

Review Problems 1

1. A scientist sets up an experiment with *Drosophila* that requires her to make a measurement on each fly every day until all the flies are dead. She knows that flies in her stock typically die at a rate of 3% per day, and she is willing to assume that probability of death is the same for each fly and is constant throughout its life. She is also willing to assume that the flies die independently of each other. She starts the experiment with 50 flies, 80 days before she plans to go on vacation.
   a. What is the probability that an individual fly survives a given day?
   b. What is the probability that an individual fly survives for 80 days?
   c. What is the probability that the experiment will be finished (i.e., that all 50 flies will have died) before her vacation?

2. A study of 6,839,854 births in the United States found a total of 6522 babies were born with a finger defect, either syndactyly (fused fingers), polydactyly (extra fingers), or adactyly (fewer than five fingers). Researchers examined 5171 of these babies with finger defects in further detail. Of these babies, 4366 had mothers that did not smoke while pregnant, and the rest had mothers that did smoke while pregnant. In a sample of 10,342 babies from the population with normal fingers, 9062 of their mothers did not smoke while pregnant, while mothers of the remaining 1280 did smoke while pregnant. Answer the following questions to establish the magnitude of the effect of smoking on the occurrence of finger defects.
   a. Why is this considered to be an observational study rather than experimental? What type of observational study is it?
   b. Use a graph to show the association between smoking during pregnancy and finger defects.

   c. What is the 95% confidence interval for the proportion of babies born in the United States with one of these finger defects? Is the defect common or rare?
   d. What is the odds ratio of these birth defects, comparing smoking to nonsmoking mothers? Provide a 95% confidence interval for the odds ratio. Based on your results, which group of mothers has the higher odds of having a child with a finger defect?
   e. Is it justified to consider your estimate of odds ratio in part (d) also to be an estimate of the relative risk? Explain.

3. The study of the spatial distribution of vegetation often makes use of random samples of "quadrats," rectangular plots of fixed size placed at random over the sampling region (e.g., a field or forest). The number of plants of each type occurring within each quadrat is then counted. In one such study, an investigator counted the number of white pine seedlings growing in eighty 10 m × 10 m quadrats to test whether the distribution of pine seedlings in the forest was random, clumped, or dispersed. She obtained the following counts:

Number of seedlings	Number of quadrats
0	47
1	6
2	5
3	8
4	6
5	6
6	2
≥ 7	0
Total	80

   a. If the null hypothesis of a random distribution of pine seedlings across the forest is correct, to what theoretical probability distribution should the observed frequencies of quadrats containing a given number of seedlings conform?

b. Carry out a formal test of the null hypothesis.

c. If the null hypothesis is rejected in part (b), determine whether the spatial distribution of seedlings is clumped or dispersed.

4. Seeds fall on a landscape that contains 70% barren rock. The remainder of the landscape is suitable habitat for the seeds to germinate. Assume that the location where a seed falls in this landscape is random with respect to the site's suitability for germination.

   a. What is the probability that a single seed lands on a suitable site for germination?

   b. If two seeds fall independently, what is the probability that both fall on suitable habitats?

   c. Suppose that three seeds fall randomly and independently onto this landscape. Use a probability tree to find the probability that exactly two of these three seeds land on suitable habitat.

5. Refer to Practice Problem 13 in Chapter 6. The researchers additionally examined the rate of increase in oxyhemoglobin flow with increasing neuronal activity in the somatosensory cortex of rat brains. They estimated a rate of increase of 1.48, with a 95% confidence interval for the population rate of $1.06 < rate < 1.91$. A rate of one is expected if the relationship between the two variables is linear, whereas a value different from one indicates a nonlinear relationship. The researchers tested the null hypothesis that the relationship between the variables is linear (H_0: $rate = 1$). Did their test reject the null hypothesis or not? Explain.

6. A researcher wished to estimate the fraction of students in a large high school who have used illegal drugs on at least one occasion. He carried out his survey through interviews with students. To satisfy legal and ethical guidelines, it was necessary to interview students only when a parent was present. Only eight students showed up to participate.

   a. Is this study likely to yield a biased estimate of illegal drug use or an unbiased estimate? Why?

   b. What effect does such a small sample size have on sampling error of an estimate, compared with a larger sample?

7. When one generation reproduces to form the next, the frequencies of alleles in the population can change by chance from generation to generation in a process called random genetic drift. An experiment was carried out using a very large laboratory population of the common fruit fly, *Drosophila melanogaster,* in which two eye-color alleles (versions of a gene) were present, one red (the norm for these flies) and the other brown. The frequency of the red allele in this base population was 0.5. The researchers created a new group of flies containing 32 alleles by randomly sampling flies from this large population (Buri 1956). They created a second new group of 32 alleles by sampling again from the base population. They repeated this procedure until there were a large number of new populations, all containing 32 alleles. The frequency distribution of the proportions of red-eyed alleles in the new groups closely matched a binomial distribution with $n = 32$ and $p = 0.5$.

   a. On average, what do we expect the mean proportion of the red-eyed allele to be in the new groups?

   b. What should be the standard deviation among groups in the proportion of red-eyed alleles?

   c. If a randomly chosen new group has 60% red-eyed alleles, what proportion of its alleles are for brown eyes?

   d. What is the probability that a randomly chosen group will have exactly 50% red-eyed alleles?

   e. What is the probability that a randomly chosen group will have 30 or more red-eyed alleles?

8. In one of the many new groups from the *Drosophila* eye-color study described in Review Problem 7, there were 9 brown-eyed alleles out of 32 alleles. In another, there were 26 brown-eyed alleles out of 32.
   a. Using the sample of alleles in each of these two new groups, separately estimate the proportion of brown-eyed alleles in the overall population of eye-color alleles. Calculate the 95% confidence intervals for the overall proportion.
   b. Do these confidence intervals overlap? How can you reconcile this with the fact that both samples came from the same population?

9. Some people who are having a heart attack do not experience chest pain, although most do. A study of people admitted to emergency rooms with heart attacks compared the death rates of people who had chest pains with those of people who did not have chest pains (Brieger et al. 2004). Of the 1763 people who had heart attacks without chest pain, 229 died, while of the 19,118 people who had heart attacks with chest pain, 822 died.
   a. Test whether the presence or absence of chest pains during a heart attack is associated with the probability of death.
   b. Give a valid estimate of the magnitude of the association between chest pain and risk of death. Put bounds on your estimate.

10. The following data are scores reflecting the number of young cod that recruited (grew to the catchable size) to the North Sea population in different years (Beaugrand et al. 2003). Scores are adjusted magnitudes (without units) rather than actual fish numbers. The measurements are listed below and arranged from low to high rather than by year.

   5.0, 5.1, 5.2, 5.2, 5.2, 5.2, 5.2, 5.2, 5.3, 5.3, 5.3, 5.3, 5.4, 5.4, 5.4, 5.4, 5.4, 5.5, 5.5, 5.5, 5.5, 5.5, 5.5, 5.5, 5.6, 5.6, 5.6, 5.7, 5.7, 5.7, 5.7, 5.7, 5.8, 5.8, 5.8, 5.9, 5.9, 6.0, 6.0

   a. What was the mean score for number of recruits over this period?
   b. What was the standard deviation in the score of number of recruits?

c. In what fraction of years did the score for the number of recruits fall within two standard deviations of the mean?

11. When the performances of individuals, or preferences for products, are judged in sequence by subjective criteria (think music audition or dance competition), does position in the sequence affect the opinion of the judges? An experiment to look for these order effects (Mantonakis et al. 2009) gave 33 volunteers four glasses of wine in sequence, one at a time. Participants were asked to say which of the four was the superior wine. Unknown to the participants, all four glasses were poured from the same bottle. Fifteen participants preferred the first glass, 5 preferred the second glass, 2 preferred the third glass, and 11 preferred the last glass. Is there evidence from these data that position in the sequence affected the preference of the volunteers?

12. Smoking is a major risk factor for a number of diseases, including strokes. Smoking is particularly dangerous for people who have already had a stroke. One way to help such people to quit smoking is to make drugs that help stop smoking free to stroke patients. Papadakis et al. (2011) investigated the effectiveness of this idea. They divided a sample of stroke patients who smoked randomly into two groups. One group of 12 patients received the normal advice and prescription for anti-smoking drugs, whereas the other group of 13 patients got the same advice and prescription but were also provided the drugs cost-free. After 6 months, five members of the cost-free group had quit smoking, while only two members of the other (control) group had quit by that time.
   a. Calculate the relative risk of smoking for the cost-free program compared with the controls (prescription only).
   b. What statistical method could be used to test for a difference between these two groups in the efficacy of the cost-free anti-smoking program?
   c. The 95% confidence interval for relative risk for this study ranges from 0.45 to 1.22. In light of this result, what do you think is the greatest weakness of this study?

13. Birds of the Caribbean islands of the Lesser Antilles are descended from immigrants originating from larger islands and the nearby mainland. The data presented here are the approximate dates of immigration, in millions of years, of each of 37 bird species now present on the Lesser Antilles (Ricklefs and Bermingham 2001). The dates were calculated from the difference in mitochondrial DNA sequences between each of the species and its closest living relative on larger islands or the mainland.

0.00, 0.00, 0.04, 0.21, 0.29, 0.54, 0.63, 0.88, 0.96, 1.25, 1.67, 1.75, 1.84, 1.96, 2.01, 2.51, 2.72, 3.30, 3.51, 4.05, 4.85, 6.94, 8.73, 10.57, 11.11, 12.45, 14.00, 17.30, 17.92, 18.05, 18.43, 22.48, 22.48, 23.48, 26.32, 26.45, 28.87

a. Plot the data in a histogram and describe the shape of the frequency distribution.

b. By viewing the graph alone, approximate the mean and median of the distribution. Which should be greater? Explain your reasoning.

c. Calculate the mean and median. Was your intuition in part (b) correct?

d. Calculate the first and third quartiles and the interquartile range.

e. Draw a box plot for these data.

14. The MathWorld web page on hypothesis testing declares that "Hypothesis testing is the use of statistics to determine the probability that a given hypothesis is true" (Weisstein 2014). Is this statement true or false? Explain.

15. *Spot the flaw.* In a newer study of "high-rise syndrome" (see Chapter 1), Vnuk et al. (2004) reported injury rates of 119 fallen cats brought to a veterinary clinic in Zagreb, Croatia. The following graph indicates the sex of the cats brought to the clinic.

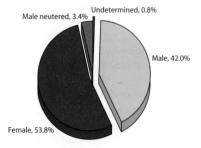

a. Identify at least two of the four principles of good graph design that are violated in this diagram. Explain your answer.

b. Pie charts are not regarded as the best way to illustrate a frequency distribution. What preferred graphical method could be used here instead?

16. For each of the following scenarios, state a good graphical method to display the data, and state the most appropriate statistical test to address the question.

a. Do whales swim past a detector with equal probability over time and independently of each other? The data are measurements of the number of whales per one-hour blocks of time.

b. Do men and women have the same probability of contracting basal carcinoma? Assume that the number of men and women in the study is very large.

c. Do men and women have the same probability of contracting basal carcinoma? Assume that the number of men and women in the study is quite small.

d. Leafcutter ants cut pieces of leaves and carry them back to their nest. Sometimes small ants called minims will ride on the leaves carried by other workers. Do minims occur on the leaves independently and with equal probability? The data are the number of minims on a sample of leaves.

e. Is the proportion of patients developing a skin rash within two weeks of treatment the same between patients taking a new drug and patients taking a placebo? Assume that 20% of patients overall get a rash and that there are 100 patients in each treatment group.

f. Each person in a sample is independently presented with two cookies that are prepared identically, except that one includes trans fats while the other is made with a healthier alternative. Is there a preference in the population of people for one or the other type of cookie? The data are the numbers of people preferring each type of cookie.

g. Does the frequency of cases of leukemia in a small town differ from the known national average?

Crab spider, *Thomisus spectabilis*

10 The normal distribution

Measurements of continuous numerical variables abound in biological data. We measure the length and weight of babies, the velocities of swallows, the times between infection and the onset of symptoms, the numbers of cones on pine trees, etc. Typically we take these measurements on a sample of individuals, but we want to be able to make inferences about the continuous variables in the population. For example, "What is a 95% confidence interval for the mean birth weight of American babies?" Or, "Which is faster, on average, an African or a European swallow?" To answer these kinds of questions, we need to understand something about the probability distributions that these data are taken from.

The normal distribution, which we introduced in Section 1.4, is the queen of all probability distributions used in the analysis of biological data. It is the ubiquitous bell-shaped curve that can be used to approximate the frequency distribution of so many biological variables. The normal distribution arguably describes more about nature than any other mathematical function; thus, it takes a preeminent role in biological statistics.

Even more important than its ability to approximate frequency distributions of data, the normal distribution can be used to approximate the *sampling distribution* of estimates, especially of sample means. Many statistical techniques have been developed for dealing with variables that have a normal sampling distribution. In the rest of this book, we focus on these techniques. This chapter describes the normal distribution and explains some of the reasons it is so important.

10.1 Bell-shaped curves and the normal distribution

Many numerical variables have frequency distributions that are bell shaped. For example, Figure 10.1-1 is a histogram of the birth weights of the 4,017,264 singleton[1] births recorded by birth certificate in the United States in 1991.

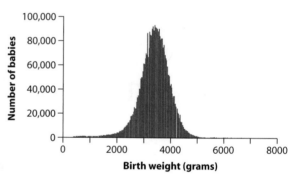

FIGURE 10.1-1
Frequency distribution of the birth weight of babies born in the United States in 1991 (Clemons and Pagano 1999).

Examine the shape of the baby weight distribution. Notice in Figure 10.1-1 that the peak (i.e., the mode) is at the center of the distribution. If we averaged all of these 4,017,264 data points, we would find the mean to be 3339 grams, which is also at the mode. The distribution is clearly shaped like a symmetrical bell. There are so many data points, and the interval widths in the histogram are so narrow, that the distribution looks almost like a smooth curve. If we collected more and more measurements and used even narrower intervals, then the graph would become even smoother.

The theoretical probability distribution describing many bell curves is called the normal distribution. The normal distribution is a continuous probability distribution, which means that it describes the probability distribution of a continuous numerical variable. It is symmetric around its mean. The further values are from the mean, the lower the probability density of observations.

---

1. "Singleton" means that the baby was not a twin, a triplet, etc.

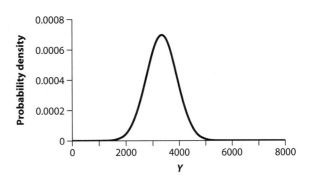

FIGURE 10.1-2
The normal distribution for a variable *Y* with mean and variance equal to that in the baby birth weight data.

The normal distribution has two parameters to describe its location and spread: the mean and the standard deviation. For example, Figure 10.1-2 shows the normal distribution having the same mean and standard deviation as the baby birth weights. It strongly resembles the frequency distribution of the real data.

The scale on the vertical axis of Figure 10.1-2 is different from that in Figure 10.1-1, because the normal distribution shows the probability density, whereas the data are expressed as counts. To find the expected relative frequency of a particular bin of the histogram, we would integrate the normal probability density from the lower bound of the bin to the upper bound. The integral of the normal distribution from minus infinity to infinity equals one, because the probabilities of all possible outcomes are accounted for.

Figure 10.1-3 includes some other examples of data whose frequency distributions resemble the normal curve: the body temperatures of adult humans, the brain sizes of undergraduate students, and the number of bristles on a fly's abdomen. In each case, we have superimposed the normal distribution with the same mean and standard deviation as the data. Bell-shaped frequency distributions appear in nature all the time, and the normal distribution is an excellent approximation to these real distributions. The number of fly bristles is actually a discrete variable; but, with a large number of possible values for this variable, it is still well approximated by a normal distribution.

Biostatistics makes great use of the normal distribution. The statistical methods in most common use assume that the data come from a normal distribution of measurements. Moreover, as we explain in Section 10.6, the normal distribution can describe some properties of samples for variables that aren't themselves normally distributed.

The *normal distribution* is a continuous probability distribution describing a bell-shaped curve. It is a good approximation to the frequency distributions of many biological variables.

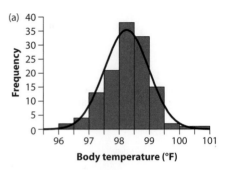

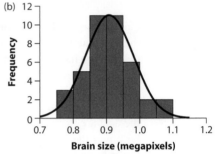

FIGURE 10.1-3
The normal distribution approximates frequency distributions in nature. (a) Human body temperature, in degrees Fahrenheit (Shoemaker 1996). (b) University undergraduate brain size (measured in number of megapixels on an MRI scan) (Willerman et al. 1991). (c) The number of bristles on the fourth and fifth segments of the abdomens of fruit flies (Falconer and MacKay 1995). The black lines show normal distributions with the same mean and standard deviation as measured in the data.

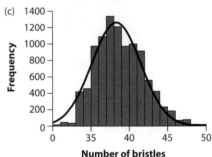

10.2 The formula for the normal distribution

You may rarely have reason to use the formula for the probability density of the normal distribution, but you will often use calculations derived from it. The formula is

$$f(Y) = \frac{1}{\sqrt{2\pi\sigma^2}}\, e^{\frac{-(Y-\mu)^2}{2\sigma^2}}.$$

This gives the probability density $f(Y)$ for a value Y. The value Y can be any real number, ranging from negative infinity to positive infinity; μ is the mean of the distribution; and σ is the standard deviation. The formula also includes the irrational constants $\pi = 3.1415\ldots$ and $e = 2.7182\ldots$ (the base of the natural logarithm).

The mean can take any real value, and the standard deviation can take any positive value. Thus, the "normal distribution" is really an infinite number of distributions, each having its own mean and standard deviation.

10.3 Properties of the normal distribution

The normal distribution has the following features that are worth remembering:

- It is a continuous distribution, so probability is measured by the area under the curve rather than the height of the curve.
- It is symmetrical around its mean.
- It has a single mode.
- The probability density is highest exactly at the mean.

This final feature, and the symmetry of the normal distribution, together imply that the mean, median, and mode are all equal to each other for the normal distribution.

There are some useful rules of thumb about areas under the normal curve. About two-thirds (68.3%, to be more exact) of the area under the normal curve lies within one standard deviation of the mean. In other words, the probability is 0.683 that the measurement of a randomly chosen observation drawn from a normal distribution falls between $\mu - \sigma$ and $\mu + \sigma$, as shown in Figure 10.3-1.

FIGURE 10.3-1
The probability that a randomly drawn measurement from a normal distribution is within one standard deviation of the mean is approximately 2/3. (More precisely, 0.683 of the observations fall within one standard deviation of the mean.)

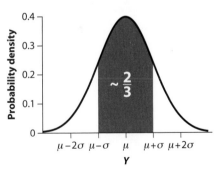

Ninety-five percent of the probability of a normal distribution lies within about two standard deviations of the mean (more precisely, within 1.96 standard deviations). In other words, the probability is 0.95 that the measurement of a randomly chosen observation drawn from a normal distribution falls between $\mu - 1.96\,\sigma$ and $\mu + 1.96\,\sigma$, as shown in Figure 10.3-2.

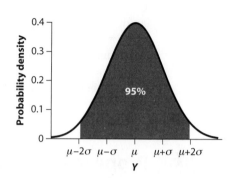

FIGURE 10.3-2
The probability is 0.95 that a randomly drawn measurement from the normal distribution is within approximately two standard deviations of the mean (more precisely, exactly 95% of measurements lie within 1.96 standard deviations of the mean).

> For a variable with normal distribution, about two-thirds of individuals are within one standard deviation of the mean, and about 95% are within two standard deviations of the mean.

10.4 The standard normal distribution and statistical tables

A normal distribution with a mean of zero and standard deviation of one is called a **standard normal distribution**.[2] Figure 10.4-1 is a plot of the standard normal. We use the symbol Z to indicate a variable having a standard normal distribution.

> The *standard normal distribution* is a normal distribution with a mean of zero and a standard deviation of one.

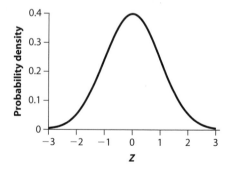

FIGURE 10.4-1
The standard normal distribution.

---

2. "Standard normal" may sound redundant and repetitive (like "bunny rabbit" or "vim and vigor"), but there are an infinite number of normal distributions, depending on what mean and variance are put into the formula. The standard normal distribution is just one of these.

Using the standard normal table

Unlike the Poisson and binomial distributions, the probability that a given event occurs when sampling from a normal distribution is difficult to compute by hand, because it requires integration of a complicated function. Instead, we use statistical tables or computers[3] to obtain probabilities under the normal curve. Statistical Table B in the back of this book gives us the probability that a random draw from a standard normal distribution is *above* a given cutoff value. For example, if we drew a number at random from a standard normal distribution, the probability is 0.025 that it would be greater than 1.96. Table 10.4-1, an excerpt from Statistical Table B, shows how we obtained this number.

TABLE 10.4-1 Probabilities of $Z > a.bc$ under the standard normal curve. The digit before and immediately after the decimal (i.e., $a.b$) are given down the first column, and the second digit after the decimal (i.e., c) is given across the top row. The answer highlighted in red shows the probability that $Z > 1.96$. Excerpted from Statistical Table B.

First two digits of $a.bc$	Second digit after decimal (c)									
	0	1	2	3	4	5	6	7	8	9
1.5	0.0668	0.0655	0.0643	0.0630	0.0618	0.0606	0.0594	0.0582	0.0571	0.0559
1.6	0.0548	0.0537	0.0526	0.0516	0.0505	0.0495	0.0485	0.0475	0.0465	0.0455
1.7	0.0446	0.0436	0.0427	0.0418	0.0409	0.0401	0.0392	0.0384	0.0375	0.0367
1.8	0.0359	0.0352	0.0344	0.0336	0.0329	0.0322	0.0314	0.0307	0.0301	0.0294
1.9	0.0287	0.0281	0.0274	0.0268	0.0262	0.0256	0.0250	0.0244	0.0239	0.0233
2.0	0.0228	0.0222	0.0217	0.0212	0.0207	0.0202	0.0197	0.0192	0.0188	0.0183
2.1	0.0179	0.0174	0.0170	0.0166	0.0162	0.0158	0.0154	0.0150	0.0146	0.0143
2.2	0.0139	0.0136	0.0132	0.0129	0.0126	0.0122	0.0119	0.0116	0.0113	0.0110

Table 10.4-1 has an unusual layout. It provides the probability that Z is greater than a given three-digit cutoff number "$a.bc$," where "a," "b," and "c" refer to digits of the number. To find the probability, begin by finding the first two digits of the cutoff number ($a.b$) in the left-hand column of the table. Then find the second digit of the number after the decimal place (c) across the top row. The probability that Z is greater than $a.bc$ is given in the corresponding cell for that row and column. To find the probability $Pr[Z > 1.96]$, for example, we would look down the left-hand column for $a.b = 1.9$ and then go across to the column that corresponds to $c = 6$ to fill in the last decimal. We find that the probability of a random draw from a standard normal distribution greater than 1.96 is 0.025 (the number shown in red in Table 10.4-1). This probability is the area under the curve to the right of a standard normal Z of 1.96 (see Figure 10.4-2).

---

3. For example, the Excel function NORMDIST(Z,0,1,TRUE) will return the probability under a standard normal distribution of getting a value less than Z. Also see the introduction to Statistical Table B.

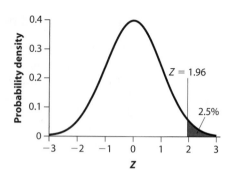

FIGURE 10.4-2
The area under the standard normal curve greater than
$Z = 1.96$ is 0.025. The area in red is given in Statistical
Table B, which is excerpted in Table 10.4-1.

Statistical Table B gives us the probability that Z is greater than a given *positive* number. Probabilities corresponding to *negative* Z values are not included. Recall, however, that all normal distributions are symmetric around their means, and the standard normal distribution is symmetric around $\mu = 0$. This means that

$$\Pr[Z < -number] = \Pr[Z > number].$$

Thus, the probability that a random observation from the standard normal distribution is *less than* -1.96 is the same as the probability that an observation is *greater than* 1.96, which is 0.025:

$$\Pr[Z < -1.96] = \Pr[Z > 1.96] = 0.025.$$

The probability that Z lies between a lower bound and an upper bound can be calculated in two steps, as shown in Figure 10.4-3. First, use Statistical Table B to find $\Pr[Z > lower\ bound]$ and $\Pr[Z > upper\ bound]$. Then, calculate the difference between these two probabilities:

$$\Pr[lower\ bound < Z < upper\ bound] = \Pr[Z > lower\ bound] - \Pr[Z > upper\ bound].$$

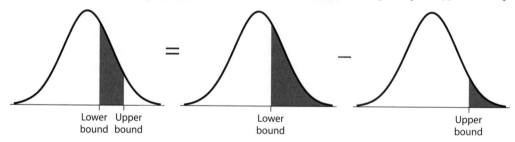

FIGURE 10.4-3 Calculating the area under the standard normal curve between a lower bound and an upper bound.

Using the standard normal to describe any normal distribution

There are an infinite number of normal distributions, but they are all similar in shape. This allows us to use a simple transformation to obtain probabilities under any normal distribution. To do so, we calculate how many standard deviations a particular value is away from the mean.

$$Z = \frac{Y - \mu}{\sigma}.$$

This standardized Z value is called a **standard normal deviate**.[4] The formula converts Y, which has a normal distribution with mean μ and standard deviation σ, to Z, which has a standard normal distribution. Look at this formula for a second. The numerator, $Y - \mu$, tells us how far away Y is from its mean, measured in the original units. If this value is negative, then Y is less than the mean; and if it is positive, then Y is greater than the mean. If we now divide this value by σ, then we can calculate how far Y is from its mean as measured by the number of standard deviations.

The probability of obtaining a measurement that is Z standard deviations from the mean is the same for all normal distributions, including the standard normal distribution. This means that we can use the table of probabilities for the standard normal distribution (Statistical Table B) to find probabilities for *any* normal distribution.

> A *standard normal deviate*, or Z, tells us how many standard deviations a particular value is from the mean.

Example 10.4 shows how the Z-standardization can be used.

EXAMPLE One small step for man?

10.4 NASA excludes anyone under 62 inches in height and anyone over 75 inches from being an astronaut pilot (NASA 2004). In metric units,[5] these values for the lower and upper height restrictions are 157.5 cm and 190.5 cm, respectively. What fraction of the young adult American population is excluded from being an astronaut pilot by these height constraints? The distribution of adult heights within a sex and age group is reasonably well approximated by a normal distribution, with the mean and standard deviation for 20- to 29-year-old males in America given by 177.6 cm and 9.7 cm, respectively (McDowell et al. 2008). For 20- to 29-year-old American females, the mean height is 163.2 cm with a standard deviation of 10.1 cm.

We can use the standard normal distribution to calculate the proportion of individuals who are made ineligible for astronaut pilot training because of their height alone. We will start with the calculation for males. Let's be very specific about what we are trying to do: we want to know the probability that a 20- to 29-year-old male individual has a height that is either less than 157.5 cm or greater than 190.5 cm:

$$\Pr\left[\textit{Height} < 157.5 \text{ or } \textit{Height} > 190.5\right].$$

---

4. Not to be confused with "everyday ordinary pervert." You don't often find a jargon term that seems to be both redundant *and* self-contradictory.

5. NASA has had infamous difficulty in converting between English and metric units. In 1999, the *Mars Climate Orbiter* missed Mars and flew off into its own independent orbit around the sun because some components of the navigational system used English units and others used metric. Oops.

It becomes much easier to answer the question if we make a sketch of what we are looking for (Figure 10.4-4).

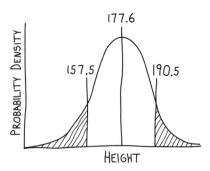

FIGURE 10.4-4
A sketch can help you keep track of the areas under the curve that you are trying to find. This rough sketch does not show the exact areas, but it properly orients the mean and the values we care about.

The maximum height (190.5 cm) is above the mean (177.6 cm), and the minimum height (157.5 cm) is below the mean. Based on the drawing in Figure 10.4-4, the outcomes "*Height* < 157.5" and "*Height* > 190.5" are mutually exclusive. By the addition rule, therefore, we can determine the fraction of American males whose height excludes them from being an astronaut pilot by summing the two parts:

$$\Pr\big[\textit{Height} < 157.5 \text{ or } \textit{Height} > 190.5\big] = \Pr\big[\textit{Height} < 157.5\big] + \Pr\big[\textit{Height} > 190.5\big].$$

Let's start with the second part first: what's the probability that an American male is too tall for NASA's restrictions? In other words, what is the probability that an American adult male is taller than 190.5 cm? The first step is to convert *Height* = 190.5 to a standard normal deviate, using the mean (177.6 cm) and standard deviation (9.7 cm) of American male height:

$$Z = \frac{190.5 \text{ cm} - 177.6 \text{ cm}}{9.7 \text{ cm}} = 1.33.$$

That is, a value of 190.5 cm occurs at 1.33 standard deviations above the mean male height.

What fraction lies above this point? In other words, what is $\Pr[Z > 1.33]$? Using Statistical Table B at the end of this book, we can see that 0.09176 of the area under the standard normal distribution lies more than 1.33 standard deviations above the mean. So, this is our answer to this part of the problem: the fraction 0.09176 of American adult males are taller than 190.5 cm.

Finding the probability of males being too short for NASA's restrictions follows a similar procedure, but with one additional step. Again we convert *Height* = 157.5 (the minimum cutoff) to a standard normal deviate:

$$Z = \frac{157.5 \text{ cm} - 177.6 \text{ cm}}{9.7 \text{ cm}} = -2.07.$$

In other words, 157.5 cm occurs at 2.07 standard deviations below the mean male height. What fraction of American adult males lies *below* this point? What is $\Pr[Z < -2.07]$? Statistical Table B gives us the probability of obtaining a value *greater than* a given number. To find the probability of getting a value *less than* a particular number, remember that the normal distribution is symmetric around its mean. Thus, the probability of being 2.07 standard deviations or more below the mean is the same as the probability of being 2.07 standard deviations or more above the mean:

$$\Pr\left[Z < -2.07\right] = \Pr\left[Z > 2.07\right].$$

When we look up the probability of being greater than 2.07 in Statistical Table B, we find that $\Pr[Z > 2.07] = 0.01923$. Thus, the fraction 0.01923 of American adult males are shorter than 157.5 cm.

Now, use the addition rule to determine the proportion of American adult men that are excluded from the astronaut pilot program by the height restrictions:

$$\Pr\left[\textit{Height} < 157.5 \text{ or } \textit{Height} > 190.5\right] = 0.01923 + 0.09176 = 0.11099.$$

The fraction 0.11099 (or 11.1%) of all 20- to 29-year-old American adult males are excluded by height. In other words, the NASA height restriction excludes a modest percentage of men.

We can make similar calculations for American adult women. Women's heights are approximated by a normal distribution with mean 163.2 cm and standard deviation 10.1 cm. These values correspond to height restrictions at 0.56 standard deviations below the mean and 2.7 standard deviations above the mean. (Check these numbers for yourself.) Using Statistical Table B, the fraction 0.28774 of the women are too short to meet NASA's guidelines and the fraction 0.00347 are too tall. In total, the height restrictions exclude 29.1% of young American women from being astronaut pilots.

10.5 The normal distribution of sample means

One of the most important facts about the normal distribution is that it can be used to describe the sampling distribution of many estimates, including the sample mean. The sampling distribution of an estimate lists all the values that we might obtain when we sample a population and describes their probabilities of occurrence (Section 4.1).

> If a variable Y has a normal distribution in a population, then the distribution of sample means $\overline{Y}$ is also normal.

For example, Figure 10.1-2 shows the normal distribution that best matches the distribution of human birth weights in the United States. This distribution is normal with mean $\mu = 3339$ g and standard deviation $\sigma = 573$ g. Suppose we took a single random sample of 10 babies from this distribution and obtained $\overline{Y} = 3084$ g. This is not equal to the true mean (3339 g) because of chance, or sampling error. The estimate based on any particular sample is influenced by who happened to get sampled and who did not, which is a random process.

Each single sample will have a mean that by chance is different from the true mean; if we took a different sample, it would differ from the truth in a different way. The different values for $\overline{Y}$ that we might have obtained, and their associated probabilities, make up the sampling distribution for $\overline{Y}$.

The fact that sample means are normally distributed, whenever the population itself is normal, is a huge time-saver. In Chapter 4, we obtained the sampling distribution for $\overline{Y}$ by going back to the population and taking a vast number of random samples, each of the same size n, calculating $\overline{Y}$ for each sample, and plotting the results. However, we are spared the trouble if Y has a normal distribution in the population, because in this case the distribution of sample means is also normally distributed, as shown in Figure 10.5-1 (we also plot the distribution of individual data points for comparison).

Figure 10.5-1 shows that the mean of the sampling distribution of $\overline{Y}$ is *also* 3339 g, the same as μ, the mean of Y itself. The mean of the sampling distribution of $\overline{Y}$ *always* equals the mean of the original distribution (μ). In other words, the sample mean based on a random sample from a normal distribution gives an unbiased estimate of μ.

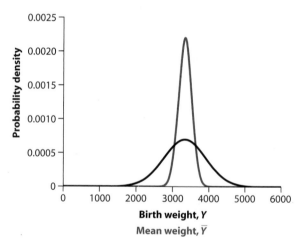

FIGURE 10.5-1
The normal distribution of sample means ($\overline{Y}$, in red) for samples of size $n = 10$ from the normal distribution describing the population of baby birth weights in Figure 10.1-2 (shown here in black for comparison). The distribution of sample means has the same mean as the individual values, but with a smaller standard deviation.

The standard deviation of the sampling distribution for $\overline{Y}$ is known as the *standard error of the mean* (Section 4.2). It is symbolized as $\sigma_{\overline{Y}}$ and is equal to the standard deviation of Y divided by the square root of the sample size (n):

$$\sigma_{\overline{Y}} = \frac{\sigma}{\sqrt{n}}.$$

This equation is correct even when Y does *not* have a normal distribution.

The standard error describes the typical amount of sampling error when estimating the mean, so it measures the precision of the estimate. Increasing sample size reduces sampling error and hence increases the precision of an estimate, as we saw in Chapter 4.

By averaging data from more babies, the sampling distribution based on samples of 100 babies is less "noisy" (has a smaller sampling error) than the sampling distribution generated from only 10 babies at a time.[6] The sampling distribution of $\overline{Y}$ would be even narrower if we used a larger sample size, as shown in Figure 10.5-2.

In later chapters, we will use the fact that the sampling distribution of $\overline{Y}$ is normal when Y is normal to calculate exact confidence intervals for population means and exact P-values when testing hypotheses about means.

FIGURE 10.5-2
The distributions of sample means based on sample sizes of $n = 10$ (*red*), $n = 100$ (*blue*), and $n = 1000$ (*black*). (Note the change in scale from the previous figures.)

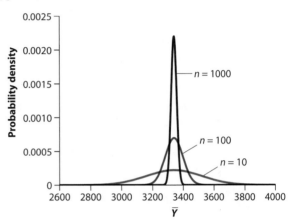

Calculating probabilities of sample means

The standard normal distribution can be used to calculate the probability of drawing a sample with a mean in a given range, assuming we know the true values of the mean and the standard deviation of the population from which the sample was taken. Given that the distribution of sample means is normal with mean μ and standard deviation $\sigma_{\overline{Y}} = \sigma/\sqrt{n}$, then

$$Z = \frac{\overline{Y} - \mu}{\sigma_{\overline{Y}}}$$

has a standard normal distribution, just like any normally distributed variable. The quantity $\sigma_{\overline{Y}}$ is the standard deviation of the sampling distribution of $\overline{Y}$, also known as the standard error of the mean. This Z tells us that the sample mean $\overline{Y}$ is Z standard errors above the true mean μ.

For example, let's take a random sample of $n = 80$ babies from the population of babies introduced in Section 10.1. In this population, weights are normally distributed with mean $\mu = 3339$ g and standard deviation $\sigma = 573$ g. What is the probability that the sample mean is at least 3370 g? In other words, what is $\Pr[\overline{Y} > 3370]$?

---

6. This may be the only time it can be said that 100 babies are less noisy than 10.

For this example, $\sigma_{\overline{Y}}$ is

$$\sigma_{\overline{Y}} = \frac{\sigma}{\sqrt{n}} = \frac{573}{\sqrt{80}} = 64.1 \text{ g.}$$

From this we calculate the Z-score:

$$Z = \frac{Y - \mu}{\sigma_{\overline{Y}}} = \frac{3370 - 3339}{64.1} = 0.48.$$

In other words, $\Pr[\overline{Y} > 3370]$ is the same as $\Pr[Z > 0.48]$. Using Statistical Table B, we find that

$$\Pr[Z > 0.48] = 0.316.$$

About 31.6% of samples of size $n = 80$ from this population will have a sample mean of 3370 g or larger.

Sampling distribution of the sample mean.

10.6 Central limit theorem

We have seen that the means of samples drawn from a normal distribution are themselves normally distributed. Another reason for the importance of the normal distribution in data analysis is that the sampling distribution of sample means $\overline{Y}$ is *approximately* normal even when the distribution of individual data points is *not* normal, provided the sample size is large enough. This astonishing fact is known as the **central limit theorem**. How large a sample is large enough depends on the shape of the distribution of observations in the population: the more similar the original distribution is to the normal, the smaller the sample size required to yield a distribution of sample means that is well approximated by the normal distribution.

> According to the *central limit theorem*, the sum or mean of a large number of measurements randomly sampled from a non-normal population is approximately normally distributed.

Example 10.6 illustrates the effect that increasing the sample size has on the sampling distribution for the mean, even when the population is highly non-normal.

EXAMPLE Young adults and the Spanish flu

10.6 Between 1918 and 1920, a strain of influenza misnamed the Spanish flu swept across the world, killing millions. This epidemic, which overlapped with World War I, caused more than twice as many deaths as the war itself. One unusual feature of the Spanish flu was that it mainly killed young adults, whereas most influenzas endanger mainly the very

young and the very old. This caused an atypical pattern in the frequency distribution of
age at death, as shown in Figure 10.6-1. The
data in the figure are from 1918 Switzerland,
which was not involved in the war, and so
represent deaths from Spanish flu in addition to all the usual causes. A total of 75,034
people are represented in this graph. This
distribution is unusual because it has three
peaks. The large spike at age 0 corresponds
to infant mortality. Another wide peak,

at approximately 75 years of age, shows elevated mortality of the elderly. In a typical
year around that time, these would have been the only two peaks. During 1918–20, the
Spanish flu caused a third peak of mortality for people in their twenties and thirties.

FIGURE 10.6-1
The frequency distribution of age at death in
Switzerland in 1918 during the Spanish flu
epidemic.

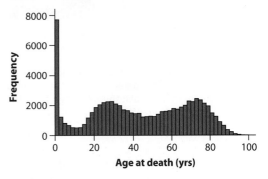

The frequency distribution in Figure 10.6-1 is extremely non-normal. It has three
peaks rather than just the one we would expect with the normal distribution, and the
distribution of the data is highly asymmetric. Let's use this distribution to illustrate
the central limit theorem. We will look at the sampling distribution of the sample
mean for a range of samples of increasing size. The distribution of sample means is
displayed in Figure 10.6-2 for the range of sample sizes $n = 1, 2, 4$, and 64.

The top left graph in Figure 10.6-2 plots the mean of random samples of size
$n = 1$, which simply re-creates the original distribution. That is the distribution of
individuals in this population. Let's now look at the distribution of sample means
when we take a sample of two or more individuals from the same population. The
top right histogram plots the sample means of many samples of size $n = 2$. The
numbers shown were generated by taking a random sample of two individuals from
the population and computing their average. This process was repeated many times,
and the distribution of the computed averages is shown. Notice how different this
frequency distribution is from that of the data. The standard deviation is lower than in
the data, and even with such a tiny sample size, the graph already looks a lot more bell
shaped. As n increases (in other words, as we take larger samples), the distribution
of sample means starts to resemble a normal distribution. For example, when $n = 4$,
the overall shape of the distribution is similar to a normal distribution, but there are

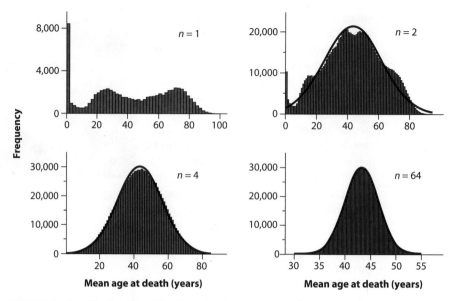

FIGURE 10.6-2 The frequency distribution of sample means for samples of increasing size. Each histogram displays the means of a large number of repeated samples of size *n* drawn from the distribution at age at death in Figure 10.6-1. The scale and range of the axes change from graph to graph.

differences in the details. The fit gets better and better as sample size increases. At some point, the sample size is large enough that the sampling distribution becomes almost identical to a normal distribution. For the Swiss 1918 age at death data, the distribution of sample averages is nearly indistinguishable from a normal distribution when $n = 64$ (the bottom right histogram in Figure 10.6-2). It is difficult to specify the sample size needed to reach this point for any particular case. For other data whose frequency distribution is even more different from the normal distribution than the Swiss mortality data, larger samples are required before the sample means converge on the normal distribution.

The most powerful statistical methods available for analyzing biological data assume that the distribution of sample means (and the distributions of some other estimates) follows a normal distribution. The beauty of the central limit theorem is that if the sample size is large enough, then it is possible to use these powerful methods even when our data are sampled from a population that is *not* normally distributed.

10.7 Normal approximation to the binomial distribution

One frequent application of the central limit theorem is the normal approximation to the binomial distribution. Recall from Section 7.1 that the binomial distribution is a discrete probability distribution. It describes the number of "successes" in n independent trials, where p is the probability of success in any one trial.

According to the central limit theorem, a binomial distribution with large n is approximated by a normal distribution. This is because the number of successes is a kind of sum: if each success is labeled as "1" and each failure is labeled as "0," then the count of the number of successes is the sum over all individuals of the ones and zeros.

The normal approximation to the binomial distribution is helpful when you are using a calculator to obtain probabilities for a binomial distribution with a large n—exactly the case when calculating exact binomial probabilities is most time-consuming. This situation comes up, for example, when carrying out a binomial test on a large sample. With a binomial test, the probability of success p_0 stated in the null hypothesis is often 0.5, in which case n needn't be too large (say 30 or more) in order for the normal approximation to be appropriate.[7] As a rule of thumb, the products np and $n(1-p)$ should both be greater than five in order to use the normal approximation if all you want to know is whether the P-value is less than 0.05. The approximation requires higher np if the goal is to provide exact P-values or to obtain critical values corresponding to significance levels smaller than 0.05 (such as when using $\alpha = 0.01$ instead).

We use the normal distribution that has the same mean and standard deviation as the binomial distribution—in other words, a mean of np and a standard deviation of $\sqrt{np(1-p)}$.

> When the number of trials n is large, the binomial probability distribution for the number of successes is approximated by a normal distribution having mean np and standard deviation $\sqrt{np(1-p)}$.

Example 10.7 applies the normal approximation to a problem requiring the binomial test.

EXAMPLE 10.7   The only good bug is a dead bug

The brown recluse spider (*Loxosceles reclusa*) often lives in human houses[8] throughout central North America. This spider is a moderate health threat, as its bite causes nasty, slow-healing wounds. Bites are rarely fatal, but the resulting wounds are so disgusting

---

7. The proportion specified by the null hypothesis is what's important, not the proportion in the data.

8. One house in Kansas had more than 2000 brown recluse spiders (Sandidge 2003).

that you will be very glad we chose to show the spider here rather than the injuries it causes. Information on the spider's diet is useful for developing effective pest management strategies. A diet-preference study (Sandidge 2003) gave each of 41 brown recluse spiders a choice between two crickets, one live and one dead. Thirty-one of the 41 spiders chose the dead cricket over the live one. Does this represent evidence for a diet preference?

A 95% confidence interval using the Agresti–Coull method (see Chapter 7) reveals that brown recluse spiders choose dead prey between about 60 and 86% of the time:

$$0.60 < p < 0.86.$$

Let's carry out a hypothesis test to determine the weight of evidence against the null hypothesis of no preference. The appropriate statistical method for analyzing these data is the binomial test with the following null and alternative hypotheses:

H_0: Brown recluse spiders show no preference for live or dead prey ($p = 0.5$).

H_A: Brown recluse spiders prefer one type of prey over the other ($p \neq 0.5$).

This is a two-sided test. The outcome of the test will be decided according to the P-value (Section 6.2), the probability of obtaining a result as extreme as or more extreme than the observed result (i.e., 31 out of 41 spiders ate the dead prey item), assuming that the null hypothesis were true. This P-value would ordinarily be calculated from the binomial distribution with $n = 41$ and $p = 0.5$:

$$P = 2 \Pr[X \geq observed]$$
$$= 2 \left[(\Pr[X = 31] + \Pr[X = 32] + \Pr[X = 33] + \cdots + \Pr[X = 41]) \right],$$

where

$$\Pr[X = i] = \binom{41}{i}(0.5)^i(0.5)^{41-i}.$$

There are 11 different probabilities to calculate in this example, which would be time-consuming to do by hand or by calculator. It is much easier to approximate the answer by using the normal distribution. The normal approximation is appropriate here because both np and $n(1-p)$ are not too small (20.5 for both in this case). Under the null hypothesis, the mean of the best-fitting normal distribution is

$$\mu = np = (41)(0.5) = 20.5,$$

and the standard deviation is

$$\sigma = \sqrt{np(1-p)} = \sqrt{41(0.5)(0.5)} = 3.20.$$

Now we can use the normal approximation to calculate the probability of getting the observed value 31 or more by chance under this distribution, and then multiply by two to yield the P-value. We will use the normal distribution having the same mean and standard deviation as our binomial distribution. In principle, we could do this by calculating

$$\Pr[X \geq Observed] \approx \Pr[Z > (Observed - np)/\sqrt{np(1 - p)}],$$

where *"Observed"* is the particular value of X that we observed in the data and the $\approx$ sign means "approximately equal to."

However, we can improve the accuracy of this approximation with a correction for continuity, as follows. Remember that the binomial distribution is a discrete distribution and the value *Observed* is an integer. To make the conversion to a continuous variable more seamless, we represent the probability of the discrete value *Observed* as the area under the continuous probability curve between *Observed* – ½ and *Observed* + ½. Therefore, if we want to use the normal distribution to approximate the probability of obtaining a value of X greater than or equal to an *Observed* value, we use the corrected formula

$$\Pr[X \geq Observed] = \Pr\left[Z > \frac{Observed - \dfrac{1}{2} - np}{\sqrt{np(1 - p)}}\right].$$

When we want to approximate the probability of obtaining a value of X *less than or equal* to an *Observed* value, we use

$$\Pr[X \leq Observed] = \Pr\left[Z < \frac{Observed + \dfrac{1}{2} - np}{\sqrt{np(1 - p)}}\right],$$

adding a half, rather than subtracting it. In both cases we are including ½ above and ½ below the specific value of *Observed* to account for the difference between the discrete and continuous distributions. This is why it is called a "continuity correction."

To calculate a two-sided P-value for our spider example, we need to calculate

$$P = 2\Pr[X \geq 31].$$

We can convert this to a Z score by using the normal approximation with the continuity correction, which effectively calculates the probability of observing a value from the normal distribution greater than 30.5:

$$P = 2\Pr\left[Z > \frac{Observed - \dfrac{1}{2} - np}{\sqrt{np(1 - p)}}\right] = 2\Pr\left[Z > \frac{31 - \dfrac{1}{2} - 41(0.5)}{\sqrt{41(0.50)(0.5)}}\right] = 2\Pr[Z > 3.12].$$

Looking to Statistical Table B, we find the probability that $Z > 3.12$ is approximately 0.0009. Multiplying this value by two to get the two-tailed P-value, we obtain

$$P = 0.0018.$$

This P-value is less than $\alpha = 0.05$, so we reject the null hypothesis. The spiders indeed show a diet preference for dead prey.

This computation is much simpler than the exact calculation based on the binomial distribution, but how good is the approximation? The exact P-value, calculated using the binomial distribution, is 0.0015 rather than 0.0018. The normal approximation is not exact. It performs best when n is large and when the population proportion p is close to 0.5. The normal approximation can make some otherwise time-consuming calculations manageable.

10.8 Summary

- The normal distribution is a bell-shaped curve, a continuous probability distribution approximating the distribution of many variables in nature.

- If a variable has a normal distribution, its mean, median, and mode are all the same. The normal distribution is symmetric around its mean.

- To describe a normal distribution, we need to know the mean μ and the standard deviation σ.

- For a normal distribution, about two-thirds (68.3%) of individuals are within one standard deviation of the mean. About 95% are within two standard deviations of the mean.

- A standard normal distribution is a normal distribution with mean of zero and standard deviation equal to one.

- All normal distributions can be converted to the standard normal distribution by computing, for each measurement, the number of standard deviations from the mean: $Z = \frac{Y - \mu}{\sigma}$, where Z is called the "standard normal deviate."

- Means of random samples drawn from a normal distribution with mean μ and standard deviation σ are also normally distributed, with the same mean μ and with standard deviation $\sigma_{\bar{Y}} = \sigma / \sqrt{n}$ where $\sigma_{\bar{Y}}$ is the standard error of the sample mean.

- According to the central limit theorem, the sum (or average) of a large number of observations randomly sampled from a non-normal distribution is approximately normally distributed.

- A binomial distribution with large n can be approximated by a normal distribution. As a rule of thumb, np and $n(1-p)$ should both be greater than five to use the normal approximation.

10.9 Quick Formula Summary

Z-standardization

What is it for? Converts values from any normal distribution with known mean μ and known variance σ into standard normal deviates.

What does it assume? The original distribution is normal with known parameters.

Formula: $Z = \dfrac{Y - \mu}{\sigma}$.

Normal approximation to the binomial distribution

What is it for? Approximates the binomial distribution using the normal distribution when *n* is large.

What does it assume? That np and $n(1 - p)$ are both five or greater.

Formula: $\Pr\left[X \geq Observed\right] = \Pr\left[Z > \dfrac{Observed - \dfrac{1}{2} - np}{\sqrt{np(1 - p)}}\right]$ and

$$\Pr\left[X \leq Observed\right] = \Pr\left[Z < \dfrac{Observed + \dfrac{1}{2} - np}{\sqrt{np(1 - p)}}\right],$$

where *n* is the number of trials, *p* is the probability of success for each trial, and *Observed* is the number of successes in the data.

PRACTICE PROBLEMS

1. **Calculation practice: Finding the probability of a range of values using the normal distribution.** The natural log of growth (change in radius per year in mm) of Engelmann spruce is approximately normally distributed with mean of 0.037 log units and standard deviation 0.385. Following these steps, determine the probability that a tree has a bad year, defined as having growth less than −0.050 log units in a year.
   a. Make a sketch of the normal distribution with mean 0.037, and mark the values that

   we are trying to determine (i.e., those values less than −0.05).
   b. Calculate the standard normal deviate (*Z*) associated with the value we are interested in here, −0.05.
   c. We are interested in the probability of getting a value less than this *Z* value. Is this probability directly shown in Statistical Table B? If not, what quantity can we use to find what we need?

d. What is the probability that a random draw from a standard normal distribution will be greater than 0.226?

e. What is the probability that a random draw from a standard normal distribution will be less than −0.226?

f. What is the probability that a tree has a bad growth year, that is, less than −0.050 in log units?

2. Calculation practice: Normal approximation to a binomial test. From 1995 to 2008 in the United States, 531 of the 648 people who were struck by lightning were men. Test whether this proportion is different from the 50% that might be expected by the proportion of men in the population as a whole (Avon 2009). Use the binomial test with a normal approximation.

a. State the null hypothesis for this binomial test.

b. Calculate the mean of the null distribution for the number of men struck by lightning under this null hypothesis.

c. What is the standard deviation of the distribution for the number of men hit by lightning under the null hypothesis?

d. Is your target value (531) above or below the value given by the null hypothesis? If above, subtract one-half from that target for the con-

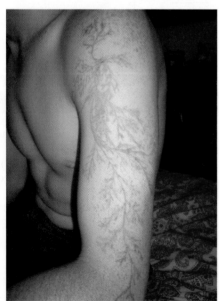

Scars from a lightning strike

tinuity correction. If below, add a half for the continuity correction.

e. What is the standard normal deviate (Z) for the continuity-corrected observed number of men, using the mean and standard deviation calculated in parts (b) and (c)?

f. What is the probability under the normal distribution of getting a result of 531 or greater (including the continuity correction)?

g. What is the two-tailed P-value for this binomial test?

h. State the conclusion from your test.

3. Assume that Z is a number randomly chosen from a standard normal distribution. Use the standard normal table to calculate each of the following probabilities:

a. $\Pr[Z > 1.34]$

b. $\Pr[Z < 1.34]$

c. $\Pr[Z > 2.15]$

d. $\Pr[Z < 1.2]$

e. $\Pr[0.52 < Z < 2.34]$

f. $\Pr[-2.34 < Z < -0.52]$

g. $\Pr[Z < -0.93]$

h. $\Pr[-1.57 < Z < -0.32]$

4. It was rumored that Britain's domestic intelligence service MI5 has an upper limit on the height of its spies, on the assumption that tall people stand out (although MI5 denies it). The rumor said that, to apply to be a spy, you can be no taller than 5 feet 11 inches (180.3 cm) if you are a man, and no taller than 5 feet 8 inches (172.7 cm) if you are a woman (supposedly to allow the spies to blend in with a crowd).

a. If the mean height of British men is 177.0 cm, with standard deviation 7.1 cm, what proportion of British men would be precluded from being spies by this hypothetical height restriction? Assume that height follows a normal distribution.

b. The mean height of women in Britain is 163.3 cm, with standard deviation 6.4 cm. Assuming a normal distribution of female height, what fraction of women meet the height standard for application to MI5?

c. Sean Connery, the original James Bond in the movies, is about 183.4 cm tall. By how many standard deviations does he exceed the height limit for spies?

5. Use the three distributions labeled *i, ii,* and *iii* to answer the following questions.

i.

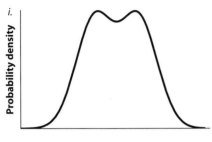

ii.

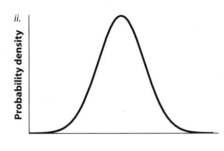

iii.

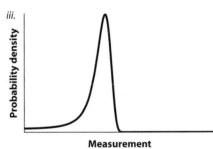

Measurement

a. Which of these distributions is most like the normal distribution? On what basis would you exclude the other two?

b. Which distribution would generate an approximately normal distribution of sample means, calculated from large random samples? Why?

6. The babies born in singleton births in the United States have birth weights that are approximately normally distributed with mean 3.296 kg and standard deviation 0.560 kg (Martin et al. 2011).

a. What is the probability of a baby being born weighing more than 5 kg?

b. What is the probability of a baby being born weighing between 3 kg and 4 kg?

c. What fraction of babies is more than 1.5 standard deviations from the mean in either direction?

d. What fraction of babies is more than 1.5 kg from the mean in either direction?

e. If you took a random sample of 10 babies, what is the probability that their mean weight $\overline{Y}$ would be greater than 3.5 kg?

7. In the accompanying pairs of graphs of normal distributions, which distribution has the highest mean? Which has the highest standard deviation? Pay careful attention to the changes in scale of the *x*-axes.

(a)

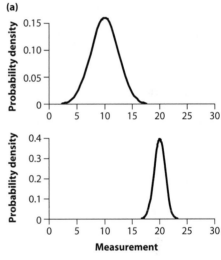

Measurement

(b)

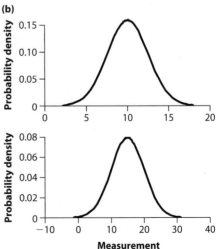

Measurement

8. In the accompanying graph, the red region accounts for 0.67 of the probability density. Estimate the standard deviation of this normal distribution.

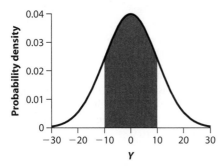

9. Suppose that a variable has a normal distribution, that the mean is 35 mm, and that 20% of the population is larger than 50 mm.
  a. What is the mode of this distribution?
  b. What is the median of this distribution?
  c. Complete the following sentence: Twenty percent of the distribution is smaller than _____.

10. Use the two distributions labeled *i* and *ii* below to answer the following questions.

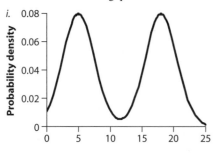

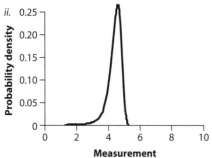

a. If we drew repeated random samples of individuals from each distribution and calculated the mean for each sample, which distribution would yield a distribution of sample means that most closely followed a normal distribution?
b. Imagine that we drew the distribution of the sum of 100 random draws from distribution *ii*. What approximate shape would this distribution have?

11. A survey of European mitochondrial DNA variation has found that the most common haplotype (genotype), known as "H", occurs in 40% of people (Roostalu et al. 2007). If we were to sample 400 individuals from the European population, what is the probability that
  a. at least 180 are haplotype H?
  b. at least 130 are haplotype H?
  c. between 115 and 170 (inclusive) are haplotype H?

12. Ninety-one out of 220 people working as cast and crew of the movie *The Conqueror*, which was filmed in 1955 downwind from nuclear bomb tests, ultimately contracted cancer (see Practice Problem 4 in Chapter 7). Based on age alone, though, only a 14% cancer rate was expected. Test the null hypothesis that the incidence of cancer in this group of people was no different than that expected, using the normal approximation to the binomial distribution.

13. The following table lists the means and standard deviations of several different normal distributions. For each, a sample of 10 individuals was taken, as well as a sample of 30 individuals. For each sample, calculate the probability that the mean of the sample was greater than the given value of *Y*.

Mean	Standard deviation	Y	n = 10: $\Pr[\bar{Y} > Y]$	n = 30: $\Pr[\bar{Y} > Y]$
14	5	15		
15	3	15.5		
−23	4	−22		
72	50	45		

ASSIGNMENT PROBLEMS

14. Assume that Z is a number randomly chosen from a standard normal distribution. Use the standard normal table to calculate each of the following probabilities:
a. $\Pr[Z > 0.24]$
b. $\Pr[Z < 0.24]$
c. $\Pr[Z > 2.01]$
d. $\Pr[Z < 1.02]$
e. $\Pr[0.60 < Z < 1.4]$
f. $\Pr[-2 < Z < 2]$
g. $\Pr[Z < 0.45]$
h. $\Pr[-0.2 < Z < 0.37]$

15. The highest recorded temperature during the month of July for a given year in Death Valley, in California, has an approximately normal distribution with a mean of 123.8°F (!) and standard deviation of 3.1°F (Weather Source 2009).
a. What is the probability for a given year that the temperature never exceeds 120°F in a given July in Death Valley?
b. What is the probability that the temperature in Death Valley goes above 128°F during July in a randomly chosen year?

16. Draw a distribution that is approximately normal, with mean equal to 10 cm and variance equal to 360. On the same graph, draw the distribution of the means of samples taken from this distribution, if each sample was a random sample of 10 individuals.

17. The following data are 40 measurements of diameter growth rate of the tropical tree *Dipteryx panamensis* from a long-term study at La Selva, Costa Rica (Clark and Clark 2012). The data are log-transformed, with the original units in millimeters.

0.0, 0.1, 0.1, 0.3, 0.4, 0.4, 0.4, 0.5, 0.5, 0.5, 0.6, 0.6, 0.7, 0.7, 0.7, 0.7, 0.8, 0.8, 0.8, 1.2, 1.2, 1.3, 1.4, 1.6, 1.9, 2.0, 2.1, 2.1, 2.2, 2.2, 2.3, 2.4, 2.5, 2.5, 2.7, 2.7, 2.7, 2.7, 2.8, 3.1

a. Make an appropriate graph of the data.
b. Examine the graph. Do the data appear to be sampled from a population having a normal

distribution? Why or why not? Identify all the features on which you based your conclusion.

18. In Europe, 53% of flowers of the rewardless orchid, *Dactylorhiza sambucina,* are yellow, whereas the remaining flowers are purple (Gigord et al. 2001). For this problem, you may use the normal approximation only if it is appropriate to do so.

a. If we took a random sample of a single individual from this population, what is the probability that it would be purple?
b. If we took a random sample of five individuals, what is the probability that at least three are yellow?
c. If we took many samples of $n = 5$ individuals, what is the expected standard deviation of the sampling distribution for the proportion of yellow flowers?
d. If we took a random sample of 263 individuals, what is the probability that no more than 150 are yellow?

19. The amount of money spent on health care per person varies enormously among countries (The World Bank 2013). In 2010, this expense ranged from 11.9 U.S. dollars per person (in Eritrea) to $8361 (in the United States). The distribution of this per capita health care expenditure is very skewed, with a long tail corresponding to countries that spend a lot on health care per capita (see the top histogram in the accompanying graph). However, if we look at the log (base 10) of each country's per capita health expenditure, it has a distribution that can be approximated by a normal distribution (see bottom histogram). On the log scale, the mean of the log expenditure is 2.47, with standard deviation equal to 0.72.

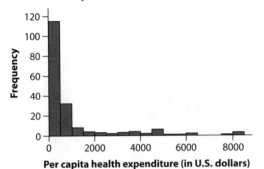

Per capita health expenditure (in U.S. dollars)

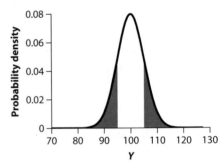

Log$_{10}$ (per capita health expenditure)

a. Assuming that log health expenditure is normal, calculate the proportion of countries that spend less than $100 per capita on health care. (This corresponds to a log expenditure of 2.)
b. Assuming that log health expenditure is normally distributed, calculate the proportion of countries that spend more than $1000 per capita on health care. (This corresponds to a log expenditure equal to 3.)
c. The true proportions of countries with per capita health expenditure less than $100 or more than $1000 are 0.30 and 0.21, respectively. Comment on why your answers from

parts (a) and (b) above do not exactly match these values.

20. Recall from Example 10.4 that more women (29.1%) than men (11.1%) are excluded from the astronaut pilot program by the minimum and maximum height restrictions. What value of minimum height for women would exclude the same total proportion of women as men, given that 0.3% of women are too tall?

21. Draw a probability distribution that isn't normal. Describe the features of your distribution that identify it as non-normal.

22. In the accompanying graph of a normal distribution, *each* of the two red areas represents one-sixth of the area under the curve. Estimate the following quantities from this graph:
a. The mean
b. The mode
c. The median
d. The standard deviation
e. The variance

23. The proportion of traffic fatalities resulting from drivers with high alcohol blood levels in 1982 was approximately normally distributed among U.S. states, with mean 0.569 and standard deviation 0.068 (U.S. Department of Transportation Traffic Safety Facts 1999).
a. What proportion of states would you expect to have more than 65% of their traffic fatalities from drunk driving?
b. What proportion of deaths due to drunk driving would you expect to be at the 25th percentile of this distribution?

24. The table at the bottom of the page lists the means and standard deviations of several different normal distributions. For each distribution, calculate the probability of drawing a single Y value greater than the given threshold and the probability of drawing a value less than that threshold.

25. In European earwigs, the males sometimes have long pincers protruding from the end of their abdomens, as shown in the accompanying photo.

In graphs (a)–(c) we have plotted three frequency distributions of sample *means* of pincer lengths (in millimeters) based on random samples from an earwig population. One of the distributions shows means of samples based on $n = 1$, one shows means of samples based on $n = 2$, and one shows means of samples based on $n = 8$. Identify which frequency distribution corresponds to each sample size. Explain the basis for your decisions.

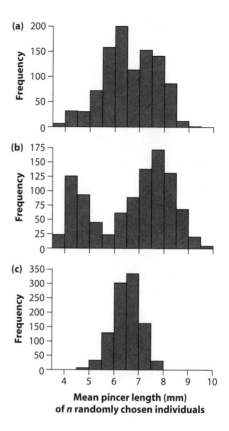

**Mean pincer length (mm)
of *n* randomly chosen individuals**

TABLE FOR PROBLEM 24

Mean	Standard deviation	Threshold	Pr[Y > threshold]	Pr[Y < threshold]
14	5	9		
15	3	18.5		
−23	4	−16		
14,000	5000	9000		

26. The crab spider, *Thomisus spectabilis*, sits on flowers and preys upon visiting honeybees, as shown in the photo at the beginning of the chapter. (Remember this the next time you sniff a wild flower.) Do honeybees distinguish between flowers that have crab spiders and flowers that do not? To test this, Heiling et al. (2003) gave 33 bees a choice between two flowers: one had a crab spider and the other did not. In 24 of the 33 trials, the bees picked the flower that had the spider. In the remaining nine trials, the bees chose the spiderless flower. With these data, carry out the appropriate hypothesis test, using an appropriate approximation to calculate *P*.

27. The following table lists the mean and standard deviation of several different normal distributions. In each case, a sample of 20 individuals was taken, as well as a sample of 50 individuals. For each sample, calculate the probability that the mean of the sample was less than the given value.

TABLE FOR PROBLEM 27

Mean	Standard deviation	Value	$n = 20$: $\Pr[\bar{Y} > \text{value}]$	$n = 50$: $\Pr[\bar{Y} > \text{value}]$
−5	5	−5.2		
10	30	8.0		
−55	20	−61.0		
12	3	12.5		

Controls in medical studies

I n 1994, researchers reported on the results of a study describing the effects of a new "wonder" drug (Lanza et al. 1994). The drug, lansoprazole, was intended to treat ulcers, and the study showed that 88% of the people who were treated with this drug got better within four weeks. Surely this was a remarkable achievement!

Another group of people were followed in the study, however. At the beginning of the study, participants who had ulcer problems were randomly sorted into two groups. One group received the new medication as planned, but the other group received a chemically inert "sugar pill" instead. The group who received no pharmacolog- ically active medication also improved over the course of the study—in fact, over 40% of the ulcer sufferers in this control group improved over the same four-week period.

In this particular study, the patients treated with the new drug did indeed get better faster than those who were not treated. But when compared with the control group, the drug was not nearly as effective as it appeared to be when looked at by itself. Of the 88% who improved after taking the drug, over 40% would have improved even without the medi- cation. How could this be?

> **I'm addicted to placebos. I could quit but it wouldn't matter.**
>
> —Steven Wright

There are several reasons. First, for most medical conditions, patients tend to get better over time anyway.[1] Most diseases are neither lethal nor permanent. We tend to go to the doctor when we are feeling at our worst, and therefore the odds are that we would soon start to improve after our worst days anyway, even without a new wonder drug. Second, humans (or at least many of them) like to please others, so there is a tendency to tell doctors that the treatment is more effective than it actually is. In both groups of the study, participants may have described an improvement in their condition, even if this was an exaggeration. Finally, being treated by a physician has benefits that go beyond the specifics of a particular drug. A doctor may suggest a new diet or advocate increased rest, for example, and participants in drug trials see doctors more often than they otherwise would and may therefore improve.

One particularly interesting form of ben- efit from treatment is psychological. In some cases, simply the knowledge of being treated may be sufficient to improve a person's con- dition. This is the so-called **placebo effect**, an improvement in a medical condition that results from the psychological effects of

---

1. Hence the old medical adage: "The common cold will go away in a week with proper treatment, but it will take seven days if left untreated."

medical treatment.[2] A sugar pill is a placebo, designed to mimic all the conditions of the medical treatment *except* for the pharmacological effects of the new drug itself.

Placebo effects are well documented for pain relief, but they are more questionable for other kinds of diseases (Hróbjartsson and Gøtzsche 2001). Placebo effects are typically larger for illnesses in which the response variable is subjectively reported by the patient. Placebo effects are smaller or nonexistent in studies that report on more objectively measured variables. The fact that the improvements are subjective does not mean that they are not real; pain, for example, is subjectively experienced but is nonetheless a real condition. In fact, MRI studies have shown that the placebo effect for pain has a neurological basis (Wager et al. 2004).

Even conditions requiring surgical interventions can show improvement without treatment. Studies of human surgery that include a "sham surgery" treatment—making surgical incisions without the specific treatment—are rare, for ethical reasons. In many cases, though, sham surgeries have been associated with real improvements in medical conditions. For example, ligation of the mammary artery

to treat anginal pain was common in the middle part of the 20th century, with improvement rates of about two-thirds after this surgery. Subsequently, however, two studies showed that improvement rates in patients with this surgery were nearly identical to those for patients who received only a sham skin incision (Shapiro and Shapiro 1997).

What we have seen is that people can improve for a wide variety of reasons, even without specific treatment. It is therefore crucial that medical studies include control treatments, in which a randomized selection of participants are treated in every way identically to those receiving a new treatment, except for the treatment itself. These controls can take the form of sugar pills, but, by preference, control patients can receive the currently most effective treatment. In this way, all patients receive care, and if the new drug or treatment is better than the old one, we have evidence in support of switching the most advisable treatment. When we report the results of a study, we want to be able to say that there is an effect, "all else being equal." The goal of careful experimental design is to make all else equal.

---

2. "Placebo" is from the Latin for "I will please." In Chaucer's time, this word was used for a sycophant or flatterer who would say whatever would please the listener rather than the truth. The word was then adopted by 19th-century physicians to refer to a treatment given to please the patient rather than to cure him or her. Slowly, doctors began to realize that in fact placebos had a therapeutic effect that had to be controlled for in medical trials.

Stalk-eyed flies

11 Inference for a normal population

We learned in Chapter 10 that many variables in nature are approximately normally distributed. Most of the methods in this book are geared toward making inferences and testing hypotheses about variables that have a normal distribution in the population. In this chapter, we begin the analysis of normally distributed measurements, a theme that will continue for much of the rest of the book.

We start by discussing estimation of the mean, including calculations for exact confidence intervals. Next we describe the simplest hypothesis test on a normal variable, the "one-sample *t*-test," which lets us ask whether the measurements of a sample of data are consistent with a hypothesized value for the population mean. We finish the chapter by discussing how to make statistical statements about the standard deviation.

303

11.1 The *t*-distribution for sample means

The sampling distribution of a statistic is the probability distribution of all the values for a statistic that we might obtain when we sample the population. In this section, we describe the sampling distribution of a statistic called *t*. This number will allow us to use data to calculate confidence limits and carry out hypothesis tests about the means of populations.

Student's *t*-distribution

Recall from Section 10.5 that the sampling distribution for the sample mean $\overline{Y}$ is a normal distribution if the variable Y is itself normally distributed. As a result, we could use the Z-standardization to calculate probabilities for $\overline{Y}$ under the normal curve:

$$Z = \frac{\overline{Y} - \mu}{\sigma_{\overline{Y}}},$$

where $\sigma_{\overline{Y}}$ is the standard error of the sample mean, the standard deviation of the sampling distribution of $\overline{Y}$.

In general, however, we can't just apply the Z-standardization to $\overline{Y}$-values calculated from real data. This is because to calculate Z we need to know $\sigma_{\overline{Y}}$, yet this is almost never possible. However, we can use the estimate of the standard error:

$$SE_{\overline{Y}} = \frac{s}{\sqrt{n}},$$

which was first discussed in Section 4.2. The numerator is the sample standard deviation (s), which is our best estimate of the true standard deviation (σ). We will usually call $SE_{\overline{Y}}$ simply "the standard error of the mean," even though it is really only an estimate of the true standard error.

Substituting $SE_{\overline{Y}}$ for $\sigma_{\overline{Y}}$ in the formula for Z leads to a related quantity called **Student's *t***:

$$t = \frac{\overline{Y} - \mu}{SE_{\overline{Y}}}.$$

This is called "Student's *t*" after its inventor,[1] but we will usually refer to it as simply *t*. While the formula for *t* resembles that for Z, the important difference is that the sampling distribution for this statistic is not the normal distribution. Instead, *t* has a *t*-distribution. $SE_{\overline{Y}}$ is not a constant like $\sigma_{\overline{Y}}$ but is a variable, varying by chance from sample to sample (because *s* itself changes from sample to sample). Therefore, the distribution of *t* is not

---

1. The statistic *t* is called Student's *t* after the nom de plume of the man who first discovered its properties, "Student." (He was more clever at statistics than pseudonyms.) "Student" in reality was William Gosset, an employee of the Guinness Brewing Company in Dublin. Gosset used a pseudonym because Guinness prohibited its employees from publishing, following the unauthorized release of some brewing secrets a few years earlier by another employee. In his honor, Guinness was appointed the official beverage of this book.

the same as *Z*. Substituting $SE_{\overline{Y}}$ for $\sigma_{\overline{Y}}$ adds sampling error to the quantity *t*. As a result, the sampling distribution of *t* is wider than the standard normal distribution. As the sample size increases, *t* becomes more like *Z*.

> The difference between the sample mean and the true mean $(\overline{Y} - \mu)$, divided by the estimated standard error $(SE_{\overline{Y}})$, has a Student's *t*-distribution with $n - 1$ degrees of freedom.

The sample size determines the number of degrees of freedom (*df*) of the *t*-distribution. The degrees of freedom specify which particular version of the *t*-distribution we need. With a *t*-distribution, we must estimate a parameter from the data—the standard deviation—before calculating *t*. As a result, the number of degrees of freedom for *t*, when applied to inference about one sample, is one less than the number of independent data points:

$$df = n - 1.$$

Consider the *t*-distribution for a sample size of five. With five individuals in a random sample, there are four degrees of freedom. Figure 11.1-1 plots the *t*-distribution with four degrees of freedom in blue; for comparison, the standard normal distribution is shown in red. In most respects, the *t*-distribution is similar to the standard normal distribution. It is symmetric around a mean of zero, is roughly bell shaped, and it has tails that fall off toward plus infinity and minus infinity.

FIGURE 11.1-1
Student's *t*-distribution with four degrees of freedom (*in blue*), compared with the standard normal distribution (*in red*). The two distributions are similar, though not identical. The tails of the *t*-distribution have more probability than the normal distribution.

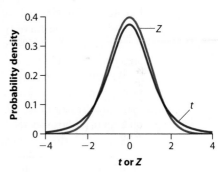

The *t*-distribution, however, is fatter in the tails than the standard normal distribution. The difference in the tails is crucial, because it is the tails that matter most when calculating confidence intervals and testing hypotheses.

For example, 95% of the area under the curve of the *t*-distribution with 4 *df* is between −2.78 and 2.78; the remaining 5% lies under the tails outside this range (we explain in the next subsection how we obtained this number). In other words, in 95% of repeated random samples of size $n = 5$ measurements from a normal population, the resulting $\overline{Y}$ will fall within 2.78 estimated standard errors of the true population mean (μ). With the standard normal (*Z*) distribution, on the other hand, 95% of the area under the curve lies between −1.96 and 1.96; the remaining 5% lies beyond these extreme values. The larger range of values of t, compared to Z, results from the uncertainty about the true value of the standard error.

Finding critical values of the t-distribution

Where did this value of 2.78 come from? The value 2.78 is the "5% critical value" of the t-distribution having $df = 5 - 1 = 4$ degrees of freedom. The 5% refers to the percentage of the area in the tails of the t-distribution (Figure 11.1-2). Every t-distribution has its own critical 5% t-value, depending on the number of degrees of freedom. Once we know the number of degrees of freedom, we can find the critical value by using a computer program or by using Statistical Table C in the back of this book.

FIGURE 11.1-2
The critical value of the t-distribution that confines a total of 5% of the area under the curve to its two tails, 2.5% to each side. With four degrees of freedom, this critical value is $t = 2.78$.

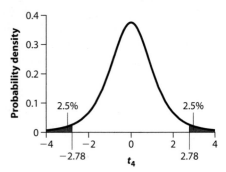

We use the symbol $t_{0.05(2),df}$ to indicate the 5% critical t-value of a t-distribution having "df" degrees of freedom. In this notation, the 0.05 stands for the fraction of the area under the curve lying in the tails of the distribution. The "(2)" indicates that the 5% area is divided between the two tails of the t-distribution—that is, 2.5% of the area under the curve lies above $t_{0.05(2),df}$ and 2.5% lies below $-t_{0.05(2),df}$.

We can use an excerpt from Statistical Table C, depicted in Table 11.1-1, to show how to find the 5% critical value for a t-distribution with four degrees of freedom. First find the row in the table corresponding to four degrees of freedom. Then find the column that corresponds to $\alpha(2) = 0.05$. The corresponding cell contains the number 2.78, which is the critical value we are looking for.

TABLE 11.1-1 Critical values of the t-distribution. Excerpted from Statistical Table C.

df	$\alpha(2) = 0.20$ $\alpha(1) = 0.10$	$\alpha(2) = 0.10$ $\alpha(1) = 0.05$	$\alpha(2) = 0.05$ $\alpha(1) = 0.025$	$\alpha(2) = 0.02$ $\alpha(1) = 0.01$	$\alpha(2) = 0.01$ $\alpha(1) = 0.005$
1	3.08	6.31	12.71	31.82	63.66
2	1.89	2.92	4.30	6.96	9.92
3	1.64	2.35	3.18	4.54	5.84
4	1.53	2.13	2.78	3.75	4.60
5	1.48	2.02	2.57	3.36	4.03
...	...	...	...	...	...

The other columns in Table 11.1-1 indicate the critical values corresponding to tail probabilities of $\alpha(2) = 0.20, 0.10, 0.02$, and 0.01 under the curve for the t-distribution. Notice, though, that the column headings in Table 11.1-1 contain $\alpha(1)$ designations, too. These values, such as $\alpha(1) = 0.025$ in the middle column of the table, indicate areas under the curve at only one tail of the t-distribution (Figure 11.1-3), and are needed when carrying out one-sided hypothesis tests.

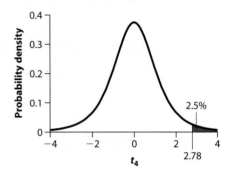

FIGURE 11.1-3
The critical value corresponding to 2.5% of the area under the curve at only one tail of the t-distribution.

In the remainder of this chapter, we use the t-distribution to compute exact confidence intervals for the population mean and to test hypotheses about the population mean.

11.2   The confidence interval for the mean of a normal distribution

As we discussed in Section 4.3, a confidence interval is a very useful way to express the precision of an estimate of a parameter. In that section, we gave an approximate confidence interval for the mean using the two standard errors (2SE) rule of thumb, but we can now do even better. We can use the t-distribution to calculate a more accurate confidence interval for the mean of a population having a normal distribution. Example 11.2 illustrates the appropriate method.

EXAMPLE   Eye to eye

11.2   The stalk-eyed fly, *Cyrtodiopsis dalmanni*, is a bizarre-looking insect from the jungles of Malaysia. Its eyes are at the ends of long stalks that emerge from its head, making the fly look like something from the cantina scene in *Star Wars*. These eye stalks are present in both sexes, but they are particularly impressive in males. The span of the eye stalk in males enhances their attractiveness to females as well as their success in battles against other males. The span, in millimeters, from one eye to the

other was measured in a random sample of nine male stalk-eyed flies.[2] The data are as follows:

$$8.69 \quad 8.15 \quad 9.25 \quad 9.45 \quad 8.96 \quad 8.65 \quad 8.43 \quad 8.79 \quad 8.63.$$

We can use these measurements to estimate the mean eye span in the fly population, and to quantify the uncertainty of our estimate using a 95% confidence interval. Assume that eye span has a normal distribution in the population.

The 95% confidence interval for the mean

The mean and standard deviation of the eye-span sample are

$$\bar{Y} = 8.778 \text{ and } s = 0.398.$$

(Review Section 3.1 if you don't remember how to do these calculations.) How precise is this estimate of the population mean? Let's describe the precision by calculating a 95% confidence interval for the mean. In Section 4.2, we used the 2SE rule of thumb to calculate this interval, but here in Section 11.2 we will use the t-distribution to give a more accurate formula. To do so, though, we must use the fact, learned in Section 11.1, that the standardized difference $(\bar{Y} - \mu)/\mathrm{SE}_{\bar{Y}}$ has a t-distribution with $df = n - 1$, assuming that Y is normally distributed. This means that in 95% of random samples from a normal distribution, the standardized difference will lie between $-t_{0.05(2), df}$ and $t_{0.05(2), df}$:

$$-t_{0.05(2), df} < \frac{\bar{Y} - \mu}{\mathrm{SE}_{\bar{Y}}} < t_{0.05(2), df}.$$

Rearranging this equation shows that, in 95% of random samples from a normal distribution, $\bar{Y} \pm t_{0.05(2), df}\, \mathrm{SE}_{\bar{Y}}$ will bracket the population mean:

$$\bar{Y} - t_{0.05(2), df}\, \mathrm{SE}_{\bar{Y}} < \mu < \bar{Y} + t_{0.05(2), df}\, \mathrm{SE}_{\bar{Y}}.$$

This is the exact 95% confidence interval for the mean. It is similar to the 2SE rule of thumb described in Section 4.2, except that we use the critical value from the t-distribution in the formula rather than "2."

> In 95% of random samples from a normal distribution, the interval from $\bar{Y} - t_{0.05(2), df}\, \mathrm{SE}_{\bar{Y}}$ to $\bar{Y} + t_{0.05(2), df}\, \mathrm{SE}_{\bar{Y}}$ will bracket the population mean. This interval is the 95% confidence interval of the mean.

Let's calculate the 95% confidence interval for mean eye span in male stalk-eyed flies. To begin, we'll need the standard error of the mean:

$$\mathrm{SE}_{\bar{Y}} = \frac{s}{\sqrt{n}} = \frac{0.398}{\sqrt{9}} = 0.133.$$

---

2. Data provided by Sam Cotton and Kevin Fowler, University College, London.

We also need the degrees of freedom to get the correct t-statistic. The sample size is $n = 9$, so the corresponding t has eight degrees of freedom. From Statistical Table C,

$$t_{0.05(2),8} = 2.31.$$

Notice that this $t = 2.31$ is greater than the 1.96 we would have gotten from the normal distribution, reflecting the fatter tails of the t-distribution. Putting all of the numbers from the eye-stalk data into the confidence interval equation, we get

$$8.778 - (2.31 \times 0.133) < \mu < 8.778 + (2.31 \times 0.133),$$

which yields

$$8.47 \text{ mm} < \mu < 9.08 \text{ mm}.$$

Thus, the 95% confidence interval for the mean of the eye span in this species, calculated from this particular random sample, is from 8.47 mm to 9.08 mm.

We do not know for certain whether the population mean eye span lies between 8.47 and 9.08. All we can say is that the 95% confidence interval for the mean will capture the population mean in 95% of random samples.

The 99% confidence interval for the mean

There's nothing special about 95%; it has just come to be adopted as the conventional level for a confidence interval. We can calculate a confidence interval for any significance level. The more general formula for a $100(1 - \alpha)\%$ confidence interval for the mean is

$$\overline{Y} - t_{\alpha(2),df}\text{SE}_{\overline{Y}} < \mu < \overline{Y} + t_{\alpha(2),df}\text{SE}_{\overline{Y}}.$$

After 95%, the 99% confidence interval is the next most popular. Its principal advantage is that it provides better coverage than the 95% interval—it covers the population mean in 99% of random samples.

Let's calculate a 99% confidence interval of the mean for the stalk-eyed fly data. All of the numbers are the same as for the 95% interval, except that now we need $t_{0.01(2),8}$, because α is now $1 - 0.99 = 0.01$. To use Statistical Table C, we go to the row for $df = 8$ and then over to the column for $\alpha(2) = 0.01$. Confirm for yourself that this yields $t_{0.01(2),8} = 3.36$. Thus, the formula for the 99% confidence interval is

$$\overline{Y} - t_{0.01(2),df}\text{SE}_{\overline{Y}} < \mu < \overline{Y} + t_{0.01(2),df}\text{SE}_{\overline{Y}}.$$

When we apply this formula to the stalk-eyed fly data, we get

$$8.778 - (3.36 \times 0.133) < \mu < 8.778 + (3.36 \times 0.133),$$

which yields

$$8.33 \text{ mm} < \mu < 9.22 \text{ mm}.$$

The 99% confidence interval is broader than the 95% interval, because we have to include more possibilities to achieve the higher probability of covering the true mean.

11.3 The one-sample *t*-test

Many methods have been developed to test hypotheses about the means of populations. Because these tests are often based on the *t*-distribution, many are called *t*-tests. In this section we learn about the **one-sample *t*-test**, which is designed to compare the mean from a sample of individuals with a value for the population mean proposed in the null hypothesis. The null hypothesis for a one-sample *t*-test is that the true mean is equal to a specific value, μ_0. The null and alternative hypothesis statements are as follows.

H_0: The true mean equals μ_0.
H_A: The true mean does not equal μ_0.

Suppose, for example, we have a null hypothesis that the mean of a variable in a population is $\mu_0 = 0$. We can then use the one-sample *t*-test to determine whether the $\overline{Y}$ that we calculate from a sample is sufficiently different from zero to warrant rejection of the null hypothesis.

> The *one-sample t-test* compares the mean of a random sample from a normal population with the population mean proposed in a null hypothesis.

The test statistic for the one-sample *t*-test is *t* (no surprise), and it is calculated by

$$t = \frac{\overline{Y} - \mu_0}{\mathrm{SE}_{\overline{Y}}},$$

where μ_0 is the population mean proposed by the null hypothesis, $\overline{Y}$ is the sample mean, and $\mathrm{SE}_{\overline{Y}}$ is the sample standard error of the mean. The sampling distribution of this test statistic under H_0 is the *t*-distribution having $n - 1$ degrees of freedom. The *P*-value for the test can be computed by comparing the observed *t* with the Student's *t*-distribution. Example 11.3 shows you how.

EXAMPLE Human body temperature

11.3 Normal human body temperature, as kids are taught in North America, is 98.6°F. But how well is this supported by data? Researchers obtained body-temperature measurements on randomly chosen healthy people (Shoemaker 1996). The data for 25 of those people are as follows:

98.4	98.6	97.8	98.8	97.9
99.0	98.2	98.8	98.8	99.0
98.0	99.2	99.5	99.4	98.4
99.1	98.4	97.6	97.4	97.5
97.5	98.8	98.6	100.0	98.4

The body temperatures are not all identical to 98.6°F, but are the measurements consistent with a population mean of 98.6°F?

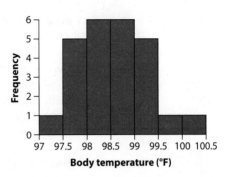

FIGURE 11.3-1
The frequency distribution of body temperatures in a sample of 25 individuals.

Body temperature is approximately normally distributed, as shown in Figure 11.3-1, so a one-sample *t*-test is appropriate. Our null and alternative hypotheses are as follows.

H_0: The mean human body temperature is 98.6°F.

H_A: The mean human body temperature is different from 98.6°F.

The null hypothesis is not arbitrary, because we are testing the common wisdom that the mean is $\mu_0 = 98.6°F$. The test is two-sided. A sample mean much higher than 98.6°F or a sample mean much lower than 98.6°F would both lead to rejection of the null hypothesis.

The sample mean body temperature is

$$\overline{Y} = 98.524,$$

and the standard deviation is

$$s = 0.678.$$

The standard error is

$$SE_{\overline{Y}} = 0.678/\sqrt{25} = 0.136.$$

We can now use the test statistic *t* to measure how different the observed mean $\overline{Y}$ is from 98.6. If the sample mean perfectly matches the hypothesized value, then *t* would equal zero. Some difference is expected just by chance, but if the null hypothesis is true, then *t* should have a *t*-distribution with $n - 1$ *df*. If the difference between $\overline{Y}$ and 98.6 is too large, then the observed *t* will lie out in one of the tails of the *t*-distribution. Hence, by comparing this observed *t* to the *t*-distribution, we assess whether 98.6 is a reasonable fit to the data. If the null hypothesis is not a good fit, then we reject H_0 and conclude that the population mean μ does not equal 98.6.

The *t*-statistic is

$$t = \frac{\overline{Y} - \mu_0}{SE_{\overline{Y}}} = \frac{98.524 - 98.6}{0.136} = -0.56.$$

Under the null hypothesis, the sampling distribution of *t* is the *t*-distribution with $n - 1$ degrees of freedom. There are 25 individuals in our sample, so $df = 25 - 1 = 24$.

A perfect match between the null hypothesis and the data would mean that $\overline{Y} = \mu_0$, and t would equal zero. Our question becomes: Is $t = -0.56$ sufficiently different from zero that we should reject the null hypothesis?

The P-value is the probability of obtaining a result as extreme as, or more extreme than, $t = -0.56$, assuming that the null hypothesis is true. This probability is the area under the t-distribution shown in Figure 11.3-2. Both tails of the t-distribution are included in the probability, because the test is two-tailed. The P-value is

$$P = \Pr[t < -0.56] + \Pr[t > 0.56].$$

Because the t-distribution is symmetric around zero, this is the same as

$$P = 2 \Pr[t > 0.56].$$

Using a computer, we find that

$$P = 0.58.$$

Since P is greater than 0.05, we do *not* reject the null hypothesis. Our results would not surprise someone who believed that mean body temperature really is 98.6°F.

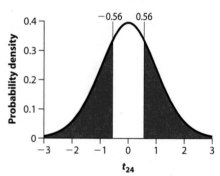

FIGURE 11.3-2
The t-distribution with 24 degrees of freedom. Shaded areas include all values less than -0.56 and greater than 0.56. These are the values as extreme as, or more extreme than, the t-statistic calculated from the data.

The same conclusion is obtained if we use Statistical Table C for the t-distribution. The probability that we need, $\Pr[t > 0.56 \text{ or } t < -0.56]$, is not given directly in Statistical Table C, but we can find the critical values of t for different α values. Remember that the critical value of a test statistic marks off the point or points in the distribution that have a certain probability α in the tails of the distribution. Values of the test statistic that are further in the tails have P-values lower than α. In other words, if the value of t is further from zero than the critical value, then we can reject the null hypothesis. If the calculated value of t is closer to zero than the critical value, then we cannot reject the null hypothesis.

For the present example, we find the row in Statistical Table C corresponding to 24 degrees of freedom and move to the right to find the critical t-value corresponding to the column $\alpha(2) = 0.05$. Confirm for yourself that this value is

$$t_{0.05(2),\,24} = 2.06.$$

This critical value defines 5% in the tails of the t-distribution, with 2.5% in each tail for a two-tailed test (Figure 11.3-3). In other words, 5% of the area under the *t*-distribution with $df = 24$ falls outside either -2.06 or 2.06. The *t*-value of -0.56 that we calculated from the data occurs inside this range. The observed *t*-statistic does not fall within one of the tails. Therefore, $P > 0.05$, and we fail to reject the null hypothesis.

FIGURE 11.3-3
Five percent of the area under the curve of the *t*-distribution with 24 degrees of freedom lies in the two tails beyond -2.06 and 2.06. The *t*-value calculated from the body temperature data ($t = -0.56$) lies closer to the value expected by the null hypothesis. Therefore, the null hypothesis is *not* rejected with these data.

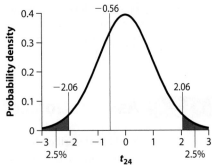

At a significance level of $\alpha = 0.05$, we would not reject the null hypothesis based on this sample of body temperatures. In other words, these data would be reasonably likely if the null hypothesis were true. With this sample, we cannot reject the view that mean body temperature in the sampled human population is 98.6°F.

Does this result imply that the common wisdom about human body temperature is correct? Well, not necessarily: the null hypothesis might still be false, but we may have lacked sufficient power to detect the difference. What range of values for mean body temperature is most plausible given the data? To answer this question, we calculated the 95% confidence interval for the mean using the methods described in Section 11.2. We obtained

$$98.24°F < \mu < 98.80°F.$$

Check the calculations for yourself. This 95% confidence interval for the mean puts fairly narrow bounds on the mean temperature in the population. The value 98.6 is within the 95% confidence interval of the mean for these data, but slightly different values for the mean body temperature between about 98.2 and 98.8 are also consistent with the data.

The effects of larger sample size: body temperature revisited

In the body-temperature study in Example 11.3, we used a sample with only 25 data points to do our calculations. A larger sample is available, that includes 130 individuals taken from the same population. Using this larger sample, the estimated mean was 98.25°F with $s = 0.733$°F. A *t*-test using this data set finds $t = -5.44$. (Check this value for yourself, for practice.) This *t* corresponds to $P = 0.000016$, so it is very unlikely that the true value of human body temperature is the canonical 98.6°F. This value is clearly rejected as the mean body temperature. The confidence interval for the mean, using the larger data set, is

$$98.12 < \mu < 98.38.$$

Why should the larger sample get a different answer than the subset? With a larger sample size, sampling error in the estimate of the mean tends to decrease. Even though the smaller sample by chance had an estimated mean not very different from the larger sample, the larger sample led to an estimate with considerably narrower bounds. Thus, a larger sample is more likely to reject a false null hypothesis.[3] Assuming that the result from the analysis of the larger sample is true, our smaller sample gave us a Type II error (i.e., a failure to detect a false null hypothesis).

11.4   Assumptions of the one-sample *t*-test

Methods described for calculating confidence intervals for the mean, and for testing a population mean using the one-sample *t*-test, make only two assumptions:

- The data are a random sample from the population. (This assumption is shared by every method of statistical inference in this book.)
- The variable is normally distributed in the population.

An excellent way to assess the assumption of normality is to examine the frequency distribution of the data using a histogram or other graphical method and look for skew, bimodality, or other departures from the normal distribution. Chapter 13 discusses methods to investigate this assumption in more detail.

Few variables in biology show an exact match to the normal distribution, but an exact match is not essential. Under certain conditions (discussed in Chapter 13), the *t*-test and the confidence interval calculations are robust to violations of the assumption of normality. A method is **robust** if the answer it gives is not sensitive to modest departures from the assumptions. Chapter 13 gives a more thorough account of this issue. Until then, we will consider only data that meet the assumption of normality reasonably well.

11.5   Estimating the standard deviation and variance of a normal population

Up to now, all of our attention has been focused on estimating and testing the population mean. But the standard deviation (or variance) is also interesting in many studies. For example, male stalk-eyed flies (see Example 11.2) often battle each other by pair-

---

3. How did common wisdom get such a basic value wrong for so long? A hint comes from the fact that 98.6°F corresponds exactly to 37°C. The original measures of human body temperature were done on the Celsius scale and rounded to the nearest degree. The estimated mean value of body temperature, 98.25°F, is 36.8°C, which when rounded off gives 37°C. Later, this rounded figure of 37°C was converted back into Fahrenheit, yielding the erroneous 98.6°F.

ing head to head and staring at each other for an extended period. The male with the longer eye stalks usually wins these staring matches, thus giving him greater access to females. When paired males have similar eye spans, the outcomes of the staring matches are difficult to predict. It is therefore interesting to know the standard deviation of eye span in the population, because it predicts the typical difference in eye span of any two males.

We must use the sample standard deviation s to estimate the population standard deviation σ. Can we say how precise our estimate s is? We can, provided that the data are from a population having a normal distribution.

Confidence limits for the variance

It is easiest first to work with the *variance*, the square of the standard deviation. The population parameter is σ^2 and the sample estimate is s^2. For the stalk-eyed fly data, the estimate of σ^2 is $s^2 = 0.1586$ mm^2.

To calculate a confidence interval for the sample variance, all we need to know is its sampling distribution. Theory tells us that if a variable Y has a normal distribution, then the sampling distribution of the quantity

$$\chi^2 = (n-1)s^2/\sigma^2$$

is the χ^2 distribution with $n - 1$ degrees of freedom. We encountered the χ^2 distribution in Chapters 8 and 9, where we used it to approximate the null sampling distribution of a goodness-of-fit test statistic. Here the same theoretical distribution is useful for calculating the precision of estimates of population variability.

A typical χ^2 distribution is shown in Figure 11.5-1. Notice that χ^2 is always zero or positive, and it extends all the way to positive infinity. This fits with the properties of a sample variance, which cannot be less than zero but can be indefinitely large. Notice also that, unlike the normal distribution, the χ^2 distribution is not symmetric around its mean.

FIGURE 11.5-1
The χ^2 distribution with critical values for a 95% confidence interval indicated. Ninety-five percent of the area under the χ^2 curve falls between the two critical values, and 5% lies beyond them.

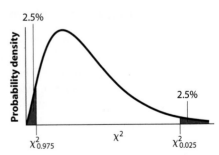

These features make it possible to determine confidence intervals for the variance. The $1 - \alpha$ confidence interval is

$$\frac{df\, s^2}{\chi^2_{\alpha/2,\, df}} < \sigma^2 < \frac{df\, s^2}{\chi^2_{1-\alpha/2,\, df}},$$

where the values $\chi^2_{\alpha/2,\,df}$ and $\chi^2_{1-\alpha/2,\,df}$ represent the critical values of the χ^2 distribution corresponding to the upper and lower tails in Figure 11.5-1. The area under the χ^2 curve to the right of $\chi^2_{\alpha/2,\,df}$ is $\alpha/2$. The other $\alpha/2$ is to the left of $\chi^2_{1-\alpha/2,\,df}$. Because the χ^2 distribution is not symmetric, we are forced to calculate the left and right tails separately. The critical values are available in Statistical Table A at the back of this book. Let's find the 95% confidence interval of the variance for eye span in male stalk-eyed flies. We have already calculated s^2 to be 0.1587. The number of degrees of freedom is just one less than the number of data points, so $df = 9 - 1 = 8$, as before. All that's left is finding the values of $\chi^2_{\alpha/2,\,df}$ and $\chi^2_{1-\alpha/2,\,df}$, which are $\chi^2_{0.025,8}$ and $\chi^2_{0.975,8}$, respectively, for the stalk-eyed fly study. Looking in Statistical Table A under eight degrees of freedom, we find the χ^2 value that has 2.5% to the right of it is $\chi^2_{0.025,8} = 17.53$, and the χ^2 value that has 2.5% probability to the left (i.e., 97.5% of the area to the right) of it is $\chi^2_{0.975,8} = 2.18$. Thus, the 95% confidence interval for the variance of this distribution is given by

$$\frac{df\, s^2}{\chi^2_{\alpha/2,\,df}} < \sigma^2 < \frac{df\, s^2}{\chi^2_{1-\alpha/2,\,df}}$$

$$\frac{8(0.1587)}{17.53} < \sigma^2 < \frac{8(0.1587)}{2.18}$$

$$0.072 < \sigma^2 < 0.582.$$

Any value of the population variance between 0.072 mm and 0.582 mm is reasonably plausible, based on these data. The true variance would be within the calculated interval in 95% of samples.

The estimate of the variance (0.1587) is closer to the lower bound (0.0724) than the upper bound (0.5824) because the χ^2 distribution is asymmetrical. However, a 95% confidence interval for the variance based on a random sample has an equal chance of falling either below or above the true variance.

Confidence limits for the standard deviation

An approximate 95% confidence interval for the standard deviation can be obtained by taking the square roots of the upper and lower limits of the 95% confidence interval for the variance. Therefore, the 95% confidence interval for the standard deviation of eye span is given by

$$\sqrt{0.0724} < \sigma < \sqrt{0.5824}$$

or

$$0.27 < \sigma < 0.76.$$

This interval surrounds the sample standard deviation of $s = 0.40$.

Assumptions

The assumptions of the method for calculating confidence intervals for the variance and standard deviation are the same as for confidence intervals for the mean—namely, the sample must be a random sample from the population, and the variable must have a normal distribution in the population. Unfortunately, the formulas for the confidence intervals for variance and standard deviation are very sensitive to the assumption of normality. The method is not robust to departures from this assumption.

11.6 Summary

- If a variable Y is normally distributed in the population with mean μ and we have random samples of n individuals, then the sample means $\overline{Y}$ are also normally distributed, with mean equal to μ (the same as the true mean of Y) and standard error $\sigma_{\overline{Y}} = \sigma / \sqrt{n}$, where σ is the true standard deviation in the population.
- The estimated standard error of the distribution of sample means is $\mathrm{SE}_{\overline{Y}} = s / \sqrt{n}$.
- If the population is normally distributed, then the standardized sample mean
  $$t = \frac{\overline{Y} - \mu}{\mathrm{SE}_{\overline{Y}}}$$ has a Student's t-distribution with $n - 1$ degrees of freedom.
- The t-distribution can be used to calculate a confidence interval for the mean.
- A one-sample t-test compares the sample mean with μ_0, a specific value for the population mean proposed in a null hypothesis. Under the null hypothesis that the population mean is equal to μ_0, the sampling distribution of the test statistic
  $$t = \frac{\overline{Y} - \mu_0}{\mathrm{SE}_{\overline{Y}}}$$ is a t-distribution with $n - 1$ degrees of freedom.
- The confidence interval for the variance is based on the χ^2 distribution. Take the square root of the limits of the confidence interval for the variance to yield an approximate confidence interval for the standard deviation.
- The confidence intervals for mean and variance, as well as the one-sample t-test, assume that the data are randomly sampled from a population with a normal distribution.

11.7 Quick Formula Summary

Confidence interval for a mean

What does it assume? Individuals are chosen as a random sample from a population that is normally distributed.

Estimate: $\overline{Y}$

Parameter: μ

Degrees of freedom: $n-1$

Formula: $\bar{Y} - t_{\alpha(2),\,df}\,\mathrm{SE}_{\bar{Y}} < \mu < \bar{Y} + t_{\alpha(2),\,df}\,\mathrm{SE}_{\bar{Y}}$, where $\mathrm{SE}_{\bar{Y}} = s/\sqrt{n}$.
   This formula gives the $1-\alpha$ confidence interval.

One-sample *t*-test

What is it for? Compares the sample mean of a numerical variable to a hypothesized value, μ_0.

What does it assume? Individuals are randomly sampled from a population that is normally distributed.

Test statistic: t

Distribution under H$_0$: The t-distribution with $n-1$ degrees of freedom.

Formula: $t = \dfrac{\bar{Y} - \mu_0}{s/\sqrt{n}}.$

Confidence interval for variance

What does it assume? Individuals are chosen as a random sample from a normally distributed population.

Estimate: s^2

Parameter: σ^2

Degrees of freedom: $df = n-1$.

Formula: $\dfrac{df\,s^2}{\chi^2_{\alpha/2,\,df}} < \sigma^2 < \dfrac{df\,s^2}{\chi^2_{1-\alpha/2,\,df}}.$

PRACTICE PROBLEMS

1. **Calculation practice: Confidence interval for a mean and one-sample *t*-test.** As the world warms, the geographic ranges of species might shift toward cooler areas. Chen et al. (2011) studied recent changes in the highest elevation at which species occur. Typically, higher elevations are cooler than lower elevations. Below are the changes in highest elevation for 31 taxa, in meters, over the late 1900s and early 2000s. (Many taxa were surveyed, including plants, vertebrates, and arthropods.) Positive numbers indicate upward shifts in elevation, and negative numbers indicate shifts to lower elevations. The values are displayed in the accompanying figure.

58.9, 7.8, 108.6, 44.8, 11.1, 19.2, 61.9, 30.5
12.7, 35.8, 7.4, 39.3, 24.0, 62.1, 24.3, 55.3
32.7, 65.3, −19.3, 7.6, −5.2, −2.1, 31.0, 69.0
88.6, 39.5, 20.7, 89.0, 69.0, 64.9, 64.8

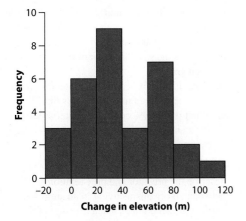

Change in elevation (m)

a. What is the sample size n?

b. What is the mean of these data points? Remember to give the units.

c. What is the standard deviation of elevational range shift? (Give the units as well.)

d. What is the standard error of the mean for these data?

e. How many degrees of freedom will be associated with a confidence interval and a one-sample t-test for the mean elevation shift?

f. What value of α is needed for a 95% confidence interval?

g. What is the critical value of t for this α and number of degrees of freedom?

h. What assumptions are necessary to use the confidence interval calculations in this chapter?

i. Calculate the 95% confidence interval for the mean using these data.

j. For the one-sample t-test, write the appropriate null and alternative hypotheses.

k. Calculate the test statistic t for this test.

l. What assumptions are necessary to do a one-sample t-test?

m. Describe the P-value for this test as accurately as you can.

n. Did species change their highest elevation on average?

2. **Calculation practice: Confidence interval for variance.** Refer to Practice Problem 1. Using the data, calculate the 95% confidence interval for the variance with the following steps.

a. What does the confidence interval refer to—the sample variance or the population variance?

b. What assumptions are necessary to apply the formula discussed in this chapter for the confidence interval for variance?

c. What is the sample variance of these data?

d. How many degrees of freedom are associated with these data for estimates of variance?

e. What is α for this analysis?

f. Find the critical value of χ^2 for $\alpha/2$?

g. What is the critical value of χ^2 for $1 - \alpha/2$?

h. Calculate the 95% confidence interval for the variance of the elevational range shift.

3. For each of the following, the mean, standard deviation, sample size, and desired confidence interval is given. Find the critical values of t required for a confidence interval of the mean.

a. $\overline{Y} = 14, s = 32, n = 12, 95\%$ confidence interval

b. $\overline{Y} = -23, s = 12, n = 32, 95\%$ confidence interval

c. $\overline{Y} = 144, s = 2.1, n = 101, 99\%$ confidence interval

d. $\overline{Y} = 3.21, s = 1.83, n = 23, 95\%$ confidence interval

e. $\overline{Y} = -152, s = 38, n = 8, 99\%$ confidence interval

4. Refer to Practice Problem 3. For each of the parts (a)–(e), find the two χ^2 critical values necessary to calculate a confidence interval of the variance.

5. Measurements of the distance between the canine tooth and last molar for 35 wolf upper jaws were made by a researcher. He found the 95% confidence interval for the mean to be 10.17 cm $< \mu <$ 10.47 cm and the 99% confidence interval to be 10.21 cm $< \mu <$ 10.44 cm. Without seeing the data, explain why he must have made a mistake.

6. Here are the data on wolf upper jaws from Practice Problem 5, in centimeters. There are 35 individuals in this data set. Assume that this variable is normally distributed.

10.2, 10.4, 9.9, 10.7, 10.3, 9.7, 10.3, 10.7, 10.1, 10.6, 10.3, 10.0, 10.2, 10.1, 10.3, 9.9,

9.7, 10.6, 10.4, 10.1, 10.6, 10.3, 10.3, 10.5, 10.2, 10.2, 10.5, 10.1, 11.2, 10.5, 10.3, 10.0, 10.3, 10.7, 11.1

a. Draw a graph to confirm that the frequency distribution of the data is roughly bell shaped.

b. What are the sample mean and the standard error of the mean for this data set? Provide an interpretation of this standard error.

c. Find the 95% confidence interval for the mean.

d. Find the 99% confidence interval for the variance.

e. Find the 99% confidence interval for the standard deviation.

7. In the data set on wolf upper jaws (Practice Problem 6), each measurement was actually the average of two measurements made on the left and right sides of the jaw of an individual wolf. Thus, a total of 70 measurements were made. Could we use $n = 70$ when calculating confidence intervals for the mean and variance? Why or why not?

8. Polyandry is the name given to a mating system in which females mate with more than one male in a breeding season. This mating system leads to competition for fertilization between sperm of different males. The prediction has been made that males in polyandrous populations should evolve larger testes than males in monogamous populations (where females mate with only one male), because larger testes produce more sperm. To test this prediction, researchers carried out an experiment on eight separate lines of yellow dung flies (Hosken and Ward 2001). In four of these lines, each female was mated with three males before laying eggs (the polyandrous populations). In the other four lines, the females mated only once. After 10 generations, the testes size (in cross-sectional area) was measured in each of these lines. The four monogamous lines had testes with areas of 0.83, 0.85, 0.82, and 0.89 mm². The polyandrous lines had testes areas of 0.96, 0.94, 0.99, and 0.91 mm².

a. Draw a graph to compare the testes areas of males in the two experimental treatments. What association is suggested?

b. Estimate the mean and standard deviation of the testes areas for both monogamous and polyandrous lines.

c. What is the standard error of the mean for each group?

d. What is the 95% confidence interval for the mean testes area in polyandrous lines?

e. What is the 99% confidence interval for the standard deviation of testes area among monogamous lines?

9. Community ecologists draw "food webs" to describe the predator and prey relationships between all organisms living in an area. A theoretical model predicts that a measure of the structure of food webs called "diet discontinuity" should be zero. Diet discontinuity is a measure of the relative numbers of predators whose prey are not ordered contiguously. Researchers have measured discontinuity scores for seven different food webs in nature. The values are given below (Cattin et al. 2004).

0.35, 0.08, 0.004, 0.08, 0.32, 0.28, 0.17

Assume that discontinuity in natural food webs has a normal distribution. Are the results consistent with the model's prediction of a zero mean discontinuity score? Carry out an appropriate hypothesis test.

10. As part of a larger study into the role of the hippocampus in memory, Fortin et al. (2004) devised a test that required rats to choose between two odors, one of which had previously been presented to them as the first in a series of odors. To validate their procedure, the researchers tested whether normal rats are able to remember and choose the odor previously presented. By chance, the rats should score only 50% on the test. But if they are able to remember, they should score better than 50%. Seven normal rats were taken through the protocol, and their scores on the memory test were on average 68.4%, with a standard deviation of 7.1%. Do the rats show the ability to perform this task at levels better than that expected by chance? State all necessary assumptions.

11. A four-year review at the Provincial Hospital in Alotau, Papua New Guinea (Barss 1984), found that about $1/40$ of their hospital admissions were injuries due to falling coconuts. If coconuts weigh on average 3.2 kg, and the upper bound of the 95% confidence interval is 3.5 kg, what is the lower bound of this confidence interval? Assume a normal distribution of coconut weights.

12. The following tables give the confidence intervals of either the standard deviation or variance from different samples. Provide the confidence interval of the other measure of spread (i.e., either standard deviation or variance).

Standard deviation	Variance
$2.22 < \sigma < 4.78$	
	$425.4 < \sigma^2 < 678.8$
$36.4 < \sigma < 59.6$	
	$185.8 < \sigma^2 < 279.0$

13. Can a human swim faster in water or in syrup? It is unknown whether the increase in the friction of the body moving through the syrup (slowing the swimmer down) is compensated by the increased power of each stroke (speeding the swimmer up). Finally, an experiment was done[4] in which researchers filled one pool with water mixed with syrup (guar gum) and another pool with normal water (Gettelfinger and Cussler 2004). They had 18 swimmers swim laps in both pools in random order. The data are presented as the relative speed of each swimmer in syrup (speed in the syrup pool divided by his or her speed in the water pool). If the syrup has no effect on swimming speed, then relative swim speed should have a mean of 1. The data, which are approximately normally distributed, are as follows:

1.08, 0.94, 1.07, 1.04, 1.02, 1.01, 0.95, 1.02, 1.08
1.02, 1.01, 0.96, 1.04, 1.02, 1.02, 0.96, 0.98, 0.99

$\Sigma(Y) = 18.21, \quad \Sigma(Y^2) = 18.4529.$

a. Draw a graph showing the frequency distribution. Why is this a good idea? What trend is suggested?

b. Test the hypothesis that relative swim speed in syrup has a mean of 1.

c. How uncertain are we about true relative swimming speed in syrup? Use the 99% confidence interval to find out.

ASSIGNMENT PROBLEMS

14. Astronauts increased in height by an average of approximately 40 mm (about an inch and a half) during the *Apollo–Soyuz* missions, due to the absence of gravity compressing their spines during their time in space. Does something similar happen here on Earth? An experiment supported by NASA measured the heights of six men immediately before going to bed, and then again after three days of bed rest (Styf et al. 1997). On average, they increased in height by 14 mm, with standard deviation 0.66 mm. Find the 95% confidence interval for the change in height after three days of bed rest.

15. Two different researchers measured the weight of two separate samples of ruby-throated hummingbirds from the same population. Each calculated a 95% confidence interval for the mean weight of these birds. Researcher 1 found the 95% confidence interval to be 3.12 g $< \mu <$ 3.48 g, while Researcher 2 found the 95% confidence interval to be 3.05 g $< \mu <$ 3.62 g.

a. Why would the two researchers get different answers?

b. Which researcher most likely had the larger sample?

c. Can you be certain about your answer in part (b)? Why or why not?

16. In the Northern Hemisphere, dolphins swim predominantly in a counterclockwise direction while sleeping. A group of researchers wanted to know whether the same was true for dolphins in the Southern Hemisphere (Stafne and Manger

4. After a rather large number of permits were applied for and granted.

2004). They watched eight sleeping dolphins and recorded the percentage of time that the dolphins swam clockwise. Assume that this is a random sample and that this variable has a normal distribution in the population. These data are as follows:

77.7, 84.8, 79.4, 84.0, 99.6, 93.6, 89.4, 97.2

a. What is the mean percentage of clockwise swimming for Southern Hemisphere dolphins?
b. What is the 95% confidence interval for the mean time swimming clockwise in the Southern Hemisphere dolphins?
c. What is the 99% confidence interval for the mean time swimming clockwise in the Southern Hemisphere dolphins?
d. What is the best estimate of the standard deviation of the percentage of clockwise swimming?
e. What is the median of the percentage of clockwise swimming?

17. Male koalas bellow[5] during the breeding season, but do females pay attention? Charlton et al. (2012) measured responses of estrous female koalas to playbacks of bellows that had been modified on the computer to simulate male callers of different body size. Females were placed one at a time into an enclosure while loudspeakers played bellows simulating a larger male on one side (randomly chosen) and a smaller male on the other side. Male bellows were played repeatedly, alternating between sides, over 10 minutes. Females often turned to look in the direction of a loudspeaker (invisible to her) during a trial. The following data measure the preference of each of 15 females for the simulated sound of the "larger" male. Preference was calculated as the number of looks toward the larger-male side minus the number of looks to the smaller-male side. Preference is positive if a female looked most often toward the larger male, and it is negative if she looked most often in the direction of the smaller male.

−2, 2, 6, 9, 13, 2, 5, 7, 2, −6, 4, 3, 2, 6, −6

a. Draw a graph to visualize the frequency distribution. What is the trend in female preference?
b. Do females pay attention to body size cues in simulated male sounds? Carry out a test, making all necessary assumptions.

18. The mating habits of threespine stickleback (a fish) have been studied intensively. One experiment examined whether fish are more likely to mate with a member of the opposite sex that was similar in body size to themselves, rather than with fish that were different in size (McKinnon et al. 2004). The mating preferences were measured in nine different fish populations, and the preference was measured by an index that is zero if the population shows no preference for mating by size, positive if the population contains fish that prefer to mate with fish of a different size, and negative if the fish mate preferentially with individuals of the same size. Notice that the independent data points here are the indices for each fish population. The nine indices are as follows:

−32.0, −29.8, −40.6, −90.8, −29.2, −28.8, −78.4, −59.2, −74.3

Assuming normality, test the hypothesis that, on average, sticklebacks do not prefer to mate differently by size. What can you conclude from these data?

19. Pit vipers (including rattlesnakes) have a pit organ located halfway between their eyes and nostrils. These organs detect the body heat of unlucky warm-blooded prey (mammals), but the snakes also use the pit organ to detect cooler spots in the environment to help in their thermoregulation. Researchers determined that western diamondback rattlesnakes had on average a 73% chance of moving into the right half of a Y-maze that was cooled to 30°C, instead of the left half which was at 40°C (Krochmal et al. 2004). Was this pattern a preference for the right side of the maze (that just happened to be cooled), or was it a direct response to the difference in heat? To test this, five snakes were put into the same maze several times individually, and the per-

5. https://soundcloud.com/new-scientist/male-koala-bellow

centage of trials in which they turned right was recorded when both sides were heated to 40°C. The average of the percentages for the five snakes was 47%, and the standard deviation was 13%.

a. Is there a preference for the right side of the maze when the temperature is equalized? Test the null hypothesis that the mean percentage of right turns is 50% in this population. Assume that the distribution of scores was normal.

b. Can you think of a way to design the experiment to test effects of temperature so that side-preferences, if present, do not affect the outcome?

20. In Seychelles warblers, young adult females known as subordinates sometimes hang around the territories of older birds. Sometimes these subordinates help feed the offspring of the older birds, and sometimes not. In one study, subordinate birds that did not help were genetically assessed to discover whether they were related to the offspring of the older birds (Richardson et al. 2003). A "relatedness" score was assigned to each subordinate, with a value of zero meaning no relationship to the offspring. Out of five subordinates examined, the mean relatedness was −0.05 with standard deviation 0.45 (relatedness can be negative if by chance sampled individuals are less related than the population as a whole). What is the 95% confidence interval for relatedness of unhelpful subordinates to the offspring?

21. Hurricanes Katrina and Rita caused the flooding of large parts of New Orleans, leaving behind large amounts of new sediment. Before the hur-

ricanes, the soils in New Orleans were known to have high concentrations of lead, which is a dangerous environmental toxin. Forty-six sites had been monitored before the hurricanes for soil lead content, as measured in mg/kg, and the soil from each of these sites was measured again after the hurricanes (Zahran et al. 2010). The data given below show the log of the ratio of the soil lead content after the hurricanes and the soil lead content before the hurricanes—we'll call this variable the "change in soil lead." (Therefore, numbers less than zero show a reduction in soil lead content after the hurricanes, and numbers greater than zero show increases.) This log ratio has an approximately normal distribution.

−0.83, −0.18, 0.14, −1.46, −0.48, −1.04, 0.25, −0.34, −0.81, −0.83, −0.60, 0.34, −0.75, 0.37, 0.26, 0.46, −0.03, −0.32, −0.53, −1.55, −0.90, −0.95, −0.13, −0.75, 0.59, −0.06, 0.39, −0.40, −0.84, −0.56, 0.44, 0.18, 0.28, −0.41, −0.26, 0.64, −0.51, −0.36, 0.49, 0.21, 0.17, 0.13, −0.63, −1.24, 0.57, −0.78

a. Draw a graph of the data, following recommended principles of good graph design (Chapter 2). What trend is suggested?

b. Determine the most-plausible range of values for the mean change in soil lead. Describe in words what the nature of that change is. Is an increase in soil lead consistent with the data? Is a decrease in soil lead consistent?

c. Test whether mean soil lead changed after the hurricanes.

22. Refer to Assignment Problem 21. In the same study, Zahran et al. (2010) also measured the lead concentration of blood (in µg/dl) of children living in the 46 areas both before and after the hurricanes. The ratio of blood lead concentration after the hurricanes to that before the hurricanes is given below. This ratio has an approximately normal distribution. A ratio of 1.0 indicates no change.

0.67, 0.57, 0.88, 0.95, 0.60, 0.98, 0.64, 0.48, 0.26, 0.77, 0.87, 0.34, 0.56, 0.88, 0.77, 0.67, 0.59, 0.56, 0.54, 0.64, 0.90, 0.00, 0.53, 0.63, 0.57, 0.70, 0.41, 0.42, 0.69, 0.45, 0.67, 0.50, 0.60, 1.16, 0.33, 0.63, 0.63, 1.00, 1.05, 0.67, 1.00, 1.00, 0.75, 1.00, 1.00, 0.63

a. Determine the most-plausible range of values for the mean change in blood lead ratio. Describe in words the nature of that change. Is a ratio greater than one consistent with the data? Is a decrease in blood lead consistent?

b. Is this ratio significantly different from one? Show the appropriate hypothesis test.

23. The evolution of blind cave forms of the fish, *Astyanax mexicanus,* is associated with large reductions in the amount of time spent sleeping. The eyed, surface forms sleep about 800 minutes per 24-hour day (about 13 hours). The accompanying graph shows the frequency distribution of sleep times per 24-hours for 23 blind individuals from a single cave population (Duboué et al. 2011). The sample mean is 129.4 and the standard deviation is 147.2. Assume that the sample is a random sample.

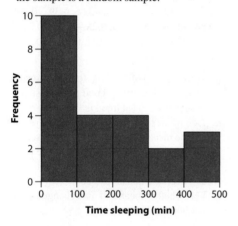

a. Using the formula in Section 11.7, calculate a 95% confidence interval for mean sleep time in the cave population.

b. Use the graph to evaluate whether the assumptions of the confidence interval are likely to be met. Explain your answer.

c. In light of your answer in part (b), can we trust the confidence interval you calculated in part (a)?

24. Without external cues such as the sun, people attempting to walk in a straight line tend to walk in circles (the accompanying image shows the paths of two participants, PS and KS, attempting to walk in a straight line in an unfamiliar forest on a cloudy day). One idea is that most individuals have a tendency to turn in one direction because of internal physiological asymmetries, or because of differences between legs in length or strength. Souman et al. (2009) tested for a directional tendency by blindfolding 15 participants in a large field and asking them to walk in a straight line. The numbers below are the median change in direction (turning angle) of each of the 15 participants measured in degrees per second. A negative angle refers to a left turn, whereas a positive number indicates a right turn.

−5.19, −1.20, −0.50, −0.33, −0.15, −0.15, −0.15, −0.07, 0.02, 0.02, 0.28, 0.37, 0.45, 1.76, 2.80

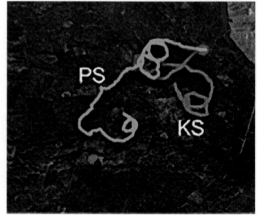

©2007 Google, Image ©2008 GeoContent, ©2008 European Space Agency, © Tele Atlas

a. Draw a graph showing the frequency distribution of the data. Is a trend in the mean angle suggested?

b. Do people tend to turn in one direction (e.g., left) more on average than the other direction (e.g., right)? Test whether the mean angle differs from zero.

c. Based on your results in part (b), is the following statement justified? "People do not have a tendency to turn more in one direction, on average, than the other direction." Explain.

25. Functionally important traits in animals tend to vary little from one individual to the next within populations, possibly because individuals that deviate too much from the mean die sooner or leave fewer offspring in the long run. If so, does variance in a trait rise after it becomes less functionally important? Billet et al. (2012) investigated this question with the semicircular canals (SC) of the inner ear of the three-toed sloth (*Bradypus variegatus*). Sloths move very slowly and infrequently, and the authors suggested that this behavior reduces the functional demands on the SC, which usually provide information on angular head movement to the brain. Indeed, the motion signal from the SC to the brain may be very weak in sloths as compared to faster-moving animals. The following numbers are measurements of the ratio of the length to the width of the anterior semicircular canals in seven sloths. Assume that this represents a random sample.

1.53, 1.06, 0.93, 1.38, 1.47, 1.20, 1.16

a. In related, faster-moving animals, the standard deviation of the ratio of the length to

the width of the anterior semicircular canals is known to be 0.09. What is the estimate of standard deviation of this measurement in three-toed sloths?

b. Based on these data, what is the most-plausible range of values for the population standard deviation in the three-toed sloth? Does this range include the known value of the standard deviation in related, faster-moving species?

c. What additional assumption is required for your answer in (b)? What do you know about how sensitive the confidence interval calculation is when the assumption is not met?

Galápagos marine iguana

12 Comparing two means

Biological data are often gathered to compare different group or treatment means. Do female hyenas differ from male hyenas in body size? Do patients treated with a new drug live significantly longer than those treated with the old drug? Do students perform better on tests if they stay up late studying or get a good night's rest? In Chapter 8, we presented methods to compare proportions of a categorical variable between different groups. In this chapter, we develop procedures for comparing means of a numerical variable between two treatments[1] or groups. We also include methods to compare two variances. All of the methods in the current chapter assume that the measurements are normally distributed in the populations.

We show analyses for two different study designs. In the paired design, both treatments have been applied to every sampled unit, such as a subject or a plot, at differ-

---

1. We will use the term "treatments" in a broad sense to refer to different states or conditions experienced by subjects, not just to formal experimental treatments.

ent times or on different sides of the body or of the plot. In the two-sample design, each group constitutes an independent random sample of individuals. In both cases, we make use of the *t*-distribution to calculate confidence intervals for the mean difference and test hypotheses about means. However, the two different study designs require alternative methods for analysis. We elaborate on the reasons in Section 12.1.

12.1   Paired sample versus two independent samples

There are two study designs to measure and test differences between the means of two treatments. To describe them, let's use an example: does clear-cutting a forest affect the number of salamanders present? Here we have two treatments (clear-cutting/no clear-cutting), and we want to know if the mean of a numerical variable (the *number of salamanders*) differs between them. "Clear-cut" is the treatment of interest, and "no clear-cut" is the control. This is the same as asking whether these two variables, *treatment* (a numerical variable) and *salamander number* (a categorical variable), are associated.

We can design this kind of a study in either of two ways: a two-sample design (see the left panel in Figure 12.1-1) or a paired design (see the right panel in Figure 12.1-1). In the *two-sample* design, we take a random sample of forest plots from the population and then randomly assign either the clear-cut treatment or the no-clear-cut treatment to each plot. In this case, we end up with two independent samples, one from each treatment. The difference in the mean number of salamanders between the clear-cut and no-clear-cut areas estimates the effect of clear-cutting on salamander number.[2]

In the *paired* design, we take a random sample of forest plots and clear-cut a randomly chosen half of each plot, leaving the other half untouched. Afterward, we count the number of salamanders in each half. The mean difference between the two sides estimates the effect of clear-cutting.

Each of these two experimental designs has benefits, and both address the same question. However, they differ in an important way that affects how the data are analyzed. In the paired design, measurements on adjacent plot-halves are not inde-

---

2. We are assuming that you have persuaded an enlightened forest company to go along with this experiment. It is more likely that the forest company has already clear-cut some areas and not others, and you must compare the two treatments after the fact with an observational study. Both the experimental approach and the observational design yield a measure of the association between clear-cutting and salamander number, but only the experimental study can tell us whether clear-cutting is the *cause* of any difference (Chapter 14). The study design issues are the same, however: whether to choose forest plots randomly from each treatment (the two-sample design) or to randomly choose forest plots that straddle both treatments, making it possible to compare cut and uncut sites that are side by side (the paired design).

FIGURE 12.1-1
Alternative designs to compare two treatments. A two-sample design is on the left; the paired design is on the right. Freestanding blocks represent sampling units, such as plots. The red and gold areas represent different treatments (e.g., clear-cut and not clear-cut). In the paired design (*right*), both treatments are applied to every unit. In the two-sample design (*left*), different treatments are applied to separate units.

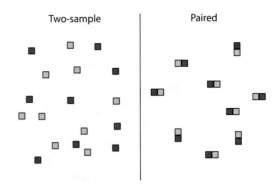

pendent. This is because they are likely to be similar in soil, water, sunlight, and other conditions that affect the number of salamanders. Because of these similarities, we can control for some of the noise between plots and see more clearly the effects of the treatment. As a result, we must analyze paired data differently than when every plot is independent of all the others, as in the case of the two-sample design.

> In the *paired* design, both treatments are applied to every sampled unit. In the *two-sample* design, each treatment group is composed of an independent, random sample of units.

Paired designs are usually more powerful than unpaired designs, because they control for a lot of the extraneous variation between plots or sampling units that sometimes obscures the effects we are looking for. It is easier to see a real difference between two treatments if nearly everything else is similar between sides of the same plot or sampling unit. Very often, though, a paired design is just not possible.

12.2   Paired comparison of means

The main advantage of the paired design is that it reduces the effects of variation among sampling units that has nothing to do with the treatment itself. This feature increases the power and the precision of estimates. For example, forest or agricultural plots likely differ greatly in their local environmental features, and this variation can make it difficult to detect a difference between the effects of two treatments applied to separate plots. The paired design reduces the impact of this variation by applying both treatments to different sides of all plots, making it easier to detect a real difference between treatments. Here are some other examples of paired study designs:

- Comparing patient weight before and after hospitalization
- Comparing fish species diversity in lakes before and after heavy metal contamination

- Testing effects of sunscreen applied to one arm of each subject compared with a placebo applied to the other arm
- Testing effects of smoking in a sample of smokers, each of which is compared with a nonsmoker closely matched by age, weight, and ethnic background
- Testing effects of socioeconomic condition on dietary preferences by comparing identical twins raised in separate adoptive families that differ in their socioeconomic conditions

The last two examples (the effects of smoking and socioeconomic condition) show that even two unique individuals can constitute a pair if they are similar due to shared physical, environmental, or genetic characteristics.

In a paired study design, the sampling unit is the pair. We must therefore reduce the two measurements made on each pair down to a single number—that is, the *difference* between the two measurements made on each sampling unit (e.g., patient, lake, subject, matched pair, or twins). This step correctly yields only as many data points as there are randomly sampled units. Thus, if 20 individuals are grouped into 10 pairs, there are 10 measurements of the difference between the two treatments. Ten would be the sample size. We can estimate and test the effect of treatment using the mean of the differences.

> Paired measurements are converted to a single measurement by taking the difference between them.

Estimating mean difference from paired data

We now describe how to estimate mean differences and calculate confidence intervals for those estimates. This method assumes that we have a random sample of pairs and that the differences between members of each pair have a normal distribution. We'll use Example 12.2 to show the concepts and the calculation.

EXAMPLE
12.2   So macho it makes you sick?

In many species, males are more likely to attract females if the males have high testosterone levels. Are males with high testosterone paying a cost for this extra mating success in other ways? One hypothesis is that males with high testosterone might be less able to fight off disease—that is, their high levels of testosterone might reduce their immunocompetence. To test this idea, Hasselquist et al. (1999) experimentally increased the testosterone levels of 13 male red-winged blackbirds by surgically implanting a small permeable tube filled with testosterone. They measured immunocompetence as the rate of antibody production in response to a nonpathogenic antigen in each bird's blood serum both before and

after the implant. The antibody production rates were measured optically, in units of log 10^{-3} optical density per minute (ln[mOD/min]).

The graph in Figure 12.2-1 shows that there is considerable variation among birds in their natural antibody production rates and that antibody production went up after the implant in some birds but went down in others. What is the mean difference between the two treatments? We can address this question by constructing a confidence interval for the mean change in antibody production.

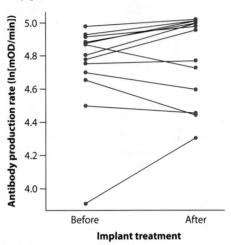

FIGURE 12.2-1
Immunocompetence of 13 red-winged blackbirds before and after testosterone implants. Immunocompetence was measured as the log of the rate of antibody production in response to an antigen (original units in mOD/min). The two measurements from each bird are connected by a line segment.

TABLE 12.2-1 Antibody production rate in blackbirds before and after testosterone implants. Each bird is represented by a single row and has a pair of antibody measurements; d is the difference ("after" minus "before") between the pair of measurements.

Male identification number	Before implant: Antibody production (ln[mOD/min])	After implant: Antibody production (ln[mOD/min])	d
1	4.65	4.44	−0.21
4	3.91	4.30	0.39
5	4.91	4.98	0.07
6	4.50	4.45	−0.05
9	4.80	5.00	0.20
10	4.88	5.00	0.12
15	4.88	5.01	0.13
16	4.78	4.96	0.18
17	4.98	5.02	0.04
19	4.87	4.73	−0.14
20	4.75	4.77	0.02
23	4.70	4.60	−0.10
24	4.93	5.01	0.08

The first step is to calculate the difference in antibody production for each male. The data, listed in Table 12.2-1, consist of a pair of measurements for each male: antibody production *before* the testosterone implant and antibody production *after* the implant. We calculated the difference between measurements within each male i as

$$d_i = \text{(antibody production of male } i \text{ after)} - \text{(antibody production of male } i \text{ before)}.$$

For bird 1, for example, $d_1 = 4.44 - 4.65 = -0.21$. It doesn't much matter whether we subtract the "after" measurement from the "before" measurement (as in Table 12.2-1) or vice versa, but it is critical that we calculate the differences the same way for each individual. A histogram of the resulting differences is shown in Figure 12.2-2.

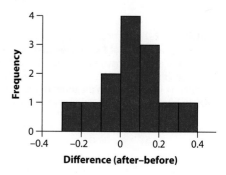

FIGURE 12.2-2
A histogram of the differences in antibody production in male blackbirds before and after testosterone implants.

We then find the sample mean difference (call it $\bar{d}$), the sample standard deviation of the differences (s_d), and the sample size (n):

$$\bar{d} = 0.056,$$
$$s_d = 0.159, \text{ and}$$
$$n = 13.$$

The trend was for immunocompetence (measured by the rate of antibody production) to go up after the testosterone implant, not down as predicted by the hypothesis. The confidence interval for the mean of a paired difference is generated in the same way as the confidence interval for any other mean (see Section 11.2). The confidence interval for the mean difference (μ_d) is

$$\bar{d} - t_{\alpha(2), df} \, SE_{\bar{d}} < \mu_d < \bar{d} + t_{\alpha(2), df} \, SE_{\bar{d}},$$

where

$$SE_{\bar{d}} = s_d/\sqrt{n}$$

is the standard error of the mean difference. Note that n is the number of *pairs*, not the total number of measurements, because pairs are the independent sampling unit. For this reason, we are carrying out the analysis on the differences.

For the blackbird data, we calculate

$$SE_{\bar{d}} = \frac{0.159}{\sqrt{13}} = 0.044.$$

With $n = 13$, we have $df = 12$, so we look in Statistical Table C to find $t_{0.05(2), 12} = 2.18$. Thus, the 95% confidence interval for the mean difference between antibody production before and after testosterone implants is

$$\bar{d} - t_{\alpha(2), df}\ SE_{\bar{d}} < \mu_d < \bar{d} + t_{\alpha(2), df}\ SE_{\bar{d}}$$

$$0.056 - 2.18(0.044) < \mu_d < 0.056 + 2.18(0.044)$$

$$-0.040 < \mu_d < 0.152.$$

In other words, the most-plausible range for the true mean difference is between -0.040 and $0.152\ \ln[\text{mOD}/\text{min}]$. While this span includes zero, it is also consistent with the possibility of a modest drop in immunocompetence following testosterone implant. A larger sample of individuals would be needed to narrow the interval further.

Paired *t*-test

The **paired *t*-test** is used to test a null hypothesis that the mean difference of paired measurements equals a specified value. The method tests differences when both treatments have been applied to every sampling unit and the data are therefore paired.

The paired *t*-test is straightforward once we reduce the two measurements made on each pair down to a single number: the difference between the two measurements. This difference is then analyzed in the same way as a regular one-sample *t*-test, as described in Chapter 11. We'll continue to analyze the blackbird testosterone data from Example 12.2 to ask whether the antibody production in blackbirds changed significantly after testosterone implants.

Because the data are paired, a paired *t*-test is appropriate, with before testosterone and after testosterone representing the two "treatments." The null and alternative hypotheses are as follows.

H_0: The mean change in antibody production after testosterone implants was zero.

H_A: The mean change in antibody production after testosterone implants was not zero.

The alternative hypothesis is two-tailed, because either greater or lesser antibody production after the testosterone implants would reject the null hypothesis. These hypothesis statements could also be written as

H_0: $\mu_d = 0$

and

H_A: $\mu_d \neq 0$,

where μ_d is the population mean difference between treatments.

Again, the first step is to calculate the difference in antibody production before and after the implants, which we have already done in Table 12.2-1. We then need the sample mean ($\bar{d} = 0.056$) and standard error ($\text{SE}_{\bar{d}} = 0.044$) of the mean differences, which we calculated in the previous subsection.

From here on, the paired t-test is identical to a one-sample t-test on the differences. We can calculate the t-statistic as

$$t = \frac{\bar{d} - \mu_{d0}}{\text{SE}_{\bar{d}}},$$

where μ_{d0} is the population mean of d proposed by the null hypothesis (0 in this example), and $\text{SE}_{\bar{d}}$ is the sample standard error of d. Under the null hypothesis, this t-statistic has a t-distribution with $df = n - 1$ degrees of freedom.

When this formula is applied to the blackbird testosterone data,

$$t = \frac{0.056 - 0}{0.044} = 1.27.$$

This test statistic has $df = 13 - 1 = 12$. The P-value for the test is

$$P = \Pr[t_{12} < -1.27] + \Pr[t_{12} > 1.27] = 2\Pr[t_{12} > 1.27].$$

Using a computer, we calculated this probability under a t-distribution with 12 df to be

$$P = 0.23.$$

P is greater than 0.05, so we do *not* reject the null hypothesis that $\mu_d = 0$ with these data.

If we did this test without a computer handy, we could use Statistical Table C instead. In that case, we would find that the critical value for a two-tailed test with $\alpha = 0.05$ and 12 degrees of freedom is

$$t_{0.05(2),\,12} = 2.18.$$

Because $t = 1.27$ does not fall outside the critical limits of -2.18 and 2.18, we know that $P > 0.05$, and we do not reject the null hypothesis.

The mean difference that we saw ($0.056\ \ln[\text{mOD}/\text{min}]$) is in the direction of higher immune system function after testosterone implants, but we cannot reject the null hypothesis that the testosterone has no effect on immunocompetence. The confidence interval that we calculated in the previous subsection indicates that a broad range of values is consistent with these data. On the basis of this data set, we do not reject the null hypothesis, but we might want to study the problem further to resolve the issue more precisely.

Assumptions

The paired t-test and the method for calculating a confidence interval for a paired difference make the same assumptions as the one-sample methods described in Chapter 11:

- The sampling units are randomly sampled from the population.
- The paired differences have a normal distribution in the population.

The analysis makes no assumptions about the distribution of either of the two measurements made on each sampling unit. These measurements can have any distribution, as long as the difference between the two measurements is approximately normally distributed.

12.3 Two-sample comparison of means

We now present methods to analyze the difference between the means of two treatments or groups in the case of a two-sample design. In a two-sample design, the two treatments are applied to separate, independent samples from two populations. We illustrate the process using Example 12.3.

EXAMPLE Spike or be spiked

12.3 The horned lizard *Phrynosoma mcallii* has many unusual features, including the ability to squirt blood from its eyes. The species is named for the fringe of spikes surrounding the head. Herpetologists recently tested the idea that long spikes help protect horned lizards from being eaten, by taking advantage of the gruesome but convenient behavior of one of their main predators—the loggerhead shrike, *Lanius ludovicianus*. The loggerhead shrike is a small predatory bird that skewers its victims on thorns or barbed wire, to save for later eating.

The researchers identified the remains of 30 horned lizards that had been killed by shrikes and measured the lengths of their horns (Young et al. 2004). As a comparison group, they measured the same trait on 154 horned lizards that were still alive and well. They compared the mean horn lengths of the dead lizards with those of the living lizards. Histograms of the horn lengths of the two groups are shown in Figure 12.3-1. Summary statistics are listed in Table 12.3-1.

TABLE 12.3-1 Summary statistics for lizard horn lengths.

Lizard group	Sample mean $\overline{Y}$ (mm)	Sample standard deviation s (mm)	Sample size n
Living	24.28	2.63	154
Killed	21.99	2.71	30

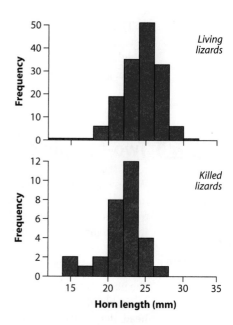

FIGURE 12.3-1
The frequency distributions of horn lengths for living and killed horned lizards. There are $n_1 = 154$ live lizards and $n_2 = 30$ killed lizards.

The lizards that were killed by shrikes are *different individuals* than the living lizards. They are not paired in any way; instead, each treatment (living or dead) is represented by a separate sample of lizards, belonging to different groups. Therefore, we must analyze the differences by using two-sample methods.

Figure 12.3-1 and Table 12.3-1 suggest that the dead lizards have shorter horns than the living lizards on average, as might be expected if shrikes avoid the longest horns. Next, we calculate a confidence interval for the difference, and we test whether the difference is real.

Confidence interval for the difference between two means

How much longer on average are the horns of the surviving lizards? The best estimate of the difference between two population means is the difference between the sample means, $\overline{Y}_1 - \overline{Y}_2$. Here, we will use $\overline{Y}_1$ to refer to the sample mean of the live lizards, and $\overline{Y}_2$ to refer to the sample mean of the dead lizards. Which group we call 1 and which we call 2 is arbitrary, but we have to be consistent with the labels throughout. We use subscripts to indicate the population or sample that a value comes from.

The method for confidence intervals makes use of the fact that, if the variable is normally distributed in both populations, then the sampling distribution for the *difference* between the sample means is also normal. Thus, the Student's t-distribution will be very helpful in describing the sampling properties of the standardized difference.

First, we will need the standard error of $\overline{Y}_1 - \overline{Y}_2$, which is given by

$$\mathrm{SE}_{\overline{Y}_1 - \overline{Y}_2} = \sqrt{s_p^2 \left(\frac{1}{n_1} + \frac{1}{n_2} \right)},$$

where

$$s_p^2 = \frac{df_1 \, s_1^2 + df_2 \, s_2^2}{df_1 + df_2}.$$

The quantity s_p^2 is called the **pooled sample variance**. It is a weighted average[3] of the sample variances s_1^2 and s_2^2 (the squared standard deviations) of the two groups. The confidence interval formula assumes that the standard deviations (and variances) of the two populations are the same. The pooled variance s_p^2 uses the information from both samples to get the best estimate of this common population variance. The df_1 and df_2 terms refer to the degrees of freedom for the variances of the two samples:

$$df_1 = n_1 - 1$$

and

$$df_2 = n_2 - 1,$$

where n_1 and n_2 are the sample sizes from the two populations.

> The *pooled sample variance* s_p^2 is the average of the variances of the samples weighted by their degrees of freedom.

Because the sampling distribution of $\overline{Y}_1 - \overline{Y}_2$ is normal, the sampling distribution of the following standardized difference has a Student's t-distribution:

$$t = \frac{(\overline{Y}_1 - \overline{Y}_2) - (\mu_1 - \mu_2)}{SE_{\overline{Y}_1 - \overline{Y}_2}}$$

with total degrees of freedom equal to

$$df = df_1 + df_2 = n_1 + n_2 - 2.$$

From these two formulas, we can calculate the confidence interval for the difference between two population means:

$$(\overline{Y}_1 - \overline{Y}_2) - t_{\alpha(2), \, df} \, SE_{\overline{Y}_1 - \overline{Y}_2} < \mu_1 - \mu_2 < (\overline{Y}_1 - \overline{Y}_2) + t_{\alpha(2), \, df} \, SE_{\overline{Y}_1 - \overline{Y}_2}.$$

Let's use the horned-lizard data to calculate the 95% confidence interval for the difference in horn length between the two groups of lizards. The place to start is with the summary statistics listed in Table 12.3-1. The difference in the means is

$$\overline{Y}_1 - \overline{Y}_2 = 24.28 - 21.99 = 2.29 \text{ mm.}$$

After that, we start from the bottom—we'll need the pooled variance to calculate the standard error of $\overline{Y}_1 - \overline{Y}_2$ and we'll need the standard error to get the confidence

---

3. A "weighted average" may count each group differently. In this case, each group is weighted by its degrees of freedom, so that the group with the larger sample size (i.e., with more information available) contributes proportionately more to the calculated average.

interval for $\mu_1 - \mu_2$. The pooled sample variance, which depends on the number of degrees of freedom ($df_1 = 153$ and $df_2 = 29$), is

$$s_p^2 = \frac{df_1\, s_1^2 + df_2\, s_2^2}{df_1 + df_2}$$

$$= \frac{153(2.63^2) + 29(2.71^2)}{153 + 29}$$

$$= 6.98.$$

The standard error of the difference between the two means is then

$$SE_{\bar{Y}_1 - \bar{Y}_2} = \sqrt{s_p^2 \left(\frac{1}{n_1} + \frac{1}{n_2}\right)} = \sqrt{6.98 \left(\frac{1}{154} + \frac{1}{30}\right)} = 0.527.$$

One common mistake is to forget that the standard error equation uses the sample sizes, n_1 and n_2, not the number of degrees of freedom, in the denominators.

There are $154 + 30 - 2 = 182$ degrees of freedom in total, so we look up the critical value of t for $df = 182$:

$$t_{0.05(2),\, 182} = 1.97.$$

(There is no row in Statistical Table C for $df = 182$, but $t_{0.05(2),\, df} = 1.97$ for both $df = 180$ and $df = 200$, and so to this order of precision, $t_{0.05(2),\, 182} = 1.97$ as well.)

By plugging these quantities into the formula, we find the 95% confidence interval for the difference in mean horn length between the living and dead lizards is

$$(\bar{Y}_1 - \bar{Y}_2) - t_{\alpha(2),\, df}\, SE_{\bar{Y}_1 - \bar{Y}_2} < \mu_1 - \mu_2 < (\bar{Y}_1 - \bar{Y}_2) + t_{\alpha(2),\, df}\, SE_{\bar{Y}_1 - \bar{Y}_2}$$

$$2.29 - 1.97\,(0.527) < \mu_1 - \mu_2 < 2.29 + 1.97\,(0.527)$$

$$1.25 < \mu_1 - \mu_2 < 3.33.$$

Thus, the 95% confidence interval for $\mu_1 - \mu_2$ is from 1.25 to 3.33 mm. We can be reasonably confident that surviving lizards have longer horns than lizards killed by shrikes, by an amount somewhere between 1.25 and 3.33 millimeters.

Two-sample *t*-test

The *two-sample t-test* is the simplest method to compare the means of a numerical variable between two independent groups. Its most common use is to test the null hypothesis that the means of two populations are equal (or, equivalently, that the difference between the means is zero):

H_0: $\mu_1 = \mu_2$

and

H_A: $\mu_1 \neq \mu_2$,

where μ_1 is the population mean for the first of the two populations and μ_2 is the mean for the second population. It doesn't matter which population we designate as population 1 and which as population 2 as long as we are consistent throughout the analysis. The two-sample t-test uses the following test statistic based on the observed difference between the sample means:[4]

$$t = \frac{(\overline{Y}_1 - \overline{Y}_2)}{SE_{\overline{Y}_1 - \overline{Y}_2}}.$$

Provided that the assumptions are met, this t-statistic has a t-distribution under the null hypothesis with $n_1 + n_2 - 2$ degrees of freedom. The denominator of this formula, $SE_{\overline{Y}_1 - \overline{Y}_2}$, is the standard error of the difference between the two sample means. This standard error is the same as that shown previously when discussing confidence intervals (p. 336):

$$SE_{\overline{Y}_1 - \overline{Y}_2} = \sqrt{s_p^2 \left(\frac{1}{n_1} + \frac{1}{n_2} \right)}.$$

The pooled sample variance, s_p^2, was defined in the previous subsection. The null hypothesis is tested by comparing the observed t-value to the theoretical t-distribution with

$$df = df_1 + df_2 = n_1 + n_2 - 2$$

degrees of freedom.

To apply the two-sample t-test to the horned lizard study, begin by writing the null and alternative hypotheses.

H_0: Lizards killed by shrikes and living lizards do not differ in mean horn length (i.e., $\mu_1 = \mu_2$).

H_A: Lizards killed by shrikes and living lizards differ in mean horn length (i.e., $\mu_1 \neq \mu_2$).

The alternative hypothesis is two-sided.

Let's apply these equations to the lizard data to perform the two-sample t-test. Once again, the difference between the sample means of the two groups of lizards is

$$\overline{Y}_1 - \overline{Y}_2 = 2.29 \text{ mm.}$$

The standard error of the difference, computed in the previous subsection, is

$$SE_{\overline{Y}_1 - \overline{Y}_2} = 0.527.$$

---

4. More generally, the null hypothesized value for the difference between the two population means can be any number: $H_0: (\mu_1 - \mu_2) = (\mu_1 - \mu_2)_0$. In this case, we would calculate the test statistic as

$$t = \frac{(\overline{Y}_1 - \overline{Y}_2) - (\mu_1 - \mu_2)_0}{SE_{\overline{Y}_1 - \overline{Y}_2}} \text{ instead.}$$

Now we can calculate the test statistic, t. According to the null hypothesis, the two means are equal; that is, $\mu_1 - \mu_2 = 0$. With this information, we can find

$$t = \frac{(\bar{Y}_1 - \bar{Y}_2)}{SE_{\bar{Y}_1 - \bar{Y}_2}} = \frac{2.29}{0.527} = 4.35.$$

We need to compare this test statistic with the distribution of possible values for t if the null hypothesis were true. The appropriate null distribution is the t-distribution, and it will have $df_1 + df_2 = 153 + 29 = 182$ degrees of freedom. The P-value is then

$$P = 2\,\Pr[t > 4.35].$$

Using a computer, we find that this t-value corresponds to

$$P = 0.000023.$$

Since $P < 0.05$, we reject the null hypothesis.

We reach the same conclusion using Statistical Table C. The critical value for $\alpha = 0.05$ for a t-distribution with 182 degrees of freedom is

$$t_{0.05(2),182} = 1.97.$$

The $t = 4.35$ calculated from these data is much further into the tail of the distribution than this critical value, so we reject the null hypothesis. In fact, the t calculated for these data is further in the tail of the distribution than all values given in Statistical Table C, including those for $\alpha = 0.0002$. From this information we may conclude that $P < 0.0002$. Based on these studies, there is a difference in the horn length of lizards eaten by shrikes, compared with live lizards. It is possible that shrikes avoid or are unable to capture the lizards with the longest horns. We can't be certain that this explains the difference in horn length, because this is an observational study. To infer causation, we would need to carry out a controlled experiment that manipulates lizard horn lengths.

Assumptions

The two-sided t-test and two-sample confidence interval for a difference in means are based on the following assumptions:

- Each of the two samples is a random sample from its population.
- The numerical variable is normally distributed in each population.
- The standard deviation (and variance) of the numerical variable is the same in both populations.

We've heard the first two assumptions before—they are required for the one-sample t-test—but the third assumption is new. The two-sample methods we have just discussed are fairly robust to violations of this assumption. With moderate sample sizes (i.e., n_1 and n_2 both greater than 30), the methods work well, even if the standard deviations in the two groups differ by as much as threefold, as long as the sample sizes of the two groups are approximately equal. If there is more than a threefold dif-

ference in standard deviations, or if the sample sizes of the two groups are very different or less than 30 with some difference in standard deviations, then the two-sample
t-test should not be used.

The two-sample *t*-test is also robust to minor deviations from the normal distribution, especially if the two distributions being compared are similar in shape. For
example, in Example 12.3, the distribution of horn size is unlikely to be perfectly
normal in either group, but in both cases the measurements do not differ greatly from
a normal distribution. The *t*-test is robust to these kinds of minor deviations from
normality. The robustness of the *t*-test improves as the sample sizes get larger. In
Chapter 13, we discuss at greater length the robustness of the *t*-test to deviations from
normality.

A two-sample *t*-test when standard deviations are unequal

An important assumption of the two-sample *t*-test is that the standard deviations of
the two populations are the same. If this assumption cannot be met, then the **Welch's
approximate *t*-test** should be used instead of the two-sample *t*-test. Welch's
t-test is similar to the two-sample *t*-test except that the standard error and degrees of
freedom are computed differently. Similarly, it is possible to calculate a confidence
interval for the difference in the means of the two groups using a formula that does
not assume equal variance in the two groups. (See the Quick Formula Summary in
Section 12.9 for the equations.) An example using the Welch's *t*-test and Welch's
confidence interval is discussed in Section 12.4.

> *Welch's t-test* compares the means of two groups and can be used even when
> the variances of the two groups are not equal.

In Chapter 13, we describe additional ways to rescue situations in which the
assumptions of normality and equal standard deviations are not met.

12.4 Using the correct sampling units

An assumption of the *t*-test, as with all other tests in this book, is that the samples
being analyzed are random samples. Quite often, repeated measurements have been
taken on each sampling unit. Because measurements made on the same sampling unit
are not independent, they require special handling. One solution we've already mentioned is to summarize the data for each sampling unit by a single measurement. One
of the biggest challenges when comparing groups is to identify correctly the independent units of replication. This decision determines not only what method is used, but
it might also affect the conclusion reached, as Example 12.4 illustrates.

EXAMPLE   So long; thanks to all the fish

12.4   One of the greatest threats to biodiversity is the introduction of alien species from outside their natural range. These introduced species often have fewer predators or parasites in the new area, so they can increase in numbers and outcompete native species. Sometimes  these species are introduced accidentally, but often they are introduced intentionally by humans. The brook trout, for example, is a species native to eastern North America that has been introduced into streams in the West for sport fishing. Biologists followed the survivorship of a native species, chinook salmon, in a series of 12 streams that either had brook trout introduced or did not (Levin et al. 2002). Their goal was to determine whether the presence of brook trout affected the survivorship of the salmon. In each stream, they released a number of tagged juvenile chinook and then recorded whether or not each chinook survived over one year. Table 12.4-1 summarizes the data.

TABLE 12.4-1 The numbers and proportion of chinook released and surviving in streams with and without brook trout. The study included 12 streams in total.

Brook trout	Number of salmon released	Number of salmon surviving	Proportion surviving
Present	820	166	0.202
Present	960	136	0.142
Present	700	153	0.219
Present	545	103	0.189
Present	769	173	0.225
Present	1001	188	0.188
Absent	467	180	0.385
Absent	959	178	0.186
Absent	1029	326	0.317
Absent	27	7	0.259
Absent	998	120	0.120
Absent	936	135	0.144
Total	9211	1865	

In all, 9211 salmon were released, of which 1865 survived and 7346 did not. A quick tally of the fish numbers by treatment yields the 2×2 table shown in Table 12.4-2.

TABLE 12.4-2 Number of salmon surviving and not surviving in each trout treatment.

	Trout absent	Trout present
Survived	946	919
Did not survive	3470	3876

We would like to test whether the proportion of salmon surviving differed between trout treatments. What method shall we use? It is tempting to carry out a χ^2 contingency test of association between treatment and survival.[5]

The problem with using the contingency test approach is that individual salmon are not a random sample. Rather, salmon are grouped by the *streams* in which they were released. If there is any inherent difference between the streams in the probability of survival, over and above any effects of brook trout, then two salmon from the same stream are likely to be more similar in their survival than two salmon picked at random. In this case, salmon from the same stream are not independent. To lump all the salmon together and analyze with a contingency test is to commit the sin of pseudo-replication (see Interleaf 4).

The key to solving the problem lies in recognizing that the stream is the independently sampled unit, not the salmon, and there are only 12 streams—six per treatment. As a result, the fates of all the salmon within a stream must be summarized by a single measurement for analysis—namely, the proportion of salmon surviving (given in the last column of Table 12.4-2). This changes the data type, because we are no longer comparing frequencies in categories, but rather differences in the means of a numerical variable. A two-sample test of the difference between the means is therefore required.

Let's label the streams with trout present as group 1 and the streams without trout as group 2. The null and alternative hypotheses are as follows:

H_0: The mean proportion of chinook surviving is the same in streams with and without brook trout ($\mu_1 = \mu_2$).

H_A: The mean proportion of chinook surviving is different between streams with and without brook trout ($\mu_1 \neq \mu_2$).

Sample means and standard deviations are listed in Table 12.4-3. The 95% confidence intervals are shown as "error bars" next to the data in Figure 12.4-1.

FIGURE 12.4-1
Proportion of chinook salmon surviving in streams with and without brook trout. Each open circle represents the measurement from a single stream. There are six streams of each type. Means and 95% confidence intervals are indicated with error bars to the right of the data.

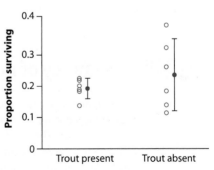

5. The results of this folly would be $\chi^2 = 7.2$, $df = 1$, and $P = 0.0071$. Thus, we would reject the null hypothesis of no association.

TABLE 12.4-3 Sample statistics for the proportion of salmon surviving in streams (Example 12.4), using the stream as the sampling unit.

Group	Sample mean	Sample standard deviation, s_i	Sample size, n_i
Brook trout present	0.194	0.0297	6
Brook trout absent	0.235	0.1036	6

Figure 12.4-1 and the summary statistics in Table 12.4-3 show that streams *without* introduced trout have a sample standard deviation more than three times that of streams *with* trout. This is a case where the Welch's approximate *t*-test is appropriate (see Section 12.3). Using Welch's *t*-test and a computer, we find that $t = 0.93$, $df = 5$, and the *P*-value for the test is $P = 0.39$. Hence, $P > 0.05$ and we cannot reject the null hypothesis. In other words, the data do *not* support the claim that the brook trout lower the survivorship of chinook salmon. The Welch's 95% confidence interval for the difference in means between these two groups ranges from -0.07 to 0.15.

The appropriate analysis, in which salmon data within streams are reduced to a single measurement per stream, might seem like a waste of hard-earned data. We started with survival data on 9211 salmon but used only six measurements per treatment. The contingency analysis rejected H_0, but the Welch's two-sample test did not! Have we thrown away data and lost power as a result?

There are two answers to this question. First, if the raw data are not randomly sampled, then it is not legitimate to analyze them as though they were a random sample. You can't lose power that you never had. But the kinder, gentler answer is that the data are not wasted. By pooling together several related data points into a single summary measure, such as a proportion, you will have an increasingly reliable measure of the true value of that measure in a given sample unit, such as a stream. As a result, little or no information is "lost."

12.5 The fallacy of indirect comparison

A common error when comparing two groups is to test each group mean separately against the same null hypothesized value, rather than directly comparing the two means with each other. The error might go something like this: "Since group 1 is significantly different from zero, but group 2 is not, group 1 and group 2 must be different from each other." We call this error the "fallacy of indirect comparison." Example 12.5 demonstrates that even papers published in scientific journals with the highest profile can make this mistake.

EXAMPLE Mommy's baby, Daddy's maybe

12.5

Do babies look more like their fathers or their mothers? The answer matters because, in most cultures, fathers are more likely to contribute to child rearing if they are convinced that a child is their biological offspring. In this case, babies who resemble their dads more closely have an evolutionary advantage, because they provide stronger assurance of paternity, which leads to greater paternal care (mothers do not face the same uncertainty of maternity). Christenfeld and Hill (1995) tested this by obtaining pictures of a series of babies and their mothers and fathers. A photograph of each baby, along with photos of three possible mothers and of three possible fathers, was shown to a large number of volunteers. Each volunteer was asked to pick which woman and which man were the parents of the baby based on facial resemblance. The percentage of volunteers who correctly guessed a parent was used as the measure of a given baby's resemblance to that parent. If there were no facial resemblance of babies to parents, then the mean resemblance should be 33.3%, the percentage of correct guesses expected by chance. If babies did resemble a parent, then the mean resemblance should be greater than 33.3%. Figure 12.5-1 shows the means for each parent and the corresponding 95% confidence intervals.

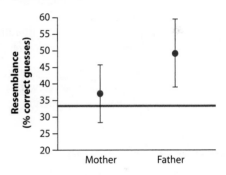

FIGURE 12.5-1
Resemblance of babies to their biological mothers and fathers, as measured by the percentage of volunteers who correctly guessed the mother and father of each baby from facial photographs. Dots are means, and vertical lines (error bars) are the 95% confidence intervals. The null expectation is 33.3% (shown with the red horizontal line). $n = 24$ babies.

The null hypothesis of no resemblance (i.e., one-third correct guesses) was soundly rejected for fathers, and the null hypothesis of no resemblance was *not* rejected for mothers. So far, so good. However, the researchers concluded from these tests that babies therefore resembled their fathers more than they resembled their mothers. This is an indirect comparison. That is, both groups were tested against the same null expectation (i.e., 33.3%), but mothers and fathers were not directly compared with each other. If they had been, no significant difference would have been found.

The problem with this sort of indirect comparison can be understood by considering a more extreme hypothetical case. In the example shown in Figure 12.5-2, the mean of group 1 is significantly different from the null expectation (indicated by the dashed line), but the mean of group 2 is not. It is false to conclude, therefore, that group 1 has a larger mean than group 2. In this example, it is group 2 that has the larger sample mean!

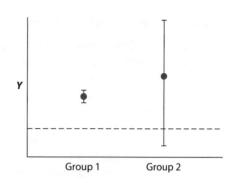

FIGURE 12.5-2
The 95% confidence intervals for the means of two
hypothetical groups. The dashed line represents the null
hypothesized value for the means of both groups.

How, then, should mothers and fathers be compared? They should be compared by testing directly whether the mean resemblance of babies to mothers is different from the mean resemblance to fathers. If the null hypothesis of no difference is rejected, then—and only then—we can conclude that babies resemble one parent more than the other.[6]

Indirectly comparing two groups by comparing them separately to the same null hypothesized value will often lead you astray. This error is extremely common in the published scientific literature (Nieuwenhuis et al. 2011). Groups should always be compared directly to each other.

> Comparisons between two groups should always be made directly, not indirectly by comparing both to the same null hypothesized value.

12.6 Interpreting overlap of confidence intervals

Scientific papers often report the means of two or more groups, along with their confidence intervals, but they might not test the difference between the means of the two groups with a two-sample t-test. How much information about whether group means differ significantly is contained in the amount of overlap between their separate confidence intervals?

It turns out that reliable conclusions can be drawn only under the two scenarios depicted in panels (a) and (b) of Figure 12.6-1.

If the 95% confidence intervals[7] of two estimates do not overlap at all (as in Figure 12.6-1, panel [a]), then the null hypothesis of equal means would be rejected

---

6. Larger follow-up studies have failed to find any difference between mothers and fathers in the degree to which their babies resemble them (Brédart and French 1999).

7. When you're reading scientific reports, interpret error bars on a graph with care. Sometimes error bars are used to show one standard error above and below the estimate, sometimes they show two standard errors, sometimes they show a confidence interval, and sometimes they just show standard deviations. Figure 12.6-1 shows confidence intervals.

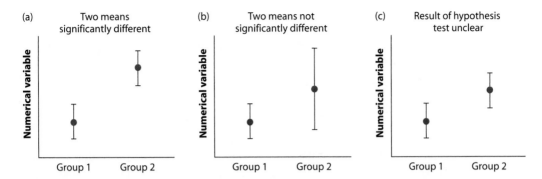

FIGURE 12.6-1 Each panel shows sample means of two groups with 95% confidence intervals. (a) The confidence intervals do not overlap; in this case, the null hypothesis of no difference between group means would be rejected. (b) The confidence interval for one group contains the sample mean of the other group; in this case, the null hypothesis of no difference would *not* be rejected. (c) The confidence intervals overlap, but neither includes the sample mean of the other group; in this case, we cannot be sure what the results of a two-sample *t*-test comparing the means would show.

at the $\alpha = 0.05$ level. If, on the other hand, the value of one of the sample means is included within the 95% confidence interval of the other mean (as in Figure 12.6-1, panel [b]), then the null hypothesis of equal means would not be rejected at $\alpha = 0.05$. Between these two extremes, where the confidence intervals overlap but neither interval includes the sample mean of the other group (as in Figure 12.6-1, panel [c]), we cannot be sure what the outcome of a two-sample *t*-test would be from a simple inspection of the overlap in confidence intervals.

12.7 Comparing variances

Up to now we have focused on comparing group means, but often we want to test whether populations differ in the variability of measurements. Several techniques are available. Here we briefly describe two of these: the *F*-test and Levene's test. In both cases, we describe the tests without much detail. The Quick Formula Summary (Section 12.9) gives the formulas, and most computer statistical programs will perform these tests.

Be warned, though: the *F*-test is highly sensitive to departures from the assumption that the measurements are normally distributed in the populations. It is not recommended for most data, therefore, because real data often show some departure from normality. Nevertheless, we present it here because many researchers continue to use it, and you will encounter it in the literature and in statistics packages on the computer. Levene's test is a popular alternative test that is more robust to the assumption of normal populations, but it has lower power.

The *F*-test of equal variances

The **F-test** evaluates whether two population variances are equal. That is, it tests the null hypothesis that

$$H_0: \ \sigma_1^2 = \sigma_2^2$$

against the alternative

$$H_A: \ \sigma_1^2 \neq \sigma_2^2,$$

where σ_1^2 is the variance (the squared standard deviation) of population 1 and σ_2^2 is the variance of population 2. The test statistic is called *F*, and it is calculated from the ratio of the two sample variances:

$$F = s_1^2 / s_2^2.$$

If the null hypothesis were true, then *F* should be near one, deviating from it only by chance. Under the null hypothesis, the *F*-statistic has an *F*-distribution with the pair of degrees of freedom $(n_1 - 1, n_2 - 1)$. The first number of the pair refers to the degrees of freedom of the top part (the numerator) of the *F*-ratio, and the second pair is for the bottom part (the denominator) of the *F*-ratio. We present more details in the Quick Formula Summary (Section 12.9). The *F*-distribution is discussed more fully in Chapter 15.

The *F*-test to compare two variances assumes that the variable is normally distributed in both populations. Unfortunately, the test is highly sensitive (i.e., not robust) to this assumption. For example, the *F*-test will often falsely reject the null hypothesis of equal variance if the distribution in one of the populations is not normal. For this reason, the *F*-test is not recommended for general use.

Levene's test for homogeneity of variances

Several alternative methods also test the null hypothesis that the variances of two or more groups are equal. One of the best is **Levene's test**, which is available in many statistical packages on the computer. Levene's test assumes that the frequency distribution of measurements is roughly symmetrical within all groups, but it performs much better than the *F*-test when this assumption is violated. Levene's test has the further advantage that it can be applied to more than two groups; in fact, it can test the null hypothesis that multiple groups all have equal variances.

Levene's test works by first calculating the absolute value of the difference between each data point and the sample mean for its group. These quantities are called "absolute deviations." The method then tests for a difference between groups in the means of these absolute deviations. The test statistic is called *W*, and it too has an *F*-distribution under the null hypothesis of equal variances. The calculations are somewhat cumbersome, so we will not detail them here, but we give the formula in the Quick Formula Summary (Section 12.9). Most modern statistical programs on the computer will do a Levene's test, however, and we recommend that you use it when comparing the variances of two or more groups. The online *Engineering Statistics Handbook*[8] is a good place to look for more information on Levene's test.

12.8   Summary

- Two study designs are available to compare two treatments. In a paired design, both treatments are applied to every randomly sampled unit. In a two-sample design, treatments are applied to separate randomly sampled units.

- Comparing two treatments in a paired design involves analyzing the mean of the differences between the two measurements of each pair. Comparing two treatments in a two-sample design involves analyzing the difference in means of two independent samples of measurements.

- A test of the mean difference between two paired treatments uses the paired *t*-test.

- Both the confidence interval for the mean difference and the paired *t*-test assume that the pairs are randomly chosen from the population and that the differences (d_i) have a normal distribution. These methods are robust to minor deviations from the assumption of normality.

- The means of a numerical variable from two separate groups or populations can be compared with a two-sample *t*-test.

- The two-sample *t*-test and the confidence intervals for the difference between the means assume that the variable is normally distributed in both populations and that the variance is the same in both populations. The methods are robust to minor deviations from these assumptions.

- The pooled sample variance is the best estimate of the variance within groups, assuming that the groups have equal variance.

- Welch's approximate *t*-test compares the means of two groups when the variances of the two groups are not equal.

- Repeated measurements made on the same sampling unit are not independent and should be summarized for each sampling unit before further analysis.

- Indirectly comparing two groups by comparing each of them separately to the same null hypothesized value will often lead you astray. Groups should always be compared directly to each other.

- For variables that are normally distributed, variances of two groups can be compared with an *F*-test. The *F*-test, however, is highly sensitive to the departures from the assumption of normal populations.

- Levene's test compares the variances of two or more groups. It is more robust than the *F*-test to departures from the assumption of normality.

---

8. http://www.itl.nist.gov/div898/handbook/eda/section3/eda35a.htm

12.9 Quick Formula Summary

Confidence interval for the mean difference (paired data)

What does it assume? Pairs are a random sample. The difference between paired measurements is normally distributed.

Parameter: μ_d

Statistic: $\bar{d}$

Degrees of freedom: $n - 1$

Formula: $\bar{d} - t_{\alpha(2),\,df}\,\mathrm{SE}_{\bar{d}} < \mu_d < \bar{d} + t_{\alpha(2),\,df}\,\mathrm{SE}_{\bar{d}}$,

where $\bar{d}$ is the mean of the differences between members of each of the pairs, $\mathrm{SE}_{\bar{d}} = s_d/\sqrt{n}$, s_d is the sample standard deviation of the differences, and n is the number of pairs.

Paired t-test

What is it for? To test whether the mean difference in a population equals a null hypothesized value, μ_{d0}.

What does it assume? Pairs are randomly sampled from a population. The differences are normally distributed.

Test statistic: t

Distribution under H_0: The t-distribution with $n - 1$ degrees of freedom, where n is the number of pairs.

Formula: $t = \dfrac{\bar{d} - \mu_{d0}}{\mathrm{SE}_{\bar{d}}}$,

where the terms are defined as for the confidence interval.

Standard error of difference between two means

Formula: $\mathrm{SE}_{\bar{Y}_1 - \bar{Y}_2} = \sqrt{s_p^2\left(\dfrac{1}{n_1} + \dfrac{1}{n_2}\right)}$,

where s_p^2 is the pooled sample variance: $s_p^2 = \dfrac{df_1 s_1^2 + df_2 s_2^2}{df_1 + df_2}$. The degrees of freedom are $df_1 = n_1 - 1$ and $df_2 = n_2 - 1$.

Confidence interval for the difference between two means (two samples)

What does it assume? Both samples are random samples. The numerical variable is normally distributed within both populations. The standard deviation of the distribution is the same in the two populations.

Degrees of freedom: $n_1 + n_2 - 2$

Statistic: $\bar{Y}_1 - \bar{Y}_2$

Parameter: $\mu_1 - \mu_2$

Formula: $(\bar{Y}_1 - \bar{Y}_2) - t_{\alpha(2),\, df}\ \mathrm{SE}_{\bar{Y}_1-\bar{Y}_2} < \mu_1 - \mu_2 < (\bar{Y}_1 - \bar{Y}_2) + t_{\alpha(2),\, df}\ \mathrm{SE}_{\bar{Y}_1-\bar{Y}_2},$

where $\mathrm{SE}_{\bar{Y}_1-\bar{Y}_2}$ is the standard error of the difference between means, as defined above.

Two-sample *t*-test

What is it for? Tests whether the difference between the means of two groups equals a null hypothesized value for the difference.

What does it assume? Both samples are random samples. The numerical variable is normally distributed within both populations. The standard deviation of the distribution is the same in the two populations.

Test statistic: t

Distribution under H$_0$: The t-distribution with $n_1 + n_2 - 2$ degrees of freedom.

Formula: $t = \dfrac{(\bar{Y}_1 - \bar{Y}_2) - (\mu_1 - \mu_2)_0}{\mathrm{SE}_{\bar{Y}_1-\bar{Y}_2}},$

where $(\mu_1 - \mu_2)_0$ is the null hypothesized value for the difference between population means, and $\mathrm{SE}_{\bar{Y}_1-\bar{Y}_2}$ is the standard error of the difference between means, as defined previously.

Welch's confidence interval for the difference between two means

What does it assume? Both samples are random samples. The numerical variable is normally distributed within both populations.

Statistic: $\bar{Y}_1 - \bar{Y}_2$

Parameter: $\mu_1 - \mu_2$

Formula: $(\bar{Y}_1 - \bar{Y}_2) \pm \dfrac{t_{\alpha(2), df}}{\sqrt{\dfrac{s_1^2}{n_1} + \dfrac{s_2^2}{n_2}}}$, where $df = \dfrac{\left(\dfrac{s_1^2}{n_1} + \dfrac{s_2^2}{n_2}\right)^2}{\left[\dfrac{(s_1^2/n_1)^2}{n_1 - 1} + \dfrac{(s_2^2/n_2)^2}{n_2 - 1}\right]}$

where df is rounded down to the nearest integer.

Welch's approximate t-test

What is it for? Tests whether the difference between the means of two groups equals a null hypothesized value when the standard deviations are unequal.

What does it assume? Both samples are random samples. The numerical variable is normally distributed within both populations.

Test statistic: t

Distribution under H_0: t-distribution. The number of degrees of freedom are fewer than in the case of the two-sample t-test. See the previous entry on Welch's confidence interval for difference between two means for the formula for df.

Formula: $t = \dfrac{(\bar{Y}_1 - \bar{Y}_2) - (\mu_1 - \mu_2)_0}{\sqrt{\dfrac{s_1^2}{n_1} + \dfrac{s_2^2}{n_2}}}.$

F-test

What is it for? Tests whether the variances of two populations are equal.

What does it assume? Both samples are random samples. The numerical variable is normally distributed within both populations.

Test statistic: F

Distribution under H_0: F-distribution with $n_1 - 1$, $n_2 - 1$ degrees of freedom. H_0 is rejected if $F \geq F_{\alpha(2), n_1-1, n_2-1}$. The quantity $F_{\alpha(1), k-1, N-k}$ is the critical value of the F-distribution corresponding to the pair of degrees of freedom ($n_1 - 1$ and $n_2 - 1$). Critical values of the F-distribution are provided in Statistical Table D. F is compared only to the upper critical value, because F, as computed below, always puts the larger sample variance in the top (the numerator) of the F-ratio.

Formula: $F = \dfrac{s_1^2}{s_2^2}$,

where s_1^2 is the larger sample variance and s_2^2 is the smaller sample variance.

Levene's test

What is it for? Testing the difference between the variances of two or more populations.

What does it assume? Both samples are random samples, and the distribution of the variable is roughly symmetrical in both populations.

Test statistic: W

Distribution under H$_0$: F-distribution with the pair of degrees of freedom $k-1$ (for the numerator of W; see the formula below) and $N-k$ (for the denominator of W), where k is the number of groups (two in the case of a two-sample test) and N is the total sample size ($n_1 + n_2$ in the case of two groups). H$_0$ is rejected if $W \geq F_{\alpha(1),\, k-1,\, N-k}$. The quantity $F_{\alpha(1),\, k-1,\, N-k}$ is the critical value of the F-distribution (see the preceding description of the F-test for an explanation of the critical value).

Formula:
$$W = \frac{(N-k)\sum_{i=1}^{k} n_i(\bar{Z}_i - \bar{Z})^2}{(k-1)\sum_{i=1}^{k}\sum_{j=1}^{n_i}(Z_{ij} - \bar{Z}_i)^2},$$

where $Z_{ij} = |Y_{ij} - \bar{Y}_i|$ is the absolute value of the deviation between individual observation Y_{ij} (symbolized as the jth data point in the ith group) and the sample mean for its group $\bar{Y}_i$ (symbolized as the sample mean for group i). $\bar{Z}_i$ is the mean of all the Z_{ij} for the ith group, and $\bar{Z}$ is the grand mean of all the Z_{ij}'s, calculated as the average of all the Z_{ij}'s regardless of group. The n_i is the number of data points in the ith group, and k is the number of groups (two in the case of a two-sample test).

PRACTICE PROBLEMS

1. **Calculation practice: Paired t-test.** Can the death rate be influenced by tax incentives?[9] Kopczuk and Slemrod (2003) investigated this possibility using data on deaths in the United States in years in which the government announced it was changing (usually raising) the tax rate on inheritance (the estate tax). The authors calculated the death rate during the 14 days before, and the 14 days after, the changes in the estate tax rates took effect. The number of deaths per day for each of these periods is given in the table at right. The data are illustrated in the strip chart on the next page (paired observations are connected by line segments).

Year	Death rate under higher estate tax rate	Death rate under lower estate tax rate
1917	22.21	24.93
1917	18.86	20.00
1919	28.21	29.93
1924	31.64	30.64
1926	18.43	20.86
1932	9.50	10.14
1934	24.29	28.00
1935	26.64	25.29
1940	35.07	35.00
1941	38.86	37.57
1942	28.50	34.79

---

9. It has been suggested that death rates can be influenced by the proximity to big events. For example, the death rate dropped a bit leading up to the new millennium and increased a bit afterward.

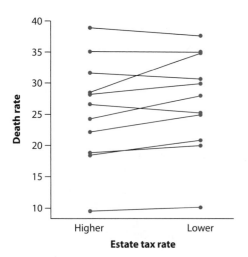

Estate tax rate

Let's use a paired *t*-test to ask whether the death rate changed significantly after the estate tax rate change.

a. State the null and alternate hypotheses for this analysis.

b. Why might a paired *t*-test be an appropriate method to apply to this comparison?

c. For each change in the estate tax, calculate the difference in death rate between the higher and lower tax regimes.

d. What is the mean of this difference?

e. What is the standard deviation of the difference?

f. What is the sample size?

g. What is the standard error of the mean difference?

h. Calculate *t* for this test.

i. How many degrees of freedom will this paired *t*-statistic have?

j. What is the critical value for the test, corresponding to $\alpha = 0.05$?

k. What is the *P*-value associated with the test statistic, and what is the conclusion from the test?

l. What scientific conclusion do you draw from these findings?

2. **Calculation practice: two-sample *t*-test.** When your authors were growing up, it was thought that humans dreamed in black and white rather than color. A recent hypothesis is that North Americans did in fact dream in black and white in the era of black-and-white television and

movies, but that we shifted back to dream in color after the introduction of color media. To test this hypothesis, Murzyn (2008) queried 30 older individuals who had grown up in the black-and-white era and 30 younger individuals who grew up with color media about their dreams. She recorded the percentage of color dreams for each individual. A mean of 68.4% of the younger peoples' dreams were in color (with a standard deviation of 31.8%). On average, 33.9% of the older individuals' dreams were in color, with a standard deviation of 36.9%. The scores were approximately normally distributed in each group. Is the difference between the two means statistically significant? We'll use a two-sample *t*-test for this comparison.

a. State the null and alternate hypotheses for this test.

b. What are the assumptions of a two-sample *t*-test? Do these data match these assumptions well enough?

c. What are the sample variances for each group?

d. How many degrees of freedom are associated with each group?

e. Calculate the pooled variance for these data.

f. What is the standard error of the difference between means?

g. Calculate *t*.

h. How many degrees of freedom does this *t*-statistic have?

i. What is the critical value of *t* with $\alpha = 0.05$ for this test?

j. What is the most precise *P*-value that you can determine for this test?

k. What can you conclude about the difference between the older and younger people in how often they dream in color? Interpret the conclusions of the test in terms of the original scientific question.

3. **Calculation practice: Confidence interval for the difference between two means.** Return to the data about the percentage of dreams in color as a function of age from Practice Problem 2. Determine a 95% confidence interval for the difference in mean percent of color dreams between the older and younger groups.

a. If you have not done so already from the previous problem, calculate the standard error of the difference between means. (This should be the same as in part (f) of Practice Problem 2.)

b. How many degrees of freedom does the t-statistic for this analysis have?

c. For a 95% confidence interval, what value of α should we use?

d. Using Statistical Table C, find the critical value of t for this α with the correct number of degrees of freedom.

e. What is the observed difference between means $(\bar{Y}_1 - \bar{Y}_2)$?

f. Calculate the 95% confidence interval for the difference between population means.

4. For each of the following scenarios, the researchers are interested in comparing the mean of a numerical variable measured in two circumstances. For each, say whether a paired t-test or two-sample t-test would be appropriate.

a. The weight of 14 patients before and after open-heart surgery.

b. The smoking rates of 14 men measured before and after a stroke.

c. The number of cigarettes smoked per day by 14 men who have had strokes compared with the number smoked by 14 men who have not had strokes.

d. The lead concentration upstream from five power plants compared with the levels downstream from the same plants.

e. The basal metabolic rate (BMR) of seven chimpanzees compared with the BMR of seven gorillas.

f. The photosynthetic rate of leaves in the crown of 10 Sitka spruce trees compared with the photosynthetic rate of leaves near the bottom of the same trees.

g. The photosynthetic rates of 10 randomly chosen Douglas-fir trees compared with 10 randomly chosen western red cedar trees.

h. The photosynthetic rate measured on 10 randomly chosen Sitka spruce trees compared with the rate measured on the western red cedar growing next to each of the Sitka spruce trees.

5. Dung beetles are one of the most common types of prey items for burrowing owls. The owls collect bits of large mammal dung and leave them around the entrance to their burrows, where they spend long hours waiting motionless for something tasty to be lured in. A research team wanted to know whether this dung actually attracted dung beetles or whether it had another use, such as to mask the odor of owl eggs from predators (Levey et al. 2004). They added dung to 10 owls' burrows, randomly chosen, and did not add dung to 10 other owl burrows. The researchers then counted the number of dung beetles consumed by the two types of owls over the next few days. The mean number of beetles consumed in the dung-addition group was 4.8, while the mean number was 0.51 in the control group. The standard deviations for the two groups were 3.26 and 0.89, respectively. What is an appropriate way to test for a difference in these two groups' beetle-capture rates?

Ronald G. Wolff

6. Practice Problem 8 in Chapter 11 described an experiment that compared the testes sizes of four experimental populations of monogamous flies to four populations of polyandrous flies. The data are as follows:

Mating system	Testes area (mm²)
Monogamous	0.83
Monogamous	0.85
Monogamous	0.82
Monogamous	0.89
Polyandrous	0.96
Polyandrous	0.94
Polyandrous	0.99
Polyandrous	0.91

a. What is the difference in mean testes size for males from monogamous populations compared to males from polyandrous populations? What is the 95% confidence interval for this difference? Assume normality and equal variances.

b. Carry out a hypothesis test to compare the means of these two groups. What conclusions can you draw?

7. In garter snakes, some males emerging from overwintering dens mimic females by producing female pheromones. Males might mimic females to warm up: males tend to emerge sooner than females and warm up in the sun while they wait for the females; the females are surrounded by males soon after emergence, and soak up their warmth. A prediction based on this idea is that males that mimic females should be covered by more males than males that don't mimic females. Observations on newly emerging garter snakes in Manitoba found that, on average, 58% of a male's body was covered by other males if he emitted female pheromones (with standard deviation 28%, measured on 49 males). In comparison, 32 males that did not emit female pheromones had, on average, 25% of their bodies covered by other males, with standard deviation 24% (Shine et al. 2001).

a. On average, how much more covered by other males are female mimics compared with nonmimics? Give a 95% confidence interval for this parameter.

b. Test the hypothesis that female mimicry has no effect on the proportion of body coverage in these garter snakes. What assumptions are you making?

8. *Spot the flaw.* Bluegill sunfish, a species of freshwater fish, prefer to feed in the open water in summer, but, in the presence of predators, they tend to hide in the weeds near shore. A study compared the growth rate of bluegills that fed in the open water with the growth rate of bluegills that fed only in nearshore vegetation. "Open-water" and "nearshore" fish were both measured in eight lakes, and the mean growth rate of open-water fish was compared to the mean growth rate of nearshore fish using a two-sample *t*-test. What was done wrong in this study?

9. The astonishing diversity of cichlid fishes of Lake Victoria is maintained by the preferences of females for males of their own species. To understand how the species arose in the first place, it is important to know the genetic basis of this preference in females. Researchers crossed two species of cichlids, *Pundamilia pundamilia* and *P. nyererei,* and raised the "F_1 hybrids" to adulthood. They measured degree of preference by the female F_1 fish for *P. pundamilia* males over *P. nyererei* males (Haeslery and Seehausen 2005). They then crossed the F_1 hybrids with each other to produce a second generation of hybrids (the "F_2"), which they also raised to adulthood and measured the same index of female preference. If a small number of genes are important in determining the preference, then the variance of the preference index will differ between these two generations (it will be highest in the F_2 hybrids). The researchers measured preference on 20 F_1 individuals and 33 F_2 individuals. The results are given below. Assume preference has a normal distribution in both populations.

F_2 hybrids: 0.380, 0.271, 0.211, 0.188, 0.157, 0.140, 0.131, 0.126, 0.126, 0.065, 0.048, 0.048, 0.024, 0.017, 0.017, 0.000, 0.000, −0.009, −0.009, −0.014, −0.032, −0.032, −0.044, −0.063, −0.068, −0.082, −0.082, −0.082, −0.143, −0.198, −0.300, −0.314, −0.348.

F_1 hybrids: 0.114, 0.101, 0.080, 0.082, 0.080, 0.067, 0.054, 0.015, 0.015, 0.007, 0.007, −0.019, −0.019, −0.019, −0.024, −0.034, −0.049, −0.058, −0.099, −0.105.

a. Choose a type of graph and compare the frequency distributions of female preference index in the F_1 and F_2 hybrids. What difference is suggested?

b. Calculate the variances of female preference index in the two hybrid crosses. Do the numbers agree with your visual estimate in (a)?

c. Test whether the variance of female preference index differs between the two crosses.

10. In most election years since 1960, a televised debate between the leading candidates for president has been influential in determining the outcome of the U.S. election. One analysis of the transcripts of these debates looked at the number of times that each candidate used the words "will," "shall," or "going to" as an indication of how many promises the candidate made. Also recorded was whether the candidate won the popular vote. (This is not always the same candidate who won the election. In 2000, for example, Al Gore won the popular vote, but George Bush attained the presidency.) The results are shown in the following table. Debates were not held in 1964, 1968, and 1972, and full transcripts were not available from 1984. Was the winner or loser significantly more likely to make promises, as measured by this index? Use an appropriate test.

Year	Candidate	Won (W) or lost (L) popular vote	Number of "will"s, "shall"s, and "going to"s
1960	Kennedy	W	163
1960	Nixon	L	122
1976	Carter	W	68
1976	Ford	L	32
1980	Reagan	W	19
1980	Carter	L	18
1988	G. Bush	W	111
1988	Dukakis	L	85
1992	Clinton	W	79
1992	G. Bush	L	75
1996	Clinton	W	56
1996	Dole	L	33
2000	Gore	W	68
2000	G. W. Bush	L	48
2004	G. W. Bush	W	176
2004	Kerry	L	149

11. Most bats are very poor at walking, but the vampire bat[10] is an exception. It is not clear why bats are poor walkers from a mechanical perspective, although a leading hypothesis has been that their hind legs are too weak. A test of this hypothesis by Riskin et al. (2005) measured and compared the strength of the hind legs between an insectivorous bat that walks poorly (*Pteronotus parnellii*) and the vampire bat (*Desmodus rotundus*). Six individual *Pteronotus* were measured, with an average hind-leg strength of 93.5 (in units of percent of body weight), with standard deviation 36.6. The mean hind-leg strength for six vampire bats was 69.3, with standard deviation 8.1.

a. Assuming that these measures of strength were normally distributed within groups, what is the appropriate test for comparing the mean hind-leg strength of these two species?

b. Look at the results closely. Without performing a hypothesis test, comment on the hypothesis that insufficient hind-leg strength is the reason that the insectivorous bat *Pteronotus* cannot walk well.

12. Ostriches live in hot environments, and they are normally exposed to the sun for long periods. Mammals in similar environments have special mechanisms for reducing the temperature of their brain relative to their body temperature. Fuller et al. (2003) tested whether ostriches could do the same. The mean body and brain temperature of six ostriches was recorded at typical hot conditions. The results, in degrees Celsius, are as follows:

Ostrich	Body temperature	Brain temperature
1	38.51	39.32
2	38.45	39.21
3	38.27	39.2
4	38.52	38.68
5	38.62	39.09
6	38.18	38.94

---

10. Vampire bats can even run. See http://www.nature.com/nature/journal/v434/n7031/extref/434292a-s2.mov.

a. Test for a mean difference in temperature between body and brain for these ostriches.

b. Compare the results to the prediction made from mammals in similar environments.

13. The following graphs are all based on random samples with more than 100 individuals. The red dots represent sample means.

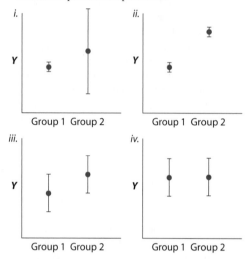

a. Assume that the error bars extend two standard errors above and two standard errors below the sample means. For which graphs can we conclude that group 1 is significantly different from group 2?

b. Assume that the error bars mark 95% confidence intervals. For which graphs can we conclude that group 1 is significantly different from group 2?

c. Assume that the error bars extend one standard error above and one standard error below the sample means. For which graphs can we conclude that group 1 is significantly different from group 2?

d. Assume that the error bars extend two standard deviations above and two standard deviations below the sample means. For which graphs can we conclude that group 1 is significantly different from group 2?

14. Vertebrates are thought to be unidirectional in growth, with size either increasing or holding steady throughout life. Marine iguanas from the Galápagos (see the photo on the first page of this chapter) are unusual because they might actually shrink during low food periods caused by El Niño events (Wikelski and Thom 2000). During these low food periods, up to 90% of the iguana population can die from starvation. The following histogram plots the changes in body length of 64 surviving iguanas during the 1992–1993 El Niño event:

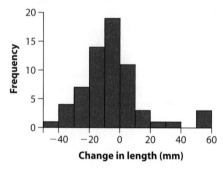

The average change in length was −5.81 mm, with a standard deviation of 19.50 mm.

a. By how much did the iguanas shrink on average? Determine the most-plausible range of values for the change in mean length of marine iguanas during the El Niño event. What assumptions are you making?

b. Using your answer in part (a), what are some of the plausible values for the mean change in weight over the El Niño event? Are the data consistent with a shrinking mean? Are they consistent with no change or even an increase in the mean?

c. How variable was the change in length among individual iguanas? Calculate the 95% confidence interval for the standard deviation of the change in length.

d. Test the hypothesis that length did not change on average during the El Niño event.

15. Red-winged blackbird males defend territories and attract females to mate and raise young there. A male protects nests and females on his territory both from other males and from predators. Males also frequently mate with females on adjacent territories. When males are successful in mating with females outside their territory, do they also attempt to protect these females and their nests from predators? An experiment measured the aggressiveness of males toward stuffed magpies placed on the territories adjacent to the males (Gray 1997). This aggressiveness was measured on a scale where larger scores were more aggressive and lower scores were less aggressive. This aggressiveness score was normally distributed. Later the researchers used DNA techniques on chicks in nests to identify whether the male had mated with the female on the neighboring territory. They compared the aggressiveness scores of the males who had mated with the adjacent female to those who had not. The results are as follows:

	Mean aggressive-ness score	Standard deviation of aggressive-ness score	Sample size (n)
Mated with neighbor	0.806	1.135	10
Did not mate with neighbor	−0.168	0.543	36

Test whether there are differences in the mean aggressiveness scores between the two groups of males. Are males aggressive to a different degree depending on whether they had mated with a neighboring female?

16. Mosquitoes find their victims in part by odor, so it makes sense to wonder whether what we eat and drink influences our attractiveness to mosquitoes. A study in West Africa (Lefèvre et al. 2010), working with the mosquito species that carry malaria, wondered whether drinking the local beer influenced attractiveness to mosquitoes. They opened a container holding 50 mosquitoes next to each of 25 alcohol-free participants and measured the proportion of mosquitoes that left the container and flew toward the participants (they called this proportion the "activation"). They repeated this procedure 15 minutes after each of the same participants had consumed a liter of beer and measured the "change in activation" (after minus before). This procedure was also carried out on another 18 human participants who were given water instead of beer. The change in activation of

mosquitoes is given for both the beer- and water-drinking groups:

Beer group: 0.36, 0.46, 0.06, 0.18, 0.25, 0.18, −0.06, −0.14, 0.12, 0.39, 0.17, −0.16, −0.05, 0.19, 0.25, 0.31, 0.17, −0.03, 0.23, −0.03, 0.26, 0.30, 0.11, 0.13, 0.21.

Water group: 0.04, 0, −0.08, −0.12, 0.201, −0.039, 0.10, 0.041, 0.02, 0.236, 0.05, 0.097, 0.122, −0.019, 0.021, −0.08, −0.165, −0.28.

a. Name three types of graphs that could be used to examine and compare the frequency distributions of the two samples. Choose one of these methods and construct a graph. What trend is suggested?

b. Test for a difference between the mean changes in mosquito activation between beer-drinking and water-drinking groups.

17. Development of an effective vaccine against HIV has proven difficult. Even though the immune systems of infected individuals produces antibodies that neutralize circulating virus, the disease destroys the CD4 T cells producing those antibodies. Balazs et al. (2011) investigated a novel HIV treatment that bypasses the immune system. They used a special strain of "humanized" mice that carry human CD4 T cells, which are susceptible to HIV. Treatment mice received human antibody-producing genes, which were injected into leg muscle using a harmless virus. Control mice were injected with a reporter gene instead (luciferase). All mice were then injected with high doses of HIV. The data below record the percentage of healthy CD4 T cells remaining in the mice five weeks later. A high value indicates that many CD4 T cells remain, and hence that HIV has been neutralized, whereas a low value indicates that the mouse has succumbed to the disease.

Antibody treatment mice: 94, 96, 92, 88, 84, 81, 76, 54.

Control treatment mice: 20, 15, 11, 7, 3, 0.

a. Plot these data using a strip chart.

b. What are the most-plausible values for the means of each of the two treatments?

Calculate 95% confidence intervals for the mean percent of both treatments.

c. Add the confidence intervals from part (b) as error bars to your plot in part (a).

d. Examine the two confidence intervals in your graph. If the null hypothesis of no difference between means were tested with these data, would the null hypothesis be rejected? How do you know?

ASSIGNMENT PROBLEMS

18. The males of stalk-eyed flies (*Cyrtodiopsis dalmanni*) have long eye stalks. The females sometimes use the length of these eye stalks to choose mates. (See Example 11.2.) Is the male's eye-stalk length affected by the quality of its diet? An experiment was carried out in which two groups of male "stalkies" were reared on different foods (David et al. 2000). One group was fed corn (considered a high-quality food), while the other was fed cotton wool (a food of substantially lower quality). Each male was raised singly and so represents an independent sampling unit. The eye spans (the distance between the eyes) were recorded in millimeters. The raw data, which are plotted as histograms at right, are as follows:

Corn diet: 2.15, 2.14, 2.13, 2.13, 2.12, 2.11, 2.10, 2.08, 2.08, 2.08, 2.04, 2.05, 2.03, 2.02, 2.01, 2.00, 1.99, 1.96, 1.95, 1.93, 1.89.

Cotton diet: 2.12, 2.07, 2.01, 1.93, 1.77, 1.68, 1.64, 1.61, 1.59, 1.58, 1.56, 1.55, 1.54, 1.49, 1.45, 1.43, 1.39, 1.34, 1.33, 1.29, 1.26, 1.24, 1.11, 1.05.

These data can be summarized as follows, where the corn-fed flies represent treatment group 1 and the cotton-fed flies represent treatment group 2.

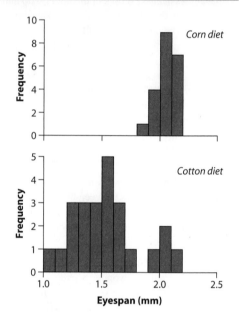

	Mean (mm)	Variance (mm²)	Sample size, *n*
Corn diet (group 1)	2.05	0.00558	21
Cotton diet (group 2)	1.54	0.0812	24

a. What is the best test to use for comparing the means of the two groups? Why?

b. Carry out the test identified in part (a), using $\alpha = 0.01$.

19. Fruit flies, like almost all other living organisms, have built-in circadian rhythms that keep time even in the absence of external stimuli. Several genes have been shown to be involved in internal timekeeping, including *per* (*period*) and *tim* (*timeless*). Mutations in these two genes, and in other genes, disrupt timekeeping abilities. Interestingly, these genes have also been shown to be involved in other time-related behavior, such as the frequency of wingbeats in male courtship behaviors. Individuals that carry particular mutations of *per* and *tim* have been shown to copulate for longer than individuals that have neither mutation. But do these two mutations affect copulation time in similar ways? The following table summarizes some data on the duration of copulation for flies that carry either the *tim* mutation or the *per* mutation (Beaver and Giebultowicz 2004):

Mutation	Mean copulation duration (min)	Standard deviation of copulation duration	Sample size, n
per	17.5	3.37	14
tim	19.9	2.47	17

a. Do these two mutations lead to different mean copulation durations? Carry out the appropriate test.

b. Do the populations carrying these mutations have different variances in copulation duration?

20. Researchers studying the number of electric fish species living in various parts of the Amazon basin were interested in whether the presence of tributaries affected the local number of electric fish species in the main rivers (Fernandes et al. 2004). They counted the number of electric fish species above and below the entrance point of a major tributary at 12 different river locations. Here's what they found:

Tributary	Upstream number of species	Downstream number of species
Içá	14	19
Jutaí	11	18
Japurá	8	8
Coari	5	7
Purus	10	16
Manacapuru	5	6
Negro	23	24
Madeira	29	30
Trombetas	19	16
Tapajós	16	20
Xingu	25	21
Tocantins	10	12

a. What is the mean difference in the number of species between areas upstream and downstream of a tributary? What is the 95% confidence interval of this mean difference?

b. Test the hypothesis that the tributaries have no effect on the number of species of electric fish.

c. State the assumptions that you had to make to complete parts (a) and (b).

21. Assignment Problem 20 from Chapter 11 discussed the relatedness of subordinate males to breeding females in the Seychelles warbler. Five subordinates that did not help feed the offspring of the older birds were measured for their relatedness to the offspring of the breeding females, with a mean relatedness of -0.05 and a standard deviation of 0.45. Another eight subordinates that *did* help feed younger offspring were also measured for their relatedness to the younger birds. For these eight, the mean relatedness was 0.27, with a standard deviation of 0.45.

a. Are helpful and unhelpful subordinates different in their mean relatedness to the younger birds? Carry out an appropriate hypothesis test.

b. Find the 95% confidence interval for the difference in mean relatedness for the two classes of subordinates.

22. In tilapia, an important freshwater food fish from Africa, the males actively court females. They have more incentive to court a female who still has eggs than a female who has already laid all of her eggs, but can they tell the difference? An experiment was done to measure the male tilapia's response to the smell of female fish (Miranda et al. 2005). Water containing feces from females that were either pre-ovulatory (they still had eggs) or post-ovulatory (they had already laid their eggs) was washed over the gills of males hooked up to an electro-olfactogram machine, which measured when the senses of the males were excited. The amplitude of the electro-olfactogram reading was used as a

measure of the excitability of the males in the two different circumstances. Six males were exposed to the scent of pre-ovulatory females; their readings averaged 1.51 with a standard deviation of 0.25. Six different males exposed to post-ovulatory females averaged readings of 0.87 with a standard deviation of 0.31. Assume that the electro-olfactogram readings were approximately normally distributed within groups.

a. Test for a difference in the excitability of the males with exposure to these two types of females.

b. What is the estimated average difference in electro-olfactogram readings between the two groups? What is the 95% confidence limit for the difference between population means?

23. A baby dolphin is born into the ocean, which is a fairly cold environment. Water has a high heat conductivity, so the thermal regulation of a newborn dolphin is quite important. It has been known for a long time that baby dolphins' blubber is different in composition and quantity from the blubber of adults. Does this make the babies better protected from the cold compared to adults? One measure of the effectiveness of blubber is its "conductance." This value was calculated on six newborn dolphins and eight adult dolphins (Dunkin et al. 2005). The newborn dolphins had an average conductance of 10.44, with a standard error of the mean equal to 0.69. The adult dolphins' conductance averaged 8.44, with the standard error of this estimate equal to 1.03. All measures are given in watts per square meter per degree Celsius.

a. Calculate the standard deviation of conductance for each group.

b. Test the null hypothesis that adults and newborns do not differ in the conductance of their blubber.

24. Weddell seals live in the Antarctic and feed on fish during long, deep dives in freezing water. The seals benefit from these feeding dives, but the food they gain comes at a metabolic cost. The dives are strenuous. A set of researchers wanted to know whether feeding per se was also energetically expensive, over and above the exertion of a regular dive (Williams et al. 2004).

They determined the metabolic cost of dives by measuring the oxygen use of seals as they surfaced for air after a dive. They measured the metabolic cost of 10 feeding dives and for each of these also measured a nonfeeding dive by the same animal that lasted the same amount of time. The data, in $(ml\ O_2\ kg^{-1})$, are as follows:

Individual	Oxygen consumption after nonfeeding dive	Oxygen consumption after feeding dive
1	42.2	71.0
2	51.7	77.3
3	59.8	82.6
4	66.5	96.1
5	81.9	106.6
6	82.0	112.8
7	81.3	121.2
8	81.3	126.4
9	96.0	127.5
10	104.1	143.1

a. Estimate the mean change in oxygen consumption during feeding dives compared with nonfeeding dives.

b. What is the 99% confidence interval for the population mean change?

c. Test the hypothesis that feeding does not change the metabolic costs of a dive.

25. Have you ever noticed that when you tear a fingernail, it tends to tear to the side and not down into the finger? (Actually, the latter doesn't bear too much thinking about.) Why might this be so? One possibility is that fingernails are tougher in one direction than another. Farren et al. (2004) compared the toughness of human fingernails along a transverse dimension (side to side) with toughness along a longitudinal direction, with 15 measurements of each. The toughness of fingernails along a transverse direction averaged 3.3 kJ/m², with a standard deviation of 0.95, while the mean toughness along the longitudinal direction was 6.2 kJ/m², with a standard deviation of 1.48 kJ/m².

a. Test for a significant difference in the toughness of these fingernails along the two dimensions. Assume that the data are from two independent samples of 15 people.

b. As it turns out, all of the fingernails in this study came from the same volunteer. How does this alter your conclusion from part (a)? Briefly, what steps would you take to design this study properly?

26. Hyenas, famously, laugh. (The technical term used by hyena biologists is "giggle.") Mathevon et al. 2010) investigated the information content of hyena giggles. In one analysis, they compared the giggles of pairs of hyenas, in which one member of each pair was the more dominant and the other socially subordinate. They measured the spectral variability of the hyena giggles using the coefficient of variation (CV) of sound spectrum features. Here are the data with these measures for each member of the pairs:

Spectral CV of dominant individual	Spectral CV of subordinate individual
0.384	0.507
0.386	0.569
0.252	0.235
0.226	0.415
0.323	0.436
0.287	0.451
0.303	0.399
0.317	0.220
0.277	0.338

Do dominant and subordinate individuals differ in the means of giggle spectral CV?

27. Refer to Practice Problem 16. The accompanying data are the values of mosquito activation (fraction of 50 mosquitos that left the container) recorded on the 25 participants before and after drinking one liter of the local beer.

a. Construct a graph illustrating the changes in mosquito activation. What trend is suggested?

b. Comment on the graph in relation to the assumptions needed to carry out a test.

c. Test whether mean activation of mosquitoes changed after consumption of beer.

d. How big is the effect of beer on activation? Construct a 95% confidence interval.

Subject	Activation before beer	Activation after beer
1	0.13	0.49
2	0.13	0.59
3	0.21	0.27
4	0.25	0.43
5	0.25	0.50
6	0.32	0.50
7	0.43	0.37
8	0.44	0.30
9	0.46	0.58
10	0.50	0.89
11	0.50	0.67
12	0.50	0.34
13	0.53	0.48
14	0.54	0.73
15	0.55	0.80
16	0.55	0.86
17	0.60	0.77
18	0.60	0.57
19	0.64	0.87
20	0.65	0.62
21	0.67	0.93
22	0.70	1.00
23	0.70	0.81
24	0.79	0.92
25	0.79	1.00

28. Does holding a weapon increase your aggressiveness afterward in other situations? Klinesmith et al. (2006) investigated this question by assigning 30 male college students to one of two groups. One group of 15 men were given a facsimile handgun to hold for 15 minutes, whereas the 15 men in the other group were given a toy instead. All men were then asked to participate in a test of taste sensitivity. Each was given a glass of water with a single drop of hot sauce to taste. Each man was also asked to add as much hot sauce as he wanted to a new glass of water to be given to the next person. (These

were not actually used on the next person.) The researchers measured how much hot sauce each man added, as a stand-in for aggression, because it correlates with the amount of pain inflicted on the next person. Do the two groups differ in the mean amount of hot sauce they add to the water? Here is a summary of the results:

Group	Sample size	Mean hot sauce added	SD hot sauce added
Gun handlers	15	13.61	8.35
Toy handlers	15	4.23	2.62

a. Assuming that the amount of hot sauce added per person is normally distributed in each group, would an ordinary two-sample t-test be an appropriate test for this analysis?

b. If not, what would be an appropriate method to use?

29. Refer to Practice Problem 17. How big is the estimated difference between the means of the antibody and control treatments? Use a confidence interval to calculate the most-plausible range of values for the difference in mean percent.

30. *Spot the flaw.* There are two types of males in bluegill sunfish. Parental males guard small territories, where they mate with females and take care of the eggs and young. Cuckolder males do not have territories or take care of young. Instead, they sneak in and release sperm when a parental male is spawning with a female, thereby fertilizing a portion of the eggs. A choice experiment was carried out on juvenile sunfish to test whether offspring from the two types of eggs (fertilized by parental male vs. fertilized by cuckolder male) are able to distinguish kin (siblings) from non-kin using odor cues. The researchers used a two-sample method to test the null hypothesis that fish are unable to discriminate between kin and non-kin. This null hypothesis was not rejected for offspring from parental males. However, the same null hypothesis was rejected for offspring from cuckolder males. The researchers concluded that offspring of cuckolder males are more able to discriminate

kin from non-kin than are offspring of parental males. What is wrong with this conclusion? What analysis should have been conducted?

31. Rutte and Taborsky (2007) tested for the existence of "generalized reciprocity" in the Norway rat, *Rattus norvegicus.* That is, they asked whether a rat that had just been helped by a second rat would be more likely itself to help a third rat than if it had not been helped. Focal female rats were trained to pull a stick attached to a tray that produced food for their partners but not for themselves. Subsequently, each focal rat's experience was manipulated in two treatments. Under one treatment, the rat was helped by three unfamiliar rats (who pulled the appropriate stick). Under the other treatment, focal rats received no help from three unfamiliar rats (who did not pull the stick). Each focal rat was exposed to both treatments in random order. Afterward, each focal rat's tendency to pull for an unfamiliar partner rat was measured. The number of pulls in a given period (in pulls/min.) by 19 focal female rats after both treatments is given below.

Focal rat	After receiving help	After recieving no help
10	0.43	0.29
11	0.86	0.14
12	0.57	0.29
20	0.86	0.57
30	0.29	0.29
31	1.14	0.71
32	0.57	0.29
33	0.86	0.86
34	1.43	0.86
40	0.86	0.86
41	0.57	0.43
42	0.86	1.00
43	0.00	0.86
50	1.00	0.43
51	0.86	1.00
52	0.86	0.00
60	1.86	1.14
61	0.86	0.71
61	0.86	0.71

a. Draw a graph to illustrate the data. What trend is evident?

b. What are the means of the two treatments, and what is the mean difference?

c. Test whether a difference was detectable between the help and no-help treatments.

d. Why is it important to apply the two treatments to the focal rats in random order?

32. Alcohol consumption is influenced by price and packaging, but what about glassware? Attwood et al. (2012) measured whether the time taken to drink a beer was influenced by the shape of the glass in which it was served. Participants were given 12 oz. (about 350 ml) of chilled lager and were told that they should drink it at their own pace while watching a nature documentary. The participants were randomly assigned to receive their beer in either a straight-sided glass or a curved, fluted glass. The data below are the total time in minutes to drink the glass of beer by the 19 women participants in the study.

Straight glass: 11.63, 10.37, 17.89, 6.96, 20.40, 20.64, 9.26, 18.11, 10.33, 23.54.

Curved glass: 7.46, 9.28, 8.90, 6.73, 8.25, 6.16, 13.09, 2.10, 6.37.

a. Show the data in a graph. What trend is suggested? Comment on other differences between the frequency distributions of the two samples.

b. Test whether the mean total time to drink the beer differs depending on beer-glass shape.

c. How much difference does it make? Provide a confidence interval of the difference.

d. A second test of the same hypotheses but using the data from male participants yielded the following results:

Straight glass: $\overline{Y}_1 = 7.987$, $s_1 = 2.459$.

Curved glass: $\overline{Y}_2 = 6.930$, $s_2 = 3.748$.

$\overline{Y}_1 - \overline{Y}_2 = 1.057$, $s_p^2 = 10.048$,
$SE_{\overline{Y}_1 - \overline{Y}_2} = 1.418$, $t = 0.746$, $df = 18$,
$P = 0.466$.

Is the following conclusion from the tests valid? "There is a significantly greater effect of beer-glass shape on mean time to drink in women than in men." Explain.

33. Spinocerebellar ataxia type 1 is a neurodegenerative disease marked by the gradual loss of motor skills and culminating in early death. It is caused by an expanded CAG repeat in the coding region of the *Ataxin-1* gene. Fryer et al. (2011) investigated the possible beneficial effects of exercise in treating the disease. They used a mild exercise regimen in a mouse model of the disease (a mouse strain in which an expanded CAG repeat was "knocked in" to the mouse version of the same gene, and that had similar symptoms). The life spans (in days) are given below for six exercised mice and six mice not given the exercise regimen. The data and 95% confidence intervals are shown in the accompanying graph.

No exercise: 240, 261, 271, 275, 276, 281.

Exercise: 261, 293, 316, 319, 324, 347.

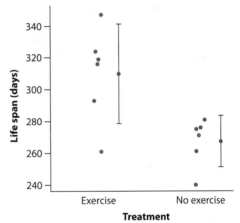

a. What type of graph is shown?

b. Using only the graph, is it possible to predict the outcome of a formal test of whether mean life span differs between the two treatments? Explain.

c. Test whether exercise affects life span in mice affected by the disease.

d. By how many days does exercise increase life span on average? Use a confidence interval to answer this question.

Which test should I use?

One of the most challenging parts of statistical analysis is deciding the right method for your particular question. In fact, with statistical computer programs so readily available, choosing the right method is usually the main challenge of data analysis left to us humans. The computer does the rest. We make choices to find the right graphical method, the right estimation approach, and the right hypothesis test. How can we choose the right method? Fortunately, the chain of logic involved in choosing the right method is similar in each case. In Section 2.7, we summarized the types of graphs available and when to use them. Here, we focus on choosing the right statistical test. We give four questions that you need to answer to help decide which test to use. The accompanying tables list information about the specific test, depending on your answers to these questions.

Does your test involve just one variable, or are you testing the association between two or more variables? Different methods apply in each case. Tests for a single variable may address whether a certain probability model fits the data or whether a population parameter (such as a mean or a proportion) equals a specified value. Tests for two variables address whether the variables are associated or whether one variable differs between groups.

Are the variables categorical or numerical? Different tests are suited to different types of data. When testing for association between two variables, it matters whether the variables are categorical, numerical, or a mixture.

Are your data paired? Two treatments can be compared either with two independent samples or with a paired design in which both treatments are applied to every unit of a single sample. Different methods are required for the two approaches.

What are the assumptions of the tests, and do your data meet those assumptions? For example, many powerful tests assume that the data are drawn from a normal distribution. If this is not true, then another approach must be found.

In this interleaf, we arrange the tests that we have already learned and several that are still to come in relation to these questions. Table 1 lists some of the common methods used for hypothesis tests involving a single variable.

Most hypothesis tests are carried out to determine whether two variables are associated or correlated. This question can be addressed when the two variables are both categorical, both numerical, or when there is one of each. Table 2 lists the most common tests used for each combination when the appropriate assumptions are met.

Many methods allow hypothesis tests of differences in a numerical response variable among different groups (see the bottom left corner of Table 2). Testing differences between groups is equivalent to a test of association between a categorical explanatory variable (group) and a response variable. Table 3 summarizes these tests and gives the particular circumstances in which each is used, along with alternatives that make fewer assumptions.

If you can organize these tests in your mind according to these classifications, it will be much easier to pick the right one. When you encounter a new test, think about whether it applies to one or more variables, whether the variables are continuous or numerical, and what assumptions it makes.

TABLE 1 Commonly used statistical tests for data on a single variable. These methods test whether a population parameter equals the value proposed in the null hypothesis or whether a specific probability model fits a frequency distribution. (Red numbers in parentheses refer to the chapter that discusses the test. Some refer to future chapters.)

Data type	Goal	Test
Categorical	Use frequency data to test whether a population proportion equals a null hypothesized value	Binomial test (7) χ^2 Goodness-of-fit test with two categories (used if sample size is too large for the binomial test) (8)
	Use frequency data to test the fit of a specific population model	χ^2 Goodness-of-fit test (8)
Numerical	Test whether the mean equals a null hypothesized value when data are approximately normal (possibly only after a transformation) (13)	One-sample t-test (11)
	Test whether the median equals a null hypothesized value when data are not normal (even after transformation)	Sign test (13)
	Use frequency data to test the fit of a discrete probability distribution	χ^2 Goodness-of-fit test (8)
	Use data to test the fit of the normal distribution	Shapiro–Wilk test (13)

TABLE 2 Commonly used tests of association between two variables. (Red numbers in parentheses refer to the chapter that discusses the test.)

		Type of explanatory variable	
		Categorical	Numerical
Type of response variable	Categorical	Contingency analysis (9)	Logistic regression (17)
	Numerical	*t*-tests, ANOVA, Mann–Whitney *U*-test, etc. [See *Table 3 for more details.*]	Linear and nonlinear regression (17) Linear correlation (16) Spearman's rank correlation (when data are not bivariate normal) (16)

TABLE 3 A comparison of methods to test differences between group means according to whether the tests assume normal distributions. (Red numbers in parentheses refer to the chapter that discusses the test.)

Number of treatments	Tests assuming normal distribution	Tests not assuming normal distributions
Two treatments (independent samples)	Two-sample *t*-test (12) Welch's *t*-test (used when variance is unequal in the two groups) (12)	Mann–Whitney *U*-test (13)
Two treatments (paired data)	Paired *t*-test (12)	Sign test (13)
More than two treatments	ANOVA (15)	Kruskal–Wallis test (15)

Sun star, *Solaster stimpsoni*

13 Handling violations of assumptions

All of the methods that we have learned about so far to estimate and test population means assume that the numerical variable has an approximately normal distribution. The two-sample *t*-test requires the further assumption that the standard deviations (and variances) are the same in the two corresponding populations. However, frequency distributions often aren't normal, and standard deviations aren't always equal. More often than we would like, our study organisms haven't read their stats books carefully enough, and the data they generate do not match the nice neat assumptions of classical statistical methods. What options are available to us when the data do not meet these assumptions?

In this chapter, we focus on four alternative options for analyzing such data:

1. *Ignore the violations of assumptions.* In some situations, we can use a procedure even if its assumptions are not strictly met. Methods for estimating and comparing *means* often work quite well when the assumption of normality is violated, especially if sample sizes are large and the violations are not too drastic.

2. *Transform the data.* For example, taking the logarithm is one way to transform data, with the result that the transformed data may better meet the assumptions. This procedure is often, but not always, effective.

3. *Use a nonparametric method.* A nonparametric method is one of a class of methods that do not require the assumption of normality. These methods can handle even badly behaved data, such as outliers that don't go away even when the data are transformed.

4. *Use a permutation test.* A permutation test uses a computer to generate a null distribution for a test statistic by repeatedly and randomly rearranging the data for one of the variables.

More alternatives are available, but these four are the most commonly used.[1] We examine each of these approaches and explain the circumstances under which they should be used. All four assume that each data point in the sample is randomly and independently chosen from the population. We begin by reviewing methods to evaluate the assumption of normality.

13.1 Detecting deviations from normality

A few techniques are available to judge whether numerical data are from a population with a normal distribution.

Graphical methods

The most convenient methods for evaluating whether data fit a normal distribution are graphical. The human eye is a powerful tool for detecting deviations from a pattern. Start by plotting a histogram of the data, separately for each group if there is more than one.

---

1. Two other options are simulation and bootstrapping, which we discuss in Chapter 19.

Data can be noisy, especially when the sample size is small, so don't expect your data to follow a perfect bell curve, even when they come from a normal population. For example, Figure 13.1-1 shows two rows of histograms of data, all sampled from a perfect normal distribution using a computer. The four in the first row are based on random samples of 10 individuals each, and the four in the second row are based on random samples of 20 individuals each. None of the eight histograms resembles a normal distribution precisely, but none is so badly behaved that it would cause us to give up our assumption of a normal population. Of course, the larger the sample size, the more likely it is for a frequency distribution to resemble that of the population it came from.

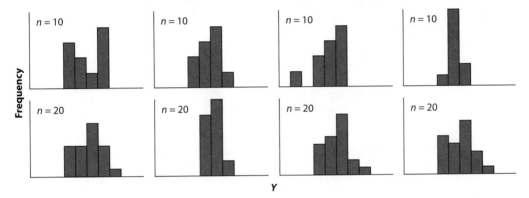

FIGURE 13.1-1 *Top row*: Histograms of four random samples of size $n = 10$ from the same normal distribution. *Bottom row*: Histograms of four random samples of size $n = 20$ from the same normal distribution.

In contrast, if the frequency distribution is strongly skewed or strongly bimodal, or if it has outliers, then the population distribution is unlikely to be normal. For example, Figure 13.1-2 shows histograms of four samples drawn from distributions that are not normal. Panels (a) and (c) are from distributions strongly skewed to the right, whereas panel (b) shows a distribution strongly skewed to the left. Panel (d) has an outlier far to the right of the rest of the data. Histograms like these indicate that the data were not from a normal population.

Besides histograms, the **normal quantile plot** is a second type of graphical technique for detecting departures from normality. The normal quantile plot compares each observation in the sample with the corresponding quantile expected from the standard normal distribution. (Remember, quantiles were introduced in Section 3.4.) An example is shown in Figure 13.1-3. Each dot on this plot represents one data point, whose measurement is indicated by its position on the horizontal axis and whose expected normal quantile is indicated on the vertical axis.[2] The points in

2. To find the normal quantile, order the observations from smallest to largest and assign each a rank, called i. The smallest data point has $i = 1$, the next smallest has $i = 2$, and so on up to the highest data point, where $i = n$. The estimated proportion of the distribution lying below an observation ranked i is $i/(n + 1)$. The corresponding normal quantile is the standard normal deviate Z having an area under the standard normal curve below it equal to $i/(n + 1)$. For example, if n is 99, the approximate fraction of the probability density lying below $i = 95$ is $95/(99 + 1) = 0.95$. The corresponding normal quantile is 1.64, which has 0.95 of the area under the normal curve below it (and 0.05 above it; see Statistical Table B).

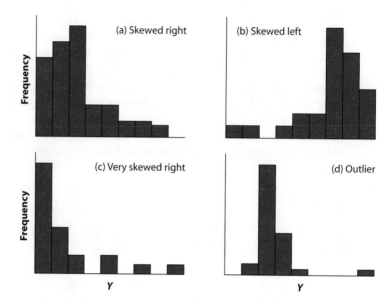

FIGURE 13.1-2 Frequency distributions of four data sets drawn from non-normal distributions. These distributions are (a) skewed right, (b) skewed left, (c) more extreme skew to the right, and (d) with an outlier.

a normal quantile plot should roughly follow a straight line if the data were sampled from a normal distribution. The dots will wiggle around a straight line even in the best of circumstances, as in Figure 13.1-3, because sampling is random. More systematic departures from a straight-line pattern, as judged by eye, indicate that the frequency distribution of the population likely deviates from the normal distribution. Substantial curvature over a large range of values or substantial jumps in the distribution indicate potential deviations from normality. We recommend using a statistics program on the computer to draw quantile plots.

FIGURE 13.1-3
A normal quantile plot for a random sample of 32 observations from a normal distribution. The points on the plot fall roughly along a straight line.

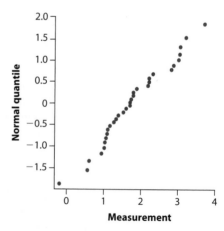

The *normal quantile plot* compares each observation in the sample with its quantile expected from the standard normal distribution. Points should fall roughly along a straight line if the data come from a normal distribution.

Example 13.1 illustrates the methods used to evaluate normality.

EXAMPLE   The benefits of marine reserves

13.1   Marine organisms do not enjoy anywhere near the same protection from human influence as do terrestrial species. However, marine reserves are becoming increasingly popular for biological conservation and the protection of fisheries. But are reserves effective in preserving marine wildlife?

Halpern (2003) matched each of 32 marine reserves to a control location, which was either the site of the reserve before it became protected or a similar unprotected site nearby. One index of protection evaluated by the study was the "biomass ratio," which is the total mass of all marine plants and animals per unit area of reserve divided by the same quantity in the unprotected control. This biomass ratio would equal one if protec-

tion had no effect. The biomass ratio would exceed one if the protection were beneficial, and it would be less than one if protection reduced biomass. The following list gives the biomass ratio for each of the 32 reserves:

1.34, 1.96, 2.49, 1.27, 1.19, 1.15, 1.29, 1.05, 1.10, 1.21, 1.31,
1.26, 1.38, 1.49, 1.84, 1.84, 3.06, 2.65, 4.25, 3.35, 2.55, 1.72,
1.52, 1.49, 1.67, 1.78, 1.71, 1.88, 0.83, 1.16, 1.31, 1.40.

Are marine reserves effective? In other words, does the mean biomass ratio differ from one?

The null and alternative hypotheses are as follows.

H_0: The mean biomass ratio is unaffected by reserve protection ($\mu = 1$).

H_A: The mean biomass ratio is affected by reserve protection ($\mu \neq 1$).

We might be tempted to use a one-sample *t*-test, but this test assumes that the biomass-ratio data are drawn from a population having a normal distribution. We must start by asking whether this assumption is valid.

The best starting point is to plot the data, as in Figure 13.1-4. The histogram shows that all is not well. The frequency distribution of the biomass ratio is skewed strongly

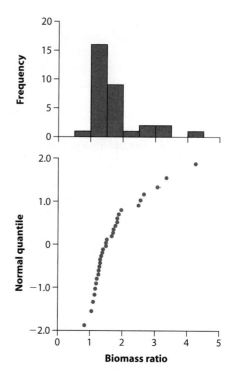

FIGURE 13.1-4
The frequency distribution of the "biomass ratio" of 32 marine reserves (*top panel*) and the corresponding normal quantile plot (*bottom panel*).

to the right. The quantile plot, moreover, shows curvature, indicating the poor fit of a normal distribution.

Formal test of normality

The "eyeball test" works quite well to detect departures from normality. However, formal goodness-of-fit tests to the normal distribution are available. These methods test the following hypotheses.

H_0: The data are sampled from a population having a normal distribution.

H_A: The data are sampled from a population not having a normal distribution.

Formal tests of normality should be used with caution. On the one hand, the test can give a false sense of security. A small sample size might not yield enough power to reject the null hypothesis of normality, even when data are drawn from a population without a normal distribution. On the other hand, a sufficiently large sample size will reject the null hypothesis of normality for many data whose departure from the normal distribution is not severe enough to warrant giving up on the methods that assume normality. The assumption of normality becomes less and less important when testing means as sample size increases, for reasons discussed in Section 13.2. As a result, we suggest that graphical methods and common sense are essential when evaluating the assumption of normality.

The **Shapiro–Wilk test** is probably the most powerful formal method for testing departures from normality. This test is carried out by most computer statistics programs, so we don't discuss the calculation details here. The Shapiro–Wilk test first estimates the mean and standard deviation of the population using the sample data. It then tests the goodness of fit to the data of the normal distribution having this same mean and standard deviation.

When we ran the Shapiro–Wilk test on the biomass-ratio data (Example 13.1), we rejected the null hypothesis that the data come from a normal distribution ($P < 0.001$), supporting what we saw from looking at the histogram and quantile plots.

> A *Shapiro–Wilk test* evaluates the goodness of fit of a normal distribution to a set of data randomly sampled from a population.

13.2 When to ignore violations of assumptions

What if the data don't meet the assumptions of normality or (for the two-sample methods) equal standard deviations? In this section, we consider the option of simply ignoring the violations. The justification is that methods for estimating and testing *means* are not highly sensitive to violations of the assumptions of normality and equality of standard deviations. Under certain conditions, these methods are robust, meaning that the answers they give are not sensitive to modest departures from the assumptions. Here we explain what these conditions are.

> A statistical procedure is *robust* if the answer it gives is not sensitive to violations of the assumptions of the method.

Violations of normality

Even though confidence intervals and *t*-tests require normally distributed data, the methods can sometimes be used to analyze data that are not normally distributed. The reason for this robustness comes from the central limit theorem (Section 10.6), which states that when a variable does not have a normal distribution, the distribution of sample means is nevertheless approximately normal when sample size is large. With large enough samples, therefore, the sampling distribution of means behaves roughly as assumed by methods based on the *t*-distribution, even when the data are not from a normally distributed population, provided that the violation of normality is not too drastic (Box and Andersen 1955). Robustness applies only to methods for means, not to methods like the F-test for testing variances, which are not robust to departures from the assumption of normality.

How large must samples be to allow us to ignore the assumption of normality? The answer depends on the shape of the distributions. If two groups are being compared and both differ from normality in different ways, then even subtle deviations from normality can cause errors in the analysis (even with fairly large sample sizes). For example, look back at the skewed distributions in Figure 13.1-2 (i.e., panels [a],[b], and [c]). If we compared two groups that both had the same skew as in panel (a) of Figure 13.1-2, we would get reasonably accurate answers from a two-sample t-test with sample sizes of about 30 in each group. However, we would require sample sizes of about 500 or more to get sufficient accuracy from a two-sample test if we were to compare one group with right-skewed measurements like those in panel (a) with another group whose data were left-skewed like those in panel (b). If the distributions are even more skewed than those in panels (a) and (b) of Figure 13.1-2, then the two-sample methods based on the t-distribution should be avoided in favor of alternative methods. The frequency distribution in panel (c) of Figure 13.1-2 is so skewed that a t-test would not give reliable answers even with extremely large sample sizes. Frequency distributions containing outliers (e.g., panel [d] of Figure 13.1-2) should never be analyzed with a t-test or a confidence interval based on the t-distribution. Methods that assume normality are very sensitive to outliers.

In the absence of definitive guidelines for every possible case, we recommend a cautious approach to data analysis. If the data show strong departures from normality, such as outliers, or if the frequency distributions in different groups are markedly different, then it is best to adopt one of the other options (i.e., data transformations, permutation tests, or nonparametric methods) rather than simply ignoring the violations of assumptions.

Unequal standard deviations

When can we ignore the assumption of equal standard deviations in two-sample methods? With moderate sample sizes (greater than 30 in each group), the two-sample methods for estimation and hypothesis testing using the t-distribution will still perform adequately with even a threefold difference between groups in their standard deviations, as long as the sample sizes of the two groups are approximately equal (Ramsey 1980). If sample sizes are not approximately equal, or if the difference in standard deviations is more than threefold, then the Welch's t-test (Section 12.3) should be used instead of the two-sample t-test, provided that the assumption of normality is met. (Indeed, Welch's t-test can be used even if the standard deviations are thought to be equal, but it is not as powerful as the two-sample t-test.) If the assumption of normality is also not met, then it is best to try data transformations, a permutation test, or one of the methods described in Chapter 19).

13.3 Data transformations

One of the best ways to handle data that don't match the assumptions of a statistical method is to see if you can **transform** the data to better meet the assumptions. Transformations can make the standard deviations more similar in different groups and improve the fit of the normal distribution to data. For example, it is often the case that a sample of data does not follow a normal distribution, but that the logarithms of the data match the normal distribution rather well. The test or estimation procedure could then be carried out on the log-transformed data instead. Our goal is to find a numeric scale on which a difference between two measurements has a similar interpretation regardless of the average measurement. For example, the difference in mass between two randomly chosen elephants is likely to be much greater than the difference in mass between two mice, simply because elephants are so much bigger. On a log scale, however, the differences among mice and among elephants might be more comparable, in which case the log scale is appropriate.

The three most frequently used transformations are the log transformation, the arcsine transformation, and the square-root transformation.[3] In this section, we examine the situations in which these three transformations are most likely to be useful. Keep in mind that, if a transformation is made to one data point, it has to be made to all data points from all samples for that variable if the data are to be compared.

> A *data transformation* changes each measurement by the same mathematical formula.

Throughout this section and beyond, we use a "prime" mark ($'$) to denote transformed data. Thus, if the original variable is Y, we would call the transformed variable Y' (pronouced "Y-prime").

Log transformation

The most common data transformation in biology is the **log transformation**. Typically, the data are converted by taking the natural log (ln) of each measurement. In mathematical terms,

$$Y' = \ln[Y].$$

Log base-10 is also sometimes used instead of the natural log, but we will use the natural log throughout this text. Again, all observations must be transformed in exactly the same way. (For example, it is not legitimate to compare the natural log of a variable from population 1 to the log base-10 of the variable in population 2.)

---

3. Other mathematical operations can also serve as valid transformations as long as there is a one-to-one correspondence between the data on its original scale and on the transformed scale (that is, it must be possible to transform back to get the original value without ambiguity).

Finally, the log transformation can be applied to data only when all the values are greater than zero. (The natural log of numbers less than or equal to zero is undefined). If the data include zero, then $Y' = \ln[Y + 1]$ can be tried instead.

In general, the log transformation is most likely to be useful under one or more of these conditions:

- The measurements are ratios or products of variables.
- The frequency distribution of the data is skewed to the right (i.e., has a long tail on the right).
- The group having the larger mean (when comparing two groups) also has the higher standard deviation.
- The data span several orders of magnitude.

For example, compare the two probability distributions in the left panel of Figure 13.3-1. Both distributions are right-skewed, and the group with the larger mean has the higher standard deviation. The log transformation fixes both problems. That is, both $\ln[Y_1]$ and $\ln[Y_2]$ are no longer skewed (they have normal distributions, instead) and both have the same standard deviation. Don't hesitate to try the log transformation in other situations as well.

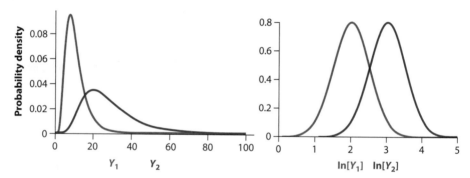

FIGURE 13.3-1 *Left panel:* Two right-skewed probability distributions having different standard deviations. In this case, the distribution with the highest standard deviation also has the highest mean. *Right panel:* The log transformations of the same variables. On the log scale, these two distributions have normal distributions with the same standard deviation.

The *log transformation* converts each data point to its logarithm.

In our experience, body measurements such as mass and length often show right-skewed frequency distributions that become more normally distributed after being log-transformed. But log transformation will not always solve the problem, even when frequency distributions are right-skewed or when the group with the larger mean also has the higher standard deviation. In these circumstances, though, the log transformation is worth a try. Just be sure to check the distribution of the transformed data to determine whether it fits the assumptions of the desired method.

Let's look again at the study on biomass ratio in marine reserves (Example 13.1). The raw data are listed in Table 13.3-1 along with the log-transformed data. Try a couple of the log calculations to make sure you get the same numbers that we did. Recall that the frequency distribution of the biomass ratio was right-skewed (Figure 13.1-4).

TABLE 13.3-1 Biomass ratios from 32 marine reserves and their log transformations.

Biomass ratio	ln[Biomass ratio]	Biomass ratio	ln[Biomass ratio]
1.34	0.29	3.06	1.12
1.96	0.67	2.65	0.97
2.49	0.91	4.25	1.45
1.27	0.24	3.35	1.21
1.19	0.17	2.55	0.94
1.15	0.14	1.72	0.54
1.29	0.25	1.52	0.42
1.05	0.05	1.49	0.40
1.10	0.10	1.67	0.51
1.21	0.19	1.78	0.58
1.31	0.27	1.71	0.54
1.26	0.23	1.88	0.63
1.38	0.32	0.83	−0.19
1.49	0.40	1.16	0.15
1.84	0.61	1.31	0.27
1.84	0.61	1.40	0.34

The effects of the log transformation on the frequency distribution of the data are shown in Figure 13.3-2. The skew has been much reduced in the log-transformed data, which now conform to a normal distribution much more closely than before (Figure 13.1-4).

FIGURE 13.3-2
Frequency distribution of the natural logarithm of the biomass ratio from 32 marine reserves.

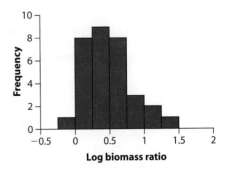

We can now apply the one-sample t-test to the transformed data, because they better meet the assumption that the sample came from a population having a normal distribution.

Let's use the data to test the following hypotheses.

H_0: The mean biomass ratio is unaffected by reserve protection ($\mu = 1$).
H_A: The mean biomass ratio is affected by reserve protection ($\mu \neq 1$).

Because we want to test the hypotheses on log-transformed data, though, we need to modify these hypotheses to reflect that transformation. Thus, a biomass ratio of 1 on the original scale is $\ln[1] = 0$ on the log-transformed scale, so our revised hypotheses are as follows.

H_0: The mean of the log biomass ratio of marine reserves is zero ($\mu' = 0$).
H_A: The mean of the log biomass ratio of marine reserves is not zero ($\mu' \neq 0$).

The prime (') denotes the transformed scale. After this, we proceed with the hypothesis test as usual. The sample mean of Y' is

$$\overline{Y}' = 0.479,$$

and the sample standard deviation is

$$s' = 0.366.$$

(Note that these are *not* the same as the log of $\overline{Y}$ and the log of the standard deviation of Y.) The corresponding one-sample t-statistic is

$$t = \frac{0.479}{0.366/\sqrt{32}} = 7.40,$$

with $df = n - 1 = 32 - 1 = 31$. Using a computer program, we find that the P-value is

$$P = 2.4 \times 10^{-8}.$$

This P-value is considerably less than 0.05, so we soundly reject the null hypothesis. Marine reserves have a higher mean log biomass than comparable unprotected areas, which implies that protected areas also have a higher biomass then unprotected areas.

Arcsine transformation

The **arcsine transformation** is used almost exclusively on data that are proportions:

$$p' = \arcsin[\sqrt{p}],$$

where p is a proportion measured on a single sampling unit. The arcsine (abbreviated, with a bizarre lack of brevity, as "arcsin") is the inverse of the sine function from trigonometry.[4] A dedicated transformation for proportions is needed because proportions tend *not* to be normally distributed, especially when the mean is close to zero or to one, and because groups differing in their mean proportion tend *not* to have equal

---

4. In other words, if $X = \sin[Y]$, then $Y = \arcsin[X]$. On many calculators, arcsine is given as $\sin^{-1}$. It is also known as the angular transformation and is measured in radians.

standard deviations. The arcsine transformation often solves both of these problems at the same time.

Don't forget the square-root component of the arcsine transformation. Note that if the original data are given as percentages, they must first be converted to proportions by dividing by 100 before applying the arcsine transformation.

The square-root transformation

The **square-root transformation** is often used when the data are counts, such as number of mates acquired, number of eggs laid, number of bacterial colonies, and so on:

$$Y' = \sqrt{Y + 1/2}.$$

This transformation, like the log transformation, sometimes helps to equalize standard deviations between groups when the group with the higher mean also has the higher standard deviation.[5]

In our experience, the effects of the square-root transformation are usually very similar to those obtained with the log transformation. It makes sense, then, to try both and see which one works best. If their effects are the same, then it doesn't matter which transformation you use.

Other transformations

When the frequency distribution of the data is skewed left, try the **square transformation**. This is done by squaring each data point:

$$Y' = Y^2.$$

If the square transformation doesn't work, try the **antilog transformation**,

$$Y' = e^Y,$$

on left-skewed data. When the data are skewed right, try the **reciprocal transformation**,

$$Y' = \frac{1}{Y}.$$

The square transformation and the reciprocal transformation are usable only if all the data points in all samples have the same sign. If all numbers in the samples are negative, then try multiplying each number by -1 before further transforming the new positive numbers.

Confidence intervals with transformations

It may often be necessary to transform the data to meet the assumptions of the methods for confidence intervals. Once the confidence interval is computed, it is usually

---

5. You will sometimes see the square-root transformation with 0 or 1 instead of $1/2$.

best to back-transform the lower and upper limits of the confidence interval before presenting the results.

For example, let's find the confidence interval for the mean of the log biomass ratio of marine reserves (Example 13.1). The original data were not normally distributed, so it is not valid to calculate a confidence interval directly from these data using the method based on the t-distribution. Recall that a log transformation fixed this problem.

The sample mean of the log-transformed data is $\overline{Y}' = 0.479$, and the sample standard deviation is $s' = 0.366$. There are a total of $n = 32$ data points in the sample. The 95% confidence interval for the mean of log of the biomass ratio is therefore

$$\overline{Y}' - t_{0.05(2),\,31}\,\mathrm{SE}_{\overline{Y}'} < \mu' < \overline{Y}' + t_{0.05(2),\,31}\,\mathrm{SE}_{\overline{Y}'}$$

$$0.479 - (2.04)\frac{0.366}{\sqrt{32}} < \mu' < 0.479 + (2.04)\frac{0.366}{\sqrt{32}}$$

$$0.347 < \mu' < 0.611$$

These limits to the confidence interval on the log scale are less easily interpreted than numbers on the original scale. However, we can back-transform the limits to yield a 95% confidence interval on the original scale. In the case of the log transformation, the back-transform is done by taking the antilog (i.e., raise e to the power of the quantity on the transformed scale). Therefore, the 95% confidence interval for the mean on the original scale (called the geometric mean)[6] is

$$e^{0.347} < \text{geometric mean} < e^{0.611}$$

or

$$1.41 < \text{geometric mean} < 1.84.$$

This 95% confidence for the geometric mean biomass ratio indicates that marine reserves have 1.41 to 1.84 times more biomass on average than the control sites do. This is a remarkably narrow interval for the mean effect of marine reserves.

Similar back-transformations can be calculated for all of the other possible transformations. Back-transformations for each transformation are given in the Quick Formula Summary (Section 13.9 at the end of the chapter).

A caveat: Avoid multiple testing with transformations

Trying multiple transformations to find which one works best to meet the assumptions of a test is an excellent plan. It is not legitimate, though, to try multiple transformations to find one that leads to a statistically significant outcome (such as a P-value smaller than 0.05). Each transformation would give a slightly different result to the statistical test, and you might be tempted to choose the result that most agreed with your preconceptions about the data. Repeated testing inflates the Type I error

---

6. The geometric mean is calculated by multiplying all data points together and the taking the nth root of the resulting product.

rate, so it should be avoided. Instead, first decide which transformation best meets the assumptions of the method, and then stick with that decision when carrying out the test.

13.4 Nonparametric alternatives to one-sample and paired *t*-tests

If ignoring violations of assumptions and transforming the data fail, sometimes you can try **nonparametric methods** instead. These methods—developed for calculating confidence intervals and hypothesis testing—make fewer assumptions about the probability distribution of the variable of interest in the population from which the data were drawn.[7] Nonparametric methods are handy when the frequency distribution of the data is not normal—such as when there are outliers. In contrast, the methods that do make assumptions about the distributions are called **parametric methods**. All of the methods we have learned about so far, including those based on the *t*-distribution, are parametric methods.

> A *nonparametric method* makes fewer assumptions than standard *parametric methods* do about the distributions of the variables.

Here we focus on nonparametric methods for hypothesis testing. We discuss nonparametric tests that address the same types of questions that we have already considered in Chapters 11 and 12. In this section, we cover alternatives to the one-sample *t*-test and the paired *t*-test. In Section 13.5, we use nonparametric methods to compare two independent groups. In subsequent chapters, we will introduce new nonparametric tests in tandem with the parametric tests designed for the same purpose.

Nonparametric tests are usually based on the **ranks** of the data points rather than the actual values of the data. In other words, the data points are ranked from smallest to largest, and the rank (first, second, third, etc.) of each data point is recorded. The actual measurements are not used again for the test. Using ranks is what frees us from making assumptions about the probability distribution of the measurements, because all distributions make similar predictions about the ranks of the measurements. Nonparametric tests are particularly useful when there are outliers in the data set, because ranks are not unduly affected by outliers.

Sign test

The **sign test** is a nonparametric method that can be used in place of the one-sample *t*-test or the paired *t*-test when the normality assumption of those tests can-

---

7. These methods are also sometimes called **distribution-free**.

not be met. The sign test assesses whether the *median* of a population equals a null hypothesized value. Measurements lying above the null hypothesized median are designated "+" and the numbers lying below are scored as "−." If the null hypothesis is correct, we expect half of the measurements to lie above the null hypothesized median and half to lie below, except for sampling error. The *P*-value can then be calculated using the binomial distribution (see Section 7.2). The sign test is simply a binomial test in which the number of data points above the null hypothesized median is compared with that expected when $p = 1/2$.

> The *sign test* compares the median of a sample to a constant specified in the null hypothesis. It makes no assumptions about the distribution of the measurement in the population.

Unfortunately, the sign test has very little power compared with the one-sample or paired *t*-test because it discards most of the information in the data. A measurement that is infinitesimally larger than the null hypothesized median and a data point that exceeds the median by several million both count only as a +. Nonetheless, the sign test is a useful tool to have in your statistical toolbox because sometimes no other test is possible.

EXAMPLE   Sexual conflict and the origin of new species

13.4   The process by which a single species splits into two species is still not well understood. One proposal involves "sexual conflict"—a genetic arms race between males and females that arises from their different reproductive roles.[8] Sexual conflict can cause rapid genetic divergence between isolated populations of the same species, leading to the formation of new species. Sexual conflict is more pronounced in species in which females mate more than once, leading to the prediction that they should form new species at a more rapid rate. To investigate this, Arnqvist et al. (2000) identified 25 insect taxa (groups) in which females mate multiple times, and they paired each of these groups to a closely related insect group in which females only mate once.

Which type of insect tends to have more species? Table 13.4-1 lists the numbers of insect species in each of the groups.

The data are paired. Thus, for each group of insects whose females mate once, there is a corresponding, closely related group of insect species in which females mate more than once. For this reason, the analysis must focus on the paired dif-

---

8. For example, in a number of insect species, male seminal fluid contains chemicals that reduce the tendency of the female to mate again with other males. However, the chemicals also reduce female survival, so females have evolved mechanisms to counter these chemicals.

TABLE 13.4-1 The number of species in 25 pairs of insect groups. Each pair matches a group of insect species in which females mate only once with a related group of insect species in which females mate multiple times.

	Number of species			
Taxon pair	Multiple-mating group	Single-mating group	Difference	Above (+) or below (−) zero
A	53	10	43	+
B	73	120	−47	−
C	228	74	154	+
D	353	289	64	+
E	157	30	127	+
F	300	4	296	+
G	34	18	16	+
H	3400	3500	−100	−
I	20	1000	−980	−
J	196	486	−290	−
K	1750	660	1090	+
L	55	63	−8	−
M	37	115	−78	−
N	100	30	70	+
O	21,000	600	20,400	+
P	37	40	−3	−
Q	7	5	2	+
R	15	7	8	+
S	18	6	12	+
T	240	13	227	+
U	15	14	1	+
V	77	16	61	+
W	15	14	1	+
X	85	6	79	+
Y	86	8	78	+

ferences. The differences listed in Table 13.4-1 were calculated by subtracting the number of species in the single-mating group from that of the corresponding multiple-mating group.

First, examine the histogram of the differences in Figure 13.4-1. These data have one outlier at 20,400, and we hardly need a Shapiro–Wilk test (Section 13.1) to tell us that the measurements are not normally distributed. At the same time, there are only 25 data points, which is too small a sample size to rely on the robustness of the paired *t*-test. There is no obvious transformation that would make these data normal, so we should pursue a nonparametric test instead. We will use the sign test to evaluate whether the median of the difference equals zero.

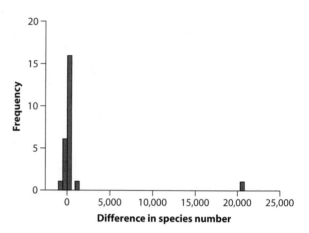

FIGURE 13.4-1
The distribution of differences in species number between single-mating and multiple-mating insect groups. There is an extreme outlier at 20,400.

Our hypotheses are as follows.

H_0: The median difference in number of species between insect groups is zero.

H_A: The median difference in number of species between these groups is not zero.

From this point on, the sign test is the same as the binomial test. If the null hypothesis is correct, then we expect half the measurements to fall above zero (+) and half to fall below zero (−). In fact, 18 out of the 25 measurements fall above zero and only seven fall below (see the last column in Table 13.4-1).

We can use the binomial distribution to calculate the P-value for the test. What is the probability of getting seven or fewer "−" observations out of 25 when the probability of a "−" observation is 0.5 under the null hypothesis? The answer is

$$\Pr[X \le 7] = \sum_{i=0}^{7} \binom{25}{i}(0.5)^i (0.5)^{25-i} = 0.02164.$$

The alternative hypothesis requires a two-sided test, so we need to double this calculated probability to account for the other tail. Thus, the P-value is

$$P = 2(0.02164) = 0.043.$$

Because this P-value is less than $\alpha = 0.05$, we reject the null hypothesis. Groups of insects whose females mate multiple times have more species than groups whose females mate singly, consistent with the sexual-conflict hypothesis.

Although it didn't come up when applying the sign test to the data in Example 13.4, what do you do about data points exactly equal to the hypothesized median? The usual approach is to drop all data points exactly equal to the median given by the null hypothesis. The test then proceeds as though those data had never existed. When this occurs, you must reduce the n used in the binomial calculations to the number of data points left after culling.

Finally, if the sample size is five or fewer, it is impossible to reject the null hypothesis from a two-sided sign test with $\alpha = 0.05$, no matter how different the true values are from the null hypothesized value. This underscores the fact that sign tests have low power and therefore require large sample sizes.

The Wilcoxon signed-rank test

The **Wilcoxon signed-rank test** is an improvement on the sign test for evaluating whether the median of a population is equal to a null-hypothesized constant. Unlike the sign test, the Wilcoxon signed-rank test retains information about magnitudes—that is, how far above or below the hypothesized median each data point lies. Unfortunately, it assumes that the distribution of measurements in the population is symmetric around the median—in other words, that no skew is present. This assumption is nearly as restrictive as the normality assumption of the one-sample *t*-test, greatly limiting its usefulness. (Skew, after all, is the usual reason the data don't fit the normal distribution, as in Example 13.4.) Because of this limitation, we do not explain the details of its calculation. Most statistics packages on the computer will carry out the Wilcoxon signed-rank test, but without highlighting its limitations.

13.5 Comparing two groups: the Mann–Whitney *U*-test

The **Mann–Whitney *U*-test** can be used in place of the two-sample *t*-test when the normal distribution assumption of the two-sample *t*-test cannot be met.[9] This method uses the ranks of the measurements to test whether the frequency distributions of two groups are the same. If the distributions of the two groups have the same shape, then the Mann–Whitney *U*-test compares the locations (medians or means) of the two groups.

> The *Mann–Whitney U-test* compares the distributions of two groups. It does not require as many assumptions as the two-sample *t*-test.

Example 13.5 shows how the Mann–Whitney *U*-test works.

EXAMPLE Sexual cannibalism in sagebrush crickets

13.5 The sage cricket, *Cyphoderris strepitans*, has an unusual form of mating. During mating, the male offers his fleshy hind wings to the female to eat. The wounds are not fatal,[10] but a male with already nibbled wings is less likely to be chosen by females he meets later. Females get some nutrition from feeding on the wings, which raises the question, "Are females more likely to mate if they are hungry?" Johnson et al. (1999) answered this ques-

---

9. Some statistical packages on the computer use an equivalent test, the Wilcoxon rank-sum test, instead of the Mann–Whitney *U*-test. The Wilcoxon rank-sum test statistic, *W*, is calculated differently from *U*, but both are based on the rank sums and give equivalent results. Don't confuse the Wilcoxon rank-sum test with the Wilcoxon signed-rank test, which is for paired data.

10. But they are disgusting.

tion by randomly dividing 24 females into two groups: one group of 11 females was starved for at least two days and another group of 13 females was fed during the same period. Finally, each female was put separately into a cage with a single (new) male, and the waiting time to mating was recorded. The data are listed in Table 13.5-1. The median time to mating was 13.0 hours for starved females and 22.8 hours for fed females.

TABLE 13.5-1 Times to mating (in hours) for female sagebrush crickets that were recently starved or fed. The measurements of fed females are in red to facilitate comparison after ranking (see Table 13.5-2).

Starved	Fed
1.9	1.5
2.1	1.7
3.8	2.4
9.0	3.6
9.6	5.7
13.0	22.6
14.7	22.8
17.9	39.0
21.7	54.4
29.0	72.1
72.3	73.6
	79.5
	88.9

We start by writing our hypotheses.

H_0: Time to mating is the same for female crickets that were starved and for those that were fed.

H_A: Time to mating differs between these two groups.

This is a two-tailed test. The frequency distributions of the two groups are shown in the histograms in Figure 13.5-1.

The data have positive skew, and a log transformation did not make them normally distributed. With the relatively small sample sizes, we cannot count on the central limit theorem to yield a normal sampling distributions of means, so the two-sample t-test is a poor choice. Therefore, we will apply a nonparametric test: the Mann–Whitney U-test.

FIGURE 13.5-1
Histograms of the time to mating for female sagebrush crickets who were starved (*upper panel*) or fed (*lower panel*).

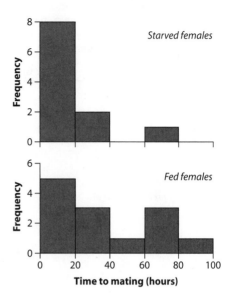

There are $n_1 = 11$ crickets that were starved and $n_2 = 13$ that were fed. For convenience, we'll label these group 1 and group 2, respectively.

To use the Mann–Whitney *U*-test, first rank all the data from smallest to largest, combining data from both groups, as shown in Table 13.5-2. The smallest measurement from either data set is from group 2, at 1.5 hours. We assign this data point a rank of 1. The second smallest data point is 1.7 hours, which we assign a rank of 2. We continue to do this, looking at data from both groups together until we have ranked them all.

Second, calculate the *rank-sum* for one of the two groups (it doesn't matter which group, but it's easier to use the group with the fewest data points). R_1 is the sum of all of the ranks in group 1 (the starved group). The ranks for group 1 are 3, 4, 7, 9, 10, 11, 12, 13, 14, 17, and 21, which added together give

$$R_1 = 3 + 4 + 7 + 9 + 10 + 11 + 12 + 13 + 14 + 17 + 21 = 121.$$

Third, use the rank-sum to calculate a new quantity, U_1, as follows:

$$U_1 = n_1 n_2 + \frac{n_1(n_1 + 1)}{2} - R_1 = 11(13) + \frac{11(11 + 1)}{2} - 121 = 88.$$

U_1 is the number of times that a data point from sample 1 is smaller than a data point from sample 2, if we compare all possible pairs of points taken one from each sample.

Also calculate U_2, as follows:

$$U_2 = n_1 n_2 - U_1 = 11(13) - 88 = 55.$$

U_2 is the quantity associated with group 2. If we applied the equation for U_1 to the data from group 2, we would get the same answer (55).

TABLE 13.5-2 Times to mating of female crickets from both groups, ordered from smallest to largest and then ranked. Data from group 2 (fed crickets) are highlighted in red to facilitate comparison.

Group	Time to mating	Rank
2	1.5	1
2	1.7	2
1	1.9	3
1	2.1	4
2	2.4	5
2	3.6	6
1	3.8	7
2	5.7	8
1	9.0	9
1	9.6	10
1	13.0	11
1	14.7	12
1	17.9	13
1	21.7	14
2	22.6	15
2	22.8	16
1	29.0	17
2	39.0	18
2	54.4	19
2	72.1	20
1	72.3	21
2	73.6	22
2	79.5	23
2	88.9	24

Fourth, choose the larger of U_1 or U_2 as our test statistic, U. In this case, U_1 is larger than U_2, so our test statistic is

$$U = U_1 = 88.$$

Finally, determine the P-value by comparing the observed U with the critical value of the null distribution for U. Critical values for cases in which sample sizes are relatively small are provided in Statistical Table E. Note that this table is based on sample sizes n rather than degrees of freedom. The null hypothesis is rejected if U equals or exceeds the critical value for U. For the cricket data, with $\alpha = 0.05$, $n_1 = 11$, and $n_2 = 13$, the critical value of U is

$$U_{0.05(2),\ 11,13} = 106.$$

Our U (88) is less than this critical value (106), so $P > 0.05$ and we cannot reject the null hypothesis. The data, therefore, provide insufficient evidence that time to mating is different between starved and fed female crickets.

Shanahan

Tied ranks

"You slept with her, didn't you?"

The data from Example 13.5 contain no observations tied for the same value, but often the same value appears more than once in a data set. When measurements are tied, we assign to all instances of the same measurement the average of the ranks that the tied points would have received. Table 13.5-3, for example, contains measurements from two hypothetical groups.

TABLE 13.5-3 Measurements from two hypothetical groups, illustrating how ties are ranked.

Group	Y	Rank
2	12	1
2	14	2
1	17	3
1	19	4.5
2	19	4.5
1	24	6
2	27	7
1	28	8

The ranks 1, 2, and 3 are assigned, as usual, to the three smallest values: 12, 14, and 17. But there are two measurements with the same value, 19. The two tied measurements, 19, would have been given ranks 4 and 5. Therefore, we assign each of the tied measurements a rank of $(4 + 5)/2 = 4.5$ (called the "midrank"). Thereafter, we continue to assign ranks to the data by assigning the next value, 24, a rank of 6, because we have already used ranks 4 and 5 for the two tied measurements (mistakes are often made at this step when ranking by hand).[11]

Data can also be in three-, four-, or even more-way ties, but we apply the same procedure. If we had three values that would otherwise have been given ranks 6, 7, and 8, then all of those individuals would be assigned a midrank of

---

11. The null distribution for the *U*-statistic is not the same when there are ties as when there are no ties, if the ties are between members of different groups. In this case the test is conservative, meaning that using the critical values in Statistical Table E yields a Type 1 error rate lower than the stated value. Ties also reduce the power of the test, but the effect is not great when there are only a few ties. Corrections for ties exist but are tedious to compute by hand.

$(6 + 7 + 8)/3 = 7$. Thereafter, we would continue by assigning the next higher data point the rank of 9.

Large samples and the normal approximation

Statistical Table E does not include critical values for the U-statistic when sample sizes are large. Fortunately, for medium and large sample sizes (n_1 and $n_2 > 10$), a transformation of the U-statistic,

$$Z = \frac{2U - n_1 n_2}{\sqrt{n_1 n_2 (n_1 + n_2 + 1)/3}},$$

has a sampling distribution that is well approximated by the standard normal distribution if the null hypothesis is true. For example, the critical value for the Z-statistic at $\alpha = 0.05$ is 1.96 (Statistical Table B).

13.6 Assumptions of nonparametric tests

While nonparametric tests do not rely on the normal distribution, they still make assumptions. Nonparametric tests assume, for example, that both samples are random samples from their populations. Without random samples, nonparametric tests are as likely as parametric tests to give erroneous answers.

As mentioned in Section 13.4, the Wilcoxon signed-rank test assumes that the distribution of measurements in the population is symmetrical. This assumption limits the utility of the method.

The Mann–Whitney U-test compares the distributions of two groups. The null hypothesis H_0 is that the distributions are the same. Rejecting H_0 therefore implies that the distributions are not the same. However, rejecting the null hypothesis does not necessarily imply that the two distributions have different locations (means or medians). Concluding that the locations are different, using a Mann–Whitney U-test, requires the additional assumption that the distributions of the two groups have the same shape. For example, the two distributions must have the same variance and the same skew. The Mann–Whitney U-test is very sensitive to differences between distributions in their shapes caused by unequal variances or different skews (Zimmerman 2003). For this reason we recommend that the Mann–Whitney U-test be used to test the null hypothesis that the distributions are the same, rather than testing the null hypothesis that the locations (means or medians) are the same. This limitation of the Mann–Whitney test is often not appreciated, and you will find many instances of misuse in the scientific literature.

13.7 Type I and Type II error rates of nonparametric methods

When the assumptions of a given test are met, the probability of making a Type I error (i.e., rejecting a true null hypothesis) is constant at α for both parametric and nonparametric tests. When the assumptions of a parametric test are not met, such as by a sharply nonnormal distribution of measurements, then the Type I error rate can become larger than the stated α-value, and sometimes this excess is extremely large (Ramsey 1980). In this case, the researchers would have false confidence in the reliability of the results. This excess Type I error rate is the main reason not to use parametric tests when the assumptions are strongly violated. Under these conditions, a nonparametric test will yield a Type I error rate equal to α, provided its less restrictive assumptions are met.

The story is different for Type II errors (i.e., failing to reject a false null hypothesis). By using only ranks, nonparametric tests use less information from the data than do parametric tests. The actual magnitudes are discarded. This loss of information causes nonparametric tests to have *less power* than the corresponding parametric test, when the assumptions of the parametric test are met. Less power means that the nonparametric test has a lower probability of rejecting a false null hypothesis, and therefore a higher Type II error rate, than does the parametric test. For this reason, most biologists prefer to use parametric tests if the assumptions can be met, and turn to nonparametric methods only after data transformation has failed to meet the assumptions of the parametric methods.

The reduced power of nonparametric tests is immaterial when the assumptions of the parametric test are strongly violated. In that case, the parametric test cannot be used at all.

> Nonparametric tests are typically less powerful than parametric tests.

How much lower is the power of the sign test and the Mann–Whitney U-test compared with their corresponding parametric tests? We can sensibly make this comparison only for cases when the assumptions of the parametric tests are met. In this case, the power of a Mann–Whitney U-test is about 95% as great as the power of a two-sample t-test *when sample sizes are large* (Mood 1954). This result is not too bad. The difference in power is greater, however, with smaller sample sizes. In the extreme case, when sample size in each of two groups is only two, the power of the Mann–Whitney U-test is zero.

The sign test has low power, much lower than the one-sample t-test and the paired t-test. Even with large samples, the sign test has only 64% of the power of a t-test (Mood 1954), and its power is even lower with smaller samples. Therefore, the sign test should be used only as a last resort, when the assumptions of the parametric tests simply cannot be met.

13.8 Permutation tests

Cheap computing has made possible new approaches for analyzing data, especially non-normal data. In this section we describe one method: permutation tests. We use the approach here to provide an alternative to the two-sample t-test and the Mann–Whitney U-test, to test for an association between a categorical explanatory variable (treatments or groups) and a numerical response variable. Permutation tests can also be used to test an association between two categorical variables (like a contingency analysis) or between two numerical variables (like a correlation analysis, described in Chapter 16). Other computer-intensive methods for estimation and hypothesis testing are covered in Chapter 19.

"Permutation" means rearrangement. In each step of a permutation test, we take the values of one of the two variables measured on the sample of individuals and randomly rearrange (permute) them. In other words, we randomly mix up the associations among the variables. This gives an idea of what values the test statistic would have if the two variables for an individual were not associated (except by chance). This method is often referred to as a randomization test, because the values for one of the two variables are repeatedly "randomized" during the test procedure. Permutation tests require us to make few assumptions about the frequency distributions of the variables and so are very versatile.

In a permutation test, a test statistic is chosen that measures the association between the two variables in the data. Here, we use the difference between sample means for two groups as our test statistic in a two-sample test.[12] This statistic is calculated on every permuted data set, each having the values of one variable randomly rearranged. Repeating the permutation procedure many times yields many values of the test statistic under the null hypothesis of no association. The frequency distribution of the test statistic is therefore used as an approximate null distribution. If the observed value of the test statistic calculated from the original data is unusual compared to the null distribution, we reject the null hypothesis of no association between the variables.

> A *permutation test* generates a null distribution for the association between two variables by repeatedly and randomly rearranging the values of one of the two variables in the data.

We illustrate the permutation test with the data from Example 13.5, in which we compared the mean time to mating of female sagebrush crickets assigned to either of two treatment groups: starved and fed. The data are reproduced in Table 13.8-1 in a slightly different format. These data are not normally distributed, as we saw from the histograms in Figure 13.5-1. As a result, we cannot use a two-sample t-test.

---

12. We can use χ^2 as our test statistic when we test association between two categorical variables. We can use the correlation coefficient when testing association between two numerical variables.

TABLE 13.8-1 Times to mating (in hours) of female sagebrush crickets that were recently starved or fed. Data from the two treatments are color-coded to more easily identifly the origin of each value later in Table 13.8-2.

Treatment	Time (hours)	Treatment	Time (hours)
Starved	1.9	Fed	1.5
Starved	2.1	Fed	1.7
Starved	3.8	Fed	2.4
Starved	9.0	Fed	3.6
Starved	9.6	Fed	5.7
Starved	13.0	Fed	22.6
Starved	14.7	Fed	22.8
Starved	17.9	Fed	39.0
Starved	21.7	Fed	54.4
Starved	29.0	Fed	72.1
Starved	72.3	Fed	73.6
		Fed	79.5
		Fed	88.9

Previously we analyzed these data using the Mann–Whitney U-test, but here we will reanalyze them with a permutation test.

There are a total of 24 data points, 11 from the starved treatment and 13 from the fed group. We are testing the following hypotheses.

H_0: Mean time to mating is the same for female crickets that were starved and for those that were fed.

H_A: Mean time to mating differs between these two groups.

This is a two-sided test, so we reject H_0 if the difference in the mean time to mating is much greater than zero or if it is much less than zero.

To carry out a permutation test of the hypotheses, we need to decide on a test statistic to describe the difference between the two treatments. The simplest statistic is the observed difference between the sample means, $\overline{Y}_1 - \overline{Y}_2$, where group 1 refers to the starved group. (We might convert this quantity to a t-statistic by dividing by the standard error of the difference, but there is little to be gained by doing so, because we won't be using the t-distribution.) From the data, the difference in the mean time to mating is

$$\overline{Y}_1 - \overline{Y}_2 = 17.73 - 35.98 = -18.26.$$

To generate the null distribution of possible values of $\overline{Y}_1 - \overline{Y}_2$, using permutation, follow these steps:

1. *Create a permuted set of data in which the values of the response variables are randomly reordered.* To do this, list all of the observations, as in Table 13.8-1. Now take all of the data values for one variable (here, time to mating) and ran-

domly rearrange them among individuals, while leaving the other variable unchanged. An example is shown in Table 13.8-2. In other words, combine all 24 values for the time to mating into a single pool. Choose one value at random, and assign it to the first individual in the starved treatment (this measurement was 3.8 in our first permutation; see Table 13.8-2). Next, choose one of the remaining 23 measurements at random and assign to the second starved individual (ours was the measurement 9.0). Continue with this process, eliminating each real data point from the sampling pool as you use it, until all 24 measurements are used up.[13] Let's call the result a "permuted sample." The size of each group in the permuted sample is the same as its size in the original data. Each original measurement is still present exactly once, but by chance it might be assigned to a different group than the group it came from. You will need a computer program to do this permutation.

TABLE 13.8-2. Outcome of a single permutation. Response measurements (time to mating) are color-coded as in Table 13.8-1 to indicate their original groups.

Treatment	Time (hrs)	Treatment	Time(hrs)
Starved	3.8	Fed	14.7
Starved	9.0	Fed	21.7
Starved	3.6	Fed	1.7
Starved	79.5	Fed	2.1
Starved	17.9	Fed	1.5
Starved	22.8	Fed	2.4
Starved	54.4	Fed	5.7
Starved	13.0	Fed	39.0
Starved	9.6	Fed	29.0
Starved	1.9	Fed	72.1
Starved	22.6	Fed	88.9
		Fed	72.3
		Fed	73.6

2. *Calculate the measure of association for the permuted sample.* For the single permuted sample shown in Table 13.8-2, the mean of the starved group is 21.65 and the mean of the fed group is 32.67. The difference is thus

$$\bar{Y}_1 - \bar{Y}_2 = 21.65 - 32.67 = -11.02.$$

This is the result from the first replicate of the permutation process.

3. *Repeat the permutation process many times—at least 1000 or more.* We repeated this permutation process a total of 10,000 times, recording the value of

---

13. In statistics jargon, this process is called "sampling without replacement." Once a value is sampled, it is removed from the pool, so the same value cannot be chosen again for the same permuted sample.

$\overline{Y}_1 - \overline{Y}_2$ each time. Figure 13.8-1 shows the resulting distribution of values for $\overline{Y}_1 - \overline{Y}_2$, from all of the 10,000 permutations. This distribution approximates the null distribution of the difference between the two groups.

To determine an approximate P-value for the test, we use the simulated null distribution for $\overline{Y}_1 - \overline{Y}_2$ in the same way that we would use a theoretical null distribution described by a formula (had one been available). To begin, find the proportion of values in the null distribution that equal or lie further in the tail than the observed value of the test statistic, calculated from the data. For our example, this is the fraction of test statistics from all of the permuted samples that are less than or equal to the observed difference between the means, $\overline{Y}_1 - \overline{Y}_2 = -18.26$. We found that 712 of the 10,000 permuted data sets, a proportion of 0.0712, yielded such an outcome (Figure 13.8-1). To obtain P, we multiply this proportion by two to take into account the equally extreme outcomes at the other tail of the null distribution (remember, this is a two-tailed test). Therefore, the P-value for the test is approximately $2(0.0712) = 0.142$. Since this value is greater than 0.05, we do not reject the null hypothesis. We are unable to conclude that starved and fed females differ in the mean time to mating.

FIGURE 13.8-1
The null distribution of the test statistic $\overline{Y}_1 - \overline{Y}_2$ from 10,000 replicates of the permutation process. Of the 10,000 permutations, only 712 had a value of $\overline{Y}_1 - \overline{Y}_2$ equal to or less than the observed value, -18.26.

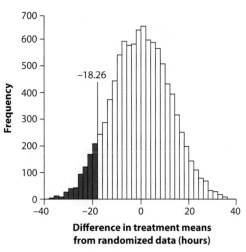

Because the permuted samples are generated by a random process, the null distribution would not be exactly the same if this test were repeated. As a result, the P-value would vary slightly from one test to another. Nevertheless, with a large number of permuted data sets, the P-value can be approximated with good precision.

Assumptions of permutation tests

Permutation tests make few assumptions and can be applied in a wide variety of circumstances, but some assumptions are required. First of all, the data must be a random sample from the population. Secondly, for permutation tests that compare means or medians between groups, the distribution of the variable must have the same shape in every population. Permutation tests are robust to violations of this assumption when sample sizes are large, more so than the Mann–Whitney U-test. Permutation tests have lower power (i.e., lower ability to reject a false null hypothesis) than parametric tests when the sample size is small, but they are more powerful than the Mann–Whitney U-test. They have similar power to parametric tests when sample sizes are large.

13.9 Summary

- Statistical methods—such as the one-sample, paired, and two-sample t-tests—that make assumptions about the distribution of variables are called parametric methods. Methods that do not make assumptions about the distribution of variables are called nonparametric methods.

- When the assumptions of parametric tests are violated, there are at least four alternative solutions: ignore the violations, if they are minor; transform the data to better meet the assumptions; use nonparametric methods, if the previous two strategies are insufficient; and use permutation tests. (More options are discussed in Chapter 19.)

- A statistical method is robust if violations of its assumptions do not greatly affect its results.

- Methods to evaluate the fit of the normal distribution to a data set include visual inspection of histograms, normal quantile plots, and the Shapiro–Wilk test.

- Parametric methods for comparing means are robust to minor violations of the assumption of normal populations, especially if the sample size is large. It is acceptable to ignore minor violations with large data sets and to proceed with the parametric tests.

- The two-sample t-test and the confidence interval for the difference between two means are robust to minor violations of the assumption of equal standard deviations between populations, if the sample sizes are approximately equal in the two groups.

- Many data can be transformed mathematically to a new scale on which the assumptions of parametric methods are met. The parametric method can then be performed on the transformed data.

- The most common data transformations are the log transformation, the arcsine transformation, and the square-root transformation. Of these, the log transformation is used most often.

- The sign test is a nonparametric alternative to the paired t-test or one-sample t-test. It is much less powerful (it is less likely to reject a false null hypothesis) than the corresponding parametric tests.

- The Wilcoxon signed-rank test is a nonparametric alternative to the paired t-test. However, it assumes a symmetrical distribution, so it should be used with caution.

- The Mann–Whitney U-test is a nonparametric alternative to the two-sample t-test. It tests the null hypothesis that the distributions are the same. Rejecting the null hypothesis allows us to conclude that the two distributions are different, but we cannot necessarily conclude that the two groups have different locations (means or medians). The U-test is less powerful than the two-sample t-test, but it is almost as powerful when the sample size is large.

- To use the Mann–Whitney U-test to test whether the locations of two distributions (medians or means) are the same, we must assume that the distributions of the variable in the two populations have the same shape.

- A permutation test is a method used to generate a null distribution for a measure of association between two variables (including a difference between groups) by randomly rearranging the observed values for one of the variables. The frequency distribution of test statistics calculated on many randomized data sets gives the null distribution of the test statistic.

- The null distribution of a test statistic generated in a permutation test is used to calculate the P-value.

13.10   Quick Formula Summary

Transformations

Log: $Y' = \ln[Y]$ or $Y' = \ln[Y + 1]$.

Arcsine: $p' = \arcsin[\sqrt{p}]$.

Square root: $Y' = \sqrt{Y + 1/2}$.

Back-transformations

Log: $Y = e^{Y'}$.

Arcsin: $p = (\sin[p'])^2$.

Square root: $Y = Y'^2 - 1/2$.

Sign test

What is it for? A nonparametric test of whether the median of a population equals a specified constant.

What does it assume? Random samples.

Test statistic: The number of measurements greater than (or less than) the median specified by the null hypothesis.

Formula: Identical to a binomial test with H_0: $p = 0.5$.

Mann–Whitney U-test

What is it for? A nonparametric test to compare the frequency distributions of two groups.

What does it assume? Random samples. When used to compare the locations (means or medians) of two distributions, the U-test additionally assumes that the distributions of the variable in the two populations have the same shape.

Test statistic: U.

Sampling distribution under H₀: The U-distribution, with sample sizes n_1 and n_2.

Formula: $U_1 = n_1 n_2 + \dfrac{n_1(n_1 + 1)}{2} - R_1$ and $U_2 = n_1 n_2 - U_1$, where R_1 is the sum of the ranks for group 1. For a two-tailed test, U is the greater of U_1 or U_2. For n_1 and $n_2 \leq 10$, use Statistical Table E. For large sample sizes, compare

$$Z = \frac{2U - n_1 n_2}{\sqrt{n_1 n_2 (n_1 + n_2 + 1)/3}}$$ to the critical value from the standard normal distribution.

PRACTICE PROBLEMS

1. **Calculation practice: Confidence interval for the mean using log transformation.** Refer to Practice Problem 19 in Chapter 10. Health spending per person from a random sample of 20 countries is given below.

Country	Per capita health expenditure in 2010
Bahrain	864
Belarus	320
Belize	239
Brunei Darussalam	882
Colombia	472
Congo, Rep.	72
Côte d'Ivoire	60
Cuba	607
Finland	3984
Germany	4668
Guinea-Bissau	47
Guyana	180
Jamaica	247
Lesotho	109
Malta	1697
Morocco	148
Namibia	361
Philippines	77
Qatar	1489
Saudi Arabia	680

We will use this sample to estimate the mean of log health expenditure, including a confidence interval.

a. Visualize the frequency distribution of the data using a histogram. What feature or features of this distribution indicate that the data are likely not from a population having normal distribution?

b. What features of this distribution make it a good candidate to try a log transformation?

c. Calculate the natural log transformation for each data point in the sample.

d. What is the sample size?

e. What is the mean of the log health expenditure?

f. What is the standard deviation of the mean log health expenditure?

g. Calculate the standard error of the mean log health expenditure.

h. Calculate the 95% confidence interval for mean log health expenditure.

i. What are the limits of this confidence interval expressed on the original (i.e., non-log) scale?

2. **Calculation practice: The sign test.** Female goldeneyes (a kind of duck) lay eggs in other

females' nests, in addition to the eggs they produce and raise in their own nests. One advantage to a female of this parasitic behavior is that other females do the work of raising her offspring. Andersson and Åhlund (2012) measured which of the eggs produced by 14 female goldeneyes were laid in other nests and which were laid in their own nests. We have converted their data into a "parasitism first" index, given below. The index is positive if a female tends to lay her earliest eggs parasitically in others' nests (and keeps the later ones for herself to raise). The index is negative if she tends to lay her last eggs in others' nests (and keeps the earlier ones). The index is zero if she alternates and her parasitic eggs are neither earlier nor later on average than eggs she keeps in her own nest.[14]

Female name	Parasitism first index
female 1	−2.3
female 2	8
female 17	10.5
female 18	4.6
female 19	5.5
female 20	6
female 37	0
female 51	5
female 55	4.5
female 58	6.5
female 70	4.6
female 76	3.8
female 80	5
female 94	5

Using the following steps, test whether goldeneyes tend to lay their eggs in other' nests before laying in their own,

a. Plot these data. Do they look normally distributed?

b. Why is a sign test suitable for these data?

c. What is the null hypothesis and the alternative hypothesis for this test?

d. Convert each data point into a positive or negative score, to express its value relative to the value stated in the null hypothesis.

e. After discarding the values that equal the value stated in the null hypothesis, what is the remaining sample size?

f. Conduct a binomial test using the observed positive and negative scores against the null expectation that half are positive and half are negative.

g. What can you conclude about the goldeneyes' behavior based on this test?

3. **Calculation practice: Mann–Whitney *U*-test.** Recycling paper has some obvious benefits, but it may have unintended consequences. For example, perhaps people are less careful about how much paper they use if they know that their waste will be recycled. Catlin and Wang (2013) tested this idea by measuring paper use in two groups of experimental participants. Each person was placed in a room alone with scissors, paper, and a trash can, and was told that he or she was testing the scissors. In the "recycling" group only, there was also a recycling bin in the room. The amount of paper used by each participant was measured in grams. The data from each person are listed below.

No recycling bin: 4, 4, 4, 4, 4, 4, 4, 5, 8, 9, 9, 9, 9, 12, 12, 13, 14, 14, 14, 14, 15, 23.

With recycling bin: 4, 5, 8, 8, 8, 9, 9, 9, 12, 14, 14, 15, 16, 19, 23, 28, 40, 43, 129, 130.

a. Make and examine histograms of these data. Are the frequency distributions of paper use in the two treatment groups similar in shape and spread?

14. Females lay just one egg per day, and so the parasitism-first behavior would have the effect of shortening the total laying period in her own nest and reducing the total span of time for all her eggs to hatch. This would get the young quickly out of the nest, where they are vulnerable to predation.

b. Based on your results in part (a), discuss your options for testing a difference between these two groups in the amount of paper used.

c. We will apply a Mann–Whitney U-test to test the hypothesis that these two treatments have the same distribution of paper use. State the null and alternative hypotheses clearly.

d. Rank all the values of paper use from smallest to largest. Properly account for ties.

e. For each treatment group, calculate the rank sum and the sample size.

f. Calculate the Mann–Whitney U_1 value for the treatment without the recycling bin.

g. Using the result from part (f), calculate U_2 for the treatment with the recycling bin.

h. What is the value of the test statistic U?

i. Calculate the P-value as accurately as you can, state the conclusion of the test, and interpret the results.

4. The four graphs shown below are normal quantile plots for four different data sets, each sampled randomly from a different population. For each graph, say whether the distribution is close to a normal distribution. If not, say how the data differs from what you might expect from a normal distribution.

FIGURE FOR PROBLEM 4

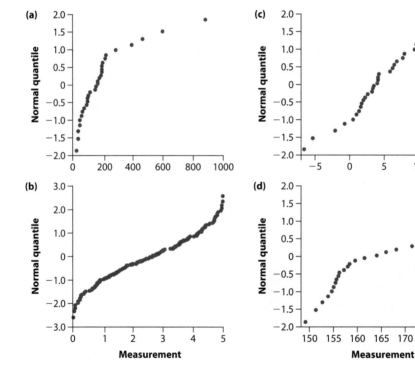

5. Below are frequency distributions of four hypothetical populations.

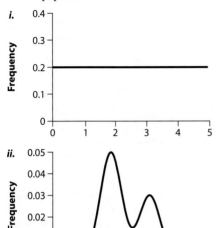

i.

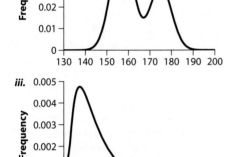

ii.

iii.

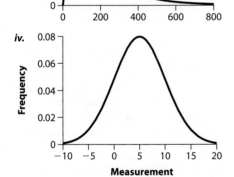

iv.

Measurement

a. For each graph (i through iv), say whether the distribution appears to be a normal distribution.

b. For each graph, imagine that you want to test the null hypothesis that the mean or median

is zero based on a random sample. What statistical method would you use? If you suggest a transformation of the data, say why you chose that transformation.

c. Match the distributions given in this problem to the quantile plots of samples in Practice Problem 4.

6. For each of the following sets of numbers, log-transform and calculate the sample mean and 95% confidence interval of the population mean. If that is impossible, say why it is impossible.

a. 10.2, 0.105, 67.3, 827

b. 2.1, 8.3, 3.2, 30.1

c. 17, −14, 37, 12

d. 12; 1.2; 125; 12,300

e. 0, 1.2, 4.5, 3.2

7. As a species becomes very rare, opportunities for mating might become reduced. This could result in low offspring numbers and further reductions in population size. Widén (1993) studied these effects in the rare *Senecio integrifolius*, a daisy-like herb, in Sweden. Below are his measurements of average percent seed set (percent of flowers producing seeds) of these plants at six different field sites in 1981:

29.8, 44.2, 58.3, 83.0, 78.2, 72.0 percent

a. Apply the arcsine transformation to these data and calculate the mean and standard deviation.

b. Calculate a 95% confidence interval for the population mean of the arcsine-transformed measurements.

c. Back-transform the upper and lower limits of the confidence interval to obtain the confidence interval on the percent scale.

8. When intruding male lions take over a pride of females, they often kill most or all of the infants in the group. This reduces the time until the females are again sexually receptive. This infanticide has many consequences for the biology of lions. It may be the reason, for example, that female lions band together in groups in the first place (to be better able to repel invading males).[15] The period after the takeover of a pride

---

15. Male infanticide has strong conservation implications, too. Most lion hunting in Africa is targeted toward the trophy males, and killing these males may provoke a takeover of their pride by other males, resulting in infanticide.

by a new group of males is an uncertain time, when the stability of the pride is unpredictable. As a result, we might predict that females will delay ovulation until this uncertainty has passed. A long-term project working on the lions of Serengeti, Tanzania, measured the time to reproduction of female lions after losing cubs to infanticide and compared this to the time to reproduction of females that had lost their cubs to accidents (Packer and Pusey 1983). The data are given below in days. Does infanticide lead to a different mean delay to reproduction in females than when cubs die from other causes? The data are not normally distributed within groups, and we have been unable to find a transformation that makes them normal. Perform an appropriate statistical test.

Accidental death: 110, 117, 133, 135, 140, 168, 171, 238, 255.

Infanticide: 211, 232, 246, 251, 275.

9. The skin of the rough-skinned newt, *Taricha granulosa*, stores an extremely poisonous neurotoxin called tetrodotoxin (TTX). In some geographical areas, garter snakes, a newt predator, have evolved some resistance to this toxin. In these areas, the newts make up a substantial part of the snakes' diet. As a first step to understanding the evolution of these traits, researchers compared resistance to TTX between two Oregon populations of garter snakes, one near Benton and the other near Warrenton (Geffeney et al. 2002). The data from 12 snakes are given in the accompanying table. Resistance is measured as the injected dose of TTX, in mass-adjusted mouse units (MAMUs), that causes a 50% reduction in crawl speed.

Locality	Resistance
Benton	0.29
Benton	0.77
Benton	0.96
Benton	0.64
Benton	0.70
Benton	0.99
Benton	0.34
Warrenton	0.17
Warrenton	0.28
Warrenton	0.20
Warrenton	0.20
Warrenton	0.37

a. Calculate summary statistics on these data and draw an appropriate graph to examine the data. Why would a two-sample *t*-test be an inappropriate method to test for differences in mean resistance?

b. List three appropriate methods that could be used to test a difference in resistance between these two populations.

c. Use a log transformation to test the hypothesis that the mean resistance does not differ between the two populations. Why might a log transformation be appropriate?

d. How big is the difference between the populations? Calculate a 95% confidence interval for the difference between populations in mean log-transformed resistance.

e. Back-transform the confidence interval from part (d) to the original scale. Provide an interpretation of this back-transformed interval.

10. When producing a 95% confidence interval for the difference between the means of two groups, under what circumstances can a violation of the assumption of equal standard deviations be ignored?

11. The following are very small data sets of human birth weights (in kg) of either singleton births or individuals born with a twin. We are interested in the difference in mean weight between singleton babies and twin babies.

Singleton: 3.5, 2.7, 2.6, 4.4.

Twin: 3.4, 4.2, 1.7.

a. Construct a valid permuted sample from these data for this difference. (Note that these samples sizes are in reality too small for an effective permutation test.)

b. Assume that we wanted to test the difference in *medians* between these two groups. Would that change the way in which the permuted sample would be created?

12. For each of the sets of samples in the accompanying figures (a)–(e), state which approach is best if the goal is to test the difference between the means of group 1 (on the left) and group 2 (on the right). Pay careful attention to differences between samples in the scales of the *x*-axis. These graphs have not been drawn according to the best practices that we outlined in Chapter 2 (histograms of two samples should be displayed one above the other and on the same scale), but they are similar to those you might obtain if you plotted them separately with a computer program.

FIGURE FOR PROBLEM 12

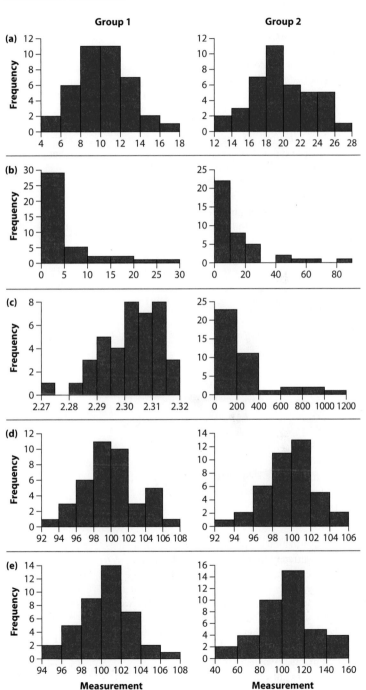

13. The nematode *Caenorhabditis elegans* is often used in studies of development and genetics. This species is an unusual animal, because most *C. elegans* individuals are hermaphrodites. That is, each worm has both ovaries and testes and acts as both a male and a female. Typically, a *C. elegans* individual will produce offspring by mating with itself. Very rarely, worms occur that have testes but not ovaries, and these males must mate with another individual to produce offspring. Therefore, it is important for the males to produce sperm that can compete well for access to eggs, and one way to do so might be to have larger sperm cells. LaMunyon and Ward (1998) measured the size of spermatids in 211 males and 700 hermaphrodites. Histograms of their findings are as follows:

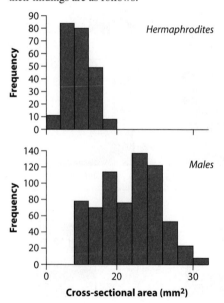

The distribution of spermatid size was significantly different from a normal distribution, so the researchers decided to perform a Mann–Whitney *U*-test. They wanted a two-sided test of the difference in spermatid size between the two types of worms. They got as far as calculating the value for *U*, which was $U = 35{,}910$.

a. What were their hypotheses?

b. Find the *P*-value for their test.

c. Why might a Mann–Whitney *U*-test with these data be inappropriate for a test of the difference between medians?

14. Only about 6% of plant species have separate male and female individuals (a syndrome called *dioecy*). The rest of the species have individuals with both male and female parts (called *monoecy*). Why are there so many more monoecious than dioecious species of plants? One possibility is that dioecious plants have low speciation rates or high extinction rates. To test this, Heilbuth (2000) compared the numbers of species in pairs of plant taxa of similar age. In each pair, one group was monoecious and the other group was the most closely related taxon that is dioecious. The data are shown in the following table.

Taxon pair	Number of species in dioecious group	Number of species in monoecious group
1	1	7000
2	1	5000
3	1150	4616
4	9	701
5	1	44
6	4	12
7	11	450
8	2	70
9	650	13
10	15	6
11	6	8
12	17	80
13	405	4
14	3	200
15	450	2770
16	4	11
17	50	53
18	370	2639
19	2	40
20	50	6
21	7	47
22	700	235
23	10	11
24	2	5
25	400	5772
26	400	72
27	1	6143
28	2	31

a. Before testing the difference, plot the data to help you decide which test to perform.

b. Carry out an appropriate test to determine whether monoecious and dioecious groups differ in the number of species.

15. The distribution of body size of mosquitoes (as measured by weight) is known to be log-normal (that is, size is normally distributed if log-transformed). The (untransformed) weights of 11 female and nine male mosquitoes are given below in milligrams (Lounibos et al. 1995). Do the two sexes weigh the same on average?

Females: 0.291, 0.208, 0.241, 0.437, 0.228, 0.256, 0.208, 0.234, 0.320, 0.340, 0.150.

Males: 0.185, 0.222, 0.149, 0.187, 0.191, 0.219, 0.132, 0.144, 0.140.

16. One of the founders of modern population genetics, J. B. S. Haldane, was once asked if he could infer anything about God from his study of nature. He replied, "An inordinate fondness for beetles," reflecting the beetles' status as the most species-rich animal group. One hypothesis for why beetles are so common is that a large number of them are plant eaters, and they may have ridden the coattails of the flowering plants (angiosperms) as the number of flowering plant species increased over the past 140 million years. A test of this hypothesis (Farrell 1998) compared the numbers of beetles in groups that feed on angiosperms to the number in the most closely related group that feeds on gymnosperms (non-flowering seed plants). Five such pairs of groups were compared in this way. The data are as follows:

Pair	Number of species in angiosperm eaters	Number of species in gymnosperm eaters
1	44,002	85
2	150	30
3	25,000	78
4	400	3
5	33,400	26

a. Are the differences in species number between these groups likely to come from a population with a normal distribution?

b. Do the groups that eat angiosperms and the groups that eat gymnosperms have different numbers of species? Carry out a formal test that does not require the assumption of a normal distribution.

c. Comment on the *P*-value and your conclusion in part (b) in light of the fact that the angiosperm-eating group had more species in all five pairs. What does this say about the power of your test?

17. Of the two scatter plots provided, one represents the real relationship between body size and brain size for 29 species of dinosaur (Hurlburt 1996). The other shows the first permuted data set created from that same data. Which do you think is the real data, and why?

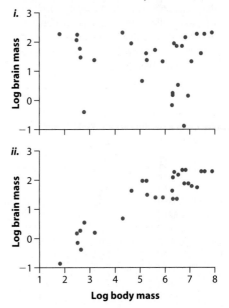

ASSIGNMENT PROBLEMS

18. Carotenoids are pigments responsible for much of the red we see in nature. For example, the bright red beak color of the male zebra finch is derived from carotenoids. Carotenoids are also important as antioxidants for humans and other animals, leading researchers to predict that carotenoids in birds may affect immune system function. To test this, a group of zebra finches was randomly divided into two groups (McGraw and Ardia 2003). Ten individual finches received supplemental carotenoids in their diet, and 10 individuals did not. All 20 birds were then measured using an assay that measures cell-mediated immunocompetence (PHA) as well as an assay that measures humoral immunity (SRBC). The data are given below. The researchers had independent reasons to believe that neither assay score would be normally distributed, so they preferred a nonparametric test.

a. Does PHA differ on average between the birds that received supplemental carotenoids and those that did not?

b. Does SRBC differ on average between the birds that received supplemental carotenoids and those that did not?

c. What assumptions did you require in parts (a) and (b)?

19. The conventional wisdom is that people who play a lot of sports have more sexual partners than those who do not. To test this, a group of French researchers asked two groups of students how many sex partners they had had in the previous year (Faurie et al. 2004). One group was composed of 250 physical education majors who regularly participated in sports; the other group was composed of 50 biology majors[16] who did not regularly do sports. The distribution of number of sex partners is shown in the graph below. Biology majors claimed an average of 1.24 sex partners in the previous year, while sports majors claimed an average of 2.4 partners in the previous year.

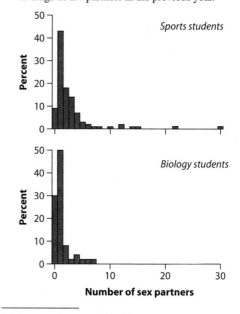

Treatment	PHA	SRBC
CAROT	1.511	2
CAROT	1.311	2
CAROT	1.460	3
CAROT	1.352	4
CAROT	1.491	5
CAROT	1.599	4
CAROT	1.653	8
CAROT	1.390	5
CAROT	1.779	7
CAROT	1.721	4
NO	1.454	4
NO	1.226	2
NO	1.198	3
NO	1.139	4
NO	1.277	3
NO	1.490	0
NO	0.912	3
NO	1.316	4
NO	1.234	2
NO	1.332	1

16. Sorry, we're not making this up.

As you can see, the distributions are very different from a normal distribution, and even with a log transformation the distributions are not normal. As a result, the researchers used a Mann–Whitney U-test to compare the median numbers of sex partners for the two groups. They found that the U_1 value corresponding to the biology majors was 8500.5, while the U_2 value corresponding to the sports group was 3999.5.

a. Finish the Mann–Whitney U-test for them. State the hypotheses and reach a conclusion.

b. Comment on your results in light of the assumptions of the Mann–Whitney U-test.

c. These researchers would like to know the *actual* number of sex partners for individuals in the two groups, rather than just the claimed number. What improvements can you suggest to their study design to achieve this goal?

20. Sockeye salmon swim sometimes hundreds of miles from the Pacific Ocean, where they grow up, to rivers for spawning. Kokanee are a type of freshwater sockeye that spend their entire lives in lakes before swimming to rivers to mate. In both types of fish, the males are bright red during mating. This red coloration is caused by carotenoid pigments, which the fish cannot synthesize but get from their food. The ocean environment is much richer in carotenoids than the lake environment, which raises the question: how do kokanee males become as red as the sockeye? One hypothesis is that the kokanee are much more efficient than the sockeye at using available carotenoids. This hypothesis was tested by an experiment in which both sockeye and kokanee individuals were raised in the lab with low levels of carotenoids in their diets (Craig and Foote 2001). Their skin color was measured electronically (as a* units on an L*a*b* standard that correlates strongly with redness). The data are as follows and are plotted in the accompanying histograms:

Kokanee: 1.11, 1.34, 1.55, 1.53, 1.50, 1.71, 1.87, 1.86, 1.82, 2.01, 1.95, 2.01, 1.66, 1.49, 1.59, 1.69, 1.80, 2.00, 2.30.

Sockeye: 0.98, 0.88, 0.97, 0.99, 1.02, 1.03, 0.99, 0.97, 0.98, 1.03, 1.08, 1.15, 0.90, 0.95, 0.94, 0.99.

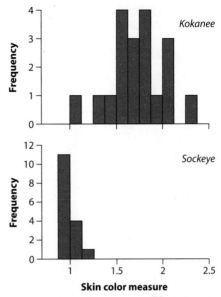

a. List two methods that would be appropriate to test whether there was a difference in mean skin color between the two groups.

b. Use a transformation to test whether there is a difference in mean between these two groups. Is there a difference in the mean of kokanee and sockeye skin color?

21. In a study of the Gouldian finch, Griffith et al. (2011) looked at stress caused by having an incompatible mate. There are two genetically distinct types of Gouldian finches, one having a red face and the other having a black face. Previous experiments have shown that female finches have a strong preference for mating with males with the same face color as themselves, and that when different face types of finch mate with one another, their offspring are less likely to survive than when both parents are the same type.

Researchers paired females sequentially with males of both types in random order. In other words, each female bred twice, once with a compatible male and once with an incompatible

With compatible male	With incompatible male
10	130
10	105
39	91
30	82
6	77
21	65
0	64
4	60
16	56
10	51
23	45
22	50
8	49
19	48
22	44
10	44
10	44
11	42
22	41
19	37
21	37
14	37
11	36
6	36
1	35
2	32
9	31
11	30
12	30
3	30
3	30
4	30
6	29
5	29
7	29
6	28
8	26
8	25
21	25
7	25
8	24
7	21
8	7

male. Each time, females produced a brood of young with the assigned male. For each pairing, the researchers measured the blood corticosterone concentration (in units of ng/ml) as an index of the amount of stress the females experienced. The corticosterone data for 43 females are given in the accompanying table.[17]

a. Plot the distribution of differences in stress levels between females with compatible and incompatible mates. What trend is suggested?

b. If we wished to carry out a test of the difference between treatment means, would a paired t-test be appropriate for these data? Why or why not?

c. Would a paired t-test be appropriate after a log transformation of the differences between treatments?

d. Would a sign test be appropriate for these data? Why or why not?

e. Test the hypothesis that stress levels are the same between females with compatible and incompatible mates. Use a sign test.

22. Using the data in Practice Problem 11, state whether each of the following sets of numbers (items a through f) is a possible permuted sample for use in testing the difference between the means of singleton and twin birth weights. If not, explain why.

	Singletons	Twins
a.	3.5, 2.7, 2.6, 4.4	3.4, 4.2, 1.7
b.	3.4, 4.2, 1.7, 3.5	2.7, 2.6, 4.4
c.	2.7, 2.6, 4.4	3.4, 4.2, 1.7, 3.4
d.	3.5, 3.5, 3.5, 3.5	3.5, 3.5, 3.5
e.	3.8, 3.8, 3.8, 3.8	3.8, 3.8, 3.8
f.	3.4, 3.5, 4.4, 3.4	4.4, 2.7, 2.6

23. One measure of the exposure of a person to tobacco smoke is the urinary cotinine-to-creatinine ratio (CCR). (Cotinine is formed in the body by breaking down nicotine.) Scientists measured this ratio in infants from smoking households (Blackburn et al. 2003). These households were divided according to their previous behavior into two groups: ones with strict controls to prevent exposure of the infant to

---

17. The data are also available at whitlockschluter.zoology.ubc.ca.

smoke (31 babies) and another group with less strict controls (133 babies). The mean (and standard deviation) of the log-transformed CCR was 1.26 (1.58) in the babies from strict households and 2.58 (1.16) from babies from less-strict households. The distribution of the log-transformed CCR was approximately normally distributed in both groups.

a. Do babies from households with strict controls differ significantly from those with less-strict controls in their exposure to smoke? Perform an appropriate test.

b. On the non-transformed scale, how much higher is the cotinine-to-creatinine ratio for babies in the less-strict group, as compared to those in the strict group?

c. Is this an observational or experimental study? Explain.

24. Use a six-sided die to make 10 permuted data sets suitable for testing the difference between the medians of the following two groups:[18]

Group A: 2.1, 4.5, 7.8.

Group B: 8.9, 10.8, 12.4.

a. Write out the 10 permuted data sets.

b. Using just these 10 permuted data sets, test the null hypothesis that the two groups have the same median, using $\alpha = 0.20$ for the significance level.

25. Researchers have observed that rainforest areas next to clear-cuts (less than 100 meters away) have a reduced tree biomass compared to rainforest areas far from clear-cuts. To go further, Laurance et al. (1997) tested whether rainforest areas more distant from the clear-cuts were also affected. They compiled data on the biomass change after clear-cutting (in tons/hectare/year) for 36 rainforest areas between 100 m and several kilometers from clear-cuts. The data are as follows:

−10.8, −4.9, −2.6, −1.6, −3, −6.2, −6.5, −9.2, −3.6, −1.8, −1.0, 0.2, 0.2, 0.1, −0.3, −1.4, −1.5, −0.8, 0.3, 0.6, 1.0, 1.2, 2.9, 3.5, 4.3, 4.7, 2.9, 2.8, 2.5, 1.7, 2.7, 1.2, 0.1, 1.3, 2.3, 0.5.

These measurements are plotted in the following histogram:

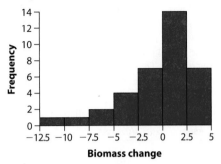

Test whether there is a change in biomass of rainforest areas following clear-cutting.

26. Male zebra finches have bright red beaks (see Assignment Problem 18), and experiments have shown that females prefer males with the reddest beaks. The red coloration in the beak comes from a class of pigments called carotenoids, which must be obtained in the diet of the birds. Pairs of brother finches were randomly assigned to alternative treatments: one was fed extra dietary carotenoids, the other was fed a diet low in carotenoids (Blount et al. 2003). Ten females were given a choice between brothers. Each female's preference was measured by the percentage of time she sat next to the carotenoid-supplemented male, standardized so that zero indicates equal time for each brother. Here are the preference data for each of the 10 pairs:

23, 27, 57, 15, 15, 54, 34, 37, 65, 12.

Choose an appropriate method and test whether females preferred one type of male over the other type.

27. The vuvuzela captured international attention during the 2010 World Cup in South Africa. In its modern incarnation, the vuvuzela is a plastic horn, about 65 cm long, that can produce a sound loud enough to cause permanent hearing damage. Blowing a vuvuzela requires a fair amount of air pressure, and Lai et al. (2011) were concerned that vuvuzela use by anyone

---

18. In this question and in some that follow, the sample sizes are often too small to provide powerful tests and are presented as exercises only.

carrying a pathogen would cause airborne contagions to be spread broadly through a crowd. They tested this idea with an experiment that compared the concentration of aerosol droplets or particles produced by people blowing vuvuzelas to that produced by the same people shouting instead. The data, measured as thousands of particles per liter, for 8 individuals are given in the table below.

	Particle concentration (1000s/liter)	
Person	Vuvuzela	Shouting
1	606	6.1
2	1077	6.4
3	220	1.3
4	396	1.8
5	1197	6.0
6	178	1.5
7	645	2.9
8	944	2.9

a. Take the log transformation of each value before finding differences. Then calculate a 95% confidence interval for the mean difference in log particle concentration between vuvuzelas and shouting.
b. Carry out an appropriate test for a mean difference in log particle concentration between the two forms of cheering.

28. The pseudoscorpion *Cordylochernes scorpioides* lives in tropical forests, where it rides on the backs of harlequin beetles to reach the decaying fig trees in which they live. Females of the species mate with multiple males over their

short lifetimes, which is puzzling because mating just once provides all the sperm she needs to fertilize her eggs. A possible advantage is that by mating multiple times, a female increases the chances of mating with at least one sperm-compatible male, if incompatibilities are present in the population. To investigate, Newcomer et al. (1999) recorded the number of successful broods by female pseudoscorpions randomly assigned one of two treatments. Females were each mated to two different males (DM treatment), or they were each mated twice to the same male (SM). This design provided the same total amount of sperm to females in both treatments, but DM females received genetically more diverse sperm than SM females did. The number of successful broods of offspring for each female is listed below. The data were not normally distributed; to test the null hypothesis of no difference between treatments in the mean number of broods, we carried out a permutation test in which the data were randomly reshuffled 10,000 times on the computer. Our test statistic was the difference between groups in the mean number of broods (SM minus DM). The observed value of this difference was –0.841. The null distribution from the 10,000 permutations is shown in the upper panel of the figure on the next page. The far left tail of the null distribution is shown in the lower panel. Numbers below each bar give the exact values of the test statistic; numbers above give the frequency of each of the values in 10,000 permutations. Using these values, carry out the permutation test.[19]

SM treatment: 0, 0, 0, 0, 0, 0, 0, 0, 0, 0, 0, 0, 0, 0, 0, 1, 1, 1, 1, 1, 1, 1, 1, 1, 1, 1, 2, 2, 2, 2, 2, 2, 2, 2, 2, 2, 2, 2, 2, 3, 3, 3, 3, 3, 3, 3, 3, 3, 3, 3, 3, 3, 4, 4, 4, 4, 4, 4, 4, 4, 4, 4, 4, 4, 5, 5, 5, 5, 5, 6, 6, 6.

DM treatment: 0, 0, 0, 0, 0, 0, 0, 1, 1, 1, 1, 1, 1, 1, 2, 2, 2, 2, 2, 2, 2, 2, 2, 2, 2, 3, 3, 3, 3, 3, 3, 3, 3, 3, 3, 3, 3, 3, 3, 3, 4, 4, 4, 4, 4, 4, 4, 4, 4, 4, 4, 4, 4, 4, 5, 5, 5, 5, 5, 5, 5, 6, 6, 6, 6, 6, 6, 7, 7.

---

19. The data are also available at whitlockschluter.zoology.ubc.ca.

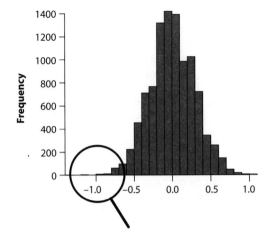

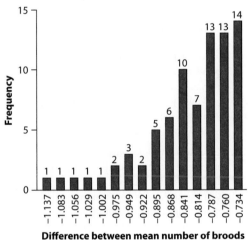

Difference between mean number of broods

the ants,[20] and isotope-labeling studies indicate that ant CHCs are transferred to the silverfish this way. Does chemical mimicry contribute to ant tolerance of silverfish? In one of a series of experiments, von Beeren et al. (2011) collected individual silverfish from ant colonies and isolated them for six days, after which most of the acquired CHCs had evaporated. The aggressive behavior of the ants toward these silverfish was then compared with behavior toward control silverfish not isolated for six days. The data below measure ant aggression toward the silverfish on a scale from 0 to 1.

Control: 0.04, 0.00, 0.22, 0.10, 0.11, 0.54.

Isolated: 0.25, 1.00, 1.00, 0.42, 0.50, 1.00, 1.00, 1.00.

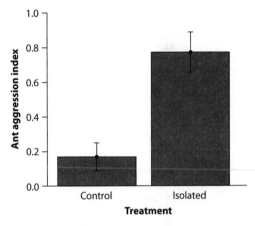

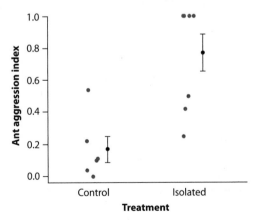

29. Despite the voracious habits of army ants, multiple species of invertebrates have managed to penetrate and exploit their societies. For example, the silverfish *Malayatelura ponerophila* is an insect that lives only in colonies of the Southeast Asia army ant, *Leptogenys distinguenda*, where it steals food brought in by the ants. If detected, the silverfish becomes ant food (see photo on the next page), but how does it usually evade detection? Ants recognize other ants from their own colony with a chemical signature, a complex blend of cuticular hydrocarbons (CHCs) on the surface of their exoskeleton. The silverfish have been observed rubbing up against

20. See movie at www.biomedcentral.com/content/download/supplementary/1472-6785-11-30-s8.mpeg.

a. Two commonly used methods for presenting the results are shown in the accompanying figure (with standard error bars). Which method is superior? Why?

b. Without transforming the data, apply an appropriate method to test whether the aggression index by ants toward silverfish was affected by isolation.

30. Dengue infects tens of millions of people annually, resulting in more than 10,000 deaths. The disease is caused by an RNA virus, which is transmitted principally by the mosquito *Aedes aegypti*. Previous work in *Drosophila* has shown that the bacterium *Wolbachia*, an endosymbiont living in the cytoplasm of the insect's cells, largely confers immunity from RNA viruses. *Wolbachia* also affects *Drosophila* sexual reproduction. By biasing its transmission to offspring, the bacterium spreads through *Drosophila* populations over multiple generations. Might it do the same to mosquitoes, and in the process rid the world of dengue?[21] *Wolbachia* does not occur naturally in *A. aegypti*, so microinjection was used to create a laboratory strain of the mosquito (called WB1) that harbors the endosymbiont. To examine the potential effects on transmission of dengue, Bian et al. (2010) infected mosquitoes from both the WB1 strain and the original wild strain with dengue. Fourteen days later, the mosquitoes were allowed to

feed on an artificial food solution for 90 min. Viral titers in the food solution were then measured (in plaque feeding units [pfu] per ml). The results are given below. Mosquitoes were tested in groups, and so each data value is an average for the group, measured in pfu/ml. The data have already been log-transformed using $\log(Y + 1)$, to better visualize the differences.

WB1: 8.0, 5.9, 4.4, 4.4, 2.4, 0.0, 0.0, 0.0.

Wild: 11.3, 10.8, 9.4, 6.5, 6.3, 5.9, 4.7, 4.2.

a. Plot the data with a strip chart. What trends are suggested? What features of the data might lead you to conclude that even after the log transformation, they do not fit a normal distribution?

b. Rank all the above data from small to large.

c. Carry out a Mann–Whitney U-test of whether WB1 differs from the wild strain in potential dengue transmission.

d. Why do you think we used $\log(Y + 1)$ rather than $\log(Y)$ when transforming the data?

e. Would your results in (c) have been different had the data not been log-transformed? Explain.

31. The cane toad (*Bufo marinus*), a large, toxic toad introduced to Australia in 1935, has been rapidly spreading across the continent at over 50 km/year. Are the fastest-moving toads at the leading edge of the expanding range morphologically different from other toads that move less rapidly? Phillips et al. (2006) compared the leg lengths of individually marked toads with the distances that they moved in a three-day period. In the following pair of graphs, the relative length of the leg is compared to the movement distance (each measurement is displayed as a deviation from the mean). One of the following two graphs is the original data, and the other shows the first permuted data set. Which is more likely to be the permuted data set? Explain your reasoning.

---

21. And advance the health of billions of mosquitoes.

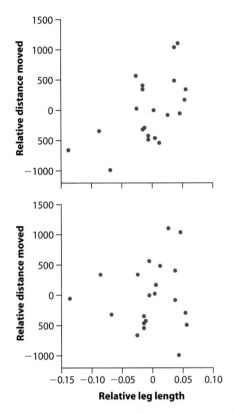

group (GF) was kept free of nonpathogenic gut microbes and all pathogens. The other (SPF) was only pathogen-free and served as controls. The following data are measurements of the percentage of T cells producing the molecule, interleukin-17, in tissue samples from 16 mice in the two groups.

SPF: 18.87, 15.65, 13.45, 12.95, 6.01, 5.84, 3.56, 3.46.

GF: 6.64, 4.51, 1.12, 0.62, 0.37, 0.61, 0.71, 0.82.

a. Use a box plot to visualize the data. What trend is suggested? In what way do the frequency distributions violate the assumptions of the two-sample t-test?

b. No transformations were effective, so we tested the difference between the medians of the two populations with a permutation test. Ten thousand randomizations were carried out, and the difference between the medians was calculated each time (SPF minus GF). Below we list the 100 largest values for the difference in median (sorted, out of 10,000). Using these values, complete the permutation test.

32. T and B lymphocytes are normal components of the immune system, but in multiple sclerosis they become autoreactive and attack the central nervous system. What triggers the autoimmune process? One hypothesis is that the disease is initiated by environmental factors, especially microbial infection. However, recent work by Berer et al. (2011) on the mouse model of the disease suggests that the autoimmune process is triggered by nonpathogenic microbes living in the gut. They compared onset of autoimmune encephalomyelitis in two treatment groups of mice from a strain that carries transgenic human CD4+ T cells, which initiate the disease. One

7.66, 7.71, 7.71, 7.71, 7.76, 7.76, 7.76, 8.425, 8.425, 8.425, 8.425, 8.425, 8.425, 8.425, 8.425, 8.425, 8.425, 8.48, 8.51, 8.51, 8.51, 8.51, 8.51, 8.51, 8.51, 8.51, 8.51, 8.51, 8.51, 8.51, 8.51, 8.51, 8.51, 8.565, 8.63, 8.63, 8.63, 8.63, 8.715, 8.715, 8.715, 8.715, 8.715, 8.825, 8.88, 8.88, 8.88, 8.88, 8.88, 8.88, 8.88, 8.88, 8.88, 9.03, 9.03, 9.03, 9.03, 9.03, 9.03, 9.03, 9.03, 9.03, 9.03, 9.03, 9.03, 9.03.

Review Problems 2

1. Women with mutations in their *BRCA1* or *BCRA2* genes ("carriers") represent about 0.5% of the U.S. population (Malone et al. 2006). Women who are carriers have an 80% chance of developing breast cancer in their lifetimes (Schubert 1997).[1] Those with normal versions of these genes have a 12% chance of developing breast cancer in their lifetime.
   a. What is the probability that a randomly sampled woman from the U.S. population is a carrier of either *BRCA1* or *BRCA2*?
   b. If 10 women are sampled from the U.S. population, what is the probability that none are carriers?
   c. If 20 women are sampled from the U.S. population, what is the probability that at least one is a carrier?
   d. What is the probability that a randomly sampled woman from the U.S. population both is a carrier of a mutant gene and develops breast cancer in her lifetime?
   e. What is the probability that a woman from the U.S. population develops breast cancer in her lifetime?
   f. What is the probability that a randomly chosen U.S. woman who develops breast cancer in her lifetime is a carrier?

2. Some researchers bought a new microgram scale. Before using it for new experiments, they wanted to ensure that the readings on the scale were accurate. They obtained a 10-μg standard weight, and then weighed this standard 30 times. Their measures were approximately normally distributed, with an average of 10.01 μg and standard deviation 0.2 μg. Test whether the scale is accurate, using these data.

3. Some genera have far more species than others. Is this just luck, or have some genera hit upon

a "key innovation" that gives them a benefit and allows more species to accumulate? One possible key innovation in plants is the ability to climb like a vine, which makes it possible to reach above other plants to compete for light. A study counted the number of species in 48 genera that had all evolved climbing (Gianoli 2004). For each of these genera, the researchers also found the most closely related non-climbing genus and counted the number of species in each. The numbers of species in these pairs of closely related genera are listed in the table below. The number of species does not

Climbing	Non-climbing	Climbing	Non-climbing
14	3	267	160
20	23	124	36
13	3	17	4
49	42	42	260
300	9	850	34
358	26	61	240
302	197	293	87
43	1	441	29
19	62	92	2
3	1	190	50
75	60	1041	262
2	3	13	8
22	11	1600	1
157	39	18	2
115	7	4	3
35	12	150	3228
130	5	350	318
15	1	142	9
650	25	525	2
845	34	24	54
6	1	180	702
30	7	320	635
307	427	625	31
2888	710	306	6

1. The actress Angelina Jolie, who is a carrier, chose to have a preventive double mastectomy to reduce the odds.

have a normal distribution in either climbing genera or nonclimbing genera, and the differences between climbing genera and their most closely related non-climbing genera are also not normally distributed.

a. Plot a histogram of the difference in number of species for each pair.

b. Carry out an appropriate test of the difference in the number of species in climbing and non-climbing genera.

4. In the early days of genetics, scientists realized that Mendel's laws of segregation could be used to predict that the second generation after a cross between two pure strains (the F_2) ought to have greater variance than the first generation after the cross (the F_1). One early test of this prediction was done by measuring flower length in a cross between two varieties of tobacco (East 1916). The following is a frequency table of the resulting individuals in both the first and second generations:

Flower length (mm)	Number of F_1 plants	Number of F_2 plants
52	0	3
55	4	9
58	10	18
61	41	47
64	75	55
67	40	93
70	3	75
73	0	60
76	0	43
79	0	25
82	0	7
85	0	8
88	0	1
n	173	444
Mean	63.53	68.76
Variance	8.62	42.37

a. Show how the means and variances were calculated for the F_1 data.

b. Test whether the variances of the two groups of plants differ, making all necessary assumptions. Is one significantly greater than the other? If so, which one?

5. Males of the Australian butterfly *Jalmenus evagoras* search for females and mate with them just as the females emerge as adults from the pupae. Multiple males might discover the same female, in which case they all attempt to mate with the same female as she emerges, forming a "mating ball." Females mate only once, and it is possible to record which male successfully mated with every female in a local population. In this way, Elgar and Pierce (1988) were able to track the mating success of 35 individual male *J. evagoras* butterflies over their complete life spans. Mating success is indicated in the accompanying table.

Lifetime number of mates	Frequency of males
0	20
1	9
2	1
3	1
4	2
5	1
6	0
7	1

a. Show these data in a graph.

b. Which probability distribution is expected to fit these data if all males have an equal probability of mating and if mating events are independent?

c. Calculate the expected frequencies of males having 0, 1, 2, ..., 7 mates under this probability distribution.

d. Add the expected frequencies to the graph you drew in part (a). Describe the pattern of differences between the observed and expected frequencies.

e. Are the differences between the observed and expected frequencies greater than we would expect by chance? Carry out a formal hypothesis test.

6. The males of some cichlid fish species are infertile until a few days after they become the socially dominant male in the presence of females. Males without a territory (and therefore without a hope of mating) have atrophied genitalia, whereas males who control a territory with females around have well-developed genitalia. They can shift from one state to the other in a matter of days. White et al. (2002) wanted to know the hormonal signal for this shift, and one candidate hormone was gonadotropin-releasing hormone (GnRH). They measured the levels of messenger RNA (mRNA) of GnRH for five territorial fish (T) and for six non-territorial fish (NT), in units of optical density relative to a known control. The data are in the accompanying table. Both distributions have positive skew (skewed to the right).

Territorial status	GnRH mRNA level
NT	0.504
NT	0.432
NT	0.744
NT	0.792
NT	0.672
NT	1.344
T	1.152
T	1.272
T	2.328
T	3.288
T	0.888

a. What procedure or procedures could be used to test for a difference in the mean GnRH levels between the T and NT populations?

b. Perform an appropriate hypothesis test for this difference.

c. Calculate a 95% confidence interval for the difference in GnRH mRNA level means between these two groups.

7. Every year Britain has a No Smoking Day, when many people voluntarily stop smoking for a day. No Smoking Day occurs on the second Wednes-

day of March each year. Waters et al. (1998) used this event to investigate the influence of stopping smoking on nonfatal injuries on the job. They compared the injury rate each year on the Wednesday of No Smoking Day to the rate for the previous Wednesday of the same year. The idea was that this comparison would control for many of the other factors that affect injury rate, such as year, time of week, and so on. The data from 1987 to 1996 (number of injuries in one day) are listed in the following table:

Year	Injuries before No Smoking Day	Injuries on No Smoking Day
1987	516	540
1988	610	620
1989	581	599
1990	586	639
1991	554	607
1992	632	603
1993	479	519
1994	583	560
1995	445	515
1996	522	556

a. How many more or fewer injuries are there on No Smoking Day, on average, compared with the normal day?

b. What is the 99% confidence interval for this difference?

c. In your own words, explain what the 99% confidence interval means.

d. Test whether the accident rate changes on No Smoking Day.

8. In the women's tennis finals at Wimbledon, the match winner is the first player to win two sets. Sets continue one after the other until there is a winner and loser; there are no ties. If one woman wins the first two sets, the match is finished at two sets. The maximum possible number of sets in a match is three.

a. Imagine that two women are equal in ability, so the probability of each woman winning any single set is 0.50. Use a probability tree to find the probability that a match lasts exactly two sets. What is the probability that the match lasts exactly three sets?

b. Imagine that one woman was better than the other, such that the probability of her victory in any set is 0.55. Use a probability tree to find the probability that the weaker woman would win the match.

9. The pitcher plant from Borneo, *Nepenthes raf-flesiana,* has two kinds of pitchers, upper and lower, that it uses to trap and digest insects. The upper pitchers use fragrance to lure mainly flying insects. The lower pitchers trap mainly ants. Di Giusto et al. (2010) tested whether odors emitted by the lower pitchers were cues for ant attraction. In one experiment, they presented individual ants, *Oecophylla smaragdina,* with air from a bag containing a lower pitcher in one arm of a Y-tube. Humidified air was provided in the other arm as a control. In 14 of 19 independent trials, ants chose the arm containing the pitcher, whereas the remaining five ants chose the control arm. Do these data indicate a preference for one arm or the other? Use an exact test.

10. For each of the following scenarios, *state the null hypothesis and identify the best statistical test to use* to answer the question stated. (Don't try to answer to the specific question.)

   a. Do stickleback fish occur with equal probability through all areas of a pond?

   b. A large number of Douglas-fir and Western hemlock trees were sampled, and the presence or absence of pine beetles on each tree was recorded. Do the tree species differ in pine beetle occurrence?

   c. A small number of Douglas-fir and Western hemlock trees were sampled, and the presence or absence of pine beetles was recorded. Do the species differ in occurrence of pine beetles? The expected number of infested fir trees is calculated as 2.3.

   d. Do patients change in mean body mass during a hospital stay?

   e. Does the amount of rainfall per day in a rainforest have a normal distribution?

   f. Which sex weighs more on average in bald eagles: males or females? Assume that the distribution of body weight in each sex is normally distributed, but that the two sexes have markedly different variances.

   g. Which sex travels more per day, on average, in sperm whales: males or females? Assume that the distribution of distance traveled is very different from a normal distribution in each sex (but similar between sexes) and that sample size is small.

   h. Do cats have greater strength in their dominant front paw (usually the right) than in their other front paw (usually the left)? The data are measurements of strength in dominant paw and other paw of a random sample of cats.

   i. Does the mean number of chirps per minute by male crickets differ when the same crickets are measured at 15°C and at 25°C?

   j. The data are water samples taken at the local beach. Does the mean number of bacteria per milliliter differ from 130 individuals per 100 ml?

11. What feature of an estimate—precision or accuracy—is most strongly affected when individuals differing in the variable of interest are not sampled independently?

12. Body size in female northern fur seals (*Callorhinus ursinus*), measured as total length, is approximately normally distributed with a mean of 124.6 cm and a standard deviation equal to 6.5 cm (Trites 1996).

   a. About what fraction of individuals have a total body length less than 110 cm?

   b. What fraction of female fur seals have a body length between 130 and 140 cm?

   c. What fraction have a body length between 120 and 125 cm?

13. Gesturing is common during human speech. Is this behavior learned via exposure? A measure was made of the number of gestures produced by each of 12 pairs of sighted individuals while talking to sighted individuals (Iverson and

Goldin-Meadow 1998). This result was compared with the number of gestures produced while talking by each of 12 pairs of people who had been blind since birth and were therefore presumably unexposed to the gestures of others. The data[2] are as follows:

Blind: 0, 1, 1, 2, 1, 1, 1, 3, 1, 0, 1, 1

Sighted: 1, 0, 1, 2, 3, 0, 1, 2, 2, 0, 3, 1

Test the hypothesis that the number of gestures is related to sightedness, using a nonparametric test.

14. Kids are often told that they should not crack their knuckles, because otherwise all sorts of terrible things may befall them. It is commonly believed that knuckle cracking leads to arthritis, which de Weber et al. (2011) recently tested in a case-control study. Of 135 patients with osteoarthritis (cases), 24 had frequently cracked their knuckles. Of 80 control patients without osteoarthritis, 19 had frequently cracked their knuckles.

    a. What type of graph is ideal for displaying these results?
    b. What is the odds ratio for osteoarthritis, comparing knuckle-crackers to non-crackers?
    c. Give a 95% confidence interval for this odds ratio.
    d. Carry out a formal hypothesis test of whether knuckle cracking is associated with osteoarthritis.

15. Have you ever had the experience that driving somewhere seems to take a really long time, but the trip back home goes faster, even though it is the same distance in reverse? Van de Ven et al. (2011) wanted to investigate how common this subjective experience was. They interviewed 69 people who had just been on trips where the outbound and inbound travel time was the same and who had been awake the whole time. They asked the people to evaluate the trips on an 11-point numeric scale, from -5 (return trip was a lot shorter) to $+5$ (return trip was a lot longer). The data are given below in a frequency table.

Return trip time score	Frequency (number of respondents)
-5	1
-4	4
-3	11
-2	9
-1	6
0	20
1	6
2	4
3	6
4	2
5	0

    a. What is the mean of the return trip time score?
    b. Calculate the 95% confidence interval of the mean return trip time score.
    c. It is also interesting to know to what extent people experience this subjective impression about travel time in the same way. What is the 95% confidence interval of the variance in the return trip time score?

16. In Chapter 3, Assignment Problem 22, you produced a box plot for the following data from Norton et al. (2011). The data are measurements of the amount of time, in seconds, that individual zebrafish with and without the *spiegeldanio* (*spd*) mutation at the *Fgfr1a* gene spent in aggressive activity over 5 minutes when presented with a mirror image of themselves. The researchers were interested in the role this gene plays in differences between individuals along the shy–bold behavioral spectrum.

Spd **mutant:** 96, 97, 100, 127, 128, 156, 162, 170, 190, 195

Wild type: 0, 21, 22, 28, 60, 80, 99, 101, 106, 129, 168

    a. With these data, estimate the magnitude of the effect of the mutation (the difference between the means) on the amount of time spent in aggressive activity. Put appropriate bounds on your estimate of the effect.
    b. What is the weight of evidence that this effect is not zero? Perform an appropriate statistical test of the difference.

---

2. Extrapolated from summary statistics in the original paper.

Frog deformities

14 Designing experiments

Two types of investigations are carried out in biology: observational and experimental. In an experimental study, the researcher assigns treatments to units or subjects so that differences in response can be compared. In an observational study, on the other hand, nature does the assigning of treatments to subjects. The researcher has no influence over which subjects receive which treatment.

What's so important about the distinction? Whereas observational studies can identify associations between treatment and response variables, properly designed experimental studies can identify the *causes* of these associations.

How do we best design an experiment to get the most information possible out of it? The short answer is that we must design to eliminate bias and to reduce the influence of sampling error. The present chapter outlines the basics on how to accomplish this feat. We also briefly discuss how to design an observational study: by taking the best features of experimental designs and incorporating as many of them as possible.

Finally, we discuss how to plan the sample size needed in an experimental or observational study.

14.1   Why do experiments?

In an experimental study, there must be at least two treatments and the experimenter (rather than nature) must assign them to units or subjects. The crucial advantage of experiments derives from the *random* assignment of treatments to units. Random assignment, or randomization, minimizes the influence of *confounding* variables (Interleaf 4), allowing the experimenter to isolate the effects of the treatment variable.

Confounding variables

Studies in biology are usually carried out with the aim of deciding how an explanatory variable or treatment affects a response variable. How are injury rates in cats with "high-rise syndrome" affected by the number of stories fallen? What is the effect of marine reserves on fish biomass? How does the use of supplemental oxygen affect the probability of surviving an ascent of Mount Everest? The easiest way to address these questions is with an observational study—that is, to gather measurements of both variables of interest on a set of subjects and estimate the association between them. If the two variables are correlated or associated, then one may be the cause of the other.

The limitation of the observational approach is that, by itself, it cannot distinguish between two completely different reasons behind an association between an explanatory variable X and a response variable Y. One possibility is that X really does cause a response in Y. For example, taking supplemental oxygen might increase the chance of survival during a climb of Mount Everest. The other possibility is that the explanatory variable X has no effect at all on the response variable Y; they are associated only because other variables affect both X and Y at the same time. For example, the use of supplemental oxygen might just be a benign indicator of a greater overall preparedness of the climbers who use it, and greater preparedness rather than oxygen use is the real cause of the enhanced survival. Variables (like preparedness) that distort the causal relationship between the measured variables of interest (oxygen use and survival) are called *confounding variables*. Recall from Interleaf 4, for example, that ice cream consumption and violent crime are correlated, but neither is the cause of the other. Instead, increases in both ice cream consumption and crime are caused by higher temperatures. Temperature is a confounding variable in this example.

> A *confounding variable* is a variable that masks or distorts the causal relationship between measured variables in a study.

Confounding variables bias the estimate of the causal relationship between measured explanatory and response variables, sometimes even reversing the apparent effect of one on the other. For example, observational studies have indicated that breast-fed babies have lower weight at six and 12 months of age compared with formula-fed infants (Interleaf 4). But an experimental study using randomization found that mean infant weight was actually *higher* in breast-fed babies at six months of age and was not less than that in formula-fed babies at 12 months (Kramer et al. 2002). The observed relationship between breast feeding and infant growth was confounded by unmeasured variables such as the socioeconomic status of the parents.

With an experiment, random assignment of treatments to participants allows researchers to tease apart the effects of the explanatory variable from those of confounding variables. With random assignment, no confounding variables will be associated with treatment except by chance. For example, if women who choose to breast-feed their babies have a different average socioeconomic background than women who choose to feed their infants formula, randomly assigning the treatments "breast feeding" and "formula feeding" to women in an experiment will break this connection, roughly equalizing the backgrounds of the two treatment groups. In this case, any resulting difference between groups in infant weight (beyond chance) must be caused by treatment.

Experimental artifacts

Unfortunately, experiments themselves might inadvertently create artificial conditions that distort cause and effect. Experiments should be designed to minimize artifacts.

> An *experimental artifact* is a bias in a measurement produced by unintended consequences of experimental procedures.

For example, experiments conducted on aquatic birds have shown that their heart rates drop sharply when they are forcibly submerged in water, compared with individuals remaining above water. The drop in heart rate has been interpreted as an oxygen-saving response. Later studies using improved technology showed that voluntary dives do not produce such a large drop in heart rate (Kanwisher et al. 1981). This finding suggested that a component of the heart rate response in forced dives was induced by the stress of being forcibly dunked underwater, rather than the dive itself. The experimental conditions introduced an artifact that for a while went unrecognized.

To prevent artifacts, experimental studies should be conducted under conditions that are as natural as possible. A potential drawback is that more natural conditions might introduce more sources of variation, reducing power and precision. Observational studies can provide important insight into what is the best setting for an experiment.

14.2   Lessons from clinical trials

The gold standard of experimental designs is the **clinical trial**, an experimental study in which two or more treatments are assigned to human participants. The design of clinical trials has been refined because the cost of making a mistake with human participants is so high. Experiments on nonhuman subjects are simply called "laboratory experiments" or "field experiments," depending on where they take place. Experimental studies in all areas of biology have been greatly informed by procedures used in clinical trials.

> A *clinical trial* is an experimental study in which two or more treatments are applied to human participants.

Before we dig into the logic of the main components of experimental design, let's look at the clinical trial in Example 14.2, which incorporates many of these features.

EXAMPLE   Reducing HIV transmission

14.2   Transmission of the HIV-1 virus via sex workers contributes to the rapid spread of AIDS in Africa. How can this transmission be reduced? In laboratory experiments, the spermicide nonoxynol-9 had shown in vitro activity against HIV-1, shown schematically at the right. This finding motivated a clinical trial by van Damme et al. (2002), who tested whether a vaginal gel containing the chemical would reduce female sex workers' risk of acquiring the disease. Data were gathered on a volunteer sample of 765 HIV-free sex workers  in six clinics in Asia and Africa. Two gel treatments were assigned randomly to women at each clinic. One gel contained nonoxynol-9, and the other contained a placebo (an inactive compound that participants could not distinguish from the treatment of interest). Neither the participants nor the researchers making observations at the clinics knew who had received the treatment and who had received the placebo. (A system of numbered codes kept track of who got which treatment.) By the end of the experiment, 59 of 376 women in the nonoxynol-9 group (15.9 %) were HIV-positive (Table 14.2-1), compared with 45 out of 389 women in the placebo group (11.6 %). Thus, the odds of contracting HIV-1 were slightly higher in the nonoxynol-9 group compared with the placebo group—which was the opposite of the expected result. The reason seems to be that repeated use of nonoxynol-9 causes tissue damage that leads to higher risk.

Design components

A good experiment is designed with two objectives in mind:

- To reduce bias in estimating and testing treatment effects
- To reduce the effects of sampling error

TABLE 14.2-1 Results of the clinical trial in Example 14.2
(n is the number of subjects).

Clinic	Nonoxynol-9		Placebo	
	n	Number infected	n	Number infected
Abidjan	78	0	84	5
Bangkok	26	0	25	0
Cotonou	100	12	103	10
Durban	94	42	93	30
Hat Yai 2	22	0	25	0
Hat Yai 3	56	5	59	0
Total	376	59	389	45

The most significant elements in the design of the clinical trial in Example 14.2 addressed these two objectives. To reduce bias, the experiment included the following elements.

1. A simultaneous control group: the study included both the treatment of interest and a control group (the women receiving the placebo).
2. Randomization: treatments were randomly assigned to women at each clinic.
3. Blinding: neither the participants nor the clinicians knew which women were assigned which treatment.

To reduce the effects of sampling error, the experiment included these elements.

1. Replication: the study was carried out on multiple independent participants.
2. Balance: the number of women was nearly equal in the two groups at every clinic.
3. Blocking: participants were grouped according to the clinic they attended, yielding multiple repetitions of the same experiment in different settings (i.e., "blocks").

> The goal of experimental design is to eliminate bias and to reduce sampling error when estimating and testing the effects of one variable on another.

In Section 14.3, we discuss the virtues of the three main strategies used to reduce bias—namely, simultaneous controls, randomization, and blinding. In Section 14.4, we explain the strategies used to reduce the effects of sampling error—namely, replication, balance, and blocking. As usual, we assume throughout that units or subjects have been randomly sampled from the population of interest.

14.3   How to reduce bias

We have seen how confounding variables in observational studies can bias the estimated effects of an explanatory variable on a response variable. The following experimental procedures are meant to eliminate bias.

Simultaneous control group

A **control group** is a group of subjects who are treated like all of the experimental subjects, except that the control group does not receive the treatment of interest.

> A *control group* is a group of subjects who do not receive the treatment of interest but who otherwise experience similar conditions as the treated subjects.

In an uncontrolled experiment, a group of subjects are treated in some way and then measured to see how they have responded. Lacking a control group for comparison, such a study cannot determine whether the treatment of interest is the cause of any of the observed changes. There are several possible reasons for this, including the following:

- Sick human participants selected for a medical treatment may tend to "bounce back" toward their average condition regardless of any effect of the treatment (Interleaf 6).

- Stress and other impacts associated with administering the treatment (such as surgery or confinement) might themselves produce a response separate from the effect of the treatment of interest.

- The health of human participants often improves after treatment merely because of their expectation that the treatment will have an effect. This phenomenon is known as the placebo effect (Interleaf 6).

The solution to all of these problems is to include a control group of subjects measured for comparison. The treatment and control subjects should be tested simultaneously or in random order, to ensure that any temporal changes in experimental conditions do not affect the outcome.

The appropriate control group will depend on the circumstance. Here are some examples:

- In clinical trials, either a placebo or the currently accepted treatment should be provided, such as in Example 14.2. A placebo is an inactive treatment that subjects cannot distinguish from the main treatment of interest.

- In experiments requiring intrusive methods to administer treatment, such as injections, surgery, restraint, or confinement, the control subjects should be perturbed in the same way as the other subjects, except for the treatment itself, as far as ethical considerations permit. The "sham operation," in which surgery is carried out without the experimental treatment itself, is an example. Sham operations are very rare in human studies, but they are more common in animal experiments.

■ In field experiments, applying a treatment of interest may physically disturb the plots receiving it and the surrounding areas, perhaps by the researchers trampling. Ideally, the same disturbance should be applied to the control plots.

Often it is desirable to have more than one control group. For example, two control groups, where one is a harmless placebo and the other is the best existing treatment, may be used in a study so that the total effect of the treatment and the improvement of the new treatment over the old may both be measured. However, using resources for multiple controls might reduce the power of the study to test its main hypotheses.

Randomization

Once the treatments have been chosen, the researcher should *randomize* their assignment to units or subjects in the sample. **Randomization** means that treatments are assigned to units at random, such as by flipping a coin. Chance rather than conscious or unconscious decision determines which units end up receiving the treatment of interest and which receive the control. A **completely randomized design** is an experimental design in which treatments are assigned to all units by randomization.

> *Randomization* is the random assignment of treatments to units in an experimental study.

The virtue of randomization is that it breaks the association between possible confounding variables and the explanatory variable, allowing the causal relationship between the explanatory and response variables to be assessed. Randomization doesn't eliminate the variation contributed by confounding variables, only their correlation with treatment. It ensures that variation from confounding variables is spread more evenly between the different treatment groups, and so it creates no bias. If randomization is done properly, any remaining influence of confounding variables occurs by chance alone, which statistical methods can account for.

Randomization should be carried out using a random process. The following steps describe one way to assign treatments randomly:

1. List all *n* subjects, one per row, in a computer spreadsheet.
2. Use the computer to give each individual a random number.[1]
3. Assign treatment A to those subjects receiving the lowest numbers and treatment B to those with the highest numbers.

---

1. The Random.org website at http://random.org/sequences will also do this.

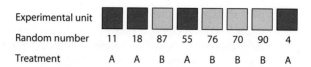

FIGURE 14.3-1 A procedure for randomization. Each of eight subjects was assigned a number between 0 and 99 that was drawn at random by a computer. Treatment A (*colored red*) was assigned to the four subjects with the lowest random numbers, whereas treatment B (*gold*) was assigned to the rest.

This process is demonstrated in Figure 14.3-1, where eight subjects are assigned to two treatments, A and B.

Other ways of assigning treatments to subjects are almost always inferior, because they do not eliminate the effects of confounding variables. For example, the following methods can lead to problems:

- Assign treatment A to all patients attending one clinic and treatment B to patients attending a second clinic. (Problem: All of the other differences between the two clinics become confounding variables. If one clinic is better than the other in general, then the difference in clinic quality would show up as a difference in treatments.)

- Assign treatments to human participants alphabetically. (Problem: This might inadvertently group individuals having the same nationality, generating unwanted differences between treatments in health histories and genetic variables.)

It is important to use a computer random-number generator or random-number tables to assign individuals randomly to treatments. "Haphazard" assignment, in which the researcher chooses a treatment while trying to make it random, has repeatedly been shown to be non-random and prone to bias.[2]

Blinding

The process of concealing information from participants and researchers about which of them receive which treatments is called **blinding**. Blinding prevents participants and researchers from changing their behavior, consciously or unconsciously, based on their knowledge of which treatment they were receiving or administering. For example, a researcher who believes that acupuncture helps alleviate back pain might unconsciously interpret a patient's report of pain differently if the researcher knows the patient was assigned the acupuncture treatment instead of a placebo. This might explain why studies that have shown acupuncture has a significant effect on back pain

---

2. What do you do if, by chance, the first four of eight units are all assigned treatment A and the last four are assigned treatment B, yielding the arrangement AAAABBBB? Some biologists might randomize again to ensure the interspersion of treatments, but that is not strictly legitimate. If the first four units are different somehow from the last four, apart from treatment, then blocking (Section 14.4) should be considered as a remedy.

are limited to those without blinding (Ernst and White 1998). Studies implementing blinding have not found that acupuncture has an ameliorating effect on back pain.

In a **single-blind experiment**, participants are unaware of the treatment they have been assigned. This requires that the treatments be indistinguishable to the participants, a particular necessity in experiments involving humans. Single-blinding prevents participants from responding differently according to their knowledge of their treatment. This is not much of a concern in nonhuman studies.

In a **double-blind experiment**, the researchers administering the treatments and measuring the response are also unaware of which subjects are receiving which treatments. This prevents researchers who are interacting with the subjects from behaving differently toward them according to their treatments. Researchers sometimes have pet hypotheses, and they might treat experimental subjects in different ways depending on their hopes for the outcome. Moreover, many response variables are difficult to measure and require some subjective interpretation, which makes the results prone to a bias in favor of the researchers' wishes and expectations. Finally, researchers are naturally more interested in the treated subjects than the control subjects, and this increased attention can itself result in improved response. Reviews of medical studies have revealed that studies carried out without double-blinding exaggerated treatment effects by 16% on average, compared with studies carried out with double-blinding (Jüni et al. 2001).

> *Blinding* is the process of concealing information from participants (sometimes including researchers) about which individuals receive which treatment.

Experiments on nonhuman subjects are also prone to bias from lack of blinding. Bebarta et al. (2003) reviewed 290 two-treatment experiments carried out on animals or on cell lines. They found that the odds of detecting a positive effect of treatment were more than threefold higher in studies without blinding than in studies with blinding.[3] Blinding can be incorporated into experiments on nonhuman subjects by using coded tags that identify the subject to a "blind" observer without revealing the treatment (and then the observer measures units from different treatments in random order).

14.4 How to reduce the influence of sampling error

Assuming we have designed our experiment to minimize sources of bias, there is still the problem of detecting any treatment effects against the background of variation between individuals ("noise") caused by other variables. Such variability creates

---

3. This result probably overestimates the effects of a lack of blinding, because the experiments without blinding also tended to have confounding problems, such as a lack of randomization (Bebarta et al. 2003).

sampling error in the estimates, reducing precision and power. How can the effects of sampling error be minimized?

One way to reduce noise is to make the experimental conditions constant. Fix the temperature, humidity, and other environmental conditions, for example, and use only participants who are of the same age, sex, genotype, and so on. In field experiments, however, highly constant experimental conditions might not be feasible. Constant conditions might not be desirable, either. By limiting the conditions of an experiment, we also limit the generality of the results—that is, the conclusions might apply only under the conditions tested and not more broadly. Until recently, a significant source of bias in medical practice stemmed from the fact that many clinical tests of the effects of medical treatments were carried out only on men, yet the treatments were subsequently applied to women as well (e.g., McCarthy 1994).

In this section, we review replication, balance, and blocking, the three main statistical design procedures used to minimize the effects of sampling error. We also review a strategy to reduce the effect of noise by using extreme treatments.

Replication

Because of variability, **replication**—the repetition of every treatment on multiple experimental units—is essential. Without replication, we would not know whether response differences were due to the treatments or just chance differences between the treatments caused by other factors. Studies that use more units (i.e., that have larger sample sizes) will have smaller standard errors and a higher probability of getting the correct answer from a hypothesis test. Larger samples give more information, and more information gives better estimates and more powerful tests.

> *Replication* is the application of every treatment to multiple, independent experimental units.

Replication is not just about the number of plants or animals used. True replication depends on the number of *independent* units in the experiment. An "experimental unit" is the independent unit to which treatments are assigned. Figure 14.4-1 shows three hypothetical experiments designed to compare the effects of two fertilizer treatments on plant growth. The lack of replication is obvious in the first design (top panel), because there is only one plant per treatment. You won't see many published experiments like it.

The lack of replication is less obvious in the second design (the middle panel of Figure 14.4-1). Although there are multiple plants per treatment in the second design, all plants in one treatment are confined to one chamber and all plants in the second treatment are confined to another chamber. If there are environmental differences between chambers (e.g., differences in light conditions or humidity) beyond those stemming from the treatment itself, then plants in the same chamber will be more similar in their responses than plants in different chambers, apart from any treatment effects. The plants in the same chamber are not independent. As a result, the chamber,

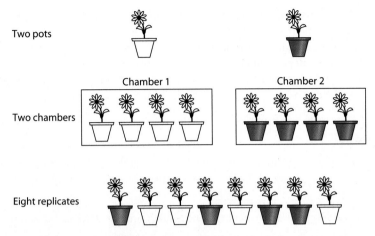

FIGURE 14.4-1 Three experimental designs used to compare plant growth under two ferti-lizer treatments (indicated by the shading of the pots). The upper ("two pots") and middle ("two chambers") designs are unreplicated.

not the plant, is the experimental unit in a test of fertilizer effects. Because there are only two chambers, one per treatment, the experiment is unreplicated.

Only the third design (the bottom panel) in Figure 14.4-1 is properly replicated, because here treatments have been randomly assigned to individual plants. A give-away indicator of replication in the third design is *interspersion* of experimental units assigned different treatments, which is an expected outcome of randomization. Such interspersion is lacking in the two-chamber design (the middle panel in Fig-ure 14.4-1), which is a clear sign of a replication problem.

An experimental unit might be a single animal or plant if individuals are randomly sampled and assigned treatments independently. Or, an experimental unit might be made up of a batch of individual organisms treated as a group, such as a field plot containing multiple individuals, a cage of animals, a household, a petri dish, or a family. Multiple individual organisms belonging to the same unit (e.g., plants in the same plot, bacteria in the same dish, members of the same family, and so on) should be considered together as a single replicate if they are likely to be more similar on average to each other than to individuals in separate units (apart from the effects of treatment).

Correctly identifying replicates in an experiment is crucial to planning its design and analyzing the results. Erroneously treating the single organism as the indepen-dent replicate when the chamber (Figure 14.4-1) or field plot is the experimental unit will lead to calculations of standard errors and *P*-values that are too small. This is pseudoreplication, as discussed in Interleaf 2.

From the standpoint of reducing sampling error, more replication is always better. As proof, examine the formula for the standard error of the difference between two sample mean responses to two treatments, $\overline{Y}_1 - \overline{Y}_2$:

$$\mathrm{SE}_{\overline{Y}_1 - \overline{Y}_2} = \sqrt{s_p^2 \left(\frac{1}{n_1} + \frac{1}{n_2} \right)}.$$

The symbols n_1 and n_2 refer to the number of experimental units, or replicates, in each of the two treatments. Based on this equation, increasing n_1 and n_2 directly reduces the standard error, increasing precision. Increased precision yields narrower confidence intervals and more powerful tests of the difference between means. On the other hand, increasing sample size also has costs in terms of time, money, and even lives. We discuss how to plan a sufficient sample size in more detail in Section 14.7.

Balance

A study design is **balanced** if all treatments have the same sample size. Conversely, a design is *un*balanced if there are *un*equal sample sizes between treatments.

> In a *balanced* experimental design, all treatments have equal sample size.

Balance is a second way to reduce the influence of sampling error on estimation and hypothesis testing. To appreciate this, look again at the equation for the standard error of the difference between two treatment means (given on page 433). For a fixed total number of experimental units, $n_1 + n_2$, the standard error is smallest when the quantity

$$\left(\frac{1}{n_1} + \frac{1}{n_2} \right)$$

is smallest, which occurs when n_1 and n_2 are equal. Convince yourself that this is true by plugging in some numbers. For example, if the total number of units is 20, the quantity $1/n_1 + 1/n_2$ is 0.2 when $n_1 = n_2 = 10$, but it is 1.05 when $n_1 = 19$ and $n_2 = 1$. With better balance, the standard error is much smaller.

To estimate the difference between two groups, we need precise estimates of the means of *both* groups. With an unbalanced design, we may know the mean of one group with great precision, but this does not help us much if we have very little information about the other group that we're comparing it with. Balance allocates the sampling effort in the optimal way.

Nevertheless, the precision of an estimate of a difference between groups always increases with larger sample sizes, even if the sample size is increased in only one of two groups. But for a fixed total number of subjects, the optimal allocation is to have an equal number in each group.

Balance has other benefits, which we discuss elsewhere in the book. For example, the methods based on the normal distribution for comparing population means are most robust to departures from the assumption of equal variances when designs are balanced or nearly so (see Chapters 12 and 15).

Blocking

Blocking is an experimental design strategy used to account for extraneous variation by dividing the experimental units into groups, called **blocks** or strata, that share common features. Within blocks, treatments are assigned randomly to experimental units. Blocking essentially repeats the same completely randomized experiment multiple times, once for each block, as shown schematically in Figure 14.4-2. Differences between treatments are evaluated only within blocks. In this way, much of the variation arising from differences between blocks is accounted for and won't reduce the power of the study.

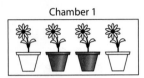 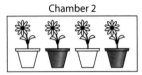

FIGURE 14.4-2 An experimental design incorporating blocking to test effects of fertilizer on plant growth (see Figure 14.4-1). Shading of the pots indicates which fertilizer treatment each plant received. Chambers might differ in unknown ways and add unwanted noise to the experiment. To remove the effects of such variation, carry out the same completely randomized experiment separately within each chamber. In this design, each chamber represents one block.

The women participating in the nonoxynol-9 HIV study discussed in Example 14.2 were grouped according to the clinic they attended. This made sense because there were age differences between women attending different clinics as well as differences in condom usage and sexual practices, all of which are likely to affect HIV transmission rates (van Damme et al. 2002). Blocking removes the variation in response among clinics, allowing more precise estimates and more powerful tests of the treatment effects.

> *Blocking* is the grouping of experimental units that have similar properties. Within each block, treatments are randomly assigned to experimental units.

The *paired* design for two treatments (Chapter 12) is an example of blocking. In a paired design, both of two treatments are applied to each plot or other experimental unit representing a block. The difference between the two responses made on each block is the measure of the treatment effect.

The **randomized block design** is analogous to the paired design, but it can have more than two treatments, as shown in Example 14.4A.

> The *randomized block design* is like a paired design but for more than two treatments.

EXAMPLE   Holey waters

14.4A   The compact size of water-filled tree holes, which can harbor diverse communities of aquatic insect larvae, makes them useful microcosms for ecological experiments. Srivastava and Lawton (1998) made artificial tree holes from plastic that mimicked the buttress tree holes of European beech trees (see image on right). They placed the plastic holes next to trees in a forest in southern England to examine how the amount of decaying leaf litter present in the holes affected the number of insect eggs deposited (mainly by mosquitoes and hover flies) and the survival of the larvae emerging from those eggs. Leaf litter is the source of all nutrients in these holes, so increasing the amount of litter might result in more food for the insect larvae. There were three different treatments. In one treatment (LL), a low amount of leaf litter was provided. In a second treatment (HH), a high level of debris was provided. In the third treatment (LH), leaf litter amounts were initially low but were then made high after eggs had been deposited. A randomized block design was used in which artificial tree holes were laid out in triplets (blocks). Each block consisted of one LL tree hole, one HH tree hole, and one LH tree hole. The location of each treatment within a block was randomized, as shown in Figure 14.4-3.

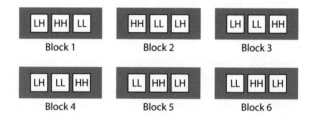

FIGURE 14.4-3 Schematic of the randomized block design used in the tree-hole study of Example 14.4A. Each block of three tree holes was placed next to its own beech tree in the woods. Within blocks, the three treatments were randomly assigned to tree holes.

As in the paired design, treatment effects in a randomized block design are measured by differences between treatments exclusively within blocks, a strategy that minimizes the influence of variation among blocks.

In the randomized block design, each treatment is applied once to every block. By accounting for some sources of sampling variation, such as the variation among trees, blocking can make differences between treatments stand out. In Chapter 18, we discuss in greater detail how to analyze data from a randomized block design.

Blocking is worthwhile if units within blocks are relatively homogeneous, apart from treatment effects, and units belonging to different blocks vary because of environmental or other differences. For example, blocks can be made up of any of these units:

- Field plots experiencing similar local environmental conditions
- Animals from the same litter
- Aquaria located on the same side of the room
- Patients attending the same clinic
- Runs of an experiment executed on the same day

One potential drawback to blocking might occur if the effects of one treatment contaminate the effects of the other in the same block. For example, watering one half of a block might raise the soil humidity of the adjacent, unwatered half. Experiments should be designed carefully to minimize contamination.

Extreme treatments

Treatment effects are easiest to detect when they are large. Small differences between treatments are difficult to detect and require larger samples, whereas larger treatment differences are more likely to stand out against random variability within treatments. Therefore, one strategy to enhance the probability of detecting differences in an experiment is to include extreme treatments. Example 14.4B shows why this might be.

EXAMPLE Plastic hormones

14.4B Bisphenol-A, or BPA, is an estrogenic compound found in plastics widely used to line food and drink containers and in dental sealants. Human daily exposures are typically in the range of 0.5–1 μg/kg body weight (Gray et al. 2004). Sakaue et al. (2001) measured sperm production of 13-week-old male rats exposed to fixed daily doses of BPA between 0 and 2000 μg/kg body weight for six days. The results are shown in a dose–response curve in Figure 14.4-4.

This experiment included doses much higher than the typical doses faced by humans at risk, a strategy that enhanced the ability to detect an effect of BPA. For example, Figure 14.4-4 shows that there was a much larger difference in mean sperm production between the 0 and 2000 μg/kg groups than between the 0 and 0.002 μg/kg treatments. If the experimenter were to design a study to compare just one of these doses with the control, using 200 or 2000 μg/kg would yield the most power, because they show the largest difference in sperm production from the control.

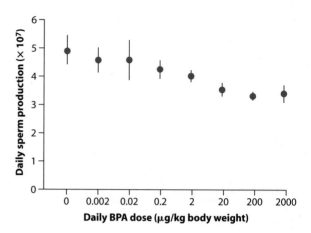

FIGURE 14.4-4
A dose-response curve showing the results of an experiment measuring the rates of sperm production of male rats exposed to fixed daily doses of bisphenol-A (BPA) (Sakaue et al. 2001). Symbols are the mean ± 1 SE.

A larger dose, or stronger treatment, can increase the probability of detecting a response. But be aware that the effects of a treatment do not always scale linearly with the magnitude of a treatment. The effects of a large dose may be qualitatively different from those of a smaller, more realistic dose. Still, as a first step, extreme treatments can be a very good way to detect whether one variable has any effect at all on another variable.

14.5 Experiments with more than one factor

Up to now, we have considered only experiments that focus on measuring and testing the effects of a single factor. A **factor** is a single treatment variable whose effects are of interest to the researcher. However, many experiments in biology investigate more than one factor, because answering two questions from a single experiment rather than just one makes more efficient use of time, supplies, and other costs.

Another reason to consider experiments with multiple factors is that the factors might interact. When operating together, the factors might have synergistic or inhibitory effects not seen when each factor is tested alone. For example, human activity has driven global increases in atmospheric CO_2 and temperature, as well as greater nitrogen deposition and precipitation. Increases in all of these factors have been shown to stimulate plant growth by experimental studies in which each treatment variable was examined separately. But what are the effects of these factors in combination? The only way to answer this is to design experiments in which more than one factor is manipulated simultaneously. If the climate variables interact when influencing plant growth, then their joint effects can be very different from their separate effects (Shaw et al. 2002).

A *factor* is a single treatment variable whose effects are of interest to the researcher.

The factorial design is the most common experimental design used to investigate more than one treatment variable, or factor, at the same time. In a **factorial design**, every combination of treatments from two (or more) treatment variables is investigated.

> An experiment having a *factorial design* investigates all treatment combinations of two or more variables. A factorial design can measure interactions between treatment variables.

The main purpose of a factorial design is to evaluate possible interactions between variables. An **interaction** between two explanatory variables means that the effect of one variable on the response depends on the state of a second variable. Example 14.5 illustrates an interaction in a factorial design.

EXAMPLE
14.5

Lethal combination

Frog populations are declining everywhere, spawning research to identify the causes. Relyea (2003) looked at how a moderate dose (1.6 mg/l) of a commonly used pesticide, carbaryl (Sevin), affected bullfrog tadpole survival. In particular, the experiment asked how the effect of carbaryl depended on whether a native predator, the red-spotted newt, was also present. The newt was caged and could cause no direct harm, but it emitted visual and chemical cues that are known to affect tadpoles. The experiment was carried out in 10-liter tubs, each containing 10 tadpoles. The four combinations of pesticide treatment (carbaryl vs. water only) and predator treatment (present or absent) were randomly assigned to tubs. For each combination of treatments, there were four replicate tubs. The effects on tadpole survival are displayed in Figure 14.5-1.

FIGURE 14.5-1
Interaction between the effects of the pesticide (carbaryl) and predator (red-spotted newt) treatments on tadpole survival. Each point gives the fraction of tadpoles in a tub that survived. Lines connect mean survival in the two pesticide treatments, separately for each predator treatment.

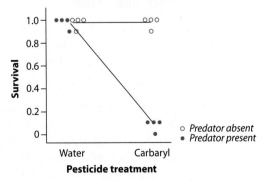

The tub, not the individual tadpole, is the experimental unit, because tadpoles sharing the same tub are not independent. The results showed that survival was high, except when pesticide was applied together with the predator—neither treatment alone had much effect (Figure 14.5-1). Thus, the two treatments, predation and pesticide, seem to have interacted—that is, the effect of one variable depends on the state of the other variable. An experiment investigating the effects of the pesticide only would have measured little effect at this dose. Similarly, an experiment investigating the effect of the predator only would not have seen an effect on survival.

> An *interaction* between two (or more) explanatory variables means that the effect of one variable depends upon the state of the other variable.

A factorial design can still be worthwhile even if there is no interaction between explanatory variables. In this case, there are efficiency advantages because the same experimental units can be used to measure the effect of two (or more) variables simultaneously.

14.6   What if you can't do experiments?

Experimental studies are not always feasible, in which case we must fall back upon observational studies. Observational studies can be very important, because they detect patterns and can help generate hypotheses. The best observational studies incorporate all of the features of experimental design used to minimize bias (e.g., simultaneous controls and blinding) and the impact of sampling error (e.g., replication, balance, blocking, and even extreme treatments), except for one: randomization. Randomization is out of the question because, in an observational study, the researcher does not assign treatments to subjects. Instead, the subjects come as they are.

Match and adjust

Without randomization, minimizing bias resulting from confounding variables is the greatest challenge of observational studies. Two types of strategies are used to limit the effects of confounding variables on a difference between treatments in a controlled observational study. One strategy, commonly used in epidemiological studies, is **matching**. With matching, every individual in the target group with a disease or other health condition is paired with a corresponding healthy individual who has the same measurements for known confounding variables, such as age, weight, sex, and ethnic background (Bland and Altman 1994).

Matching is often used when designing case-control studies. Recall from Chapter 9 that in a case-control study, exposure to one or more possible causal factors is compared between a sample of individuals having a disease (the cases) and a second sample of participants not having the disease (the controls). Matching ensures that the cases and controls are otherwise similar. For example, Dziekan et al. (2000) investigated possible causes of a hospital outbreak of antibiotic-resistant *Staphylococcus*.

The 67 infected cases were each paired with a control individual matched for age, sex, hospital admission date, and admission department.

Matching reduces bias by limiting the contribution of suspected confounding variables to differences between treatments. Unlike randomization in an experiment, matching in an observational study does not account for all confounding variables, only those explicitly used to match participants. Thus, while matching reduces bias, it does not eliminate bias. Matching also reduces sampling error by grouping experimental units into similar pairs, analogous to blocking in experimental studies. It is with such a matched case-control design that the link between smoking and lung cancer was convincingly demonstrated.

> With *matching*, every individual in the treatment group is paired with a control individual having the same or closely similar values for the suspected confounding variables.

In a weaker version of this approach, a comparison group is chosen that has a frequency distribution of measurements for each confounding variable that is similar to that of the treatment group, but no pairing takes place. For example, attention deficit/hyperactivity disorder (ADHD) is often treated with stimulants, such as amphetamines. Biederman et al. (2009) carried out an observational study to examine the psychiatric impacts later in life of stimulant treatment. A sample of ADHD youths who had been treated with stimulants was compared with a control sample of untreated ADHD individuals that was similar to the treated group in the distribution of ages, sex, ethnic background, sensorimotor function, other psychiatric conditions, and IQ.

The second strategy used to limit the effects of confounding variables in a controlled observational study is adjustment, in which statistical methods such as analysis of covariance (Chapter 18) are used to correct for differences between treatment and control groups in suspected confounding variables. For example, LaCroix et al. (1996) compared the incidence of cardiovascular disease between two groups of older adults: those who walked more than four hours per week and those who walked less than one hour per week. The ages of the adults were not identical in the two groups, and this could affect the results. To compensate, the authors examined the relationship between cardiovascular disease and age within each exercise group, so that they could compare the predicted disease rates in the two groups for adults of the same age. This approach is discussed in more detail in Chapter 18.

14.7 Choosing a sample size

A key part of planning an experiment or observational study is to decide how many independent units or participants to include. There is no point in conducting a study whose sample size is too small to detect the expected treatment effect. Equally, there is no point in making an estimate if the confidence interval for the treatment effect is expected to be extremely broad because of a small sample size. Using too many participants is also undesirable, because each replicate costs time and money, and adding one more might put yet another individual in harm's way, depending on the study. If the treatment is unsafe, as the spermicide nonoxynol-9 appears to be (Example 14.2), then we want to injure as few people or animals as possible in coming to this conclusion. Ethics boards and animal-care committees require researchers to justify the sample sizes for proposed experiments. How is the decision made? Here in Section 14.7, we answer this question for two objectives: when the goal is to achieve a predetermined level of *precision* of an estimate of treatment effect, or when we want to achieve predetermined *power* in a test of the null hypothesis of no treatment effect. We focus here on techniques for studies that compare the means of two groups. Formulas to help plan experiments for some other kinds of data are given in the Quick Formula Summary (Section 14.9).

> An important part of planning an experiment or observational study is choosing a sample size that will give sufficient power or precision.

Plan for precision

A frequent goal of studies in biology is to estimate the magnitude of the treatment effect as precisely as possible. Planning for precision involves choosing a sample size that yields a confidence interval of expected width. Typically, we hope to set the bounds as narrowly as we can afford.

By way of example, let's develop a plan for a two-treatment comparison of means. Let the unknown population mean of the response variable be μ_1 in the treatment group of interest and μ_2 in the control group. When the results are in, we will compute the sample means $\overline{Y}_1$ and $\overline{Y}_2$ and use them to calculate a 95% confidence interval for $\mu_1 - \mu_2$, the difference between the population means of the treatment and control groups. To simplify matters somewhat, we will assume that the sample sizes in both treatments are the same number, n. Let's also assume that the measurement in the two populations is normally distributed and has the same standard deviation, σ.

In this case, a 95% confidence interval for $\mu_1 - \mu_2$ will take the form

$$\overline{Y}_1 - \overline{Y}_2 \pm \text{margin of error},$$

where "margin of error" is half the width of the confidence interval. Planning for precision involves deciding in advance how much uncertainty we can tolerate. Once we've decided that, then the sample size needed in each group is approximately

$$n = 8\left(\frac{\sigma}{\text{margin of error}}\right)^2.$$

This formula is derived from the 2SE rule of thumb that was introduced in Section 4.3.[4] According to this formula, a larger sample size is needed if σ, the standard deviation within groups, is large than if it is small. Additionally, a larger sample size is needed to achieve a high precision (a narrow confidence interval) than to achieve a low precision.

A major challenge in planning sample size is that key factors, like σ, are not known. Typically, a researcher makes an educated guess for these unknown parameters based on pilot studies or previous investigations. (If no information is available, then consider carrying out a small pilot study first, before attempting a large experiment.)

For example, let's plan an experiment to measure the effect of diet on the eye span of male stalk-eyed flies (Example 11.2). The planned experiment will randomly place individual fly larvae into cups containing either corn or spinach. The target parameter is the difference between mean eye spans in the two diet treatments, $\mu_1 - \mu_2$. Assume that we would like to obtain a 95% confidence interval for this difference whose expected margin of error is 0.1 mm (i.e., the desired full width of the confidence interval is 0.2 mm). How many male flies should be used in each treatment to achieve this goal?

Our sample estimate for σ was about 0.4, based on the sample of nine individuals in Example 11.2. Using these values gives

$$ n = 8\left(\frac{\sigma}{\text{margin of error}}\right)^2 = 8\left(\frac{0.4}{0.1}\right)^2 = 128. $$

This is the sample size in each treatment, so the total number of male flies would be 256. At this point, we would need to decide whether this sample size is feasible in an experiment. If not, then there might be no point in carrying out the experiment. Alternatively, we could revisit the desired width of the 95% confidence interval. That is, could we be satisfied with a higher margin of error? If so, then we should decide on this new width and then recalculate n.

After all this planning, imagine that the experiment is run and we now have our data. Will the confidence interval we calculate have the precision we planned for? There are two reasons that it probably won't. First, 0.4 was just an educated guess for the value of σ to help our planning, and it was based on only nine individuals. The true value of σ in the population might be larger or smaller. Second, even if we were lucky and the true value of σ really is close to 0.4, the within-treatment standard deviation s from the experiment will not equal 0.4 because of sampling error. The resulting confidence interval will be narrower or wider accordingly. The probability that the width of the resulting confidence interval is less than or equal to the desired width is only about 0.5. To increase the probability of obtaining a confidence interval no wider than the desired interval width, we would need an even larger sample size.

Figure 14.7-1 shows the general relationship between the expected precision of the 95% confidence interval and n, the sample size in each of two groups. The

---

4. The margin of error is approximately twice the standard error of the difference between sample means (2SE), or $2\sqrt{\sigma^2(\frac{1}{n} + \frac{1}{n})} = 2\sqrt{2\sigma^2/n} = \sqrt{8\sigma^2/n}$. Solving for n gives the rule in the text.

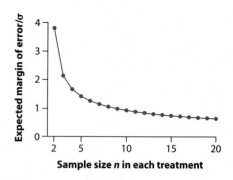

FIGURE 14.7-1
Expected precision of the 95% confidence interval for the difference between two treatment means depending on sample size n in each treatment. The vertical axis is given in standardized units, (margin of error)$/\sigma$. We calculated the expected confidence interval using the t-distribution, rather than with the 2SE approximation.

variable on the vertical axis is standardized as margin of error divided by σ. The effect of sample size from $n = 2$ to $n = 20$ is shown.

The graph shows that very small sample sizes lead to very wide interval estimates of the difference between treatment means. More data gives better precision. Note also that interval precision initially declines rapidly with increasing sample size (e.g., from $n = 2$ to $n = 10$), but it then declines more slowly (e.g., from $n = 10$ to $n = 20$). Precision is 0.63 at $n = 20$, but it drops to 0.40 by $n = 50$, to 0.28 by $n = 100$, and to 0.20 by $n = 200$. Thus, we get diminishing returns by increasing the sample size past a certain point.

Plan for power

Next we consider choosing a sample size based on a desired probability of rejecting a false null hypothesis—that is, planning a sample based on a desired power. Imagine, for example, that we want to test the following hypotheses on the effect of diet on eye span in stalk-eyed flies.

H_0: $\mu_1 - \mu_2 = 0$.
H_A: $\mu_1 - \mu_2 \neq 0$.

The null hypothesis is that diet has no effect on mean eye span. The power of this test is the probability of rejecting H_0 if it is false. Planning for power involves choosing a sample size that would have a high probability of rejecting H_0 if the absolute value of the difference between the means, $|\mu_1 - \mu_2|$, is at least as great as a specified value D. The value for D won't be the true difference between the means; it is just the minimum we care about. By specifying a value for D in a sample size calculation, we are deciding that we aren't much interested in rejecting the null hypothesis of no difference if $|\mu_1 - \mu_2|$ is smaller than D.

A conventional power to aim for is 0.80. That is, if H_0 is false, we aim to demonstrate that it is false in 80% of the experiments (the other 20% of experiments would fail to reject H_0 even though it is false). If we aim for a power of 0.80 and a conventional significance level of $\alpha = 0.05$, then a quick approximation to the planned sample size n in each of two groups is

$$n \approx 16\left(\frac{\sigma}{D}\right)^2$$

(Lehr 1992). This formula assumes that the two populations are normally distributed and have the same standard deviation (σ), which we are forced to assume is known. A more exact formula is provided in the Quick Formula Summary (Section 14.9), which also allows you to choose other values for power and significance level.

For a given power and significance level, a larger sample size is needed when the standard deviation σ within groups is large, or if the minimum difference that we wish to detect is small.

Let's return to our experiment to test the effect of diet on the eye span of male stalk-eyed flies. We would like to reject H$_0$ at $\alpha = 0.05$ with probability 0.80 if the absolute value of the difference between means were truly $D = |\mu_1 - \mu_2| = 0.2$ mm. How many males should be used in each treatment?

Let's assume again that $\sigma = 0.4$. Using this value in the equation for power gives

$$n = 16\left(\frac{0.4}{0.2}\right)^2 = 64.$$

This is the number in each treatment, so the total number of males needed in the experiment would be 128.

These power calculations assume that we know the standard deviation (σ), which is stretching the truth. For this and other reasons, we must always view the results of power calculations with a great deal of caution. The calculations provide useful guidelines, but they do not give infallible answers.

We have explored only the sample sizes needed to compare the means of two groups, but similar methods are available for other kinds of statistical comparisons as well. Sample sizes for desired precision and power are available for one- and two-sample means, proportions, and odds ratios in the Quick Formula Summary (Section 14.9). A variety of computer programs are available to calculate sample sizes when planning for power and precision. A good place to start investigating these programs is http://www.divms.uiowa.edu/~rlenth/Power/.

Plan for data loss

The methods given here in Section 14.7 for planning sample sizes refer to sample sizes still available at the *end* of the experiment. But some experimental individuals may die, leave the study, or be lost between the start and the end of the study. The starting sample sizes should be made even larger to compensate.

14.8 Summary

- In an experimental study, the researcher assigns treatments to subjects.
- The purpose of an experimental study is to examine the causal relationship between an explanatory variable, such as treatment, and a response variable. The virtue of experiments is that the effect of treatment can be isolated by randomizing the effects of confounding variables.

- A confounding variable masks or distorts the causal relationship between an explanatory variable and a response variable in a study.
- A clinical trial is an experimental study involving human participants.
- Experiments should be designed to minimize bias and limit the effects of sampling error.
- Bias in experimental studies is reduced by the use of controls, by randomizing the assignment of treatments to experimental units, and by blinding.
- In a completely randomized experiment, treatments are assigned to experimental units by randomization. Randomization reduces the bias caused by confounding variables by making nonexperimental variables equal (on average) between treatments.
- The effect of sampling error in experimental studies is reduced by replication, by blocking, and by balanced designs.
- A randomized block design is like a paired design but for more than two treatments.
- The use of extreme treatments can increase the power of the experiment to detect a treatment effect.
- Observational studies should employ as many of the strategies of experimental studies as possible to minimize bias and limit the effect of sampling error.
- Although randomization is not possible in observational studies, the effects of confounding variables can be reduced by matching and by adjusting for differences between treatments in known confounding variables.
- A factorial design is used to investigate the interaction between two or more treatment variables. The factorial design includes all possible combinations of the treatment variables.
- When planning an experiment, the number of experimental units to include can be chosen so as to achieve the desired width of confidence interval for the difference between treatment means.
- Alternatively, the number of experimental units to include when planning an experiment can be chosen so that the probability of rejecting a false H_0 (power) is high for a specified magnitude of the difference between treatment means.
- Compensate for possible data loss when planning sample sizes for an experiment.

14.9   Quick Formula Summary

Planning for precision

Planned sample size for a 95% confidence interval of a proportion

What is it for? To set the sample size of a planned experiment to achieve approximately a specified half-width ("margin of error") of a 95% confidence interval for a proportion p.

What does it assume? The population proportion p is not close to zero or one, and n is large.

Formula: $n \approx \dfrac{4p(1 - p)}{(\text{margin of error})^2}$, where p is the proportion being estimated and

"margin of error" is the half-width of the confidence interval for the proportion p. For the most conservative scenario, set $p = 0.50$ when calculating n. The symbol $\approx$ stands for "is approximately equal to."

Planned sample size for a 95% confidence interval of a log-odds ratio

What is it for? To set the sample size n in each of two groups for a planned experiment to achieve approximately a specified half-width ("margin of error") of a 95% confidence interval for a log-odds ratio, $\ln(OR)$.

What does it assume? Sample size n is the same in both groups.

Formula: $n \approx \dfrac{4}{(\text{margin of error})^2}\left(\dfrac{1}{p_1} + \dfrac{1}{1 - p_1} + \dfrac{1}{p_2} + \dfrac{1}{1 - p_2} \right)$, where "margin

of error" is the half-width of the confidence interval for $\ln(OR)$, and p_1 and p_2 are the probabilities of success in the two treatment groups.

Planned sample size for a 95% confidence interval of the difference between two proportions

What is it for? To set the sample size n in each of two groups for a planned experiment to achieve approximately a specified half-width ("margin of error") of a 95% confidence interval for a difference between two proportions, $p_1 - p_2$. This is an alternative approach to the one that uses a log-odds ratio to compare the proportion of successes in two treatment groups.

What does it assume? Sample size n is the same in both groups.

Formula: $n \approx \dfrac{8\bar{p}(1 - \bar{p})}{(\text{margin of error})^2}$, where "margin of error" is the half-width of

the confidence interval for $p_1 - p_2$. p_1 and p_2 are the probabilities of success in the two treatment groups, and $\bar{p}$ is the average of the two proportions—that is, $\bar{p} = (p_1 + p_2)/2$.

Planned sample size for a 95% confidence interval of the mean

What is it for? To set the sample size of a planned experiment to achieve approximately a specified half-width ("margin of error") of a 95% confidence interval for a mean μ.

What does it assume? The population is normally distributed with known standard deviation σ.

Formula: $n \approx 4\left(\dfrac{\sigma}{\text{margin of error}}\right)^2$, where n is the planned sample size, and "margin

of error" is the half-width of the confidence interval for the mean μ.

Planned sample size for a 95% confidence interval of the difference between two means

What is it for? To set the sample size of a planned experiment so as to achieve approximately a specified half-width ("margin of error") of the 95% confidence interval for $\mu_1 - \mu_2$.

What does it assume? Populations are normally distributed with equal standard deviations σ. The value of σ is known. Sample size n is the same in both groups.

Formula: $n \approx 8\left(\dfrac{\sigma}{\text{margin of error}}\right)^2$, where n is the planned sample size within each

group, and "margin of error" is the half-width of the confidence interval for the difference between means.

Planning for power

Planned sample size for a binomial test of 80% power at $\alpha = 0.05$

What is it for? To set the sample size n of a planned experiment to achieve approximately a power of 0.80 in a binomial test at $\alpha = 0.05$.

What does it assume? The proportion p_0 under the null hypothesis is not close to zero or one, and n is not small. Sample size n is the same in both groups.

Formula: $n \approx \dfrac{8p_0(1 - p_0)}{D^2}$, where p_0 is the proportion under the null hypothesis, and $D = p - p_0$ is the predetermined difference we wish to be able to detect between the population parameter p and that specified under the null hypothesis.

Planned sample size for 2 × 2 contingency test of 80% power at $\alpha = 0.05$

What is it for? To set the sample size of a planned experiment so as to achieve approximately a power of 0.80 at $\alpha = 0.05$ in a contingency test of the difference between the proportion of successes in two treatment groups (or, equivalently, a test that the odds ratio equals one).

What does it assume? The average probability of success in the two treatment groups is known. Sample size n is the same in both treatment groups.

Formula: $n \approx \dfrac{8\bar{p}(1 - \bar{p})}{D^2}$, where $\bar{p}$ is the average of the two probabilities of success [i.e., $\bar{p} = (p_1 + p_2)/2$], and $D = p_1 - p_2$ is the predetermined difference we wish to be able to detect between the two proportions.

Planned sample size for a one-sample or paired t-test of 80% power at $\alpha = 0.05$

What is it for? To set the sample size of a planned experiment so as to achieve approximately a power of 0.80 in a one-sample or paired t-test at $\alpha = 0.05$.

What does it assume? The population is normally distributed with standard deviation σ. The value of σ is known.

Formula: $n \approx 8\left(\dfrac{\sigma}{D}\right)^2$, where n is the sample size within each group, and $D = \mu$ is the predetermined value of the mean (or the mean difference in the case of a paired test) that we wish to detect.

Planned sample size for a two-sample t-test of 80% power at $\alpha = 0.05$

What is it for? To set the sample size of a planned experiment so as to achieve approximately a power of 0.80 in a two-sample t-test at $\alpha = 0.05$.

What does it assume? Populations are normally distributed with equal standard deviation σ. Sample size n is the same in both groups.

Formula: $n \approx 16\left(\dfrac{\sigma}{D}\right)^2$, where n is the sample size within each group, and

$D = |\mu_1 - \mu_2|$ is the predetermined difference between means we wish to detect.

PRACTICE PROBLEMS

1. Identify which goal of experimental design (i.e., reducing bias or limiting sampling error) is aided by the following procedures:
   a. Using a genetically uniform animal stock to test treatment effects
   b. Using a completely randomized design
   c. Grouping related experimental units together
   d. Taking the response measurements while unaware of the treatments assigned to experimental units
   e. Using a computer to randomly assign treatments to experimental units within each block

2. Using a coin toss for each unit, assign two hypothetical treatments to eight experimental units.
   a. Write the sequence of eight assignments you ended up with.
   b. Did you end up with an equal number of units in each treatment?
   c. What is the probability of an unbalanced design using this approach?
   d. Recommend a procedure for randomly assigning treatments to units that always results in a balanced design.

3. A series of plots were placed in a large agricultural field in preparation for an experiment to investigate the effects of three fertilizers differing in their chemical composition. Before assigning treatments, it was noticed that plots differed along a moisture gradient. What strategy would you suggest the researchers implement to minimize the impact of this gradient on the ability to measure a treatment effect? Explain with an illustration the experimental design you would recommend.

4. You read the following statement in a journal article: "On the basis of an alpha level of 0.05 and a power of 80%, the planned sample size was 129 subjects in each treatment group." State in plain language what this means.

5. Example 12.4 described a study in which salmon were introduced to 12 streams with and without brook trout to investigate the effect of brook trout on salmon survival. Is this an experimental study or an observational study? Explain the basis for your reasoning.

6. Identify the consequences (i.e., increase, decrease, or none) that the following procedures are likely to have on both bias and sampling error in an observational study.
   a. Matching sampling units between treatment and control
   b. Increasing sample size
   c. Ensuring that the frequency distribution of subject ages is the same in the two treatments
   d. Using a balanced design

7. In 1899, the *British Medical Journal* (page 933) reported the results of a medical procedure involving the subcutaneous infusion of a salt solution for the treatment of extremely severe pneumonia: "Dr. Clement Penrose has tried the effect of subcutaneous salt infusions as a last extremity in severe cases of pneumonia. He continues this treatment with inhalations of oxygen. He has had experience of three cases, all considered hopeless, and succeeded in saving one. In the other two the prolongation of life and the relief of symptoms were so marked that Dr. Penrose regretted that the treatment had not been employed earlier."
   a. Is this an experimental study? Why or why not?
   b. What design components might Dr. Penrose have included in an experiment to test the effectiveness of his treatment?

8. In a study of the effects of marijuana on the risk of cancer in oral squamous cells, Rosenblatt et al. (2004) examined 407 recent cases of the cancer from western Washington state. They also randomly sampled 615 healthy people from the same region having similar frequency distributions of age and sex as the cancer cases. They found that a similar proportion of the cancer cases (25.6%) and healthy participants (24.4%) reported ever having used marijuana (odds ratio = 0.9; 95% confidence interval, 0.6 < OR < 1.3).

   a. What name is given to this type of study? Is it an experimental study or an observational study? Explain.

   b. Does this study include a control group? Explain.

   c. What was the purpose of ensuring that the healthy participants were similar in age and sex to the cancer cases?

   d. Can we conclude that marijuana does not cause cancer in oral squamous cells in this population?

9. After stinging its victim, the honeybee leaves behind the barbed stinger, poison sac, and muscles that continue to pump venom into the wound. Visscher et al. (1996) compared the effects of two methods of removing the stinger left behind: scraping off with a credit card or pinching off with thumb and index finger. A total of 40 stings were induced on volunteers. Twenty were removed with the credit card method, and 20 were removed with the pinching method. The size of the subsequent welt by each sting was measured after 10 minutes. All 40 measurements came from two volunteers (both authors of the study), each of whom received one treatment 10 times on one arm and the other treatment 10 times on the other arm. Pinching led to a slightly smaller average welt, but the difference between methods was not significant.

   a. All 40 measurements were combined to estimate means, standard errors, and the P-value for a two-sample t-test of the difference between treatment means. What is wrong with this approach?

   b. How should the data be analyzed? Describe how the quantities would be calculated and what type of statistical test would be used.

   c. Suggest two improvements to the experimental design.

10. What is the justification for including extreme doses well outside the range of exposures encountered by people at risk in a dose–response study on animals of the effects of a hazardous substance? What are the problems with this approach?

11. A strain of sweet corn has been genetically modified with a gene from the bacterium *Bacillus thuringiensis* (Bt) to express the protein Cry1Ab, which is toxic to caterpillars that eat the leaves. Unfortunately, the pollen of transformed corn plants contains the toxin, too. Corn pollen dusts the leaves of other plants growing nearby, where it might have negative effects on non-pest caterpillars. You are hired to conduct a study to measure the effects on monarch butterfly caterpillars of ingesting Bt-modified pollen that has landed on the leaves of milkweed, a plant commonly growing in or near cornfields. You decide to use a completely randomized design to compare the effect of two treatments on monarch pupal weight. In one treatment, you place potted milkweed plants in plots of Bt-modified corn, where their leaves receive pollen carrying the toxin. In the other treatment, you place milkweed plants in plots with ordinary corn that has not been transformed with the Bt gene. You place a monarch larva on every milkweed plant. Previous studies have estimated that the standard deviation of pupal weight in monarch butterflies is about 0.25 g.

a. Suppose your goal at the end of the experiment is to calculate a 95% confidence interval for the difference between treatments in mean monarch pupal weights. How many plots would you plan in each treatment if your goal was to produce a confidence interval for the difference in mean pupal weights between treatments having a total width of 0.4 g?

b. What sample size would you need if you decided that 0.4 was not precise enough, and that you wished to halve this interval to 0.2?

c. Imagine that your permits allow you to plant only five plots of Bt-transformed corn, so that the only way you can increase the total sample size for the whole experiment is to increase the number of plots in the ordinary corn treatment. To achieve the same width of confidence interval as in part (a), would the total sample size needed (both treatments combined) likely be greater, smaller, or no different from that calculated in part (a)? Explain.

d. In designing the experiment, why would you not simply place all the milkweed plants for one treatment at random locations in a single large Bt-transformed corn field, and all the milkweed plants for the other treatment at random locations in a single large normal corn field?

12. In the Bt and monarch study described in Practice Problem 11, how many plots would you plan per treatment if your goal were to carry out a test having 80% power to reject the null hypothesis of no treatment effects when the difference between treatments means is at least 0.25 g?

13. Consider the results of a six-year observational study that documented health changes related to homeopathic care (Spence and Thompson 2005). Homeopathic treatment was defined as "stimulating the body's autoregulatory mechanisms using microdoses of toxins." Every one of the 6544 patients in the study was assigned to a hospital outpatient unit for homeopathic treatment. Of these, 4627 patients (70.7%) reported positive health changes following treatment.

Suggest a major improvement to the design of this study.

14. The fish species *Astyanax mexicanus* includes blind, cave-inhabiting populations whose eyes degenerate during embryonic development. To understand how eye degeneration worked, Yamamoto and Jeffery (2000) replaced the lens of the degenerate eye on one side (randomly chosen) of a blind cave fish embryo with a lens from the embryo of a "normal," sighted fish. This procedure was repeated on all individuals in a sample of blind cave fish. Final eye size was measured on both sides of each experimental fish, after embryonic development was complete. Remarkably, a normal-sized eye was restored on the transplant eyes of blind cave fish but not on the unmanipulated side. Based on the preceding description of a laboratory experiment, identify which of the six main strategies of experimental design (listed in Section 14.2) were incorporated.

15. Blaustein et al. (1997) used a field experiment to investigate whether increased UV-B radiation was a cause of amphibian deformities (see the photo at the beginning of this chapter). They measured long-toed salamanders either exposed to or shielded from natural UV-B radiation. It was not possible to carry out all replicates simultaneously, so the researchers carried them out over several days. They made sure that both treatments were included on each day. In their analysis, they grouped replicates together that were carried out on the same day.

a. By grouping experiments carried out on the same day, what experimental procedure were they using?

b. What is the main reason for adopting this procedure in an experimental study?

16. In 1976, Ewan Cameron and Linus Pauling (the only person to have won two unshared Nobel Prizes) published a paper showing that vitamin C was an effective treatment for some kinds of cancer. They measured the life spans of a sample of 100 patients who were given extra doses of vitamin C. As a control, they pulled the records of several hundred patients from the same clinic who had died from the same types of terminal

cancer, and who were matched to the vitamin C patients for their age, sex, and type of cancer. They found that the patients with extra vitamin C lived on average 2.7 times longer than the controls. A later study by Moertel et al. (1985) randomly assigned two treatments to cancer patients, supplemental vitamin C and control, and followed the patients with a double-blind study. This later study found no difference between the two groups for their life spans.

a. Give plausible reasons why the two studies might have found different results.

b. From the information given, which study is expected to give the most reliable results? Why?

ASSIGNMENT PROBLEMS

17. Identify the consequences (i.e., increase, decrease, or none) that the following procedures are likely to have on both bias and sampling error in an experimental study.

  a. Assigning treatment to subjects alphabetically, not randomly
  b. Increasing sample size
  c. Calculating power
  d. Applying every treatment to every experimental unit in random order
  e. Using a sample of convenience instead of a random sample
  f. Testing only one treatment group, without a control group
  g. Using a balanced design
  h. Informing the human participants which treatment they will receive

18. The experiment described in Example 12.2 compared antibody production in 13 male red-winged blackbirds before and after testosterone implants. The units of antibody levels were log 10^{-3} optical density per minute ($\ln[\text{mOD}/\text{min}]$). The mean change in antibody production was $\bar{d} = 0.056$, and the standard deviation was $s_d = 0.159$. If you were assigned the task of repeating this experiment to test the hypothesis that testosterone changed antibody levels, what sample size (i.e., number of blackbirds) would you plan to ensure that a mean change of 0.05 units could be detected with probability 0.8? Explain the steps you took to determine this value.

19. Two clinical trials were designed to test the effectiveness of laser treatment for acne. Seaton et al. (2003) randomly divided participants into two groups. One group received the laser treatment, whereas the other group received a sham treatment. Orringer et al. (2004) used an alternate design in which laser treatment was applied to one side of the face, randomly chosen, and the sham treatment was applied to the other side. The number of facial lesions was the response variable.

  a. Identify the main component of experimental design that differs between the two studies. Give the statistical term identifying this component in experimental design.
  b. Under what circumstances would there be an advantage to using the "divided-face" design over the completely randomized (two-sample) design?
  c. Assuming that the advantage identified in part (b) is met, can you think of a disadvantage of the divided-face design?[5]

20. Identify the consequences (i.e., increase, decrease, or none) that the following procedures are likely to have on both bias and sampling error in an observational study.

  a. Planning for data loss
  b. Taking measurements of the subjects while unaware of which subjects belong to which group
  c. Including only one sex and age group in the study

---

5. Other than a possible social dilemma.

d. Adjusting for body size using analysis of covariance

21. Identify which goal of experimental design (i.e., reducing bias or limiting effects of sampling error) is aided by the following procedures:
   a. Including extreme treatment levels
   b. Using a paired design
   c. Keeping room temperature constant in an experiment designed to test the effects of a pesticide on insect survival
   d. Eliminating artifacts when designing the treatment of interest
   e. Adding a sham operation group

22. Identify the particular feature that defines each of the following experimental designs, and list the specific advantages provided by the feature you identify.
   a. Factorial design
   b. Randomized block design
   c. Completely randomized design

23. Kirsch (2010) argues that in double-blind clinical trials to test the effects of antidepressants, a large fraction of patients figure out whether they have been given the antidepressant or the placebo by noticing the presence or absence of known side effects of the antidepressant. Doctors evaluating the patients are also able to determine which treatment patients are receiving. How might this situation affect the results of the clinical trial? Specifically, is the treatment effect (difference between the means of the antidepressant and placebo treatments) likely to be overestimated, underestimated, or unaffected by this knowledge?

24. Michalsen et al. (2003) conducted a study to examine the effects of "leech therapy" for pain resulting from osteoarthritis of the knee. Two treatments were randomly assigned to 51 patients with osteoarthritis of the knee. Patients in the leech treatment received 4–6 medicinal leeches applied to the soft tissue of the affected knee in a single session. The animals were left to feed ad libitum until they detached themselves, on average 70 minutes later. Patients in the control treatment were given diclofenac gel and were told to apply it twice daily to the affected area. Pain was assessed by a questionnaire given by per-

sonnel unaware of the treatments applied to each patient. The results showed that seven days after the start of treatment, pain was significantly lower in the leech group.
   a. Does this study include a control group? Explain.
   b. Is this an experimental study or an observational study? Explain.
   c. Is this a completely randomized design or a randomized block design? Explain.
   d. Which strategy for reducing bias was not adopted in this study? How might its absence have affected the results?

25. Design a study to compare the reaction times of the left and right hands of right-handed people using a computer mouse. Two design choices are available to you. In the first, a sample of right-handed participants are randomly divided into two groups. Reaction time with the left hand is measured in one group, and reaction time with the right hand is tested in the other group.
   a. What is the second design choice available to you?
   b. Under what circumstances would the second choice be the preferred choice?
   c. Assume that you decide to go with the completely randomized design and that, at the end of the experiment, you aim to calculate a 95% confidence interval for the difference between the mean reaction times of left and right hands. To achieve a confidence interval of specified width, what information would you require to plan an appropriate sample size?

26. Identify the single most significant flaw in each of the following experimental designs. Use statistical language to identify what's missing.
   a. In a test of the effectiveness of acupuncture in treating migraine headaches, a random sample of patients at a migraine clinic were provided with a novel acupuncture treatment daily for six months. The patients were interviewed at the start of the study and at the end to determine whether there had been any change in the severity of their migraines.
   b. In a modified study, a second sample of patients were chosen after the acupuncture

treatment was completed on the first set of patients. This second sample of patients received a placebo in pill form for six months. At the end of the study, perceived pain levels in the two groups were compared.

c. In a modified study, the sample of patients was divided into two groups according to gender. The women received the acupuncture treatment, and the men received the placebo medication in pill form. At the end of the experiment, perceived pain levels in the two groups were compared.

d. In a modified study, patients were randomly divided into two groups. One group received the acupuncture treatment, and the other received a fixed dose of placebo medication in pill form. At the end of the experiment, perceived pain levels in the two groups were compared.

27. Young et al. (2006) took measurements of subordinate female meerkats to determine the changes in reproductive physiology experienced by females that are evicted from their social groups. They compared evicted females and those not evicted in their level of plasma luteinizing hormone following a GnRH hormone challenge. They found that nine evicted females had a sample average of 6.2 mIU/ml (milli-International Units per milliliter) of plasma luteinizing hormone compared with 12.1 mIU/ml in 18 females that had not been evicted. The pooled sample variance was 28.4.

a. Is this an experimental study or an observational study? Explain.

b. The sample size was unequal between the two groups of females compared. How would this affect the power of a hypothesis test of the difference between group means compared with a more balanced design? Explain.

c. How would the imbalance of the sample sizes affect the width of the confidence interval for the difference between group means compared with a more balanced design? Explain.

d. If you were planning to repeat the comparison of plasma luteinizing hormone between these two groups of females, what sample size would you plan to achieve an expected half-width of 3 mIU/ml for a 95% confidence interval of the difference between means? Explain the steps you took to determine this value.

28. Diet restriction is known to extend life and reduce the occurrence of age-related diseases. To understand the mechanism better, you propose to carry out a study to look at the separate effects of age and diet restriction, and the interaction between age and diet restriction, on the activity of liver cells in rats. What experimental design should you consider employing? Why?

Data dredging

In the spoof journal *Annals of Improbable Research*, a satirical article reported on a study of the so-called butterfly effect (Inaudi et al. 1995). This effect, a mainstay of the popular representation of chaos theory, says that small initial causes, like the flapping of a butterfly's wings, can ultimately have large effects, like a hurricane, on the other side of the world. The fearless researchers set out to measure this effect by capturing several dozen butterflies and holding them in captivity in Switzerland. Each day, they checked the butterflies and recorded whether or not they flapped their wings. Then, using the lab's phone, they called their girlfriends in Paris each day to ask whether or not it was raining.[1] At the end of the study, the students tested each butterfly for an association between its daily flapping behavior and the daily weather in Paris. They found that the flapping days of one of the butterflies closely matched the rainfall days in Paris ($P < 0.05$). They exulted, "Not only have we proven that the butterfly effect exists, we have found the butterfly."

These guys were clearly joking, but statistically speaking, where did they go wrong? The answer is that they went "data dredging."

They performed many statistical tests and eventually one of them was significant. Data dredging (also called "data snooping" or "data fishing") is the carrying out of many statistical tests in hope of finding at least one statistically significant result.

The problem with data dredging is that the probability of making *at least one* Type I error (i.e., of obtaining a false positive) is greater than the significance level α when many tests are conducted, if the null hypothesis is true (as it surely is in the butterfly example). Each hypothesis test has some chance of error, and these errors are compounded over multiple tests. There is a much larger probability of getting an error out of several tries than in any one try. By analogy, we might get away with playing Russian roulette once, but we would be unlikely to survive a month of playing once a night.

It's useful to do a few calculations to see how big the problem might be. The probability of making no Type I errors in N independent tests is $(1 - \alpha)^N$. Thus, the chance of making at least one Type I error from N independent tests is $1 - (1 - \alpha)^N$. This means that, if we use $\alpha = 0.05$ and carry out 20 independent tests of true null hypotheses, the probability that at least one of these tests will falsely reject the null hypothesis is about 65%. If we carry out 100

---

1. They continued the experiment "until the first phone bill reached our Office of Financial Services."

tests, then the chance of rejecting at least one of the null hypotheses becomes 99.4%, even if all the null hypotheses are true. With data dredging, a false positive result is almost inevitable.

Nevertheless, multiple testing is common in biology, and for good reasons. A dedicated experimentalist on human participants might measure many conceivable responses (e.g., blood pressure, body temperature, red blood cell count, white blood cell count, speed of recovery, appetite, and weight change) and perhaps even a few extra variables that might be long shots. The result is that the clinician might end up carrying out 10 or 20 tests of treatment effects, raising the probability of a false positive result. This level of multiple testing pales next to that seen in gene mapping. Locating a gene for a single trait, such as a genetic disease, typically involves thousands of statistical tests (one for each section of the genome). What should be done about the soaring Type I error rates resulting from so much testing?

The answer to this question depends on your goals. If your goal is simply to *explore* the data, to discover the possibilities but not to provide rigorous tests, then you need do nothing special about multiple testing except report the number of tests that you carried out and note which ones yielded a significant result. If you admit that you dredged the data, your results can still be useful. New hypotheses and unexpected discoveries can emerge from a thorough fishing expedition. However, the individual significant results that pop up from data dredging cannot yet be taken seriously, due to the high probability of one or more Type I errors. Some of the significant results might indeed be real, but it will be difficult to establish which ones. Rather, a new study must be carried out with new data to test

any promising results that emerged from the exploratory approach. Another strategy sometimes used when exploring data is to divide the data randomly into two independent parts. One part is used for data dredging, and the other part is used to confirm any positive results suggested by the dredging.

If your goals from multiple testing are more rigorous (e.g., you want to determine which variable really did respond to treatment in a clinical trial, or which location in the genome really does contain a gene for a heritable disease), then steps must be taken to *correct* for the inflation of Type I error rates that occurs with multiple testing. The simplest way to accomplish this is to use a more stringent significance level—that is, one smaller than the usual $\alpha = 0.05$.

The most common correction for multiple comparisons is the **Bonferroni correction**. In the simplest version of this method, each test uses a significance level α^* rather than α, where

$$\alpha^* = \frac{\alpha}{\text{number of tests}}$$

For example, if we typically adopt the significance level $\alpha = 0.05$ when carrying out a single test, then to carry out 12 separate tests we should use the significance level $\alpha^* = 0.05/12 = 0.00417$ instead. In this case, we would reject H_0 in each test only if P were less than or equal to 0.00417. With the Bonferroni correction, the probability of getting at least one Type I error during the course of carrying out all 12 tests is approximately equal to the initial α-value (i.e., 0.05 in this case).

Keep in mind, though, that applying the Bonferroni correction greatly reduces the power of single tests. This is the price paid for asking many questions of the data. More than ever, we should be mindful not to "accept the null hypothesis." It is okay to be skeptical when a null hypothesis is not rejected and power is so limited, but there is little to do about it except to repeat the study and look again.

Another, increasingly popular approach to correct for multiple comparisons is called the **false discovery rate (FDR)**. To use this approach, carry out all of the multiple tests at a fixed significance level α (e.g., the usual 0.05). Gather all of the tests that yield a statistically significant result (i.e., all of the tests for which $P \leq \alpha$). We can call this subset of tests the "discoveries." The FDR estimates the proportion of discoveries that are false positives. In other words, the FDR is the proportion of tests for which the null hypothesis was rejected yet the null hypothesis was true. For example, Brem et al. (2005) carried out hundreds of statistical hypothesis tests of interactions between pairs of yeast genes. Of these tests, 225 yielded a statisti-cally significant result (the "discoveries"). Using the false discovery rate method, they estimated that 12 of these 225 tests were false positives, leaving 213 "true" discoveries.

An extension of the FDR calculates a quantity called the q-value for each discovery. The q-value is analogous to a P-value, providing a measure of the strength of support from the data that the null hypothesis is false in a specific test. The smaller the q-value, the stronger is the evidence that H_0 is false and should be rejected. Unlike the P-value, the q-value takes into account other tests carried out at the same time. The idea is that, by choosing to reject H_0 only if the q-value is 0.05 or less, we reject the null hypothesis falsely in only 5% of tests. FDR and q-values are a more powerful approach to dealing with multiple comparisons, and we expect their use to increase in biological applications over the next decade. Consult Benjamini and Hochberg (1995) or Storey and Tibshirani (2003) for more details.

Egyptian vulture

15 Comparing means of more than two groups

How would we analyze the results of a clinical trial that randomly assigns not two but *three* treatments to a sample of patients? Two of the treatments might be different medications and the third a placebo control. Such a design can answer more questions than a two-treatment experiment because more comparisons can be made in the same experiment. For example, are both medications better than the placebo? If they are, then by how much? Is one medication superior to the other? If so, how much better is it?

How do we compare the means of the three groups? At first glance, it might seem reasonable to compare them two at a time: first compare the means of groups 1 and 2, then compare groups 2 and 3, and finally, compare groups 1 and 3. This analysis-by-twos quickly runs into problems, because testing multiple pairs of

459

means inflates the probability of committing at least one Type I error (recall the data dredging discussed in Interleaf 8). The danger is modest when comparing only three groups, but it escalates rapidly with an increasing number of groups. Comparing five groups would require 10 tests, which would give as much as a 40% chance of falsely rejecting at least one of those null hypotheses if they were all true.

The best solution is the **analysis of variance**, or **ANOVA**, which compares means of multiple groups simultaneously in a single analysis. Analysis of *variance* might seem like a misnomer, given our intention to compare *means*, but testing for variation among groups is equivalent to asking whether the means differ. ANOVA was originally developed by the biologist and statistician R. A. Fisher, who was first mentioned in Interleaf 1.

This chapter discusses one-way or single-factor analysis of variance, which investigates the means of several groups differing by one explanatory variable or factor. Two-way or two-factor ANOVA is discussed in Chapter 18.

15.1 The analysis of variance

Analysis of variance is the most powerful approach known for simultaneously testing whether the means of k groups are equal. It works by assessing whether individuals chosen from different groups are, on average, more different than individuals chosen from the same group. Example 15.1 introduces the method.

EXAMPLE
15.1

The knees who say night

Traveling to a different time zone can cause jet lag, but people adjust as the schedule of light to their eyes in the new time zone gradually resets their internal, circadian clock. This change in their internal clock is called a phase shift. Campbell and Murphy (1998) reported that the human circadian clock can also be reset by exposing the back of the *knee* to light, a finding met with skepticism by some, but hailed as a major discovery by others. Aspects of the experimental design were subsequently challenged.[1] The data in

Table 15.1-1 are from a later experiment by Wright and Czeisler (2002) that re-examined the phenomenon. The new experiment measured circadian rhythm by the daily cycle of melatonin production in 22 people randomly assigned to one of three light treatments. Participants were awakened from sleep and subjected to a single three-hour episode of

---

1. In the 1998 experiment, participants' eyes had been exposed to low levels of light while their knees were being illuminated.

bright lights applied to the eyes only, to the knees only, or to neither (the control group). Effects of treatment on the circadian rhythm were measured two days later by the magnitude of phase shift in each participant's daily cycle of melatonin production. Results are plotted in Figure 15.1-1. A negative measurement indicates a delay in melatonin production, which is the predicted effect of light treatment; a positive number indicates an advance. Does light treatment affect phase shift?

TABLE 15.1-1 Raw data and descriptive statistics of phase shift, in hours, for the circadian rhythm experiment.

Treatment	Data (h)	$\overline{Y}$	s	n
Control	0.53, 0.36, 0.20, −0.37, −0.60, −0.64, −0.68, −1.27	−0.3088	0.6176	8
Knees	0.73, 0.31, 0.03, −0.29, −0.56, −0.96, −1.61	−0.3357	0.7908	7
Eyes	−0.78, −0.86, −1.35, −1.48, −1.52, −2.04, −2.83	−1.5514	0.7063	7

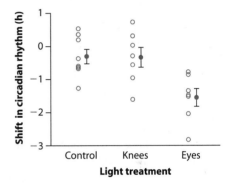

FIGURE 15.1-1
Strip chart showing the phase shift in the circadian rhythm of melatonin production in 22 experimental participants given alternative light treatments (*open circles*). Filled dots and vertical lines (*error bars*) are group means ±1 standard error.

Hypotheses

The null hypothesis of ANOVA is that the population means μ_i are the same for all treatments. (Throughout this chapter, we'll use subscripts on various quantities to indicate the group that each refers to. Generically, we'll refer to population i, which will have mean μ_i; for example, the mean of group 3 will be written as μ_3.)

Under the null hypothesis, the sample means $\overline{Y}_i$ differ from each other solely because of random sampling error. The alternative hypothesis is that the mean phase shift is not the same in all three light-treatment populations.

H_0: $\mu_1 = \mu_2 = \mu_3$.

H_A: At least one μ_i is different from the others.

The alternative hypothesis does not state that every mean is different from all the others, but only that at least one mean stands apart.

> Rejecting H_0 in ANOVA is evidence that the mean of at least one group is different from the others.

ANOVA in a nutshell

Even if all the groups in a study had the same true mean, the data would likely show a different sample mean for each group. This is because of sampling error—the chance difference between a sample estimate and the true value of a population parameter caused by random sampling. Thus, we expect to see variation among sample means taken from different groups even when the null hypothesis is true and the groups all have the same mean.

The key insight of ANOVA is that we can estimate how much variation among group means *ought* to be present from sampling error alone if the null hypothesis is true. If there really are no differences among the populations, then taking a random sample from each population is equivalent to taking the same number of random samples from a single population. As we learned in Chapter 4, when we talked about the sampling distribution of the mean, the amount of variation we expect to see *among* the sample means of repeated random samples from a population is directly related to sample size and amount of variation we see among subjects within the samples.[2]

In contrast, if the null hypotheis is *not* true, then there are real differences in means among the groups. There are still differences among sample means caused by chance, but on top of that there are differences among sample means caused by real variation among the population means. ANOVA lets us determine whether there is more variance among the sample means than we would expect by chance alone. If so, then we can infer that there are real differences among the population means.

That's the basic idea behind ANOVA. To take it one step further, we introduce the two measures of variation that are calculated from the data and compared in a test of the null hypothesis. In the terminology of this chapter, both quantities are called "mean squares."[3]

The *group mean square* (MS_{groups}) is proportional to the observed amount of variation among the group sample means. You can think of this quantity as representing the variation among the sampled subjects that belong to different groups. The *error mean square* (MS_{error}) estimates the variance among subjects that belong to the same group. It is analogous to the pooled sample variance in two-sample comparisons.

---

2. If several samples of the same size are taken from a population, the variance among sample means $\sigma_{\bar{Y}}^2$ is σ^2/n, where σ^2 is the variance within populations and n is the number of subjects in each sample (sample size). Remember from Chapter 4 that the standard deviation of the sampling distribution for $\bar{Y}$ is $\sigma/\sqrt{n}$. This is the standard error of the mean, and squaring it gives the variance, σ^2/n.

3. Unfortunately, analysis of variance has its own special jargon. You need to learn it so that the terms are familiar when you see them used in published articles and in the output of statistical packages on the computer.

Under the null hypothesis that the true means of groups do not differ, individuals belonging to different groups will on average be no more different from one another than individuals belonging to the same groups. The group mean square and the error mean square should be equal (except by chance). But if the null hypothesis is false, we expect the group mean square to *exceed* the error mean square. In this case the variation among individuals belonging to different groups is expected to be *greater than* the variation among subjects belonging to the same group.

The comparison of mean squares is done with an *F*-ratio.[4]

$$F = \frac{\text{group mean square}}{\text{error mean square}} = \frac{\text{MS}_{\text{groups}}}{\text{MS}_{\text{error}}}.$$

If the null hypothesis is true, and the means do not differ, the group and error mean squares on average will be similar and *F* should be close to 1. If the null hypothesis is false, the real differences among group means should inflate the group mean square and *F* is expected to exceed 1. Those are the only two possibilities. In the following sections, we'll show how to calculate the mean squares and carry out the formal test using the *F*-distribution.

ANOVA tables

If you run an analysis of variance on a data set using the computer, the results will likely include an **ANOVA table**. In Table 15.1-2, we show the ANOVA table for the circadian rhythm experiment discussed in Example 15.1.

TABLE 15.1-2 ANOVA table for the results of the circadian rhythm experiment (Example 15.1).

Source of variation	Sum of squares	df	Mean squares	F-ratio	P
Groups (treatment)	7.224	2	3.6122	7.29	0.004
Error	9.415	19	0.4955		
Total	16.639	21			

The table organizes all the computations leading to a test of the null hypothesis of no differences among population means. It includes the group and error mean squares as well as their ratio, *F*. The mean squares are computed from other quantities that we'll learn about in the next section: sums of squares and their degrees of freedom (*df*). We'll see in the next two sections how these numbers are calculated and interpreted.

---

4. Continuing the logic of footnote 2, if the null hypothesis is true and we have several samples of size *n*, then the variance among sample means $\sigma_{\bar{Y}}^2$ equals σ^2/n, where σ^2 is the variance within populations. Rewriting, $n\sigma_{\bar{Y}}^2 = \sigma^2$. Under the null hypothesis, $\text{MS}_{\text{groups}}$ provides an estimate of $n\sigma_{\bar{Y}}^2$, and MS_{error} is an estimate of σ^2. In this case, $\text{MS}_{\text{groups}}$ will on average be the same as MS_{error}, and the *F*-ratio $\text{MS}_{\text{groups}}/\text{MS}_{\text{error}}$ should equal 1, except by chance.

ANOVA tables also show the *P*-value for the test of the null hypothesis. For the circadian rhythm data, the *P*-value is 0.004, allowing us to reject the null hypothesis that the mean phase shift is the same for all three treatment populations.

Partitioning the sum of squares

We'll show you the ANOVA calculations in stages, beginning with the *sums of squares*. This step separates the two sources of variation in the data: within groups and among groups. In the formulas that follow, we refer to an individual observation as Y_{ij}. Here, i refers to the group to which the individual belongs, and j refers generally to the *j*th individual in the group. As you can see in Figure 15.1-2, the deviation between each observation Y_{ij} and the grand mean $\overline{Y}$ (the mean of all the observations) can be split into its two parts,

$$Y_{ij} - \overline{Y} = (Y_{ij} - \overline{Y}_i) + (\overline{Y}_i - \overline{Y}).$$

The first part of the split, $(Y_{ij} - \overline{Y}_i)$, is the deviation between each observation and its group mean. These deviations are illustrated with vertical lines in the right panel of Figure 15.1-2. The second part of the split, $(\overline{Y}_i - \overline{Y})$, is the deviation between the mean of the group to which the observation belongs and the grand mean. These deviations are illustrated for the circadian rhythm data in the middle panel of Figure 15.1-2.

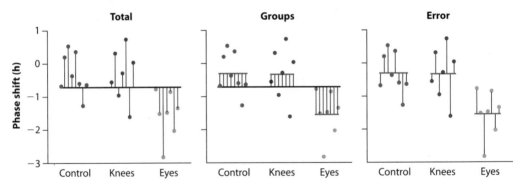

FIGURE 15.1-2 Depiction of the portions of variation in the circadian rhythm data (Example 15.1). Each dot is the measurement of phase shift in a single subject. The long horizontal line in black is the grand mean, $\overline{Y}$. Short horizontal lines in red are the sample means of the three light-treatment groups. Vertical lines represent deviations.

Repeating this split for all observations, and then squaring the deviations and taking their sum, partitions the total variation in the data set into its within- and among-group components,

$$SS_{total} = \sum_i \sum_j (Y_{ij} - \overline{Y})^2 = \sum_i \sum_j (Y_{ij} - \overline{Y}_i)^2 + \sum_i n_i (\overline{Y}_i - \overline{Y})^2,$$

where n_i is the number of observations in group i.[5] The term on the far left of the equation is the *total sum of squares*, symbolized as SS_{total}. The two parts on the right of the equal sign are the *error sum of squares*, SS_{error}, and the *group sum of squares*, SS_{groups}. In other words,

$$SS_{total} = SS_{error} + SS_{groups}.$$

To begin our calculations we also need the grand mean $\overline{Y}$ (the mean of all the data from all groups combined):

$$\overline{Y} = \frac{\sum\limits_{i} n_i \overline{Y}_i}{N},$$

where N is the total sample size, $N = \sum n_i$. This equation for the grand mean is equivalent to adding up all the individual observations from all the groups and dividing by the total number of measurements. The grand mean is not the same as the average of the group sample means if the sample size is not the same in every group. For the circadian rhythm data,

$$\overline{Y} = \frac{8(-0.3087) + 7(-0.3357) + 7(-1.5514)}{22} = -0.7127.$$

The group sum of squares is then calculated as

$$SS_{groups} = \sum_{i} n_i (\overline{Y}_i - \overline{Y})^2.$$

For the circadian rhythm data, this is

$$SS_{groups} = 8[-0.3087 - (-0.7127)]^2 + 7[-0.3357 - (-0.7127)]^2 + 7[-1.5514 - (-0.7127)]^2$$
$$= 7.224.$$

The error sum of squares can be calculated as

$$SS_{error} = \sum_{i} \sum_{j} (Y_{ij} - \overline{Y}_i)^2 = \sum_{i} s_i^2 (n_i - 1),$$

where s_i is the sample standard deviations from group i. For the circadian rhythm data,

$$SS_{error} = (0.6176)^2(8 - 1) + (0.7908)^2(7 - 1) + (0.7063)^2(7 - 1)$$
$$= 9.415.$$

The value of SS_{total} is then calculated from the other two quantities by addition:

$$SS_{total} = SS_{error} + SS_{groups} = 9.415 + 7.224 = 16.639.$$

These sums of squares go into the second column of the ANOVA table (Table 15.1-2).

---

5. The n_i in the second term comes from the fact that we're adding up over all the n_i individuals in group i.

Calculating the mean squares

The *group mean square,* symbolized by MS_{groups}, is calculated from the deviations of group sample means $(\overline{Y}_i)$ around the grand mean of all the measurements $(\overline{Y})$:

$$MS_{groups} = \frac{SS_{groups}}{df_{groups}},$$

where df_{groups} is the number of degrees of freedom for the groups. The degrees of freedom for groups is one less than the number of groups:

$$df_{groups} = k - 1,$$

where k is the number of groups.

MS_{groups} measures the observed amount of variation among the sample means from different groups. It represents the variation among subjects that belong to different groups. For the circadian rhythm data,

$$MS_{groups} = \frac{SS_{groups}}{df_{groups}} = \frac{7.224}{3 - 1} = 3.6122.$$

> The *group mean square* of ANOVA represents variation among the sampled individuals belonging to different groups. It will on average be similar to the error mean square if population means are equal.

The *error mean square,* symbolized as MS_{error}, measures variance within groups. ANOVA assumes that σ^2 (i.e., the variance of Y) is the same in every population. In other words, we assume that $\sigma^2 = \sigma_1^2 = \sigma_2^2 = \ldots = \sigma_k^2$ for all k groups. The best estimate of this variance within groups is the pooled sample variance, just as in the two-sample *t*-test (Chapter 12). In ANOVA, the pooled sample variance is called the error mean square, and it is calculated as

$$MS_{error} = \frac{SS_{error}}{df_{error}}.$$

We showed how to calculate the error sum of squares (SS_{error}) on p. 465. The denominator of this formula is the number of degrees of freedom for error, and it is just the sum of the degrees of freedom for the different groups:

$$df_{error} = \sum (n_i - 1) = N - k.$$

N is the total number of data points in all groups combined:

$$N = \sum n_i$$

For the circadian rhythm data in Table 15.1-1,

$$MS_{error} = \frac{SS_{error}}{df_{error}} = \frac{9.415}{22 - 3} = 0.4955.$$

The degrees of freedom and the mean squares go into the ANOVA table next to the sums of squares (Table 15.1-2).

> The *error mean square* of ANOVA is the pooled sample variance, a measure of the variation among individuals within the same groups.

The variance ratio, F

Under the null hypothesis that the population means of all groups are the same, the variation among individuals belonging to different groups (represented by MS_{groups}) will on average be the same as the variation among individuals belonging to the same group (estimated by MS_{error}). In ANOVA, therefore, we test for a difference by calculating the ratio of MS_{groups} over MS_{error}:

$$F = \frac{MS_{groups}}{MS_{error}}.$$

This F-ratio is the test statistic in analysis of variance. Under the null hypothesis, F will on average lie close to one, differing from it only because of sampling variation in the numerator and denominator. If the null hypothesis is false, however, and the alternative hypothesis is correct, then MS_{groups} should *exceed* MS_{error} and we expect F to be greater than one.

For the circadian rhythm data,

$$F = \frac{3.6122}{0.4955} = 7.29.$$

To calculate the P-value, we need the sampling distribution for the F-statistic under H_0. This null distribution for the F-statistic is called the F-distribution. The F-distribution has a pair of degrees of freedom, one for the numerator (top) of the F-ratio and a second for the denominator (bottom). The numerator (MS_{groups}) has $k-1$ degrees of freedom, and the denominator (MS_{error}) has $N-k$ degrees of freedom.

For the circadian rhythm data, $N = 22$ and $k = 3$, so there are $k-1 = 3-1 = 2$ degrees of freedom for the numerator and $N-k = 22-3 = 19$ degrees of freedom for the denominator. The F-distribution with 2 and 19 df is therefore the appropriate null distribution for F. The number of degrees of freedom for the numerator is always presented first when specifying the F-distribution. This is important because the F-distribution with 2 and 19 degrees of freedom is *not* the same as the F-distribution having 19 and 2 degrees of freedom.

Figure 15.1-3 illustrates the F-distribution with 2 and 19 degrees of freedom. The distribution ranges from zero to positive infinity. The right tail of the curve is the part we are interested in. This is because if H_0 is false, then MS_{groups} should *exceed* MS_{error} and lead to an F-ratio in the right tail of the F-distribution. F-ratios less than one might occur, but only by chance.

FIGURE 15.1-3
The F-distribution with 2 and 19 degrees of freedom. The value of F ranges from zero to positive infinity. The area under the curve to the right of the critical value $F = 3.52$ (*shaded*) is 0.05. (See Statistical Table D.)

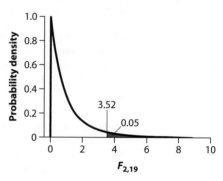

The critical value corresponding to the area of 0.05 in the right tail of the F-distribution is found in Statistical Table D. An excerpt is shown in Table 15.1-3.

TABLE 15.1-3 An excerpt from Statistical Table D, with critical values of the F-distribution corresponding to the significance level $\alpha(1) = 0.05$.

Denominator df	Numerator df									
	1	2	3	4	5	6	7	8	9	10
10	4.96	4.10	3.71	3.48	3.33	3.22	3.14	3.07	3.02	2.98
11	4.84	3.98	3.59	3.36	3.20	3.09	3.01	2.95	2.90	2.85
12	4.75	3.89	3.49	3.26	3.11	3.00	2.91	2.85	2.80	2.75
13	4.67	3.81	3.41	3.18	3.03	2.92	2.83	2.77	2.71	2.67
14	4.60	3.74	3.34	3.11	2.96	2.85	2.76	2.70	2.65	2.60
15	4.54	3.68	3.29	3.06	2.90	2.79	2.71	2.64	2.59	2.54
16	4.49	3.63	3.24	3.01	2.85	2.74	2.66	2.59	2.54	2.49
17	4.45	3.59	3.20	2.96	2.81	2.70	2.61	2.55	2.49	2.45
18	4.41	3.55	3.16	2.93	2.77	2.66	2.58	2.51	2.46	2.41
19	4.38	3.52	3.13	2.90	2.74	2.63	2.54	2.48	2.42	2.38
20	4.35	3.49	3.10	2.87	2.71	2.60	2.51	2.45	2.39	2.35

To find the critical value of the F-distribution having 2 and 19 degrees of freedom, locate the cell in the table corresponding to 2 df in the numerator and 19 df in the denominator. This value, 3.52, is highlighted in Table 15.1-3. We write it as

$$F_{0.05(1), 2, 19} = 3.52.$$

In this formula, "(1)" indicates that we are looking only at the right tail of the F-distribution. In other words, the area under the curve in Figure 15.1-3 to the right of

3.52 is 0.05. Because our observed value of F (i.e., 7.29) is larger than 3.52, it lies farther out in the right tail of the F-distribution, so P must be less than 0.05. Therefore, we reject the null hypothesis.

The exact P-value is the area under the curve of the F-distribution to the *right* of the observed F-value:

$$P = \Pr[F > 7.29].$$

If you analyzed the data with a statistics package on the computer, you would obtain this probability directly: $P = 0.004$. Again, $P < 0.05$, so we reject H_0. We have evidence that mean phase shift differs among light treatments.

Rejecting H_0 indicates only that at least one of the population means μ_i is different from the others, not that every μ_i is different from all of the others. We show in Section 15.4 how to take this process further to decide which means are different.

Variation explained: R^2

The R^2 **value** ("R-squared") is used in ANOVA to summarize the contribution of group differences to the total variation in the data. The quantity is based on the fact that the total sum of squares can be split into its two parts, the error sum of squares and the group sum of squares:

$$\mathrm{SS}_{total} = \mathrm{SS}_{error} + \mathrm{SS}_{groups}.$$

R^2 is the group portion of variation expressed as a fraction of the total:

$$R^2 = \frac{\mathrm{SS}_{groups}}{\mathrm{SS}_{total}}.$$

It can be thought of, loosely, as the "fraction of the variation in Y that is explained by groups." It is a reflection of how much narrower the scatter of measurements is around the group means compared with the scatter around the grand mean (compare the first and third panels of Figure 15.1-2).

R^2 takes on values between zero and one. When R^2 is close to zero, the group means are all very similar and most of the variability is within groups—that is, the explanatory variable defining the groups explains *very little* of the variation in Y. Conversely, an R^2 close to one indicates that little variation in Y is left over after the different group means are taken into account—that is, the explanatory variable explains *most* of the variation in Y.

For the circadian rhythm data,

$$R^2 = \frac{7.224}{16.639} = 0.43.$$

In other words, 43% of the total sum of squares among subjects in the magnitude of phase shift is explained by differences among them in light treatment. The remaining 57% of the variability among subjects is "error"—variance unexplained by the explanatory variable, light treatment.

R^2 measures the fraction of the variation in Y that is explained by group differences.

ANOVA with two groups

The analysis of variance works even when $k = 2$. ANOVA and the two-sample t-test give identical results[6] when testing the null hypothesis H_0: $\mu_1 = \mu_2$. An advantage of the two-sample t-test is that it generalizes more easily to other hypothesized differences between means, such as H_0: $\mu_1 - \mu_2 = 10$. Welch's t-test (Section 12.3) is additionally useful when the variances are very different between groups, whereas ANOVA requires more similar variances.

15.2 Assumptions and alternatives

The assumptions of analysis of variance are the same as those of the two-sample t-test, but they must hold for all k groups. To review:

- The measurements in every group represent a random sample from the corresponding population.
- The variable is normally distributed in each of the k populations.
- The variance is the same in all k populations.

Methods to evaluate these assumptions were discussed in Chapter 13.

The robustness of ANOVA

The ANOVA is surprisingly robust to deviations from the assumption of normality, particularly when the sample sizes are large. This robustness stems from a property of sample means described by the central limit theorem (Section 10.6)—that is, within each group, the sampling distribution of means is approximately normal when the sample size is large, even if the variable itself does not have a normal distribution (Chapter 13).

ANOVA is also robust to departures from the assumption of equal variance in the k populations, but only if the samples are all large, about the same size, and if there is no more than about a tenfold difference among the variances.

---

6. The test statistics and null distributions are basically the same. The F-ratio for two means (having one degree of freedom for groups and "df" degrees of freedom for error) is the same as the square of the two-sample t-statistic having "df" degrees of freedom.

Data transformations

If the data do not conform to the assumptions of ANOVA, they can be transformed as described in Chapter 13. Any of the transformations discussed there (e.g., log transformations and arcsine transformations) can be applied to ANOVA. If you get lucky, transforming the data will simultaneously make the data more normal and make the variances more equal, but this does not always happen.

Nonparametric alternatives to ANOVA

If the normality assumption of ANOVA is not met and transformations are unsuccessful, then there is a nonparametric alternative to single-factor ANOVA. The **Kruskal–Wallis test**, a nonparametric method based on ranks, is the equivalent of the Mann–Whitney U-test (Section 13.5) when there are more than two groups. It is sometimes referred to as analysis of variance based on ranks. It makes all the same assumptions as the Mann–Whitney U-test, but it is applied to more than two groups:

- All group samples are random samples from the corresponding populations.
- To use Kruskal–Wallis as a test of differences among populations in means or medians, the distribution of the variable must have the same shape in every population.

As in the Mann–Whitney U-test, the Kruskal–Wallis test begins by ranking the data from all groups together (employing the strategy for tied observations first described in Section 13.5). The sum of the ranks for each group, R_i, is then used to calculate the Kruskal–Wallis test statistic, H. The formula is given in the Quick Formula Summary (Section 15.8). In general, we recommend using a computer program for this procedure. Under the null hypothesis of no difference among populations, the sampling distribution of H is approximately χ^2 with $k-1$ degrees of freedom, where k is the number of groups. The null hypothesis is rejected if H is greater than or equal to the critical value from the appropriate χ^2 distribution, $\chi^2_{\alpha, k-1}$.

As with the Mann–Whitney U-test, the Kruskal–Wallis test has little power when sample sizes are very small. Remember that power is the probability of rejecting a false null hypothesis, so more power is better. Therefore, ANOVA is preferred if its assumptions can be met. The Kruskal–Wallis test has nearly the same power as ANOVA when sample sizes are large.

15.3 Planned comparisons

Analysis of variance is the start, but not necessarily the end, of efforts to compare the means of more than two groups. Researchers might want to answer two additional questions, such as "Which means are different?" and "What is the magnitude of the difference between means?" ANOVA by itself answers neither of these questions.

There are two approaches to figuring out which means are different and by how much—namely, planned and unplanned comparisons of means. A **planned comparison** is a comparison between means identified as being of crucial interest during the design of the study, identified *prior* to obtaining the data. A planned comparison must have a strong prior justification, such as an expectation from theory or a prior study.[7] Only one or a small number of planned comparisons is allowed, to minimize inflating the Type I error rate. An **unplanned comparison** is one of multiple comparisons, such as between all pairs of means, carried out to help determine where differences between means lie. Unplanned comparisons represent a kind of data dredging (Interleaf 8), so it is necessary to protect against rising Type I errors. Here we briefly describe an example of a planned comparison. Unplanned comparisons are covered in Section 15.4.

> A *planned comparison* is a comparison between means planned during the design of the study, identified before the data are examined.

Planned comparison between two means

A good example of a planned comparison between two means comes from the circadian rhythm experiment (Example 15.1). The main point of that experiment was to contrast the mean phase shift of melatonin production between the knee-treatment group and the control group, which received no extra light. Because the experiment was built around this contrast, we are justified in using methods for planned comparisons to ask, "How big is the difference between the knee-treatment and control groups?" and "Is the difference between these two groups statistically significant?"

The method for a planned comparison between two means is almost the same as the two-sample comparison based on the *t*-distribution that we learned about in Chapter 12. Only the standard error is calculated differently: the planned comparison uses the pooled sample variance (the error mean square) based on all *k* groups (and the corresponding error degrees of freedom), rather than that based only on the two groups being compared. This step increases precision and power. The planned comparison method assumes, just as in ANOVA, that the variance is the same within all groups. Details of the modified formulas for a planned comparison are provided in the Quick Formula Summary (Section 15.8).

For example, let's examine the confidence interval for the difference between the means of the "knee" and "control" treatment groups. The difference between the sample means (knee treatment minus control) is small:

$$\overline{Y}_2 - \overline{Y}_1 = (-0.336) - (-0.309) = -0.027 \, \text{h}.$$

---

7. For this reason, they are also called a priori comparisons.

The standard error for this difference is

$$SE = \sqrt{MS_{error}\left(\frac{1}{n_1} + \frac{1}{n_2}\right)}.$$

For the knee versus control comparison, the standard error of the difference between these two means is

$$SE = 0.364,$$

which has $N - k = 22 - 3 = 19$ degrees of freedom, the same as that for the error mean square (Table 15.1-2). The critical value from the t-distribution is $t_{0.05(2),\,19} = 2.09$ (Statistical Table C), which leads to the planned 95% confidence interval for the difference between the population means:

$$(-0.027) - 0.364(2.09) < \mu_2 - \mu_1 < (-0.027) + 0.364(2.09)$$

or

$$-0.788 < \mu_2 - \mu_1 < 0.734,$$

where the units are in hours. If we avoided using the planned-comparison method and simply used the two-sample method for a confidence interval introduced in Chapter 12, we would obtain the following 95% confidence interval instead:

$$-0.813 < \mu_2 - \mu_1 < 0.759.$$

Thus, the planned-comparison method has a slightly higher precision. Similarly, the planned-comparison method for testing the null hypothesis of no difference between these two means has slightly higher power. These are the main advantages of using the planned-comparison methods, provided that the assumption of equal variance is met.

Planned comparisons make all of the same assumptions as ANOVA—namely, random samples from populations, a normal distribution of the variable in every population, and equal variances in all populations. Because each comparison typically involves only one pair of means, planned comparisons are not as robust as ANOVA to violations of the assumptions.

15.4 Unplanned comparisons

The formulas for planned comparisons are not valid for unplanned comparisons, because unplanned comparisons need to make adjustments for the inflated false-positive rate (Type I errors) that accompanies multiple testing (Interleaf 8). The number of possible comparisons involving k groups is potentially large. Unplanned comparisons basically represent data dredging or "snooping" because we are poring through data to find differences. This is not inherently bad, provided that the method used corrects for the number of comparisons.

Testing all pairs of means to find out which groups stand apart from the others is the most common type of unplanned comparison, and the **Tukey–Kramer test** is the most commonly used procedure for accomplishing this.[8] The method assumes that we have already carried out an ANOVA and that the null hypothesis of no differences among means has been rejected. Example 15.4 illustrates unplanned comparisons with the Tukey–Kramer method.

EXAMPLE
15.4

Wood wide web

Most plants have underground associations with fungi, called mycorrhizae, that provide minerals and antibiotics to the plant in exchange for sugars. The fungi's hyphae (branching filaments) extend through the soil and may connect to other plants, even to other plant species, creating an underground network that might cause a trickle of nutrients to flow from plant to plant. Simard et al. (1997) measured the flow of carbon between seedlings of birch and Douglas-fir and tested whether the carbon flow rate depended on shading. Shaded trees may draw more carbon via the mycorrhizae than trees in full sun. In each of three shade treatments, five pairs of birch and Douglas-fir seedlings were planted and allowed to grow for one year. Then,

each birch was covered in a sealed bag for two hours and supplied with carbon dioxide (CO_2) whose carbon consisted entirely of the carbon-13 isotope (atmospheric CO_2 is made up almost entirely of carbon-12). The same was done to its partnered fir seedling, except that CO_2 with carbon-14 was used. The amounts of carbon-13 and carbon-14 present in the tissues of both plants of each pair were measured nine days later. Because different isotopes of carbon were used on birch and Douglas-fir, it was possible to calculate the amount of carbon transferred from each plant to the other. Most transfer occurred from birch to Douglas-fir. Descriptive statistics for the average net carbon gain by Douglas-fir, in milligrams, are given in Table 15.4-1.

TABLE 15.4-1 Summary of the net amount of carbon transferred from birch to Douglas-fir (Example 15.4).

Shade treatment	Sample mean $\overline{Y}_i$ (mg)	Sample standard deviation, s_i	n_i
Deep shade	18.33	6.98	5
Partial shade	8.29	4.76	5
No shade	5.21	3.00	5

---

8. Testing all pairs of means is not the only kind of unplanned comparison. For example, the Scheffé method tests any linear combination of means, but it is too conservative when applied only to pairs of means. Three other methods to test all pairs of means—Duncan's multiple-range test, Fisher's least-significant-difference test, and the Newman–Keuls test—provide less protection than the Tukey–Kramer test against inflated Type I error rates. The Tukey–Kramer method is also called "Tukey's honestly significant difference (HSD)" test. Unplanned comparisons are also called "post hoc tests," "a posteriori tests," or simply "multiple comparison tests."

The ANOVA results for these data, testing the null hypothesis for no differences among treatment means, are shown in Table 15.4-2. The different shade treatments indeed led to differences in the mean net carbon gain by Douglas-fir seedlings, as shown by the high F-ratio and low P-value. It remains to be seen, though, which means are different. That is, are all of the means detectably different from each other? If not, which of the means are different from the others?

TABLE 15.4-2 ANOVA table summarizing results of the Douglas-fir carbon-transfer data (Example 15.4).

Source of variation	Sum of squares	df	Mean squares	F-ratio	P
Groups (treatments)	470.704	2	235.352	8.784	0.004
Error	321.512	12	26.793		
Total	792.216	14			

Testing all pairs of means using the Tukey–Kramer method

The Tukey–Kramer method works like a series of two-sample t-tests, but it uses a larger critical value to limit the Type I error rate. With the Tukey–Kramer test, the probability of making at least one Type I error throughout the course of testing *all pairs* of means is no greater than our significance level α.

To carry out the Tukey–Kramer procedure, the means must first be ordered from smallest to largest:

No shade	Partial shade	Deep shade
$\overline{Y}_3$	$\overline{Y}_2$	$\overline{Y}_1$
5.21	8.29	18.33

Then, compare every pair of means in turn. For example, the hypotheses for the comparison between the means for "deep shade" (call it μ_1) and "partial shade" (call it μ_2) are as follows.

H_0: $\mu_1 - \mu_2 = 0$.
H_A: $\mu_1 - \mu_2 \neq 0$.

The test statistic (q) is calculated using a standard error based on the MS_{error}. This test statistic is then compared with the q-distribution having k and $N - k$ degrees of freedom. Critical values are provided in Statistical Table F. Calculation details are given in the Quick Formula Summary (Section 15.8).

The results for all the pairwise tests in the carbon-transfer data are given in Table 15.4-3.

TABLE 15.4-3 Summary of Tukey–Kramer tests carried out on the results of Example 15.4.

Group i	Group j	$\bar{Y}_i - \bar{Y}_j$	SE	Test statistic q	Critical value $q_{0.05,3,12}$	Conclusion
Deep	No	13.12	3.2737	4.008	2.67	Reject H_0
Deep	Partial	10.04	3.2737	3.067	2.67	Reject H_0
Partial	No	3.08	3.2737	0.941	2.67	Do not reject H_0

The results are unambiguous. The mean of the deep-shade group is different from that of both the partial-shade and no-shade groups, whereas the means of the partial-shade and no-shade groups are not significantly different from each other.

The results of the Tukey–Kramer procedure are often indicated using symbols, such as those shown in Figure 15.4-1 for displaying the means. Groups in the figure are assigned the same symbol if their means are not significantly different, based on the unplanned comparisons (e.g., the partial-shade and no-shade treatments are both assigned the symbol "b" in Figure 15.4-1). Sample means in the figure are assigned a unique symbol if they are different from all other means (e.g., the deep-shade treatment is given the symbol "a" in Figure 15.4-1).[9]

FIGURE 15.4-1
Using symbols to indicate the outcome of Tukey–Kramer tests of all pairs of means. Dots and error bars indicate means ±1 SE of the three treatment groups in Example 15.4. Two means are assigned a different symbol if they are significantly different ("a" vs. "b"), whereas means are assigned the same symbol ("b" in this case) if they are not significantly different.

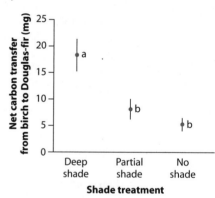

The Tukey–Kramer results do not imply that the partial-shade and no-shade treatments have the same mean carbon transfer rates. Assuming that they do would be making the mistake of accepting a null hypothesis. The results merely indicate that a test of their differences did not reject H_0.

> With the Tukey–Kramer method, the probability of making at least one Type I error throughout the course of testing all pairs of means is no greater than the significance level α.

---

9. One or more groups with intermediate means might be assigned two symbols if the Tukey-Kramer results were ambiguous. For example, if it had turned out that the partial-shade mean was not significantly different from the means of either the deep-shade or no-shade groups, yet the deep-shade and no-shade means were different from each other, then partial shade would have been assigned both symbols "a" and "b" to indicate this ambiguity.

Assumptions

The Tukey–Kramer method makes all of the same assumptions as ANOVA—namely, random samples, a normal distribution of the variable in every population, and equal variances in all groups. Because each comparison tests only one pair of means at a time, rather than all means simultaneously, the method is not as robust as ANOVA to violations of these assumptions.

The P-value for the Tukey–Kramer test is exact when the experimental design is balanced—that is, when the sample size is the same in every group ($n_1 = n_2 = \ldots = n_k$). If the sample sizes are different, then the Tukey–Kramer test is conservative, which means that the real probability of making at least one Type I error, when testing all pairs of means, is smaller than the stated α. This makes it harder to reject H_0, which is why the test is deemed "conservative."

15.5 Fixed and random effects

Up to now, we have been analyzing *fixed* groups—namely, studies in which the different categories of the explanatory variable are predetermined, of direct interest, and repeatable. ANOVA on fixed groups is called **fixed-effects** analysis of variance. Other examples of fixed effects include

- Alternative medical treatments in a clinical trial
- Fixed doses of a toxin
- Different heights above low tide in the intertidal zone
- Different sexes or age categories of individuals

Any conclusions reached about differences among fixed groups apply only to those fixed groups. If a difference among drug treatments was found in some response variable, for example, we could not generalize the results to other drugs not included in the study.

By contrast, there is a second type of ANOVA in which the groups are not fixed, but instead are *randomly chosen*. Randomly chosen groups are not predetermined, but instead are randomly sampled from a much larger "population" of groups. ANOVA applied to random groups is called **random-effects** ANOVA.[10] Examples of random effects include

- Family, in a study of resemblance among relatives in IQ scores
- Subject or individual, in a study involving repeated measurements of individuals

Because random effects are randomly sampled from a population, conclusions reached about differences among groups *can* be generalized to the whole population of groups.

---

10. Some authors refer to fixed-effects ANOVA as "Model 1" and random-effects ANOVA as "Model 2."

With random effects, the specific groups included in a study are ephemeral and would not typically be reused. For example, consider a study to investigate whether families differ from one another in the mean IQ scores of their children. This study would begin with a random sample of families from a population of families. "Family" would be used as the group variable in an ANOVA, and the replicates would be the different children making up each family. A later study attempting to address the same question in the same population would not attempt to relocate the same families used in the previous study, because the population, not the families themselves, is the target of study. Rather, a new study would begin with a new random sample of families and would again be able to generalize the results to the whole population. In contrast, a new study of the effects of light treatment—a fixed effect—on circadian rhythm could quite easily use the same light treatments as previous studies.

> An explanatory variable is called a *fixed effect* if the groups are predefined and are of direct interest. An explanatory variable is called a *random effect* if the groups are randomly sampled from a population of possible groups.

F-tests of differences among group means for fixed effects are changed when ANOVA is expanded to examine the effects of more than one explanatory variable simultaneously. We discuss some of these issues in Chapter 18.

15.6 ANOVA with randomly chosen groups

Because the groups are not of specific interest in a random-effects ANOVA, planned and unplanned comparisons aren't particularly useful. The main use of random-effects ANOVA is to estimate *variance components*—the amount of the variance in the data that is among random groups and the amount that is within groups. Among other uses, variance components are employed in animal and plant breeding to identify the contributions of genes and the environment to variance in traits. Variance components are also useful for quantifying measurement error in data, as shown in Example 15.6.

EXAMPLE   Walking-stick limbs

15.6   The walking stick *Timema cristinae* is a wingless herbivorous insect that lives on plants in chaparral habitats of California. In a study of the insect's adaptations to different plant species, Nosil and Crespi (2006) measured a variety of traits using digital photographs of specimens collected from a study site in California. They used a

femur length

computer to measure various traits on the photographs. Because the researchers were concerned about measurement error, they took two separate photographs of each specimen. After taking the first photo, they returned the insect to storage and then retrieved it again for the second photograph. After measuring traits on one set of photographs, they repeated the measurements on the second set. Very often the result was different the second time around, indicating measurement error. How large was the measurement error compared with real variation among individuals in the trait? Figure 15.6-1 illustrates the two measurements of femur length made on 25 specimens. The data are listed in Table 15.6-1.

FIGURE 15.6-1
Strip chart showing the pair of femur length measurements (connected by a line segment) obtained from separate photographs of each of 25 walking sticks.

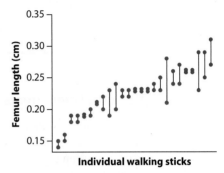

TABLE 15.6-1 Femur length, in centimeters, measured from separate photographs of 25 walking sticks. Two measurements were made per specimen to evaluate measurement error.

Specimen	Femur length (cm)	Specimen	Femur length (cm)
1	0.26, 0.26	14	0.27, 0.24
2	0.23, 0.19	15	0.23, 0.29
3	0.25, 0.23	16	0.23, 0.23
4	0.26, 0.26	17	0.14, 0.15
5	0.23, 0.22	18	0.19, 0.19
6	0.23, 0.23	19	0.31, 0.27
7	0.22, 0.23	20	0.23, 0.24
8	0.21, 0.28	21	0.16, 0.15
9	0.24, 0.26	22	0.22, 0.20
10	0.24, 0.20	23	0.19, 0.18
11	0.29, 0.25	24	0.21, 0.21
12	0.23, 0.23	25	0.19, 0.20
13	0.18, 0.19		

ANOVA calculations

The ANOVA results, which are needed to calculate the variance components of the walking-stick data, are presented in Table 15.6-2. With only one explanatory variable, the calculations for random-effects ANOVA are the same as for fixed effects. The "groups" are the individual insects, and the replicates are the repeat measurements made of each insect. Because our goal here is to estimate the variance components, we refrain from hypothesis testing.

TABLE 15.6-2 ANOVA table with results of repeat femur-length measurements of 25 walking sticks.

Source of variation	Sum of squares	df	Mean squares
Groups (individual insects)	0.059132	24	0.002464
Error	0.008900	25	0.000356
Total	0.068032	49	

Variance components

Single-factor ANOVA with random effects differs from fixed-effects ANOVA by having *two* levels of random variation[11] in the response variable Y. The first level is variation within groups, and the second level is variation between groups. In the stick insect study (Example 15.6), the first level (variation within groups) is the variance among repeat measurements made on the same individual, which is exclusively due to measurement error in the study. ANOVA assumes that the "true" variance between measurements is the same in every group. In Example 15.6, this means that measurement error is the same for every individual insect, except by chance. We use the symbol σ^2 to indicate the value of the variance within groups in the population. The single best estimate of σ^2 is MS_{error}, the error mean square. In the walking sticks, $MS_{error} = 0.000356$ (Table 15.6-2).

To evaluate the second level of random variation (i.e., variation between groups), each group is assumed to have its own mean. For the walking sticks, the mean femur length of an individual insect is its "true" femur length—the length we would obtain if we measured its femur a great many times and took the average measurement. Random-effects ANOVA assumes that group means have a normal probability distribution in the population with a grand mean μ_A (the mean of all the group means) and a variance σ_A^2 (the variance among the group means in the population of groups). In our study of the walking sticks, we assume that true femur length varies between individual insects in the population according to a normal distribution. The parameter σ_A^2 is the variance among insects in the population in their average femur length.

In random-effects ANOVA, the parameters σ^2 and σ_A^2 are called **variance components**. Together they describe all the variance in the response variable Y—that is, σ^2 describes the variance within groups (e.g., the measurement error in the walking-stick study), and σ_A^2 describes the variance among groups (e.g., the differences between the true femur lengths of individual insects).

---

11. Fixed-effects ANOVA also has two levels of variation, but only variation within groups is random. Differences among fixed groups are not random.

> The *variance components* in a random-effects ANOVA make up all the variance in the response variable: variance within groups (σ^2) and variance among groups (σ_A^2).

If sample size is the same in all groups (the design is balanced), the group mean square from the ANOVA results (Table 15.6-2) can be used to estimate σ_A^2. Using the symbol s_A^2 to indicate the estimate of σ_A^2,

$$s_A^2 = \frac{\text{MS}_{\text{groups}} - \text{MS}_{\text{error}}}{n},$$

where n is the number of measurements taken within each group.[12]

There were $n = 2$ measurements for each insect specimen in the data for Example 15.6, so

$$s_A^2 = \frac{0.002464 - 0.000356}{2} = 0.00105 \, \text{cm}^2.$$

Thus, the best estimate of the variance in femur length among individuals in this population is 0.00105.

Repeatability

Repeatability is the fraction of the summed variance that is present among groups:

$$\text{Repeatability} = \frac{s_A^2}{s_A^2 + \text{MS}_{\text{error}}}.$$

The denominator of the repeatability equation (i.e., $s_A^2 + \text{MS}_{\text{error}}$) estimates the total amount of measurement variance in the population, summing the variance among groups and the variance within groups.

Repeatability measures the overall similarity of repeat measurements made on the same group.[13] A repeatability near zero, the lowest possible value, indicates that nearly all of the variance in the response variable results from differences between separate measurements made on the same group. In our example, this would mean that measurement error greatly dominates the variation found in the data. In contrast, a repeatability near one, the maximum value, indicates that repeated measurements on the same group give nearly the same answer every time.

Calculated on the walking-stick data,

$$\text{Repeatability} = \frac{0.00105}{0.00105 + 0.000356} = 0.75.$$

---

12. Methods based on likelihood (Chapter 20) are required if the number of measurements is not identical in every group (see Pinheiro and Bates 2000). Also, s_A^2 can sometimes be negative by chance, even though σ_A^2 can be only zero or positive. When this happens, s_A^2 is set to zero.

13. The repeatability is also called the "intraclass correlation."

In other words, an estimated 75% of the total variance in femur length measurements in the population is the result of differences in true femur length between individual insects. The remaining 25% is the result of measurement error.

The repeatability of femur length is not one, because the walking sticks are small and their femurs are even smaller. Slight variation in the position of the insect when a photograph is taken can lead to different length measurements. It is important, therefore, to report repeatability estimates in your research papers. If measurement error is present, then it is best to take several measurements of each specimen and take their average.

Don't confuse repeatability with the R^2 value (Section 15.1). Repeatability reflects the magnitudes of variance components, which estimate specific population parameters. It applies only to random effects. R^2, on the other hand, isn't an estimate of any population parameter. R^2 just measures the reduction in the amount of scatter around the group means compared with that around the grand mean. R^2 is based on the sums of squares rather than variances, and it can be applied to fixed or random effects.

Assumptions

Random-effects ANOVA makes all of the same assumptions as fixed-effects ANOVA, but adds two more. It assumes that groups are randomly sampled. Also, random-effects ANOVA assumes that the group means have a normal distribution in the population.

15.7 Summary

- The analysis of variance (ANOVA) tests differences among the means of multiple groups.
- ANOVA works by comparing the variance among subjects within groups (the error mean square, MS_{error}) with the variation among the sampled individuals belonging to different groups (represented by the group mean square, MS_{groups}).
- The test statistic in ANOVA is the F-ratio, where $F = MS_{groups}/MS_{error}$.
- Under H_0, the F-ratio should be close to one except by chance. Under H_A, the F-ratio is expected to exceed one. H_0 is rejected if F is much larger than one. When F is significantly greater than one, it implies that there is real variance among group means, more than is expected by sampling error alone.
- The quantities needed to test hypotheses or estimate variance components in single-factor ANOVA are summarized in an ANOVA table.
- R^2 measures the fraction of the variability explained by the explanatory (group) variable in analysis of variance. It measures the reduction in the amount of

scatter of the measurements around their group means compared with that around the grand mean.

- ANOVA assumes that the Y-variable has a normal distribution in each population, with equal variance in all populations.
- ANOVA is robust to departures from the assumption of normal populations, especially if the sample size is large. ANOVA is also robust to moderate departures from the assumption of equal standard deviations, if the study design is balanced with large samples.
- The Kruskal–Wallis test is an alternative, nonparametric method used to test the null hypothesis that the distributions of the variable in different groups are the same. It can be used if the normality assumption of ANOVA cannot be met.
- When used to test the null hypothesis that population means or medians differ between groups, the Kruskal–Wallis test assumes that the frequency distributions of measurements in different groups have the same shape (i.e., the same assumption as the Mann–Whitney U-test).
- Planned comparisons between means are few in number and represent only comparisons identified as crucial before the data are collected and analyzed. Unplanned comparisons are a more comprehensive set of comparisons done in search of interesting patterns. Unplanned comparisons require special methods to protect against high Type I error rates.
- The Tukey–Kramer method to compare all pairs of means is the most commonly used method for unplanned comparisons.
- In fixed-effects ANOVA, the groups are predetermined, repeatable categories of direct interest. The results of ANOVA apply only to those groups included in the study.
- In random-effects ANOVA, groups are a random sample from a population of groups. The results of random-effects ANOVA can be generalized to the population of groups.
- The repeatability estimates the fraction of the total variance that is present among groups in random-effects ANOVA. Repeatability is frequently used to evaluate the importance of measurement error.

15.8 Quick Formula Summary

Analysis of variance (ANOVA)

What is it for? Testing the difference among means of k groups simultaneously.

What does it assume? The variable is normally distributed with equal standard deviations (and therefore equal variances) in all k populations. Each sample is a random sample.

Test statistic: F

Sampling distribution under H$_0$: F-distribution with $k - 1$ and $N - k$ degrees of freedom. Use the right tail of the F-distribution in ANOVA.

Formulas:

Source of variation	Sum of squares	df	Mean squares	F-ratio
Groups	$SS_{groups} = \sum_i n_i (\bar{Y}_i - \bar{Y})^2$	$k - 1$	$\dfrac{SS_{groups}}{df_{groups}}$	$\dfrac{MS_{groups}}{MS_{error}}$
Error	$SS_{error} = \sum_i s_i^2 (n_i - 1)$	$N - k$	$\dfrac{SS_{error}}{df_{error}}$	
Total	$SS_{groups} + SS_{error}$	$N - 1$		

In these formulas, n_i is the sample size in group i, k is the number of groups,

$$\bar{Y} = \frac{\sum_i n_i (\bar{Y}_i)}{N}$$ is the grand mean, and $N = \sum_i n_i$ is the total sample size.

R squared (R²)

What is it for? Measuring the fraction of the variation in Y that is "explained" by differences among groups.

Formula: $R^2 = \dfrac{SS_{groups}}{SS_{total}}$,

where SS_{groups} is the sum of squares for groups and SS_{total} is the total sum of squares.

Kruskal–Wallis test

What is it for? Testing differences among k populations in the means or medians of their distributions.

What does it assume? Random samples. The frequency distributions of measurements in the different groups have the same shape.

Test statistic: H

Sampling distribution under H$_0$: Approximately χ^2 with $k - 1$ degrees of freedom.

Formula: $H = \dfrac{12}{N(N + 1)} \sum_i \dfrac{R_i^2}{n_i} - 3(N + 1)$,

where N is the total sample size and R_i is the sum of the ranks for group i.

Observations are ranked from small to large, as described in Section 13.5 for the Mann–Whitney U-test.

Planned confidence interval for the difference between two of k means

What is it for? Estimating the difference between means of two out of k populations when the comparison is *planned* before the experiment.

What does it assume? Random samples. The variable is normally distributed with equal variances in all k populations.

Estimate: $\bar{Y}_i - \bar{Y}_j$, where i and j are the two populations, with $i \neq j$.

Parameter: $\mu_i - \mu_j$

Formula: $(\bar{Y}_i - \bar{Y}_j) - \text{SE}\, t_{0.05(2),N-k} < \mu_i - \mu_j < (\bar{Y}_i - \bar{Y}_j) + \text{SE}\, t_{0.05(2),N-k}$,

where $\text{SE} = \sqrt{\text{MS}_{\text{error}}\left(\dfrac{1}{n_i} + \dfrac{1}{n_j}\right)}$.

Planned test of the difference between two of k means

What is it for? Testing the difference between two out of k population means when the comparison is *planned*.

What does it assume? Random samples. The variable is normally distributed with equal variances in all k populations.

Test statistic: t

Distribution under H_0: The t-distribution with $N - k$ degrees of freedom.

Formulae: $t = \dfrac{(\bar{Y}_i - \bar{Y}_j)}{\text{SE}}$,

where $\bar{Y}_i - \bar{Y}_j$ is the observed difference between the means of two groups i and j (with $i \neq j$), and $\text{SE} = \sqrt{\text{MS}_{\text{error}}\left(\dfrac{1}{n_i} + \dfrac{1}{n_j}\right)}$ is the standard error of the difference between the two means.

Tukey–Kramer test of all pairs of means

What is it for? Testing the differences of all pairs of k means. These tests are *unplanned*.

What does it assume? Random samples. The variable is normally distributed with equal variances in all k populations. An ANOVA has already rejected the null hypothesis of equal means for all k groups.

Test statistic: q

Sampling distribution under H$_0$: The q-distribution with k means and $N - k$ degrees of freedom.

Formula: $q = \dfrac{(\bar{Y}_i - \bar{Y}_j)}{SE}$,

where $\bar{Y}_i - \bar{Y}_j$ is the observed difference between the means of two groups i and j (with $i \neq j$), $SE = \sqrt{MS_{error}\left(\dfrac{1}{n_i} + \dfrac{1}{n_j}\right)}$, and n_i and n_j are the sample sizes for the two groups i and j, respectively.

Repeatability and variance components

What is it for? Repeatability is the fraction of the total variance that is among groups in a random-effects ANOVA.

What does it assume? Groups are a random sample from a population of groups. Repeat measurements within groups are randomly sampled and normally distributed, with an equal standard deviation (and variance) in all groups. Group means have a normal distribution in the population.

Formula: Repeatability $= \dfrac{s_A^2}{s_A^2 + MS_{error}}$,

where $s_A^2 = \dfrac{MS_{groups} - MS_{error}}{n}$, and n is the sample size within each group (assumed to be equal for our formulas).

PRACTICE PROBLEMS

1. **Calculation practice: Analysis of variance.** Many humans like the effect of caffeine, but it occurs in plants as a deterrent to herbivory by animals. Caffeine is also found in flower nectar, and nectar is meant as a reward for pollinators, not a deterrent. How does caffeine in nectar affect visitation by pollinators? Singaravelan et al. (2005) set up feeding stations where bees were offered a choice between a control solution with 20% sucrose or a caffeinated solution with 20% sucrose plus some quantity of caffeine. Over the course of the experiment, four different concentrations of caffeine were provided: 50, 100, 150, and 200 ppm. The response variable was the difference between the amount of nectar consumed from the caffeine feeders and that removed from the control feeders at the same station (in grams). Here are the data and strip chart, including standard error bars:

   50 ppm caffeine: –0.4, 0.34, 0.19, 0.05, –0.14

100 ppm caffeine: 0.01, –0.39, –0.08, –0.09, –0.31

150 ppm caffeine: 0.65, 0.53, 0.39, –0.15, 0.46

200 ppm caffeine: 0.24, 0.44, 0.13, 1.03, 0.05

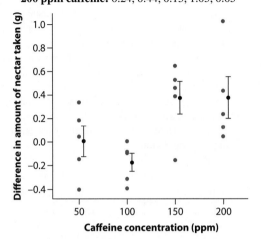

Does the mean amount of nectar taken depend on the concentraton of caffeine in the nectar? We will carry out an analysis of variance to find out.

a. State the null and alternate hypotheses appropriate for this question.
b. Calculate the following summary statistics for each group (i.e., for each caffeine treatment): n_i, $\overline{Y}_i$, and s_i.
c. Set up an ANOVA table to keep track of your results. Add to this table through the remaining steps.
d. What is the mean square error MS_{error}?
e. How many degrees of freedom are associated with error?
f. Calculate the estimate of the grand mean.
g. Calculate the group sum of squares.

h. Calculate the group degrees of freedom and the group mean square.
i. What is F for this example?
j. Use Statistical Table D (or a computer) to find the P-value for this test.

2. **Calculation practice: Analyze results from a Tukey–Kramer test.** Using the same data as in Practice Problem 1, use the results from a Tukey–Kramer test to illustrate which of the pairs of groups are significantly different in their means. The Tukey–Kramer test computations are given in the table at the bottom of this page.
a. What null hypotheses are being tested by the Tukey–Kramer procedure?
b. On a single line of text, write the sample mean of the four treatment groups, from smallest to largest.
c. Begin from the largest sample mean on the right. Draw a line underneath this sample mean, and continue the line to the left until it underlines all sample means whose group means are not significantly different from the largest mean.
d. Now move to the next largest mean, second from right. Draw a line underneath this sample mean to the left until it underlines all sample means whose group means are not significantly different. Keep this line if it extends farther to the left than the line from part (c); otherwise, discard it because it contains no new information.
e. Now move to the next largest mean, third from right. Repeat the procedure outlined in part (d). Continue the procedure in (c) and (d) until it has been carried out on all the means. When you reach the smallest mean, create a new underline if this group mean has not yet been underlined in a previous step.

TABLE FOR PROBLEM 2

Group i	Group j	$\overline{Y}_i - \overline{Y}_j$	SE	q	$q_{0.05,4,12}$	Conclusion
200	100	0.550	0.190	2.89	2.86	Reject
200	50	0.370	0.190	1.94	2.86	Do not reject
200	150	0.002	0.190	0.01	2.86	Do not reject
150	100	0.548	0.190	2.88	2.86	Reject
150	50	0.368	0.190	1.93	2.86	Do not reject
50	100	0.180	0.190	0.95	2.86	Do not reject

f. Assign each unique line a symbol such as "a," "b," and so on. Under each of the four sample means, write the symbols for all the lines that underline it. Give a group more than one symbol (e.g., "a,b") if it is underlined by more than one unique underline.

g. Use the underlines and symbols from part (f) to summarize in words the overall results of the Tukey–Kramer test. That is, state which means group together on the basis of statistical significance.

3. Calculation practice: Repeatability. The following anonymized data show the midterm and final exam grades (%) for eight undergraduate students from a biostatistics class at a major university. The partial ANOVA table provides sums of squares and mean squares. What is the repeatability of grade performance?

Individual	Midterm grade	Final exam grade
1	78	81
2	84	65
3	94	75
4	82	62
5	58	60
6	62	86
7	81	92
8	80	89

Source	Sum of squares	df	Mean squares
Individual	1145.94	7	163.71
Error	956.50	8	119.56
Total	2102.44		

a. Calculate s_A^2 from the mean squares and the sample size. What variance component does this quantity estimate?

b. Calculate the repeatability using s_A^2 and MS_{error}.

c. Interpret the repeatability you just calculated. What fraction of the variance among test scores is estimated to reflect true differences among students in performance, and what fraction is measurement variance from test to test within students?

d. What assumptions are you making when estimating repeatability to test scores?

4. An important issue in conservation biology is how dispersal among populations influences the persistence of species in a fragmented landscape. Molofsky and Ferdy (2005) measured this in the annual plant *Cardamine pensylvanica*, a weed that produces explosively dispersed seeds. Four treatments were used to manipulate seed dispersal by changing the distance among experimental plant populations. These treatments were adjacent (continuous treatment), separated by 23.2 cm (medium), separated by 49.5 cm (long), or separated by partitions that blocked all seed dispersal among populations (isolated). Treatments were randomly assigned to plant populations. The data below are the number of generations that the populations persisted in four replicates of each treatment.

Treatment	Generations persisted
Isolated	13, 8, 8, 8
Medium	14, 12, 16, 16
Long	13, 9, 10, 11
Continuous	9, 13, 13, 16

a. What is the explanatory variable in this analysis? What is the response variable?

b. Is this an experimental study or an observational study? Explain.

c. Calculate the sample means for each group, and then calculate a confidence interval for the mean of each group. What assumptions have you made?

d. Display the data along with the means and confidence intervals in a graph.

5. Analysis of variance carried out on the *Cardamine* data of Practice Problem 4 yielded the following results:

Source of variation	Sum of squares	df	Mean squares	F-ratio	P
Groups	63.188		0.035		
Error	63.250				
Total					

a. What are the null and alternative hypotheses being tested?

b. Fill in the rest of the table.

c. What is the sampling distribution of the F-statistic under the null hypothesis?

d. What does P measure?

e. In words, explain what each "sum of squares" measures.

f. If you wanted to determine the fraction of the variation in generations persisted that is "explained" by treatment, what quantity would you use?

g. Calculate the quantity identified in part (f).

6. The Tukey–Kramer procedure was carried out on the results of Practice Problems 4 and 5, yielding the results at the bottom of this page.

a. What are the null and alternative hypotheses being tested?

b. Are these comparisons considered planned or unplanned? Why?

c. Only the largest pairwise difference between means, that between the "medium" and "isolated" treatments, is statistically significant. How is this possible, given that neither of these two means is significantly different from the means of the other two groups?

d. Using symbols, summarize the results of the Tukey–Kramer tests on the graph you created in part (d) of Practice Problem 4.

e. Explain in words why the critical value for each test (2.97) is larger than the critical value of the t-distribution having 12 degrees of freedom (2.18).

7. Imagine a hypothetical experiment with multiple treatments but relatively small sample sizes within treatments. The goal is to test whether the treatment means are equal. Calculations and graphical analysis indicate that the data differ markedly from a normal distribution.

a. What other two options are available to carry out a test?

b. Which of the other two options should be attempted first? Why?

8. Finding the causes of the major mental illnesses, schizophrenia and bipolar disorder, is a major activity in medical research. Tkachev et al. (2003) compared expression levels of several genes involved in the production and activity of myelin, a tissue important in nerve function, in the brains of 15 persons with schizophrenia, 15 with bipolar illness, and 15 control individuals. The results presented in the accompanying table summarize the expression levels of the proteolipid protein 1 gene *(PLP1)* in the 45 brains.

a. The main objective of the study was to compare *PLP1* gene expression in persons having schizophrenia with that of control individuals. Using a planned comparison approach, compute a 95% confidence interval for the difference between the means of these two groups.

TABLE FOR PROBLEM 6

Group i	Group j	$\bar{Y}_i - \bar{Y}_j$	SE	q	$q_{0.05,4,12}$	Conclusion
Medium	Isolated	5.25	1.623	3.234	2.97	Reject H_0
Medium	Long	3.75	1.623	2.310	2.97	Do not reject H_0
Medium	Continuous	1.75	1.623	1.078	2.97	Do not reject H_0
Continuous	Isolated	3.50	1.623	2.156	2.97	Do not reject H_0
Continuous	Long	2.00	1.623	1.232	2.97	Do not reject H_0
Long	Isolated	1.50	1.623	0.924	2.97	Do not reject H_0

TABLE FOR PROBLEM 8

Group	Raw data (normalized units)	Mean	Standard deviation
Control	−0.02, −0.27, −0.11, 0.09, 0.25, −0.02, 0.48, −0.24, 0.06, 0.07, −0.30, −0.18, 0.04, −0.16, 0.25	−0.004	0.218
Schizophrenia	−0.1, −0.31, −0.05, 0.11, −0.38, 0.23, −0.23, −0.28, −0.36, −0.22, −0.40, −0.19, −0.34, −0.29, −0.12	−0.195	0.182
Bipolar	−0.34, −0.39, −0.22, −0.32, −0.32, −0.05, −0.43, −0.33, −0.41, −0.36, −0.25, −0.29, 0.06, −0.30, 0.01	−0.263	0.151

b. Is a planned comparison appropriate in part (a)? Explain.

c. What are your assumptions in part (a)?

9. Use the data from Practice Problem 8 to solve this problem.

a. Test whether mean *PLP1* gene expression differs among the schizophrenia, bipolar, and control groups.

b. What are your assumptions in part (a)?

c. Is the analysis in part (a) a random-effects or fixed-effects ANOVA? Explain.

d. What quantity would you use to describe the fraction of the variation in expression levels explained by group differences? Calculate this quantity for the data from Practice Problem 8.

e. What method would you use after ANOVA to determine which group means were different from each other?

10. The bright yellow head of the adult Egyptian vulture (see the photo at the beginning of the chapter) requires carotenoid pigments. These pigments cannot be synthesized by the vultures, so they must be obtained through their diet. Unfortunately, carotenoids are scarce in rotten flesh and bones, but they are readily available in the dung of ungulates. Perhaps for this reason, Egyptian vultures are frequently seen eating the droppings of cows, goats, and sheep in Spain, where they have been studied.[14] Ungulates are common in some areas but not in others. Negro et al. (2002) measured plasma carotenoids in wild-caught vultures at four randomly chosen locations in Spain as part of a study to determine the causes of variation among sites in Spain in carotenoid availability.

Site	Mean concentration (μg/ml)	Standard deviation	n
1	1.86	1.22	22
2	5.75	2.46	72
3	6.44	3.42	77
4	11.37	1.96	11

a. Use the data provided in the table to test whether the mean plasma concentration of

carotenoids in wild Egyptian vultures differs among sites.

b. What are the assumptions of your analysis in part (a)?

11. One way to assess whether a trait in males has a genetic basis is to determine how similar the measurements of that trait are among his offspring born to different, randomly chosen females. In a lab experiment, Kotiaho et al. (2001) randomly sampled 12 male dung beetles, *Onthophagus taurus*, and mated each of them to three different virgin females. The average body-condition score of offspring born to each of the three females is listed for each male in the following table.

Male	Offspring body-condition scores
1	0.82, 0.44, 0.92
2	0.35, 0.19, 1.39
3	0.12, 0.84, 0.16
4	0.49, 0.59, −0.23
5	0.44, 0.33, 0.07
6	0.00, 0.29, 0.30
7	0.69, −0.49, −0.60
8	0.13, −0.43, 0.66
9	0.21, −0.34, −0.09
10	−0.35, −1.40, −0.45
11	−1.04, −0.34, −0.82
12	−1.27, −0.75, −0.86

ANOVA was used to test whether males differed in the mean condition of their offspring using the three measurements for each male. The results were $SS_{groups} = 9.940$ and $SS_{error} = 4.682$.

---

14. Spaniards have nicknamed the Egyptian vulture "moñiguero," which politely translates as "dung-eater."

a. Is this a random-effects or a fixed-effects ANOVA? Explain the reason behind your answer.

b. What is the repeatability of the offspring condition of males mated to different females?[15]

12. Imagine that you are a statistical consultant and a researcher comes to you with a problem. The researcher has a random sample of data from each of six groups, and she wants to test whether the means of the six groups differ. In each of the following situations, recommend the best procedure.

a. The data are approximately normally distributed with equal standard deviations in all groups. Sample size is small.

b. The data are *not* normally distributed and do not have equal standard deviations in all groups. Sample size is small.

c. The data are *not* normally distributed. Groups have nearly equal standard deviations. Sample size is large and equal among groups.

d. The null hypothesis of no difference among group means was rejected. Now the researcher wants to find out which means are different. The data are approximately normally distributed with equal standard deviations in all groups, and the sample size is small.

13. Pea aphids, *Acyrthosiphon pisum*, can be red or green. Weirdly, red aphids make carotenoids (red pigments) with genes that jumped from a fungus into the aphid genome some time during recent evolutionary history. What's more, some red aphids start out red and then change to green later in life. Observation suggested that color changers were infected with a bacterium, *Rickettsiella*. To test whether *Rickettsiella* was the cause of color change, Tsuchida et al. (2010) experimentally injected the bacterium into a sample of red aphids. The data below are color measurements of genetically identical and bacteria-free red aphids that were either

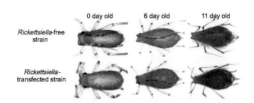

uninjected (original), injected successfully with *Rickettsiella* (infected), or injected but the bacterium failed to establish (uninfected). Color was measured as hue angle in degrees; for these data, small angles indicate red, whereas larger angles represent green.

Original: 30.7, 25.4, 26.2, 23.0, 20.9, 20.7, 15.8, 17.4, 17.6, 17.0, 16.5, 15.3

Uninfected: 25.2, 22.3, 18.5, 15.4, 15.3, 17.0, 16.6, 18.6, 19.0

Infected: 43.3, 42.3, 40.7, 41.2, 39.6, 39.5, 36.2, 36.2, 34.4, 30.7, 31.9

a. Show the data in a graph. What trend is suggested?

b. By eye, describe how the data might depart from the assumptions of ANOVA.

c. The data were analyzed using a Kruskal–Wallis test. What are the null and alternative hypotheses for this test?

d. The results of the test were as follows: $H = 21.1$. What is the conclusion?

e. H is not calculated directly from the original measurements. What does it use instead?

f. Under what assumption are we able to use the results of the Kruskal–Wallis test to conclude that the means differ among the three groups? Is this assumption met here?

14. Tsetse flies are the vectors of human sleeping sickness and animal trypanosomiasis in Africa. The tsetse fly species *Glossina palpalis* feeds on the blood of a variety of animals, including humans, and an important question is whether the feeding preferences of individuals can be affected by learning. To investigate this, Bouyer

---

15. The repeatability of a trait among the offspring of males born from different females helps estimate the "heritability" of that trait, the fraction of variation in the trait in the population that is genetic rather than environmental. Differences among males indicate a genetic component, because they were randomly mated and their offspring were raised in a common (lab) environment.

et al. (2007) provided cohorts of male tsetse flies with a first blood meal of either cows or lizards. After two days, the flies were offered a second blood meal of cows only. The data below measure the proportion of flies in each cohort that took a meal from the cows (the remaining individuals chose not to feed).

Treatment: first blood meal	Proportion of flies taking second meal from cow
Lizard	0.66, 0.58, 0.52, 0.37, 0.35, 0.34, 0.29
Cow	1.00, 0.98, 0.97, 0.96, 0.87, 0.83

a. Display the results of the study in a graph.
b. What assumptions must be met before using ANOVA to test for differences between the two treatment groups in the mean proportion of flies taking the second blood meal from cows? In view of your results in part (a), do you see a problem meeting these assumptions?
c. Consider using a transformation to fix the problems identified in part (b). Given the data, what transformation should be

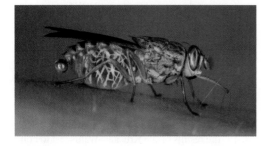

attempted first? Try this transformation on the data. Did it fix the problems?
d. Using the transformed data, test whether the means of the two blood-meal groups are different.

15. In a field experiment designed to investigate the role of genetic diversity in ecosystems, Reusch et al. (2005) planted eelgrass in plots in a shallow estuary in the Baltic Sea. Eighteen eelgrass shoots were planted in every plot. Some plots (randomly chosen) were planted with only one eelgrass genotype, others were planted with three genotypes, and others were planted with six different eelgrass genotypes. At the end of the experiment, the total number of shoots was counted in each plot. The results from 32 plots are given in the following table:

Treatment (number of genotypes)	Number of shoots at end of experiment
1	11, 14, 21, 27, 28, 30, 32, 36, 38, 49, 61, 71
3	20, 35, 36, 41, 46, 47, 52, 53, 58, 67
6	31, 45, 45, 47, 48, 62, 64, 69, 84, 86

a. What are the hypotheses for the test?
b. Carry out the test using ANOVA. Summarize your results in an ANOVA table.
c. What assumptions have you made?
d. Is this a fixed-effects or random-effects ANOVA? Explain your reasoning.

ASSIGNMENT PROBLEMS

16. Tukey–Kramer tests carried out on the results in Practice Problem 15 yielded the accompanying table of results. Groups refer to the number of genotypes in the corresponding treatment.
a. Fill in the conclusions in the table.
b. Write the sample means of the three treatment groups, from smallest to largest, and use symbols to indicate the groups to which

the means belong. Summarize the result in words.

Group i	Group j	$\bar{Y}_i - \bar{Y}_j$	SE	q	$q_{0.05, k, N-k}$	Conclusion
6	1	23.26	7.13	3.26	2.47	
6	3	12.60	7.45	1.69	2.47	
3	1	10.67	7.13	1.50	2.47	

c. Are these planned or unplanned comparisons? Explain.

d. Why not use a series of two-sample *t*-tests instead of the Tukey–Kramer method?

e. In the preceding analysis, what is the probability of making at least one Type I error during the course of carrying out all of the pairwise tests of differences between means?

17. Dormant eggs of the zooplankton *Daphnia* survive in lake sediments for decades, making it possible to measure their physiological traits in past years. Hairston et al. (1999) extracted *Daphnia* eggs from sediment cores of Lake Constance in Europe to examine trends in resistance to dietary cyanobacteria, a toxic food type that has increased in density since 1960 in response to increased nutrients in the lake. The data and accompanying histogram give the resistance level of 32 *Daphnia* clones, each initiated from single eggs extracted from deposits laid down during years of low, medium, and high cyanobacteria density between 1962 and 1997. Resistance is the average growth rate of individuals fed cyanobacteria divided by the growth rate when individuals from the same clone are fed a high-quality algal food instead. We wish to test whether resistance differs among *Daphnia* clones from the three cyanobacteria density groups.

Cyanobacterium density	Measurements of resistance
Low	0.56, 0.57, 0.58, 0.62, 0.64, 0.65, 0.67, 0.68, 0.74, 0.78, 0.85, 0.86
Medium	0.70, 0.74, 0.75, 0.76, 0.78, 0.79, 0.80, 0.82, 0.83, 0.86
High	0.65, 0.73, 0.74, 0.76, 0.81, 0.82, 0.85, 0.86, 0.88, 0.90

a. Examine the histograms of the data. Give the two main reasons why caution is warranted before using ANOVA to test for differences among group means.

b. The data were analyzed using a Kruskal–Wallis test. What are the null and alternative hypotheses for this test?

c. The results of the test were as follows: $H = 8.20$. What is the conclusion?

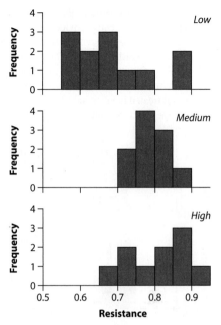

d. *H* is not calculated directly from the original measurements. What does it use instead?

e. Under what assumption would we be able to use the results of the test to draw a conclusion about whether the means or medians are the same in the three groups? Is this assumption met here?

18. When using analysis of variance, what are the main advantages of the following factors?
   a. Large sample size
   b. Balanced design

19. Examine Figure 15.4-1 (p. 476), which shows means and standard errors for three treatment groups. If this graph was to be the only graph to display the results in a scientific report, what additional feature(s) would you recommend to ensure it follows the principles of good graph design?

20. As the "baby boom" generation ages, interest in finding treatments that extend life span has surged. Experimental research mainly uses nonhuman animals. In a recent experiment, Evason et al. (2005) tested the influence of the anticonvulsant medication trimethadione on the life span of the nematode worm, *Caenorhabditis*

elegans. The study compared the effects of three trimethadione treatments (provided at the larval stage, at the adult stage, and at both stages) and a water treatment (the control). The resulting life spans are shown in the following histograms for 50 worms in each treatment. Assume that each worm was treated independently. The data are available at whitlockschluter.zoology.ubc.ca.

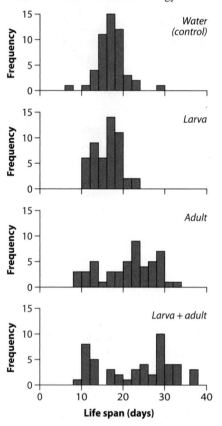

Life span (days)

a. ANOVA would be the preferred method to compare the means of each group. What problem or problems do you see in applying this method to the data shown in the preceding histograms?

b. The data were analyzed with a Kruskal–Wallis test. What are the null and alternative hypotheses?

c. The Kruskal–Wallis test yielded the following rank sums.

	Life stage of trimethadione treatment		
None (water)	Larval stage	Adult stage	Both stages
4201	3672	6003.5	6223.5

$H = 29.27$. Can we conclude from this result that the means of the treatment groups are unequal? Explain.

21. An observational study gathered data on the rate of progression of multiple sclerosis in patients diagnosed with the disease at different ages. Differences in the mean rate of progression were tested among several groups that differed by age-of-diagnosis using ANOVA. The results gave $P = 0.12$. From the following list, choose all of the correct conclusions that follow from this result (Borenstein 1997). Explain the basis of your answers.

   a. The mean rate of progression does not differ among age groups.

   b. The study has failed to show a difference among means of age groups, but the existence of a difference cannot be ruled out.

   c. If a difference among age groups exists, then it is probably small.

   d. If the study had included a larger sample size, it probably would have detected a significant difference among age groups.

22. Head width was measured twice, in cm, on the random sample of 25 walking-stick insects described in Example 15.6. The data are as follows:

Specimen	Head width (cm)	Specimen	Head width (cm)
1	0.15, 0.15	14	0.15, 0.18
2	0.18, 0.18	15	0.19, 0.21
3	0.17, 0.18	16	0.18, 0.19
4	0.21, 0.21	17	0.15, 0.15
5	0.15, 0.16	18	0.17, 0.17
6	0.19, 0.19	19	0.20, 0.21
7	0.18, 0.17	20	0.18, 0.18
8	0.18, 0.23	21	0.18, 0.16
9	0.18, 0.17	22	0.17, 0.18
10	0.19, 0.21	23	0.16, 0.13
11	0.17, 0.20	24	0.18, 0.16
12	0.19, 0.21	25	0.17, 0.17
13	0.16, 0.18		

a. Use ANOVA calculations to estimate the variance within groups for head width.

b. Calculate the estimate of the variance among groups.

c. What is the repeatability of the head-width measurements?

d. Compare your result in part (c) with that for femur length analyzed in Example 15.6. Which trait has higher repeatability? Which trait is more affected by measurement error?

23. The accompanying table presents mean cone size (mass) of lodgepole pine in 16 study sites in three types of environments in western North America (Edelaar and Benkman 2006). The three environments were islands of lodgepole pines in which pine squirrels were absent (an "island" here refers to a patch of lodgepole pine surrounded by other habitat and separated from the large tracts of contiguous lodgepole pine forests), islands with squirrels present, and sites within the large areas of extensive lodgepole pines ("mainland") that all have squirrels.

Habitat type	Raw data (g)	Mean	SD
Island, squirrels absent	9.6, 9.4, 8.9, 8.8, 8.5, 8.2	8.90	0.53
Island, squirrels present	6.8, 6.6, 6.0, 5.7, 5.3	6.08	0.62
Mainland, squirrels present	6.7, 6.4, 6.2, 5.7, 5.6	6.12	0.47

The main comparison of interest in this study, identified before the data were gathered, was the comparison between islands with and without squirrels, because this comparison controls for any effects of forest isolation on the mass of lodgepole pine cones.

a. What do we label this type of comparison?

b. Taking into account the type of comparison identified in part (a), calculate a 95% confidence interval for the difference in cone mass between islands with and without squirrels. Assume that sites were randomly sampled.

c. Using these data, carry out a test of the differences among the means of all three groups.

24. People with an autoimmune disease like lupus produce antibodies that react to their own tissues. Research has shown that lupus-prone strains of mice have B cells with reduced expression levels of the receptor gene $Fc\gamma RIIB$, suggesting that low expression of this gene might contribute to the autoimmune reaction. To test this, McGaha et al. (2005) experimentally enhanced expression of the $Fc\gamma RIIB$ gene in bone marrow taken from a lupus-prone mouse strain and transplanted it back into irradiated mice of the same strain. Other mice of the same strain were subjected to the same procedures but received bone marrow that was not enhanced for $Fc\gamma RIIB$ expression (i.e., the sham treatment). Mice in a third group were left untreated. Autoimmune reactivity was measured six months later. The following is a frequency table indicating the highest dilution of blood serum, in a fixed series of dilutions, at which reactivity could be detected (a high dilution reflects a high autoimmune reactivity).

	Number of mice		
Dilution measured	Enhanced	Sham-treated	Untreated
100	3	0	0
200	4	0	0
400	2	3	2
800	0	2	4
Total	9	5	6

a. Is this a balanced design? Explain.

b. What distinct purposes do the sham-treated and untreated groups serve in this experiment?

c. Calculate the mean and standard deviation of dilution measurements in each group.

d. We would like to test whether the mean dilutions of the three groups are different. Based on your answers to parts (a) and (c), why should we be cautious about employing ANOVA?

e. Choose a transformation that overcomes the main difficulty in part (d). Display your resulting sample means and standard deviations.

f. Using the transformed data, test whether there is a difference among treatment groups in the mean of dilution measurements.

g. What method would we use next to help decide which group means differed from the others?

25. Huey and Dunham (1987) measured the running speed of fence lizards, *Sceloporus merriami*, in Big Bend National Park in Texas. Individual lizards were captured and placed in a 2.3-meter raceway, where their running speeds were measured. Lizards were then tagged and released. The researchers returned to the park the following year, captured many of the same lizards again, and measured their sprint speed in the same way as previously. The pair of measurements for 34 individual lizards is as follows:

Lizard	Sprint speed (m/s)	Lizard	Sprint speed (m/s)
1	1.43, 1.37	18	2.28, 2.05
2	1.56, 1.30	19	2.44, 1.92
3	1.64, 1.36	20	2.23, 2.12
4	2.13, 1.54	21	2.53, 2.11
5	1.96, 1.82	22	2.20, 2.22
6	1.89, 1.79	23	2.16, 2.27
7	1.72, 1.72	24	2.25, 2.39
8	1.80, 1.80	25	2.42, 2.33
9	1.87, 1.87	26	2.61, 2.33
10	1.61, 1.88	27	2.62, 2.39
11	1.60, 1.98	28	3.09, 2.17
12	1.71, 2.08	29	2.13, 2.54
13	1.83, 2.16	30	2.44, 2.63
14	1.92, 2.08	31	2.76, 2.69
15	1.90, 2.01	32	2.96, 2.64
16	2.06, 2.03	33	3.13, 2.81
17	2.06, 1.97	34	3.27, 2.88

a. With these data, calculate the repeatability of running speed.

b. What does repeatability measure?

26. Mosquitoes contribute to more human deaths than any other organism, because they transmit diseases such as malaria, dengue fever, and yellow fever. Some of these diseases develop or grow inside the mosquito—a process that can take some time. Therefore, one possible strategy to reduce transmission of disease is to cause mosquitos to die slightly sooner, leaving insufficient time for the disease to develop. Fang et al. (2011) tested the idea by infecting mosquitos with a fungus (*Metarhizium anisopliae*) that reduces the life span of the insect. In addition, they developed a transgenic strain of fungus that carries a gene for scorpine, a protein from scorpion venom known to inhibit the gamete stages of malaria. They compared three groups of mosquitoes: a "control" group that was not treated with fungus, a "wild type" group that was infected with unmodified fungus, and a "scorpine" group that was infected with the transgenic fungus. Each mosquito was infected with malaria. The response variable was the log number of sporozoites (infectious cells of malaria) in the salivary glands of the mosquitoes. Here are the data:

Control: 7.2, 7.4, 7.4, 7.7, 7.9, 7.9, 8.0, 8.2, 8.3, 8.4, 8.4, 8.5, 9.1, 9.2, 9.2

Wild type: 5.6, 6.5, 6.7, 7.0, 7.5, 7.9, 7.9, 8.0, 8.0, 8.2, 8.4, 9.0, 9.1, 9.0, 9.1

Scorpine: 0.0, 4.4, 5.3, 5.6, 4.1, 5.3, 5.9, 6.0, 6.0, 6.1, 6.2, 7.0, 7.5

a. Show the data in a graph. What pattern is suggested?

b. Examine the frequency distributions of the data. What statistical approach would be the most appropriate to determine whether these treatments vary in their number of sporozoites? Why?

27. Does adding math to a scientific paper make readers think that it has more value? Eriksson (2012) sent two abstracts of scientific papers to 200 people with postgraduate degrees. For each participant, one of the abstracts was randomly chosen and had a meaningless sentence inserted describing an unrelated mathematical model, while the other had no mathematical addition. The sentence had no conceptual connection to the subject matter of the abstract; it was just meaningless mathematics in that context. Participants were asked to rate the quality of the research in each abstract on a scale from 1 to 100, and the differences between the scores of their two abstracts—score of the abstract with math minus score of abstract without math—were recorded. Participants were also asked for the subject matter of their postgraduate degree: math, science, technology (MST); medicine (M): humanities, social science (HS); or other (O). A box plot of the data and summaries of the results for each group are given below; the full data set can be found at whitlockschluter.zoology.ubc.ca.

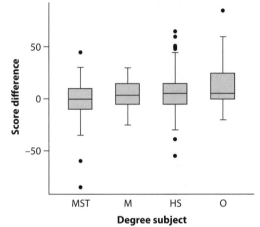

Degree subject	Mean score difference	SD score difference	n
MST	−1.28	19.24	69
M	3.06	15.99	16
HS	6.60	21.15	84
O	13.90	23.31	31

a. Examine the graph and judge by eye how well the data likely fit the assumptions of ANOVA.

b. Test whether the subject background of the participants affected how much the added math changed their views of the abstracts on average.

c. Is the relationship between degree subject and score difference strong? Answer using R^2.

28. The parasitoid wasp, *Leptopilina heterotoma*, injects eggs into young larvae of fruit flies, *Drosophila melanogaster*. One reaction by the flies is to self-medicate by consuming alcohol (ethanol), which is naturally present in the decaying fruits where they live. The ethanol reduces oviposition by wasps, and it increases death rates of wasp larvae within parasitized flies. Kacsoh et al. (2013) investigated whether the presence of the wasp influences where female fruit flies prefer to lay their eggs. They presented female flies in cages with two dishes of fly food, one having 6% ethanol and the other with 0% ethanol. They recorded the proportion of eggs laid in the 6% ethanol dish when females were placed with female wasps, with male wasps, or with no wasps. The data below give the proportion of eggs laid in the ethanol dish for multiple replicates of each wasp treatment.

No wasp: 0.25, 0.40, 0.46, 0.44

Male wasp: 0.42, 0.47, 0.31, 0.52

Female wasp: 0.89, 0.83, 0.92, 0.93

a. Proportion data often show differences in standard deviations between groups that differ in the mean proportion (tending to be smaller in groups whose means are close to 0 or 1). Do these data show such a trend? To answer, make a table of the means and standard deviations of groups (include sample sizes).

b. Carry out a transformation suitable for proportion data and then make a new table. Does the transformation reduce heterogeneity among groups in the standard deviation?

c. Test the null hypothesis of no differences among group means using the transformed data.

d. What fraction of the variation in the response variable is explained by treatment?

29. Refer to Assignment Problem 28.

a. Illustrate the (transformed) data in a graph. Add means and error bars showing standard errors of means.

b. The table at the bottom of the page shows partial results of Tukey–Kramer multiple comparisons of means. Complete the table by adding the test conclusions.

c. Use symbols to illustrate the results of the Tukey–Kramer test. Add the symbols to your graph in part (a) to show which means group together on the basis of statistical significance.

30. Fiddler crabs are so called because males have a greatly enlarged "major" claw, which is used to attract females and to defend a burrow. Darnell and Munguia (2011) recently suggested that this appendage might also act as a heat sink, keeping males cooler while out of the burrow on hot days. To test this, they placed four groups of crabs into separate plastic cups and supplied a source of radiant heat (60-watt light bulb) from above. The four groups were intact male crabs; male crabs with the major claw removed; male crabs with the other (minor) claw removed (control); and intact female fiddler crabs. They measured body temperature of crabs every 10 minutes for 1.5 hours. These measurements were used to calculate a rate of heat gain for every individual crab in degrees C/log minute. Rates of heat gain for all crabs are provided below.

Female: 1.9, 1.6, 1.4, 1.1, 1.6, 1.8, 1.9, 1.7, 1.5, 1.8, 1.7, 1.7, 1.8, 1.7, 1.8, 2.0, 1.8, 1.7, 1.6, 1.6, 1.5

Intact male: 1.9, 1.2, 1.0, 0.9, 1.4, 1.0, 1.3, 1.4, 1.1, 1.0, 1.4, 1.2, 1.4, 1.4, 1.5, 1.5, 1.1, 1.4, 1.3, 1.3, 1.3

Male minor removed: 1.2, 1.0, 0.9, 0.8, 1.2, 0.9, 1.1, 1.1, 1.3, 1.3, 1.3, 1.1, 1.4, 1.5, 1.4, 1.4, 1.2, 1.4, 1.3, 1.2, 1.4

Male major removed: 1.2, 0.9, 1.4, 1.2, 1.2, 1.6, 1.9, 1.4, 1.4, 1.4, 1.6, 1.4, 1.7, 1.3, 1.5, 1.2, 1.3, 1.6, 1.5, 1.5, 1.5

a. Show these data in a graph. What trends are suggested?

b. Use ANOVA to test whether mean rate of heat gain differs among groups.

31. Refer to Assignment Problem 30 on fiddler crab claws.

a. The main comparison of interest, which was identified before carrying out the experiment, was to test for a difference between the two male groups "Major removed" and "Minor removed." What test method is justified in this case?

b. The table at the bottom of the page shows partial results of Tukey–Kramer multiple comparisons of means. In what way does this method differ from the method identified in part (a)?

c. Complete the table by adding the test conclusions.

d. Use symbols to illustrate the results of the Tukey–Kramer test. Describe in words which

TABLE FOR PROBLEM 29

Group i	Group j	$\bar{Y}_i - \bar{Y}_j$	SE	q	$q_{0.05,4,12}$	Conclusion
Female	No	0.572	0.0626	9.13	2.79	
Female	Male	0.527	0.0626	8.42	2.79	
Male	No	0.044	0.0626	0.71	2.79	

TABLE FOR PROBLEM 31

Group i	Group j	$\bar{Y}_i - \bar{Y}_j$	SE	q	$q_{0.05,4,12}$	Conclusion
Female	Minor rem.	0.4667	0.0642	7.263	2.62	
Female	Intact male	0.3905	0.0642	6.077	2.62	
Female	Major rem.	0.2619	0.0642	4.076	2.62	
Major rem.	Minor rem.	0.2048	0.0642	3.187	2.62	
Major rem.	Intact male	0.1286	0.0642	2.001	2.62	
Intact male	Minor rem.	0.0762	0.0642	1.186	2.62	

population means are grouped together based on statistical significance.

32. The graphs at the right are from a study investigating hippocampal volume loss in 107 patients with drug-resistant epilepsy (Cook et al. 1993). The graphs depict the association between hippocampal volume loss (measured using MRI as the volume of the smaller half of the hippocampus divided by the volume of the larger half, expressed as a percentage) and patient history. Patients were grouped on the basis of whether they had a record of childhood febrile seizures (CFS), childhood non-febrile seizures (no CFS) and no childhood seizures.

a. Which accompanying graph, the box plot (top) or the bar graph (bottom; indicating means and SEs), best depicts the patterns in the data? Why?

b. Which statistical method would you recommend to test whether groups differed in hippocampal volume loss? Why?

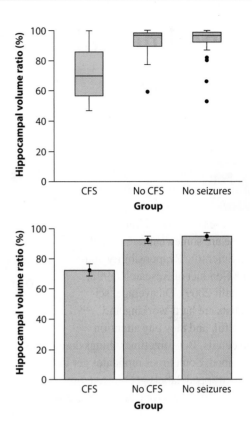

Experimental and statistical mistakes

Science is mostly done by intelligent people who want to get it right. There are cases of outright fraud, but these are somewhat rare (Panel on Scientific Responsibility and the Conduct of Research, 1992; Fanelli 2009). On average, scientists are hard-working and careful, and they pay attention to details. But sometimes things don't go as planned. Sometimes replicates get lost, vials get swapped, labels get blurred, tired hands write down the wrong numbers, the wrong button is pushed, or the wrong statistical test is applied. Many of these mistakes are caught by the researchers or by the peer review process, but not all. Sometimes, as Richard Nixon once said, "Mistakes were made."

"Experimental mistakes" are errors made during the process of an experiment, in which the protocol actually followed was not the one intended. No one knows how frequently important mistakes are made in research, but we do know that they happen. For example, a study of the dopamine neurotoxicity of MDMA ("ecstasy") found significant toxic effects in the brains of 15 monkeys (Ricaurte et al. 2002). Before this study, this sort of brain damage produced by MDMA was thought to be unlikely, whereas brain damage was a well-known side effect of using methamphetamine. The Ricaurte et al. (2002)

> Life is like a sewer. What you get out of it depends on what you put into it.
>
> —Tom Lehrer

paper was published in *Science*, a high-profile scientific weekly. But as it turned out, the lab that did the study had received its shipment of MDMA at the same time that it had received some methamphetamine as well, and the labels on the two bottles were swapped! Unbeknownst to them, the entire experiment had been done with methamphetamine rather than MDMA (Ricaurte 2003). The mistake was caught by careful follow-up work by this lab, but most mistakes of this type would not have been caught. Fundamental experimental errors sometimes make their way into the literature, even in the most prominent journals.

It is difficult, if not impossible, to know how often experimental mistakes are made or how important they are. It is a little easier to investigate how often statistical mistakes are made when analyzing the data. A few studies have looked at the rates of statistical errors in published papers. The number of mistakes that are found is sobering. Surveys of papers in medical journals routinely find that one-third to one-half of papers that use statistics make at least one minor mistake (Gore et al. 1977; Kanter and Taylor 1994; McGuigan 1995). This is likely to be an underestimate, since not all statistical mistakes made are detectable in the actual published paper. More importantly, though, about 8% of medical papers make statistical mistakes important enough to alter the conclusions of the paper (Gore et al. 1977). These statistical errors are not limited to the medical literature. Even in a field like ecology, which prides itself on its sophisticated use of statistics, a survey found statistical mistakes in about half of the papers (Hurlbert and White 1993). These mistakes are sometimes jokingly referred to as "Type III errors."

It is interesting to compare the 8% rate at which conclusions are changed by statistical mistakes to the 5% rate of Type I error that we normally tolerate. If we demand this level of confidence against errors of chance, we should also keep in mind the relatively high rates of other errors that might affect the result. It is one more reason why all studies should be repeated.

The moral is, be careful when doing science and when reading about science. Trust the authors to have done a good job, but don't expect their work to be perfect. Watch for mistakes. Very often, great scientific advances come from spotting, and fixing, the mistakes of our predecessors.

Western trillium

16 Correlation between numerical variables

When two numerical variables are associated, we say that they are **correlated**. For example, brain size and body size are positively correlated across mammal species, as the graph[1] on the next page demonstrates. Large-bodied species tend to have large brains, and small-bodied species tend to have small brains.

The correlation coefficient is a quantity that describes the strength and direction of an association between two numerical variables measured on a sample of subjects or units. Correlation reflects the amount of "scatter" in a scatter plot of the two variables. Unlike linear regression (Chapter 17), correlation fits no line to the data. It does not measure how steeply one variable changes when the other is varied. In this chapter, we show how to measure a correlation, put confidence bounds on estimates of a correlation coefficient, and test hypotheses about correlation.

---

1. Modified from Jerison (2006).

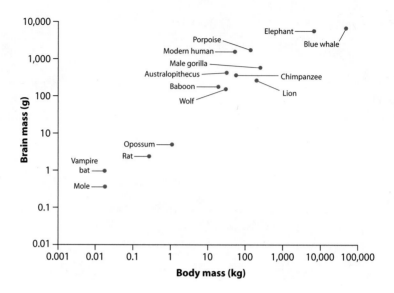

16.1 Estimating a linear correlation coefficient

The linear **correlation coefficient**[2] measures the tendency of two numerical variables (call them X and Y) to co-vary—that is, to change together along a line. We use the lowercase Greek letter ρ (rho, pronounced "row"[3]) to represent the correlation between X and Y in the population. We use r to represent the correlation between X and Y in a sample taken from the population.

> The *correlation coefficient* measures the strength and direction of the association between two numerical variables.

The correlation coefficient

The formula for the sample correlation coefficient r has three parts, two of which may look familiar and one of which is new:

$$r = \frac{\sum_i (X_i - \bar{X})(Y_i - \bar{Y})}{\sqrt{\sum_i (X_i - \bar{X})^2}\sqrt{\sum_i (Y_i - \bar{Y})^2}}$$

2. The linear correlation coefficient is usually called simply the "correlation coefficient." It is also called the Pearson's correlation coefficient after Karl Pearson, who first defined it. Pearson was one of the founders of the modern field of statistics (see Interleaf 1).

3. Be careful: Don't mix up the Greek letter ρ with the roman letter p, which we use to represent proportion (Chapter 7).

The term in the numerator of the formula is called the sum of products,[4] and it measures how deviations in X and Y vary together. A deviation is the difference between an observation and its mean (Figure 16.1-1). If an observation i is above the mean for both X and Y (upper right corner of the figure), then its deviations are both positive and so the product of its deviations $(X_i - \overline{X})(Y_i - \overline{Y})$ is a positive number. If an observation is below the mean in both X and Y (lower left corner of the figure), then both its deviations are negative, and so the product of the deviations $(X_i - \overline{X})(Y_i - \overline{Y})$ is again positive. Observations lying in the other two corners of the plane have a positive deviation for one variable and a negative deviation for the other, so they have a negative product of deviations. The sum of products adds all the products of deviations. This sum will be positive if most of the observations are in the lower left and upper right corners of the plane, like those shown in Figure 16.1-1. The sum will be negative if most observations lie in the upper left and lower right corners. If the scatter of observations fills all four corners of the plane, then the positive and negative values cancel in the sum, yielding a sum of products close to zero.

FIGURE 16.1-1
The position of X and Y observations in relation to the means, $\overline{X}$ and $\overline{Y}$ (indicated by an open square at the intersection of the two dashed lines). Observations lying in the upper right corner are above both $\overline{X}$ and $\overline{Y}$, and so have positive deviations in X and Y. Those lying in the lower left corner fall below both $\overline{X}$ and $\overline{Y}$, and so have negative deviations in X and Y. In the other two corners, observations have a positive deviation for one trait and a negative deviation for the other.

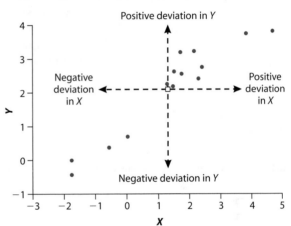

The denominator in the formula for r includes the sums of squares for X and for Y (under square-root signs). You will have calculated quantities like these in Section 3.1 as part of calculating a standard deviation.[5]

The population correlation coefficient, ρ, is calculated using the same formula as r except it is measured on all individuals in the population. The correlation coefficient ρ and its sample estimate r lie between -1 and 1.

The correlation coefficient has no units, which means it is readily interpretable whatever the variables (provided they are numerical). A negative correlation means

---

4. We provide a shortcut formula for this quantity in the Quick Formula Summary (Section 16.8 at the end of the chapter) to help when you are doing calculations by hand.

5. Another way to write the equation for the correlation coefficient is $r = \dfrac{\text{Covariance}(X,Y)}{s_X s_Y}$, where the covariance of X and Y is in the numerator, and the standard deviations of X and Y are in the denominator. The covariance of X and Y is a measure of their relationship, analogous to the variance of a single variable. Its formula is provided in the Quick Formula Summary (Section 16.8).

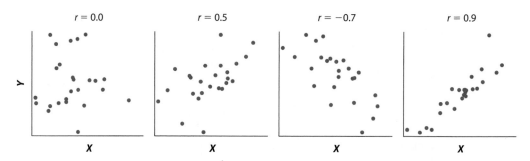

FIGURE 16.1-2 Scatter plots illustrating correlations between two numerical variables. In each case, $n = 25$.

that one variable decreases as the other increases, whereas a positive correlation means that both variables increase and decrease together (Figure 16.1-2).

The strongest correlations (i.e., $r = 1.0$ or $r = -1.0$) occur when all points lie along a straight line (e.g., the left panel in Figure 16.1-3), which is why we refer to the correlation as linear. Two variables might be strongly associated yet have no correlation (i.e., $r = 0$) if the relationship between them is nonlinear (the right panel in Figure 16.1-3). Example 16.1 illustrates the use of the correlation coefficient.

FIGURE 16.1-3
Correlation between two variables that are strongly associated. On the left, measurements of X and Y lie along a straight line, producing the maximum correlation possible ($r = 1$). On the right, the relationship is nonlinear and exhibits no correlation ($r = 0$).

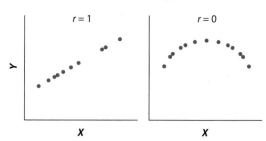

EXAMPLE   Flipping the bird

16.1   Adults who mistreat children were often the target of maltreatment themselves when they were young. Does a similar association occur in nonhuman animals, where the causes might be more readily studied? Müller et al. (2011) investigated this possibility in the Nazca booby (*Sula granti*), a colonial nesting seabird of the Galápagos Islands. Unattended chicks in nests frequently receive visits from unrelated adults, who behave mainly aggressively toward them. The researchers counted the number of such visits to nests of 24 booby chicks. These chicks were given unique numbered rings on their legs, which allowed the researchers to observe their behavior years later when they had become adults. The first variable in Table 16.1 gives the number of non-parent adult visits experienced by the 24 focal birds while they were growing up in the nest. The second variable measures the number of visits to nests

of unrelated chicks by these same birds when they were adults. The second variable has been corrected for other variables measured by the researchers and so is not on the same scale as the first variable.

TABLE 16.1 Number of non-parent adult visits experienced by boobies as nestlings compared to the number of similar behaviors performed by the same birds when an adult. $n = 24$.

Number of visits	Future aggressive behavior	Number of visits	Future aggressive behavior
1	−0.80	13	−0.10
7	−0.92	13	0.04
15	−0.80	14	0.13
4	−0.46	12	0.19
11	−0.47	13	0.25
14	−0.46	9	0.23
23	−0.23	8	0.15
14	−0.16	18	0.23
9	−0.23	22	0.31
5	−0.23	22	0.18
4	−0.16	23	0.17
10	−0.10	31	0.39

A scatter plot of the data, shown in Figure 16.1-4, suggests that previous experience of nestlings at the hands of adult boobies is positively associated with the behavior of the same birds when they become adults themselves, although the association does not seem strong. The larger the number of visits received as nestlings, the more the birds perform such events toward unrelated nestlings when they are adults. Correlation is appropiate to measure the strength of this association, since the data consist of two numerical variables measured on a sample of individuals. By itself this association does not imply that one variable is the cause of the other, however—correlation is not causation (Interleaf 4). Further studies including experiments would be needed to establish causation.

FIGURE 16.1-4
Scatter plot of the relationship between the number of visits experienced by nestling boobies and the future behavior of the same individuals as adults. $n = 24$.

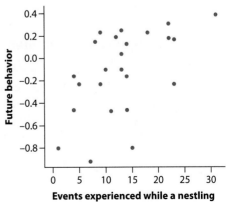

How strong is the association between past experience and future booby behavior? The correlation coefficient r quantifies the strength of the association between the two variables. Let's use X to refer to the variable "number of visits" by the boobies, and Y to refer to their future aggressive behavior. To calculate r, we obtained the following three quantities:

$$\sum_i (X_i - \bar{X})(Y_i - \bar{Y}) = 33.086$$

$$\sum_i (X_i - \bar{X})^2 = 1194.625$$

$$\sum_i (Y_i - \bar{Y})^2 = 3.217,$$

yielding

$$r = \frac{33.086}{\sqrt{1194.625}\sqrt{3.217}} = 0.534.$$

The sample correlation coefficient r between the two variables is 0.534.

Standard error

The data are merely a sample taken to estimate the correlation between the same two variables in a population, ρ. The standard error of r—that is, the standard deviation of its sampling distribution—is one way to assess how close our estimate is likely to be to the population parameter ρ. The standard error is

$$SE_r = \sqrt{\frac{1 - r^2}{n - 2}}.$$

For our example data set, the standard error is

$$SE_r = \sqrt{\frac{1 - (0.534)^2}{24 - 2}} = 0.180.$$

It is not ideal to use this quantity to calculate a confidence interval, because the sampling distribution of r is not normal. However, we will use SE_r later in hypothesis testing. Calculating a confidence interval requires the modified method shown next.

Approximate confidence interval

The 95% confidence interval for ρ puts bounds on our estimate of the population correlation, identifying the range of values that are compatible with the data. Fisher discovered an approximate confidence interval for ρ. The approximation is best when sample size is large. With this method, we convert r to a new quantity called z that approximately follows a normal sampling distribution. The Fisher's z-transformation is

$$z = 0.5 \ln\left(\frac{1 + r}{1 - r}\right),$$

where ln is the natural logarithm. The z-transform of ρ is symbolized as ζ (the lower-case Greek letter zeta), so z represents the value in a sample, and ζ is the true value in the population. The standard error of the sampling distribution for z is approximately

$$\sigma_z = \sqrt{\frac{1}{n-3}}.$$

For the booby aggression ratio data (Example 16.1),

$$z = 0.5 \ln\left(\frac{1+0.534}{1-0.534}\right) = 0.595,$$

and

$$\sigma_z = \sqrt{\frac{1}{24-3}} = 0.218.$$

The sampling distribution of the statistic z is approximately normal, so we can use the standard normal distribution to generate the 95% confidence interval for ζ:

$$z - 1.96\,\sigma_z < \zeta < z + 1.96\,\sigma_z.$$

The quantity 1.96 is the value of Z_{crit}, the two-tailed critical value of the standard normal distribution corresponding to $\alpha = 0.05$ (Statistical Table B; $\Pr[Z > 1.96] = 0.025$). More generally, to obtain a $1 - \alpha$ confidence interval, find the value of Z_{crit} such that $\Pr[Z > Z_{crit}] = \alpha/2$.

For the booby aggression ratio data, the 95% confidence interval for ζ is

$$0.595 - 1.96\,(0.218) < \zeta < 0.595 + 1.96\,(0.218)$$

$$0.168 < \zeta < 1.023.$$

To complete the analysis, we convert the lower and upper bounds of this confidence interval back to the original correlation scale using the inverse of Fisher's transformation,

$$r = \frac{e^{2z} - 1}{e^{2z} + 1},$$

where e is the base of the natural logarithm. For our example, this yields

$$\frac{e^{2(0.168)} - 1}{e^{2(0.168)} + 1} < \rho < \frac{e^{2(1.023)} - 1}{e^{2(1.023)} + 1}$$

or

$$0.166 < \rho < 0.771,$$

which, when rounded to two digits, is

$$0.17 < \rho < 0.77.$$

This calculation shows that the data are consistent with a fairly broad range of values for the population correlation between the number of visits experienced as a nestling and future aggressive behavior. At the same time, we can be reasonably confident that ρ is greater than zero and is well below one.

The formula for σ_z gives only an approximation to the standard deviation of the sampling distribution for z, so the 95% confidence interval for ζ (and therefore ρ) is also an approximation.

16.2 Testing the null hypothesis of zero correlation

The most common use of hypothesis testing in correlation analysis is to test the null hypothesis that the population correlation ρ is exactly zero:[6]

H_0: $\rho = 0$.
H_A: $\rho \neq 0$.

Example 16.2 shows how the method works.

EXAMPLE   **What big inbreeding coefficients you have**
16.2

By 1970, the wolf (*Canis lupus*) had been wiped out in Norway and Sweden. Around 1980, two wolves immigrated from farther east and founded a new population. By 2002, the new population totaled approximately 100 wolves. A new population started by so few individuals, however, might be expected to suffer problems caused by inbreeding. Liberg et al. (2005) compiled observations on reproduction  between 1983 and 2002 and constructed the pedigree of the wolves in the small population. The data listed in Table 16.2-1 show the inbreeding coefficients of litters produced by mated pairs and the number of pups of each litter surviving their first winter. An inbreeding coefficient is zero if parents of the litter were unrelated, 0.25 if parents were brother and sister whose own parents were unrelated, and greater than 0.25 if inbreeding had continued for more generations.

Are inbreeding coefficients of the litters associated with the number of pups surviving their first winter? A scatter plot of the data, shown in Figure 16.2-1, suggests a negative association between inbreeding coefficient and number of surviving pups.

---

6. Testing the more general null hypothesis H_0: $\rho = \rho_0$, where ρ_0 is a number other than zero, requires a different method that makes use of the Fisher's z-transformation. We do not present it here (see, e.g., Sokal and Rohlf 2008).

TABLE 16.2-1 Inbreeding coefficients of litters of mated wolf pairs and the number of pups surviving their first winter. $n = 24$ litters.

Inbreeding coefficient	Number of pups	Inbreeding coefficient	Number of pups
0.00	6	0.24	3
0.00	6	0.24	2
0.13	7	0.24	2
0.13	5	0.25	6
0.13	4	0.27	3
0.19	8	0.30	5
0.19	7	0.30	3
0.19	4	0.30	2
0.22	4	0.30	1
0.24	3	0.36	3
0.24	3	0.37	2
0.24	3	0.40	3

FIGURE 16.2-1
The number of surviving wolf pups in a litter and inbreeding coefficient. Overlapping points have been offset slightly to render them visible. The total number of litters, $n = 24$.

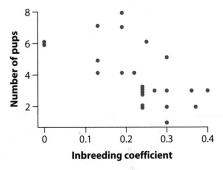

The correlation coefficient between inbreeding coefficient and number of pups can be calculated from the following quantities:

$$\sum_i (X_i - \overline{X})(Y_i - \overline{Y}) = -2.612$$

$$\sum_i (X_i - \overline{X})^2 = 0.228$$

$$\sum_i (Y_i - \overline{Y})^2 = 80.958.$$

Putting these into the formula for r gives

$$r = \frac{-2.612}{\sqrt{0.228}\sqrt{80.958}} = -0.608.$$

The observed correlation coefficient is less than zero, but we want to test whether this correlation is sufficiently strong to warrant rejection of the null hypothesis that the population correlation ρ is zero. Our two hypotheses are as follows:

H_0: There is no relationship between the inbreeding coefficient and the number of pups ($\rho = 0$).

H_A: Inbreeding depression and the number of pups are correlated ($\rho \neq 0$).

To test the hypotheses, we calculate the t-statistic,

$$t = \frac{r}{\text{SE}_r},$$

where the standard error (SE_r) is calculated as

$$\text{SE}_r = \sqrt{\frac{1 - r^2}{n - 2}}.$$

Under the null hypothesis of zero correlation, the sampling distribution of the t-statistic is a Student's t-distribution with $n - 2$ degrees of freedom.[7]

For the wolf data,

$$\text{SE}_r = \sqrt{\frac{1 - (-0.608)^2}{24 - 2}} = 0.169,$$

and so

$$t = \frac{-0.608}{0.169} = -3.60.$$

The P-value for this t-statistic is $P = 0.002$, obtained using a computer. Using Statistical Table C, instead, the critical value for the t-distribution having 22 degrees of freedom is $t_{0.05(2),\,22} = 2.075$ with $\alpha = 0.05$. Since $t = -3.60$ is less than -2.075 and farther from the value expected by the null hypothesis, P must be less than 0.05. With $P = 0.002$, we can be confident that inbreeding coefficient and number of pups are negatively related, and we reject the null hypothesis of zero correlation.

16.3 Assumptions

The methods used to estimate and test a population correlation assume that the sample of individuals is a random sample from the population. In addition, correlation analysis assumes that the measurements have a **bivariate normal distribution** in the population. A bivariate normal distribution is a bell-shaped probability distribution in two dimensions rather than one (Figure 16.3-1).

---

7. There are n independent data points when calculating a correlation coefficient, but there are two fewer degrees of freedom because we have to use two summaries of the data, $\overline{X}$ and $\overline{Y}$, when we calculate r.

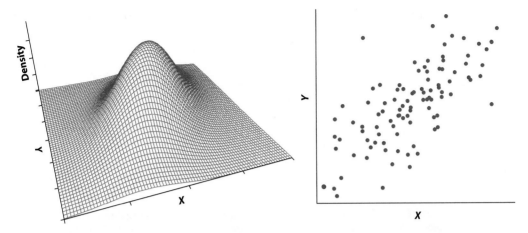

FIGURE 16.3-1 The left panel shows a bivariate normal distribution with a correlation of 0.7 between X and Y. Height above the plane represents the probability density of each pair of values of X and Y. The right panel shows a random sample of 100 observations from the bivariate normal distribution shown on the left.

A bivariate normal distribution has the following features:

- The relationship between X and Y is linear.
- The cloud of points in a scatter plot of X and Y has a circular or elliptical shape.
- The frequency distributions of X and Y separately are normal.

This is not a complete list of the features of the bivariate normal distribution, but it is the set that matters most and is easiest to evaluate with data. Inspecting the scatter plot of the data is probably the best way to check the assumption of bivariate normality. The right panel of Figure 16.3-1, a random sample from a bivariate normal distribution having a correlation of $\rho = 0.70$, shows an example of what a scatter of points should look like when the assumption of bivariate normality is met. All of the examples shown in the scatter plots of Figure 16.1-2 are also random samples from bivariate normal distributions.

Figure 16.3-2 shows the most frequent types of departures from bivariate normality. These are also the types of departures that most seriously affect correlation analysis:

- A cloud of points that is funnel shaped (i.e., wider at one end than at the other)
- The presence of outliers
- A relationship between X and Y that is not linear

Histograms depicting the frequency distributions of X and Y separately are also helpful. If either X or Y has a decidedly skewed distribution, then the frequency distribution of the two variables is not bivariate normal.

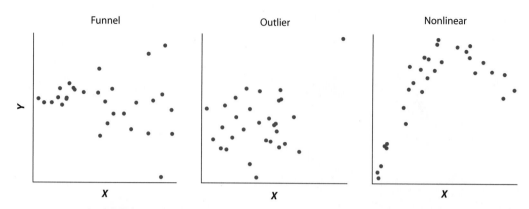

FIGURE 16.3-2 Data from three distributions that differ from bivariate normality. Scatter plots show a funnel shape (*left*), an outlier (*middle*), and a nonlinear relationship between *X* and *Y* (*right*).

What do we do if the assumption of bivariate normality is not met? Two strategies are available—namely, transforming the data and nonparametric methods. It is best to try first to transform *X, Y,* or both variables to see if the assumptions are better met on a new scale. Here are the usual transformations, first described in Chapter 13:

- The log transformation [an all-purpose transformation, as long as the data are not negative; use $\log(X + 1)$ or $\log(Y + 1)$ if there are zeros]
- The square-root transformation (which often works for data that are counts)
- The arcsine transformation (for data that are proportions)

Log transformations are good to try if the relationship between the two variables is nonlinear or if the variance in one variable seems to increase with the value of the other variable.

If transforming the data is unsuccessful, use a nonparametric method instead, such as Spearman's rank correlation. This method is explained in Section 16.5.

16.4 The correlation coefficient depends on the range

The correlation between two variables *X* and *Y* depends on the range of values included. For this reason, we must be cautious when comparing correlations between studies that use a different range. For example, the top panel of Figure 16.4-1 shows that the population density of different species of stream invertebrates (*Y*) is strongly correlated with their body mass (*X*) when both variables are log-transformed (Schmid et al. 2000). Now imagine a second study of the same two variables in streams whose invertebrates have a smaller range of values of body mass. To see what effect this

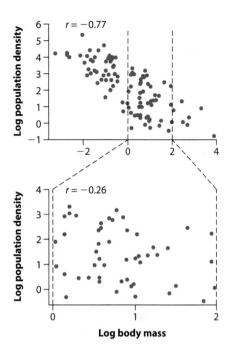

FIGURE 16.4-1
The correlation between two variables depends on the range of X-values included. The bottom graph plots those data points lying between the two dashed lines in the top graph. The data drawn from a smaller range of X have a reduced correlation coefficient. The data are log (base 10) of population density (individuals/m²) and body mass (μg) of different species of stream invertebrates (Schmid et al. 2000).

has, we've taken the points falling between the two dashed lines in the top panel and replotted them in the lower panel. The correlation is weaker, because the amount of scatter relative to total variation is increased.

This effect means that the correlation coefficient between the same two variables is not comparable between separate studies unless they include a comparable range of values.

16.5   Spearman's rank correlation

Many situations require a test of zero correlation between variables that do not meet the assumption of bivariate normality, even after data transformation. For these cases, the nonparametric **Spearman's rank correlation** is used. Spearman's rank correlation uses the ranks of both the X and Y variables to calculate a measure of correlation. It does not make assumptions about the distribution of the variables, but it still assumes that the individuals are randomly chosen from the population. Example 16.5 illustrates the method.

> The *Spearman's rank correlation* measures the strength and direction of the linear association between the ranks of two variables.

EXAMPLE   The miracles of memory

16.5

How reliable are people's recollections of having witnessed "miracles"? One way to investigate is to compare
different accounts of extraordinary magic tricks. For example, of the many illusions performed by magicians, none
is more renowned than the Indian rope trick. In the most
sensational version of the trick, a magician tosses into the
air one end of a rope, which forms into a rigid pole. A boy
climbs up the rope and disappears at the top. The magician scolds the boy to return but gets no reply, whereupon
he grabs a knife, climbs the rope, and also disappears.
The boy's body then falls in pieces from the sky into a
basket on the ground. The magician descends the rope
and retrieves the boy from the basket, revealing him to be
unharmed and in one piece.

Wiseman and Lamont (1996) tracked down 21 firsthand, written accounts of the Indian rope trick. They gave
a score to each description according to how impressive it
was. For example, a score of 1 was given if the observer
saw only that "boy climbs up rope, then climbs down
again." The most impressive accounts in the sample, "boy
climbs rope, vanishes at the top, reappears in basket in full view of audience," were given
a score of 5, the highest possible. For each account, the researchers also recorded the
number of years that had elapsed between the date that the trick was witnessed and the
date the memory of it was written down. The measurements of impressiveness score and
number of years elapsed are shown in the scatter plot in Figure 16.5-1. Is there an association between the impressiveness of eyewitness accounts and the time elapsed until the
writing of the description? If so, then it might indicate a tendency of human memories to
become more exaggerated and less accurate with time.

FIGURE 16.5-1
Impressiveness of written accounts of the Indian rope
trick by firsthand observers and the number of years
elapsed between witnessing the event and writing the
account. $n = 21$.

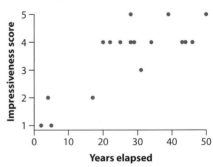

A test of the null hypothesis of zero correlation is what we would like to carry
out. But the assumption of bivariate normality is clearly violated because the impressiveness score is a discrete ordered score. A test of zero correlation is nevertheless

possible using a nonparametric method, because the different categories for the impressiveness score can be ranked as shown in Table 16.5-1. The Spearman's rank correlation measures the association between the ranks of the two variables. Spearman's rank correlation is measured by the parameter ρ_S, which is estimated by r_S.

TABLE 16.5-1 Raw data from Example 16.5-1, and their ranks. Each variable is ranked separately. Midranks are assigned when there are ties. $n = 21$.

Years elapsed	Rank years	Impressiveness score	Rank impressiveness
2	1	1	2
5	3.5	1	2
5	3.5	1	2
4	2	2	5
17	5.5	2	5
17	5.5	2	5
31	13	3	7
20	7	4	12.5
22	8	4	12.5
25	9	4	12.5
28	10.5	4	12.5
29	12	4	12.5
34	14.5	4	12.5
43	17	4	12.5
44	18	4	12.5
46	19	4	12.5
34	14.5	4	12.5
28	10.5	5	19.5
39	16	5	19.5
50	20.5	5	19.5
50	20.5	5	19.5

The Spearman's correlation coefficient is the linear correlation coefficient computed on the *ranks* of the data. The two variables must be ranked separately, from low to high.

The data from Figure 16.5-1 are listed in Table 16.5-1. The table also includes the separate rankings of each variable. As first discussed in Section 13.5, we assign midranks when there are ties. The midrank is the average of the ranks associated with a set of tied observations. For example, the three measurements with the lowest impressiveness score (1) were all assigned the midrank 2, which is the average of the three ranks associated with the three lowest values: 1, 2, and 3. The 10 values having impressiveness score 4 are associated with the ranks 8 through 17, and so were assigned the midrank 12.5. In the following calculations, R refers to the rank of years elapsed, and S refers to the rank of the impressiveness score:

$$\sum_i (R_i - \overline{R})(S_i - \overline{S}) = 566$$

$$\sum_i (R_i - \overline{R})^2 = 767.5$$

$$\sum_i (S_i - \overline{S})^2 = 678.5,$$

yielding

$$r_S = \frac{566}{\sqrt{767.5}\sqrt{678.5}} = 0.784.$$

The hypotheses for the test are

H_0: $\rho_S = 0$

H_A: $\rho_S \neq 0$,

where ρ_S refers to the Spearman's correlation in the population. To determine the P-value for the test, compare r_S with the critical value[8] given in Statistical Table G corresponding to a sample size of 21:

$$r_{S(0.05,\,21)} = 0.435.$$

Since $r_S = 0.784$ is greater than 0.435, $P < 0.05$, and so we reject the null hypothesis (a computer program gave $P = 0.0003$). We conclude that there is a positive correlation between the impressiveness score of the eyewitness accounts of the Indian rope trick and the number of years elapsed between viewing the trick and the retelling of it in writing. The likely explanation for these findings is that eyewitness accounts of the Indian rope trick are exaggerated,[9] becoming more so with time.

Procedure for large n

Statistical Table G provides critical values for the Spearman's rank correlation for sample sizes up to $n = 100$. For larger n, use the procedure for the linear correlation coefficient, but applied to the ranks. Calculate the t-statistic,

$$t = \frac{r_S}{SE[r_S]},$$

---

8. Different computer programs might yield slightly different P-values for the same data, depending on the approximation used to compute it.

9. According to the authors of the report, one eyewitness claimed to have seen the trick performed and had taken a photograph. Examination of the photograph showed only a boy balancing on the end of a long pole.

where

$$SE[r_S] = \sqrt{\frac{1 - r_S^2}{n - 2}}.$$

Under the null hypothesis of no Spearman's rank correlation in the population ($\rho_S = 0$), t is approximately t-distributed with $n - 2$ degrees of freedom. Reject the null hypothesis of zero rank correlation if $t \geq t_{0.05(2), n-2}$ or $t \leq -t_{0.05(2), n-2}$.

Assumptions of Spearman's correlation

The Spearman's rank correlation assumes that observations are a random sample from the population. It also assumes that the relationship between the two numerical variables is monotonic. That is, as one variable increases, the other variable either (1) increases or does not change; or (2) decreases or does not change. Essentially, this means that the relationship between the ranks of the two numerical variables is linear.

16.6 The effects of measurement error on correlation

When a variable is not measured perfectly, we say that there is **measurement error**. Measurement error is difficult to avoid. Some biological traits are extremely challenging to measure, and measurement error can sometimes be an important component of variation (Chapter 15). For example, behavioral traits are notorious for having low repeatability: a behavior measured on one individual might be quite different the next time it is measured on the same individual.

> *Measurement error* is the difference between the true value of a variable for an individual and its measured value.

Measurement error in either X or Y tends to weaken the observed correlation between the variables. The same thing happens if there is measurement error in both X and Y, if the errors in X and Y are uncorrelated. With measurement error, r will tend to underestimate the magnitude of ρ (it will tend to be closer to zero on average than the true correlation), a bias called **attenuation** (Figure 16.6-1).

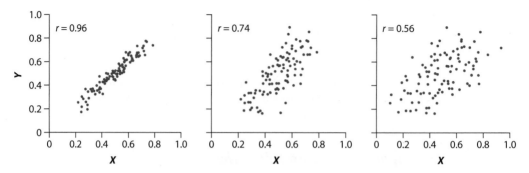

FIGURE 16.6-1   Attenuation. In the left panel, *X* and *Y* are measured without error and are highly correlated. In the middle panel, *Y* is measured with error. In the right panel, both *X* and *Y* are measured with
error, and these errors are uncorrelated. In all cases, the true correlation is very strong, but the correlation
appears weaker when the variables are measured with error.

In the Quick Formula Summary (Section 16.8), we include an equation that corrects the estimate of correlation for the effects of measurement error. The method
requires that repeated measurements have been made on the same individuals. This
corrected correlation, r^*, can't be used in place of r in confidence intervals for ρ or in
hypothesis testing. However, it might be useful for comparison with the uncorrected
correlation to evaluate how measurement error is affecting the observed correlation
between two variables. In general, measurement error can be reduced by taking precise measurements. If this is not possible, then it is best to measure each individual
multiple times and use the average measurement in subsequent analysis.

16.7 Summary

- The correlation coefficient (r) measures the strength and direction of the association between two numerical variables.
- Correlation implies association, not causation. It is appropriate when two variables are measured on a sample of individuals whether or not there is a causal
  connection between the variables.
- The correlation coefficient ranges from −1 (the maximum negative correlation)
  to 0 (no correlation) to 1 (the maximum positive correlation).
- Analysis using the correlation assumes that the two numerical variables have a
  bivariate normal distribution and that the individuals are randomly sampled.
- With a bivariate normal distribution, the relationship between *X* and *Y* is linear,
  the cloud of points in a scatter plot of *X* and *Y* is circular or elliptical in shape,
  and the frequency distributions of *X* and *Y* separately are normal. (This is just a
  partial list of its features.)

- The scatter plot is a useful tool for examining the assumption of bivariate normality. Histograms of X and Y should both appear normal.
- The Spearman's rank correlation measures the linear correlation between the ranks of two variables, where each variable is ranked separately from low to high.
- The correlation between two variables is expected to be weaker when only a narrow range of X values is represented.
- Measurement error biases the estimate of a correlation coefficient toward zero.

16.8 Quick Formula Summary

Shortcuts

Sum of Products: $\displaystyle\sum_i (X_i - \overline{X})(Y_i - \overline{Y}) = \sum_i (X_i Y_i) - \frac{\left(\sum_i X_i\right)\left(\sum_i Y_i\right)}{n}$

Sum of Squares: $\displaystyle\sum_i (X_i - \overline{X})^2 = \sum_i (X_i^2) - \frac{\left(\sum_i X_i\right)^2}{n}$

$\displaystyle\sum_i (Y_i - \overline{Y})^2 = \sum_i (Y_i^2) - \frac{\left(\sum_i Y_i\right)^2}{n}$

Covariance

What is it for? Measuring the strength of an association between two numerical variables.

Estimate: Covariance(X, Y)

Formula: $\displaystyle\text{Covariance}(X, Y) = \frac{\sum_i (X_i - \overline{X})(Y_i - \overline{Y})}{n - 1}$

Correlation coefficient

What is it for? Measuring the strength of a linear association between two numerical variables.

What does it assume? Bivariate normality and random sampling.

Parameter: ρ

Estimate: r

Formula: $r = \dfrac{\sum\limits_{i}(X_i - \bar{X})(Y_i - \bar{Y})}{\sqrt{\sum\limits_{i}(X_i - \bar{X})^2}\sqrt{\sum\limits_{i}(Y_i - \bar{Y})^2}}$

Standard error: $\mathrm{SE}_r = \sqrt{\dfrac{1 - r^2}{n - 2}}$

Degrees of freedom: $n - 2$

Alternate formula: $r = \dfrac{\mathrm{Covariance}(X,Y)}{s_X s_Y}$, where $\mathrm{Covariance}(X,\,Y)$ is the covariance between X and Y, and where s_X and s_Y are the sample standard deviations of X and Y, respectively.

Confidence interval (approximate) for a population correlation

What does it assume? The sample is a random sample. The numerical variables X and Y have a bivariate normal distribution in the population. Sample size is not too small for the approximation.

Parameter: ρ

Estimate: r

Formula: $z - Z_{\mathrm{crit}}\,\sigma_z < \zeta < z + Z_{\mathrm{crit}}\,\sigma_z,$

where $z = 0.5\ln\left(\dfrac{1 + r}{1 - r}\right)$ is Fisher's z-transformation of r, $\sigma_z = \sqrt{\dfrac{1}{n - 3}}$ is the approximate standard error of z, and Z_{crit} is the critical value of the standard normal distribution for which $\Pr[Z > Z_{\mathrm{crit}}] = \alpha/2$. To obtain the confidence interval for ρ, back-transform the limits of the resulting confidence interval using the inverse of Fisher's transformation, $r = \dfrac{e^{2z} - 1}{e^{2z} + 1}$.

The *t*-test of zero linear correlation

What is it for? To test the null hypothesis that the population parameter (ρ) is zero.

What does it assume? Bivariate normality and random sampling.

Test statistic: t

Distribution under H$_0$: t-distributed with $n - 2$ degrees of freedom.

Formula: $t = \dfrac{r}{SE_r}$,

where $SE_r = \sqrt{\dfrac{1 - r^2}{n - 2}}$ is the standard error of r.

Spearman's rank correlation

What is it for? To measure correlation between the two variables, when the variables do not meet the assumptions of correlation.

What does it assume? A linear relation between the ranks of X and Y, and random sampling.

Parameter: ρ_S

Estimate: r_S

Formula: Same as for linear correlation but calculated on ranks.

Spearman's rank correlation test

What is it for? To test the null hypothesis that the rank correlation in the population (ρ_S) is zero.

What does it assume? A linear relation between the ranks of X and Y, and random sampling.

Test statistic when $n \leq 100$: r_S

Distribution under H$_0$: Distribution of the Spearman's rank correlation (Statistical Table G).

Test statistic when $n > 100$: t

Distribution under H$_0$: t-distributed with $n - 2$ degrees of freedom.

Formula: $t = \dfrac{r_S}{SE[r_S]}$,

where $SE[r_S] = \sqrt{\dfrac{1 - r_S^2}{n - 2}}$ is the standard error of r_S.

Correlation corrected for measurement error

What does it assume? That X and Y have been measured two or more times independently on all individuals, and that measurement error in X is uncorrelated with measurement error in Y. The following formula assumes that the correlation r between X and Y is measured using the average of the repeat measurements for every individual.

Parameter: ρ

Estimate: r^*

Formula: $r^* = \dfrac{r}{\sqrt{R_X R_Y}}$,

where r is calculated from the average of the repeat measurements for every individual (Adolph and Hardin 2007). R_X and R_Y are repeatabilities of X and Y, respectively. Repeatability here is similar to that described in Chapter 15, except that here the repeatability is for the average values of repeat measurements made on each individual rather than single measurements. The formula for R_X is

$$R_X = \frac{s_A^2}{s_A^2 + (\mathrm{MS}_{error}/m)},$$

where m is the number of repeat measurements made on each individual and

$$s_A^2 = \frac{\mathrm{MS}_{groups} - \mathrm{MS}_{error}}{m}.$$

MS_{groups} and MS_{error} are calculated from a random effects ANOVA on the repeat measurements for X, as explained in Section 15.6. The formula for R_Y is calculated in the same way using measurements of the Y-variable. Confidence intervals for the corrected correlation coefficients are discussed in Charles (2005).

PRACTICE PROBLEMS

1. **Calculation practice. Estimate a correlation coefficient.** In their study of hyena laughter, or "giggling" (see Chapter 12, Assignment Problem 26), Mathevon et al. (2010) asked whether sound spectral properties of hyenas' giggles are associated with age. The accompanying figure and data show the giggle fundamental frequency (in hertz) and the age (in years) of 16 hyenas. What is the correlation coefficient in the data, and what is most-plausible range of values for the population correlation?

Age (years)	Fundametal frequency (Hz)
2	840
2	670
2	580
6	470
9	540
10	660
13	510
10	520
14	500
14	480
12	400
7	650
11	460
11	500
14	580
20	500

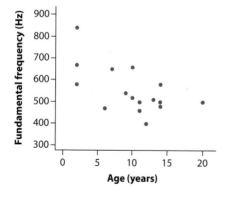

a. What type of graph is this? Does it suggest a positive or negative association?

b. Calculate the sum of squares for age.

c. Calculate the sum of squares for fundamental frequency.

d. Calculate the sum of products between age and frequency.

e. Compute the correlation coefficient, r.

f. Compute the Fisher's z-transformation of the correlation coefficient.

g. Calculate the approximate standard error of the z-transformed correlation.

h. What is Z_{crit}, the two-tailed critical value of the standard normal distribution corresponding to $\alpha = 0.05$?

i. Calculate the lower and upper bounds of the 95% confidence interval for ζ, the z-transformed population correlation.

j. Transform the lower and upper bounds of the confidence interval for ζ, yielding the 95% confidence interval for p.

2. **Calculation practice: Standard error and hypothesis testing for a correlation.** Refer to Practice Problem 1. Test whether there is a correlation in the population between giggle fundamental frequency and age.

a. State the null and alternative hypotheses.

b. Calculate the standard error of the sample correlation r.

c. Calculate the t-statistic.

d. What is the sample size? What are the degrees of freedom?

e. Obtain the critical value for t corresponding to $\alpha = 0.05$.

f. What is the conclusion from the test?

3. **Calculation practice: Spearman rank correlation.** As human populations became more urban from prehistory to the present, disease

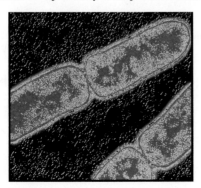

transmission between people likely increased. Over time, this might have led to the evolution of enhanced resistance to certain diseases in settled human populations. For example, a mutation in the *SLC11A1* gene in humans causes resistance to tuberculosis. Barnes et al. (2011) examined the frequency of the resistant allele in different towns and villages in Europe and Asia and compared it to how long humans had been settled in the site ("duration of settlement"). If settlement led to the evolution of greater resistance to tuberculosis, there should be a positive association between the frequency of the resistant allele and the duration of settlement. The data are below (durations based on dates BC have been rounded). The relationship appears curvilinear, so we will use the Spearman's correlation to test an association.

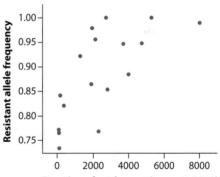

Settlement	Date	Duration (years)	Allele frequency
Çatal Höyük	6000 BC	8010	0.990
Susa, other	3250 BC	5260	1.000
Harappa	2725 BC	4735	0.948
Sanxingdui	2000 BC	4010	0.885
Knossos, other	1700 BC	3710	0.947
Carthage	800 BC	2810	0.854
Tarquinia, other	720 BC	2730	1.000
Angkor Borei	300 BC	2310	0.769
Tong'gorou	100 BC	2110	0.956
Colchester	55 AD	1955	0.979
Aksum	100 AD	1910	0.865
Nara	710 AD	1300	0.922
Yakutsk	1632 AD	378	0.821
Bathurst	1816 AD	194	0.842
Blantyre	1880 AD	130	0.734
Kiruna	1900 AD	110	0.766
Juba	1919 AD	91	0.772

a. Rank duration from low to high.
b. Rank allele frequency from low to high, assigning midranks to ties.
c. Calculate the sum of squares for the ranks of duration, allele frequency, and the sum of products.
d. Compute the Spearman's correlation coefficient.
e. What is the sample size?
f. State the null and alternative hypotheses.
g. Obtain the critical value for the Spearman's correlation corresponding to $\alpha = 0.05$.
h. Draw the appropriate conclusion.

4. Visually estimate the value of the correlation coefficient in each of the four following scatter plots.

(a)

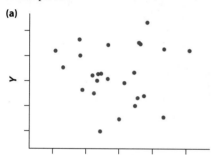

(b)

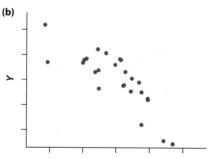

(c)

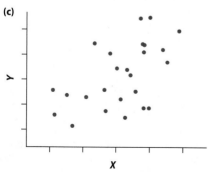

(d)

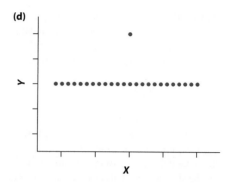

5. Birds of many species retain the same breeding partner year after year. In some of these species, male and female partners migrate separately and spend the winter in different places, often thousands of kilometers apart. Yet they manage to find one another again each spring. In a field study of individually banded pairs of black-tailed godwits, Gunnarsson et al. (2004) recorded spring arrival dates of males and females on the breeding grounds in the year after they were observed breeding together. The data for 10 pairs are provided in the accompanying table. Arrival date is measured as the number of days since March 31.

Female arrival date	Male arrival date
24	22
36	35
35	35
35	44
38	46
50	50
55	55
56	56
57	56
69	59

a. Display the relationship between arrival dates of males and females in a graph. What type of graph did you use?
b. Describe the pattern in part (a) briefly. Is there a relationship? Is it positive or negative? Is it linear or nonlinear? Is it weak or strong?

c. Calculate the correlation coefficient between arrival dates of male and female godwits. Include a standard error for your estimate.

d. What does the standard error in part (c) refer to?

e. Calculate an approximate 95% confidence interval for ρ.

6. Answer the following questions using the data for the godwits in Practice Problem 5.
   a. Adding 30 to each of the observations for males converts arrival dates to "days since March 1" rather than March 31. How is the correlation coefficient between arrival dates of males and females affected? What can you conclude about the effects on the correlation coefficient of adding a constant to one or both of the variables?
   b. Dividing female arrival dates by seven converts their arrival dates to "weeks since March 31" rather than days. How does this affect the correlation between male and female arrival dates? What can you conclude about the effects on the correlation coefficient of multiplying one or both of the variables by a constant?

7. Use the godwit data in Practice Problem 5 to test whether the mean arrival dates of male and female partners differ significantly. What assumptions are required?

8. When measuring a correlation between two variables, under what circumstances would it be best to make and average several repeat measurements of each subject for a given variable rather than measure the variable only once on each subject?

9. In large wolf populations, most inbreeding coefficients of litters are close to zero, and very few are as high or higher than 0.25. What effect is this narrower range of inbreeding coefficients expected to have on the correlation between the number of pups surviving and inbreeding coefficient, compared with that measured in the Scandinavian population of Example 16.2? Explain.

10. Large males of the European earwig, *Forficula auricularia*, develop abdominal forceps, which are used in fighting and courtship. Smaller males do not develop the forceps. Tomkins and Brown (2004) compared the proportion of males having forceps on islands in the North Sea with the population density of earwigs, measured as number caught per trap. Their data are listed in the following table.

Islands	Earwig density (number per trap)	Proportion of males with forceps
1	0.3	0.04
2	5.2	0.02
3	2.5	0.07
4	25.6	0.06
5	8.1	0.13
6	0.3	0.15
7	4.7	0.20
8	3.3	0.24
9	7.8	0.25
10	20.0	0.19
11	31.0	0.22
12	25.0	0.32
13	43.3	0.30
14	33.9	0.44
15	33.9	0.52
16	32.7	0.55
17	33.8	0.62
18	12.7	0.66
19	57.0	0.46
20	52.5	0.38
21	64.0	0.38
22	70.4	0.46

a. The distribution of the two variables is not bivariate normal, and transforming the data does not improve matters. Choosing the most appropriate method, test whether the two variables are correlated.

b. What are your assumptions in part (a)?

11. Earwig density on an island and the proportion of males with forceps are estimates, so the measurements of both variables include sampling error. In light of this fact, would the true correlation between the two variables tend to be larger, smaller, or the same as the measured correlation?

12. According to the immunocompetence handicap hypothesis, males of a species evolve high reproductive effort to the point that they divert resources away from immune function. To test this, Simmons and Roberts (2005) measured sperm viability and immune function in lab-raised male crickets to test whether male reproductive effort affects immune function. The data in the following graph show male sperm viability and lysozyme activity, an important defense against bacterial infection. Each point is the average of the males in a single family of crickets. Lysozyme activity is measured as the area of clear region around an inoculation of 2 µl of hemolymph onto an agar plate containing bacteria. The total sample size n is 41.

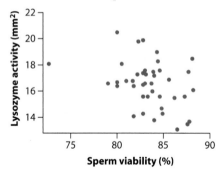

a. What assumption of linear correlation analysis is violated by these data? Explain.

b. Assuming that transforming the data doesn't help, what is the most appropriate method to test the null hypothesis of no correlation between male sperm viability and male lysozyme activity?

c. When we applied the most appropriate method to these data, we obtained the fol-

lowing numbers based on the ranks of sperm viability and lysozyme activity:

Sum of products: −1744.5
Sum of squares (sperm viability): 5726.0
Sum of squares (lysozyme activity): 5729.5

Using these figures, test the hypothesis that sperm viability and male lysozyme activity are correlated.

d. What assumptions have you made in part (c)?

13. Left-handed people have an advantage in many sports, and it has been suggested that left-handedness might have been advantageous in hand-to-hand fights in early societies. (Left-handed people can get a lot of practice against right-handed opponents, whereas right-handers are less experienced against lefties.) To explore this potential advantage, Faurie and Raymond (2005) compared the frequency of left-handed individuals in traditional societies with measures of the level of violence in those societies. The following table lists the data for one index of violence, the rate of homicide, measured in number / 1000 people / year.

Society	Percent left-handed	Homicide rate
Dioula	3.5	0.01
Ntumu	8.2	0.02
Kreyol	6.6	0.03
Inuit	6.4	0.17
Baka	10.2	0.50
Jimi	13.0	5.37
Eipo	20.4	3.02
Yanomamo	22.7	3.98

a. Use a graph to illustrate the association between the two variables.

b. What assumption of linear correlation analysis is violated by these data? Explain.

c. Before resorting to a nonparametric method, what strategy is available to test for a correlation between percent left-handedness and homicide rate?

d. Carry out this strategy, using a scatter plot to assess your success in meeting assumptions.

e. Using your results from part (d), test whether the two variables are correlated.

14. Does stress age you? As part of an investigation, Epel et al. (2004) measured telomere length in blood mononuclear cells of healthy premenopausal women, each of whom was the biological mother and caregiver of a chronically ill child. Telomeres are complexes of DNA and protein that cap chromosomal ends. They tend to shorten with cell divisions and with advancing age. A scatter plot of the data is shown below. Telomere length is measured as a ratio compared to a standard. Chronicity is the number of years since the child's diagnosis. The data in the scatter plot can be summarized as follows:

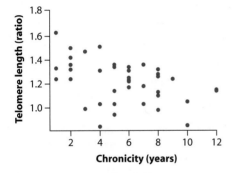

> **Sum of products:** −8.636
> **Sum of squares for chronicity:** 327.436
> **Sum of squares for telomere length:** 1.228
> **Total sample size (n):** 38

a. Describe the pattern in the scatter plot briefly in words. Is there a relationship? Is it positive or negative? Is it linear or nonlinear? Is it weak or strong?

b. Calculate the linear correlation between telomere length and the chronicity of caregiving.

c. Calculate a 95% confidence interval for the population correlation.

d. Provide an interpretation for the interval in part (c). What does the interval represent?

e. What are your assumptions in part (c)? Does the scatter plot support these assumptions? Explain.

ASSIGNMENT PROBLEMS

15. Does learning a second language change brain structure? Mechelli et al. (2004) tested 22 native Italian speakers who had learned English as a second language. Proficiencies in reading, writing, and speech were assessed using a number of tests whose results were summarized by a proficiency score. Gray-matter density was measured in the left inferior parietal region of the brain using a neuroimaging technique, as mm³ of gray matter per voxel. (A voxel is a picture element, or "pixel," in three dimensions.) The data are listed in the accompanying table.

a. Display the association between the two variables in a scatter plot.

b. Calculate the correlation between second language proficiency and gray-matter density.

c. Test the null hypothesis of zero correlation.

d. What are your assumptions in part (c)?

e. Does the scatter plot support these assumptions? Explain.

Proficiency score for second language	Gray-matter density (mm³/voxel)	Proficiency score for second language	Gray-matter density (mm³/voxel)
0.26	−0.070	2.75	−0.008
0.44	−0.080	3.25	−0.006
0.89	−0.008	3.85	0.022
1.26	−0.009	3.04	0.018
1.69	−0.023	2.55	0.023
1.97	−0.009	2.50	0.022
1.98	−0.036	3.11	0.036
2.24	−0.029	3.18	0.059
2.24	−0.008	3.52	0.062
2.58	−0.023	3.59	0.049
2.50	−0.006	3.40	0.033

DATA TABLE FOR PROBLEM 16

Site	Attachment	Area (ha)	Number of butterfly species	Number of bird species	ln (number of plant species)
A	4.4	23.8	6	12	5.1
B	4.5	16.0	14	18	5.5
C	4.7	6.9	8	8	6.4
D	4.5	2.3	10	17	4.7
E	4.3	5.7	6	7	5.3
F	3.8	1.2	5	4	4.6
G	4.4	1.4	5	8	4.5
H	4.6	15.0	7	22	5.5
I	4.1	3.1	9	7	5.2
J	4.2	3.8	5	4	4.6
K	4.6	7.6	10	11	4.5
L	4.2	12.9	9	11	5.0
M	4.3	4.0	12	13	5.0
N	4.4	5.6	11	16	5.6
O	4.2	4.9	7	7	5.4

f. Do the results demonstrate that second language proficiency affects gray-matter density in the brain? Why or why not?

16. In an increasingly urban world, are there psychological benefits to biodiversity? Fuller et al. (2007) measured the number of plant, bird, and butterfly species in 15 urban green spaces of varying size in Sheffield, England, a city of more than a half-million people. They also interviewed 312 green-space users and asked a series of questions related to the degree of psychological well-being obtained from green-space use. From the answers, the researchers obtained a measure of user "attachment" to green spaces (strength of emotional ties). Their results are in the table at the top of this page.
   a. Which of the three measures of green-space biodiversity (number of butterfly species, number of bird species, and ln number of plant species) is most strongly correlated with the "attachment" variable? Provide a standard error with each of your correlations.
   b. Provide an approximate 95% confidence interval for each of your correlations in part (a).

17. Use the data in Assignment Problem 16 to calculate a 95% confidence interval for the correlation between attachment and green-space area.

18. Pacific salmon return from the ocean to streams to spawn and die, bringing with them a lot of nutrients from one ecosystem to another. Bears kill and bring onto land up to half of the salmon present in the river, where the fish remains decompose, fetilizing the forest at the stream edge. Hocking and Reynolds (2011) measured the association between the density of salmon in streams (kg/m) and the abundance of the aptly named salmonberry, *Rubus spectabilis* (measured as the proportion of herb cover that is salmonberry). The graph on the next page shows the relationship between the square-root-transformed salmon density ($Y' = \sqrt{Y + 1/2}$) and the arcsine square root of salmon abundance. The data are available at http://whitlockschluter.zoology.ubc.ca.

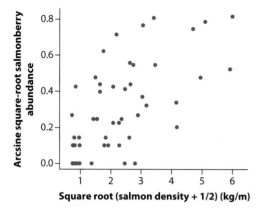

Square root (salmon density + 1/2) (kg/m)

Here are some intermediate computations for square root of salmon density (X) and arcsine square root of salmonberry density (Y):

$$\sum_{i=1}^{n} X_i = 113.86 \qquad \sum_{i=1}^{n} Y_i = 15.14$$

$$n = 50 \qquad \sum_{i=1}^{n} X_i^2 = 353.82$$

$$\sum_{i=1}^{n} Y_i^2 = 7.65 \qquad \sum_{i=1}^{n} (X_i Y_i) = 45.42$$

a. Why is it a good idea to transform these data?

b. What other transformations might have been attempted instead? Try one of them (using the data on whitlockschluter.zoology.ubc.ca) and see if the results is as effective. Describe your actions.

c. Before calculating the correlation coefficient, predict from the figure whether the correlation coefficient will be positive or negative.

d. Calculate the correlation coefficient between salmon density and salmonberry density, transformed as in the figure. Include the standard error for the coefficient.

e. Test the null hypothesis that this correlation is zero.

f. In what way would these data support a hypothesis that nutrients from salmon are good for salmonberries?

19. The following data are from a laboratory experiment by Smallwood et al. (1998) in which liver preparations from five rats were used to measure the relationship between the administered concentration of taurocholate (a salt normally occurring in liver bile) and the unbound fraction of taurocholate in the liver.

Rat	Concentration (μM)	Unbound fraction
1	3	0.63
2	6	0.44
3	12	0.31
4	24	0.19
5	48	0.13

a. Calculate the correlation coefficient between the taurocholate unbound fraction and the concentration.

b. Plot the relationship between the two variables in a graph.

c. Examine the plot in part (b). The relationship appears to be maximally strong, yet the correlation coefficient you calculated in part (a) is not near the maximum possible value. Why not?

d. What steps would you take with these data to meet the assumptions of correlation analysis?

20. If you are having trouble solving homework problems, should you sleep on it and try again in the morning? Huber et al. (2004) asked 10 participants to perform a complex spatial learning task on the computer just before going to sleep. EEG recordings were then taken of the electrical activity of brain cells during their sleep. The magnitude of the increase in their "slow-wave" sleep after learning the complex task, compared to baseline amounts, is listed in the following table for all 10 participants. Also provided is the increase in performance recorded when the participants were challenged with the same task upon waking.

Increase in slow-wave sleep (%)	Improvement in task performance (%)
8	8
14	3
13	0
15	0
17	8
18	15
31	14
32	10
44	27
54	26

a. Calculate the correlation coefficient between the magnitude of the increase in slow-wave sleep and the magnitude of the improvement in performance upon waking.

b. What is the standard error for your estimate in part (a)?

c. Provide an interpretation of the quantity you calculated in part (b). What does it measure?

d. Test the hypothesis that the two variables are correlated in the population.

e. Is this an observational or an experimental study? Explain.

21. Both of the variables in Assignment Problem 20 are measurements that include some measurement error.

a. How would this measurement error affect the correlation between the two variables?

b. What steps could be taken in the design of the study to minimize the effect of measurement error?

c. For a given variable, what quantity is used to estimate the proportion of the variance among subjects not attributable to measurement error?

22. Logging in western North America impacts populations of western trillium, a long-lived perennial that inhabits conifer forests (*Trillium ovatum*; see the photo at the beginning of the chapter). Jules and Rathcke (1999) measured attributes of eight local populations of western trillium, confined to forest patches of varying size created by logging in southwestern Oregon. Their data, presented in the following table, compare estimates of recruitment (the density of new plants produced in each population per year) at each site with the distance from the site to the edge of the forest fragment.

Local population	Distance to clear-cut edge (m)	Recruitment
1	67	0.0053
2	65	0.0021
3	61	0.0069
4	30	0.0006
5	84	0.0124
6	97	0.0045
7	16	0.0028
8	332	0.0182

a. Display these data in an appropriate graph. Examine the graph and describe the shape of the distribution. What departures from the assumption of correlation analysis do you detect?

b. Choose a transformation and transform one or both of the two variables. Plot the results. Did the transformation solve the problem? If not, try a different transformation.

c. Using the transformed data, estimate the correlation coefficient between the two variables. Provide a standard error with your estimate.

d. Calculate an approximate 95% confidence interval for the correlation coefficient.

23. Cocaine is thought to affect the brain by blocking the dopamine transporter, increasing the amount of dopamine in the nerve synapse. To investigate this idea, Volkow et al. (1997) administered intravenous doses of 0.3 to 0.6 mg/kg of cocaine to volunteers. They used PET scans to compare the magnitude of the perceived "high" of regular cocaine users with the percentage of dopamine receptors blocked. The results for 34 subjects are illustrated below. Full data are available at http://whitlockschluter. zoology.ubc.ca.

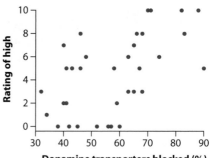

a. Using the following quantities, calculated from these data, estimate the correlation between the percentage of dopamine receptors blocked and subjects' ratings of the cocaine high. Provide a standard error with your estimate.

Sum of products: 957.5

Sum of squares (receptors blocked): 8145.441

Sum of squares (rating of high): 372.5

b. Calculate a 99% confidence interval for the correlation in the population.

c. What are your assumptions in part (b)?

d. Imagine the following scenario: A second team of researchers carried out a similar study using the same population and sample size. They used a narrower range of intravenous doses of cocaine in their experiment, which led to a smaller range of values than in the first study for the percentage of dopamine receptors blocked. When they analyzed their results, they found only a low correlation between percentage dopamine receptors blocked and perceived high. In their published report, they concluded that the true correlation between these variables is much lower than estimated in the Volkow et al. study. Who is right? Explain.

24. In a study of the relationship between personality and humor appreciation, Mobbs et al. (2005) measured two dimensions of personality, neuroticism and extroversion, in 17 healthy volunteers. Scores along the personality dimensions were based on a 60-item, self-report questionnaire. What is the association between these two personality dimensions?

Subject	Extroversion	Neuroticism
1	43	49
2	46	53
3	48	67
4	48	57
5	48	56
6	50	48
7	51	60
8	51	41
9	53	51
10	58	47
11	62	41
12	63	51
13	63	30
14	63	28
15	67	55
16	67	47
17	67	39

a. Plot the data in a graph.

b. What is the correlation coefficient in the data, and what is most-plausible range of values for the population correlation? Use the 95% confidence interval.

25. Refer to Problem 24. Carry out a formal test of null hypothesis of zero correlation in the population between the two behavioral dimensions.

26. There is evidence that higher consumption of foods containing chemicals called flavonols—including cocoa, red wine, green tea, and some fruits—increases brain function in several ways. Messerli (2012) asked whether chocolate consumption in a country is correlated with the number of Nobel Prizes for the country over all time. The data are below. Both chocolate consumption and number of Nobel Prizes are scaled to the number of people in each country.

Country	Chocolate consumption (kg/person/year)	Nobel Prizes (per 100 million)
Australia	4.5	5.5
Austria	8.5	24.4
Belgium	4.4	8.6
Brazil	2.9	0
Canada	3.9	6
China	0.7	0
Denmark	8.5	25.3
Finland	7.3	7.6
France	6.3	9
Germany	11.6	12.7
Greece	2.5	1.9
Ireland	8.8	12.8
Italy	3.7	3.2
Japan	1.8	1.4
Netherlands	4.5	11.5
Norway	9.4	23.4
Poland	3.5	3.1
Portugal	1.9	2.2
Spain	3.6	1.7
Sweden	6.4	31.9
Switzerland	11.9	32.8
United Kingdom	9.7	18.8
United States	5.3	10.6

a. Plot and examine these data. What challenges do you anticipate if your goal is to test whether chocolate consumption and number of Nobel Prizes are correlated? Describe any issue you identify.

b. Without transforming the data, test for an association between the two variables using an appropriate method.

c. Interpret the findings of the study appropriately. Does chocolate consumption increase the probability of winning a Nobel Prize? Should it be recommended as a national priority?

27. Biopsy is often used to distinguish cancerous from harmless tumors before resorting to surgery. Ridgway et al. (2004) investigated the ability of MIB-1 monoclonal antibodies, which detect rapidly proliferating cells with staining, to distinguish known breast tumor types from biopsies on a postoperative sample. The following measurements were taken to determine whether the MIB-1 index measured on biopsy is associated with whole tumor size. MIB-1 index was measured double-blind on histological sections of tumor tissue by the number of stained cells counted at a particular microscope magnification.

   a. Examine the association in a graph. What is the trend? Do the data look bivariate normal?

   b. Using an appropriate method, and without transforming the data, test whether there is an association between MIB-1 index and tumor size.

Tumor size (mm)	MIB-1 index
10	1
13	39
15	7
20	154
20	141
20	26
21	41
23	1
25	7
25	24
26	67
30	1
30	27
35	1
35	19
35	42
40	37
45	2
47	1
70	23
130	93
130	32

Publication bias

When we read an article in a journal we respect, we tend to believe what we read. The paper has been carefully vetted by expert referees, and the decision to publish it was made by an editor who is likely among the best scientists in the field. We might look at the authors of the papers and at the methods section, and we often find that both are above reproach. Very often the authors will have used statistical analysis, and usually it seems to be done correctly. We see that P is less than 0.05, or even less than 0.01, and we interpret that to mean that there was a low probability of getting results as extreme as those observed if the null hypothesis were true. When we've finished reading, we've just learned something. Haven't we?

It turns out that the papers that are actually published, especially those published in the "better quality," more widely read journals, are not a random sample of all studies done. This is true in at least two ways. First, the papers that get published are, on average, reporting on science that is done better than those submitted to journals but rejected by the editors. This is as it should be, for far too much science is done poorly and journal space is limited.

Second, papers that are more "interesting" get published more often than boring papers. Again, there is nothing intrinsically wrong

The odds of publishing a study whose main outcome was a *P*-value less than 0.05 are about 2.5 times higher than those of studies obtaining *P* > 0.10.

with this, but it can raise a serious problem: papers that do *not* reject the null hypothesis of no effect are usually thought, by most editors and even most authors, to be less interesting than those that do reject the null hypothesis. Thus, papers that reject the null hypothesis are more likely to be published than those that do not. Moreover, papers that describe large effects of an experiment are more interesting than those showing smaller effects, so published papers tend to show larger effects than unpublished studies.

As a result, the science that gets published is a biased selection of all science done. The difference between the true effect and the average effect published in journals is called **publication bias**. Publication bias can seriously skew our perception of nature. It is difficult to quantify, though, because we don't know much about the research that isn't published, such as how much there is and what results were obtained. After all, publication is how we normally find out about research.

One way to detect publication bias takes advantage of the fact that scientists must file for permits from an ethics review board to do research on humans in research hospitals. Records of these reviews allow people studying publication bias to know how many studies were carried out, so that they can follow up on the results and publication outcome of each study. Such reviews consistently find that the odds of publishing a study whose main outcome obtained a P-value less than 0.05 are about 2.5 times higher than those of studies obtaining $P > 0.10$ (Easterbrook et al. 1991; Dickersin et al. 1992).[1] Moreover, researchers are slower to publish nonsignificant results ($P > 0.05$) when they do publish them (Stern and Simes 1997). Studies with $P < 0.05$ are more likely to get into more widely read journals (Ioannidis et al. 1997), and subsequently, they are more likely to be cited by other papers (Gøtzsche 1987). These findings are disturbing—the papers that we read are not necessarily representative of the truth.

Another, perhaps more troubling, source of evidence about the possibility of publication bias comes from statistical analyses of drug trials according to their funding source. Most analyses of this sort have found that studies funded by drug-manufacturing companies are about 3.5 times more likely to yield a result favorable to the company than are publicly funded studies (Melander et al. 2003; Leopold et al. 2003; Bekelman et al. 2003). The implication is that

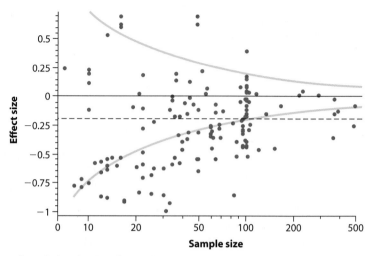

A funnel plot showing the results of 140 studies that measured the association between left–right asymmetry and male mating success. The effect size is the correlation coefficient between asymmetry and mating success. Adapted from Palmer (1999).

research is unlikely to be published if it reflects poorly on the interests of the company funding it.

Small studies finding minor or nonsignificant effects are more likely to be left unpublished than are large studies with similarly weak results. Perhaps scientists who have done a large study are more determined to publish whatever the result, to get some payback for all their work. On the other hand, a researcher who has carried out a small study and gets inconclusive results is quite likely to assume that the study had insufficient power and leave it unpublished. If publication bias is present, therefore, then there should be a relationship between publication, the size of the effect, and the sample size, as depicted in a **funnel plot**. A funnel plot is a scatter plot of the magnitude of the effect detected in published studies and their corresponding sample sizes.

The figure in this interleaf is a funnel plot showing the results of 140 published studies, each of which examined the relationship

---

1. These studies also show that the main source of the bias is the authors themselves, not the journal editors. People are less likely to go to all the trouble of writing a paper if they are not as interested in the results, or if they fear that it will not be published anyway.

between the mating success of individual males in a study species and the degree of left–right asymmetry in a male trait (Palmer 1999). In studies of the mating success in flies, for example, asymmetry might measure the absolute value of the difference between the lengths of the left and right wings. In human studies, asymmetry might measure the difference in proportions of the two sides of the face. The 140 studies devoted solely to one male feature might seem excessive, but the causes of romantic success in nature are of great interest to biologists. Most of the 140 studies followed on the heels of claims that the symmetry of traits may be even more important than the traits themselves in explaining why, in nature and in human societies, some males get more than their fair share of mates and others get less. But does asymmetry really matter?

Each point in the figure is from a different study. The x-axis gives the sample size of each study, whereas the y-axis gives the "effect size," which in this case is the estimated correlation coefficient between asymmetry and mating success. A negative effect size means that males with greater asymmetry had lower mating success than males with less asymmetry. The plot combines studies of many types of animals carried out by many researchers in many jungles, shopping malls, and laboratories.

The solid horizontal line in the funnel plot represents the null hypothesis tested in every study that the true correlation coefficient between male asymmetry and mating success is zero. The gold curves mark the critical values for tests of the null hypothesis. Points falling outside these bounds are statistically significant at $\alpha = 0.05$. The dashed horizontal line marks the average of the observed effect sizes of all 140 published studies.

This funnel plot is highly revealing. For example, note that the range of published correlation coefficients is broad when the sample size is small and narrow when the sample size is large. This is expected, though, because larger sample sizes should yield more precise estimates (see Chapter 4). This expectation gives the funnel plot its name.

Other features of the funnel plot are unexpected and are cause for concern. In the first place, very few small studies yielded estimates close to the average effect size. Instead, most small studies found very large effects, in contrast to the larger studies, which tended to find smaller effects. Even more disturbing, many results for the smaller studies are statistically significant, clumping outside the lower critical value for significance (indicated in the figure by the lower gold curve). Again, this finding contrasts with the largest studies, most of which found no statistically significant effect. What is behind these unexpected patterns?

The probable answer is that most small studies—those finding weak and nonsignificant effects—were never published. As a result, the published papers are not representative of all studies done. There must be many studies with weaker, nonsignificant effects still sitting in file drawers and on hard drives in universities around the world, never to see the light of day. Another implication is that the average effect size of all of the published studies is overestimated, so reading just the published papers gives a biased view of the strength of the relationship between asymmetry and male mating success. Symmetry apparently gives the hopeful male at most a slight edge in romance.

As mentioned previously, the more interesting a result is, the more likely it is to be published in a high-profile journal. One of the things that makes a result interesting is an effect that is stronger than previously believed. This can occur because our previous assumptions were wrong, but it can also occur because the effect was overestimated. As a consequence, dramatic claims in published papers often turn out later to be exaggerated or even false. This is not always or even usually the result of bad science, but rather is due to publication bias.

The lesson to take home about publication bias is that flashy new results of published studies should always be repeated, preferably by different researchers working with larger sample sizes and bent upon publishing no matter what the outcome. In this way, the excesses of publication bias can be detected and corrected, yielding more accurate views of the patterns in nature.

Red campion, *Silene dioica*

17 Regression

Regression is the method used to predict values of one numerical variable from values of another. For example, the scatter plot on the following page shows how genetic diversity in a local contemporary human population is predicted by its dispersal distance from East Africa by fitting a straight line to the data points.[1] Modern humans emerged from Africa around 60,000 years ago, and our ancestors lost some genetic variation at each step as they spread to new lands.

The line fitted to the data is the regression line. The line can be used to *predict* the genetic diversity of a local human population (the response variable), even for a locale not included in this study, based on its dispersal distance from East Africa (the explanatory variable). The slope or steepness of the regression line indicates the *rate of change* of genetic diversity with distance. It shows that humans lose 0.076 units of

---

1. Modified from Prugnolle et al. (2005).

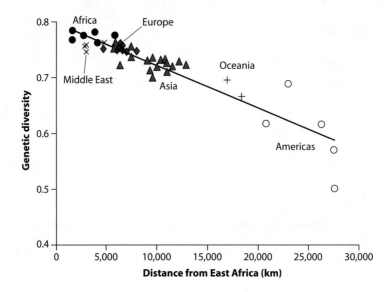

genetic diversity, about 10% of the maximum, with every 10,000-km distance from East Africa. Both features of the relationship are captured in the equation for the line.

In this chapter, we show how to estimate the regression line, how to put bounds on its predictions, and how to test hypotheses about the slope. Our focus is linear regression, but we also introduce some general principles of nonlinear regression.

> *Regression* is a method that predicts values of one numerical variable from values of another numerical variable.

Like correlation (Chapter 16), linear regression measures aspects of the linear relationship between two numerical variables. However, there are important differences. Regression fits a line through the data to predict one variable from another and to measure how steeply one variable changes with changes in the other. Correlation does none of these things. It measures strength of association between two variables, reflecting the amount of scatter in the data.

Regression is used on data from either of two study designs. In the first, individuals are randomly sampled from a population. Two variables are measured on the sample, and one of them (deemed the explanatory variable) is used to predict the other (response) variable. In the second design, the researcher fixes or chooses values of the explanatory variable, which represent treatments or doses. The response variable is then measured on one or more individuals assigned to each treatment. The calculations are the same in both cases. Examples 17.1 and 17.3 in this chapter illustrate these two designs.

17.1 Linear regression

The most common type of regression is **linear regression**, which draws a straight line through the data to predict the response variable (Y, shown on the vertical axis) from the explanatory variable (X, shown on the horizontal axis). One important assumption of the linear regression method is that the relationship between the two variables really is linear. Example 17.1 shows how to use linear regression to predict the value of a response variable.

EXAMPLE The lion's nose

17.1 Managing the trophy hunting of African lions is an important part of maintaining viable lion populations. Knowing the ages of the male lions helps, because removing males older than six years has little impact on lion social structure, whereas taking younger males is more disruptive. Whitman et al. (2004) showed that the amount of black pigmentation on the nose of male lions increases as they get older and so might be used to estimate the age of unknown lions. The relationship between age and the proportion of black pigmentation on the noses of 32 male

lions of known age in Tanzania is shown in the scatter plot in Figure 17.1-1. The raw data are listed in Table 17.1-1. We can use these data to predict a lion's age from the proportion of black on his nose.

FIGURE 17.1-1
Scatter plot of the known ages of 32 male lions (Y, vertical axis) and the proportion of black on their noses (X, horizontal axis).

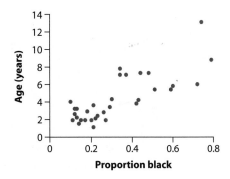

The scatter plot in Figure 17.1-1 puts age as the response variable (the vertical axis) and the proportion of black on the nose as the explanatory variable (the horizontal axis), rather than the reverse, because we want to predict age from the proportion of black, not the other way around.

TABLE 17.1-1 The proportion of black on the noses of 32 male lions of known age.

Proportion black	Age (years)	Proportion black	Age (years)
0.21	1.1	0.30	4.3
0.14	1.5	0.42	3.8
0.11	1.9	0.43	4.2
0.13	2.2	0.59	5.4
0.12	2.6	0.60	5.8
0.13	3.2	0.72	6.0
0.12	3.2	0.29	3.4
0.18	2.9	0.10	4.0
0.23	2.4	0.48	7.3
0.22	2.1	0.44	7.3
0.20	1.9	0.34	7.8
0.17	1.9	0.37	7.1
0.15	1.9	0.34	7.1
0.27	1.9	0.74	13.1
0.26	2.8	0.79	8.8
0.21	3.6	0.51	5.4

The method of least squares

Many straight lines can be drawn through a scatter of points, so how do we find the "best" one? Ideally, we would find a line that leads to the most accurate predictions of Y from X. This is the line that has the smallest possible deviations in Y (the vertical axis) between the data points and the regression line (Figure 17.1-2).

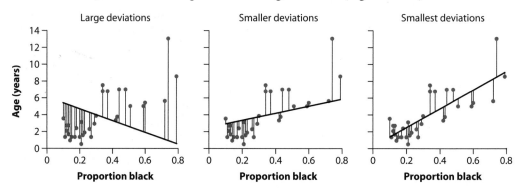

FIGURE 17.1-2 Illustration of the deviations between the data and several possible regression lines (the heavy black lines) drawn through the data points originally plotted in Figure 17.1-1. Vertical lines are the deviations in Y between each point and the regression line. The line in the right panel is the least-squares regression line.

The **least-squares regression** line is the line for which the sum of all the *squared* deviations in *Y* is smallest. We square the deviations from the regression line for the same reason that we square deviations from the mean when calculating an ordinary variance—to overcome the fact that some deviations are positive (the points above the regression line) and others are negative (the points below the regression line), which would cancel each other out in a simple average.

Formula for the line

The regression line through a scatter of points is described mathematically by the following equation:

$$Y = a + bX.$$

The symbol *Y* is the response variable (displayed on the vertical axis in a scatter plot), and *X* is the explanatory variable (the horizontal axis). The formula has two coefficients, *a* and *b*. The coefficient *a* is the *Y*-intercept, or just the **intercept**. Mathematically, *a* is the value of *Y* when *X* is zero (hence, it is the *Y*-value where the regression line "intercepts" the *y*-axis). Its units are the same as the units of *Y*. The right panel in Figure 17.1-3 shows two regression lines that have different intercepts.

The coefficient *b* is the **slope** of the regression line. It measures how much *Y* changes per unit change in *X*. Its units are the ratio of the units of *Y* and *X*. If *b* is positive, then larger values of *X* predict larger values of *Y*. If *b* is negative, then larger values of *X* predict smaller values of *Y*. The first three panels of Figure 17.1-3 show the slope of the line when *b* is positive, negative, and equal to zero.

> The *slope* of a linear regression is the rate of change in *Y* per unit of *X*.

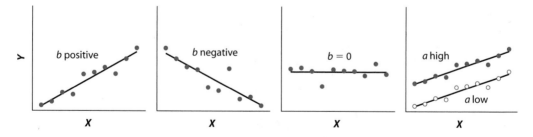

FIGURE 17.1-3 Comparing the slope of a line when *b* is positive (*far left*), negative (*left*), and zero (*right*); comparing a line with a high intercept *a* and one with a low intercept *a* (*far right*).

Calculating the slope and intercept

Typically, you would use a computer to calculate the regression line, but we provide the formulas here for use with a calculator. The slope of the least-squares regression line is computed as[2]

$$b = \frac{\sum_i (X_i - \overline{X})(Y_i - \overline{Y})}{\sum_i (X_i - \overline{X})^2},$$

where $\overline{X}$ and $\overline{Y}$ are the sample means of the two variables, and X_i and Y_i refer to the X and Y measurements of individual i. The top of this formula is the sum of products, something first encountered in Section 16.1. The bottom is the sum of squares for X. Shortcut formulas for these sums are given in the Quick Formula Summary (Section 17.10).

Once we have the slope b, getting the intercept is relatively straightforward, because the least-squares regression line always goes through the point $(\overline{X}, \overline{Y})$. As a result, we know that

$$\overline{Y} = a + b\overline{X}.$$

So, we find a by simple algebra:

$$a = \overline{Y} - b\overline{X}.$$

We can now use these formulas to calculate the coefficients of the least-squares regression line for the lion data in Example 17.1. First, though, we need the following quantities calculated from the data in Table 17.1-1:

$$\overline{X} = 0.3222 \qquad\qquad \overline{Y} = 4.3094$$

$$\sum_i (X_i - \overline{X})^2 = 1.2221 \qquad\qquad \sum_i (Y_i - \overline{Y})^2 = 222.0872$$

$$\sum_i (X_i - \overline{X})(Y_i - \overline{Y}) = 13.0123.$$

The slope is then

$$b = \frac{\sum_i (X_i - \overline{X})(Y_i - \overline{Y})}{\sum_i (X_i - \overline{X})^2} = \frac{13.0123}{1.2221} = 10.647.$$

The slope b measures the change in age of male lions per unit increase in the proportion of black on the nose. Its units are years per unit proportion black.

---

2   Another way to write the formula for the slope is $b = \dfrac{\text{Covariance}(X,Y)}{s_X^2}$. The term in the numerator is the covariance between X and Y and the term in the denominator is the variance in X. The formula for the covariance can be found in the Quick Formula Summary of the correlation chapter (Section 16.8).

The intercept, in years, is

$$a = \overline{Y} - b\overline{X} = 4.3094 - 10.647(0.3222) = 0.879.$$

The formula for the line that predicts age from the proportion of black pigmentation on the nose in these lions can be written by putting all of this together, with appropriate rounding:

$$Y = 0.88 + 10.65X.$$

This equation could also be written as

$$Age = 0.88 + 10.65(\text{proportion black}).$$

Figure 17.1-4 shows what this line looks like when it is plotted on the scatter plot[3] shown originally in Figure 17.1-1. The slope of the line indicates that on average, lion age increases by 10.65 years per unit change in the proportion of the nose that is black. We can say, equivalently, that age goes up by 1.065 years for each 0.1 increase of black on the nose.

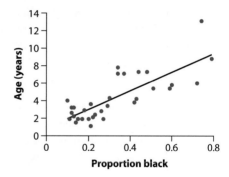

FIGURE 17.1-4
The regression line for the lion data from Example 17.1.

Populations and samples

The regression line is not just calculated for the sake of the data. It is typically used to estimate the true regression of Y on X in the population from which the data are a sample. The regression equation for the population is

$$Y = \alpha + \beta X,$$

where β is the slope in the population, and α is the intercept. The quantities α and β are population parameters, whereas a and b are their sample estimates.

Under one sampling scenario for linear regression, we have a random sample of (X, Y) pairs of measurements from a population. The lion data correspond to this scenario. Or, under a second scenario, the researcher fixes or chooses values of X to

---

3. To draw the line by hand, plot two points given by the regression equation and use a ruler to connect them. Two convenient points are the intercept (the predicted Y at $X = 0$) and the mean, because the regression line always passes through the point $(\overline{X}, \overline{Y})$. Don't extend the actual line to pass through the Y-intercept at $X = 0$ if this point lies beyond the range of X in the data (as in Figure 17.1-4).

include in the study, and Y is then measured on a sample of one or more individuals for each X-value included in the study. In either case, regression assumes that there is a population of *possible* Y-values for every possible value of X. The mean Y-value for each value of X lies on the true regression line.

For example, one of the lions in the data for Example 17.1 has a value of 0.6 for the proportion of black on its nose. We assume that there is a population of lions having the value $X = 0.6$ for the proportion of black on their noses, even though the data includes just one lion with the value. The mean age of all lions in the population having $X = 0.6$ is assumed to lie on the true regression line.

Predicted values

Now that we have the regression line, we can use it to determine points on the line that correspond to specified values of X. These points on the regression line are called **predictions**. We will symbolize predictions as $\hat{Y}$ ("Y-hat") to distinguish them from values of Y (i.e., actual data points), which lie above or below the line but not usually on it. The predicted value of Y for a given value of X estimates the mean of Y for the whole population of individuals having that value of X. For example, to predict the age of a male lion corresponding to a proportion of 0.50 black on the nose, plug the value $X = 0.50$ into the regression formula:

$$\hat{Y} = a + b(0.50) = 0.88 + 10.65(0.50) = 6.2.$$

In other words, the regression line predicts that lions with a proportion of black $X = 0.50$ will be 6.2 years old on average. If we observed a lion with 0.5 proportion black on its nose, we could predict its age, even though we had never seen a lion exactly like that before.

According to Table 17.1-1, the value $X = 0.50$ was not represented in the sample, although it falls within the range of observed X-values (i.e., 0.10 to 0.79). For reasons that are explained in Section 17.2, we can reliably make predictions only by using values of X that lie within the range of values in the sample.

> The predicted value of Y from a regression line estimates the mean value of Y for all individuals having a given value of X.

Residuals

Residuals measure the scatter of points above and below the least-squares regression line. They are crucial for evaluating the fit of the line to the data. Each observation in the data has a corresponding residual, measuring the vertical deviation from the least-squares regression line (see the right panel in Figure 17.1-2). The point on the regression line used to calculate the residual for individual i is $\hat{Y}_i$, the value predicted when its corresponding value for X_i is plugged into the regression formula:

$$\hat{Y}_i = a + bX_i.$$

For example, the 31st lion in the sample ($i = 31$) has a proportion $X_{31} = 0.79$ of black on its nose (see Table 17.1-1). The corresponding age $\hat{Y}_{31}$ predicted for a lion with this much black on the nose is

$$\hat{Y}_{31} = 0.88 + 10.65(0.79) = 9.29.$$

The actual age of the lion was 8.8 years, which is below the predicted value. The residual is the observed value minus the predicted value:

$$\text{residual}_{31} = (Y_{31} - \hat{Y}_{31}) = (8.8 - 9.3) = -0.49 \text{ years.}$$

The variance of the residuals, symbolized as $\text{MS}_{\text{residual}}$, quantifies the spread of the scatter of points above and below the line. In regression jargon, this variance is called the "residual mean square":

$$\text{MS}_{\text{residual}} = \frac{\sum_i (Y_i - \hat{Y}_i)^2}{n - 2}.$$

The $\text{MS}_{\text{residual}}$ is like an ordinary variance, but it has $n - 2$ degrees of freedom[4] rather than $n - 1$. It is analogous to the error mean square in the analysis of variance (Section 15.1). The following alternate formula is easier to use, though, because you don't need to calculate each $\hat{Y}_i$:

$$\text{MS}_{\text{residual}} = \frac{\sum_i (Y_i - \overline{Y})^2 - b \sum_i (X_i - \overline{X})(Y_i - \overline{Y})}{n - 2}.$$

Shortcuts for calculating the sum of squares and products are provided in the Quick Formula Summary (Section 17.10).

All of the quantities needed to determine $\text{MS}_{\text{residual}}$ for the lion data have been calculated previously on page 544. Inserting these values into the equation for $\text{MS}_{\text{residual}}$ yields

$$\text{MS}_{\text{residual}} = \frac{222.0872 - 10.647(13.0123)}{32 - 2} = 2.785.$$

Standard error of slope

Like any other estimate, there is uncertainty associated with the sample estimate b of the population slope β. Uncertainty is measured by the standard error, the standard deviation of the sampling distribution of b. The smaller the standard error, the higher the precision and the lower the uncertainty of the estimate of the slope. If the assumptions of linear regression are met (Section 17.5), then the sampling distribution of b

---

4. The degrees of freedom are $n - 2$ rather than $n - 1$, because we couldn't calculate the predicted values $\hat{Y}_i$ without first calculating *two* other quantities using the data—namely, the slope and the intercept of the regression line.

is a normal distribution having a mean equal to β and a standard error estimated from data as

$$SE_b = \sqrt{\frac{MS_{residual}}{\sum_i (X_i - \bar{X})^2}}.$$

The quantity on top of the fraction under the square-root sign is the residual mean square, and the quantity on the bottom is the sum of squares for X.

The standard error of b for the lion data is

$$SE_b = \sqrt{\frac{MS_{residual}}{\sum_i (X_i - \bar{X})^2}} = \sqrt{\frac{2.785}{1.2221}} = 1.510.$$

The standard error of the slope has the same units as the slope itself (i.e., years per unit of proportion black for the lion data in Example 17.1).

Confidence interval for the slope

A confidence interval for the parameter β is given by

$$b - t_{\alpha(2),df}SE_b < \beta < b + t_{\alpha(2),df}SE_b,$$

where $t_{\alpha(2),df}$ is the two-tailed critical value of the t-distribution having $df = n - 2$ degrees of freedom. For a 95% confidence interval, $\alpha = 0.05$, and for a 99% confidence interval, $\alpha = 0.01$. For the lion data, $t_{0.05(2),30} = 2.042$ (Statistical Table C), so the 95% confidence interval for the slope is

$$10.647 - 2.042(1.510) < \beta < 10.647 + 2.042(1.510)$$
$$7.56 < \beta < 13.73.$$

This is a modest range of most-plausible values for the slope. The mean age of lions increases by as little as 7.6 years per unit proportion of black on the nose, or by as much as 13.7 years.

17.2 Confidence in predictions

The regression line calculated from data predicts the mean value of Y for any specified value of X lying between the smallest and largest X in the data. This line is calculated with error, however, which affects how precise the predictions are. Here in Section 17.2 we quantify the precision of predictions. We also discuss the hazards of *extrapolating*—making predictions when the values of X lie beyond the range of X-values in the data.

Confidence intervals for predictions

Two subtly different types of predictions can be made using the regression line. The first predicts the *mean* Y for a given X. What, for example, is the mean age of all male lions in the population whose noses are 60% black (i.e., X = 0.60)? The second type predicts a *single* Y for a given X. (For example, how old is that lion over there, given that 60% of its nose is black?) Usually we just want to predict the mean Y for each X (i.e., the first prediction) because we are interested in the overall trend. In special situations, though, we also want to predict an individual Y-value (i.e., the second prediction). This is especially true in the lion study (Example 17.1). A hunter who encounters a male lion would want to know the age of that specific lion if he or she wishes to avoid shooting a young male.

Both types of predictions generate the same value for $\hat{Y}$. They differ in the precision of the predictions. In the case of lions with 60% black on their noses,

$$\hat{Y} = a + bX = 0.88 + 10.65(0.60) = 7.27 \text{ years.}$$

Regardless of the prediction goals, this is the best prediction of age. The precision of the prediction is lower, however, if the goal is to predict the age of an individual lion rather than the mean age of lions having the specified proportion of black on their noses. This is because the prediction for a single Y-value includes uncertainty stemming from variation in Y among the individuals in the population having the same value of X (i.e., not all male lions having 60% black noses are the same age). The two graphs in Figure 17.2-1 illustrate these differences in precision of predictions using confidence intervals.

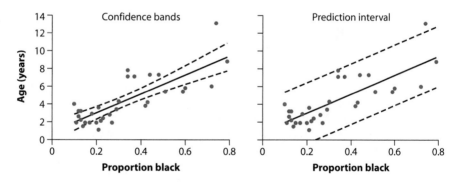

FIGURE 17.2-1 *Left:* 95% confidence bands for the predicted mean age of male lions at every value of proportion of black on their noses. *Right:* 95% prediction intervals for the predicted age of single lions. $n = 32$.

The left panel of Figure 17.2-1 shows the 95% confidence intervals for the predicted mean lion age at every X. The upper curve connects the upper bounds of all of the 95% confidence intervals for the predicted mean Y-values, one for every X between the smallest and largest X in the data. The lower curve connects the lower bounds of these same confidence intervals. Together the upper and lower curves showing the confidence intervals for the mean Y are called the 95% **confidence bands**. These bands are narrowest in the vicinity of $\overline{X}$, the mean value for proportion of black on the nose, and they flare outward toward the extremes of the range of data. The uncertainty of predictions always increases the farther the X-value is from the mean X in the data. In 95% of samples, the confidence bands will bracket the true regression line in the population.

The right panel of Figure 17.2-1 shows the 95% **prediction intervals**. The upper and lower curves connect the upper and lower limits of the 95% prediction intervals for a *single Y* over the range of X-values in the data. These are much wider than the confidence bands because predicting an individual lion's age from the color of its nose is more uncertain than predicting the mean age of all lions having the same proportion of black on their noses. Prediction intervals bracket most of the individual data points in the sample, because they incorporate the variability in Y from individual to individual at a given X.

> *Confidence bands* measure the precision of the predicted mean Y for each value of X. *Prediction intervals* measure the precision of the predicted single Y-values for each X.

Most statistical packages on the computer will calculate and display confidence bands and prediction intervals. We haven't given calculation details, but we provide the formulas in the Quick Formula Summary (Section 17.10).

Extrapolation

We've stressed that regression can be used to predict Y for any value of X lying between the smallest and largest values of X in the data set. Regression cannot be used to predict the value of the response variable when an X-value lies well outside the range of the data. This is because there is no way to ensure that the relationship between X and Y continues to be linear beyond the range of the data. Predicting Y for X-values beyond the range of the data is called **extrapolation**. The graph in Figure 17.2-2 illustrates the problem.

FIGURE 17.2-2
Ear lengths of 206 adults 30 years old or more as a function of their ages. Modified from Heathcote (1995).

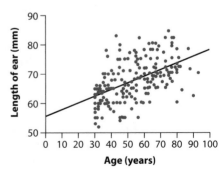

The data are measurements of ear length taken on a sample of adults at least 30 years old (Heathcote 1995). The linear regression equation calculated from these data (in millimeters) is

$$\text{ear length} = 55.9 + 0.22(\text{age}).$$

The results suggest that our ears grow longer by about 0.22 mm per year on average as we age. The intercept of this equation, which predicts the ear length at birth (i.e., when age is zero), is 56 mm. This makes no sense, though. To quote the authors of the study (Altman and Bland 1998), "A baby with ears 5.6 cm long would look like Dumbo." The relationship between ear length and age is not linear from birth, but we wouldn't know this unless we took measurements over the complete range of ages.

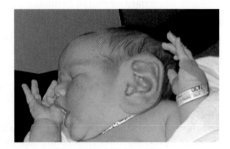

> *Extrapolation* is the prediction of the value of a response variable outside the range of X-values in the data.

17.3 Testing hypotheses about a slope

Hypothesis testing in regression is used to evaluate whether the population slope equals a null hypothesized value, β_0, which is typically (but not always) zero. The test statistic t is

$$t = \frac{b - \beta_0}{\text{SE}_b},$$

where b is the estimate of the slope in the sample and SE_b is the standard error of b. Under the null hypothesis, this test statistic has a t-distribution with $n - 2$ degrees of freedom. Example 17.3 shows how to use this test.

EXAMPLE   Prairie Home Campion

17.3   Human activity is reducing species numbers in many of the world's ecosystems. Does this decrease affect basic ecosystem properties? Or are different plant species largely substitutable, with lost species compensated by those species remaining? To find out, Tilman et al. (2006) seeded 161 plots, each measuring 9 × 9 meters, at the Cedar Creek Reserve in Minnesota. They used a varying number of prairie plant species

and measured plant biomass production over 10 subsequent years. Treatments of either 1, 2, 4, 8, or 16 plant species (randomly chosen from a set of 18 perennials) were randomly assigned to plots. After 10 years of measurement, the researchers measured the "stability" of plant biomass production in every plot as mean biomass divided by the standard deviation in biomass over the 10 years (the reciprocal of the coefficient of variation; Section 3.1). Results are plotted in Figure 17.3-1. Stability has been log-transformed to reduce skew. The data are available at whitlockschluter.zoology.ubc.ca.

FIGURE 17.3-1
Stability of plant biomass production over 10 years in 161 plots and the initial number of plant species assigned to plots. Stability was log-transformed (natural log) to better meet the assumptions of regression. The line is the least-squares regression line. Data are from Tilman et al. (2006).

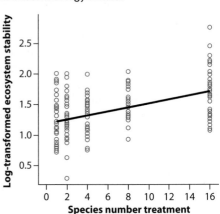

Unlike the previous example, involving lions, the data in Example 17.3 are from an experiment in which the values of the explanatory variable were fixed treatments. In contrast to correlation, regression does not require the explanatory variable to follow a normal distribution.

The *t*-test of regression slope

The null hypothesis is that the measure of ecosystem stability cannot be predicted from the species number treatment—that is, the slope of the linear regression of ecosystem stability on number of species is zero. The alternative hypothesis is that stability either increases or decreases with increasing number of species. This is a two-tailed test.

H_0: The slope of the regression of log ecosystem stability on species number is zero ($\beta = 0$).

H_A: The slope of the regression of log ecosystem stability on species number is not zero ($\beta \neq 0$).

The following quantities calculated from the data are needed for the *t*-test of zero slope:

$$\bar{X} = 6.3168 \qquad\qquad \bar{Y} = 1.4063$$

$$\sum_i (X_i - \bar{X})^2 = 5088.8447 \qquad\qquad \sum_i (Y_i - \bar{Y})^2 = 24.8149$$

$$\sum_i (X_i - \bar{X})(Y_i - \bar{Y}) = 167.5548 \qquad\qquad n = 161.$$

The best estimate of the slope is

$$b = \frac{\sum_i (X_i - \bar{X})(Y_i - \bar{Y})}{\sum_i (X_i - \bar{X})} = \frac{167.5548}{5088.8447} = 0.03293.$$

Calculating the intercept as before, and rounding, we get the least-squares regression line as

$$Y = 1.20 + 0.033X,$$

which can also be written as

$$\text{Log stability} = 1.20 + 0.033(\text{number of species}).$$

This line has a positive slope, as shown in Figure 17.3-1. The estimate of slope ($b = 0.033$) indicates that log stability of biomass production rises by the amount 0.033 for every species added to plots.

To calculate the standard error of the slope, we need the mean square residual:

$$\begin{aligned} \text{MS}_{\text{residual}} &= \frac{\sum_i (Y_i - \bar{Y})^2 - b \sum_i (X_i - \bar{X})(Y_i - \bar{Y})}{n - 2} \\ &= \frac{24.8149 - 0.03293(167.5548)}{161 - 2} \\ &= 0.12137. \end{aligned}$$

Thus, the standard error of b is

$$\text{SE}_b = \sqrt{\frac{\text{MS}_{\text{residual}}}{\sum_i (X_i - \bar{X})^2}} = \sqrt{\frac{0.12137}{5088.8447}} = 0.004884.$$

We now have all of the elements needed to calculate the t-statistic:

$$t = \frac{b - \beta_0}{\text{SE}_b} = \frac{0.03293 - 0}{0.004884} = 6.74.$$

We must compare this t-statistic with the t-distribution having $df = n - 2 = 161 - 2 = 159$ degrees of freedom. Using a computer, we find that $t = 6.74$ corresponds to $P = 2.7 \times 10^{-10}$, so we reject the null hypothesis. We reach the same conclusion if we use the critical value for the t-distribution with $df = 159$ (Statistical Table C):

$$t_{0.05(2), 159} = 1.97.$$

Since $t = 6.74$ is greater than 1.97, $P < 0.05$, and we reject H$_0$. In other words, increasing the number of plant species in plots increases the stability of plant biomass production of the ecosystem. A 95% confidence interval for the population slope, calculated from the formula in Section 17.1, is

$$0.0233 < \beta < 0.0426.$$

indicating that the estimate of slope has fairly tight bounds.

The ANOVA approach

In the literature, and in the output of regression analyses conducted on the computer, you will encounter tests of regression slopes that use an F-test rather than a t-test. Just as ANOVA can be used to compare two population means in place of the two-sample t-test, ANOVA can be used to test for a significant slope in place of the t-test of slope. The resulting P-values are identical. Table 17.3-2 shows the ANOVA table for the ecosystem stability data. Formulas for the quantities are given in the Quick Formula Summary (Section 17.10).

TABLE 17.3-2 ANOVA table testing the effect of plant species number on the stability of biomass production.

Source of variation	Sum of Squares	df	Mean Squares	F-ratio	P
Regression	5.5169	1	5.5169	45.45	2.73×10^{-10}
Residual	19.2980	159	0.1214		
Total	24.8149	160			

The basic idea behind the ANOVA approach in regression is similar to that when testing differences among means of multiple groups (Chapter 15). If the null hypothesis is true, then the population regression line is flat with a slope of 0. In this case, the amount of variation in Y among individual data points having the same value for X (represented by the residual mean square) is expected to equal the amount of variation among data points having different X values (represented by the regression mean square), except by chance. If the null hypothesis is false, we expect the regression mean square to exceed the residual mean square. The comparison of mean squares is done with an F-ratio.

The first step to estimating these two sources of variation in the data is to take the deviation between each Y-measurement Y_i and the grand mean $\overline{Y}$ and break it into two parts. The residual part is the deviation between Y_i and its predicted value on the regression line (i.e., $Y_i - \hat{Y}_i$, analogous to the "error" component in ANOVA). The regression, on the other hand, is the difference between the predicted value for each point and $\overline{Y}$ (i.e., $\hat{Y}_i - \overline{Y}$, analogous to the "groups" component in ANOVA). The sum of squared deviations corresponding to each of these two sources of variation and their total are computed and used to calulate the mean squares. The test statistic is an F-ratio of the two mean squares (the mean square regression divided by the mean square residuals). If the null hypothesis is true, and the slope of the population regression β is zero, then the F-ratio is expected to be 1 (except by chance). If the slope of the regression is not zero, however, then F is expected to be greater than 1.

The ANOVA approach can be used when the test is two-sided and the null hypothesized slope is zero.

Using R^2 to measure the fit of the line to data

We can measure the fraction of variation in Y that is "explained" by X in the estimated linear regression with the quantity R^2:

$$R^2 = \frac{SS_{regression}}{SS_{total}}.$$

R^2 is calculated from the sums of squares in the ANOVA table, and it is analogous to the R^2 in analysis of variance (Section 15.1), which measures the fraction of variation in the sample of Y-values accounted for by differences between groups.[5] If R^2 is close to one (i.e., its maximum possible value), then X predicts most of the variation in the values of Y. In this case, the Y-observations will be clustered tightly around the regression line with little scatter. If R^2 is close to zero (i.e., its minimum value), then X does not predict much of the variation in Y, and the data points will be widely scattered above and below the regression line.

For the ecosystem stability study, R^2 can be calculated from the quantities in the ANOVA table, Table 17.3-2:

$$R^2 = \frac{5.5169}{24.8149} = 0.222.$$

Thus, number of plant species explained 22% of the variation in log-transformed ecosystem stability, a moderate percentage.

17.4 Regression toward the mean

Suppose a study measured the cholesterol levels of 100 men randomly chosen from a population. After their initial measurement, the men were put on a new drug therapy, designed to reduce cholesterol levels. After one year, the cholesterol level of each man was measured again and compared with the first measurement. Figure 17.4-1 shows a scatter plot of the results. The researchers were delighted to find that cholesterol levels had dropped on average in the men who had previously had the highest levels. Their excitement dimmed, though, when they realized that cholesterol levels had *increased* on average in the men who had previously had low levels of cholesterol. What happened? Had they discovered a drug with complex effects?

In this hypothetical example, there was no effect at all due to the drug—the trend resulted entirely from a general phenomenon known as **regression toward the mean**. If two variables measured on a sample of individuals, such as consecutive measures of cholesterol, have a correlation less than one, then individuals that are far from the mean on the first measure will on average lie closer to the mean for the

---

5. R^2 is sometimes written as r^2. In regression with only one explanatory variable, the R^2 value is the same as the square of the correlation coefficient r.

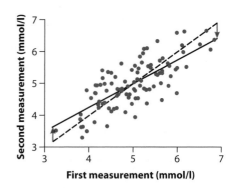

FIGURE 17.4-1
Regression toward the mean. These hypothetical data are two cholesterol measurements taken on the same 100 men. The dashed line is the one-to-one line with a slope of one. The solid line is the regression line predicting the second measure from the first. It has a slope less than one, as indicated by the blue arrows.

second measure. Even without an effect of the drug treatment, average cholesterol levels of the men with the highest levels on the first measure were expected to drop by the second measurement, and average levels of the men who originally had low levels were expected to rise.[6]

> *Regression toward the mean* results when two variables measured on a sample of individuals have a correlation less than one. Individuals that are far from the mean for one of the measurements will, on average, lie closer to the mean for the other measurement.

Regression toward the mean is a tricky concept, which is perhaps why it is often overlooked. Think of it this way: each of the men in the study has a "true," underlying cholesterol value, but his "measured" cholesterol value varies randomly with time and circumstance around the true value. The subset of men who scored highest on the first measurement therefore likely included a disproportionate number of men whose cholesterol measurement was higher than its true value the first time. The second measurement made on each of these men is expected to be closer to its true value on average, bringing down the average for the subset of men as a whole. Similarly, the subset of men who initially scored lowest likely included a disproportionate number of men whose measured values were lower than their true values, so on the second measurement they would seem to improve.[7]

Regression toward the mean is potentially a large problem in any study that tends to focus on individuals in one tail of the distribution. In many medical studies, for example, only sick people are included in the research, as indicated by their initial assessment before the study. Because of regression toward the mean, many of these

---

6. Regression toward the mean happens even when the overall mean changes from the first to the second measurement.

7. Regression was named for this property by Galton (1886), who studied heights of parents and their grown children. He noticed that tall fathers tended to have sons shorter than their fathers on average, while the reverse was true for short fathers. Galton was concerned at first that this tendency would eventually eliminate variation in height in the population and prevent evolution. He later realized that, in each generation, very tall (and very short) individuals are still present, but they are not necessarily the offspring of the tallest (or shortest) parents.

people will appear to improve even if the treatment has no effect. Interpreting this improvement as if it were a response to the treatment, instead of a mathematical fact of regression, is called the **regression fallacy**. It is one of the reasons that experiments should always include a control group for comparison.

Regression toward the mean is an issue only in observational studies, not in randomized experiments, where the value of the explanatory variable (X) is set by the experimenter. Kelly and Price (2005) discuss ways to disentangle biologically meaningful trends from the effect of regression toward the mean.

17.5 Assumptions of regression

When using linear regression, the following assumptions must be met for confidence intervals and hypothesis tests to be accurate:

- At each value of X, there is a population of possible Y-values whose mean lies on the true regression line (this is the assumption that the relationship must be linear).
- At each value of X, the distribution of possible Y-values is normal.
- The variance of Y-values is the same at all values of X.
- At each value of X, the Y-measurements represent a random sample from the population of possible Y-values.

Figure 17.5-1 illustrates the first three of these assumptions. In the next few sections, we explore some of the ways to examine deviations from these assumptions. We also discuss methods to try when the assumptions are not supported by the data.

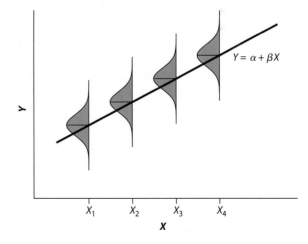

FIGURE 17.5-1
Illustration of the assumptions of linear regression. At each value of X, there is a normally distributed population of Y-values with the mean on the true regression line. The variance of the Y-values is assumed to be the same for every value of X.

Unlike correlation analysis, no assumptions are made about the distribution of X when using regression. In regression, for example, it is not necessary that the distribution of X-values is normal. It is not even necessary that X-values are randomly sampled—they might be fixed by the experimenter, instead, as they were in the ecosystem stability study (Example 17.3).

Outliers

Besides creating a non-normal distribution of Y-values at the corresponding value of X, and violating the assumption of equal variance in Y, outliers disproportionately affect estimates of the regression slope and intercept. If an outlier is present, biologists usually examine and report its influence on the results by comparing the regression line produced with and without the outlier. For example, Figure 17.5-2 shows how the average amount of white in the tails of dark-eyed juncos varies with latitude. One outlier is present, however, indicated in red (a population that formed in 1983 on the campus of the University of California, San Diego). Without the outlier, the estimate of slope is $b = -0.37$ (black line in Figure 17.5-3), and the null hypothesis of zero slope is rejected ($t = -2.66, P = 0.024$). Including the outlier changes the slope substantially (red line; $b = -0.18$), and the null hypothesis of zero slope is not rejected ($t = -0.81, P = 0.43$).

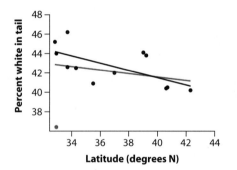 

FIGURE 17.5-2 Graph showing the effect of an outlier on an estimate of the regression line. The data are the percentage of white in the tail feathers of the dark-eyed junco at sites at different latitudes in California (Yeh 2004). The black regression line was calculated after excluding the red point on the lower left, whereas the red regression line included it.

Outliers are especially likely to be influential if they occur at or beyond the range of X-values in the rest of the data. If the outlier has a large effect, then alternative approaches might also be sought. One approach is to transform X or Y, such as by taking logarithms, to see if this brings the outlier closer to the rest of the distribution. Further solutions include robust regression methods (Rousseeuw and Leroy 2003) and permutation testing (Chapter 13).

Detecting nonlinearity

Visually inspecting the scatter plot is a useful method for detecting departures from the assumption of a linear relationship between Y and X. Often, this approach is enough to conclude that the relationship between X and Y is not linear. Forcing a linear regression through the scatter plot can sometimes make the nonlinearity even more obvious (see the left panel of Figure 17.5-3).

Scatter-plot "smoothing," a method discussed in greater detail in Section 17.8, can also aid the eye in detecting a nonlinear relationship (see the right panel of Figure 17.5-3). Most statistics packages on the computer are able to carry out scatter-plot smoothing.

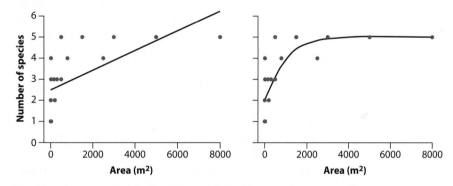

FIGURE 17.5-3 A scatter plot showing the relationship between the number of fish species and the surface area of 20 desert pools (Kodric-Brown and Brown 1993). The left panel fits a linear regression to the data to highlight how poorly a straight line matches the data. The right panel adds a "smoothed" fit to the same data (see Section 17.8).

Detecting non-normality and unequal variance

It is often difficult to decide whether the assumptions of normally distributed residuals and equal variance of residuals are met. Visual inspection of a **residual plot** can help. In a residual plot, the residual for every data point $(Y_i - \hat{Y}_i)$ is plotted against X_i, the corresponding value of the explanatory variable. This plot is best made with the aid of a computer. Two examples of residual plots are shown in Figure 17.5-4.

If the assumptions of normality and equal variance of residuals are met, then the residual plot should have all of the following features:

- A roughly symmetric cloud of points above and below the horizontal line at zero, with a higher density of points close to the line than away from the line
- Little noticeable curvature as we move from left to right along the x-axis
- Approximately equal variance of points above and below the line at all values of X

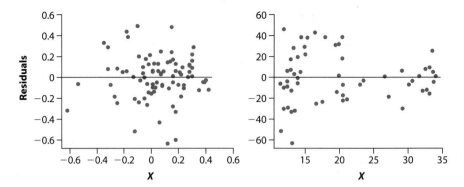

FIGURE 17.5-4 Two examples of residual plots. Data on the left are from a linear regression of the cap color of offspring on that of their parents in the blue tit, a British bird (Hadfield et al. 2006). Those on the right are from a linear regression of firing rates of cockroach neurons on temperature (Murphy and Heath 1983).

The blue tit data in the left panel of Figure 17.5-4 fit these requirements reasonably well. The density of observations peaks near the horizontal line and spreads outward above and below in a fairly symmetrical fashion. The spread of points above and below the line is similar across the range of X-values. (The spread may seem low at the extreme left end, but we can't tell because there are only two data points). The cockroach data in the right panel of Figure 17.5-4 do not fit these requirements as well. The spread of points above and below the horizontal line is considerably higher at low values of X than at high values of X.

A *residual plot* is a scatter plot of the residuals $(Y_i - \hat{Y}_i)$ against the X_i, the values of the explanatory variable.

Normal quantile plots (Section 13.1) and histograms of the residuals are yet other ways to evaluate the assumption that the residuals are normally distributed.

17.6 Transformations

Some (but not all) nonlinear relationships can be made linear with a suitable transformation. The most versatile transformation in biology is the log transformation (Section 13.3). The power and exponential relationships, which are commonly encountered in biology, are two nonlinear relationships that can be made linear with a log transformation, as shown in Figure 17.6-1.

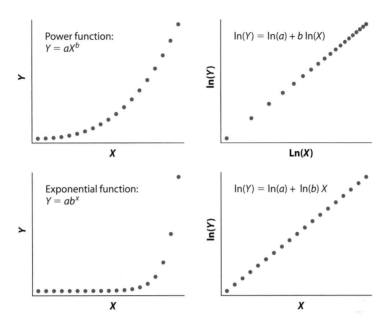

FIGURE 17.6-1 The power function (*upper left*) and the exponential function (*lower left*) are two types of nonlinear relationships that can be made linear using the log transformation. Plot log of Y against log of X if the two variables are described by a power function (*upper right*). Plot the log of Y against X if the relationship between these two variables is described by an exponential function (*lower right*).

For example, the relationship between the number of fish species in 20 desert pools and the surface area of pools (Figure 17.5-3) looks like it might fit a power curve or an exponential curve. We tried a log-transformation of both the number of species and the surface area, and we obtained the graph shown in Figure 17.6-2. The straight line fits the transformed data much better than the untransformed data. We can now proceed to estimate parameters and test hypotheses about this relationship using the methods of linear regression, setting Y to be the log of the number of fish species and X to be the log of the surface area of the pools.

FIGURE 17.6-2
A scatter plot of the log-transformed number of fish species and surface area of desert pools. This relationship is more linear than the one in Figure 17.5-3, which is based on the untransformed data.

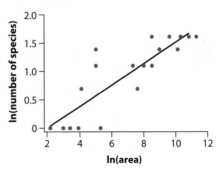

Transformation can also be used to help meet other assumptions of linear regression. For example, if a residual plot reveals that the variance of Y increases with increasing X, then transforming Y can often improve matters (it may be necessary, though, to transform X as well to keep the relationship linear). For example, the number of pollen grains received by flowers of the iris *Lapeirousia anceps*, which is pollinated by long-proboscid flies, increases with increasing flower tube length (Pauw et al. 2009). In a residual plot from a regression using the untransformed data (see the left panel of Figure 17.6-3), the variance of residuals increases from the smallest X-values to larger X-values. This problem goes away when Y is square-root transformed (see the right panel of Figure 17.6-3).

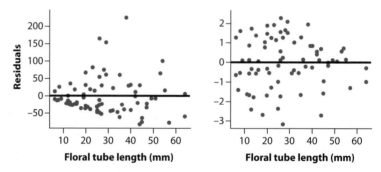

FIGURE 17.6-3 The effect of a square-root transformation on the residuals from a linear regression of number of pollen grains received on floral tube length of an iris species (Pauw et al. 2009). Residuals from a linear regression calculated on the original data (*left panel*) do not fit the equal-variance assumptions of linear regression, but residuals from a regression using the square root of the number of pollen grains (*right panel*) have more equal variances.

The square-root transformation, originally described in Section 13.3, often resolves unequal variance problems when the data are counts, as in Figure 17.6-3. The log transformation can also be effective when the variance in Y increases with increasing X. Arcsine transformation is often effective when Y is a proportion. When analyzing data that violate the assumptions of regression, try simple transformations of X and/or Y to see if they help to meet the assumptions of linear regression.

17.7 The effects of measurement error on regression

Recall from Section 16.6 that measurement error occurs when a variable is not measured with complete accuracy. Many biological traits, such as behavior or aspects of physiology, can be difficult to measure accurately, so measurement error can be an important component of variation (Section 15.6).

The effects of measurement error on regression differ from the effects on correlation (Section 16.6). The effect depends on the variable. Measurement error in Y

increases the variance of the residuals, as shown in Figure 17.7-1 when you compare the scatter in the middle panel (measurement error in Y) with that in the left panel (no measurement error). This increases the sampling error of the estimate of the slope and of the predictions but has no effect on expected slope.

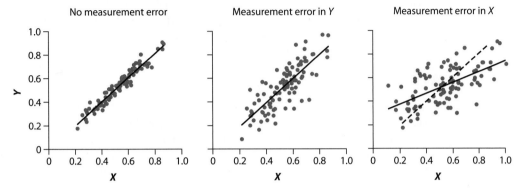

FIGURE 17.7-1 The effects of measurement error on the estimate of regression slope. X and Y are measured without error in the left panel. Y is measured with error in the middle panel, which has little effect on expected slope but increases the variability in the residuals. X is measured with error in the right panel, which causes the expected estimate of the slope to decline (*solid line*) compared with the slope in the absence of measurement error (*dashed line*). The variability of the residuals also increases.

Measurement error in X (the right panel in Figure 17.7-1) also increases the variance of the residuals, and in addition it causes bias in the expected estimate of the slope. With measurement error in X, b will tend to lie closer to zero on average than the population quantity β. On average, the largest values of X in the data will include disproportionately many measurements that were erroneously overestimated. Since the true X-values of these points are smaller than their measured values, they predict Y-values that on average lie closer to the mean than would the same X-values in the absence of measurement error. Conversely, the smallest values of X will tend to include disproportionately many measurements that were erroneously underestimated. These underestimated X-values will be associated with Y-values that lie closer to the mean on average than in the absence of measurement error.

17.8 Nonlinear regression

Transformations won't always successfully convert a nonlinear relationship into one that can be analyzed using linear regression. Nonlinear regression methods, however, are readily available in most statistics packages on the computer. Here in Section 17.8, we outline some basic principles for nonlinear regression.

The assumptions of nonlinear regression are almost the same as those of linear regression (Section 17.5), except here we usually assume that the true relationship between X and Y has a specific nonlinear form.

The immediate problem when turning to nonlinear regression is the nearly unlimited number of options. There are so many mathematical functions to choose from that it can be difficult to know where to begin. The appropriate choice depends on the data, but a few guidelines can help you make a sensible choice.

A curve with an asymptote

The best advice we can offer is to keep things simple, unless the data suggest otherwise. Figure 17.8-1 illustrates this principle. The left panel shows a nonlinear regression model fitted to data on the population growth rate of a species of phytoplankton and the concentration of iron, a limiting nutrient. The mathematical function fit to the data passes through every data point. The resulting $MS_{residual}$ is zero.

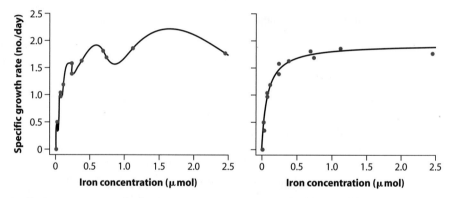

FIGURE 17.8-1 Population growth rate of a species of phytoplankton in culture in relation to the concentration of iron in the medium (data from Sunda and Huntsman 1997). The curve in the left panel is an arbitrarily complex function that passes through all of the data points. The curve in the right panel is a Michaelis–Menten curve that fits the data more simply.

This is hardly the best possible outcome, however, even though each data point fits precisely. The problem with the curve in the left panel of Figure 17.8-1 is that it would probably do a terrible job of predicting any *new* observations obtained from the same population, because the curve does not describe the general trend. Such a complicated curve is also difficult to justify biologically—is there good reason to think that all the peaks and dips in this curve truly reflect the effects of iron on the growth of phytoplankton?

Greater simplicity, as demonstrated by the fitted curve in the right panel of Figure 17.8-1, solves both of these problems. The data are the same as in the left panel, but this time we've fit the much simpler function,

$$Y = \frac{aX}{b + X}.$$

This is the *Michaelis–Menten* equation used frequently in biochemistry. The curve rises from a *Y*-intercept at zero and increases at a declining rate with increasing *X*, eventually reaching a saturation point, or asymptote. The asymptote is represented in the formula by the constant *a*, whereas *b* determines how fast the curve rises to the asymptote. Reminiscent of linear regression, the Michaelis–Menten equation has only two parameters to estimate (the true asymptote α and the true rate parameter β). These parameters are very different, however, from those of linear regression. We obtained the curve on the right of Figure 17.8-1 with a statistics package on the computer that used least squares to find the best fit.

The virtue of linear regression is simplicity. We should strive to retain this property as we look at the wide range of nonlinear functions available.

Quadratic curves

The *quadratic* curve is often used in biology to fit a humped-shape curve to data, such as the relationship shown in Figure 17.8-2 between the number of plant species present in ponds (*Y*) and pond productivity (*X*). The curve is a symmetric parabola described by the quadratic (second-degree polynomial) equation,

$$Y = a + bX + cX^2.$$

This equation is similar to the formula for a straight line except that one more term has been added for the square of *X*, and another regression coefficient *c* must be computed. When *c* is negative, the curve is humped, as in Figure 17.8-2. When *c* is positive, the parabola curves upward in a U-shape.

Asymptotic and quadratic curves are just two of several nonlinear functions commonly used in biology. The choice between them must depend on the data: Is the relationship asymptotic or humped? If humped, is the hump symmetric or does it fall more steeply on one side than the other? If it falls more steeply on one side, then we must search for another function altogether. A good guide to a variety of curves used in biology can be found in Motulsky (1999).

FIGURE 17.8-2
A quadratic curve fit to the relationship between the number of plant species present in ponds and pond productivity (Chase and Leibold 2002).

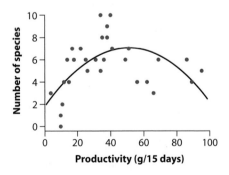

Formula-free curve fitting

With the aid of a computer, it is possible to fit curves to data without specifying a formula. Often called **smoothing**, this approach gathers information from groups of nearby observations to estimate how the mean of Y changes with increasing values of X. There are several methods, including "kernel," "spline," and "loess" smoothing. We bypass the technical details here, but Example 17.8 illustrates the utility of the approach.

EXAMPLE   The incredible shrinking seal

17.8   Trites (1996) amassed a large set of measurements of the ages and sizes of northern fur seals (*Callorhinus ursinus*) in the Pacific Ocean, gathered over decades by many researchers. Most measurements were taken between spring and fall. In summer, females spend a lot of time on land giving birth and nursing young. The graph in Figure 17.8-3 shows the measurements for nearly 10,000 adult females. In a preliminary analysis, a line was fitted to the data, but the relationship appeared nonlinear. Average length increased with age but appeared to taper off by about 4500 days (i.e., about 12 years old).

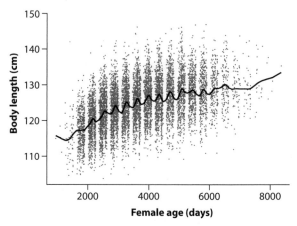

FIGURE 17.8-3
Measurements of body length as a function of age for female fur seals. The spline fit is in black.

To fit the data, we used a spline technique to calculate a smoothed fit of body length on female age. The result is plotted in Figure 17.8-3. Astonishingly, the curve indicates that average female body length does not rise steadily with age, but oscillates each year. Female fur seals become longer each summer and then shorter again by winter (keep in mind that these changes are in length, not weight). Elongation results in part because the seals are heavier on land than in water, and the added weight stretches the skeleton during summer breeding. The skeleton shrinks back after the seals return to the water in the winter. It would have been difficult to come up with a formula that captured this complex relationship.

The seal example illustrates that it is not always easy to anticipate or see the type of relationship present in the data by visual inspection alone. Smoothing techniques can help. Example 17.8 also demonstrates that we don't always need a mathematical formula if all we want to do is fit the data and use it to improve our understanding of a biological system.

The fit obtained by smoothing is controlled by a smoothing coefficient that determines how bumpy the curve is. You can adjust this coefficient in statistical computer programs so that you can explore its effects. A low value for the coefficient results in a bumpy curve that, in the extreme, would pass through all the data points. A larger value of the coefficient gives a smoother fit. Computer programs usually use rules of thumb to choose the best value for the smoothing parameter, but it is wise to try alternatives to see what effect varying the smoothing coefficient has on the curve.

17.9 Logistic regression: fitting a binary response variable

Logistic regression is a special type of nonlinear regression developed for a *binary* response variable—that is, when the Y-variable measured on independent units is either zero or one. A common use for logistic regression is to fit a dose–response curve, where mortality (or survival) of individuals (the "response") is plotted against the concentration of a drug, toxin, or other chemical (the "dose").[8] Here we provide a quick description with a minimum of calculations.

Our example data set is shown in Figure 17.9-1, from an experiment in which 160 guppies were exposed to cold temperatures (5°C) for different durations (3, 8, 12, or 18 minutes). Each treatment, or dose, was assigned to 40 fish. The study was by Pitkow (1960), who carried out the experiment to identify the physiological mechanism causing fish death at cold temperatures. Mortality is the binary response variable (Y) measured on each independent fish (1 = died, 0 = survived), and duration is the explanatory variable. We used logistic regression to fit the curve predicting the probability of fish mortality from duration of exposure (Figure 17.9-1). Because the data are binary, the curve describes the proportion of fish that die (the mean of Y) at levels of exposure X. Table 17.9-1 gives a frequency table of the data.

Linear regression of Y on X is unsuitable for these data because the binary response variable violates three of its assumptions. The relationship between Y and X

8. Not all dose–response data are binary. For example, the response variable Y is often the fraction of individuals in a unit that die, where a unit refers to a randomly sampled group of subjects such as members of a family, petri dish, aquarium, or forest plot. Such data might be analyzed with ordinary linear regression after making a suitable transformation.

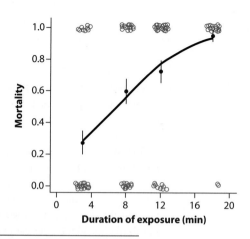

FIGURE 17.9-1
Mortality of guppies in relation to duration of exposure to a temperature of 5°C (data from Pitkow 1960). Treatments were 3, 8, 12, or 18 minutes of exposure, with 40 fish in each of the four treatments. Each point (red circle) indicates a different individual (points were offset using a random perturbation to reduce overlap). $Y = 1$ if the individual died, whereas $Y = 0$ if the individual survived. Black dots indicate the proportion of deaths (± 1 SE) in each treatment. The curve is the logistic regression predicting the probability of death.

TABLE 17.9-1 Number of fish (out of 40) in two mortality groups at each of four cold-temperature treatments.

	Duration of exposure (min)			
	3	8	12	18
Died ($Y = 1$)	11	24	29	38
Survived ($Y = 0$)	29	16	11	2

is not linear, because the predicted values $\hat{Y}$ cannot fall outside the interval between 0 and 1. The residuals $Y - \hat{Y}$ are not normally distributed; they are binary—each point is either $0 - \hat{Y}$ or $1 - \hat{Y}$. The variance of the residuals is not constant: variance is expected to be highest when the predicted Y is near 0.5, and lowest when the prediction of Y is close to zero or one. (This is because when the prediction is zero or one, most of the data are zeros or ones (respectively) and the variance among individuals is therefore small.) These problems are not fixed with a simple transformation of the data.

All three problems are solved by logistic regression. The method fits a curve constrained to lie between 0 and 1. Instead of the normal distribution, it assumes that outcomes at every X have a binomial distribution (Section 7.1) being either one or zero. The probability of an event (in this case, dying) is given by the corresponding predicted value on the regression curve. Finally, to correct for differences in the variance of residuals at different values of X, logistic regression weights each residual by its estimated variance obtained from the binomial distribution.

Logistic regression fits the following equation to binary data:

$$\text{log-odds}(Y) = \alpha + \beta X.$$

The log-odds refers to the natural log of the odds of Y (Section 9.2).[9] The right side of the equation ($\alpha + \beta X$) is the formula for a straight line, with α the intercept and β

---

9. Log-odds$(Y) = \ln[Y/(1 - Y)]$. See Section 9.2, which used the symbol p in place of Y. Log-odds(Y) is also known as the logit function.

the slope. In other words, an ordinary line is used to fit the log-odds of the proportion of individuals dying. The curve shown in Figure 17.9-1 is based on estimates of these two parameters: a for intercept and b for slope,

$$\text{log-odds}(Y) = a + bX.$$

Methods to calculate this regression curve from data are available in most computer statistical packages. When we analyzed the guppy data with such a statistics package, we obtained the output in Table 17.9-2, showing estimates of regression coefficients, and Table 17.9-3, showing the results of a test of the null hypothesis of zero slope (H$_0$: $\beta = 0$). We will explain the contents of these two tables in turn.

TABLE 17.9-2 Logistic regression output. The values shown are the estimates for a and b of the intercept (α) and the slope (β) of the logistic regression curve for the cold-fish data. SE refers to the standard error of estimates.

	Estimate	SE
Intercept	−1.66	0.41
Slope	0.24	0.04
Number of iterations: 4		

From Table 17.9-2 we see that the best estimate for the intercept is given by $a = -1.66 \pm 0.41$ SE, and the best estimate for the slope is $b = 0.24 \pm 0.04$ SE.

Computer programs might additionally provide Wald statistics (symbolized by z) and corresponding P-values for approximate tests of the null hypotheses H$_0$: $\alpha = 0$ and H$_0$: $\beta = 0$. However, the Wald method is inaccurate, and we do not present these results. The log-likelihood ratio test (Table 17.9-3) should be used instead.

Remember: the estimates a and b are not intercept and slope of a linear regression of Y on X. Instead, they describe the linear relationship between X and the predicted log odds of Y [which we'll call log-odds($\hat{Y}$)]. To obtain the predicted values ($\hat{Y}$), we need to convert the log-odds to ordinary proportions:

$$\hat{Y} = \frac{e^{\text{log-odds}(\hat{Y})}}{1 + e^{\text{log-odds}(\hat{Y})}}$$

For example, to predict the proportion of fish dying for a cold-temperature duration of 10 minutes, we calculate

$$\text{log-odds}(\hat{Y}) = a + bX = -1.66 + 0.24(10) = 0.74,$$

$$\hat{Y} = \frac{e^{0.74}}{1 + e^{0.74}} = 0.68.$$

In other words, a duration of 10 minutes at 5°C is predicted to cause 68% mortality.

Another useful quantity for the regression curve is the *LD*50 (lethal dose 50), the estimated dose predicting 50% mortality:

$$LD50 = -\frac{a}{b}.$$

For the fish data,

$$LD50 = -\frac{-1.66}{0.24} = 6.92 \text{ minutes.}$$

Computer output for logistic regression will typically also include the number of iterations used in the calculations (Table 17.9-2). Logistic regression uses maximum likelihood (Chapter 20) to fit the curve to the data, and no formula exists to calculate the estimates. Instead, the computer uses a series of iterations to search for the best-fit curve. The search ceases when there are no further improvements in the fit from successive iterations.

Table 17.9-3 is an *analysis of deviance* table, with the results of a log-likelihood ratio test (see Section 20.5) of the null hypothesis that there is no relationship between mortality and duration (H_0: $\beta = 0$ vs. H_A: $\beta \neq 0$). The analysis of deviance table is analogous to the ANOVA table in ordinary linear regression. The method fits two models to the data, one in which the variable duration is absent, and one in which it is present.

Null model: log-odds(Y) = α.
Regression model: log-odds(Y) = $\alpha + \beta X$.

The null model is a restatement of the null hypothesis that $\beta = 0$, whereas the regression model is a restatement of the alternative hypothesis that $\beta \neq 0$. Analysis of deviance compares the fit of the two models to the data. If the improvement in fit of the regression model over the null model is too large to be explained by chance, the null hypothesis is rejected.

TABLE 17.9-3 Analysis of deviance table, containing results of the log-likelihood ratio test for the cold-fish data.

Model	df	Deviance	Residual df	Residual deviance	P
Null			159	209.55	
Duration	1	44.86	158	164.69	2.12×10^{-11}

The key quantity in the table is the *improvement* in fit when $\beta \neq 0$, which is here labeled simply "Deviance." This quantity is the difference in fit between the two models, where "fit" is the discrepancy between the observed values of Y and the values predicted, $\hat{Y}$, from each model. For the fish data, the improvement in fit is $209.55 - 164.69 = 44.86$. Residual deviance is analogous to residual mean square in ordinary linear regression.

Under the null hypothesis that $\beta = 0$, deviance has a χ^2 distribution with 1 *df*. The critical value for $\chi^2_{0.05,1} = 3.84$. Since $44.86 > 3.84$, $P < 0.05$, and we reject the null hypothesis. An approximation to the exact *P*-value is given in the table.

17.10 Summary

- Regression is a method used to predict the value of a numerical response variable Y from the value of a numerical explanatory variable X.

- Linear regression fits a straight line through a scatter of points. The equation for the regression line is $Y = a + bX$, where b is the slope of the line and a is the intercept.

- The least squares regression line is found by minimizing the sum of the squared differences between the observed Y-values and the values predicted by the line.

- The residuals are the differences between the observed values of Y and the values predicted by the least-squares regression line, $\hat{Y}$. The variance of the residuals, $MS_{residual}$, measures the spread of points above and below the regression line.

- Linear regression calculated on a sample of points estimates the straight-line relationship between the two variables in the population. The formula for the population regression line is $Y = \alpha + \beta X$, where β is the slope of the line and α is the intercept.

- If the assumptions of regression are met, then the sampling distribution of b is normal with mean β and standard deviation estimated by the standard error of the estimate of the slope, SE_b.

- The confidence interval for the slope β is based on the t-distribution.

- There are two types of prediction in regression: the mean Y at a given X, and a single Y observation at X. Both predictions generate the same value $\hat{Y}$, but they have very different precision. Precision is lower when predicting an individual Y because it includes the variability between individuals having the same X-value.

- Confidence intervals for predicted mean Y-values at each X are represented by confidence bands. Analogous intervals for the predicted Y-values of a single individual are called prediction intervals.

- Extrapolation is the prediction of Y at values of X beyond the range of X-values in the sample. Extrapolation is problematic, though, because there is no way to ensure that the relationship between X and Y continues to be linear beyond the data.

- If the null hypothesis is correct that the slope β of a population regression line is zero, then the test statistic $t = b/SE_b$ has a t-distribution with $n - 2$ degrees of freedom.

- An ANOVA table and F-test can also be used to test the null hypothesis that the population slope $\beta = 0$.

- R^2 is a measure of the fit of a regression line to the data. It measures the "fraction of the variation in Y that is explained by X."

■ Regression toward the mean results when two imperfectly correlated variables are compared by regression. Individuals that are far from the mean for one of the measurements will on average lie closer to the mean for the other measurement.

■ Methods for regression assume that the relationship between X and Y falls along a straight line, that the Y-measurements at each value of X are a random sample from a population of Y-values, that the distribution of Y-values at each value of X is normal, and that the variance of Y-values is the same at all values of X.

■ The scatter plot and the residual plot are graphical devices for detecting departures from the assumptions of linear regression.

■ Transformations of X and/or Y can be used to render a nonlinear relationship linear and to correct violations of the assumption of equal variance of residuals at every X.

■ The log-transformation is the most versatile transformation. It is useful when Y is related to X by a power function or by an exponential function.

■ If transformations do not work, nonlinear regression is an option.

■ Nonlinear regression curves should be kept as simple as the data warrant. Overly complex curves may be biologically unjustified and have low predictive power.

■ Smoothing methods make it possible to fit nonlinear curves to data without specifying a formula.

■ Logistic regression allows us to use a numerical variable to predict the probability that an individual has a particular value of a binary response variable.

17.11 Quick Formula Summary

Shortcuts

Sum of products: $\sum_i (X_i - \overline{X})(Y_i - \overline{Y}) = \sum_i (X_i Y_i) - \dfrac{\left(\sum_i X_i\right)\left(\sum_i Y_i\right)}{n}$.

Sum of squares for X: $\sum_i (X_i - \overline{X})^2 = \sum_i (X_i^2) - \dfrac{\left(\sum_i X_i\right)^2}{n}$.

Regression slope

What is it for? Estimating the slope of the linear equation $Y = \alpha + \beta X$ between an explanatory variable X and a response variable Y.

What does it assume? The relationship between X and Y is linear; each Y-measurement at a given X is a random sample from a population of Y-measurements; the distribution of Y-values at each value of X is normal; and the variance of Y-values is the same at all values of X.

Parameter: β

Estimate: b

Formula:
$$b = \frac{\sum_i (X_i - \bar{X})(Y_i - \bar{Y})}{\sum_i (X_i - \bar{X})^2}.$$

Standard error:
$$SE_b = \sqrt{\frac{MS_{residual}}{\sum_i (X_i - \bar{X})^2}},$$

where $MS_{residual}$ is the mean squared residual (the estimated variance of the residuals);

$$MS_{residual} = \frac{\sum_i (Y_i - \hat{Y}_i)^2}{n - 2}.$$

A quicker formula for $MS_{residual}$, not requiring you to calculate the $\hat{Y}$ first, is

$$MS_{residual} = \frac{\sum_i (Y_i - \bar{Y})^2 - b \sum_i (X_i - \bar{X})(Y_i - \bar{Y})}{n - 2}.$$

Regression intercept

What is it for? Estimating the intercept of the linear equation $Y = \alpha + \beta X$.

What does it assume? Same as the assumptions for the regression slope.

Parameter: α

Estimate: a

Formula: $a = \bar{Y} - b\bar{X}$.

Standard error: Set $X = 0$ in the formula for the standard error of the predicted mean Y at a given X [see "Confidence interval for the predicted mean Y at a given X (confidence bands)" on p. 574]. This is valid only if the value $X = 0$ falls within the range of X-values in the data.

Confidence interval for the regression slope

What is it for? An interval estimate of the population slope.

What does it assume? Same as the assumptions for the regression slope.

Statistic: b

Parameter: β

Formula: $b - t_{\alpha(2), df}\text{SE}_b < \beta < b + t_{\alpha(2), df}\text{SE}_b$, where SE_b is the standard error of the slope (see the formula given previously with the regression slope), and $t_{\alpha(2), n-2}$ is the two-tailed critical value of the t-distribution having $df = n - 2$.

Confidence interval for the predicted mean Y at a given X (confidence bands)

What is it for? An interval estimate of the predicted *mean Y* of all individuals in the population having the given value of X.

What does it assume? Same as the assumptions for the regression slope.

Statistic: $\hat{Y}$

Formula: $\hat{Y} - t_{\alpha(2), n-2}\,\text{SE}[\hat{Y}] < \text{predicted } Y < \hat{Y} + t_{\alpha(2), n-2}\,\text{SE}[\hat{Y}]$,

where $\text{SE}[\hat{Y}] = \sqrt{\text{MS}_{\text{residual}}\left(\dfrac{1}{n} + \dfrac{(X_i - \overline{X})^2}{\sum\limits_i (X_i - \overline{X})^2}\right)}$ is the standard error of the

prediction, $\text{MS}_{\text{residual}}$ is the mean square residual (see the formula given previously under "Regression slope"), and $t_{\alpha(2), n-2}$ is the two-tailed critical value of the t-distribution having $df = n - 2$.

Confidence interval for the predicted individual Y at a given X (prediction intervals)

What is it for? An interval estimate of the predicted Y for a *single individual* having the given value of X.

What does it assume? Same as the assumptions for the regression slope.

Statistic: $\hat{Y}$

Formula: $\hat{Y} - t_{\alpha(2), n-2}\,\text{SE}_1[\hat{Y}] < \text{predicted } Y < \hat{Y} + t_{\alpha(2), n-2}\,\text{SE}[\hat{Y}]$,

where $\text{SE}_1[\hat{Y}] = \sqrt{\text{MS}_{\text{residual}}\left(1 + \dfrac{1}{n} + \dfrac{(X_i - \overline{X})^2}{\sum\limits_i (X_i - \overline{X})^2}\right)}$ is the standard error of

the prediction, $\text{MS}_{\text{residual}}$ is the mean square residual (see the formula given previously under "Regression slope"), and $t_{\alpha(2),\, n-2}$ is the two-tailed critical value of the t-distribution having $df = n - 2$.

The t-test of a regression slope

What is it for? To test the null hypothesis that the population parameter β equals a null hypothesized value β_0.

What does it assume? Same as the assumptions for the regression slope.

Test statistic: t

Distribution under H$_0$: t-distributed with $n - 2$ degrees of freedom.

Formula: $t = \dfrac{b - \beta_0}{\text{SE}_b}$, where SE_b is the standard error of b (see the formula given previously under "Regression slope").

The ANOVA method for testing zero slope

What is it for? To test the null hypothesis that the slope β equals zero, and to partition sources of variation.

What does it assume? Same as the assumptions for the regression slope.

Test statistic: F

Distribution under H$_0$: F distribution. F is compared with $F_{\alpha(1),1,\, n-2}$.

Source of variation	Sum of squares	df	Mean squares	F-ratio
Regression	$\text{SS}_{\text{regression}} = \sum_i (\hat{Y}_i - \overline{Y})^2$	1	$\dfrac{\text{SS}_{\text{regression}}}{df_{\text{regression}}}$	$\dfrac{\text{MS}_{\text{regression}}}{\text{MS}_{\text{residual}}}$
Residual	$\text{SS}_{\text{residual}} = \sum_i (Y_i - \hat{Y}_i)^2$	$n - 2$	$\dfrac{\text{SS}_{\text{residual}}}{df_{\text{residual}}}$	
Total	$\text{SS}_{\text{total}} = \sum_i (Y_i - \overline{Y})^2$	$n - 1$		

R squared (R^2)

What is it for? Measuring the fraction of the variation in Y that is "explained" by X.

Formula: $R^2 = \dfrac{\text{SS}_{\text{regression}}}{\text{SS}_{\text{total}}}$,

where $\text{SS}_{\text{regression}}$ is the sum of squares for regression and SS_{total} is the total sum of squares.

PRACTICE PROBLEMS

1. **Calculation problem: Regression lines.** Men's faces have higher width-to-height ratios than women's, on average. This turns out to reflect a difference in testosterone expression during puberty. Testosterone is also known to predict aggressive behavior. Does face shape predict aggression? To test this, Carré and McCormick (2008) compared the face width-to-height ratio of 21 university hockey players with the average number of penalty minutes awarded per game for aggressive infractions like fighting or cross-checking. Their data are below along with some partial calculations. We will calculate the equation for the line that best predicts penalty minutes from face width-to-height ratio.

Face width: height ratio (X)	Penalty minutes per game (Y)	Face width: height ratio (X)	Penalty minutes per game (Y)
1.59	0.44	1.81	1.06
1.67	1.43	1.83	1.20
1.65	1.57	1.83	1.23
1.72	0.14	1.84	0.80
1.79	0.27	1.87	2.53
1.77	0.35	1.92	1.23
1.74	0.85	1.95	1.10
1.74	1.13	1.98	1.61
1.77	1.47	1.99	1.95
1.78	1.51	2.07	2.95
1.76	1.99		

a. Plot the data in a scatter plot.

b. Examine the graph. Based on this graph, do the assumptions of linear regression appear to be met?

c. Calculate the means of the two variables. (While you're doing so, record the sum of all X-values and the sum of all Y-values.)

d. Calculate the sum of X^2, the sum of Y^2, and the sum of the product XY.

e. Calculate the sum of products

$$\left(\sum_i (X_i - \bar{X})(Y_i - \bar{Y}) \right)$$ and the sum of

squares $$\left(\sum_i (X_i - \bar{X})^2 \right)$$ of the explanatory

variable, face ratio.

f. How steeply does the number of penalty minutes increase per unit increase in face ratio? From the sum of products and sum of squares for face ratio, calculate the estimate b of the slope. Double-check that the sign of the slope matches your impression from the scatter plot.

g. Calculate the estimate of the intercept, a, from the variable means and b.

h. Write the result in the form of an equation for the line. Add your line to the graph in (a).

2. **Calculation problem: Standard error and confidence intervals of the slope.** How uncertain is our estimate of slope? Using the face ratio and hockey aggressive penalty data from Practice Problem 1, calculate the standard error and confidence interval of the slope of the linear regression.

a. Calculate the total sum of squares for the response variable, penalty minutes.

b. Calculate the residual mean square $MS_{residual}$, using the total sum of square for Y, the sum of products, and the slope b.

c. With the sum of squares for X and $MS_{residual}$, calculate the standard error of b.

d. How many degrees of freedom does this analysis of the slope have?

e. Find the two-tailed critical t-value for a 95% confidence interval ($\alpha = 0.05$) for the appropriate df.

f. Calculate the confidence interval of the population slope, β.

3. **Calculation problem: Testing the null hypothesis that the slope equals zero.** Can the relationship be explained by chance? Using the face ratio and hockey penalty data from Practice Problem 1, test the null hypothesis that the slope of the regression line is zero.

a. State the null and alternate hypotheses.

b. What is β_0 for this null hypothesis?

c. Calculate the test statistic t from b, β_0, and the standard error of b.

d. Find the critical value of t appropriate for the degrees of freedom, at $\alpha = 0.05$.

e. Is the absolute value of the *t* for this test greater than the critical value?

f. Using a computer or the statistical tables, be as precise as possible about the *P*-value for this test. Draw conclusions from the test.

g. What fraction of the variation in average penalty minutes per game is accounted for by face ratio? Calculate the value of R^2.

4. Golomb et al. (2012) looked at whether higher chocolate consumption predicts higher body mass in humans. They fitted the data using a linear regression having chocolate consumption (number of times consumed per week) as the explanatory variable and the body mass index (BMI) as the response variable. BMI measures body mass relative to height, with high BMI typically meaning an overweight person. The slope of the regression was −0.142, with a standard error of 0.053 and a *P*-value of 0.008. Is this evidence that people who eat more chocolate have higher BMI? Why or why not?

5. What is the formula for each of the following four regression lines?

(a)
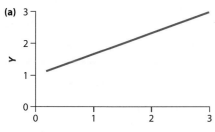

(b)
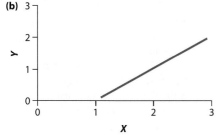

(c)
(d)
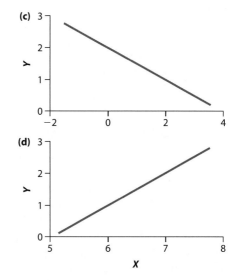

6. Some species seem to thrive in captivity, whereas others are prone to health and behavior difficulties when caged. Maternal care problems in some captive species, for example, lead to high infant mortality. Can these differences be predicted? The following data are measurements of the infant mortality (percentage of births) of 20 carnivore species in captivity along with the log (base-10) of the minimal home-range sizes (in km²) of the same species in the wild (Clubb and Mason 2003).

Log₁₀ home-range size	Captive infant mortality (%)	Log₁₀ home-range size	Captive infant mortality (%)
−1.3	4	0.1	29
−0.5	22	0.2	33
−0.3	0	0.4	33
0.2	0	1.3	42
0.1	11	1.2	33
0.5	13	1.4	20
1.0	17	1.6	19
0.3	25	1.6	25
0.4	24	1.8	25
0.5	27	3.1	65

a. Draw a scatter plot of these data, with the log of home-range size as the explanatory variable. Describe the relationship between the two variables in words.

b. Estimate the slope and intercept of the least-squares regression line, with the log of home-range size as the explanatory variable. Add this line to your plot.

c. Does home-range size in the wild predict the mortality of captive carnivores? Carry out a formal test. Assume that the species data are independent.[10]

d. Outliers should be investigated because they might have a substantial effect on the estimates of the slope and intercept. Recalculate the slope and intercept of the regression line from part (c) after excluding the outlier at large home-range size (which corresponds to the polar bear). Add the new line to your plot. By how much did it change the slope?

7. The following two graphs display data gathered to test whether the exercise performance of women at high elevations depends on the stage of their menstrual cycle (Brutsaert et al. 2002). In the upper panel, the explanatory variable is the progesterone level and the response variable is the ventilation rate at submaximal exercise levels. The line is the least-squares regression. The lower panel is the corresponding residual plot.

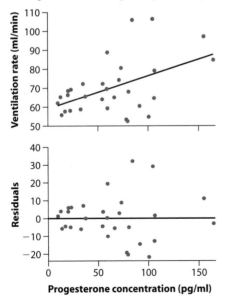

a. What is a "least-squares" regression line?

b. What are residuals?

c. Assume that the random sampling assumption is met. By viewing these plots, assess whether each of the three other main assumptions of linear regression is likely to be met in this study.

8. The slopes of the regression lines on the following graph show that the winning Olympic 100-m sprint times for men and women have been getting shorter and shorter over the years, with a steeper trend in women than in men (the graph is modified from Tatem et al. 2004). If trends continue, women are predicted to have a shorter winning time than men by the year 2156. What cautions should be applied to this conclusion? Explain.

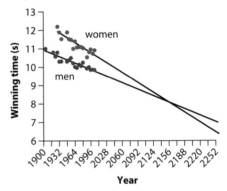

9. In an analysis of the performance of Major League Baseball players, Schaal and Smith (2000) found that the batting scores of the top 10 players in the 1997 baseball season dropped on average in 1998. What is the best interpretation of this finding?

a. Players who did well in 1997 reduced their effort the following year, realizing that they didn't need to work as hard to get an above-average result.

b. Players performing above average in 1997 were older and more worn out by 1998.

c. Regression toward the mean.

d. Possibly (a) and (b), but (c) is likely and cannot be ruled out.

---

10. Independence is a perilous assumption when comparing species (or other taxa) because species are related to one another to varying degrees in the phylogeny (see Interleaf 11).

10. Hybrid offspring of parents of different species are often sterile. How different must the parent species be, genetically, to produce this effect? The accompanying table (Moyle et al. 2004) lists the proportion of pollen grains that are sterile in hybrid offspring of crosses between pairs of species of *Silene* (bladder campions—see the photo on the first page of this chapter). Also listed is the genetic difference between the pair of species, as measured by DNA sequence divergence. Assume that different species pairs are independent.

Silene species pair	Genetic distance	Proportion of pollen that is sterile
1	0.00	0.02
2	0.00	0.06
3	0.00	0.14
4	0.00	0.24
5	0.00	0.30
6	0.03	0.62
7	0.02	0.28
8	0.03	0.23
9	0.04	0.15
10	0.04	0.45
11	0.05	0.84
12	0.11	0.65
13	0.12	0.77
14	0.12	1.00
15	0.13	1.00
16	0.13	0.93
17	0.13	0.91
18	0.14	0.93
19	0.13	0.96
20	0.13	1.00
21	0.15	1.00
22	0.16	0.97
23	0.18	1.00

a. We would like to predict the proportion of hybrid pollen that is sterile (Y) from the genetic distance between the species (X). Since the response variable is a proportion, what transformation would be your first choice to help meet the assumptions of linear regression?

b. Transform the proportions and then produce a scatter plot of the data. Estimate and draw the regression line.

c. Calculate the 95% confidence interval for the slope of the line.

11. Rattlesnakes often eat large meals that require significant increases in metabolism for efficient digestion. Snakes are known to adjust their thermoregulatory behavior after feeding, seeking out warmer spots to increase their metabolic rates. Can snakes increase body temperature, though, even without this behavior? Tattersall et al. (2004) measured the change in body temperature of snakes after meals of various sizes, and we have used their data in an inappropriate way in the following graph, fitting a nonlinear mathematical function indicated by the curve.

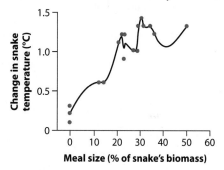

a. Why is the nonlinear fit shown inappropriate?

b. What alternative procedure would you recommend to achieve the goal of predicting snake body-temperature change from meal size?

12. Male lizards in the species *Crotaphytus collaris* use their jaws as weapons during territorial interactions. Lappin and Husak (2005) tested whether weapon performance (bite force) predicted territory size in this species. Their measurements for both variables are listed in the following table for 11 males.

Bite force (N)	Territory area (m²)
28.2	437
33.9	589
29.5	871
39.8	977
41.7	1288
44.7	2138
46.8	2455
47.9	3548
36.3	2692
35.5	2042
33.9	3020

a. How rapidly does territory size increase with bite force? Estimate the slope of the regression line. Provide a standard error for your estimate.

b. How uncertain is our estimate of slope? Provide a 99% confidence interval for β.

c. Provide an interpretation for the 99% confidence interval in part (b). What does it measure?

d. Bite force is difficult to measure accurately, and so the values shown probably include some measurement error. Is the slope of the true regression line most likely to be underestimated, overestimated, or unaffected as a result?

e. Territory area is difficult to measure accurately, so the values shown probably include some measurement error. Is the slope of the true regression line most likely to be underestimated, overestimated, or unaffected as a result?

13. An ANOVA carried out to test the null hypothesis of zero slope for the regression of lizard territory area on bite force (see Practice Problem 12) yielded the following results.

Source of variation	Sum of squares	df	Mean squares	F-ratio
Regression	3758539	1		
Residual	7303662	9		
Total				

a. Complete the ANOVA table.

b. Using the F-statistic, test the null hypothesis of zero slope at the significance level $\alpha = 0.05$.

c. What are your assumptions in part (b)?

d. What does the $MS_{residual}$ measure?

e. Calculate the R^2 statistic. What does it measure?

14. James et al. (1997) demonstrated that the chemical hypoxanthine in the vitreous humour (the colorless jelly filling the eye) shows a postmortem linear increase in concentration with time since death. This suggests that hypoxanthine concentration might be useful in predicting time of death when it is unknown. The following graph shows measurements collected by the researchers on 48 subjects whose time of death was known. The regression line, the 95% confidence bands, and the 95% prediction interval are included on the graph.

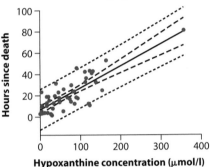

a. The data set depicted in the graph includes one conspicuous outlier on the far right. If you were advising the forensic scientists who gathered these data, how would you suggest they handle the outlier?

b. What do the confidence bands measure?

c. Are the inner dashed lines the confidence bands or the prediction interval?

d. If the regression depicted in the graph was to be used to predict the time of death in a murder case, which bands would provide the most relevant measure of uncertainty, the confidence bands or the prediction interval? Why?

15. Social spiders live together in kin groups that build communal webs and cooperate in gathering prey. The following web measurements were

gathered on 17 colonies of the social spider *Cyrtophora citricola* in Gabon (Rypstra 1979).

Colony	Height of web aboveground (cm)	Number of spiders
1	90	17
2	150	32
3	270	96
4	320	195
5	180	372
6	380	135
7	200	83
8	120	36
9	240	85
10	120	20
11	210	82
12	250	95
13	140	59
14	300	89
15	290	152
16	180	62
17	280	64

a. Use these data to draw a scatter plot of the relationship between the colony height aboveground (explanatory variable) and the number of spiders in the colony (response variable).

b. Examine the scatter plot and determine any impediments that might make it difficult to use linear regression to predict number of spiders in a colony from colony height.

c. In view of what you discerned in part (b), what method would you recommend to test whether colony height predicts the number of spiders?

16. Identify the assumption(s) of linear regression that is (are) violated in each of the following residual plots.

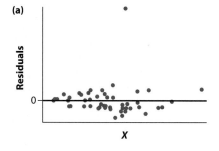

17. The forests of the northern United States and Canada have no native terrestrial earthworms, but many exotic species (including those used as bait when fishing) have been introduced. These immigrant species are dramatically changing the soil. The following data were gathered to predict the nitrogen content of mineral soils of 39 hardwood forest plots in Michigan's Upper Peninsula from the number of earthworm species found in those plots (Gundale et al. 2005).

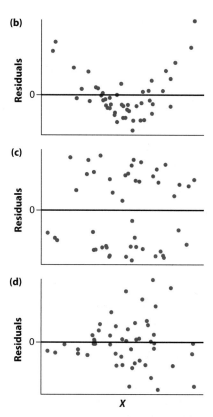

Earthworm species	Nitrogen content (%)
0	0.22, 0.19, 0.16, 0.08, 0.05
1	0.33, 0.30, 0.26, 0.24, 0.20, 0.18, 0.14, 0.13, 0.11, 0.09, 0.08
2	0.27, 0.24, 0.23, 0.18, 0.16, 0.13
3	0.32, 0.32, 0.29, 0.24, 0.22, 0.12, 0.40
4	0.34, 0.33, 0.23, 0.21, 0.18, 0.17, 0.15, 0.14
5	0.20, 0.54

a. Draw a scatter plot of these data, using the number of earthworm species as the explanatory variable.

b. Using the following intermediate calculations, calculate the regression line to predict the total nitrogen content of the soil from the number of earthworm species present. Add the line to your plot.

$$\bar{X} = 2.205 \qquad \bar{Y} = 0.215$$

$$\sum_i (X_i - \bar{X})^2 = 86.359$$

$$\sum_i (Y_i - \bar{Y})^2 = 0.366$$

$$\sum_i (X_i - \bar{X})(Y_i - \bar{Y}) = 2.453.$$

c. What are the units of your estimate of slope, b?

d. What is the predicted nitrogen content of soil having five earthworm species?

e. Calculate a standard error of the slope.

f. Produce a 95% confidence interval for the slope.

18. Is the scaling of respiratory metabolism to body size in plants similar to that found in animals, where an approximate 3/4-power relation seems to hold? The data below are measurements of aboveground mass (in g) and respiration rate (in nmol/s) in 10 individuals of Japanese cypress trees (*Chamaecyparis obtusa*). They were obtained from a larger data set amassed by Reich et al. (2006). Respiratory metabolism (Y) is expected to depend on body mass (X) by the power law, $Y = \alpha X^\beta$, where β is the scaling exponent.

Aboveground mass (g)	Respiration rate (nmol/s)
453	666
1283	643
695	1512
1640	2198
1207	2535
2096	4176
2804	3196
3528	3494
5940	7386
10,000	10,363

a. Use linear regression to estimate β for Japanese cypress. Include a standard error of your estimate.

b. Plot your line and the data in a scatter plot.

c. Use the 95% confidence interval to determine the range of most-plausible values for β based on these data. Does this range include the value $3/4$?

d. Carry out a formal test of the null hypothesis that $\beta = 3/4$.

e. It is a challenge to estimate mass and respiration rate of a living tree in the field, and both measurements are likely subject to measurement error. How is measurement error in each of these two traits likely to affect the estimate of the exponent?

ASSIGNMENT PROBLEMS

19. You might think that increasing the resources available would elevate the number of plant species that an area could support, but the evidence suggests otherwise. The data in the accompanying table are from the Park Grass Experiment at Rothamsted Experimental Station in the U.K., where grassland field plots have been fertilized annually for the past 150 years (collated by Harpole and Tilman 2007). The number of plant species recorded in 10 plots is given in response to the number of different nutrient types added

Plot	Number of nutrients added	Number of plant species
1	0	36
2	0	36
3	0	32
4	1	34
5	2	33
6	3	30
7	1	20
8	3	23
9	4	21
10	4	16

in the fertilizer treatment (nutrient types include nitrogen, phosphorus, potassium, and so on).

a. Draw a scatter plot of these data. Which variable should be the explanatory variable (X), and which should be the response variable (Y)?

b. What is the rate of change in the number of plant species supported per nutrient type added? Provide a standard error for your estimate.

c. Add the least-squares regression line to your scatter plot. What fraction of the variation in the number of plant species is "explained" by the number of nutrients added?

d. Test the null hypothesis of no treatment effect on the number of plant species.

20. Heusner (1991) assembled the following data on the mass and basal metabolic rate of 17 species of primates, including the potto shown in the accompanying photo.

Species	Mass (g)	Basal metabolic rate (watts)
Alouatta palliata	4670.0	11.6
Aotus trivirgatus	1020.0	2.6
Arctocebus calabarensis	206.0	0.7
Callithrix jachus	190.0	0.9
Cebuella pygmaea	105.0	0.6
Cheirogaleus medius	300.0	1.1
Euoticus elegantilus	261.5	1.2
Galago crassicaudatus	1039.0	2.9
Galago demidovii	61.0	0.4
Galago elegantulus	261.5	1.2
Homo sapiens	70,000.0	82.8
Lemur fulvus	2330.0	4.2
Nycticebus coucang	1300.0	1.7
Papio anubis	9500.0	16.0
Perodicticus potto	1011.0	2.1
Saguinus geoffroyi	225.0	1.3
Saimiri sciureus	800.0	4.4

Previous research has indicated that basal metabolic rate (R) of mammal species depends on body mass (M) in the following way: $R = \alpha M^{\beta}$, where α and β are constants.

a. Use linear regression to estimate β for primates. Call your estimate b.

b. Plot your line and the data in a scatter plot.

c. How precise is the estimate of β? Provide a standard error for b and a 95% confidence interval for β. Assume that the species data are independent.

21. Previous evidence and some theory predict that the exponent β describing the relationship between metabolic rate and mass should equal 3/4. Using the data from Assignment Problem 20, test whether the exponent differs from the expected value of 3/4.

22. The white forehead patch of the male collared flycatcher is important in mate attraction. Griffith and Sheldon (2001) found that the length of the patch varied from year to year. They measured the forehead patch on a sample of 30 males in two consecutive years, 1998 and 1999, on the Swedish island of Gotland. The scatter plot provided gives the pair of measurements for each male. The solid regression line predicts the 1999 measurement from the 1998 measurement. The dashed line is drawn through the means for 1998 and 1999, but it has a slope of one. The difference between the two lines indicates that males with the longest patches in 1998 had smaller patches in 1999, relative to the other birds. Similarly, the males with the smallest patches in 1998 had larger patches, on average, in 1999, relative to other birds.

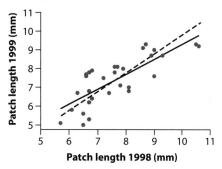

a. The following table summarizes the data. Use these numbers to calculate the regression slope.

	Mean	Sum of squares	Sum of products
Patch length 1998	7.62	45.43	36.26
Patch length 1999	7.40	47.03	

b. Now let the patch length in 1998 be the response variable (Y). Use the patch length in 1999 to predict patch length in 1998. What is the slope of this new regression?

c. What is the most likely reason that the slope is less than one in both regressions?

23. Seedlings of understory trees in mature tropical rainforests must survive and grow using intermittent flecks of sunlight. How does the length of exposure to these flecks of sunlight (fleck duration) affect growth? Leakey et al. (2005) experimentally irradiated seedlings of the Southeast Asian rainforest tree *Shorea leprosula* with flecks of light of varying duration while maintaining the same total irradiance to all the seedlings. Their data for 21 seedlings are listed in the following table.

Tree	Mean fleck duration (min)	Relative growth rate (mm/mm/week)
1	3.4	0.013
2	3.2	0.008
3	3.0	0.007
4	2.7	0.005
5	2.8	0.003
6	3.2	0.003
7	2.2	0.005
8	2.2	0.003
9	2.4	0.000
10	4.4	0.009
11	5.1	0.010
12	6.3	0.009
13	7.3	0.009
14	6.0	0.016
15	5.9	0.025
16	7.1	0.021
17	8.8	0.024
18	7.4	0.019
19	7.5	0.016
20	7.5	0.014
21	7.9	0.014

$$\bar{X} = 5.062 \qquad \bar{Y} = 0.0111$$

$$\sum_i (X_i - \bar{X})^2 = 100.210$$

$$\sum_i (Y_i - \bar{Y})^2 = 0.001024$$

$$\sum_i (X_i - \bar{X})(Y_i - \bar{Y}) = 0.2535.$$

a. What is the rate of change in relative growth rate per minute of fleck duration? Provide a standard error for your estimate.

b. Using these data, test the hypothesis that fleck duration affects seedling growth rate.

c. Calculate a 99% confidence interval for the slope of the population regression.

d. What are your assumptions in parts (a)–(c)?

e. What is the main procedure you would employ to evaluate those assumptions?

24. How do we estimate a regression relationship when each subject is measured multiple times over a series of X-values? The easiest approach is to use a *summary* slope for each individual and then calculate the average slope. Green et al. (2001) dealt with exactly this type of problem in their study of macaroni penguins exercised on treadmills. Each penguin was exercised at a range of speeds, and its oxygen consumption was measured in relation to its heart rate (a proxy for metabolic rate). The graph provided shows the relationship for just two individual penguins.

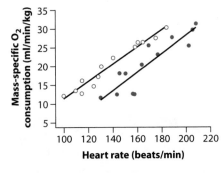

The following table lists the estimated regression slopes for each of 24 penguins in three categories.

Group	Regression slope
Breeding males	0.31, 0.34, 0.30, 0.38, 0.35, 0.33, 0.32, 0.32, 0.37
Breeding females	0.30, 0.32, 0.23, 0.38, 0.31, 0.26, 0.42, 0.28, 0.35
Molting females	0.25, 0.41, 0.32, 0.34, 0.27, 0.23

a. Calculate the mean, standard deviation, and sample size of the slope for penguins in each of the three groups. Display your results in a table.

b. Test whether the means of the slopes are equal between the three groups.

25. Many species of beetle produce large horns that are used as weapons or shields. The resources required to build these horns, though, might be diverted from other useful structures. To test this, Emlen (2001) measured the sizes of wings and horns in 19 females of the beetle species *Onthophagus sagittarius*. Both traits were scaled for body-size differences and hence are referred to as relative horn and wing sizes. Emlen's data are shown in the following scatter plot along with the least squares regression line ($Y = -0.13 - 132.6X$).

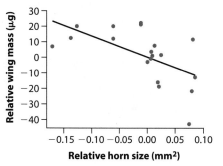

We used this regression line to predict the horn lengths at each of the 19 observed horn sizes. These are given in the following table along with the raw data.

Relative horn size (mm²)	Relative wing mass (μg)	Predicted relative wing mass (μg)
0.074	−42.8	−9.9
0.079	−21.7	−10.6
0.019	−18.8	−2.6
0.017	−16.0	−2.4
0.085	−12.8	−11.4
0.081	11.6	−10.9
0.011	7.6	−1.6
0.023	1.6	−3.2
0.005	3.7	−0.8
0.007	1.1	−1.1
0.004	−0.8	−0.7
−0.002	−2.9	0.1
−0.065	12.1	8.5
−0.065	20.1	8.5
−0.014	21.2	1.7
−0.014	22.2	1.7
−0.132	20.1	17.4
−0.143	12.5	18.8
−0.177	7.0	23.3

a. Use these results to calculate the residuals.

b. Use your results from part (a) to produce a residual plot.

c. Use the graph provided and your residual plot to evaluate the main assumptions of linear regression.

d. In light of your conclusions in part (c), what steps should be taken?

26. Can the songs of extinct species be predicted? Gua et al. (2012) used measurements of living species of katydid to predict the call frequency, or "pitch," of the extinct *Archaboilus musicus* based on a 165-million-year-old fossil. Male katydids call by stridulating—rubbing forewings together so that a scraper on one wing rubs against a "file" on the other. Call frequency is predicted by file length (see accompanying graph; the data are available at whitlockschluter. zoology.ubc.ca). File length of a single well-preserved fossil of the extinct *Archaboilus musicus* was 9.34 mm. What was its call frequency?

Summary for log-transformed data are as follows:

$$n = 58, \quad \sum_i X_i = 33.241, \quad \sum_i Y_i = 183.936,$$

$$\sum_i X_i^2 = 42.615, \quad \sum_i Y_i^2 = 609.994,$$

$$\sum_i X_i Y_i = 86.720.$$

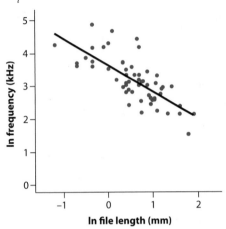

a. Calculate the regression line from the summary numbers provided. Assume for the purpose of this exercise that the data points are independent.[11]

b. On the basis of this regression, what is the predicted log-transformed call frequency of *Archaboilus musicus*? The log file length for this species is 2.23.

c. What is the most-plausible range of values for the stridulation frequency of the 165-million-year-old katydid? Give the appropriate 95% confidence interval or prediction interval to determine this.

d. Calls with a frequency above about 20 kHz [or ln(frequency) of about 3.0] are ultrasonic and inaudible by most humans. How confident can we be that the calls of *Archaboilus musicus* were audible to humans? Answer based on your confidence or prediction interval in part (c).

27. The parasitic bacterium *Pasteuria ramosa* castrates and later kills its host, the crustacean *Daphnia magna*. The length of time between infection and host death affects the number of spores eventually produced and released by the parasite, as the following scatter plot reveals. The *x*-axis measures age at death for 32 infected host individuals, and the response variable is the square-root-transformed number of spores produced by the infecting parasite (Jensen et al. 2006).

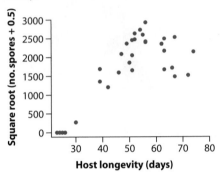

a. Describe the shape of the relationship between the number of spores and host longevity.

b. What equation would be best to try first if you wanted to carry out a nonlinear regression of Y on X?

28. Human brains have a large frontal cortex with excessive metabolic demands compared with the brains of other primates. However, the human brain is also three or more times the size of the brains of other primates. Is it possible that the metabolic demands of the human frontal cortex are just an expected consequence of greater brain size? Sherwood et al. (2006) investigated this question in a number of ways. Their data in the accompanying table and scatter plot shows the relationship between the glia–neuron ratio (an indirect measure of the metabolic requirements of brain neurons) and the log-transformed brain mass in nonhuman primates. A linear regression is drawn through these data.

---

11. The authors of the study used a method to correct for non-independence introduced by phylogeny (see Interleaf 11), but their prediction was similar.

Species	Brain mass (g)	ln (brain mass)	Glia–neuron ratio
Homo sapiens	1373.3	7.22	1.65
Pan troglodytes	336.2	5.82	1.20
Gorilla gorilla	509.2	6.23	1.21
Pongo pygmaeus	342.7	5.84	0.98
Hylobates muelleri	101.8	4.62	1.22
Papio anubis	155.8	5.05	0.97
Mandrillus sphinx	159.2	5.07	1.02
Macaca maura	92.6	4.63	1.09
Erythrocebus patas	102.3	4.53	0.84
Cercopithecus kandti	71.6	4.27	1.15
Colobus angolensis	74.4	4.31	1.20
Trachypithecus francoisi	91.2	4.51	1.14
Alouatta caraya	55.8	4.02	1.12
Saimire boliviensis	24.1	3.18	0.51
Aotus trivirgatus	13.2	2.58	0.63
Saguinus oedipus	10.0	2.30	0.46
Leontopithecus rosalia	12.2	2.50	0.60
Pithecia pithecia	30.0	3.40	0.64

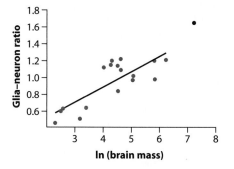

a. Determine the equation of the regression line for nonhuman primates.

b. Using the nonhuman primate relationship, what is the predicted glia–neuron ratio for humans, given their brain mass?

c. Determine the most-plausible range of values for the prediction. Which confidence interval is relevant for your prediction of human glia–neuron ratio in (b): the confidence interval for the predicted mean glia–neuron ratio at the given brain mass, or the interval for the prediction of a single new observation?

d. Carry out the calculation of the 95% confidence interval chosen in part (c). (See the Quick Formula Summary for the method.) Assume for the purpose of this exercise that the species data are independent.

e. On the basis of your result in part (d), does the human brain have an excessive glia–neuron ratio for its mass compared with other primates? Explain.

f. Considering the position of human data point relative to those data used to generate the regression line (see accompanying figure), what additional caution is warranted? Why?

29. Golenda et al. (1999) carried out a human clinical trial to investigate the effectiveness of a formulation of DEET (*N,N*-diethyl-*m*-toluamide) in preventing mosquito bites. The study applied DEET to the underside of the left forearm of volunteers. Cages containing 15 fresh mosquitoes were then placed over the skin for five minutes, and the number of bites was recorded. This was repeated four times at intervals of three hours. The scatter plot provided displays the total number of bites (square-root transformed) received by 52 women in the study in relation to the dose of DEET they received.

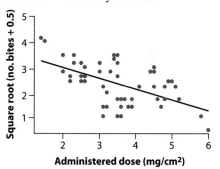

a. What are the uses of the square-root transformation in linear regression?

b. What feature of this study justifies our calling it an *experimental* study rather than just an observational study?

c. Complete the ANOVA table for these data.

Source of variation	Sum of squares	df	Mean squares	F-ratio
Regression	9.97315			
Residual				
Total	32.0569			

d. Use the *F*-statistic to test the null hypothesis of zero slope.

e. Calculate the R^2 statistic. What does it measure?

30. Calculating the year of birth of cadavers is a tricky enterprise. One method proposed is based on the radioactivity of the enamel of the body's teeth. The proportion of the radioisotope ^{14}C in the atmosphere increased dramatically during the era of aboveground nuclear bomb testing between 1955 and 1963. Given that the enamel of a tooth is non-regenerating, measuring the ^{14}C content of a tooth tells when the tooth developed, and therefore the year of birth of its owner. Predictions based on this method seem quite accurate (Spalding et al. 2005), as shown in the accompanying graph. The *x*-axis is $\triangle^{14}C$, which measures the amount of ^{14}C relative to a standard (as a percentage).

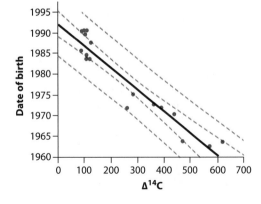

There are three sets of lines on this graph. The solid line represents the least-squares regression line, predicting the actual year of birth from the estimate based on amount of ^{14}C. One pair of dashed lines shows the 95% confidence bands and the other shows the 95% prediction interval.

a. What is the approximate slope of the regression line?

b. Which pair of lines shows the confidence bands? What do these confidence bands tell us?

c. Which pair of lines shows the prediction interval? What does this prediction interval tell us?

31. A lot of attention has been paid recently to portion size in restaurants, and how it may affect obesity in North Americans. Portions have grown greatly over the last few decades. But is this phenomenon new? Wansink and Wansink (2010) looked at representations of the Last Supper in European paintings painted between about 1000 AD and 1700 AD. They scanned the images and measured the size of the food portions portrayed (relative to the sizes of heads in the painting). (For example, the painting reproduced here was painted by Ugolino di Nerio in 1234 AD.) They reported the year of the painting and the portion size as follows:

Portion size	Year	Portion size	Year
3.08	999	3.99	1520
2.70	1004	4.24	1542
2.14	1050	4.80	1515
2.91	1098	5.40	1522
3.69	1314	5.27	1568
4.41	1314	5.44	1554
3.51	1350	5.44	1544
2.44	1309	5.70	1544
3.21	1398	3.04	1561
2.78	1400	3.30	1573
3.39	1467	3.47	1618
3.21	1458	3.56	1626
3.17	1486	5.87	1707
2.78	1494	1.93	1153
2.57	1479	1.76	1434
3.30	1527	1.84	1426
3.81	1525		

a. Calculate a regression line that best describes the relationship between year of painting and

the portion size. What is the trend? How rapidly has portion size changed in paintings?

b. What is the most-plausible range of values for the slope of this relationship? Calculate a 95% confidence interval.

c. Test for a change in relative portion size painted in these works of art with the year in which they were painted.[12]

d. Draw a residual plot of these data and examine it carefully. Can you see any cause for concern about using a linear regression? Suggest an approach that could be tried to address the problem.

32. Scarlet king snakes (left photo) are relatively harmless snakes from the southeastern United States. Most individuals have a conspicuous color pattern very similar to the extremely venomous coral snake (right photo). The king snake mimics are thought to gain a survival advantage when coral snakes are present, because predators have learned to avoid coral snakes. However, king snakes also live well outside the range of coral snakes, where the conspicuous colors of these mimics should make them more vulnerable than non-mimic king snakes, because the predators have not learned to avoid coral snakes. To test this, Harper and Pfennig (2008) compared predation rates on mimic and non-mimic king snake color patterns at locations with varying distance from the boundary of the range of coral snakes. The results are given in the table. The first variable is the distance in km between each study location and the boundary of the area where coral snakes are present. Negative numbers mean locations inside the range of coral snakes, and positive numbers mean locations outside the range. At each location, plasticine dummies of king snakes were set out in the habitat, with half the dummies painted to look like mimics and the other half like less-conspicuous non-mimics. The second variable in the table is the proportion of attacks by predators on the mimics at each location.

Distance from boundary	Proportion of attacks on mimics
−97	0
−47	0.01
−33	0
−23	0
−72	0.33
−23	0.5
152	0.4
−15	0.67
97	0.66
113	0.66
105	1
80	1
138	1
148	1
152	1
49	0.4
48	0

a. Give the equation for the line that best predicts proportion of attacks on mimics from the distance to the boundary. What is the trend? Assume that the relationship is linear over the range of the data (being a proportion, the true relationship cannot extend below zero or above one).

b. Test the hypothesis that distance to the boundary predicts the proportion of attacks on mimics.

---

12. A reply to this study by an art historian (Rich 2011) pointed out that what may have been changing was head size, not portion size. Apparently artists at that time used size as an indication of importance, so they depicted things like Jesus's head in the Last Supper as larger than life. However, later artists drew head sizes more realistically.

33. The warm temperatures of spring and summer arrive earlier now at high latitudes than they did in the past, as a result of human-caused climate change. One consequence is that many organisms start breeding earlier in the year than in previous years, often at suboptimal times. For example, historically the great tit *Parus major* (a well-studied European bird; see Example 2.3A) laid its eggs on dates that resulted in the chicks hatching around the time that caterpillars, a major source of food, became abundant. Currently, a shift in breeding date has led to a mismatch between hatching date and the dates when the caterpillars appear. Does this mismatch affect the growth rate of the bird population? To test this, Reed et al. (2013) used multiple years of study to examine the average timing mismatch (X), in days, with the growth rate of the bird population (Y), expressed as log of the ratio of the number of birds in one year over the number of birds in the previous year. A growth rate greater than zero indicates that the population is increasing, whereas a negative value indicates that the population is declining. Their data are summarized below (available at whitlockschluter. zoology.ubc.ca).

$$\sum_i X_i = 4.923885, \ \sum_i Y_i = 41.12394, \ n = 38,$$

$$\sum_i X_i^2 = 2005.83430, \ \sum_i Y_i^2 = 50.15243,$$

$$\sum_i X_i Y_i = -3.97524.$$

a. Find the formula of the line that best predicts population growth rate from mismatch. What is the trend in growth rate with timing mismatch?

b. What is the confidence interval for the slope of this line?

c. Are the data consistent with a "substantial" effect on population growth rate (where "substantial" refers to a decline of 0.1 or more in growth per 10-day mismatch, which would be enough to cause extinction with expected climate change)?

d. Is there a significant relationship between mismatch and growth rate of the population? Carry out a formal test.

34. Dads transmit many more new mutations than do mothers to their babies at conception. These mutations occur from copying errors during sperm production. There is increasing interest in the effect of father age on this process. As part of a larger study into the genetics of mental illness, Kong et al. (2012) used complete genome squencing of 21 father-child pairs to tally the total number of new mutations inherited from each father (in this particular sample, all the offspring were afflicted with schizophrenia). These counts are listed in the following table along with fathers' ages at offspring conception.

Age of father (years)	Number of new mutations
16	39
18	41
20	39
19	49
22	50
24	54
24	55
24	61
25	57
28	52
29	54
30	57
32	61
37	67
36	70
34	77
30	83
29	67
33	68
26	54
33	65

a. Graph the relationship between number of new mutations (Y) and father's age (X). Add the regression line to your plot.

b. Based on these data, how rapidly does the number of new mutations increase with father's age? Provide a standard error for your estimate.

c. What is the predicted mean number of new mutations from fathers 36 years of age? How does this compare with the predicted number for fathers only 18 years old?

d. Use the ANOVA approach to test the null hypothesis of no relationship between father's age and number of new mutations. Include an ANOVA table with your results.

e. What fraction of the variation among fathers in the number of new mutations is explained by father's age?

35. The threat of bioterrorism makes it necessary to quantify the risk of exposure to infectious agents such as anthrax (*Bacillus anthracis*). Hans (2002) measured the mortality of rhesus monkeys in an exposure chamber to aerosolized anthrax spores of varying concentration. The data are available at whitlockschluter.zoology. ubc.ca and are tabulated below.

Anthrax concentration (spores/l)	Survived	Died
29,300	7	1
32,100	4	4
45,300	3	5
57,300	2	6
64,800	3	5
67,000	5	3
100,000	0	8
125,000	1	7
166,000	0	8

a. Graph the relationship between mortality (Y) and anthrax concentration (X).

b. We would like to use these data to predict the probability of death based on anthrax concentration. Which assumptions of linear regression are violated by these data? Explain.

c. Which method could be used instead to predict mortality from anthrax concentration?

d. An analysis of these data using the method in part (c) yielded the following results. Using these results, what is the predicted mortality from a concentration of 100,000 spores/l?

	Estimate	SE
Intercept	−1.7445	0.6206
Slope	0.00003643	0.00001119

Model	df	Deviance	Residual df	Residual deviance
Null			71	92.982
Duration	1	19.02	70	73.962

e. Based on these results, add the regression curve to your plot in (a).

f. Based on these data, what is the concentration predicting a 50% mortality (include units)?

g. Using the results in part (d), test the null hypothesis of zero slope.

36. Do individual differences in stress physiology influence survival or reproduction in natural populations? Blas et al. (2007) investigated this question in a Spanish population of European white stork (*Ciconia ciconia*). The accompanying data display stress-induced corticosterone levels circulating in the blood of 34 storks, measured once when they were nestlings, and their survival over the subsequent five years of study. "Stress" involved restraining each stork for 45 minutes and then taking a blood sample.

Corticosterone (ng/ml)	Survival	Corticosterone (ng/ml)	Survival
26	1	31.2	0
28.2	1	34.9	0
29.8	1	35.9	0
34.9	1	41.8	0
34.9	1	43	0
35.9	1	45.1	0
37.4	1	46.8	0
37.6	1	46.8	0
38.3	1	47.4	0
39.9	1	47.4	0
41.6	1	47.7	0
42.3	1	47.8	0
52	1	50.7	0
26.6	0	51.6	0
27	0	56.4	0
27.9	0	57.6	0
31.1	0	61.1	0

a. Graph the relationship between survival (Y) and stress-induced corticosterone levels (X).

b. Give three reasons that linear regression would not be suitable for these data.

c. What regression method could be used instead to predict stork survival from corticosterone levels? How does the method overcome the problems noted in part (b)?

d. An analysis of these data using the method in (c) yielded the following results. Based on these results, add the regression curve to your plot in part (a).

	Estimate	SE
Intercept	2.70304	1.74725
Slope	−0.07980	0.04368

Model	df	Deviance	Residual df	Residual deviance
Null			33	45.234
Duration	1	3.84	32	41.396

e. Based on these data, what is the estimated concentration predicting a 50% mortality (include units)?

f. Using the results in part (d), test the null hypothesis of zero slope.

Using species as data points

Many types of studies in biology use species measurements as data points. We've encountered several cases in this book. In Chapter 2, for instance, we looked at the frequency distribution of bird abundances using the abundance measurements of different bird species as data points (Example 2.2B). In Chapter 16, we illustrated the association between brain and body mass in mammals by using measurements of different mammal species as data points. What we haven't told you yet is how tricky it can be to analyze such data. The trouble is that species data are not usually *independent*. The reason is that species share an evolutionary history. Here we explain the situation and what can be done about it.

The following study of lilies illustrates the problem created by shared evolutionary history. Patterson and Givnish (2002) found that lily species flowering in the low-light environment of the forest understory, such as the bluebead lily (*Clintonia borealis*; below left),

tend to have small and inconspicuous flowers that are whitish or greenish in color. Lilies that live in sunny, open habitats, or that live in deciduous woods but flower before the tree leaves come out, such as the Turk's-cap lily (*Lilium superbum*; below right), tend to have large, showy flowers. Data from 17 lily species, shown in the branching figure on the next page, indicate an almost perfect association between habitat and flower type. All 10 species flowering in open habitats had large and showy flowers. Six of the seven species flowering in shaded habitats had relatively small and inconspicuous flowers. A χ^2 contingency test with these data soundly rejects the null hypothesis of no association ($\chi^2 = 13.24$, $df = 1$, $P = 0.0003$).

However, this contingency test assumes that the data from the 17 species of lilies are independent. The figure indicates that this assumption is likely false. The branching tree in the figure is a *phylogeny,* indicating the

INTERLEAF

11

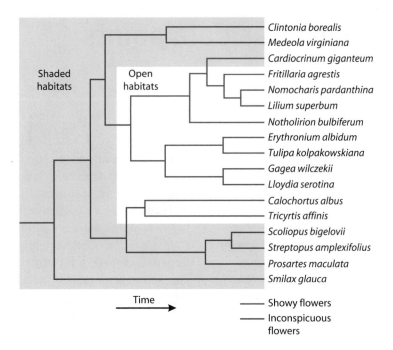

Shaded habitats

Open habitats

Clintonia borealis
Medeola virginiana
Cardiocrinum giganteum
Fritillaria agrestis
Nomocharis pardanthina
Lilium superbum
Notholirion bulbiferum
Erythronium albidum
Tulipa kolpakowskiana
Gagea wilczekii
Lloydia serotina
Calochortus albus
Tricyrtis affinis
Scoliopus bigelovii
Streptopus amplexifolius
Prosartes maculata
Smilax glauca

Time

——— Showy flowers
——— Inconspicuous flowers

ancestor-descendant relationships among the 17 lily species in the data set, which are at the tips of the tree. Branching points, or nodes, in the tree represent points in history when a single ancestor split into two descendant species. Two lily species at the tips of the tree are relatively closely related if they have a recent ancestor in common, such as *Nomocharis pardanthina* and *Lilium superbum*. Two species are more distantly related if their common ancestor is deeper in the tree, such as *Lilium superbum* and *Prosartes maculata*. The color of the branches in the figure indicates flower type, and the background shading indicates habitat type. (We can only guess about the habitat types and flower types of the ancestors, because they are not alive any more. The colors and shading used in the figure represent just one of the more likely possible scenarios for the transitions in habitat and flower type through history.)

The crucial insight from the tree is that closely related lily species tend to have the same flower type. They also tend to have the same habitat type. Both attributes were likely inherited from their common ancestor. In all, *closely related species are more similar on average than species picked at random.* This means that the species data points are not independent. The situation is like that confronted when pollsters interview more than one person from the same household, or when a bird researcher measures more than one chick from the same nest. However, in the present case, the non-independence is not the fault of the way the species were sampled by the researcher. It is generated by the process of evolution. This is a problem unique to biology.

The preceding example is about discrete variables, but the same issue arises when species data are numerical. The following

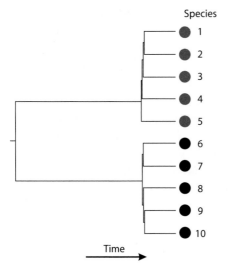

Species

1
2
3
4
5
6
7
8
9
10

Time

extreme case makes the point. The branching tree in the figure above shows a phylogeny for 10 hypothetical species. In this example, the species fall into two lineages that split from a common ancestor a long time ago (at the first branching point at the far left of the figure). The lineages have been evolving separately ever since. More recently, each lineage split into five new species more or less simultaneously. We've used distinct colors to represent species in the two groups.

Now consider two numerical variables, *X* and *Y*, measured on all 10 hypothetical species. It is often the case that species in the same group (indicated by red or black) will be more similar to each other in these traits

than to species in the alternative group, just because they share a more recent common ancestor. The scatter plot at bottom left shows example measurements for the 10 species. The symbols in the scatter plot indicate the main lineage to which the species belong.

If we paid no attention to the evolutionary relationships among the 10 species represented by the different symbols, and if we assumed that the species data were independent, we would conclude that *X* and *Y* were positively correlated ($r = 0.69$, $df = 8$, $P = 0.016$). However, it is clear from the scatter plot that closely related species have similar values of *X* and of *Y*, a likely outcome of the common history they shared until very recently. Within the two groups, there appears to be no correlation between *X* and *Y*.

What can be done about the non-independence of species data resulting from shared evolutionary history? Happily, methods have been developed that correct for the problem. They make it possible to test whether an association is present and to put confidence limits on the strength of the association. The most widely used method for analyzing associations between continuously varying species traits is known as **phylogenetically independent contrasts**, invented by Felsenstein (1985). His explanation of how and why it works is very clear, and we refer you to his original paper for details. Analogous methods have been developed for categorical species traits.[1]

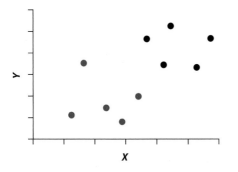

Y

X

---

1. Specialized computer programs are available to carry out phylogenetic comparative methods for continuously varying traits (such as phylogenetically independent contrasts) and discrete traits, such as MESQUITE (Maddison and Maddison 2011). Several contributed packages are available for the R statistical computing language (see the topic "Trait Evolution" at http:cran.r-project.org/web/views/Phylogenetics.html).

None of the methods that correct for the problem of non-independence of species traits are foolproof, because all make assumptions that can be difficult to verify. For example, they all assume that the process of trait evolution through time can be adequately mimicked by a simple mathematical model of a "random walk." If the mathematical model badly describes the process of evolution in a specific instance, then using the method can be even worse than ignoring the problem of shared history altogether and just using conventional statistical methods. Biologists nowadays tend to cover all bases and analyze their data both ways. Another strategy is to begin an analysis of species data by examining whether closely related species really are more similar on average than species picked at random. If not, then conventional statistical methods are adequate. If so, then the more specialized methods are used, often along with the results from conventional statistical methods, so that the outcomes can be compared.

Review Problems 3

1. The early movies by Eadweard Muybridge in the late 19th century showed for the first time the exact positions and movements of the legs during walking by horses and other large mammals. How much has this scientific analysis affected the representation of such animals in art? And how well do modern images of quadrupeds depict walking compared to images made by prehistoric humans? Horvath et al. (2012) examined a large number of images of horses and other animals in art created after Muybridge, in art made by modern humans before Muybridge, and in art from prehistoric humans depicted in cave paintings. For each image, they assessed whether the animal was presented in a biologically realistic posture. The data are at the bottom of the page.

   a. Draw a graph of these data. What is the pattern?

   b. Is there a statistically significant difference in modern images before and after Muybridge in the probability of getting the posture correct?

   c. What assumptions are you making in (b)?

   d. Calculate a confidence interval for the proportion of prehistoric paintings that depicted walking posture correctly.

2. Assume that the few remaining hairs on a balding man's head occur independently of each other and with equal probability for each square centimeter (cm) of scalp. Imagine that this man has 2.3 hairs per square cm on average. What is the probability that a randomly chosen square cm of his scalp has exactly four hairs?

3. Many species have "assortative mating," meaning that a female is more likely to mate with a male that is similar to her in some particular feature, such as body size, than with a dissimilar male. Imagine a female butterfly weighing 0.4 g in a population where the weight of males is normally distributed with mean 0.3 g and standard deviation 0.08 g. Assume that the female encounters males independently of his body weight.

   a. What is the probability that the first male she encounters is within 0.1 g of her own weight?

   b. What is the probability that the first five males she encounters are all more than 0.1 g different from her in body weight?

4. *Spot the flaw.* In their more recent study of "high-rise syndrome" (see Chapter 1), Vnuk et al. (2004) reported injury scores (0–4) of 119 fallen cats brought to a veterinary clinic in

TABLE FOR PROBLEM 1

Period	Correct walking posture	Incorrect walking posture
Prehistoric	21	18
Modern (pre-Muybridge)	45	227
Modern (post-Muybridge)	289	397

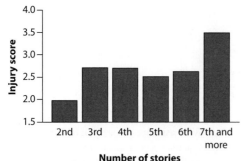

Number of stories

Zagreb, Croatia. The graph illustrates the average injury scores by the number of stories fallen. Identify the two principles of good graph design that are violated in this figure.

5. Most of us would like science to find ways to extend our lives. Genetic research has found some promising variants of genes in other organisms that greatly increase life span. Some mutations in the gene *daf-2* cause the worm *Caenorhabditis elegans*[1] to live almost three times as long as normal worms. But does this greater life span come at a cost? Below are data for the average life span (in days) of worms having one of 14 different mutations at the *daf-2* gene, along with data on the number of offspring produced during their lives (expressed as a percentage of number of offspring of normal

daf-2 mutation	Life span (days)	Relative number of offspring
e1365	28.2	101
m577	25.8	95
sa193	33.5	96
e1371	31.8	99
e1368	33.2	85
m41	27.0	87
m212	48.4	95
e1369	52.8	88
m120	32.3	93
e1370	33.8	70
m596	36.8	75
m579	44.3	73
e1391	63.4	61
e979	50.3	70

unmutated worms) (Gems et al. 1998). Here we wish to calculate the correlation between these variables.

a. Both variables have positive skew. Find an appropriate transformation to reduce the skew (you will find that skew can be reduced but not altogether eliminated).

b. Include a plot of the transformed data. What trend is suggested?

c. Is life span significantly correlated with number of offspring? Answer using the transformed data.

6. Mendel famously discovered the basics of genetics with garden peas. He proposed a law of independent assortment, that the inheritance of different genes should be independent. We now know that this "law" is erroneous because genes that are linked on the same chromosome tend to be inherited together. Mendel (1866) used the following data from a cross of peas to test the predictions made by independent assortment.

Yellow smooth: 315

Yellow wrinkled: 101

Green smooth: 108

Green wrinkled: 32

The traits yellow/green and smooth/wrinkled are determined by different genes. If independent assortment were true, then the traits of the offspring of the cross should have the following proportions: 9/16 yellow smooth peas, 3/16 yellow wrinkly peas, 3/16 green smooth peas, and 1/16 green wrinkly pea. Test whether Mendel's prediction about the proportions is consistent with these data.

7. Van Hylckama Vlieg et al. (2009) investigated the relationship between oral contraceptive use and thrombosis in women. In a sample of 1524 adult female patients who had thrombosis, 103 had taken oral contraceptives regularly. In a second sample of 1760 women from the same population who did not have thrombosis, 658 had taken oral contraceptives regularly.

---

1. The lowly worm *C. elegans* is a popular study organism in aging research. See also Chapter 15, Assignment Problem 20. The *daf-2* protein is an insulin receptor.

a. What type of study design was used?
b. Graph the data. Which treatment condition (oral contraceptive use) had the higher proportion of women with thrombosis?
c. Test for an association between oral contraceptive use and thrombosis.
d. What is the odds ratio of thrombosis in women taking oral contraceptives compared to women not taking oral contraceptive? Include a confidence interval for the population odds ratio.
e. Under what circumstances can we say that the odds ratio estimated in part (d) is a reasonable estimate of the relative risk of thrombosis?

8. Studying the influence of metabolic differences among individuals on survival or reproductive success in nature requires that an individual's metabolism doesn't vary too much from time point to time point. To investigate, Hayes and O'Connor (1999) measured repeatability of thermogenic capacity (ability to generate heat) by recording maximal rate of oxygen consumption (VO_2max) in high-altitude deer mice exposed to cold temperatures in a wind tunnel. A sample of 34 mice were measured twice about 68 days apart. The two measurements on each mouse, in ml/min, are given below.

Mouse	VO_2max	Mouse	VO_2max
1	5.62, 5.84	18	5.75, 5.92
2	5.13, 4.75	19	4.41, 5.15
3	5.00, 5.65	20	4.63, 4.82
4	5.76, 6.07	21	5.59, 6.42
5	6.10, 5.22	22	5.14, 5.31
6	5.11, 5.68	23	5.18, 5.30
7	5.63, 6.10	24	5.13, 5.21
8	5.40, 6.20	25	4.80, 4.64
9	4.62, 5.54	26	5.69, 6.17
10	5.06, 5.56	27	5.00, 3.70
11	4.29, 5.41	28	4.98, 5.32
12	5.24, 5.58	29	5.33, 5.86
13	4.81, 5.04	30	5.15, 5.50
14	5.84, 5.69	31	4.79, 5.46
15	5.34, 5.66	32	5.95, 5.85
16	5.53, 5.89	33	5.91, 6.04
17	5.59, 4.80	34	4.78, 4.83

a. Calculate the variance components of VO_2max within and among deer mice.
b. What is the repeatability of thermogenic capacity, as measured using VO_2max under cold exposure?
c. What are your assumptions in parts (a) and (b)?

9. Collins and Bell (2004) investigated the impacts of elevated carbon dioxide (CO_2) concentrations on plant evolution. They raised separate lines of the unicellular algae *Chlamydomonas* under normal and high CO_2 levels. After 1000 generations, they measured the growth rate of all of the experimental lines in a high CO_2 environment. The results for 14 experimental lines are presented in the following table. Growth rate is measured relative to the starting strain and has no units. Use these data to test whether the mean growth rate is associated with the CO_2 treatment.

CO_2 treatment	Growth rate
Normal	2.31
Normal	1.95
Normal	1.86
Normal	1.59
Normal	1.55
Normal	1.30
Normal	1.07
High	2.37
High	1.89
High	1.55
High	1.49
High	1.26
High	1.20
High	0.98

10. To investigate whether subcutaneous fat provides insulation in humans, Sloan and Keatinge (1973) measured the rate of heat loss by boys swimming for up to 40 min in water at 20.3°C and expending energy at about 4.8 kcal/min. Heat loss was measured by the change in body temperature, recorded using a thermometer under the tongue, divided by time spent swimming, in minutes. The authors measured an index of body "leanness" on each boy as the reciprocal of the skin-fold thickness adjusted for

total skin surface area (in meters squared) and body mass (in kg). Their data are listed in the following table.

Body leanness (m/kg)	Heat-loss rate (°C/min)
7.0	0.103
7.0	0.097
6.2	0.090
5.0	0.091
4.4	0.071
3.3	0.024
3.6	0.014
2.8	0.041
2.4	0.031
2.1	0.010
2.1	0.006
1.7	0.002

a. Draw a scatter plot of these data, showing the relationship.

b. Does body leanness predict heat-loss rate? Using the following intermediate calculations, calculate the regression line and add it to your plot in part (a). Carry out a formal test.

$$\bar{X} = 3.96667 \qquad \bar{Y} = 0.04833$$

$$\sum_i (X_i - \bar{X})^2 = 41.14667$$

$$\sum_i (Y_i - \bar{Y})^2 = 0.01696$$

$$\sum_i (X_i - \bar{X})(Y_i - \bar{Y}) = 0.78053.$$

c. How uncertain is the estimate of slope? Calculate a 95% confidence interval.

d. What are your assumptions in parts (b) and (c)?

e. What fraction of the variation in heat-loss rate is predictable from body leanness?

11. For each of the following scenarios, state what statistic would be used to estimate the effect of interest.

a. How different are the two hospitals X and Y in the frequency of doctors who wash and do not wash their hands before medical procedures?

b. How different is the number of bacteria on hands between people who wash for one

minute and people who do not wash their hands?

c. How different are athletes and nonathletes in the mean number of mitochondria per muscle cell?

d. How different is the number of mitochondria per cell between the muscles of people's dominant arm and the muscles in their other arm?

e. What fraction of individuals in an elephant population are male?

f. How different are the frequencies of males in two populations of elephants?

g. How much variation in weight is there among individuals in an elephant population?

h. How strong is the association between the number of mitochondria per cell in arm muscles and leg muscles?

12. For each of the following scenarios or questions, say which method for hypothesis testing would be most appropriate to best answer the scientific question. Unless otherwise stated, make any necessary assumptions. Be as specific as possible. (*Do not try to answer the biological question posed; just say what statistical technique would be best.*)

a. Do Hospital A and Hospital B differ in the frequency of doctors who wash their hands before medical procedures?

b. Does the mean rate at which doctors wash their hands before medical procedures vary among the three hospitals X, Y, and Z?

c. Does the rate of hand washing at hospitals predict the proportion of patients catching infections?

d. Does washing hands for five minutes leave a different number of bacteria on people's hands than washing for one minute, on average?

e. Which group washes hands for the greatest mean number of minutes each time, doctors or nurses?

f. Does whether or not doctors wash their hands before the first examination of patients have an effect on the lengths of patient stays in the hospital?

g. Do athletes have more mitochondria per muscle cell on average than nonathletes? (Assume that mitochondria per cell is normally distributed among individuals.)

h. Do athletes have a greater mean number of mitochondria per cell than nonathletes? (Assume that the number of mitochondria is not normally distributed but they have the same shape of distribution in the two groups.)

i. Is the mean number of mitochondria per cell in the muscles of people's dominant arm different from the number in their other arm?

j. Is the proportion of males in an elephant population equal to 0.50?

k. Do two populations of elephants have the same proportion of males?

l. Are the left tusks of elephants on average longer than their right tusks?

m. Are elephants spread out over the savanna independently and with equal probability everywhere?

n. Is the length of elephants' trunks normally distributed?

o. Are male elephants more variable in weight than are females?

p. Do male and female elephants differ in their mean growth rates? (Assume that elephant growth rate is not normally distributed but males and females have the same shape of distribution.)

q. Does the thickness of an elephant's first left molar predict its age?

r. Does the thickness of an elephant's first left molar predict whether it lives at least to 5 years of age?

13. The naked mole rat is a very unusual creature. For one thing, it is the only known mammal that is eusocial, with most individuals forgoing reproduction and instead helping to raise the offspring of the "king" and "queen" of their colony. They also live many times longer than other animals their size, and even up to twice as long as their two closely related species, the blind mole rat and the Damaraland mole rat (Edrey et al. 2012). It is possible that this difference is in part caused by differential expression of a tran-

scription factor called HIF1-α, which regulates proteins called neuregulins (neural growth factors thought to be involved in maintaining nerve function). The data below give measurements of HIF1-α expression in several individuals from each of the three species (expressed as a percentage of the expression of actin, a common protein used as a reference). Does the expression of HIF1-α differ among these three species?

Naked mole rat: 3.5, 3.8, 5.6, 12.9, 13.9, 28.2

Blind mole rat: 5.2, 8.7, 8.9, 11.4, 12.6

Damaraland mole rat: 4.3, 5.2, 8.4, 10.2, 10.2, 20.6

a. Show the data in a graph. What trend is suggested?

b. Do the species differ significantly in their mean amount of HIF1-α? (Use a log transformation to improve the fit to assumptions.)

14. Previous studies have shown that the antibody titers in obese people are lower after vaccination than in people of normal weight. One suggested reason is that the vaccines may not effectively penetrate the layer of subcutaneous fat in obese individuals. To test this, Middleman et al. (2010) compared the response to hepatitis B virus vaccine in obese participants in two different groups. The researchers vaccinated one group of 10 individuals with standard 1-inch (2.5 cm) needles. They used 1.5-inch (3.8 cm) needles instead for a second group of 14 individuals. They later measured the antibody titers (in units of mIU/ml) of each participant. Greater num-

bers indicate a more successful response to the vaccine. These results are as follows.

Short-needle group: 51.6, 87.4, 143.6, 144.6, 189.7, 189.8, 208.9, 324.7, 368, 383.9

Long-needle group: 28.0, 181.6, 203.9, 243, 249.6, 274.3, 341.2, 349.6, 393.0, 429.2, 464.2, 473.1, 492.9, 647.0

a. What is the most-plausible range of values for the difference in mean antibody titers between the long- and short-needle groups? Use the 95% confidence interval to answer this question.

b. Use an appropriate hypothesis test to compare the means of the two groups. What can you conclude about the effectiveness of the vaccine as a function of the length of the needle?

c. What is the 95% confidence interval for antibody titer in the long-needle group?

15. *Spot the flaw.* Refer to Problem 14. Grover, a student who skipped reading Chapter 2 of this book, made the following graph when presenting the results of the needle-length study to his epidemiology class. The points are means, and the error bars are 95% confidence intervals. What is the biggest weakness of the graph?

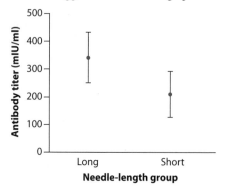

16. We are accustomed to thinking that the proportion of males at birth is fixed by genetics in birds and mammals to be close to 50%. Some have suggested, however, that the sex of offspring can be adjusted by females, such as in response to

the quality of her mate or the number of helpers she has. West and Sheldon (2002) found a total of 15 studies, all done on birds, that have measured changes in the sex ratio in response to such social factors, and each has expressed its results in terms of a coefficient ranging from −1 to 1. The coefficient is positive if the change in the sex ratio of offspring is in agreement with evolutionary theory, and negative if the data disagree with the theory. A coefficient of zero indicates no association with social factors in the data. The measures are as follows:

−0.160, −0.037, 0.034, 0.144, 0.137, 0.118, 0.395, 0.363, 0.350, 0.376, 0.253, 0.440, 0.453, 0.460, 0.563.

The frequency distribution of the coefficients is shown in the following histogram.

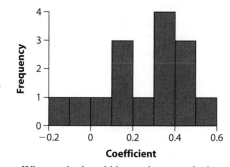

a. What method could be used to test whether these data are consistent with a mean or median coefficient of zero? Discuss why you would use this method in contrast to specific other methods.

b. Apply the best method to these data to test whether the mean coefficient differs from zero.

17. Cellulose from the butts of smoked cigarettes is commonly used by urban birds in nest construction. In an observational study of the house finches in Mexico City, Suárez-Rodríguez et al. (2013) discovered that nests with more cellulose from smoked cigarette butts contained fewer nest-dwelling ectoparasites of birds (such as mites) than nests with less cellulose from smoked cigarettes. In a separate experimental

study, the researchers placed thermal traps in 28 active house finch nests (the parasites hiding in the nest are drawn to the warmth and become trapped). Smoked Marlboro cigarette butts[2] were placed in the trap in about half the nests, randomly chosen. In the other nests, filters from unsmoked cigarettes (lacking tobacco residues) were used as control. At the end of the experiment, the researchers counted the number of ectoparasites caught in each trap. The traps containing smoked butts had fewer ectoparasites than traps with unsmoked filters.

a. Which study provided the stronger evidence that the chemical contents of smoked cigarette butts deters ectoparasites: the observational study or the experimental study? Explain your reasoning.

b. Which of the six commonly used components of experimental design were not incorporated in the experimental study described above? What benefit might result from including them?

18. We often assume that the mapping between words and their meanings is completely arbitrary. Maurer et al. (2006) tested whether this was completely true. College students were shown the following two shapes, and asked to say which was "bouba" and which was "kiki".

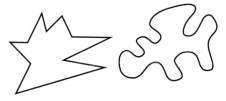

Eighteen of 20 students called the angular shape on the left kiki, while the other two called that shape bouba.

a. Calculate a confidence interval for the proportion of adults who would call the left shape kiki and the other bouba.

b. Test whether kiki and bouba are used with equal probability in the student population.

19. Sex, with its many benefits, also brings risk. For example, individuals that are more promiscuous are exposed to more sexually transmitted diseases. This is true for other primates as well as for our own species. Different species of primates vary widely in the mean number of sexual partners per individual, and this raises the question, are the immune systems of more promiscuous species different from those of less promiscuous species? Researchers approached this question by comparing pairs of closely related primate species, in which one species of the pair was more promiscuous and the other less promiscuous (Nunn et al. 2000). They measured the mean white blood cell (WBC) count in cells per nanoliter for each species. The results are listed in the following table.

WBC count: Less promiscuous species	WBC count: More promiscuous species
5.7	10.4
7.2	10.4
7.4	9.9
8.1	9.1
8.4	9.2
9.2	11.9
9.1	9.3
9.1	8.9
10.6	12.5

a. What is the mean difference in WBC count between less and more promiscuous species? Which type of species (more promiscuous or less promiscuous) has the higher WBC count on average?

b. What is the 99% confidence interval for this difference?

c. Test the null hypothesis that there is no mean difference in WBC count between more and less promiscuous species.

d. What are your assumptions in (b) and (c)?

---

2. Cigarettes were consumed using an artificial smoking machine and contained chemical residues from the smoked tobacco.

Polyphemus pediculus

18 Multiple explanatory variables

Up to this point in the book, we have discussed methods to predict a response variable from at most one explanatory variable. Many biological studies, however, investigate more than one explanatory variable at the same time. One reason for this is efficiency: for the same amount of work, or just a little more, we can obtain answers to more than one question if we include more than one explanatory variable in a study. For example, a study of the causes of nearsightedness in children might want to examine the effects of genetic factors *and* the amount of time subjects spend reading at the same time, rather than just one of these variables alone. The same sample of subjects can be used to address both questions.

The analysis of data from a study having more than one explanatory variable follows the purpose and design of the study. We considered several study designs in

Chapter 14, and three of these will serve as the basis of the present chapter. The first design is an experiment that includes *blocking* (Section 14.4) to improve the detection of treatment effects. The second design is the *factorial* experiment (Section 14.5), which is carried out to investigate the effects of two or more treatment variables (often called factors) and their interaction. The third design adjusts for the effects of known confounding variables (also called *covariates*; see Interleaf 4 and Section 14.1) when comparing two or more groups.

In this chapter, we introduce an all-purpose method that can be used to analyze data from designs fulfilling all of these objectives and more. The method is called *general linear models*. General linear models is a large topic. It includes two-factor ANOVA, multiple regression, analysis of covariance, and other methods you might have encountered already in the scientific literature. Our purpose in this chapter is to introduce the basic elements by example, using visual displays of data and models. Each analysis begins with a model statement. We will show you how model statements are constructed, and how the results are interpreted, when analyzing block and factorial designs and when adjusting for a covariate. Our goal is to give you an overview of what is possible when there are multiple explanatory variables, but we don't have space here to describe all of the many important complications that arise when analyzing these kinds of study designs. To go further or to apply this approach to your own data, consult more advanced textbooks, such as Grafen and Hails (2002) or Quinn and Keough (2002).

18.1 ANOVA and linear regression are linear models

ANOVA, linear regression, and more complicated analyses having multiple explanatory variables all involve a response variable Y that can be represented by a linear model plus random error. By **model**, we mean a mathematical representation of the relationship between a response variable Y and one or more explanatory variables. The scatter of Y measurements around the model is random error, which results from chance and various effects not included in the model. Linear regression is an obvious example of a linear model (Chapter 17). Let's quickly review key elements of regression and then see how they can be generalized.

> A *model* is a mathematical representation of the relationship between a response variable Y and one or more explanatory variables.

Modeling with linear regression

The model of linear regression is a straight line,

$$Y = \alpha + \beta X,$$

where α and β are the intercept and slope, respectively. Individual values of the response variable are scattered above and below the regression line, representing random error. In Example 17.3, for example, we fitted a linear regression to the relationship between the natural logarithm of stability of plant biomass production in prairie plot (Y) and treatments of either 1, 2, 4, 8, or 16 plant species (X). If we call the response variable LOGSTABILITY and the explanatory variable SPECIESNUMBER, then the equation with the variable names is

$$\text{LOGSTABILITY} = \alpha + \beta\,(\text{SPECIESNUMBER}).$$

Least squares yielded the best fit of this model to the data, which is drawn in the right panel of Figure 18.1-1. The data points are scattered above and below the best-fit line, representing random error.

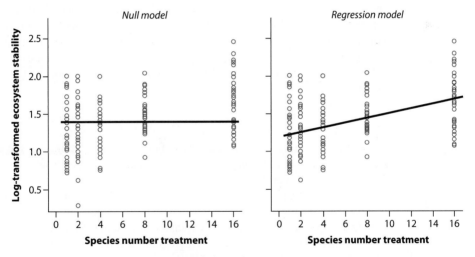

FIGURE 18.1-1 Comparison of the fits of the null model (*left panel*) and the linear regression model (*right panel*) to the plant biomass stability data of Example 17.3.

When we tested the null hypothesis that $\beta = 0$, we were really comparing the fit of two linear models to the data: the linear regression model and the *null model*, in which the slope β is set to zero. The null model is a simplified linear model because setting β equal to zero removes the variable species number (SPECIESNUMBER) from the equation, yielding the model statement

$$\text{LOGSTABILITY} = \alpha',$$

where α' is just a constant. Fitting this null model to the data using least squares yields the flat line depicted in the left panel of Figure 18.1-1; the best estimate of the constant is the sample mean log stability of plant biomass production.

By visually comparing the two panels of Figure 18.1-1, you can see that the full regression model—the one that includes the explanatory variable SPECIESNUM-BER—is the superior fit. The residuals are smaller than for the null model: data points on average lie closer to the line fitted through the data (right panel) than to the flat line having zero slope (left panel). Even if the null hypothesis were true, though, we expect the data to have a slope that departs from zero just by chance. An important question is whether the regression model fits the data *significantly* better than the null model. The *F*-ratio (Section 17.3) is used to evaluate this: if *F* is sufficiently large, then the reduction in the magnitudes of the residuals represents a significant improvement in fit, and the null model is rejected. Another important question is whether the coefficients of the linear model, such as the slope in this example, are large enough to be interesting biologically. Graphical displays of model fits are a valuable tool for evaluating this question, and we emphasize this approach in the current chapter.

Generalizing linear regression

The method of general linear models takes this regression approach and extends it in two key ways. First, it allows more explanatory variables to be included in the model. Second, the method incorporates *categorical* explanatory variables in the same framework, not just numerical explanatory variables as in regression.

For example, the linear model for single-factor ANOVA (one categorical variable) is

$$Y = \mu + A.$$

The constant μ is the grand mean, the average of all the observations combined, and A stands for the group or treatment effect. For each observation, A is the difference between the mean of its group and the grand mean. The data themselves are scattered above and below the group means, representing random error.

The resemblance between the model for a single categorical variable and the model for linear regression highlights their fundamental similarity. Both models include a response variable and an explanatory variable. Both also have a constant term—namely, the intercept in linear regression, and the grand mean in the case of a categorical variable. The only real difference is that the explanatory variable is categorical for one model and numerical for the other.[1]

In Example 15.1, we compared phase shift of the circadian rhythms of subjects assigned one of three different light treatments. We can analyze these data with a general linear model as follows. Phase shift (call it SHIFT) is our response variable, and the light treatment (TREATMENT) is the explanatory variable of interest. The model statement can be written as

$$\text{SHIFT} = \text{CONSTANT} + \text{TREATMENT}.$$

---

1. To analyze categorical data with a linear model, your computer package converts the categories to numerical variables that indicate the groups to which every subject belongs. This behind-the-scenes trick allows categorical variables to be analyzed in the same way as numerical variables.

This word statement is similar to the way most statistical packages on the computer require you to enter your model statement if you want to analyze your data with a general linear model.[2] The CONSTANT term stands for the grand mean. TREATMENT represents the light treatment variable, indicating the group to which individuals belong. If there is a treatment effect, the TREATMENT group means will differ. The hypotheses are as follows.

H_0: Treatment means are all the same.

H_A: Treatment means are not all the same.

As we showed previously for linear regression, the significance of a treatment variable in a general linear model is tested by comparing the fit of *two* models to the data using an *F*-test. The two models correspond to the null and alternative hypotheses. The model that includes the TREATMENT term represents the alternative hypothesis— that there is an effect of the treatment on circadian rhythm. The other model is the null hypothesis. Under the null hypothesis, all the group means are the same (i.e., there is no treatment effect). Under the null hypothesis, the TREATMENT variable contributes nothing and is removed from the model, yielding

$$SHIFT = CONSTANT.$$

Figure 18.1-2 shows the fits of the null (left panel) and alternative (right panel) models to the circadian rhythm data. Horizontal lines give the predicted values (analogous to the $\hat{Y}$ in linear regression) under the two models.[3] The plot on the right suggests that the effect of the "eyes" treatment is to shift circadian rhythm by an hour or more.

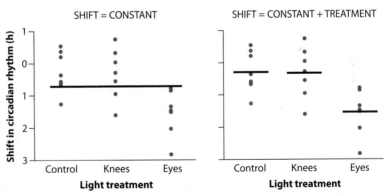

FIGURE 18.1-2 The fit of a general linear model to the circadian rhythm data (*right panel*) compared with the fit of the null model (*left panel*). Horizontal lines indicate the values predicted by the model.

---

2. You would provide the real variable names from your data spreadsheet when generating the model statement on the computer. For example, if the treatment variable was named "LightTreatmentPosition" in your data set, then this is what you should use in the model statement instead of TREATMENT. Some packages might not require you to type the CONSTANT term in the word statement for the sake of brevity, but a constant is nevertheless included in the analysis.

3. In single-factor ANOVA, the predicted value for every data point under the null model is just the grand mean, $\overline{Y}$. The predicted value for an observation under the alternative model is the mean for its group $\overline{Y}_i$.

Even if the null hypothesis were true, we expect the full model—the one including the treatment variable TREATMENT—to fit the data best, because the sample means of the different groups will not be identical simply by chance. An important question is whether the model that includes the TREATMENT variable fits the data *significantly* better than the null model.

The *F*-ratio is used to test whether including the treatment variable in the model results in a significant improvement in the fit of the model to the data, compared with the fit of the null model lacking the treatment variable. The test was summarized in Chapter 15 (Table 15.1-2). The *P*-value (0.004) indicated that the improvement in fit was sufficiently large to warrant rejection of H_0 in favor of the alternative model.

> *F* measures the improvement in the fit of a general linear model to the data when the term describing a given explanatory variable is included in the model, compared with the null model in which the term is absent.

General linear models

A unified model encompassing both linear regression and single-factor ANOVA can therefore be stated using the following generic format:

$$\text{RESPONSE} = \text{CONSTANT} + \text{VARIABLE}.$$

RESPONSE is the response variable, which is numerical. CONSTANT refers to a constant and can represent the mean or intercept, depending on whether the explanatory variable (VARIABLE) is categorical or numerical. To keep the format as generic as possible, the model statement leaves out the coefficients for the explanatory variable—namely, the slope in the case of a numerical explanatory variable, and group effects in the case of a categorical explanatory variable. Henceforth, we will use simple word statements like this to describe all types of general linear models. Model statements will be recognizably different mainly by the explanatory variables that are, or are not, included.

Linear models having more than one explanatory variable differ from single-variable models in two respects. They have one more variable, but more importantly they might include an *interaction* between variables, as indicated by the last term in the following word statement:

$$\text{RESPONSE} = \text{CONSTANT} + \text{VARIABLE1} + \text{VARIABLE2} + \text{VARIABLE1} * \text{VARIABLE2}.$$

An interaction between two explanatory variables means that the effect of one variable on the response depends on the value of the second variable (Chapter 14). With an interaction, the average response for a particular combination of the two variables differs from that expected simply from adding the average effect of each variable separately.

A few example models for two explanatory variables are listed in Table 18.1-1. As we show, the different models are best suited to analyzing data from particular experimental designs or observational studies. In the following sections we introduce several of these models for two explanatory variables, emphasizing key concepts rather than computational details.

TABLE 18.1-1 Linear models having one or two explanatory variables, with examples of study designs analyzed. Symbols refer to "words" in the model statement: μ is a constant (mean or intercept); Y is the numerical response variable; X is a numerical explanatory variable; A and B are fixed, categorical variables, whereas b indicates a blocking or other random-effect categorical variable. Models covered in this chapter are indicated in bold.

Linear model	Other name	Example study design
$Y = \mu + X$	**Linear regression**	Dose–response
$Y = \mu + A$	**One-way (single factor) ANOVA**	Completely randomized
$Y = \mu + A + b$	**Two-way ANOVA, no replication**	Randomized block
$Y = \mu + A + B + A * B$	**Two-way, fixed-effect ANOVA**	Factorial experiment
$Y = \mu + A + b + A * b$	Two-way, mixed-effects ANOVA	Factorial experiment
$Y = \mu + X + A$	**Analysis of covariance (ANCOVA)**	Observational study
$Y = \mu + X_1 + X_2 + X_1 * X_2$	Multiple regression	Dose–response

18.2   Analyzing experiments with blocking

Here we show how to analyze an experiment designed with one treatment variable plus a blocking variable. Blocking (Section 14.4) results in an additional variable (block) that must be included in the analysis. We begin with the randomized block design because it is the simplest experiment incorporating blocking.

Analyzing data from a randomized block design

A randomized block design is like a paired design, but for more than two treatments. Data from such a design is analyzed with the following linear model:

RESPONSE = CONSTANT + BLOCK + TREATMENT.

In the typical randomized block design, such as Example 18.2, every treatment is replicated exactly once within each block. It yields exactly one data point from each combination of treatment and block, and so no interaction term is included in the linear model.[4]

EXAMPLE
18.2

Zooplankton depredation

Svanbäck and Bolnick (2007) set up a field experiment in the shallows of a small lake on Vancouver Island (British Columbia) that allowed them to measure how fish abundance affects the abundance and diversity of prey species. They used a randomized block design to minimize noise caused by background variation in prey availability between locations in the lake. Three fish abundance treatments—Low, High, and Control—were set up at each of five locations in the lake. In the Low treatment, 30 small fish were added to a mesh cage having a surface area of 3 m $\times$ 3 m. In the High treatment, 90 fish were added to a second cage nearby. An unenclosed space of equal area adjacent to each pair of enclosures served as the Control. Table 18.2-1 shows the diversity of zooplankton prey in each treatment at the five locations after 13 days. Diversity is measured using an index called Levins' D, which takes both the number of species and their rarity into account (e.g., common species count more than rare species). Does treatment affect zooplankton diversity?

Table 18.2-1 Zooplankton diversity D in three fish abundance treatments.

Abundance treatment	Location (block)				
	1	2	3	4	5
Control	4.1	3.2	3.0	2.3	2.5
Low	2.2	2.4	1.5	1.3	2.6
High	1.3	2.0	1.0	1.0	1.6

To analyze these data, we can't simply combine all 15 measurements into three treatment groups, because the three measurements made at each location in the lake are not independent. We ran into the same issue in Section 12.2 when analyzing paired measurements. Instead, we need to designate each location as a distinct category of a blocking variable and include it in the analysis. If the blocking variable accounts for some of the variation in the data, then it can improve our ability to detect an effect of the treatment of interest. As in a paired t-test, treatment effects are assessed by the differences in response to different treatments within each block.

Model formula

Let's call the abundance treatment ABUNDANCE in our statement of the general linear model. BLOCK is the individual location. Finally, let's call the response variable DIVERSITY. The full linear model, including all terms, is

$$\text{DIVERSITY} = \text{CONSTANT} + \text{BLOCK} + \text{ABUNDANCE}.$$

---

4. At least two data points are needed for each combination of two categorical variables before an interaction between the variables can be fitted.

This linear model resembles one for single-factor ANOVA, except that we've added a blocking variable. Fish abundance treatment is the factor we are interested in, so the null and alternative hypotheses are as follows.

H_0: Mean zooplankton diversity is the same in every abundance treatment.

H_A: Mean zooplankton diversity is not the same in every abundance treatment.

The linear model, including all of the terms, represents the alternative hypothesis, whereas the model representing the null hypothesis (the null model) leaves out ABUNDANCE:

$$DIVERSITY = CONSTANT + BLOCK.$$

Fitting the model to data

The graphs in Figure 18.2-1 visually compare the fits of the null (left panel) and alternative (right panel) models. The horizontal lines indicate predicted values for each model. The residual is the difference between each data point and the corresponding predicted value. The predicted values under the full model (right panel) suggest that the diversity of zooplankton was lowest in the High treatment and highest in the Control. The predicted values under the null model (left panel) lie on five horizontal lines rather than one line, as in Figure 18.1-2, because the null model for a test of the effect of ABUNDANCE still includes the blocking variable, and there are five blocks, each with its own average value for the response, D.

The ANOVA table provides the results of the test (Table 18.2-2). Adding ABUNDANCE significantly improves the fit of the linear model ($F = 16.37$; $df = 2,14$; $P = 0.001$).

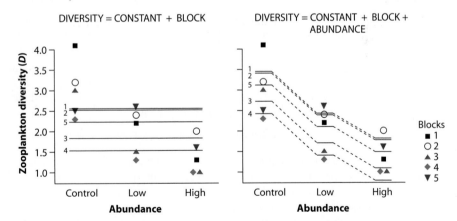

FIGURE 18.2-1 Comparison of the fits of two linear models to the zooplankton diversity data. Symbols indicate blocks (locations). Horizontal lines are predicted values, with numbers indicating blocks. The left panel shows the fit of the null model, including the CONSTANT and BLOCK terms. The right panel shows the fit of the "full" model, which includes the ABUNDANCE treatment variable.

TABLE 18.2-2 ANOVA results of fitting the linear model to the zooplankton diversity data.

Source of variation	Sum of squares	df	Mean square	F	P
BLOCK	2.3400	4	0.5850		
ABUNDANCE	6.8573	2	3.4287	16.37	0.001
Residual	1.6760	8	0.2095		
Total	10.8733	14			

Computer output from programs fitting linear models to data typically also include coefficients estimating magnitudes of effects with standard errors. We do not provide interpretations of these coefficients here. Instead, we emphasize graphical displays of model fits to evaluate effects. The model predicted values shown in the right panel of Figure 18.2-1 suggest that fish abundance had relatively large effects on zooplankton diversity in this experiment.

The effect of BLOCK is not tested here, because it is of less interest than the effect of ABUNDANCE, the treatment variable. Blocking is included in the analysis only to reduce the effect of variation between locations in zooplankton diversity when testing the effect of the fish abundance treatment. BLOCK should still be retained in the analysis whether or not it is statistically significant, because it is there by design and because either way it can still improve the ability to detect a treatment effect.

18.3 Analyzing factorial designs

Here we illustrate the analysis of data from a factorial experiment, an experiment in which all combinations of the values of two (or more) explanatory variables are investigated (Section 14.5). The explanatory variables are called factors; they represent treatments of direct interest. (In contrast, a blocking variable is not considered a factor because it is not of direct interest—a block is included only to improve the detection of treatment effects.) The linear model for an experiment having two treatment variables (call them A and B, for short) is

$$\text{RESPONSE} = \text{CONSTANT} + A + B + A*B.$$

A and B terms in the model represent the main effects, whereas $A*B$ is the interaction term. Figure 18.3-1 illustrates the interpretation of these model terms. Main effects are so named because they represent the effects of each factor alone, when averaged over the categories of the other factor.

Remember that all subjects in the experiment will have a value for both the A and B variables. A main effect of A is present in the data when the mean value of the response variable differs among subjects belonging to different A treatment groups, when averaged over subjects' values for the B variable (top left panel of Figure 18.3-1). Similarly, a main effect of B is present in the data when the mean

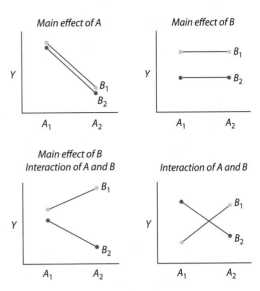

FIGURE 18.3-1 Interaction plots of effects in a hypothetical experiment with two factors (variables) *A* and *B*, each having two treatment categories. The title of each panel indicates which effects are present. Dots represent means. Lines connect means of each *B* group between different *A* groups. An interaction between *A* and *B* is present in the data if the lines are not parallel.

response differs among subjects in different *B* treatment groups, averaging over their values for the *A* variable (top right and bottom left panels of Figure 18.3-1). An interaction is present in the data if the magnitude of the difference between *A* groups differs according to which *B* group subjects belong, as indicated by nonparallel lines in the interaction plot (bottom left and right panels of Figure 18.3-1). An example will make these ideas more concrete.

We restrict ourselves to an example involving *fixed* factors. As discussed in Section 15.5, the different categories of a fixed factor are predetermined, of direct interest, and repeatable. The analysis is different when one or more of the factors is *random*, instead. A random factor is a variable whose groups are not predetermined, but instead are randomly sampled from a "population" of groups (see Section 15.5).

EXAMPLE   Interaction zone

18.3   Harley (2003) investigated how herbivores affect the abundance of plants living in the intertidal habitat of coastal Washington using field transplants of a red alga, *Mazzaella parksii*. The experiment also examined whether the effect of herbivores on the algae depended on where in the intertidal zone the plants were growing. Thirty-two study plots were established just above the low-tide mark, and another 32 plots were set at mid-height between the low- and high-tide marks. The plots were cleared of all organisms, and then a constant amount of new algae was glued onto the rock surface in the center of each plot. Using copper fencing, herbivores (mainly limpets and snails) were excluded from a randomly chosen half of the plots at each height. The remaining plots were left accessible to herbivores. The design was balanced and included every combination of the height and

herbivore treatments (the data are summarized in Table 18.3-1, page 618). At the end of the experiment, the surface area covered by the algae (in cm²) was measured in each plot. The data were square-root transformed to improve the fit to the assumptions of normal distributions with equal variance. Means and standard errors are shown in Figure 18.3-2. Figure 18.3-3 shows the data.

The experiment had two factors, herbivory treatment and height. Figure 18.3-2 suggests that an interaction between these factors might be present, because herbivore treatment had a strong effect on algal cover at low height, but little or even an opposite effect at mid-height. We can fit a linear model to the data to decide which effects are statistically significant.

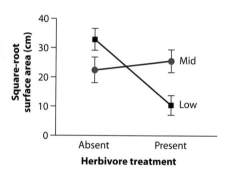

FIGURE 18.3-2
Mean surface area of algae at every combination of treatments. Original units are cm², and $n = 16$ in each treatment combination. The data are shown in Figure 18.3-3.

Model formula

We'll use ALGAE to indicate the response variable, the square-root-transformed algal cover. HERBIVORY and HEIGHT refer to the two factors of interest. Finally, HERBIVORY*HEIGHT will indicate the interaction between HERBIVORY and HEIGHT. The full general linear model is then written as

ALGAE = CONSTANT + HERBIVORY + HEIGHT + HERBIVORY*HEIGHT.

This formula captures every effect that can be examined from the data. A significant HERBIVORY term would indicate that the growth rate of algae differs between HERBIVORY treatments, when averaged over HEIGHT categories. A significant HEIGHT term would indicate that algal growth differs between HEIGHT levels, when averaged over the HERBIVORY treatments. HEIGHT and HERBIVORY are known as the **main effects** because each represents the effects of that factor alone, when averaged over the categories of the other factor. However, the overall effect of herbivory and of height treatments also includes their contribution to the interaction. The HERBIVORY*HEIGHT term in the model represents the differences in slope between line segments in the interaction plot (Figure 18.3-2). The interaction term will be different from zero if the effect of HERBIVORY on algal growth is different for different values of HEIGHT.

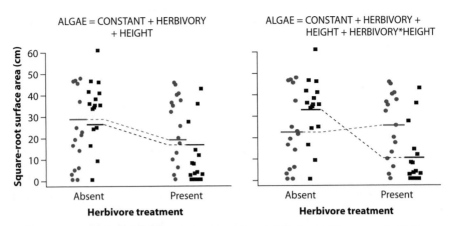

FIGURE 18.3-3 Visual depiction of the fit of the full model (*right panel*) compared with the fit of the null model (*left panel*) lacking HERBIVORY*HEIGHT, the interaction term. Horizontal lines are predicted values under each model. Red circles indicate mid-height above low tide, whereas black squares indicate low height. Points in each combination of herbivory and height treatments are spread out to reduce overlap.

Testing the factors

An *F*-test from an analysis of variance is used to examine the improvement in fit of the model to the data when each main effect or interaction is present in the model, compared to when it is absent. There are three sets of null and alternative hypotheses to be tested.

HERBIVORY (main effect):

H_0: There is no difference between herbivory treatments in mean algal cover.

H_A: There is a difference between herbivory treatments in mean algal cover.

HEIGHT (main effect):

H_0: There is no difference between height treatments in mean algal cover.

H_A: There is a difference between height treatments in mean algal cover.

HERBIVORY*HEIGHT (interaction effect):

H_0: The effect of herbivory on algal cover does not depend on height in the intertidal zone.

H_A: The effect of herbivory on algal cover depends on height in the intertidal zone.

Each of these sets of hypotheses is tested by comparing the fit of the full model to the data with the fit when the term of interest is deleted from the model. To test the interaction term, for example, the fit of the full model is compared with a null model in which the main effects are present but the interaction is absent:

$$\text{ALGAE} = \text{CONSTANT} + \text{HERBIVORY} + \text{HEIGHT}.$$

TABLE 18.3-1 ANOVA results of fitting the two-factor model to the herbivory data.

Source of variation	Sum of squares	df	Mean square	F	P
HERBIVORY	1,512.18	1	1512.18	6.36	0.014
HEIGHT	88.97	1	88.97	0.37	0.543
HERBIVORY*HEIGHT	2,616.96	1	2616.96	11.00	0.002
Residual	14,270.52	60	237.84		
Total	18,488.63	63			

The fits of these two models are depicted in Figure 18.3-3. The model including the interaction term (right panel of Figure 18.3-3) suggests that herbivory treatment has little effect at mid-height, but a substantial effect at low height.

Each combination of herbivory and height treatment represents a separate group in Figure 18.3-3. Under the full model (right panel), which includes the HERBIVORY*HEIGHT interaction term, the predicted values are the group sample means. The null model (Figure 18.3-3, left panel) lacks the interaction term, which constrains the difference between predicted values at low and mid-heights to be the same in both herbivory treatments. The fit is the best possible in which the lines connecting means from different height treatments are constrained to be parallel. This leads to greater residuals (i.e., greater vertical distances between points and corresponding predicted values) under the null model than under the full model. According to an F-test (Table 18.3-1), the interaction term is indeed significant, so the null hypothesis of no interaction is rejected ($F = 11.00$; $df = 1, 60$; $P = 0.002$).

The other two F-ratios in Table 18.3-1 test the significance of the main effects, HERBIVORY and HEIGHT. Each test compares the fit of the full model with that of a null model in which the term of interest is removed (but all other terms are still included).[5] The results show that the null hypothesis of no HERBIVORY main effect is also rejected ($F = 6.36$; $df = 1, 60$; $P = 0.014$). This effect is evident in the interaction plot by a higher overall mean algal cover in the absence of herbivores than in their presence averaged over height categories (Figure 18.3-3). No significant main effect of HEIGHT was detected ($F = 0.37$; $df = 1, 60$; $P = 0.543$). This is just the main effect, however; HEIGHT still interacted with HERBIVORY to influence algal growth (Figure 18.3-2). Height in the intertidal zone has its effects by changing the way herbivores affect algal cover.

---

5. There is difference of opinion on the appropriate null model for testing a main effect. The method we use here (called "Type 3") tests each main effect against a null model that includes all other terms in the model, including any interactions. Some statisticians recommend an alternative method ("Type 1") in which the null model for testing a given effect includes only those terms that appear before it in the model statement, with interactions always appearing last. In this case the order in which terms appear in your model statement might affect the outcome of F-tests for main effects. Different statistics packages on the computer adopt one or the other of these approaches as their default without necessarily highlighting this fact or making it obvious from the output (e.g., JMP uses Type 3, whereas R uses Type 1). None of this affects the fit of the full model to the data or the test of interaction effects.

The importance of distinguishing fixed and random factors

F-tests to compare the fits of null and alternative models to the data are straightforward when all factors are fixed, as in Example 18.3. When one or more factors are random, however, a subtle change occurs that affects how *F*-ratios are calculated. The change is required because the groups are randomly sampled in the case of the random factor, which contributes extra sampling error to the design. This sampling error adds noise to the measurement of differences between group means for other factors that interact with the random factor. The *F*-ratios must be calculated differently to compensate. Most statistics packages assume that all factors are fixed until instructed otherwise. Designating factors as random takes some extra work on your part (you might need to consult the manual of your statistics package to figure out how to do this). If factors are not properly identified as fixed or random, the results given by the computer program will be wrong. Consult more advanced statistics references, such as Sokal and Rohlf (2012), for details on how random factors influence the expected mean squares for treatment effects.

18.4 Adjusting for the effects of a covariate

Our last application is a general linear model having one categorical explanatory variable and one numeric explanatory variable, with a numerical response variable. The method is also called analysis of covariance (ANCOVA). The method is often used to investigate whether linear regressions fitted to data sampled from two or more groups have the same slope. Here we use the method to adjust the effects of a categorical treatment variable to account for a known numerical confounding variable, often called the covariate. Confounding variables bias estimates of treatment effects, as we described in Interleaf 4 and Chapter 14. Experimental studies eliminate such biases by randomly assigning treatments to experimental units, but often experiments are not feasible. A frequent strategy in an observational study is to include known confounding variables in the analysis and to "correct" for their distorting influence on the estimation of the treatment effect.

The model that we would like to fit is

$$\text{RESPONSE} = \text{CONSTANT} + \text{COVARIATE} + \text{TREATMENT}.$$

This model contains no term for the interaction between the covariate and the treatment or factor. Adjusting for the effect of a covariate is simpler when no interaction is present. The effect of treatment changes with the value of the covariate when an interaction is present, complicating the effort to adjust for the covariate. The usual strategy is then to fit the data in a two-stage process. In the first round, the interaction between the treatment and the covariate is included in the linear model and tested. If no significant interaction is detected, the interaction term is dropped from the model in the second round, in which the treatment effect is estimated and tested. Failure to detect an interaction does not confirm that an interaction is truly absent, but the result is often used to justify using a linear model without an interaction term. As usual, graphing the data can help to decide whether this strategy is a sensible one in each particular case.

EXAMPLE

18.4

Mole-rat layabouts

Mole rats are the only known mammals with distinct social castes. A single queen and a small number of males are the only reproducing individuals in a colony. Remaining individuals, called workers, gather food, defend the colony, care for the young, and maintain the burrows. Recently, it was discovered that there might be two worker castes in the Damaraland mole rat (*Cryptomys damarensis*). "Frequent workers" do almost all of the work in the colony, whereas "infrequent workers" do little work except on rare occasions after rains, when they extend the burrow system. To assess the physiological differences between the two types of workers, Scantlebury et al. (2006) compared daily energy expenditures of wild mole rats during a dry season. Energy expenditure appears to vary with body mass in both groups (Figure 18.4-1), but infrequent workers are heavier than frequent workers. How different is mean daily energy expenditure between the two groups when adjusted for differences in body mass?

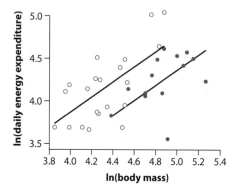

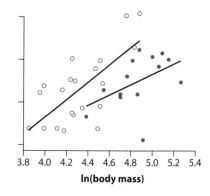

FIGURE 18.4-1 Log-transformed daily energy expenditure and body mass of "frequent workers" (*open circles*, $n_1 = 21$) and "infrequent workers" (*red-filled circles*, $n_2 = 14$) of Damaraland mole rats in a dry season. Original units for the two measurements are kJ/day and g. Predicted values in the right panel include the CASTE*MASS interaction term, whereas the null model (the left panel) lacks the interaction term.

Testing interaction

To analyze these data, we used the log transformation of daily energy expenditure and body mass to improve the fit to the assumptions of general linear models. We'll call the log-transformed response variable ENERGY and the log-transformed body mass MASS. The factor of interest is CASTE, whereas MASS is the covariate. All together, the full general linear model is

$$\text{ENERGY} = \text{CONSTANT} + \text{CASTE} + \text{MASS} + \text{CASTE*MASS}.$$

CASTE*MASS indicates the interaction. The MASS variable is numerical, whereas the CASTE variable is categorical. The model describes a linear regression of ENERGY on MASS, separately for each category of CASTE, as shown by the predicted values for each worker caste in Figure 18.4-1 (right panel).

The question of interest is whether energy expenditure differs between worker castes after adjusting for their differences in body size. The answer is easiest to obtain if we can assume that the regression lines predicting energy expenditure from body mass have the same slope in the two castes (left panel of Figure 18.4-1). Therefore, the usual first step when adjusting for a covariate is a test of equal slopes. This is a test of the interaction term in the general linear model.

> A general linear model with one numerical and one categorical explanatory variable fits separate regression lines to each group of the categorical variable. A test of the interaction term between the two variables is a test of whether the slopes of the regression lines is the same for all groups of the categorical variable.

The hypotheses for the interaction term are as follows.

H_0: There is no interaction between caste and body mass.

H_A: There is an interaction between caste and body mass.

To test the hypotheses, we compared the fit of the full model, which contains the CASTE*MASS interaction term, with that of the null model (ENERGY = CONSTANT + CASTE + MASS), which lacks the interaction.

The fit of the null model to the data is illustrated in the left panel of Figure 18.4-1. Without an interaction term, the regression lines for worker castes have the same slope. To compare the two models, we focus exclusively on the F-test of the interaction term (CASTE*MASS) in Table 18.4-1, the ANOVA table of results. The interaction term is not statistically significant ($F = 1.02$; $df = 1, 31$; $P = 0.321$). The null hypothesis of equal slopes (no interaction) is therefore *not* rejected by the data.

The test result does not mean that the slopes are truly equal—it is not wise to "accept" a null hypothesis just because you have failed to reject it. But the *assumption* that the slopes are equal seems reasonable to make at this point, and the data do not contradict it. Figure 18.4-1 suggests that the assumption is reasonable.

TABLE 18.4-1 ANOVA table for the general linear model fitted to the mole-rat data. We test only the interaction term in this round.

Source of variation	Sum of squares	df	Mean square	F	P
CASTE	0.0570	1	0.0570		
MASS	1.3618	1	1.3618		
CASTE*MASS	0.0896	1	0.0896	1.02	0.321
Residual	2.7249	31	0.0879		
Total	4.2333	34			

Fitting a model without an interaction term

If we can assume that there is no difference between the regression slopes, then we can fit a model without an interaction term:

$$ENERGY = CONSTANT + CASTE + MASS.$$

The fit of this model is illustrated in Figure 18.4-1 (left panel). Now, for the second round of the analysis, the hypotheses are as follows.

H_0: Castes do not differ in energy expenditure.

H_A: Castes differ in energy expenditure.

The test involves comparing the fit of the "full" model (ENERGY = CONSTANT + CASTE + MASS) with that of the null model lacking the CASTE term (ENERGY = CONSTANT + MASS).

This null model is fitted by a single linear regression of energy on mass, calculated on all of the data combined. Both models include MASS, so the test of differences between castes is "adjusted" for mass differences. The ANOVA results are listed in Table 18.4-2.

The F-ratio for CASTE is significant ($F = 7.25$; $df = 1, 32$; $P = 0.011$), confirming that the two worker castes differ in their mean daily energy expenditure after adjusting for body mass. The magnitude of the difference is reflected by the vertical gap between the regression lines for the two castes (Figure 18.4-1, left panel). Infrequent workers expend less energy than frequent workers during the dry season. The results for MASS are also statistically significant ($F = 21.39$; $df = 1, 32$; $P < 0.001$),

TABLE 18.4-2 ANOVA table for the general linear model without an interaction term fitted to the mole-rat data.

Source of variation	Sum of squares	df	Mean square	F	P
MASS	1.8815	1	1.8815	21.39	<0.001
CASTE	0.6375	1	0.6375	7.25	0.011
Residual	2.8145	32	0.0880		
Total	5.3335	34			

indicating that energy expenditure changes with body mass—the regression slope in Figure 18.4-1 is significantly different from zero.

Keep in mind that this was an observational study. Energy expenditure was statistically "adjusted" using naturally occurring variation in body mass, rather than experimentally induced variation in mass, which might yield a different regression slope and hence a different value for the effect of the factor. The analysis of covariance is still prone to bias resulting from *other* confounding variables not included in the model. The justification for using the method in observational studies is that bias is reduced by including one (or more) important covariates in the model, but it is not necessarily eliminated.

How would we have proceeded had the data indicated that the regression lines did not have equal slopes and the interaction term should not be dropped from the model? In this event, the difference between castes is not a constant but changes with mass (as illustrated in the right panel of Figure 18.4-1). Adjusting for mass would then require that we specify a value of body mass at which to estimate caste differences.

Leaving out an interaction term in a linear model to adjust for a confounding variable does not imply that we should usually drop nonsignificant terms from general linear models fitted to data. Leaving model terms such as an interaction out of a linear model should be done with great caution. A sensible rule is that the analysis should follow the design and purpose of the study. In general, variables that are part of the study design should be retained in the general linear model fitted to the data.

18.5 Assumptions of general linear models

The assumptions of general linear models are the same as those for regression and ANOVA.

- The measurements at every combination of values for the explanatory variables (e.g., every block and treatment combination) are a random sample from the population of possible measurements.
- The measurements for every combination of values for the explanatory variables have a normal distribution in the corresponding population.
- The variance of the response variable is the same for all combinations of the explanatory variables.

As in linear regression, the residual plot is a useful technique to evaluate these assumptions in general linear models. In Figure 18.5-1, we present the residual plot for the mole-rat data fitted with the general linear model after dropping the interaction term. Residuals in a general linear model have the same interpretation as in linear regression. Each residual is the difference between an observed Y-value and the value of Y predicted by the model. The residual plot has the predicted values along the horizontal axis and the residuals along the vertical axis. Statistical packages on the computer can compute predicted values and residuals for you.

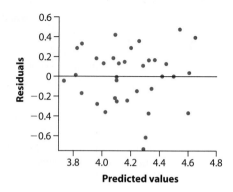

FIGURE 18.5-1
Residual plot for the general linear model without an interaction term fitted to the mole-rat data in Example 18.4.

If the assumptions of general linear models are met, then the residual plot should have the following features:

- A roughly symmetric cloud of points above and below the horizontal line at zero, with a higher density of points close to the line than away from the line
- Little noticeable curvature as we move from left to right along the horizontal axis
- Approximately equal variance of points above and below the horizontal line at all predicted values

The residual plot for the mole-rat example meets these criteria reasonably well (Figure 18.5-1), although one or two data points with predicted ENERGY of about 4.3 have very low residuals and might represent outliers. In general, if assumptions are violated, then a log or other transformation of the response variable can sometimes improve the situation, just as in single-factor analysis of variance. When an outlier is present, it is advisable to determine how its presence influences the results. In the present example, deleting the most extreme outlier with predicted ENERGY of 4.3 did not perceptibly change the results.

18.6 Summary

- Some experiments and observational studies have more than one explanatory variable. These usually can be analyzed by using a general linear model approach, in which the response variable is represented by a linear model plus random error.
- A model is a mathematical representation of the relationship between a response variable Y and one (or more) explanatory variables.
- Linear regression is an example of a linear model. General linear models extend the regression approach to include multiple explanatory variables that may be numerical or categorical.

- The general linear model approach begins with a statement of the model to be fitted to the data.
- The F-ratio is used to test whether including a term of interest in the general linear model results in a significant improvement in the fit of the model to the data, compared with the fit of the null model lacking the term.
- Graphical displays of model fits and the data are a valuable tool for evaluating the magnitude of effects.
- General linear models assume that every combination of values of the explanatory variables has a random sample of Y-values from a population having a normal distribution with equal variance.
- After fitting a model to the data, a plot of the residuals against the predicted values (i.e., a residual plot) is a useful method to evaluate whether the assumptions of general linear models are met.
- Experiments with blocking include the block as an explanatory variable in the general linear model statement.
- Model statements for factorial designs include the main effects of the factors and their interaction.
- F-ratios are calculated differently when a factorial design includes one or more random factors, compared with that when only fixed factors are present. It is important to make sure that, when using statistical programs on the computer, the correct designation of fixed or random is associated with each factor.
- A general linear model with one numerical and one categorical explanatory variable (also called analysis of covariance, or ANCOVA) fits separate regression lines to each group of the categorical variable. An interaction between the variables means that the regression slopes differ among the groups.
- Analysis of covariance is often used to adjust for known confounding variables, or covariates, when testing treatment effects. The procedure is simplest if we can assume that no interaction term is present. Testing the interaction term is usually the first step in this analysis.
- Leaving model terms such as an interaction out of a linear model should be done with great caution. Variables that are part of the study design should usually be retained in the general linear model fitted to the data.

PRACTICE PROBLEMS

1. Examine the accompanying interaction plots. Each is based on hypothetical data from a factorial experiment with two factors *A* and *B*. In each case, indicate which of the main effects and interaction are likely to be present, and which are likely to be weak or absent.

(a)

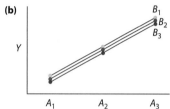

(b)

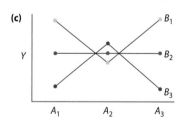

(c)

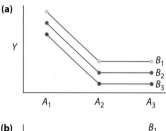

2. Rest in fruit flies, *Drosophila melanogaster*, has many features in common with mammalian sleep. Its study might lead to a better understanding of sleep in mammals, including humans. Hendricks et al. (2001) examined the role of the signaling molecule cyclic AMP (cAMP) by comparing the mean number of hours of resting in six different lines of mutant or transgenic fly lines having different levels of cAMP expression. Measurements are hours per 24 hours divided by the mean of "wild type" flies. Means ($\pm$ SE) are shown for the different fly lines in the accompanying graph.

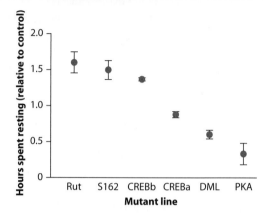

a. Write the statement of the general linear model to be fit to these data to compare means between groups. Indicate what each term in the model represents.

b. Write the corresponding statement for the null hypothesis of no differences between mutant lines.

c. Using a ruler, add the predicted values for each model to the figure (approximate positions will suffice).

d. What test statistic should be used to test whether the null model should be rejected in favor of the alternative?

3. A study of the Magellanic penguin (*Spheniscus magellanicus*) measured stress-induced levels of the hormone corticosterone in chicks living in either tourist-visited areas or undisturbed areas of a breeding colony in Argentina (Walker et al. 2005). Chicks at three stages of development were included in the study—recently hatched, midway through growth, and close to fledging.

Penguin chicks were stressed (captured) by the researchers and their corticosterone concentrations were measured 30 minutes later. The following graph diagrams the mean hormone concentrations for the three age groups of chicks from tourist-visited (*filled circles*) and undisturbed (*open circles*) areas of the colony.

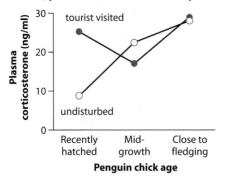

Penguin chick age

a. What is the response variable?
b. What are the explanatory variables?
c. The line segments in this plot are not parallel. What does this suggest?
d. Is this an observational or experimental study? Explain.
e. Did the study use a factorial design? Explain.

4. Refer to Practice Problem 3.
   a. Write a complete model statement for a general linear model to fit to the penguin data. Indicate what each term in the model represents.
   b. What are the null hypotheses tested in the ANOVA table for the general linear model?
   c. What are the assumptions of your analysis?

5. Give three reasons that studies in biology sometimes have more than one explanatory variable.

6. Evidence is mounting that a part of the brain known as the hippocampus is crucial for tasks that depend on relating information from multiple sources, such as tasks requiring spatial memory. Broadbent et al. (2004) tested this experimentally by surgically inducing lesions of different extent in the hippocampus of rats and measuring the subsequent memory performance of the rats in a maze. Their data are plotted in the accompanying graph.

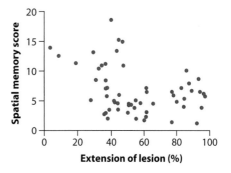

Extension of lesion (%)

a. Write a model statement for a general linear model to fit to these data. Indicate what each term in the model represents.
b. Write the corresponding statement for the null model.
c. Using a ruler, add the predicted values for each model to the figure (approximate positions will suffice).

7. The ejaculate of male *Drosophila* contains a protein, called SP, that reduces the life span of mated females. How the protein does this remains mysterious. One possibility is that the protein manipulates females to produce more eggs, and the extra effort reduces her life span. To investigate, Barnes et al. (2008) housed young female flies with fertile young males either intermittently (low-cost treatment) or continuously (high-cost treatment) for the rest of their lives—up to 56 days. Males in the high-cost treatment were replaced every four days with fresh young males to ensure continued, frequent mating. The same happened in the low-cost treatment, except that the fertile males were present on only one day of each four-day cycle; during the other three days, females were given mutant non-mating males, so that male density was always the same in both treatments. Two strains of females were used: fertile and sterile. Sterile females do not produce eggs, and so do not bear a cost of extra egg production. Sample size was 210 females in each combination of treatment (low-cost vs. high-cost) and fertility (fertile vs. sterile). (Sample size was 212 in the low-cost, sterile combination.) The data are shown in the figure on the next page. Lines connect mean life span between the two treatments (fertile females indicated by solid line and filled

circles, sterile females by dashed line and open circles).

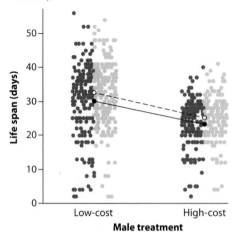

a. What type of experimental design was used?
b. What was the response variable, and what were the explanatory variables?
c. Provide the word statement of the linear model employed.
d. Examine the graph, and state which effects are likely to be present. Does there appear to be a main effect of treatment and a main effect of female fertility? Which main effect is likely to be larger? Is there an interaction between the explanatory variables?
e. The ANOVA results of the general linear model analysis are shown in the table at the bottom of the page. What are the null hypotheses being tested?
f. Using this table, determine which main effects were found to be statistically signif-

icant. Do the results agree with your assessment in part (d) based on the graph?
g. From the ANOVA table alone, is it possible to determine which females live longer—fertile or sterile females?
h. Based on these results, does the cost of more frequent mating (more frequent in the high-cost treatment than in the low-cost treatment) affect sterile and fertile females similarly? Comment on whether these results support or fail to support the hypothesis that SP reduces female life span by causing females to produce more eggs.

8. The foraging gene (*for*) has been found to underlie variation in foraging behavior in several insect species. Ben-Shahar et al. (2002) examined whether the gene might influence behavioral differences in the honeybee (*Apis mellifera*). Worker bees perform tasks in the hive such as brood care ("nursing") when they are young, but switch to foraging for nectar and pollen outside the hive as they age. The authors compared *for* gene expression in nurse and

TABLE FOR PROBLEM 7

Source	Sum of squares	df	Mean square	F	P
Treatment	10244.6	1	10244.6	150.54	< 0.0001
Fertility	1001.3	1	1001.3	14.71	0.0001
Treatment * fertility	24.8	1	24.8	0.36	0.547
Error	57,027.8	838	68.1		
Total	62027.5	841			

foraging worker bees in three bee colonies. The results are compiled in the accompanying table. Gene expression is measured in arbitrary units.

Worker type	Colony	*for* gene expression
Nurse	1	0.99
Forager	1	1.93
Nurse	2	1.00
Forager	2	2.36
Nurse	3	0.24
Forager	3	1.96

a. Draw an interaction plot for these data.
b. Treating COLONY as a blocking variable, write the statement of a general linear model to fit to these data. Indicate what each term in the model represents.
c. Write the corresponding null model for a test of whether worker types differ in their mean gene expression.
d. Is worker type a random effect or a fixed effect? Explain.
e. What is the purpose of a blocking variable in experimental design?

9. The following table lists the ANOVA results for the general linear model fit to the data in Practice Problem 8.

Source of variation	Sum of squares	df	Mean square	F	P
BLOCK	0.342	2	0.171		
WORKERTYPE	2.693	1	2.693	35.34	0.03
Residual	0.152	2	0.076		
Total	3.187	5			

a. Explain in words what the F-ratio for WORKERTYPE measures.
b. Explain in words what the F-ratio for BLOCK measures.
c. The term for BLOCK is not statistically significant. Should it be dropped from the general linear model? Explain.
d. Explain in words what the residuals are.
e. Explain what is plotted along each axis in the following residual plot for the bee data.

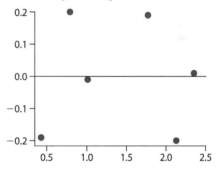

ASSIGNMENT PROBLEMS

10. Examine the accompanying plots. Each is based on hypothetical data from an experiment with a categorical factor A and a continuous covari- ate X. In each case, indicate which of the main effects and interaction are likely to be present, and which are likely to be weak or absent.

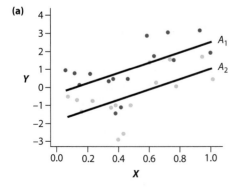

(a)

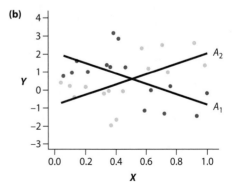

(b)

(c)

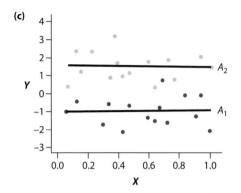

11. Langford et al. (2006) investigated whether lab mice experiencing discomfort "empathize" with familiar mice also in discomfort. They conducted an experiment in which individual mice were given an injection of 0.9% acetic acid into the abdomen, causing mild discomfort. These mice were placed in one of three different treatments: (1) isolation, (2) with a familiar companion mouse (a cage mate) that was not injected, or (3) with a familiar companion also injected and exhibiting behaviors associated with discomfort. The response variable was the percentage of time that each treated mouse exhibited a characteristic "stretching" behavior (measured by abdominal constriction) indicative of discomfort. The data on 42 male mice are below.

Isolated: 46.7, 38.9, 65.6, 35.6, 32.2, 30.0, 41.1, 63.3, 0.0, 53.3, 22.2, 48.9, 5.6, 14.4, 46.7, 45.6, 42.2

Companion not injected: 56.7, 51.1, 50.0, 51.1, 44.4, 2.2, 41.1, 33.3, 25.6, 22.2, 14.4, 3.3, 64.4

Companion injected: 36.7, 81.1, 66.7, 66.7, 44.4, 54.4, 63.3, 62.2, 58.9, 50.0, 54.4, 57.8

a. Write a model statement for a general linear model fit to these data. Indicate what each term in the model represents.

b. Write the corresponding statement for the null model.

c. Plot the data and add the predicted values for the null model and the "full" model. If mice empathize, focal mice should stretch most often, on average, when the companion

mouse is injected. Is this the pattern in the data?

d. Is the fit of the full model significantly better than that of the null model? Carry out the appropriate hypothesis test.

12. Were Neanderthals smaller-brained than modern humans? Estimates of cranial capacity from fossils indicate that Neanderthals had large brains, but also that they had a large body size. The accompanying graph shows the data from Ruff et al. (1977) on estimated log-transformed brain and body sizes of Neanderthal specimens (*filled circles*) and early modern humans (*open circles*). The goal of the analysis was to determine whether humans and Neanderthals have different brain sizes once their differences in body size are taken into account. The ANOVA results of the model are listed in the following table.

Source of variation	Sum of squares	df	Mean square	F	P
SPECIES	0.00547	1	0.00547	1.24	0.274
MASS	0.09810	1	0.09810	22.15	<0.001
SPECIES *MASS	0.00485	1	0.00485	1.09	0.303
Residual	0.15503	35	0.00443		
Total	0.26345	38			

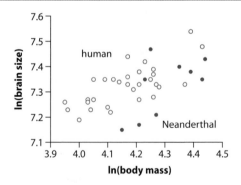

a. Examine the ANOVA table and the graph and write the statement of the general linear model that was fit to these data.

b. What does the SPECIES*MASS term represent?

c. What does the *F*-ratio corresponding to the SPECIES*MASS term represent?

d. What null and alternative hypotheses are tested with the SPECIES*MASS term?

e. What can you conclude from the *F*-ratio and *P*-value listed in the table for the SPECIES*MASS term?

f. In view of the goals of the study, what steps would you recommend next to test whether the brain sizes of Neanderthal and early modern humans differ after adjusting for differences in body size?

13. Using a ruler, add the predicted values from the analysis recommended in part (f) of Assignment Problem 12 to the scatter plot (approximate positions will suffice).

14. For each of the following scenarios, draw an interaction plot (like that in Figure 18.3-2) showing the results of a hypothetical experiment having two factors, A and B, each having two groups, in which there is
a. a main effect of A, no main effect of B, and no interaction between A and B;
b. a main effect of A, a main effect of B, and an interaction between A and B;
c. no main effect of A or B, and an interaction between A and B.

15. In a study of the effects of commercial fishing on fish populations, Hsieh et al. (2006) measured the year-to-year coefficient of variation (CV) of larval population sizes of exploited and unexploited fish species in the California current system. They compared the two groups of fish species by using a general linear model that adjusted for differences between exploited and unexploited species in the age of maturation (MATURATION), which also seemed to influence the coefficient of variation. Data for 13 exploited species and 15 unexploited species are shown in the following graph, along with the predicted values of the model for the two groups of fish.

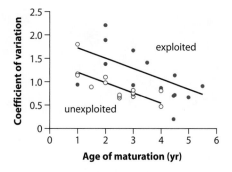

The model was

$$CV = CONSTANT + MATURATION + EXPLOITATION,$$

where EXPLOITATION is the explanatory variable representing the two groups of fish species (exploited and unexploited). The ANOVA results are listed in the following table.

Source of variation	Sum of squares	df	Mean square	F	P
MATURATION	1.7313	1	1.7313	17.65	0.0003
EXPLOITATION	1.5924	1	1.5924	16.23	0.0005
Residual	2.4518	25	0.0981		
Total	5.7755	27			

a. Explain the steps that likely led to the authors using the model analyzed above.

b. What are the assumptions of this analysis?

c. Is there a significant difference between exploited and unexploited fish in their year-to-year coefficients of variation? Explain the basis for your conclusion.

16. The tortoise beetle *Deloyala guttata* feeds and lays eggs on leaves of the two morning glory species *Ipomea pandurata* and *I. purpurea*.

TABLE FOR PROBLEM 16

Family number	Mean development time on *I. pandurata* (days)	Mean development time on *I. purpurea* (days)
18	15.1	14.1
19	14.8	14.5
25	15.9	14.0
50	16.9	17.1
65	14.7	14.7
66	15.6	14.4

Rausher (1984) investigated whether there was genetic variation in the population in the relative abilities of beetles to exploit the two plant species. To test this, he randomly sampled six beetle families from a local population by crossing randomly sampled males and females. He then raised half the offspring from each family on leaves of *I. pandurata* and the other half on *I. purpurea*. Here we analyze development time (the days from hatching to formation of the pupa) of female offspring from each family. If genetic variation is present in the relative abilities to exploit the two plant species, then there should be an interaction between family and plant species. Means are given in the table on page 631; standard errors were about 0.7.

a. Draw an interaction plot for these results. Briefly describe (in words) the patterns revealed. Is an interaction present? How can you tell?

b. Write a model statement for a "full" general linear model to fit to these data. Explain what each term in the model represents.

c. Which factors in the general linear model are random and which are fixed? Explain.

d. What are the null hypotheses to test in the corresponding ANOVA table?

e. What assumptions are required to test these hypotheses?

f. What does the F-statistic for any given term in the model signify?

17. Females of the yellow dung fly, *Scathophaga stercoraria*, mate with multiple males, and the sperm of different males "compete" to fertilize her eggs. The last male to mate usually gains a disproportionate number of fertilizations. In a laboratory experiment on male and female dung flies from two populations, one in Switzerland

and the other in the United Kingdom (U.K.), Hosken et al. (2002) found that fertilizations by the last male (assessed by DNA fingerprinting) depended on the population of origin of both the male and female. Females were mated to two males in turn, one from each population, and the father of each egg laid was then determined. On average, males from the same population as the female fared worse than foreign males. The following graph, redrawn from Hosken et al. (2002), shows the mean percentage of offspring sired by the second male ± SE.

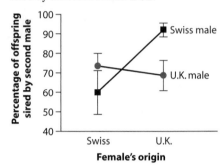

a. What do we call the type of experimental design that was carried out?

b. Write the statement of a general linear model to fit to these data.

c. Based on the graph, which F-ratios are likely to be greater than one?

18. Does light environment have an influence on the development of color vision? The accompanying data, from Fuller et al. (2010), are measurements of the relative abilities of bluefin killifish from two wild populations to detect short wavelengths of light (blue light in our own visible color spectrum). One population was from a swamp, whose tea-stained water filters out blue wavelengths, whereas the other population was from a clear-water spring. Fish were crossed and raised in the lab under two light conditions simulating those in the wild: clear and tea-stained. Sensitivity to blue light was measured as the relative expression of the *SWS1* opsin gene in the eyes of the fish (as a proportion of total expression of all opsins). Opsin proteins in eyes detect light of specific wavelengths; SWS1 is so named because it is shortwave sensitive. The data are from a single individual from each of 33 families. Because the fish were raised in a common lab environment, population differences are likely to be genetically based, whereas differences between fish under different water clarity conditions are environmentally induced.

Population	Water clarity	Relative expression of SWS1
Spring	Clear	0.16, 0.11, 0.12, 0.11, 0.08, 0.09, 0.14, 0.16
Swamp	Clear	0.08, 0.13, 0.07, 0.12, 0.12, 0.05, 0.06, 0.11
Spring	Tea	0.09, 0.09, 0.08, 0.10, 0.12, 0.06, 0.13, 0.08, 0.11, 0.07
Swamp	Tea	0.06, 0.07, 0.08, 0.08, 0.10, 0.03, 0.03

a. How many factors are included in this experiment? Identify them.
b. What type of experimental design was used?
c. Draw an interaction plot of the data (remember to show the data also).
d. Provide a word statement of a full linear model to fit to the data.
e. Examine your graph from part (c), and state which effects are likely to be present in the results. Say how you reached your conclusions. Explain whether the genetic and enviromentally induced effects on *SWS1* opsin expression appear to be in the same direction.
f. The ANOVA results of the general linear model analysis are shown in the table below. What null hypotheses are tested?
g. Using this table, indicate which main effects were found to be statistically significant. Do the results agree with your assessment in part (e) based on your graph?

TABLE FOR PROBLEM 18

Source	Sum of squares	df	Mean square	F	P
Population	0.00670	1	0.00670	8.98	0.006
Water treatment	0.00647	1	0.00647	8.68	0.006
Population * water treatment	0.000000	1	0.000000	0.00	0.999
Error	0.021619	29	0.000745		
Total	0.03479	32			

Red panda

19 Computer-intensive methods

The advent of fast and cheap computers has changed the way statistics is done. Most obviously, graphics and the tedious calculations required for classical statistical techniques can be done at the touch of a button, so that the large amount of time that was once spent on such tasks could be used to collect more data or drink more coffee. But the computer has allowed more than just speedier calculations of what could already be done. Computers have also made possible new approaches for analyzing data that were not feasible before. This chapter describes two of these methods: simulation and bootstrapping.

Simulation is primarily a method for hypothesis testing, whereas bootstrapping is a method designed to calculate the precision of estimates. These methods are particularly useful when the assumptions of standard statistical methods cannot be met or when no standard method exists. Simulation and bootstrapping both require a large number of calculations, and they are impractical except with the aid of a computer. The value of these techniques is that they can be applied to almost any type of statistical problem.

In this chapter, we describe simulation and bootstrapping using relatively simple examples. We emphasize the conceptual basis of each method without assuming that the reader is adept at computer programming.

19.1 Hypothesis testing using simulation

Simulation is a computer-intensive method used in hypothesis testing, where the major challenge is to determine the null distribution—that is, the sampling distribution of the test statistic when the null hypothesis is true. In many situations, this distribution can be calculated from probability theory and simulation is not needed. In some situations, however, the null distribution is too difficult to calculate from theory. Computer simulation of the sampling process can be an excellent way of getting an approximation of the null distribution in these cases. **Simulation** uses a computer to mimic—or "simulate"—sampling from a population under the null hypothesis.

With simulation, we use a computer to create an imaginary population whose parameter values are those specified by the null hypothesis. We then use the computer to simulate sampling from this imaginary population, using the same protocol as was used to collect the real data. Each time we take a simulated sample, we use it to calculate the test statistic. We repeat this simulation process a large number of times. The frequency distribution of values obtained for the test statistic from all of the simulations is an approximate null distribution for our hypothesis test.

> *Simulation* uses a computer to imitate the process of repeated sampling from a population to approximate the null distribution of a test statistic.

We have used simulation already without saying so. In Example 6.2, we used computer simulation to generate the null distribution for the number of right-handed toads in a random sample of 18 toads under the null hypothesis that the proportion of right-handed toads in the population was 0.5. Simulation was unnecessary in that case because, as we now know, the binomial test is faster, easier, and gives an exact *P*-value. For Example 19.1, however, simulation is the best or only option.

EXAMPLE 19.1 How did he know? The non-randomness of haphazard choice

Some stage performers claim to have real telepathic powers—the ability to read minds. However, these powers can be convincingly faked. In one example, the performer asks every member of the audience to think of a two-digit number. After a show of pretending to read their minds, the performer states a number that a surprisingly large fraction of the audience was thinking of. This feat would be surprising if the people thought of all two-digit numbers with equal probability, but not if people everywhere tend to pick the same few numbers. Figure 19.1-1 shows the numbers chosen independently by 350 volunteers (Marks 2000). Are all two-digit numbers selected with equal probability?

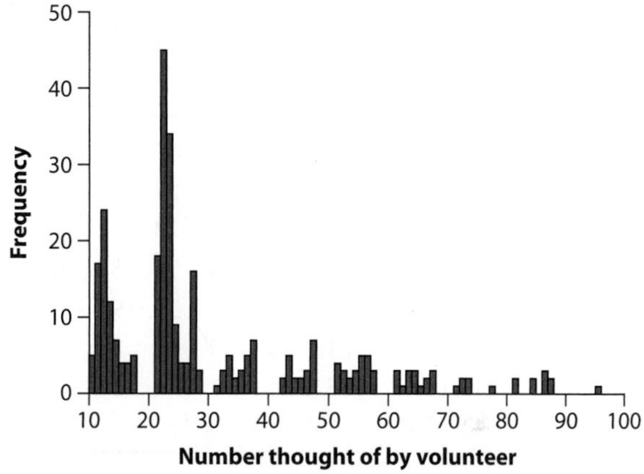

FIGURE 19.1-1
The distribution of two-digit numbers chosen by volunteers.

According to the histogram in Figure 19.1-1, the two-digit numbers chosen don't seem to occur with equal probability at all. A large number of people chose numbers in the teens and twenties, with few choosing larger numbers. Almost nobody chose multiples of ten. Let's use these data to test the null hypothesis that every two-digit number occurs with equal probability. The hypotheses are as follows.

H_0: Two-digit numbers are chosen with equal probability.

H_A: Two-digit numbers are not chosen with equal probability.

To analyze this problem, our first thought might be to use a χ^2 goodness-of-fit test (Chapter 8). There are 90 categories of outcome (each of the 90 integers between 10 and 99). The expected frequency of occurrence of each category is $350/90 = 3.89$, and the observed frequencies are those shown in Figure 19.1-1. The resulting value of the test statistic is $\chi^2 = 1111.4$. If this number exceeds the critical value for the null distribution at $\alpha = 0.05$, then we can reject H_0.

But here we run into a problem. The expected frequency of 3.89 for each category violates the requirements of the χ^2 test that no more than 20% of the categories should have expected values less than five. As a result, the null distribution of the test statistic χ^2 is not a χ^2 distribution, so we cannot use that distribution to calculate a P-value. How do we determine the null distribution of our test statistic so that we can get a P-value?

One possible solution is to use computer simulation to generate the null distribution for the test statistic. Here's how it's done, in five steps:

1. *Use a computer to create and sample an imaginary population whose parameter values are those specified by the null hypothesis.* In the case of the

TABLE 19.1-1 A subset of the results of a simulation. Each row has the first 12 of 350 numbers randomly sampled from an imaginary population in which all numbers between 10 and 99 occur with equal probability. The last column has the χ^2 statistic calculated on each simulated sample of 350 numbers.

Simulation number	Simulated samples of 350 numbers (the first 12 numbers of each sample are shown)	Test statistic χ^2
1	32 42 27 74 78 86 98 71 28 50 41 54 ...	86.1
2	38 63 98 36 88 10 74 35 62 52 90 48 ...	80.5
3	52 45 59 44 44 25 94 29 27 64 24 47 ...	95.4
4	69 55 31 42 35 48 52 37 91 40 67 67 ...	78.4
5	11 66 32 11 76 96 73 20 64 40 37 49 ...	87.7
6	46 15 87 23 92 18 91 26 31 23 40 51 ...	96.9
7	61 35 58 33 58 82 67 95 16 64 59 64 ...	95.4
8	47 54 16 39 91 68 49 57 10 21 79 51 ...	106.7
9	21 84 30 66 81 13 16 18 81 91 52 95 ...	92.3
10	88 26 48 44 34 72 89 14 98 35 99 54 ...	100.5
...		...

mentalist's numbers, simulating a single sample involves drawing 350 two-digit numbers between 10 and 99 at random and with equal probability. Each simulated sample must have 350 numbers, because 350 is the sample size of the real data. The first row of Table 19.1-1 lists the first 12 numbers of our first simulated sample of 350 numbers.

2. *Calculate the test statistic on the simulated sample.* For the number-choosing example, we have decided to use χ^2 as the test statistic. This statistic does not necessarily have a χ^2 distribution under H_0, because of the small expected frequencies, but χ^2 is still a suitable measure of the fit between the data and the null hypothesis. In our first simulated sample, the χ^2 value turned out to be 86.1 (Table 19.1-1).

3. *Repeat steps 1 and 2 a large number of times.* We repeated the simulated sampling process 10,000 times, calculating χ^2 each time. (Typically a simulation should involve at least 1000 replicate samples.) Table 19.1-1 shows a subset of outcomes for the first 10 of our simulated samples.

4. *Gather all of the simulated values for the test statistic to form the null distribution.* The distribution of simulated test statistics can be used as the null distribution of the estimate. The frequency distribution of all 10,000 values for χ^2 that we obtained from our example simulations is plotted in Figure 19.1-2. This is our approximate null distribution for the χ^2 statistic.

5. *Compare the test statistic from the data to the null distribution.* We use the simulated null distribution to get an approximate *P*-value. In an ordinary χ^2 goodness-of-fit test, the *P*-value for the χ^2 statistic is the probability under the null distribution of obtaining a χ^2 statistic as large or larger than the observed

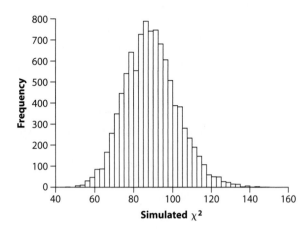

FIGURE 19.1-2
The null distribution for the χ^2 statistic based on 10,000 simulated random samples from an imaginary population conforming to the null hypothesis (Example 19.1). The test statistic calculated from the data, $\chi^2 = 1111.4$, is far greater than for any of the simulated samples.

value of the test statistic. The same is true with a simulated null distribution for χ^2: the *P*-value is approximated by the fraction of simulated values for χ^2 that equal or exceed the observed value of χ^2 (in our case, $\chi^2 = 1111.4$). According to Figure 19.1-2, *none* of the 10,000 simulated χ^2 values exceeded the observed χ^2 statistic. This means that the approximate *P*-value is less than 1 in 10,000 (i.e., $P < 0.0001$).[1] To be more precise, we would have to run more simulations.

These results show that when people choose numbers haphazardly, the outcome is highly non-random. This can make mentalists appear to have telepathic powers when the only power they possess is that of statistics.

19.2 Bootstrap standard errors and confidence intervals

The **bootstrap** is a computer-intensive procedure used to approximate the sampling distribution of an estimate. Bootstrapping creates this sampling distribution by taking new samples randomly and repeatedly *from the data themselves*. Unlike simulation, the bootstrap is not directly intended for testing hypotheses. Instead, the bootstrap is used to find a standard error or confidence interval for a parameter estimate. The bootstrap is especially useful when no formula is available for the standard error or when the sampling distribution of the estimate of interest is unknown.

Recall from Section 4.1 that the sampling distribution is the probability distribution of sample estimates when a population is sampled repeatedly in the same way. The standard error is the standard deviation of this sampling distribution. In principle,

---

1. We can't say $P = 0$, because we might find a more extreme value from the null distribution if we ran more simulations.

therefore, we might obtain a standard error of an estimate by taking repeated samples from the population, calculating the sample estimate each time, and then taking the standard deviation of the many sample estimates. In reality, however, we can't do repeated sampling; collecting data is expensive, and it is best to put all individuals collected into one sample if we had more data. However, if the size of our sample from the population is large, then we do have easy access to a part of the population—namely, the part that was already sampled. Bootstrapping is a kind of repeated sampling, but instead of taking individuals from the population directly, we use a computer to draw the samples from the *data*, a procedure called "resampling." If the data set is large enough, then bootstrap samples drawn in this way will have statistical properties very similar to the distribution of possible sample estimates obtained from the population itself.

> *Bootstrapping* uses resampling from the data to approximate the sampling distribution of an estimate.

The bootstrap is therefore a bit strange: we resample from the data itself to generate many new data sets, and from these we infer the sampling distribution of the estimate. If you think about it, this is almost cheating, because we use the one and only data set to infer the distribution of estimates from all possible data sets. Hence the name "bootstrap," coming from the idea of picking yourself up by your own bootstraps.[2] The method was proposed by Bradley Efron in 1979, when desktop computers started to become available. The bootstrap is now commonly used in biology and other sciences.

Example 19.2 shows how to calculate a standard error and a confidence interval using the bootstrap. This particular example estimates a median, a simple quantity for which it is otherwise difficult to calculate a sampling distribution. Bootstrapping, however, can be applied to essentially any type of estimate.[3]

EXAMPLE The language center in chimps' brains

19.2 One of the things that makes humans different from other organisms is our well-developed capacity for complex speech. Chimps and gorillas can learn some rudimentary language, but with a capacity far below that of humans. Speech production in humans is associated with a part of the brain called "Brodmann's area 44," which is part of Broca's area. In humans, this area is larger in the left hemisphere of the brain than in the right,

---

2. The term supposedly derives from the adventures of Baron Munchausen, who found himself at the bottom of a hole in the ground, lost until he had the idea of picking himself up by his own bootstraps. (Rumored to be from *The Travels and Surprising Adventures of Baron Munchausen Illustrated with Thirty-seven Curious Engravings from the Baron's Own Designs and Five Illustrations*, by G. Cruikshank.) Efron (1979) mentions that he was tempted to call this method the "shotgun" because it could "blow the head off any problem if the statistician can stand the resulting mess."

3. The most common use of the bootstrap in biology is to calculate the uncertainty of estimates of phylogenies—the evolutionary relationships of a sample of species (see Practice Problem 6 for an example).

and this asymmetry has been shown to be important for language development. With the advent of magnetic resonance imaging (MRI), it is possible to ask whether this area is asymmetric in other apes' brains as well. A sample of 20 chimpanzees were scanned with MRI, and the asymmetry of their Brodmann's area 44 was recorded (Cantalupo and Hopkins 2001). This asymmetry score is left measurement minus the

right, divided by the average of the two sides. The raw data are listed in Table 19.2-1. The sample median asymmetry score was 0.14. We want to quantify the uncertainty of this estimate of the population median by calculating its standard error.

TABLE 19.2-1 Asymmetry scores for Brodmann's area 44 in 20 chimpanzees.

Name of chimp	Asymmetry score
Austin	0.30
Carmichael	0.16
Chuck	−0.24
Dobbs	−0.25
Donald	0.36
Hoboh	0.17
Jimmy Carter	0.11
Lazarus	0.12
Merv	0.34
Storer	0.32
Ada	0.71
Anna	0.09
Atlanta	1.12
Cheri	−0.22
Jeannie	1.19
Kengee	0.01
Lana	−0.24
Lulu	0.24
Mary	−0.30
Panzee	−0.16

The frequency distribution of asymmetry scores shown in Figure 19.2-1 is skewed to the right and might even be bimodal. A transformation of these data would be difficult to find because the range of values includes negative numbers. What to do? Bootstrapping provides a suitable approach.

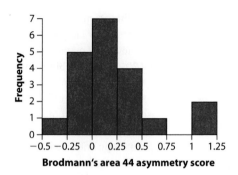

FIGURE 19.2-1
The frequency distribution of asymmetry scores for Brodmann's area 44 in 20 chimpanzees. A negative score indicates that the area is larger on the right side of the chimp's brain, while chimps with positive scores show a larger area in the left hemisphere.

Bootstrap standard error

To generate a bootstrap standard error, there are four steps to follow. First, we list the steps all at once here, and then we go through them again with the data.

1. *Use the computer to take a random sample of individuals from the original data.* Each individual in the data has an equal chance of being sampled. The bootstrap sample should contain the same number of individuals as the original data. Each time an observation is chosen, it is left available in the data set to be sampled again, so the probability of it being sampled remains unchanged.[4]

2. *Calculate the estimate using the measurements in the bootstrap sample from step 1.* This is the first **bootstrap replicate estimate**.

3. *Repeat steps 1 and 2 a large number of times* (10,000 times is reasonable). The frequency distribution of all bootstrap replicate estimates approximates the sampling distribution of the estimate.

4. *Calculate the sample standard deviation of all the bootstrap replicate estimates obtained in steps 1–3.* The resulting quantity is called the **bootstrap standard error**.

The last point is worth repeating: the standard error is the standard deviation of the sampling distribution of estimates.[5]

We can now apply these four steps to the chimp data. There are 20 data points in the sample, so each bootstrap sample must also have 20 measurements. Each of the 20 measurements in the bootstrap sample is chosen with equal probability from the values in the original data. Applying step 1, the following is the first bootstrap replicate that we obtained:

0.24	0.36	0.30	0.16	0.34	−0.24	0.30	1.19	0.32	0.32
0.36	0.01	0.01	0.11	0.11	−0.25	0.12	0.32	−0.24	0.17

---

4. In statistics jargon, this is called "sampling with replacement."

5. A common mistake is to calculate the standard error of the bootstrap estimates by dividing the standard deviation by the square root of the number of bootstrap replicates, by analogy with the standard error of the mean. This results in a quantity that is much too small to be the bootstrap standard error.

Each of the measurements in this first bootstrap sample is present in the original data set. By chance, some of the original data points are present more than once in the bootstrap sample. For example, the score 0.32 (from the chimp named Storer) is present three times. Also by chance, some of the original data points are absent from this first bootstrap sample. For example, the score 0.71 (from the chimp named Ada) was not sampled. The sample median of this bootstrap sample is 0.205, so this is our first bootstrap replicate estimate of the median asymmetry score (step 2).

We repeated this process 10,000 times, calculating the sample median of the measurements each time (step 3). Figure 19.2-2 shows the frequency distribution of the bootstrap replicate estimates of the sample median.

FIGURE 19.2-2
The distribution of 10,000 bootstrap replicate estimates for the median asymmetry of Brodmann's area 44 in chimpanzees.

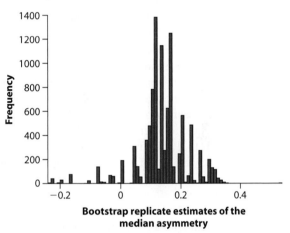

The mean of the bootstrap replicate estimates is 0.142, which is very close to the estimated median from the original data (0.14). Remember that the bootstrap procedure is calculating a sampling distribution for an estimate, not a null distribution for a hypothesis test. As such, the overall mean of the bootstrap replicate estimates should be close to the estimate first calculated on the original data.[6]

The standard deviation of these bootstrap replicate estimates is 0.085 (step 4). This is the bootstrap standard error of our sample median: SE = 0.085.

> The *bootstrap standard error* is the standard deviation of the bootstrap replicate estimates obtained from resampling the data.

Because the bootstrap samples come from the data, which generally do not represent the full population, the bootstrap standard error tends to be slightly smaller than the true standard error. This effect is negligible when the sample size is large.

---

6. The exception is when the estimate itself is biased. The bootstrap can be used to estimate and correct for bias, but we don't present the method here.

Confidence intervals by bootstrapping

The approximate sampling distribution generated by the bootstrap can also be used to calculate an approximate confidence interval for the population parameter. We present the most commonly used method here.[7] The bootstrap $1 - \alpha$ confidence interval can be obtained from the bootstrap sampling distribution by finding the points that separate $\alpha/2$ of the distribution into each of the left and right tails. In other words, an approximate 95% confidence interval ranges from the 0.025 quantile to the 0.975 quantile of the bootstrap sampling distribution.

For example, let's compute a 95% confidence interval for the population median asymmetry using the bootstrap sampling distribution displayed in Figure 19.2-2. To determine the lower bound of the 95% confidence interval, we must find the 0.025 quantile—the value that was greater than or equal to the 250th (i.e., 0.025 × 10,000) sorted bootstrap replicate estimate. In our example, after sorting the 10,000 bootstrap replicate estimates from small to large, the 250th sorted value was −0.075. To determine the upper bound of the confidence interval, we must find the value that is less than or equal to 2.5% of the sorted bootstrap estimates—in other words, the 9751st measurement in the sorted bootstrap estimates (note that 250 out of 10,000 values are equal to or greater than the 9751st estimate). For the chimp brain data, this number is 0.31. As a result, the 95% bootstrap confidence interval for the population median asymmetry of Brodmann's area 44 in chimps is

$$-0.075 < \text{median} < 0.31.$$

This is a relatively wide confidence interval, indicating that the data are consistent with a broad range of possible values for the population median asymmetry of Brodmann's area 44 in chimps. This asymmetry of brain structure, thought to be so important for language development in humans, might have a median as low as zero (or very slightly larger on the left side than on the right side of the brain) or as large as 0.31 in our closest relative.[8]

Bootstrapping with multiple groups

Bootstrapping can be used for just about any statistic that can be estimated and for which conventional methods are unavailable. For example, bootstrapping may be used to obtain standard errors and confidence intervals for measures of the difference between groups. Here we show a bootstrap method to compare two groups whose frequency distributions are not normally distributed and might not have the same shape. The data from each group is resampled separately, and the bootstrap sampling uses the original sample sizes for each group. This bootstrap procedure generates "new" data sets that mimic sampling repeatedly from the original populations.

---

7. Consult the excellent book *An Introduction to the Bootstrap*, by Efron and Tibshirani (1993), for more details and other options.

8. The 95% confidence interval for *mean* asymmetry is similarly broad, but it doesn't overlap zero.

Let's revisit the data from Example 13.5, which was problematic because the normality assumption was not met and transforming the data didn't fix the problem. In this example, we examined the relationship between female starvation treatment (female crickets were either starved or fed before the experiment) and the number of hours females waited before mating with a male ("time to mate"). (Remember that in this species the females sometimes munch on the males' wings during mating, so hunger may play a role in willingness to mate.) The frequency distribution of mating times was skewed in both treatments (see Figure 13.5-1), and it is useful to ask how experimental treatment affected the *median* time to mate. We can use the bootstrap to determine a confidence interval for the difference between the medians of the two populations (starved and fed females).

Using the data in Table 13.5-1, we first calculate an estimate of the difference between the median mating times of starved and fed crickets. This estimate is

$$\text{median}_{\text{starved}} - \text{median}_{\text{fed}} = 13.0 - 22.8 = -9.8 \text{ hours.}$$

To find a bootstrap confidence interval for the difference between the population medians, we resample the measurements of individual females separately for each group for each bootstrap replicate. There were $n_1 = 11$ females in the starved treatment and $n_2 = 13$ females in the fed group; so for each bootstrap replicate, we resample with replacement 11 females from the fed group and 13 from the starved crickets. These sample sizes are probably too small to yield an accurate confidence limit for the difference in medians between the two populations. However, the small sample sizes makes it easier to show how resampling works. For example, here is one bootstrap sample from these data:

Starved: 17.9, 9.6, 9.0, 9.0, 14.7, 21.7, 72.3, 2.1, 13.0, 9.0, 1.9

Fed: 39.0, 3.6, 2.4, 22.6, 1.7, 79.5, 1.5, 3.6, 54.4, 72.1, 22.8, 3.6, 22.6

Compare these values to the original data in Table 13.5-1. For each case, the values for *hunger status* and the *time to mating* in the bootstrap sample come from the same individual in the original data set. In each treatment, some of the original females were, by chance, sampled more than once in this bootstrap replicate (for example, the individual from the starved group with time to mating of 9.0). Some of the original individuals were not sampled at all in this bootstrap sample (for example, the individual from the starved goup with time to mating of 3.8). The median female time to mate in the starved treatment in this bootstrap sample was 9.6. The median in the fed treatment in this bootstrap sample is 22.6. Using this bootstrap sample, we get a bootstrap replicate estimate for the difference between treatment medians equal to $9.6 - 22.6 = -13$ hours.

We repeated the steps of this bootstrap procedure a total of 1000 times. The frequency distribution of bootstrap replicate estimates is shown in Figure 19.2-3. The 2.5% quantile of this distribution is -62.5, and the 97% quantile is at 12.2. The 95% confidence interval for the difference between population medians is therefore

$$-62.5 < \text{difference between medians} < 12.2.$$

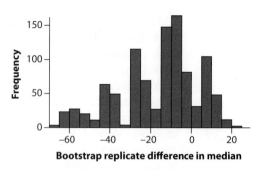

FIGURE 19.2-3
The distribution of bootstrap replicate estimates of the difference in median time to mating, in hours, between starved and fed female crickets.

These data yield a fairly wide confidence interval for the difference in median time to mating between starved and fed female crickets.

Assumptions and limitations of the bootstrap

The main assumption of the bootstrap is that each sample is a random sample from its corresponding population. Additionally, each sample must be large enough that the frequency distribution of the measurements in the sample is a good approximation of the frequency distribution in the population. The larger the sample, the greater the resemblance between the frequency distribution of measurements in the sample and that in the population. Bootstrap analyses based on small samples will, on average, produce standard errors that are too small and confidence intervals that are too narrow, overestimating the precision of the estimate.

19.3 Summary

- Simulation is a method for hypothesis testing in which a computer is used to mimic repeated sampling from an imaginary population whose properties conform to those stated in the null hypothesis. The frequency distribution of test statistics calculated on the simulated samples gives a null distribution of the test statistic.

- The simulated null distribution of a test statistic is used to calculate *P*-values for hypothesis testing.

- Bootstrapping is a method for calculating standard errors of estimates and confidence intervals for parameters. It uses resampling from the data to approximate the sampling distribution for an estimate.

- The standard deviation of the bootstrap sampling distribution for an estimate is the bootstrap standard error of the estimate.

- Bootstrap confidence intervals can be calculated from quantiles of the distribution of bootstrap replicate estimates.

- Bootstrapping requires large sample sizes to generate reliable estimates of the sampling distribution.

PRACTICE PROBLEMS

1. The following are very small data sets of birth weights (in kg) of either singleton births or individuals born with a twin.

   Singleton: 3.5, 2.7, 2.6, 4.4

   Twin: 3.4, 4.2, 1.7

   We are interested in the difference in mean weight between singleton babies and twin babies.
   a. Construct a valid bootstrap sample from these data for this difference.
   b. Assume that we wanted to estimate and test the difference in *medians* between these two groups. Would that change the way in which the bootstrap sample should be created?

2. Using the data from Practice Problem 1, state whether each of the following data sets (a through f) is a possible bootstrap replicate sample for use in determining a bootstrap confidence interval for the difference in mean birth weight. If not, explain why.

	Singletons	Twins
a.	3.5, 2.7, 2.6, 4.4	3.4, 4.2, 1.7
b.	3.4, 4.2, 1.7, 3.5	2.7, 2.6, 4.4
c.	2.7, 2.6, 4.4	3.4, 4.2, 1.7, 3.4
d.	3.5, 3.5, 3.5, 3.5	3.4, 3.4, 3.4
e.	3.8, 3.8, 3.8, 3.8	3.4, 4.2, 1.7
f.	3.5, 3.5, 4.4, 2.7	4.2, 1.7, 4.2

3. The following table lists 100 bootstrap replicate estimates for an estimate of the mean length (in cm) of timber wolf jaws. (The values have been sorted into ascending order for your convenience.) Use the numbers to approximate a 90% confidence interval for the mean length.

10.15	10.17	10.17	10.18	10.20
10.20	10.20	10.20	10.21	10.21
10.21	10.21	10.22	10.22	10.23
10.23	10.23	10.23	10.23	10.24
10.24	10.24	10.24	10.24	10.25
10.25	10.25	10.25	10.25	10.26
10.26	10.26	10.26	10.26	10.26
10.27	10.27	10.27	10.27	10.27
10.27	10.27	10.28	10.28	10.28
10.28	10.28	10.29	10.29	10.29
10.29	10.29	10.29	10.30	10.30
10.30	10.30	10.31	10.31	10.31
10.31	10.32	10.32	10.32	10.32
10.33	10.33	10.33	10.34	10.34
10.34	10.35	10.35	10.35	10.35
10.35	10.35	10.35	10.36	10.36
10.36	10.37	10.37	10.37	10.37
10.37	10.38	10.38	10.38	10.38
10.38	10.39	10.40	10.40	10.40
10.40	10.41	10.41	10.44	10.48

4. Prairie voles are monogamous and social, whereas their close relative the meadow voles are polygamous and solitary. The species differ in their expression of a vasopressin receptor gene (*V1aR*) in their forebrains (expression measures the rate of protein production by a gene). The scientific hypothesis is that this receptor of vasopressin, an important neurotransmitter, might influence the voles' social behavior. Geneticists were able to experimentally increase the expression of the *V1aR* gene in a sample of meadow voles, and they compared the behavior of the resulting individuals with that of control individuals without excess *V1aR* (Lim et al. 2004). They measured the time each vole spent huddling with a partner, under

the assumption that greater huddling time is indicative of a more social animal, as described in Assignment Problem 14 in Chapter 3. The following graph shows the bootstrap sampling distribution of the difference in median huddling time between the two groups, based on 1000 bootstrap replicates. Estimate by eye the bootstrap standard error of the mean difference in median huddling time.

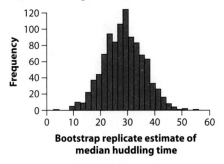

Bootstrap replicate estimate of median huddling time

5. The "broken stick" model is often used in ecology to represent how species should divide up space or resources if these are divided randomly. For example, Waldron (2007) compared the geographic range sizes of closely related bird species in North America. For each of 65 pairs of bird species, he used maps to calculate the total range sizes of the two species (in km²), and then he calculated the ratio of the size of the smaller of the two ranges over the larger range size. This ratio can range from nearly zero, if one species of a pair has a much larger range than the other, to one, if both species have the same range sizes. He observed that the average ratio for all the pairs was 0.48. He then used simulation to test the null hypothesis that the mean of this ratio was no different than that expected if species divided the region "randomly" according to a randomly broken stick, with one species of every pair getting the short end and the other species the long end. Using a computer, he took a stick of unit length, randomly broke it into two parts, and then calculated the ratio of the length of the smaller piece to the length of the larger piece. He did this 65 times, once for every pair of closely related species, and then took the average ratio. This process was repeated 10,000 times. The resulting 10,000 values for the ratio are shown in the accompanying graph.

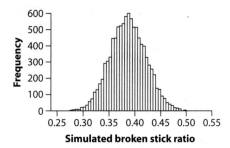

Simulated broken stick ratio

a. What does this frequency distribution of simulated values for the ratio estimate?
b. Only 42 of the 10,000 simulated values for the ratio were greater than or equal to 0.48, the observed ratio in the bird data. Using this information, test the null hypothesis that the observed mean ratio of the bird range sizes is no different than expected if the species divided the region randomly as a broken stick.

6. The most common use of the bootstrap in biology is to estimate the uncertainty in trees of evolutionary relationships (phylogenies) estimated from DNA sequence data. For example, the tree provided for this problem is a phylogeny of the carnivores, a group of mammals, based on their gene sequences (Flynn et al. 2000). The arrangement of branches on the tree indicates how species are related by descent. Pick any two named species on the right (representing the present time) and follow their branches

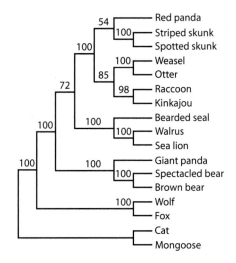

backward in time (to the left) until the branches meet at a node representing their common ancestor. Species are closely related if they share a recent common ancestor. In this tree, for example, the cat is more closely related to the mongoose than either of these two species is to the wolf. The tree is not necessarily the true phylogeny, but an estimate based on the assumption that species sharing a recent common ancestor will have had less time to become different in their DNA sequences than more distantly related species. The bootstrap is used to calculate the uncertainty in the arrangement of the branches of the estimated evolutionary tree. To do this, the bootstrapping procedure resamples the DNA sequence data and recalculates a phylogenetic tree from each new bootstrap sample (we'll spare you the details). This is repeated a large number of times. The method then examines every bootstrap replicate tree and compares it to the tree estimated from the original data. Every node of the carnivore tree shown here gives a number between 0 and 100 that refers to the percentage of bootstrap replicate trees that have exactly the same set of named species descended from that node. For example, the number 100 on the branch leading to the two skunks means that 100% of the bootstrap replicate trees had a node just like it leading to the same two skunks. The number 54 leading to the trio of red panda (pictured on the first page of this chapter), striped skunk, and spotted skunk indicates that just 54% of the bootstrap replicate trees had a node containing only those same three species. The remaining 46% of the bootstrap replicate trees had different arrangements (e.g., the red panda might have been grouped with the weasel or raccoon).

a. What percentage of the bootstrap replicate trees showed the raccoon as the species most closely related to the kinkajou?

b. What percentage of the bootstrap replicate trees grouped the giant panda with the two bear species?

c. From the results of this bootstrap, which of the following statements can be made with greater confidence, that "red pandas are most closely related to skunks" or that "giant pandas are most closely related to bears"?

7. Use a six-sided die to generate 10 bootstrap replicates of the following data set, which has three numbers from each of two groups:[9]

Group A: 2.1, 4.5, 7.8

Group B: 8.9, 10.8, 12.4

a. Write out the bootstrap replicate data sets.

b. Use these 10 bootstrap replicates to get an approximate 80% confidence interval for the difference between the medians of Group A and Group B.

8. Suppose we want to calculate the bootstrap standard errors for the difference between two groups in their medians. We would also like to do the same for the difference between the two groups in their interquartile ranges. Could we use the same procedure to generate the bootstrap replicate samples for both standard errors?

9. In a hypothetical study of a predatory fish, the bootstrap was used to help generate confidence intervals for the mean waiting time between meals. A waiting-time measurement was obtained for 79 fish in a random sample from the predatory fish population. The mean waiting time was 109 seconds. Ten thousand bootstrap replicates were used to calculate a sampling distribution for the estimate. The mean of the bootstrap replicates was 108.6 seconds, and the standard deviation of the bootstrap replicate estimates was 10.4 seconds. What is the bootstrap standard error of the estimated mean waiting time?

10. Assume that you have two moderately small samples, both drawn from separate populations that are normally distributed but with unequal variances. You want to compare the means of these two populations. What technique would

---

9. In this question and in ones that follow, the sample sizes and numbers of bootstrap replicates are often too small to provide accurate bootstrap calculations and are presented as exercises only. This question calls for the 80% confidence interval only to prevent you from having to roll the die too many times. In general, results based on bootstraps use the same levels of confidence as more traditional statistics.

you use to test the null hypothesis that the means are equal?

11. The snail *Helix aspersa* has an unusual mating system. Each individual is a hermaphrodite, producing both sperm and eggs, but the truly bizarre part is that these snails have evolved a "love dart" with which they try to stab their mate. (If you look closely at the photo of the snails, you can see the love dart protruding through the head of the lower snail.) This love dart is coated with a drug that enhances the amount of sperm that the stabbed snail receives and stores. This storage effect is expected to be greater when the recipient is smaller, because the relative dose would therefore be larger. Rogers and Chase (2001) measured the size of stabbed snails (i.e., the volume of their shells) and the number of stabber sperm that were stored. Is the size of the recipient snail correlated with the amount of sperm it stores? The data are listed in the table provided. The table also gives two computer-generated data sets. One of the computer-generated data sets was created by a bootstrap procedure that resampled the observed data, and the other was made by a permutation procedure (see Chapter 13).

a. Could "computer-generated data A" have been created for a bootstrap estimate? Describe how you determined this.

b. Could "computer-generated data B" have been created for a bootstrap estimate? Describe how you determined this.

c. What statistic should be used on these data to measure whether snail size is correlated with the number of sperm stored?

Original data		Computer-generated data A		Computer-generated data B	
Shell volume (cm³)	Number of sperm stored	Shell volume (cm³)	Number of sperm stored	Shell volume (cm³)	Number of sperm stored
2.2	2474	2.2	1807	6.0	260
2.5	2897	2.5	2897	3.7	2516
2.6	2658	2.6	1843	2.5	2897
2.6	2471	2.6	2474	3.3	2009
2.7	2250	2.7	260	3.3	2009
2.9	2606	2.9	440	2.5	1843
3.2	2978	3.2	2332	3.0	1552
3.7	2516	3.7	2009	2.5	2897
3.3	2332	3.3	2250	4.0	1138
3.3	2009	3.3	1158	2.5	1843
2.9	1807	2.9	2516	2.7	2250
3.0	1552	3.0	2606	4.2	440
2.5	1843	2.5	1138	2.5	2897
3.5	1158	3.5	2978	2.5	2897
4.2	440	4.2	2658	4.2	440
6.0	260	6.0	1552	3.7	2516
5.9	788	5.9	788	2.7	2250
4.0	1138	4.0	2471	3.3	2009

ASSIGNMENT PROBLEMS

12. Using the data in Practice Problem 11 on shell volume and the number of sperm stored, we carried out a bootstrap analysis of the correlation coefficient r between the two variables. The cumulative frequency distribution provided here summarizes the 1000 bootstrap replicates of the estimate of the correlation coefficient. The correlation coefficient calculated from the original data was −0.78. From this graph, determine the approximate 95% bootstrap confidence interval for the correlation between shell volume and the number of sperm stored.

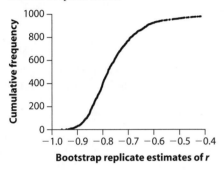

Bootstrap replicate estimates of r

13. The surface of plant leaves is a thriving ecosystem, home to many microorganisms such as bacteria. These bacteria can affect the health of the plant, because some cause disease and some serve as ice-nucleation sites, thus causing frost damage. Hirano et al. (1982) were interested in the probability distribution of the number of bacteria on corn plant leaves. The frequency distribution of the number of bacteria on a random sample of single leaves is shown in the accompanying figure. The number of bacteria is calculated as the number per gram of leaf.

  a. Imagine that we wished to estimate the mean number of bacteria per gram of leaf and the uncertainty of our estimate. Would a confidence interval for the mean based on the t-distribution be appropriate? Explain your answer.

  b. Name two appropriate methods to calculate a confidence interval for the mean with these data.

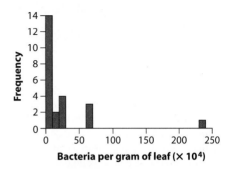

Bacteria per gram of leaf ($\times\ 10^4$)

  c. The authors also measured the frequency distribution of the number of bacteria on soybean leaves. The distribution was shaped similarly to the one shown here for corn leaves. List two valid methods that could be used to test for a difference in the median numbers of bacteria on corn and soybean leaves.

14. In Example 13.1, we described the results of a study that measured the change in biomass in marine areas after an increase in protection. The distribution of biomass ratios (i.e., the biomass with protection divided by the biomass without protection) was highly skewed, so we performed a log transformation to overcome this problem when estimating the mean. Here we take a different approach, and we use the bootstrap to estimate the *median* biomass ratio. The distribution of bootstrap replicate estimates of the median are shown in the following histogram. The mean and standard deviation of this distribution are 1.484 and 0.134, respectively, and there are 1000 bootstrap replicates.

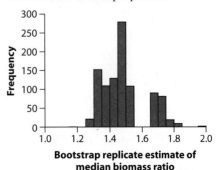

Bootstrap replicate estimate of median biomass ratio

a. What does this frequency distribution estimate?

b. What is the bootstrap standard error of the median biomass ratio?

c. What is the 95% confidence interval for the median biomass ratio? Interpret the histogram to give an approximate answer.

15. Outliers can occur in a sample of data for several reasons, including instrument failure, measurement error, and real variation in the population. The "trimmed mean" is a method developed for estimating a population mean when a measurement is prone to producing outliers. A trimmed mean is an ordinary sample mean calculated on data after the most extreme measurements have been dropped according to a percentile criterion. A 5% trimmed mean drops the measurements below the 5th percentile (0.05 quantile) and above the 95th percentile. In 1882, Simon Newcomb estimated the speed of light by measuring the time it took for light to return to his lab after being bounced off a mirror, a round trip of 7442 meters. The following is the frequency distribution of 66 measurements Newcomb made of this time, in microseconds (i.e., in millionths of a second) (Stigler 1977).

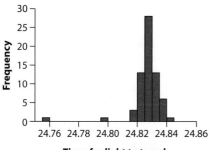

Time for light to travel 7442 m (microseconds)

The distribution includes outliers, and the trimmed mean is an objective way to increase the precision of the estimate. For the light data, the 5% trimmed mean drops the three smallest and three largest values, yielding the value 24.8274 for the mean of the remaining measurements. Bootstrapping is an excellent way to calculate the uncertainty of the trimmed mean. The results of 1000 bootstrap replicate estimates of the trimmed mean are shown in the following

histogram. Of the bootstrap estimates, 2.5% were below 24.8254, 5% were below 24.8260, 95% were below 24.8285, and 97.5% were below 24.8288.

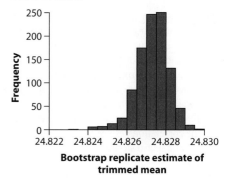

Bootstrap replicate estimate of trimmed mean

a. Calculate a 95% confidence interval for the trimmed mean.

b. The ordinary, "untrimmed" mean of the 66 data points, including the outliers, is 24.8262. Is the ordinary sample mean contained in the 95% confidence interval for the trimmed mean?

16. The variance-to-mean ratio is a useful measurement of how clumped or dispersed events are in space or time relative to the random expectation (see Section 8.6). A high ratio indicates that events are clumped, whereas a low ratio indicates "overdispersion" of events. Davis et al. (2009) used this approach to examine the dispersion of "compensatory mutations" affecting protein sequences. Most mutations to genes are harmful, but compensatory mutations occasionally occur that counteract some of the damage caused by other mutations. The authors gathered data on 77 harmful mutations of varying lengths and a total of 328 compensatory mutations that have been discovered in the same genes. These data come from organisms ranging from viruses to fruit flies. The authors recorded the number of compensatory mutations at each amino acid position in the genes. They calculated the mean and variance in the number of compensatory mutations per amino acid position and calculated variance/mean = 2.64. They used simulation to test the null hypothesis that mutations were randomly and independently located in the genes. Compensatory mutations

were placed at random and independently. After each simulation, the computer calculated the resulting variance-to-mean ratio in the number of compensatory mutations per amino acid position. The results of 10,000 such simulations are plotted in the accompanying histogram.

a. What does this frequency distribution estimate?

b. Using the results shown in the frequency distribution and the observed variance/mean ratio of 2.64, test the null hypothesis that the true variance/mean ratio is as expected from the random placement of compensatory mutations.

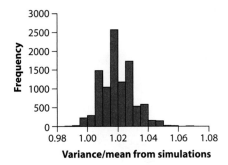

Chimpanzee

20 Likelihood

Biologists gather data in hopes of discovering the correct value of a population parameter among its many possible values. Consider, for example, the relationship between the metabolic rate of organisms and their body mass. According to one theory, the rate should increase in proportion to body mass raised to the power of 3/4, whereas another theory predicts the power should be 2/3, instead (Savage et al. 2004). Which of these alternatives is correct? We must look to the data to find out.

Likelihood measures how well alternative values of a parameter fit the data. The approach is based on the idea that the best choice of a parameter value, among all the possibilities, is the one with the highest likelihood—the one for which the data have the highest probability of occurring. If the probability of obtaining the data is much higher for one possible value of the parameter than for another, then the first is probably closer to the truth.

Likelihood is a very general approach that can be applied to every type of problem encountered so far in this book, though we have not made much direct use of it until now. One advantage of likelihood techniques is that they can be applied to data that are not normally distributed.

Here we introduce the concept of likelihood and identify some of its applications in biology. Our goals are to review the key features of the approach so that its uses in the scientific literature may be more readily understood.

20.1 What is likelihood?

Likelihood measures how well a set of data supports a particular value for a parameter. The likelihood of a specific value for a parameter is the probability of obtaining the observed data if the parameter were equal to that specific value. Using the likelihood method involves calculating the probability of obtaining the observed data for each possible value of the parameter and then comparing this probability between the different possible values.

> Likelihood measures how well the data support a particular value for a parameter. It is the probability of obtaining the observed data if the parameter equaled that value.

The likelihood of a particular value tells us little by itself. Rather, the likelihood of a value gains meaning when compared with the likelihoods of other possible values. The likelihood should be relatively high for those values close to the true population parameter and relatively low for values that are far from it. The parameter value gaining greatest support among all possible values is called the **maximum likelihood estimate**. It is the value for which the probability of obtaining the observed data is highest. With the likelihood approach, the maximum likelihood estimate is our best estimate of the true value of the parameter.

> The *maximum likelihood estimate* is the value of the parameter for which the likelihood is highest. It is our best estimate of the parameter.

Nearly all of the estimation techniques we have learned about in this book (for example, the mean and proportion) yield maximum likelihood estimates, although we haven't referred to them in that way.

20.2 Two uses of likelihood in biology

Here we showcase two of the most frequent uses of likelihood in biology—namely, phylogeny estimation and gene mapping. We don't present any of the computations or details. Our goal is to illustrate how likelihood is applied.

Phylogeny estimation

Thomas Henry Huxley[1] was right when he said that either the chimpanzee or the gorilla was the closest living relative of the human species. But which ape is our closest cousin? At least three possible relationships between humans and other apes have been proposed, as shown by the tree diagrams in Figure 20.2-1. According to the hypothesis represented by the leftmost tree, we share our most recent common ancestor with the chimpanzees (the bonobo is the pygmy chimp). According to the tree on the right, however, we are closest to the gorilla, instead. Finally, the tree in the middle represents the hypothesis that we stand apart from the rabble, equally related to chimp and gorilla, who are each other's closest relatives.

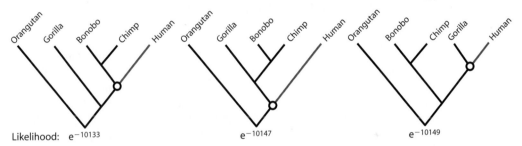

FIGURE 20.2-1 Three proposed trees of ancestor-descendant relationships between humans and the other great apes. The human branch and our shared ancestor with the other apes is highlighted. Numbers at the bottom are the likelihoods of each proposal based on gene sequence data (Rannala and Yang 1996). The likelihood of the leftmost tree is the highest at e^{-10133}, where e is the base of the natural logarithm.

   Likelihood is frequently used to estimate trees of ancestor-descendant (phylogenetic) relationships using DNA sequence data. Rannala and Yang (1996) determined the likelihood of each of the three proposed trees in Figure 20.2-1 by using gene sequence data from five apes. The likelihood of a given tree refers to the probability of obtaining the observed gene sequences for the five species if that tree were the correct tree. This probability is based on a probability model of gene sequence evolution. In this model, the gene sequences for any pair of species were identical at the moment they split from their most recent common ancestor, after which each species accumulated differences gradually and randomly over time. Under this model, the most

---

1. Huxley was known as "Darwin's bulldog" for his vigorous support of Darwin's theory of evolution.

closely related species today should have the most similar gene sequences, and the most distantly related species should have the most different sequences.

The tree on the left in Figure 20.2-1 had the highest likelihood, making it the maximum likelihood estimate. Humans are most closely related to the chimps, justifying our nickname, the "third chimpanzee" (Diamond 1992).

Gene mapping

Huge efforts are under way to find the genes that underlie inherited forms of human diseases. One approach tests whether genetic "markers" in the human genome differ between individuals having the disease of interest and those not having the disease. A marker is a unique, easily identifiable site in the genome whose gene sequence ("state") tends to vary among individuals in the population. Many markers throughout the human genome are available for gene-mapping studies.

Hamshere et al. (2005) used this approach to find a gene responsible for human schizoaffective disorder. This debilitating mental illness is known to run in families, and it is therefore likely to have a genetic component. The data were the marker states of a sample of individuals afflicted with the disease and their healthy family members. If a gene involved with the disease were present, then the states of markers closest to this gene should show differences in frequency between diseased and healthy individuals.

Their approach is illustrated in Figure 20.2-2 with results from chromosome 1. At each marker along the chromosome, the researchers calculated *two* likelihoods, one for each of two hypotheses. One hypothesis was that a gene increasing risk of schizoaffective disorder was really present at that site. The second hypothesis was the null hypothesis that no gene affecting the disorder was present at that site. The likelihood in each case is the probability of obtaining the observed data (the differences between healthy and diseased individuals in the frequency of marker states) if the given hypothesis is correct. The log of the ratio of these two likelihoods (the first likelihood divided by the second likelihood) measures the strength of evidence that

FIGURE 20.2-2
Evidence for a gene affecting schizo-affective disorder on human chromosome 1. Redrawn from Hamshere et al. (2005), with permission.

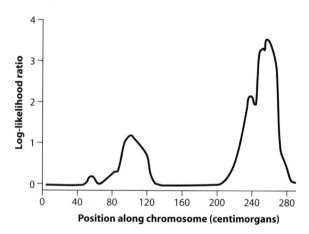

a gene is located at that site.[2] This procedure was repeated at every marker along the chromosome, yielding the curve shown in Figure 20.2-2.

The highest point on the curve is located approximately at position 260, which indicates where, along chromosome 1, a gene for schizoaffective disorder is most likely to be present.

20.3 Maximum likelihood estimation

Likelihood is a probability. The likelihood L of a particular value for a parameter is the probability of obtaining the data if the parameter equaled that value. Using a vertical bar to represent *given* or *conditional upon* (see Chapter 5), this definition of L can be written as

$$L[value \mid data] = \Pr[data \mid parameter = value].$$

Maximum likelihood estimation involves finding the parameter value that has the largest L. Here we show how this is accomplished using Example 20.3, which comes from a study that investigated a population proportion.

EXAMPLE   Unruly passengers

20.3   The tiny wasp, *Trichogramma brassicae*, parasitizes eggs of the cabbage white butterfly, *Pieris brassicae*. The wasp rides on a female butterfly (the arrow in the photo at the right points to a small wasp on a butterfly's leg). When the butterfly lays her eggs, the wasp climbs down and parasitizes the freshly laid eggs. Fatouros et al. (2005) tested whether the wasps could distinguish mated female butterflies (with fertilized eggs) from unmated females. They carried out a series of trials in which a single wasp was presented simultaneously with two female butterflies, one of them a virgin female and the other recently mated. Of the 32 wasps that rode on females, 23 chose the mated female, whereas nine chose the unmated female. We can use these data to provide an interval estimate of the population proportion.

You learned in Chapter 7 how to estimate a population proportion and to calculate its confidence interval. The probability model is relatively simple for this case, however, so we can use it to demonstrate how to estimate a parameter using maximum likelihood.

---

2. The arbitrary convention in gene-mapping studies is to use the base 10 logarithm rather than the natural log. The resulting quantity is called the LOD score.

Probability model

Maximum likelihood estimation requires a probability model that specifies the probabilities of different outcomes of the data-sampling process depending on the parameter being estimated. In the wasp study (Example 20.3), the outcome measured was the number of wasps that chose the mated female butterfly. The parameter of interest is p, the unknown proportion of wasps in the population that would choose the mated female. Let's assume that each of the n wasps tested represented a single random trial and that separate trials were independent. Under these conditions, the number of wasps choosing the mated female should follow a binomial distribution with the probability of "success" in any one trial equal to p. In this case, the probability that exactly Y females choose mated females depends on p as follows:

$$\Pr[Y \ choose \ mated \,|\, p] = \binom{n}{Y} p^Y (1 - p)^{n-Y}.$$

This formula for the binomial distribution[3] was first introduced in Chapter 7, except that here we have written the probability of Y successes as conditional on the unknown p. We use this formulation because we need to vary p to see how this affects the calculated probability of obtaining the observed data.

The likelihood formula

We use the symbols $L[p \,|\, Y \ choose \ mated]$ to indicate "the likelihood of a particular value of p, given that Y wasps chose the mated female." This likelihood is defined as "the probability that Y wasps choose the mated female in n trials, given that the population parameter equals the particular value for p." Thus, the formula for the likelihood is

$$L[p \,|\, Y \ choose \ mated] = \Pr[Y \ choose \ mated \,|\, p] = \binom{n}{Y} p^Y (1 - p)^{n-Y}.$$

To calculate the likelihood of a specific value for p, set Y equal to 23, the observed number of wasps choosing mated females. Also, plug in the total number of trials, $n = 32$, yielding

$$L[p \,|\, 23 \ choose \ mated] = \binom{32}{23} p^{23} (1 - p)^9.$$

The likelihood of $p = 0.5$ is

$$L[p = 0.5 \,|\, 23 \ choose \ mated] = \binom{32}{23} (0.5)^{23} (1 - 0.5)^9$$

$$= 0.00653.$$

---

3. Remember that $\binom{n}{Y}$ is called "n choose Y" and is shorthand for $\dfrac{n!}{Y!(n - Y)!}$. The symbol $n!$ represents "n factorial."

This likelihood represents the support for the possibility that exactly half of the wasps in the population would choose the mated female, given that 23 of 32 sampled wasps did so. We cannot interpret this number in isolation, however, because likelihoods are informative only when compared with the likelihoods of other values for the parameter.

It is usually easier to work with the log of the likelihood rather than the likelihood itself.[4] The **log-likelihood** is the natural log of the likelihood. The formula for the log-likelihood of p, given the observed Y, is

$$\ln L[p \mid Y \text{ choose mated}] = \ln\left[\binom{n}{Y}\right] + Y \ln[p] + (n - Y)\ln[1 - p].$$

To understand how to use this formula, plug in the values $n = 32$ and $Y = 23$. The log-likelihood of $p = 0.5$, given the data, is

$$\ln L[p \mid 23 \text{ choose mated}] = \ln\left[\binom{32}{23}\right] + 23 \ln[0.5] + 9 \ln[1 - 0.5]$$
$$= -5.03125.$$

This value is the same as the natural log of $L[0.5 \mid 23 \text{ choose mated}] = 0.00653$ calculated previously (except for rounding errors).

> The *log-likelihood* of a value for the population parameter is the natural log of its likelihood.

The maximum likelihood estimate

The maximum likelihood estimate of a parameter is the specific value having the highest likelihood, given the data. The value of the parameter that maximizes the likelihood is also the one that maximizes the log-likelihood, so we can work with the log-likelihood to find the maximum likelihood estimate.

A straightforward way to find the maximum likelihood estimate is to use the computer to calculate the log-likelihood across the range of possible parameter values and then pick the highest one. This is the approach we use here. Alternatively, we could use calculus to find the maximum.

The work of finding the maximum likelihood value for p is much reduced by using a program on the computer, rather than your calculator. In either case, start by evaluating the log-likelihood for several values of p over a broad range to see its overall shape. In Table 20.3-1, for example, we calculate the log-likelihood of values of p

---

4. One reason is that some likelihoods can be so small as to push the lower limits of your calculator and even your computer. Using logs also makes it easier to evaluate quantities such as $\binom{32}{23}$, which might push the upper limits of computer memory. With logs, multiplication becomes addition: $\ln[A \times B] = \ln[A] + \ln[B]$, and powers become multiples: $\ln[A^B] = B \ln[A]$.

TABLE 20.3-1 The log-likelihood calculated for a range of values of p using a spreadsheet program on the computer.

Proportion p	Log-likelihood
0.1	− 36.758
0.2	− 21.876
0.3	− 13.752
0.4	− 8.523
0.5	− 5.031
0.6	− 2.846
0.7	− 1.890
0.8	− 2.468
0.9	− 5.997

between 0.1 and 0.9 in increments of 0.1. You can see that the log-likelihood is highest when the proportion p is about 0.7 and that it declines at larger and smaller values.

Next, narrow the search by using a finer sequence of values for p. Table 20.3-2 shows the results for values of p between 0.52 and 0.88 in increments of 0.01. The log-likelihood reaches its maximum when p is 0.72. This value is therefore the maximum likelihood estimate.

In Figure 20.3-1, we plot the log-likelihood of all possible values of p between 0.4 and 0.9. The resulting curve is called the **log-likelihood curve**. The maximum[5] of the log-likelihood curve occurs at the value $\hat{p} = 0.72$, so this is the maximum

FIGURE 20.3-1
The log-likelihood curve for p, the estimated proportion of wasps choosing the mated female butterfly. The log-likelihood is maximized at $\hat{p} = 0.72$.

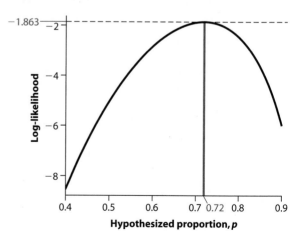

---

5. Log-likelihood curves based on simple probability models usually have only one peak, but there can be more than one peak in more complex models. It is important to check the full range of possible values for the parameter to be sure you find the highest peak.

TABLE 20.3-2 The log-likelihood calculated for a narrower range of values for p. The log-likelihood is maximized at $\hat{p} = 0.72$. The third column calculates the difference between each log-likelihood and the maximum log-likelihood.

Proportion, p	Log-likelihood	Log-likelihood $-$ maximum
0.52	-4.497	-2.634
0.53	-4.248	-2.385
0.54	-4.012	-2.149
0.55	-3.787	-1.924
0.56	-3.575	-1.712
0.57	-3.375	-1.512
0.58	-3.187	-1.324
0.59	-3.010	-1.147
0.60	-2.846	-0.983
0.61	-2.694	-0.831
0.62	-2.554	-0.691
0.63	-2.426	-0.563
0.64	-2.310	-0.447
0.65	-2.207	-0.344
0.66	-2.117	-0.254
0.67	-2.039	-0.176
0.68	-1.976	-0.113
0.69	-1.926	-0.063
0.70	-1.890	-0.027
0.71	-1.869	-0.006
0.72	-1.863	0.000
0.73	-1.873	-0.010
0.74	-1.900	-0.037
0.75	-1.944	-0.081
0.76	-2.007	-0.144
0.77	-2.089	-0.226
0.78	-2.192	-0.329
0.79	-2.318	-0.455
0.80	-2.468	-0.605
0.81	-2.644	-0.781
0.82	-2.848	-0.985
0.83	-3.084	-1.221
0.84	-3.354	-1.491
0.85	-3.663	-1.800
0.86	-4.014	-2.151
0.87	-4.416	-2.553
0.88	-4.873	-3.010

likelihood estimate of the population proportion. We give this value the symbol $\hat{p}$ to indicate that 0.72 is our maximum likelihood estimate.[6]

Likelihood-based confidence intervals

The log-likelihood curve also allows us to calculate a confidence interval for the population parameter. The range of values for p whose log-likelihood lies within $\chi^2_{\alpha,1}/2$ units of the maximum constitutes the $1 - \alpha$ **likelihood-based confidence interval** (Meeker and Escobar 1995). For example, an approximate 95% confidence interval for p is the range of values whose log-likelihood lies within 1.92 units of the maximum, since $\chi^2_{0.05,1}/2 = 3.84/2 = 1.92$. The 95% confidence interval is therefore determined directly from the log-likelihood curve. Figure 20.3-2 shows that the limits of the 95% confidence interval for the proportion p are $0.55 < p < 0.86$. A table of calculation results (e.g., Table 20.3-2) can also help to find these limits.

FIGURE 20.3-2
The likelihood-based 95% confidence interval for p. The top horizontal line indicates the highest log-likelihood (-1.863). The line immediately below corresponds to 1.92 units below the maximum. The 95% confidence interval, indicated by the red lines, is the range of values for the parameter whose log-likelihoods fall within 1.92 units of the maximum. This interval ranges from 0.55 to 0.86 for the wasp–butterfly data.

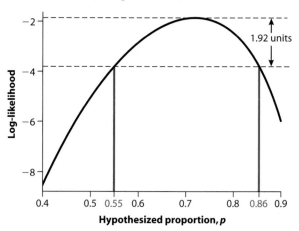

Based on these data, we conclude that the proportion of wasps in the population that would choose the mated female butterfly lies between 0.55 and 0.86 (which is very similar to the confidence interval we could calculate using the Agresti–Coull method (Section 7.3). This most-plausible range for the population proportion is relatively broad and includes relatively weak preference values (but still greater than 0.5) as well as relatively strong preference values. A larger sample size would be needed to narrow the range further.

This confidence-interval method based on the likelihood curve is more accurate than other methods used to generate confidence intervals for the maximum likelihood estimates (Meeker and Escobar 1995). It also requires no formula other than that needed to calculate the log-likelihood curve.

---

6. Notice that the maximum likelihood estimate of $\hat{p} = 0.72$ is the same as the conventional estimate for the proportion, $Y/n = 23/32 = 0.72$. This is no coincidence: Y/n is the formula for the maximum likelihood estimate of a population proportion. This shortcut could have spared us work, except that we wanted to show the general approach for finding a maximum likelihood estimate.

The *likelihood-based confidence interval* for a population parameter is calculated directly from the log-likelihood curve.

20.4 Versatility of maximum likelihood estimation

The great advantage of maximum likelihood estimation is its versatility. It can be applied to almost any situation in which it is possible to write a model describing the probability of different outcomes. Using Example 20.4, we demonstrate a less familiar problem than estimating a proportion, in which likelihood methods have proved invaluable.

EXAMPLE Conservation scoop

20.4 Counting elephants is more challenging than you might think, at least when they live in dense forest and feed at night. Eggert et al. (2003) used "capture–recapture" to estimate the total number of forest elephants inhabiting Kakum National Park in Ghana without having to see a single one. The researchers spent about two weeks in the park collecting elephant dung, from which they were able to extract pure elephant DNA. Using five genes, the researchers generated a unique DNA fingerprint for every elephant "encountered" via dung deposits. By using this fingerprint method, they identified 27 elephant individuals over the first seven days of sampling. We call this the first sample, and refer to these 27 elephants as "marked." Over

the next eight days, they sampled 74 individuals, of which 15 had already been detected in the first sample. We'll refer to these 15 elephants as "recaptured." Based on the number of recaptures in the second sample, what is the total population size of elephants in the park?

This kind of data can be used to estimate population size because the first sample tells us how many have been marked, and the second sample tells us the proportion of the total population that has been marked. By dividing the number marked by the proportion marked, we can estimate the total number of individuals in the population, N, which is the parameter of interest.

Probability model

The simplest probability model for population size estimation makes the following assumptions:

- The population of elephants in the park is constant—there were no births, deaths, immigrants, or emigrants while the study was being carried out.
- Random sampling—the dung of every elephant, marked or unmarked, had an equal and independent chance of being sampled.

Our data are the number of "recaptures" in the second sample—the individuals that had already been marked in the first sample. We know that 15 individuals were recaptured in the data, but, if we could go back and take another "second" sample of 74 individuals under identical conditions, we would probably obtain a different value for the number recaptured. In other words, the number recaptured has a probability distribution that depends on how many individuals there are in the population (N) and on the sample sizes.

To use the likelihood approach, we need to know the probability of obtaining the observed number of recaptures for different possible values of the parameter N. Finding this probability can be one of the biggest challenges of using the likelihood method. Often, though, someone has already investigated similar data and has published the needed formula in the scientific literature. For mark–recapture data, the formula is already available—we found it in Gazey and Staley (1986). Use n_1 to represent the number of individuals captured and marked in the first seven days of study. In this case, $n_1 = 27$. Let n_2 equal the size of the second sample of individuals obtained over the next eight days. In this case, $n_2 = 74$. The probability of Y recaptures, given the population size N, is[7]

$$\Pr[\text{number recaptures} = Y \mid N] = \frac{\binom{n_1}{Y}\binom{N - n_1}{n_2 - Y}}{\binom{N}{n_2}}.$$

The term on the right of the equal sign represents the proportion of all possible random samples of n_2 elephants that include exactly Y recaptures (previously marked) and $n_2 - Y$ new individuals (not previously marked).

The likelihood formula

From this point on, the steps are the same as those we followed in Example 20.3 when estimating a proportion. Let $L[N \mid Y \ recaptures]$ indicate the likelihood of a particular value for N, given that there were Y recaptures. The likelihood is defined

---

7. This probability distribution is called the hypergeometric distribution, which gives the probability of a given number of individuals of a particular type from a sample of known size and a population of known size, assuming that the individuals are sampled without replacement.

as the probability of obtaining Y recaptures if N equals the specified value—namely, $\Pr[\textit{number recaptures} = Y|N]$. Thus, the formula for the likelihood is

$$L[N|Y\ \textit{recaptures}] = \frac{\binom{n_1}{Y}\binom{N-n_1}{n_2-Y}}{\binom{N}{n_2}}.$$

It is easier to work with the log-likelihood rather than the likelihood itself:

$$\ln L[N|Y\ \textit{recaptures}] = \ln\left[\binom{n_1}{Y}\right] + \ln\left[\binom{N-n_1}{n_2-Y}\right] - \ln\left[\binom{N}{n_2}\right].$$

To calculate the log-likelihood of a specific value of N, substitute $Y = 15$ into the preceding equation. Plugging in the fixed sample sizes, $n_1 = 27$ and $n_2 = 74$, as well, yields

$$\ln L[N|15\ \textit{recaptures}] = \ln\left[\binom{27}{15}\right] + \ln\left[\binom{N-27}{74-15}\right] - \ln\left[\binom{N}{74}\right].$$

We are now ready to calculate the log-likelihood for specific values of N.

The maximum likelihood estimate

We plugged the formula for the log-likelihood into a computer program and calculated the log-likelihood of all possible values of N between 90 and 250. Try this yourself to see if you obtain the same answers we did. Our resulting log-likelihood curve is shown in Figure 20.4-1. The maximum of the log-likelihood curve occurs at the value $\hat{N} = 133$, so this is the maximum likelihood estimate of the elephant population size.

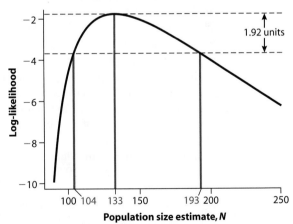

FIGURE 20.4-1
The log-likelihood curve for N, the total number of elephants in Kakum National Park in Ghana. The log-likelihood is maximized at $\hat{N} = 133$. The values 104 and 193 are the limits of the likelihood-based 95% confidence interval for N.

We also used the computer to determine the likelihood-based 95% confidence interval for N. This confidence interval includes all parameter values whose log-likelihood is within $\chi^2_{0.05,1}/2 = 3.84/2 = 1.92$ units below the maximum log-likelihood. These limits occurred at the N values 104 and 193, as shown in Figure 20.4-1. Based on these data, the population size of elephants in Kakum National Park, Ghana, is likely to be in the range

$$104 < N < 193.$$

Such a broad interval estimate for the population size estimation is not unusual for capture–recapture methods.

Bias

The maximum likelihood estimate should be relatively close to the true population parameter, compared with other values having lower likelihood. Nevertheless, maximum likelihood estimates are often biased. On average, that is, a maximum likelihood estimate tends to fall to one side of the population parameter that it is intended to estimate. For example, the maximum likelihood method for estimating population size using capture–recapture techniques tends to underestimate population size, on average, even when all the assumptions are met (Krebs 1999). In some cases, corrections are known that will compensate for bias. The bias is usually small, however, compared with the breadth of the 95% confidence interval. Bias diminishes as sample size increases (Edwards 1992).

20.5 Log-likelihood ratio test

The log-likelihood ratio test uses likelihood to compare how well two probability models fit the data. In one of the models, the parameter or parameters of interest are constrained to match that specified by a null hypothesis. The constraint is relaxed in the second probability model, which corresponds to the alternative hypothesis. This second model is fit using parameter values that best match the data. If the likelihood of the second model, in which the parameters best match the data, is significantly higher than the likelihood of the first model, in which the parameters are constrained to match H_0, then we reject the null hypothesis. Importantly, one of the hypotheses has to be a special case of the other; for example, a parameter that is allowed to vary in one hypothesis is set to a specific value in the other.

In this section, we introduce the log-likelihood ratio test. We analyze the wasp choice data from Example 20.3 (Fatouros et al. 2005) and test the null hypothesis that the population proportion p is 0.5. We use the binomial distribution to calculate the log-likelihood of $p = 0.5$, and we compare it with the log-likelihood of the alternative model, in which the parameter p is set to the maximum likelihood estimate.

Likelihood ratio test statistic

The test statistic for the likelihood ratio test, G, is twice the natural log of the ratio of the likelihoods of the two hypotheses. For the simplest case of only a single parameter,

$$G = 2 \ln\left(\frac{L[\,maximum\ likelihood\ value\ of\ parameter\,|\,data]}{L[\,parameter\ value\ under\ H_0\,|\,data\,]}\right).$$

The formula is similar when there is more than one parameter to be estimated. The remarkable feature of the log-likelihood ratio is that, if H_0 is true, then G is approximately χ^2 distributed. This means that we can use the χ^2 distribution to calculate a P-value and thus decide whether the null hypothesis should be rejected. The approximation is reliable only when the sample sizes are large.[8]

The degrees of freedom for G equal the difference between the two hypotheses in the number of parameters that require estimation using the data. In the case of only a single parameter, there is one degree of freedom.

Testing a population proportion

In the wasp experiment of Example 20.3, more wasps chose the mated female butterfly (23) than the unmated female (9). Is this by itself evidence that wasps in the population prefer mated female butterflies? We will use the log-likelihood ratio test to answer this question. The hypotheses are as follows.

H_0: Wasps choose mated and unmated females with equal probability ($p = 0.5$).

H_A: Wasps prefer one type of female over the other ($p \neq 0.5$).

We could use the binomial test (Chapter 7) or the χ^2 goodness-of-fit test (Chapter 8) to test these hypotheses. The probability models are relatively simple for this case, though, so we will use it to demonstrate the log-likelihood ratio test. This is a two-tailed test because a preference for either the mated or the unmated female is possible if the null hypothesis is false.

Our test statistic is the log-likelihood ratio,

$$G = 2 \ln\left(\frac{L[\hat{p}\,|\,Y\ chose\ mated\ female\,]}{L[\,p_0\,|\,Y\ chose\ mated\ female\,]}\right),$$

where $L[\,p_0\,|\,Y\ chose\ mated\ female\,]$ is the likelihood of p_0, the value of p specified by the null hypothesis. $L[\hat{p}\,|\,Y\ chose\ mated\ female]$ is the likelihood of $\hat{p}$, the maximum likelihood estimate of p. It is generally easiest to work with log-likelihoods, so we rewrite the previous formula as

$$G = 2(\ln L[\hat{p}\,|\,Y\ chose\ mated\ female\,] - \ln L[\,p_0\,|\,Y\ chose\ mated\ female\,]).$$

---

8. For small samples, simulation can be used to find the null distribution of G (see Chapter 19).

Under the null hypothesis, $p_0 = 0.5$, and in the data $Y = 23$ wasps chose the mated butterflies. Plugging these values into the likelihood formula for p, we get

$$\ln L\left[p_0 \mid 23 \text{ chose mated female}\right] = \ln\left[\binom{32}{23}(0.5)^{23}(1 - 0.5)^9\right]$$

$$= \ln\left[\binom{32}{23}\right] + 23 \ln(0.5) + 9 \ln(0.5)$$

$$= -5.031.$$

Under the alternative hypothesis, p is unknown and is estimated using maximum likelihood. We already determined in Section 20.3 that the maximum likelihood estimate is $\hat{p} = 0.72$. The corresponding log-likelihood of this $\hat{p}$ was found to be -1.863 (Figure 20.3-2).

We now have what we need to calculate the test statistic:

$$G = 2(\ln L\left[\hat{p} \mid 23 \text{ chose mated female}\right] - \ln L\left[p_0 \mid 23 \text{ chose mated female}\right]$$

$$= 2\left[-1.863 - (-5.031)\right]$$

$$= 6.336.$$

Under the null hypothesis, the test statistic G is approximately χ^2 distributed. The difference between the two hypotheses in the number of parameters needing estimation from the data was one. The test statistic G therefore has $df = 1$.

The critical value for the χ^2 distribution having one degree of freedom and $\alpha = 0.05$ is available in Statistical Table A:

$$\chi^2_{0.05,1} = 3.84.$$

Because $G > 3.84$, $P < 0.05$. Using a computer to calculate the exact probability for the χ^2 distribution, we find that $P = 0.012$. We reject the null hypothesis based on these data. The wasps indeed prefer mated female butterflies over unmated females.[9]

20.6 Summary

- Likelihood measures the level of support provided by data for a particular value of a population parameter. The likelihood of a specified value is the probability of obtaining the observed data if the parameter equaled that value.

- The maximum likelihood estimate of a parameter is the value having highest support among all possible values. It is the parameter value for which the probability of obtaining the observed data (the likelihood) is highest.

- Maximum likelihood estimation depends on a probability model that specifies the probabilities of different outcomes from the process that produced the data.

---

9. The researchers later determined that the chemical cue that the wasps use to distinguish mated from unmated females is benzyl cyanide, which the male butterfly passes to the female during mating. The compound is an "anti-aphrodisiac," rendering the mated female less attractive to other male butterflies (Fatouros et al. 2005).

- The log-likelihood is the natural log of the likelihood. The hypothesis that maximizes the log-likelihood is the same as the one maximizing the likelihood.
- The log-likelihood curve describes the log-likelihood of a range of values of the parameter. The maximum likelihood estimate can be found from the value that gives the highest point on the curve.
- The likelihood-based confidence interval for a single parameter is the range of values of the parameter whose log-likelihoods lie within $\chi^2_{\alpha,1}/2$ units of the maximum. The range for an approximate 95% confidence interval includes all values for the parameter whose log-likelihoods are within $3.841/2 = 1.92$ units of the maximum.
- The log-likelihood ratio test uses likelihood to compare the fits of two probability models to the data. In the case of a single parameter, one of the models constrains the parameter to equal that specified by a null hypothesis. The alternative probability model is fit using the maximum likelihood estimate of the parameter.

20.7 Quick Formula Summary

Likelihood

Formula: $L[value \mid data] = \Pr[data \mid parameter = value]$.

Likelihood-based confidence interval for a single parameter

What is it for? To obtain a confidence interval estimate for a parameter.

What does it assume? Data are randomly sampled from a population. The specific assumptions of the probability model are correct.

Formula: Obtained directly from the log-likelihood curve. To do so, find the range of parameter values whose log-likelihood lies within $\dfrac{\chi^2_{\alpha,1}}{2}$ units of the maximum.

Log-likelihood ratio test for a single parameter

What is it for? To compare the fit of two probability models to data using likelihood. In one model, the parameter is fixed to that specified by the null hypothesis. The other model is fit using the maximum likelihood estimate of the parameter.

What does it assume? Data are randomly sampled from a population. The specific assumptions of the probability model are correct.

Test statistic: G

Distribution under H$_0$: Approximately a χ^2 distribution with degrees of freedom equal to the difference between the two hypotheses in the number of parameters requiring estimation from the data.

Formula: $G = 2 \ln\left(\dfrac{L\left[\textit{maximum likelihood value of parameter} \mid \textit{data}\right]}{L\left[\textit{parameter value under } H_0 \mid \textit{data}\right]}\right).$

PRACTICE PROBLEMS

1. Albino animals are well known, but what about plants? Apiron and Zohary (1961) found a recessive mutation in orchard grass (*Dactylis glomerata*) that causes a chlorophyll deficiency. Individuals with two deficient copies of the gene lack the green pigment entirely (and die), but heterozygous individuals, which have one normal copy of the gene and one deficient copy, appear to produce normal levels of chlorophyll. Using progeny testing, the authors found that 12 of 47 plants sampled from the Matzleva Valley in Israel were heterozygotes. Assume that the plants were randomly sampled.

   a. What probability distribution should we use to calculate the probability of obtaining a specific number of heterozygous individuals (*p*) in a random sample of size 47?

   b. Write the formula for the likelihood of *p*, given the data. What does the likelihood measure?

   c. Write the formula for the log-likelihood of *p*.

   d. Calculate the log-likelihood of the value $p = 0.50$, given the data.

2. Refer to Practice Problem 1. The accompanying table lists log-likelihoods of values of *p* between 0.1 and 0.4, in increments of 0.02.

   a. Using only the table calculations, identify the maximum likelihood estimate of *p* to within 0.02 units.

   b. Using these same calculations, generate a likelihood-based 95% confidence interval for *p*.

Proportion, *p*	Log-likelihood	Log-likelihood − maximum
0.10	−6.639	−4.615
0.12	−5.238	−3.214
0.14	−4.193	−2.169
0.16	−3.414	−1.390
0.18	−2.844	−0.820
0.20	−2.444	−0.420
0.22	−2.186	−0.162
0.24	−2.051	−0.027
0.26	−2.024	0.000
0.28	−2.094	−0.070
0.30	−2.252	−0.228
0.32	−2.492	−0.468
0.34	−2.809	−0.785
0.36	−3.201	−1.177
0.38	−3.663	−1.639
0.40	−4.195	−2.171

3. Noonan et al. (2006) used DNA sequence data and the method of maximum likelihood to add the extinct Neanderthal species to the tree of humans and other apes (see Figure 20.2-1). The accompanying graph is a likelihood curve[10] for the time of the split between humans and Neanderthals, which is more recent than that between humans and the other living apes.

   a. What does the likelihood curve measure in this example?

   b. With the aid of a ruler, use the likelihood curve to find the maximum likelihood estimate of divergence time.

   c. Using this same curve, approximate the likelihood-based 95% confidence interval for divergence time.

---

10. The variable on the vertical axis is the log-likelihood after adding a constant to each log-likelihood value, so that the maximum is zero. This doesn't change any of the ensuing calculations.

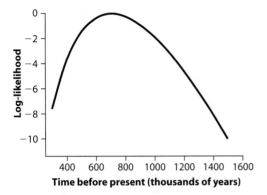

Time before present (thousands of years)

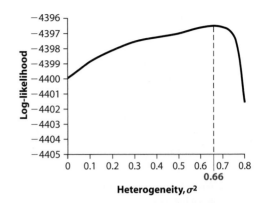

Heterogeneity, σ^2

d. What does the interval in part (c) measure? Explain.

4. Yashin et al. (2000) compared a sample of 197 centenarians (people 100 years old or older) with a group of 465 younger people (aged between 5 and 80) to examine whether the two groups differed in the frequencies of genetic markers. As part of an attempt to estimate which genetic markers are associated with mortality, the researchers tested for the presence of "hidden heterogeneity," intrinsic differences in mortality rate among individuals within groups. To accomplish this, they generated a likelihood model for a heterogeneity parameter, σ^2, and calculated the log-likelihood for all values of σ^2 between 0 and 0.8, given the data. Their methods are too complicated to describe fully here, but their results are summarized in the accompanying graph. The maximum likelihood estimate for σ^2 was 0.66. The null and alternative hypotheses are as follows.

H_0: Heterogeneity in mortality is absent $(\sigma^2 = 0)$.

H_A: Heterogeneity in mortality is present $(\sigma^2 \neq 0)$.

a. Using the log-likelihood curve, find the approximate value of the log-likelihood under the null hypothesis that heterogeneity is zero. Assume that the probability model used depends only on the single parameter shown.

b. With the aid of a ruler, use the log-likelihood curve to find the log-likelihood corresponding to the maximum likelihood estimate.

c. Using your results from parts (a) and (b), calculate the test statistic for a log-likelihood ratio test.

d. Under H_0, what is the approximate null distribution for the log-likelihood ratio statistic?

e. Using your results from parts (a)–(d), carry out the log-likelihood ratio test. Report your conclusion.

5. "Have you ever stolen something worth more than $10?" Anyone asked this question in a survey might be reluctant to answer truthfully, especially if he or she did not wish to make known a past misbehavior. The respondent might be more willing to tell the truth if the questioner doesn't know which of two questions, randomly chosen, the responder is answering (Warner 1965). This approach was used to estimate the true fraction of thieves among the third-year biology undergraduate population on a university campus in 2006. A total of 185 students participated. Each student was instructed to flip a coin and conceal the outcome. He or she was to respond with a yes if the outcome was heads. If the outcome was tails, the student was to answer the theft question truthfully with a yes or a no. The result: 113 of the 185 students responded with a yes, whereas the remaining 72 answered no. Assume that students answered truthfully and independently, that the probability of heads was 0.5, and that the sample of students was a random sample. Use likelihood to estimate the fraction of thieves in the student population.

a. Construct a probability tree (Chapter 5) to show that the probability of a student answering yes is $(1 + s)/2$, where s is the fraction of thieves in the population.

b. If the assumptions are met, the number of yes answers in the survey of n students, Y, should follow a binomial distribution with the probability of a yes equal to $(1 + s)/2$:

$$\Pr\left[Y \text{ yeses} \mid s\right] = \binom{n}{Y}\left(\frac{1 + s}{2}\right)^{Y}\left(\frac{1 - s}{2}\right)^{n-Y}.$$

Write the formula for the likelihood of s, given the data.

c. Write the formula for the log-likelihood of s, given the data.

d. Calculate the log-likelihood that the fraction of thieves s is zero.

6. Refer to Practice Problem 5.

a. Using a spreadsheet or other program, calculate the log-likelihood of values of s between 0.0 and 0.5 in increments of 0.01. Using this information, determine the maximum likelihood estimate of the fraction of thieves.

b. Using the same approach as in part (a), calculate the likelihood-based 95% confidence interval for the parameter s.

c. Are there truly thieves among us? Using the values for the likelihood calculated in your answers to Practice Problem 5, use the log-likelihood ratio test to test the null hypothesis that s is zero.

7. A regulatory gene controls the expression of other genes—it turns them on and off. Many of these targets are themselves regulatory genes, instructing genes to turn on or off. Guelzim et al. (2002) determined the number of regulatory genes controlled by a sample of genes in yeast. Their data are listed in the following table.

Number of regulatory genes controlled, i	Frequency, f_i
0	72
1	18
2	10
3	5
4	1
5	3
Total	109

The shape of this frequency distribution suggests that it might be approximated by a probability distribution known as the geometric distribution. Under a geometric distribution, the fraction of genes controlling no regulatory genes (here, $i = 0$) is p. The fraction controlling exactly one regulatory gene ($i = 1$) is $(1 - p)p$. The fraction controlling exactly two regulatory genes ($i = 2$) is $(1 - p)^2 p$, and so on, yielding

$$\Pr[i] = (1 - p)^i p,$$

where i is 0, 1, 2, or more. Maximum likelihood methods can be used to estimate the parameter p, the fraction controlling no regulatory genes, from data. The log-likelihood formula for p is

$$\ln\left[L[p \mid data]\right] = \ln\left[1 - p\right]\left(\sum_i i f_i\right) + n \ln[p],$$

where f_i is the frequency of observations corresponding to $i = 0, 1, 2$, and so on, and n is the total sample size.

a. Using the data in the table and the preceding formula, calculate the log-likelihood of values of p between 0.1 and 0.9 in increments of 0.01. Use a computer. Draw the relationship between the log-likelihood and p with a log-likelihood curve.

b. What is the maximum likelihood estimate $\hat{p}$, to an accuracy of two decimal places?

c. What is the value of the log-likelihood at the maximum likelihood estimate $\hat{p}$?

d. Using the formula for the geometric distribution, $\Pr[i] = (1 - p)^i p$, calculate the predicted proportion of genes regulating $i = 0$ to 5 genes based on $\hat{p}$. Plot these values on a histogram with the observed proportions. Based on the result, do the data appear to follow a geometric distribution?

e. Identify one method you might use to test the null hypothesis that a geometric distribution fits the data.

8. Phylogenetic trees like those in Figure 20.2-1, if dated, permit us to estimate the rate at which new species have formed over time. For example, the following tree indicates the timing of events in the history of the living species of *Dubautia*, the Hawaiian silverswords (Baldwin and Sanderson 1998), a famously diverse group of plants found only on the Hawaiian islands.

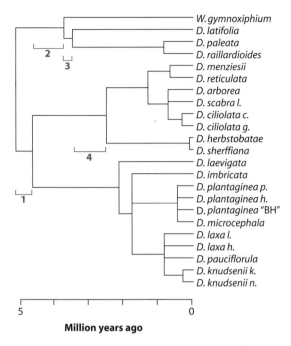

Million years ago

Assume that the number of *Dubautia* species has increased exponentially over time and has suffered no extinctions. In this case, the time interval between successive branching events on the tree provides information about the rate of increase of the number of species, λ (lambda). The earliest four intervals are indicated in red on the tree; with 23 species, there are $n = 21$ such intervals in total. Let Y_i measure the duration of each interval i multiplied by the number of species alive at that time. We'll call these Y_i the "waiting times." The $n = 21$ values of Y_i from the above tree are

1.0, 3.0, 0.8, 5.0, 2.4, 2.1, 0.8, 3.6, 5.0, 0.0, 0.0, 0.0, 1.4, 0.0, 1.6, 3.4, 0.0, 0.0, 2, 0.0, 4.4.

Under the assumptions given, the Y_i values should follow an *exponential* distribution (Hey 1992), with probability highest at zero and declining smoothly with increasing Y. To illustrate, an exponential distribution is superimposed on the following histogram of the Y_i values.

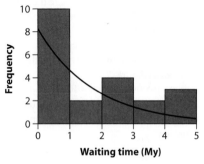

Waiting time (My)

Use the waiting-time data to estimate the rate λ. The formula for the log-likelihood of λ, given the Y_i, is

$$\ln L\left[\lambda \mid observed \; waiting \; times\right] = n \ln\left[\lambda\right] - \lambda \sum_i Y_i.$$

a. Using a computer program, calculate the log-likelihood of values of λ between 0.1 and 0.9 in increments of 0.01. Draw the relationship between the log-likelihood and λ with a log-likelihood curve.

b. Find the maximum likelihood estimate of λ to two decimal places. This estimates the number of new species produced per species per million years in *Dubautia*.

c. Using a similar approach as in part (b), find the likelihood-based 95% confidence interval for λ.

ASSIGNMENT PROBLEMS

9. Huntington disease is an inherited neurodegenerative disease caused by a mutation in the *huntingtin* gene. When did this mutation arise? García-Planells et al. (2005) found that most Spanish cases of the disease share the same ancestral mutation. Using gene sequence data from patients at genetic loci linked to the mutation, the authors applied a likelihood method similar to that used to date events in the human–ape lineage (Figure 20.2-1). Their calculations produced the following log-likelihood curve for the date of origin of the *huntingtin* mutation, measured as the number of generations before the present.

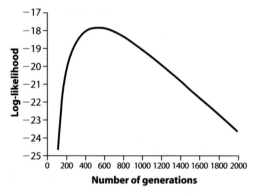

Number of generations

a. With the aid of a ruler, use this curve to find the maximum likelihood estimate for the date of origin of the mutation shared by most Spanish cases of Huntington disease.

b. Using this same curve, approximate a 95% confidence interval for the date of origin.

c. What does the interval in part (b) measure? Explain.

10. The Mediterranean shrub *Thymelaea hirsute* has five sexual types, the most curious of which is "gender labile." Such individuals change their predominant sex from year to year. Ramadan et al. (1994) found 13 gender-labile individuals in a sample of 68 shrubs from a single habitat in Egypt.

a. What probability distribution would we use to calculate the probability of a specific number of gender-labile individuals in a sample of size 68?

b. Write the formula for the likelihood of p, given the data. What does this likelihood measure?

c. What are your assumptions in part (b)?

d. Write the formula for the log-likelihood of p.

e. Calculate the log-likelihood of the value $p = 0.40$, given the data.

11. Refer to Assignment Problem 10. The following table lists log-likelihoods for a range of values of p between 0.1 and 0.3, in increments of 0.01.

Proportion, p	Log-likelihood	Log-likelihood − maximum
0.10	−4.652	−2.550
0.11	−4.027	−1.925
0.12	−3.517	−1.415
0.13	−3.105	−1.003
0.14	−2.778	−0.676
0.15	−2.524	−0.422
0.16	−2.336	−0.234
0.17	−2.207	−0.105
0.18	−2.130	−0.028
0.19	−2.102	0.000
0.20	−2.119	−0.170
0.21	−2.176	−0.074
0.22	−2.272	−0.170
0.23	−2.404	−0.302
0.24	−2.570	−0.468
0.25	−2.768	−0.666
0.26	−2.996	−0.894
0.27	−3.254	−1.152
0.28	−3.539	−1.437
0.29	−3.853	−1.751
0.30	−4.192	−2.090

a. Using only our calculations based on these increments, identify the maximum likelihood estimate.

b. Using these same calculations, generate a likelihood-based 95% confidence interval for p.

12. Sacktor et al. (2000) measured the neuropsychological performance of 33 HIV-positive patients undergoing antiretroviral therapy. Hand-use performance improved in 23 of the 33 patients but deteriorated in the remaining 10 patients. The graph below shows the log-likelihood curve for the population proportion p of patients whose hand-use performance improved after antiretroviral therapy.

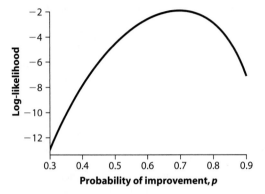

a. Using only the information in the log-likelihood curve, test the null hypothesis that the probability p is 0.5 using the log-likelihood ratio test.

b. Using the same curve, provide an approximate 95% confidence interval for the proportion of patients improving in the population.

13. Ants are capable of chemically discriminating between nestmates and non-nestmates. Ozaki et al. (2005) discovered sensory cells in the antennae of ants that respond only to the cuticular hydrocarbons (CHCs) of non-nestmates. They showed that 42 of 48 head-attached antenna preparations responded to chemical extracts of CHC compounds obtained from non-nestmates. Use likelihood to estimate the proportion of preparations p responding to non-nestmate CHC extracts.

a. Using a computer program, calculate the log-likelihood of a series of values of p between 0.50 and 0.99 in increments of 0.01. Draw the relationship between the log-likelihood and p with a log-likelihood curve.

b. Find the maximum likelihood estimate for p to two decimal places.

c. Calculate the likelihood-based 95% confidence interval for the parameter p.

14. Although we used the log-likelihood ratio test to analyze the data on the preference of parasitic

wasps for female butterflies (Example 20.3), we could have used the binomial test (Chapter 7) or the χ^2 goodness-of-fit test (Chapter 8). The data are reproduced in the following frequency table.

Preference	Observed frequency
Chose mated female	23
Chose virgin female	9
Total	32

Analyze the data again, but this time use the G-test, a goodness-of-fit test that we learned about in Chapter 9. The formula for the G-statistic is given in the following equation, where $Observed_i$ and $Expected_i$ refer to the observed and expected frequencies in category i, respectively. Under the null hypothesis, G is approximately χ^2 distributed with $k - 1$ degrees of freedom, where k is the number of categories in total.

$$G = 2\sum_i Observed_i \ln\left[\frac{Observed_i}{Expected_i}\right].$$

a. Restate the hypotheses for the parasitic wasp and butterfly data, and provide the expected frequencies under the null hypotheses.

b. Calculate the G-statistic using the formula provided.

c. Compare your result in part (b) with the critical value from the χ^2 distribution with the appropriate degrees of freedom, and state your conclusion.

d. Compare the value of G that you calculated in part (b) with the value of the log-likelihood ratio statistic that we calculated in Section 20.3. Based on your comparison, what can you infer about the G-test?

15. Dispersal and the movement distance of organisms are sometimes described using a probability distribution known as the geometric distribution (see Practice Problem 7). For example, the following histograms show the number of home ranges separating the locations where individual male and female field voles, *Microtus agrestis*, were first trapped, and the locations they were caught in a subsequent trapping period (Sandell et al. 1991). Most individuals didn't move at all, a few moved just one home

range away, and a small fraction moved two home ranges.

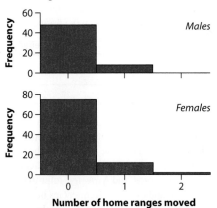

Number of home ranges moved

If these frequencies are described by a geometric distribution, then the fraction of observations in the first category (here, $i = 0$) is given by the parameter p. The fraction in the second category ($i = 1$) is $(1 - p)p$. The fraction in the third category ($i = 2$) is $(1 - p)^2 p$, and so on, yielding

$$\Pr[i] = (1 - p)^i p,$$

where i is 0, 1, 2, or more. Maximum likelihood methods can be used to estimate the parameter p from the data. The log-likelihood formula for p is

$$\ln L[p \mid data] = \ln[1 - p]\left(\sum_i i f_i\right) + n \ln[p],$$

where f_i is the frequency of observations corresponding to $i = 0, 1, 2$, and so on.

a. Forty-eight of 56 male voles stayed put ($i = 0$), whereas the remaining eight moved one home range. Using a computer program, calculate the log-likelihood of values of p between 0.70 and 0.99 with intervals of 0.01. Draw the relationship between the log-likelihood and p with a log-likelihood curve.

b. Find the maximum likelihood estimate $\hat{p}$ for male voles to two decimal places.

c. Obtain a likelihood-based 95% confidence interval for p.

16. Refer to Assignment Problem 15. Seventy-five of 89 female voles remained where they were first caught ($i = 0$), 12 moved just one home range away ($i = 1$), and two moved two home ranges away ($i = 2$).

a. Repeating the procedures described in Assignment Problem 15, calculate the maximum likelihood estimate of p for females.

b. Similarly, find a likelihood-based 95% confidence interval for p. Does it overlap that for males? (See Assignment Problem 15.)

17. Is the perception of eye-gazing in humans acquired through experience, or is it innate? To test this, Farroni et al. (2002) presented 17 infants (3–5 days old) with paired photographs of faces. The face in one photograph had a direct gaze, whereas the other had an averted gaze (see image). Fifteen of 17 infants spent more time gazing at the face with the direct gaze. Use the log-likelihood ratio test to test the null hypothesis of no preference. Show the steps of your calculations.

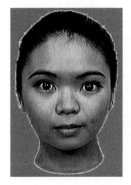 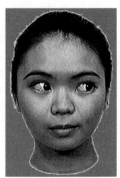

18. Life spans of individuals in a population often approximate an exponential distribution, a continuous probability distribution having probability density $f(Y) = \lambda e^{-\lambda Y}$, where λ is the mortality rate. To estimate the mortality rate of foraging honey bees, Visscher and Dukas (1997) recorded the entire foraging life span of 33 individual worker bees in a local bee population in a natural setting. The 33 life spans (in hours) are listed as follows and in a histogram.

2.3, 2.3, 3.9, 4.0, 7.1, 9.5, 9.6, 10.8, 12.8, 13.6, 14.6, 18.2, 18.2, 19.0, 20.9, 24.3, 25.8, 25.9, 26.5, 27.1, 30.0, 33.3, 34.8, 34.8, 35.7, 36.9, 41.3, 44.2, 54.2, 55.8, 65.8, 71.1, 84.7

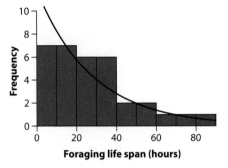

Foraging life span (hours)

If life span follows an exponential distribution, then the log-likelihood of λ, given the data is

$$\ln L[\lambda \mid observed\ life\ spans] = n \ln(\lambda) - \lambda \sum_i Y_i,$$

where Y_i is the life span of individual i, and n is the sample size.

a. Estimate λ, the mortality rate of bees per hour. Using a computer, find the maximum likelihood estimate of λ to two decimal places.

b. What is the value of the log-likelihood at the maximum likelihood estimate $\hat{\lambda}$?

c. What is likelihood-based 95% confidence interval for λ?

Blood red moon
© 2014, Fred Espenak, www.MrEclipse.com

21 Meta-analysis
combining information
from multiple studies

Most papers in scientific journals present the outcome of just one experiment or of one observational study. Each study is typically based on a single random sample or on a small number of related samples. Compelling issues in biology, however, deserve more than one study.

You will rarely find only one published study on a topic, unique in addressing an interesting problem. The first test of a truly original idea is almost always followed by studies that duplicate it to some extent. Most new studies typically address a question that has been asked and answered before, often many times. For example, by 1985 at least 118 data sets had been published testing whether the phase of the moon affects human behavior (Rotton and Kelly 1985). At least 24 studies have investigated whether acupuncture can help you stop smoking (White et al. 2006). Studies on a topic accumulate in the literature, and at some point these must be summarized to yield an overall conclusion. How do we combine the information that we get from multiple studies?

The review article is the traditional answer to this question. An expert in the field assembles the studies published on a topic, thinks about them carefully and (hopefully) fairly, and then writes a review article summarizing the overall conclusions reached. There is much to be said for this approach. A first-rate review article advances a field far beyond a mere summary. Such a review will propose new hypotheses, uncover previously unnoticed relationships, and point to new paths of research. A strong review provides a solid indication of the state of thought and knowledge about a particular topic.

But the traditional review usually lacks a quantitative methodology, which might lead to two problems. First, the conclusions of one reviewer are often partly subjective, perhaps weighing studies that support the author's preferences more heavily than studies with opposing views. An extreme example of this is Linus Pauling's (1986) book *How to Live Longer and Feel Better*, in which the author cited 30 studies supporting his own idea that taking large daily doses of vitamin C reduces the risk of contracting the common cold. Pauling[1] cited no studies opposing this idea, even though a number of such studies had been published (Knipschild 1994). Not all reviews are blatantly biased, but complete objectivity is difficult for any reviewer to achieve when evaluating a diverse set of published studies.

The second problem is that it is extremely difficult to balance multiple studies by intuition alone without quantitative tools. The combined strength of statistical support and the average magnitude of an effect simply cannot be determined without some calculations. "Vote counting" is a step in the right direction—that is, count the number of published studies that have rejected the null hypothesis of zero effect and compare this count with the number of studies failing to reject the null hypothesis. Yet, even this approach ignores important information on the size of the study, the strength of an effect, and the magnitude of our uncertainty.

A set of techniques called *meta-analysis* provide quantitative tools for the synthesis of information from multiple studies. In this chapter, we give an overview of the approach using examples, and we point out advantages and limitations. We

---

1. Linus Pauling is the only person to have won two unshared Nobel Prizes, Chemistry in 1954 and Peace in 1962, so his behavior does not represent that of a crank.

also indicate how best to report the results of your own studies so that your results will be useful to future meta-analyses.

21.1   What is meta-analysis?

Meta-analysis is the "analysis of analyses." It is not a single statistical test but rather a set of techniques that allows us to address statistical questions about collections of data obtained from the published literature and other sources. The researcher begins a meta-analysis by gathering all known studies about a particular topic. Typically, these studies report a broad range of results, with some rejecting the null hypothesis and others not rejecting the null hypothesis. Given such a broad range of results, how can we decide what the truth is? Meta-analysis provides ways to combine information from independent studies.

With meta-analysis, we can address questions like these:

- How strong is the effect we are studying, on average?
- Does the collection of studies reject a null hypothesis?
- How variable is the effect? What factors might explain this variability?
- Is there any bias resulting from the publication of some studies and not others?
- Are there enough unpublished studies to change our conclusions?

The quantitative details of meta-analysis lie beyond the scope of this book. If you want to know more about the techniques, have a look at Borenstein et al. (2009).

> A *meta-analysis* compiles all known scientific studies estimating or testing an effect and quantitatively combines them to give an overall estimate of the effect.

Why repeat a study?

Why would anyone bother to test an idea that has already been addressed in one or more previous studies that have already provided answers? Repetition is necessary in science, because errors—both statistical errors and methodological flaws—can be made in individual studies. A single study might yield a significant result when the null hypothesis is true (a Type I error) or it may fail to detect a true effect (a Type II error). Moreover, what is true in one circumstance may not be true in another. For example, one kind of animal might tend to evolve to larger size on islands (think of the three-meter Komodo dragon on the Indonesian islands of Komodo and Flores), whereas another kind of animal may do the opposite (for example, the five-foot pygmy elephants living on Flores, thought to be the original diet of the dragons).

Moreover, we all make mistakes sometimes, so experimental designs can be flawed or errors can be made in recording data. In short, we need to have more than one study on any phenomenon that we care about.

21.2 The power of meta-analysis

One of the most useful aspects of meta-analysis is that it can greatly increase the statistical power of our collective statistical tests. Any one study will have limitations, sometimes severe, on the amount of time and money available to do research. As a result, the sample size of any given study is limited, yet the power of a study depends on sample size. We can combine the power of individual studies when we address the scientific question with meta-analysis. As a result, weak but important effects can sometimes be detected. This power can be particularly important in medical studies, when even a small reduction in mortality rates from an inexpensive and relatively harmless treatment can result in thousands of lives saved.

EXAMPLE Aspirin and myocardial infarction

21.2 People at risk of stroke or myocardial infarction (a "heart attack") are often advised to take aspirin or another antiplatelet medication. These kinds of medication are known to reduce the risk of future stroke, myocardial infarction, and death (all of which we will refer to as "vascular events"). But how do we know this is true? It turns out that the effect of aspirin and other antiplatelet agents was confirmed by a large meta-analysis, conducted by a large collaboration of researchers (Antiplatelet Trialists' Collaboration 1994). They combined information from 142 randomized trials of antiplatelet treatments on patients who had previously had a stroke or similar disease. In total, the trials included more than 70,000 patients. The results of these trials are summarized in Figure 21.2-1.

FIGURE 21.2-1
The vascular event rate for patients in antiplatelet therapy compared with patients in the corresponding control group. Each dot is a separate study, except at the origin, where 28 studies overlap. The one-to-one line marks equal vascular event rates in the two treatments.

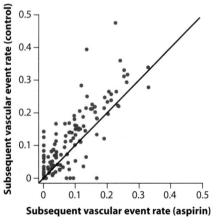

Before the meta-analysis, whether aspirin helped prevent vascular events was far from clear. Of the 142 studies reviewed, only 87 showed a better result for patients on antiplatelet therapy than for the control patients. Moreover, only 19 of these 87 showed significantly better results for the treated patients than the control patients at $\alpha = 0.05$, while 68 had nonsignificant results. Two of the 142 studies even showed a significantly worse rate of vascular events with aspirin treatment! A simple vote-counting procedure does not give overwhelming support for aspirin.

When the results are combined in a meta-analysis, however, the answer is surprisingly robust. It was relatively straightforward to combine the results, because all of the experiments involved the same kind of data: frequencies of control and treated patients who either did or did not have subsequent vascular events. Each of these studies had roughly the same numbers of treated individuals as control individuals. For display purposes only, we added up the frequencies of patients in each category and produced the contingency table shown in Table 21.2-1.

TABLE 21.2-1 Numbers of patients in each treatment group and outcome category in 142 studies combined.

	Treated with antiplatelet therapy	Control (no antiplatelet therapy)	Row totals
Vascular event (stroke, infarction, death)	4,183	5,400	9,583
No vascular event	32,353	31,311	63,664
Column Totals	36,536	36,711	73,247

In total, 14.7% (5400/36,711) of the patients in the control groups had subsequent vascular events, compared with 11.4% (4183/36,536) in the treated group. This difference may seem small, but it corresponds to the control patients having a 30% higher chance of serious problems than the treated patients. Put another way, if just the individuals in the controls of these experiments had been given antiplatelet treatment, there would have been about 1200 fewer strokes, infarctions, and deaths.

Since the side effects associated with antiplatelet treatment are low, this result (if statistically supported) would justify using aspirin. We cannot do a simple odds-ratio (OR) calculation on the combined table, because the data points are not all independent. We can, however, combine the odds ratios for each study using the Mantel–Haenszel confidence interval (see the Quick Formula Summary in Section 21.8). The Mantel–Haenszel procedure is specifically designed to combine information across multiple contingency analyses. Doing so, the researchers found that taking aspirin had a clear beneficial effect ($0.71 < OR < 0.75$).

Without the analysis of the data across all available studies, there would have been little chance of confirming the value of this relatively important treatment. In fact, the meta-analysis came 10 years after it would have been possible to show an effect by combining across studies that already existed at the time. It took another six years before the results of this meta-analysis were accepted by the medical community, perhaps in part due to the relative newness of meta-analysis techniques (Hunt 1997). The number of extra deaths and suffering in the intervening years that could have been avoided by earlier application and acceptance of these meta-analysis techniques is sobering to think about.

21.3   Meta-analysis can give a balanced view

Studies that find dramatic results are sometimes given more attention in the press and scientific literature than better studies with less "newsworthy" results. As a consequence, we require fair summaries of the breadth of knowledge on a question to give us a balanced view. Meta-analysis attempts to cover all studies fairly.

EXAMPLE   The Transylvania effect

21.3   Many people believe that a full moon can affect human behavior. The word "lunacy" is derived from the Latin *luna*, moon, and legends of strange beings, such as werewolves and vampires, have been connected to full moons for centuries. Some studies have shown that abnormal human behavior increases during full moons, as measured by such variables as homicide rates, psychiatric hospital admissions, suicide rates, crisis calls, etc. This phenomenon is called the "Transylvania effect." Other studies, however, have reported no significant effects. A meta-analysis of these studies (Rotton and Kelly 1985) found no statistically significant change in "lunacy" during the full moon. Moreover, the best estimate of the average effect was that less than 1% of the variation in these events was explained by moon phase. Even if the moon did affect human abnormal behavior, the average effect is so small as to be unimportant. The results are shown in the funnel plot in Figure 21.3-1. (See Interleaf 10 for a discussion of funnel plots.)

The literature on the Transylvania effect includes some studies with apparently very strong relationships between moon phase and human behavior. The most

FIGURE 21.3-1
A funnel plot of the effect sizes from the meta-analysis of the "Transylvania effect." The horizontal line marks zero correlation between moon phase and unusual behavior.

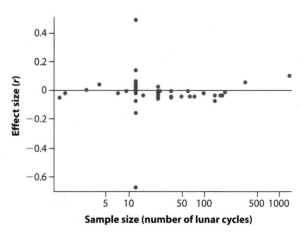

extraordinary of these papers have gotten a great deal of media attention and word of mouth among police and hospital workers. Thus, there is a tendency to believe that the sensational studies are typical of all those performed. This meta-analysis shows that these are just the tips of the sampling iceberg; when all data sets are examined, there is little or no effect of moon phase on aberrant human behavior.

21.4   The steps of a meta-analysis

In this section, we describe the process of combining information using meta-analysis. We focus on some of the issues that arise when summarizing data, but we don't provide mathematical details.

Define the question

A key step in meta-analysis is defining the question and the breadth of studies to which it applies. Some meta-analyses apply a question to a very narrow set of studies, all of which are expected to estimate the same true value. Meta-analyses in medical research are usually of this type. For example, in the aspirin/heart attack meta-analysis of Example 21.2, the question was whether aspirin affected the risk of future myocardial events. To answer this question, the authors reviewed studies that had very similar properties. They included only studies that were all done on human participants who had suffered a stroke or heart attack. All of the studies were randomized clinical trials comparing an aspirin treatment with a control treatment, and all measured the future risks to the participants in terms of significant health problems—namely, heart attacks, strokes, or death. Different studies, therefore, estimated very similar effects, and all should give more or less the same answer, except for sampling error. The goal of the meta-analysis was to combine the multiple studies—in effect, to create one large study with more power than any of the single studies. A meta-analysis of such a homogeneous set of studies is called a fixed-effect meta-analysis.

At the other extreme, a meta-analysis might address a question whose answer requires reviewing a broad and heterogeneous mix of studies. Meta-analyses in non-medical areas of biology are often of this type. For example, Kingsolver et al. (2001) conducted a meta-analysis to estimate the average strength of natural selection in the wild. Answering this question required combining studies carried out on different species living in different kinds of environments. The variables measured were not the same from study to study. A meta-analysis of such a heterogeneous set of studies is called random-effect meta-analysis. In this case, the goal is to obtain an average over the separate effects of all the studies. If the studies in the meta-analysis are similar enough, and if we can standardize their responses in some way, then the average effect may be informative.

Example 21.4 describes a meta-analysis of a heterogeneous mix of studies, even though all were carried out on the same species.

EXAMPLE   Testosterone and aggression

21.4   Using a meta-analysis, Book et al. (2001) asked the question, "Are testosterone levels and aggression correlated in humans?" The studies reviewed for this meta-analysis were all done on humans, but the ways aggression was measured were extremely diverse. Some compared the levels of testosterone in prisoners convicted of violent crimes with those of prisoners convicted of property

crimes. One correlated the levels of testosterone in university students with their answers to questionnaires that asked them for levels of agreement to statements like "If somebody hits me, I hit back." Others were less prosaic. One measured the levels of aggression in !Kung San males by counting "their scars and sometimes still open wounds in the head region." Another compared drunken Finnish spouse-abusers with drunken Finns drinking quietly in a bar. Another compared members of "rambunctious" fraternities with "responsible" frats.[2]

In this example, "aggression" was defined in a variety of ways. They tested the null hypothesis: "Testosterone has no average effect on human aggression levels." The authors wanted to ask a rather broad question with a variety of types of people and types of aggression, so many studies were relevant. These studies do not repeat themselves; they each address different *specific* questions, but they all address the same *general* question. Meta-analyses in biology can be even broader, often including results from a large number of species. This is appropriate, if the question is general enough, but it should always be remembered that the different subgroups in the meta-analysis may actually have different responses.

Review the literature

Once the question is defined, the hardest part of a meta-analysis ensues. We must collect all of the available information that pertains to the question at hand. In principle, this is simple, but in practice, collecting even a fraction of the pertinent literature can be a Herculean task. Meta-analysts often take years to collect the information for their reviews, although computerized databases are making the task easier. Most libraries do not carry subscriptions to all journals, so getting copies of some articles can be difficult. More importantly, many good studies are not published in journals at all and therefore are not in computer databases. Sometimes they are reported in the "gray literature"—monographs and in-house publications of various foundations, institutes, and government agencies. Lots of science is reported only in doctoral dissertations and master's theses. Finally, many studies are not published at all, so an

---

2. One of the "rambunctious" frat houses was described as "only standing because it was made out of steel and concrete." At one of the "responsible" frats, they "talked a lot about computers and calculus."

effort has to be made to contact researchers in the field to discover what information exists but is not widely available.

It is crucial to find everything, to minimize the effects of "publication bias." As discussed in Interleaf 10, studies that find large and significant effects are more likely to be published, more likely to be in "first-rate" journals, and more likely to be referenced in other articles. All of this means the studies that we can find easily are *different* from those that we cannot so easily find. Some statistical techniques exist to partially account for such biases (see the discussion of funnel plots in Interleaf 10 and the "file-drawer problem" in Section 21.5), but they are not nearly as reliable as a proper literature review.

Meta-analyses contain from as few as a handful of studies to as many as thousands, depending on the amount of information available, the amount of effort expended by the meta-analysts themselves, and the breadth of the question asked. The more information there is available for analysis, the more reliable the final results are likely to be.

One key issue that arises when compiling studies is whether to include studies of apparently poor quality. An unfortunate fact of science is that not all studies are done well; in spite of long training, large amounts of money spent on research, and the peer review process, some bad science is still published. What do we do about these studies? We could just ignore them. But this raises a thorny point: we may be more likely to reject a study as "bad" if its conclusions do not agree with our own opinions about the scientific question at hand. Good meta-analyses circumvent this problem by having the methods (alone, separated from the results) of each study rated by independent scientists and then weighting the study's contribution to the meta-analysis by the judges' scores. Others set standards a priori, such as collecting only studies that experimentally and randomly assigned individuals to treatments and controls. This question of how to deal with varying quality of science is not fully resolved, and it represents one of the greatest challenges for the future of meta-analysis.

Compute effect sizes

The core quantity of meta-analysis is the **effect size**, which measures how strong the association is between the explanatory variable X and the response variable Y. The magnitude is important—for example, we might need to know whether the effect of a drug is large enough to exceed any negative consequences or costs the treatment might have.

> The *effect size* is a standardized measure of the magnitude of the relationship between two variables.

In meta-analysis, we are combining across multiple studies, and usually these studies do not measure their results on the same yardstick. Some may even be measuring completely different things. How do we combine measures of "aggression" that range from the number of scars (a numerical variable) to whether or not someone is imprisoned for murder (a categorical variable)? How do we combine information that comes from a correlation analysis with results from comparisons of means and contingency tables? Fortunately, methods have been developed for putting quantities obtained from different kinds of data onto a common, standardized scale.

Most meta-analyses use one of three common measures of effect size. These are the odds ratio (*OR*; see Section 9.2), the correlation coefficient (*r*; see Section 16.1), and the standardized mean difference (SMD; see below). Which measure of effect size is used depends on the variables. The odds ratio compares the frequencies of "success" and "failure" between two groups or treatments. The correlation coefficient typically measures the relationship between two numerical variables, although it can also be generalized to describe many kinds of data (Rosenthal 1991). The standardized mean difference is useful when most studies being combined compare the mean of a numerical response variable between two groups. SMD is defined as

$$\text{SMD} = \frac{\overline{Y}_1 - \overline{Y}_2}{s_{\text{pooled}}},$$

where s_{pooled} is the square root of the pooled sample variance. In other words, SMD is the difference in the sample means of two groups divided by the pooled estimate of their standard deviation (the square root of the pooled sample variance). SMD measures the difference between groups in units of standard deviation.[3]

Each of these effect size measures allows the results of multiple studies to be compared with one another. The effect sizes from the different studies included in a meta-analysis must be calculated in the same way. For example, when using SMD, the mean of the control group should be subtracted from the mean of the treatment group in every study. If a positive effect size indicates improvement in one study, then every other study that showed an improvement associated with the treatment ought to have a positive effect size, too.

Table 21.4-1 lists results from nine of 54 studies in the meta-analysis of testosterone and aggression (Book et al. 2001; only nine are shown here to keep the table brief). Of these nine studies, the original measure of association was either a correlation coefficient (*r*), or a difference of group means. Finally, all measures of association were converted to the equivalent correlation coefficient *r* to put them on the same scale. (See Rosenthal 1991 for computational details.) Degrees of freedom (*df*) are also listed in Table 21.4-1, reflecting the sample size of each study, along with the *P*-value.

Several of these nine studies rejected the null hypothesis that there is no association between aggression and testosterone, but not all. The funnel plot in Figure 21.4-1 shows how variable the results are among all 54 studies. Some studies show a strong positive relationship between testosterone and aggression, whereas others show no significant effect. One study even shows a significant negative correlation. This variability stems in part from sampling error, but it also likely reflects differences between studies in the types of variables measured and the populations sampled.

---

3. The SMD is often called Cohen's *d* (Borenstein et al. 2009). A similar measure, Hedges' *g*, includes a correction for small sample size.

TABLE 21.4-1 A subset of the results of a meta-analysis on the relationship between testosterone and aggression in humans.

References	Type of study	Effect size (r)	df	P
Gray et al. (1991)	Aging men (39–70); questionnaires	0.02	1677	0.21
Houser (1979)	20-something males; questionnaires	0.086	3	0.45
Olweus et al. (1980)	Swedish male adolescents; self-reports of aggression	0.24	56	0.035
Banks and Dabbs (1996)	College students compared with "Americans who belonged to a deviant and delinquent urban subculture"	0.30	61	0.008
Harris et al. (1996)	University students; questionnaires	0.36 men 0.41 women	153 149	1.7×10^{-6} 5.8×10^{-8}
Christiansen and Winkler (1992)	!Kung San men; number of scars from fights	0.06	112	0.27
Lindman et al. (1992)	Drunken Finns arrested for spousal abuse vs. drunken Finns in bars	0.18	34	0.15
Dabbs et al. (1996)	"Rambunctious" frats vs. "responsible" frats	0.26	93	0.006

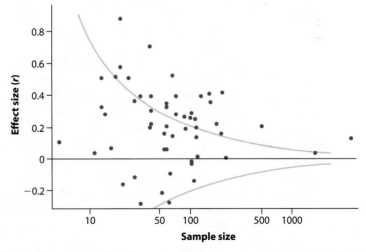

FIGURE 21.4-1 A funnel plot of studies comparing human aggression to levels of testosterone. The curves show the approximate boundaries of the critical regions that would reject the null hypothesis in any one study with $\alpha = 0.05$.

Determine the average effect size

Once we have all the effect sizes in comparable units, we can calculate the average effect over all of the studies. Typically, however, we do not take just a simple average of the individual effect sizes. Instead, we acknowledge that some studies give us more information than others, and we weight studies according to the precision of their estimates. The most common reason that some studies have more information is simply that their sample sizes were larger. It is also possible to weight studies by their quality: studies done with superior methodology might be weighted more heavily than those with more suspect technique.

Calculate confidence intervals and test hypotheses

Meta-analysis techniques make it possible to calculate confidence intervals and test hypotheses for the mean effect size, accounting for the input of all studies included. Typically, the confidence intervals for the effect size averaged over all studies is much smaller than that obtained from any one study, and hypothesis tests are much more powerful than in any of the individual studies. Different techniques are used for fixed-effects and random-effects meta-analyses. See Gurevitch and Hedges (1999) and Cooper and Hedges (1994) for further discussion.

Our example data show an average correlation between testosterone and aggression of $\bar{r} = 0.096$, with a 95% confidence interval for the mean correlation of $0.055 < \bar{\rho} < 0.136$. In other words, the evidence points to a real effect, but it is quite small. Most of the individual studies were not powerful enough to detect an effect this small, because their sample sizes were too small.

The mean effect size has to be understood in the context of the studies that are used to calculate it. Averaging over systematically biased estimates will still give a biased estimate. Moreover, the results can only be interpreted relative to the population of individuals addressed in the studies. If the studies are all about humans, then the interpretation can be reasonably applied only to humans. If the studies come from a diversity of species, then the conclusions can be applied to a mix of species like those in the studies. It is rare—well, to be honest, impossible—that the group of studies available in the literature truly reflects a random selection of all possible studies that may have been done on a given topic. A mixture of studies to determine the effects of a particular agent of human mortality, for example, might be heavily biased toward studies done in first-world hospitals, so we would have to interpret the results according to this bias. Medical studies are done disproportionately in large research hospitals, and it is certainly possible that the standard of care in these hospitals is systematically different from other, less studied health-care settings. The efficacy rate of a treatment could be different in these other settings, meaning that the meta-analysis could be systematically biased.

Look for effects of study quality

Not all studies in a meta-analysis are the same. They vary in sample size and in their methodologies. As part of a meta-analysis, we can and should ask whether these differences among studies influence the outcome.

Larger published studies (those based on larger sample sizes) are more likely to give more reliable estimates of effect size, because effect of publication bias is likely to be lower. Small studies not yielding a statistically significant outcome are less likely to be published than those yielding a significant result (see Interleaf 10), inflating the average effect size for those small studies that make it to publication. Large studies would be less affected by this problem, because the larger sample sizes are more likely to detect an effect if it is present. Large studies also require much more effort and funds than small studies; therefore, researchers are more likely to publish their conclusions whatever the results. One way to address this possibility is to look across all studies included in the meta-analysis for a correlation between sample size and effect size. For example, there is a negative relationship between sample size and effect size in the meta-analysis comparing morphological asymmetry to mating success discussed in Interleaf 10 (Spearman $r_S = -0.30$, $df = 138$, $P = 0.003$). The effect size is much smaller for large studies than for small studies. (See the funnel plot in Interleaf 10.) As a result, we should distrust the smaller studies in that meta-analysis. In contrast, in the meta-analysis of the effects of aspirin on vascular events (Example 21.2), there was no detectable relationship between sample size and effect size. In that meta-analysis, we have no reason to think that the smaller studies were influenced by publication bias.

Moreover, there are often differences in average effect size between studies of high quality and those of low quality. For example, studies without blinding (Section 14.3) have systematically larger effect sizes than those studies including this method to reduce bias (Jüni et al. 2001). Meta-analyses commonly find differences in average effect size between observational and experimental studies, between studies that did and did not include randomization, and so on. When such differences are found, we should use only the better-quality studies to draw our conclusions.

Look for associations

Another advantage of meta-analysis is that the different studies can be used to examine the effects of methodological or other differences between studies. That is, we are looking for **moderator variables**—variables that can explain some of the variation in effect size.

Scientifically interesting factors can be responsible for the heterogeneity of effect size across studies. Meta-analysis can suggest relationships that were not even addressed in any of the component studies. For example, a meta-analysis of the efficacy of homework showed that homework stimulated learning, despite claims to the contrary in the education literature (Cooper and Valentine 2001). More surprising, homework seemed to have little benefit for elementary schoolchildren but large benefits for high school students. This effect was detected in the meta-analysis only; none of the individual studies being analyzed tested the effect of grade level directly, and no one had predicted it theoretically. A potentially important conclusion was reached by combining information across studies and looking at the effect of a key moderator variable that had not been explicitly addressed in any one study.

21.5 File-drawer problem

All of the meta-analysis techniques we have described assume that the studies being reviewed are a random sample of all possible studies on that topic. But in one very important way, the studies available for synthesis are not random: they tend to be drawn mainly from the published literature, and this literature usually suffers from publication bias. As mentioned in Interleaf 10, publication bias is the difference in mean effect size between published studies and all studies on the topic.

In meta-analysis, the difficulties caused by publication bias are called the **file-drawer problem**, in reference to the unknown studies sitting unavailable in researchers' file drawers or hidden in obscure journals.

> The *file-drawer problem* is the possible bias in estimates and tests caused by publication bias.

Meta-analysis can increase power and reduce the Type II error rate—that is, combining across studies makes it easier to reject a false null hypothesis. In some cases, if the available data are biased, meta-analysis may also have a higher chance of rejecting a true null hypothesis than expected (Type I error).

A few methods partially address these problems. Funnel plots (see Figure 21.4-1 and Interleaf 10), for example, can give some indication of the bias resulting from small studies. (If you haven't done so already, please read Interleaf 10. We won't repeat that material here.)

Another partial solution to the file-drawer problem is to calculate how many missing studies would be needed to change the overall result of the meta-analysis. A standard technique assumes that all of these missing studies failed to reject the null hypothesis of no effect. The method then calculates the number of missing studies required to reach the point at which the null hypothesis is no longer rejected by the meta-analysis. This number is called the **fail-safe number**.[4] If the fail-safe number is small (i.e., roughly the same as the number of published studies included in the initial meta-analysis), then the results of that meta-analysis would be regarded as unreliable. If the fail-safe number is very large (e.g., in the millions), then we can be more

---

4. Calculation details of the fail-safe number, also called the "file-drawer method," can be found on p. 405 of Cooper and Hedges (1994).

certain that the meta-analysis is giving us the right answer—it is simply too unlikely for a million unpublished studies on the subject to exist.

The fail-safe number calculation assumes that the direction of the effect detected in a study does not influence the likelihood of publication. Unfortunately, studies that detect an effect opposite to that of most published studies might also be lost from publication. For example, studies that detect significant harm to patients taking a drug may be less likely to be published than those that find an advantage to taking the drug. If studies in the opposite direction from that desired also go missing from a meta-analysis database, then the fail-safe number cannot be reliably interpreted.

21.6 How to make your paper accessible to meta-analysis

Many published papers do not report enough information for meta-analysts to extract the numbers that they need. As a result, many otherwise relevant papers have to be discarded. This difficulty can be avoided by a few simple changes in the way information is presented. Here are some suggestions.

- **Always give estimates of the sizes of the effects and provide their standard errors.** Much too often, the size of the effect (e.g., the odds ratio, the correlation coefficient, etc.) is not given, and an author presents only a P-value. Also, give estimates of the mean and standard deviation of the important variables, as both may be needed for some effect-size calculations. It is surprising how often estimates of means and the sizes of effects are not given, even though we care far more about the size of the effect than about the P-value.

- **Give the values of your test statistics and the number of degrees of freedom.** The degrees of freedom are essential for most of meta-analysis. In particular, the degrees of freedom are needed for calculating a weighted average of effect size across studies. Similarly, the values of the test statistics are needed for some methods used to calculate effect sizes.

- **Make the data accessible.** Publish the raw data in the paper or on an online archive, such as datadryad.org. If the data are available for scrutiny by others, the information in the study can be faithfully transcribed into future meta-analyses.

21.7 Summary

- Most scientific questions have been addressed by more than one published study. Meta-analysis is a general name for methods that quantitatively combine information from multiple studies that address the same question.

- Meta-analysis increases power and decreases the Type II error rate.
- An effect size is a standardized measure of the results of a study, used for all studies included in a meta-analysis. Common effect sizes include odds ratios, correlation coefficients, and standardized mean differences.
- The studies easily found in the literature are not a random sample of all studies done. Because of publication bias, the mean effect size in collections of published studies will often be larger than the true mean. Because meta-analysts have more difficulty finding unpublished studies than published studies, the values estimated via meta-analysis can be biased. This difficulty is called the file-drawer problem.
- One measure of the possible effect of publication bias is the fail-safe number. The fail-safe number is the number of hypothetical, unpublished studies failing to reject the null hypothesis that would need to be added to the meta-analysis to change its results.
- The confidence intervals of estimates of effect sizes are smaller, usually much smaller, for a meta-analysis than for its component studies, because the meta-analysis uses more information. On the other hand, meta-analysis can give estimates of effect size that are more biased than those from the larger studies they summarize, due to publication bias.
- Meta-analysis can also investigate the influence of study methodology and other variables not studied in the original articles.
- To make your study amenable to meta-analysis, provide exact calculations of effects, standard errors, standard deviations, test statistics, and degrees of freedom.

21.8 Quick Formula Summary

Mantel–Haenszel estimate of odds ratios from combined studies

What is it for? To estimate a confidence interval for an odds ratio based on data from multiple studies.

What does it assume? Data in each study are from random samples, there is no publication bias, and the odds ratio in the population is the same for all studies.

Degrees of freedom: 1

Formula: $OR_{MH} = \dfrac{\sum\limits_{i=1}^{S} a_i d_i / n_i}{\sum\limits_{i=1}^{S} b_i c_i / n_i},$

where a, b, c, and d are defined for each study according to the following table, s is the number of studies, and $n_i = a_i + b_i + c_i + d_i$.

		Explanatory variable	
		Treatment	Control
Response	Success	a_i	b_i
variable	Failure	c_i	d_i

The confidence interval for a Mantel–Haenszel estimate should be calculated with the aid of a computer.

Mantel–Haenszel test

What is it for? To test the null hypothesis that the odds ratio estimated in multiple studies equals one.

What does it assume? Data in each study are from random samples, there is no publication bias, and the odds ratio in the population is the same for all studies.

Test statistic: χ^2_{MH}

Distribution under H_0: χ^2 distribution with one degree of freedom.

Formula:
$$\chi^2_{MH} = \frac{\left[\left|\sum_{i=1}^{S}\left(a_i - \frac{(a_i + b_i)(a_i + c_i)}{n_i}\right)\right| - 0.5\right]^2}{\sum_{i=1}^{S}\frac{(a_i + b_i)(a_i + c_i)(b_i + d_i)(c_i + d_i)}{n_i^2(n_i - 1)}},$$

where the terms are defined as for the Mantel–Haenszel estimate of odds ratios and $||$ denotes the absolute value of its contents.

PRACTICE PROBLEMS

Note: Some of these problems address topics related to publication bias, as discussed in Interleaf 10.

1. Give two reasons why researchers are more likely to publish results when $P < 0.05$ than when $P > 0.05$.

2. The accompanying funnel plot shows the results of studies that estimated the heritability of various traits in a large number of species. Heritability is the proportion of the variation among

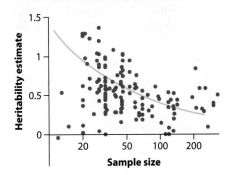

individuals in a population that can be explained by genetic differences among individuals. Because it is a proportion, heritability is a unitless number between zero and one. The methods of estimating heritability, however, allow estimates outside of these limits. The curve on this graph marks the critical values with $\alpha = 0.05$ for the null hypothesis that the population heritability equals zero.

a. Look carefully at this funnel plot. Is there any evidence for publication bias?

b. What kinds of decisions made by researchers and editors are likely to be behind the pattern?

3. The meta-analysis on the relationship between testosterone and aggression mentioned in Example 21.2 also included several studies on whether success in certain sports, such as tennis, was correlated with testosterone levels. Discuss the value of including these studies in this particular meta-analysis.

4. The file *selection.csv*, available online,[5] contains the effect sizes and sample sizes from a meta-analysis of studies measuring the strength of natural selection on morphological traits in different wild animal and plant populations (Kingsolver et al. 2001). Make and examine a funnel plot of these results. (You don't need to draw the curve showing statistical significance.)

5. Choose a paper from the scientific literature on a topic that you are interested in. Identify an interesting question that the paper addresses statistically, and attempt to extract from the paper the necessary information for a meta-analysis. What is the sample size? What is the effect size? What test statistic was used, and what was its value?

6. A hypothetical meta-analysis on the effects of a new drug on the reduction in the size of ovarian cancer tumors finds that no published studies failed to find a statistically significant effect of the drug and that the average size of the effect of the drug is small in the largest studies but larger in the smaller studies. The authors found the mean effect size was large, and they interpreted this as evidence for a large benefit of the drug. Why might you doubt the results of this meta-analysis?

7. Why might meta-analysis increase the Type I error rate, relative to individual studies?

ASSIGNMENT PROBLEMS

8. A meta-analysis on 25 studies testing the effects of a new enzymatic treatment on pityriasis (a common skin disease) finds a significant effect of the treatment. The fail-safe number was calculated to be 5.

a. How much confidence should you give the results of this meta-analysis?

b. If the fail-safe number had been 1500, how would you feel about the results? Explain your reasoning.

9. In what ways is meta-analysis an improvement over a traditional review of the literature on a research topic?

10. What measure of effect size would most likely be used in each of the following meta-analyses?

a. A meta-analysis on the effect of the drug ibuprofen on the frequency of individuals contracting kidney disease.

b. A meta-analysis of whether excluding parasites from a population affects mean growth rate.

c. A meta-analysis of the association between human height and life span.

d. A meta-analysis comparing the survival of parasitized and unparasitized birds.

e. A meta-analysis comparing the difference in body size of males and females across species.

11. A meta-analysis of the effects of predators on their prey populations collected the results of 45

---

5. whitlockschluter.zoology.ubc.ca

studies (Salo et al. 2007). The effect size used to describe these studies was Hedges' *g*, similar to the SMD. As one part of the meta-analysis, the researchers discovered that predators that had been artificially introduced by humans ("alien predators") had larger effects on prey than did native predators. None of the 45 individual studies had compared alien and native predators.

a. What kind of variable is the classification of predators as alien or native? (Answer using the specific language of meta-analysis.)

b. Australia has had a disproportionately large number of alien species introduced with great ecological damage. As a result, most of the studies compared in this meta-analysis on alien predators were done in Australia. Comment on how this should affect our understanding of the results.

c. None of the original studies compared alien and native predators. Explain how meta-analysis made it possible to answer a novel question based on these studies, even though none of the individual studies addressed this question.

12. In most meta-analyses, some studies find answers that are opposite to those found by the average of all studies. Give two reasons for this phenomenon.

Statistical tables

This appendix gives probabilities and critical values for a few of the most commonly used probability distributions. More can be found in such references as Rohlf and Sokal (1994), *Statistical Tables*.

Using statistical tables

Statistical tables are used to calculate probabilities and to put bounds on the P-value for a test statistic. They are also used to find the critical values for calculating confidence intervals. With easy access to computers becoming more common, statistical tables are becoming less necessary. One advantage of using computer programs is that they provide precise P-values, whereas statistical tables often only bracket a P-value (such as "$0.025 < P < 0.05$"). Moreover, the tables in this book only give critical values corresponding to the most commonly used significance levels, such as 0.05 and 0.01. Statistical tables are useful when computers are unavailable, or when putting bounds on P is satisfactory. For some of the simpler distributions, we give instructions for calculating more exact probabilities in both the free statistical package R and the spreadsheet program Excel, on the introductory page of each table.

Some probability distributions depend on the degrees of freedom. Examples include the χ^2 distribution, the Student's t-distribution, and the F-distribution. In such cases, the statistical tables in this book provide quantities for many but not all values of the degrees of freedom. What should you do when the probability distribution corresponding to a specific number of degrees of freedom is missing from the tables? For example, you might be looking for critical values of the t-distribution having 149 degrees of freedom, but Statistical Table C provides such quantities only for t-distributions having 140 and 160 degrees of freedom. In this case, there are two approaches to obtaining the missing critical values, as follows:

1. *The conservative approach*: Find the two critical values in the table corresponding to distributions having degrees of freedom just above and just below

the desired degrees of freedom. Use the critical value that makes your confidence interval widest, or that makes it most difficult to reject the null hypothesis of your test.

2. *Interpolation*: Find the two critical values in the table that correspond to the degrees of freedom just above and just below the desired number of degrees of freedom. Use linear interpolation to estimate the critical value for your desired degrees of freedom.

For example, let's say that you want to find the critical value for the F-distribution having five degrees of freedom in the numerator and 132 degrees of freedom in the denominator, $F_{0.05(1),5,132}$. Statistical Table D only gives $F_{0.05(1),5,100} = 2.31$ and $F_{0.05(1),5,200} = 2.26$. Because 132 lies between 100 and 200, the critical value must lie between 2.31 and 2.26. The following figure shows how to find the critical value by linear interpolation.

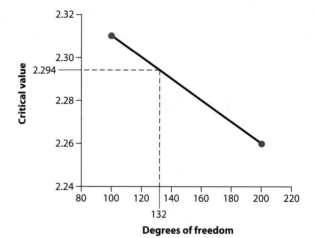

The desired critical value (C) is obtained as

$$C = 2.31 + (2.26 - 2.31)\left(\frac{132 - 100}{200 - 100}\right)$$
$$= 2.294.$$

The general formula for calculating a critical value from interpolation is

$$C = C_{low} + (C_{high} - C_{low})\left(\frac{df - df_{low}}{df_{high} - df_{low}}\right),$$

where df_{low} and df_{high} are the degrees of freedom just below and just above df, the desired number of degrees of freedom, and C_{low} and C_{high} are their corresponding critical values.

Statistical Table A: The χ^2 distribution

This table gives critical values of the χ^2 distribution. The critical value $\chi^2_{\alpha, df}$ defines the area α in the right tail of the χ^2 distribution with df degrees of freedom. To find a critical value in the table, select the desired α along the top row and the number of degrees of freedom df in the far left column. For example, if $df = 5$ and $\alpha = 0.05$, then the critical value is 11.07; that is, the probability of a value greater than or equal to 11.07 is 0.05.

The critical value for a χ^2 distribution having $df = 5$ when $\alpha = 0.05$. The area in red indicates the tail probability corresponding to 0.05, or 5%. The boundary of the red section is $\chi^2 = 11.07$. The probability of a value greater than or equal to 11.07 is 0.05.

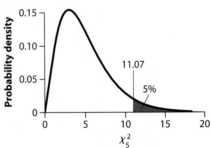

Exact probabilities under the right tail of the χ^2 distribution for any value of χ^2 can be obtained using one of many computer programs. For example, in the free statistical package R, to find the P-value corresponding to a χ^2 value of 11.07 with 5 df, enter the following text into the R console:

```
1 - pchisq(df = 5, q = 11.07)
```

More generally, replace 5 and 11.07 with the appropriate df and χ^2 calculated for your data.

The common spreadsheet program Excel can calculate exact P-values for χ^2 as well. In a cell, write

```
= 1 - CHISQ.DIST(11.07, 5, TRUE)
```

where the first number in the parentheses is the observed χ^2 and the second number is the df.

					α					
df	0.999	0.995	0.99	0.975	0.95	0.05	0.025	0.01	0.005	0.001
1	0.0000016	0.000039	0.00016	0.00098	0.00393	3.84	5.02	6.63	7.88	10.83
2	0.002	0.01	0.02	0.05	0.10	5.99	7.38	9.21	10.60	13.82
3	0.02	0.07	0.11	0.22	0.35	7.81	9.35	11.34	12.84	16.27
4	0.09	0.21	0.30	0.48	0.71	9.49	11.14	13.28	14.86	18.47
5	0.21	0.41	0.55	0.83	1.15	11.07	12.83	15.09	16.75	20.52
6	0.38	0.68	0.87	1.24	1.64	12.59	14.45	16.81	18.55	22.46
7	0.60	0.99	1.24	1.69	2.17	14.07	16.01	18.48	20.28	24.32
8	0.86	1.34	1.65	2.18	2.73	15.51	17.53	20.09	21.95	26.12
9	1.15	1.73	2.09	2.70	3.33	16.92	19.02	21.67	23.59	27.88
10	1.48	2.16	2.56	3.25	3.94	18.31	20.48	23.21	25.19	29.59
11	1.83	2.60	3.05	3.82	4.57	19.68	21.92	24.72	26.76	31.26
12	2.21	3.07	3.57	4.40	5.23	21.03	23.34	26.22	28.30	32.91
13	2.62	3.57	4.11	5.01	5.89	22.36	24.74	27.69	29.82	34.53
14	3.04	4.07	4.66	5.63	6.57	23.68	26.12	29.14	31.32	36.12
15	3.48	4.60	5.23	6.26	7.26	25.00	27.49	30.58	32.80	37.70
16	3.94	5.14	5.81	6.91	7.96	26.30	28.85	32.00	34.27	39.25
17	4.42	5.70	6.41	7.56	8.67	27.59	30.19	33.41	35.72	40.79
18	4.90	6.26	7.01	8.23	9.39	28.87	31.53	34.81	37.16	42.31
19	5.41	6.84	7.63	8.91	10.12	30.14	32.85	36.19	38.58	43.82
20	5.92	7.43	8.26	9.59	10.85	31.41	34.17	37.57	40.00	45.31
21	6.45	8.03	8.90	10.28	11.59	32.67	35.48	38.93	41.40	46.80
22	6.98	8.64	9.54	10.98	12.34	33.92	36.78	40.29	42.80	48.27
23	7.53	9.26	10.20	11.69	13.09	35.17	38.08	41.64	44.18	49.73
24	8.08	9.89	10.86	12.40	13.85	36.42	39.36	42.98	45.56	51.18
25	8.65	10.52	11.52	13.12	14.61	37.65	40.65	44.31	46.93	52.62
26	9.22	11.16	12.20	13.84	15.38	38.89	41.92	45.64	48.29	54.05
27	9.80	11.81	12.88	14.57	16.15	40.11	43.19	46.96	49.64	55.48
28	10.39	12.46	13.56	15.31	16.93	41.34	44.46	48.28	50.99	56.89
29	10.99	13.12	14.26	16.05	17.71	42.56	45.72	49.59	52.34	58.30
30	11.59	13.79	14.95	16.79	18.49	43.77	46.98	50.89	53.67	59.70
31	12.20	14.46	15.66	17.54	19.28	44.99	48.23	52.19	55.00	61.10
32	12.81	15.13	16.36	18.29	20.07	46.19	49.48	53.49	56.33	62.49
33	13.43	15.82	17.07	19.05	20.87	47.40	50.73	54.78	57.65	63.87
34	14.06	16.50	17.79	19.81	21.66	48.60	51.97	56.06	58.96	65.25
35	14.69	17.19	18.51	20.57	22.47	49.80	53.20	57.34	60.27	66.62
36	15.32	17.89	19.23	21.34	23.27	51.00	54.44	58.62	61.58	67.99
37	15.97	18.59	19.96	22.11	24.07	52.19	55.67	59.89	62.88	69.35
38	16.61	19.29	20.69	22.88	24.88	53.38	56.90	61.16	64.18	70.70
39	17.26	20.00	21.43	23.65	25.70	54.57	58.12	62.43	65.48	72.05
40	17.92	20.71	22.16	24.43	26.51	55.76	59.34	63.69	66.77	73.40
41	18.58	21.42	22.91	25.21	27.33	56.94	60.56	64.95	68.05	74.74
42	19.24	22.14	23.65	26.00	28.14	58.12	61.78	66.21	69.34	76.08
43	19.91	22.86	24.40	26.79	28.96	59.30	62.99	67.46	70.62	77.42
44	20.58	23.58	25.15	27.57	29.79	60.48	64.20	68.71	71.89	78.75
45	21.25	24.31	25.90	28.37	30.61	61.66	65.41	69.96	73.17	80.08
46	21.93	25.04	26.66	29.16	31.44	62.83	66.62	71.20	74.44	81.40
47	22.61	25.77	27.42	29.96	32.27	64.00	67.82	72.44	75.70	82.72
48	23.29	26.51	28.18	30.75	33.10	65.17	69.02	73.68	76.97	84.04
49	23.98	27.25	28.94	31.55	33.93	66.34	70.22	74.92	78.23	85.35
50	24.67	27.99	29.71	32.36	34.76	67.50	71.42	76.15	79.49	86.66

					α					
df	**0.999**	**0.995**	**0.99**	**0.975**	**0.95**	**0.05**	**0.025**	**0.01**	**0.005**	**0.001**
51	25.37	28.73	30.48	33.16	35.60	68.67	72.62	77.39	80.75	87.97
52	26.07	29.48	31.25	33.97	36.44	69.83	73.81	78.62	82.00	89.27
53	26.76	30.23	32.02	34.78	37.28	70.99	75.00	79.84	83.25	90.57
54	27.47	30.98	32.79	35.59	38.12	72.15	76.19	81.07	84.50	91.87
55	28.17	31.73	33.57	36.40	38.96	73.31	77.38	82.29	85.75	93.17
56	28.88	32.49	34.35	37.21	39.80	74.47	78.57	83.51	86.99	94.46
57	29.59	33.25	35.13	38.03	40.65	75.62	79.75	84.73	88.24	95.75
58	30.30	34.01	35.91	38.84	41.49	76.78	80.94	85.95	89.48	97.04
59	31.02	34.77	36.70	39.66	42.34	77.93	82.12	87.17	90.72	98.32
60	31.74	35.53	37.48	40.48	43.19	79.08	83.30	88.38	91.95	99.61
61	32.46	36.30	38.27	41.30	44.04	80.23	84.48	89.59	93.19	100.89
62	33.18	37.07	39.06	42.13	44.89	81.38	85.65	90.80	94.42	102.17
63	33.91	37.84	39.86	42.95	45.74	82.53	86.83	92.01	95.65	103.44
64	34.63	38.61	40.65	43.78	46.59	83.68	88.00	93.22	96.88	104.72
65	35.36	39.38	41.44	44.60	47.45	84.82	89.18	94.42	98.11	105.99
66	36.09	40.16	42.24	45.43	48.31	85.96	90.35	95.63	99.33	107.26
67	36.83	40.94	43.04	46.26	49.16	87.11	91.52	96.83	100.55	108.53
68	37.56	41.71	43.84	47.09	50.02	88.25	92.69	98.03	101.78	109.79
69	38.30	42.49	44.64	47.92	50.88	89.39	93.86	99.23	103.00	111.06
70	39.04	43.28	45.44	48.76	51.74	90.53	95.02	100.43	104.21	112.32
71	39.78	44.06	46.25	49.59	52.60	91.67	96.19	101.62	105.43	113.58
72	40.52	44.84	47.05	50.43	53.46	92.81	97.35	102.82	106.65	114.84
73	41.26	45.63	47.86	51.26	54.33	93.95	98.52	104.01	107.86	116.09
74	42.01	46.42	48.67	52.10	55.19	95.08	99.68	105.20	109.07	117.35
75	42.76	47.21	49.48	52.94	56.05	96.22	100.84	106.39	110.29	118.60
76	43.51	48.00	50.29	53.78	56.92	97.35	102.00	107.58	111.50	119.85
77	44.26	48.79	51.10	54.62	57.79	98.48	103.16	108.77	112.70	121.10
78	45.01	49.58	51.91	55.47	58.65	99.62	104.32	109.96	113.91	122.35
79	45.76	50.38	52.72	56.31	59.52	100.75	105.47	111.14	115.12	123.59
80	46.52	51.17	53.54	57.15	60.39	101.88	106.63	112.33	116.32	124.84
81	47.28	51.97	54.36	58.00	61.26	103.01	107.78	113.51	117.52	126.08
82	48.04	52.77	55.17	58.84	62.13	104.14	108.94	114.69	118.73	127.32
83	48.80	53.57	55.99	59.69	63.00	105.27	110.09	115.88	119.93	128.56
84	49.56	54.37	56.81	60.54	63.88	106.39	111.24	117.06	121.13	129.80
85	50.32	55.17	57.63	61.39	64.75	107.52	112.39	118.24	122.32	131.04
86	51.08	55.97	58.46	62.24	65.62	108.65	113.54	119.41	123.52	132.28
87	51.85	56.78	59.28	63.09	66.50	109.77	114.69	120.59	124.72	133.51
88	52.62	57.58	60.10	63.94	67.37	110.90	115.84	121.77	125.91	134.75
89	53.39	58.39	60.93	64.79	68.25	112.02	116.99	122.94	127.11	135.98
90	54.16	59.20	61.75	65.65	69.13	113.15	118.14	124.12	128.30	137.21
91	54.93	60.00	62.58	66.50	70.00	114.27	119.28	125.29	129.49	138.44
92	55.70	60.81	63.41	67.36	70.88	115.39	120.43	126.46	130.68	139.67
93	56.47	61.63	64.24	68.21	71.76	116.51	121.57	127.63	131.87	140.89
94	57.25	62.44	65.07	69.07	72.64	117.63	122.72	128.80	133.06	142.12
95	58.02	63.25	65.90	69.92	73.52	118.75	123.86	129.97	134.25	143.34
96	58.80	64.06	66.73	70.78	74.40	119.87	125.00	131.14	135.43	144.57
97	59.58	64.88	67.56	71.64	75.28	120.99	126.14	132.31	136.62	145.79
98	60.36	65.69	68.40	72.50	76.16	122.11	127.28	133.48	137.80	147.01
99	61.14	66.51	69.23	73.36	77.05	123.23	128.42	134.64	138.99	148.23
100	61.92	67.33	70.06	74.22	77.93	124.34	129.56	135.81	140.17	149.45

Statistical Table B: The standard normal (Z) distribution

This table gives probabilities under the right tail of the standard normal distribution. To determine the probability of sampling a value greater than or equal to a specific value of Z, find the first two digits of Z (called a and b) in the far left column of the table, and then find the last digit (c) in the top row of the table. For example, to find the probability of sampling a value greater than or equal to $Z = 1.96$, use $a.b = 1.9$ and $c = 6$. This probability is 0.025, or 2.5%.

An example of a probability under the standard normal curve: the probability of sampling a value greater than or equal to the value 1.96 is 0.025, or 2.5%.

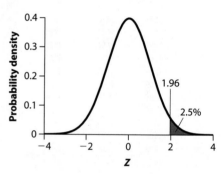

Exact probabilities under the right tail of the standard normal distribution can be obtained in any one of various computer packages. In R, the probability under the standard normal distribution of obtaining a value of Z greater than 1.96 is found using

```
1 - pnorm(q = 1.96)
```

Replace 1.96 with your desired value of Z.

In Excel, this probability can be calculated by entering the following into a spreadsheet cell:

```
= 1 - NORM.DIST(1.96, 0, 1, TRUE)
```

First two digits of *a.bc*	Second digit after decimal (*c*)									
	0	**1**	**2**	**3**	**4**	**5**	**6**	**7**	**8**	**9**
0.0	0.5	0.49601	0.49202	0.48803	0.48405	0.48006	0.47608	0.47210	0.46812	0.46414
0.1	0.46017	0.45620	0.45224	0.44828	0.44433	0.44038	0.43644	0.43251	0.42858	0.42465
0.2	0.42074	0.41683	0.41294	0.40905	0.40517	0.40129	0.39743	0.39358	0.38974	0.38591
0.3	0.38209	0.37828	0.37448	0.37070	0.36693	0.36317	0.35942	0.35569	0.35197	0.34827
0.4	0.34458	0.34090	0.33724	0.33360	0.32997	0.32636	0.32276	0.31918	0.31561	0.31207
0.5	0.30854	0.30503	0.30153	0.29806	0.29460	0.29116	0.28774	0.28434	0.28096	0.27760
0.6	0.27425	0.27093	0.26763	0.26435	0.26109	0.25785	0.25463	0.25143	0.24825	0.24510
0.7	0.24196	0.23885	0.23576	0.23270	0.22965	0.22663	0.22363	0.22065	0.21770	0.21476
0.8	0.21186	0.20897	0.20611	0.20327	0.20045	0.19766	0.19489	0.19215	0.18943	0.18673
0.9	0.18406	0.18141	0.17879	0.17619	0.17361	0.17106	0.16853	0.16602	0.16354	0.16109
1.0	0.15866	0.15625	0.15386	0.15151	0.14917	0.14686	0.14457	0.14231	0.14007	0.13786
1.1	0.13567	0.13350	0.13136	0.12924	0.12714	0.12507	0.12302	0.12100	0.11900	0.11702
1.2	0.11507	0.11314	0.11123	0.10935	0.10749	0.10565	0.10383	0.10204	0.10027	0.09853
1.3	0.09680	0.09510	0.09342	0.09176	0.09012	0.08851	0.08691	0.08534	0.08379	0.08226
1.4	0.08076	0.07927	0.07780	0.07636	0.07493	0.07353	0.07215	0.07078	0.06944	0.06811
1.5	0.06681	0.06552	0.06426	0.06301	0.06178	0.06057	0.05938	0.05821	0.05705	0.05592
1.6	0.05480	0.05370	0.05262	0.05155	0.05050	0.04947	0.04846	0.04746	0.04648	0.04551
1.7	0.04457	0.04363	0.04272	0.04182	0.04093	0.04006	0.03920	0.03836	0.03754	0.03673
1.8	0.03593	0.03515	0.03438	0.03362	0.03288	0.03216	0.03144	0.03074	0.03005	0.02938
1.9	0.02872	0.02807	0.02743	0.02680	0.02619	0.02559	0.02500	0.02442	0.02385	0.02330
2.0	0.02275	0.02222	0.02169	0.02118	0.02068	0.02018	0.01970	0.01923	0.01876	0.01831
2.1	0.01786	0.01743	0.01700	0.01659	0.01618	0.01578	0.01539	0.01500	0.01463	0.01426
2.2	0.01390	0.01355	0.01321	0.01287	0.01255	0.01222	0.01191	0.01160	0.01130	0.01101
2.3	0.01072	0.01044	0.01017	0.00990	0.00964	0.00939	0.00914	0.00889	0.00866	0.00842
2.4	0.00820	0.00798	0.00776	0.00755	0.00734	0.00714	0.00695	0.00676	0.00657	0.00639
2.5	0.00621	0.00604	0.00587	0.00570	0.00554	0.00539	0.00523	0.00508	0.00494	0.00480
2.6	0.00466	0.00453	0.00440	0.00427	0.00415	0.00402	0.00391	0.00379	0.00368	0.00357
2.7	0.00347	0.00336	0.00326	0.00317	0.00307	0.00298	0.00289	0.00280	0.00272	0.00264
2.8	0.00256	0.00248	0.00240	0.00233	0.00226	0.00219	0.00212	0.00205	0.00199	0.00193
2.9	0.00187	0.00181	0.00175	0.00169	0.00164	0.00159	0.00154	0.00149	0.00144	0.00139
3.0	0.00135	0.00131	0.00126	0.00122	0.00118	0.00114	0.00111	0.00107	0.00104	0.00100
3.1	0.00097	0.00094	0.00090	0.00087	0.00084	0.00082	0.00079	0.00076	0.00074	0.00071
3.2	0.00069	0.00066	0.00064	0.00062	0.00060	0.00058	0.00056	0.00054	0.00052	0.00050
3.3	0.00048	0.00047	0.00045	0.00043	0.00042	0.00040	0.00039	0.00038	0.00036	0.00035
3.4	0.00034	0.00032	0.00031	0.00030	0.00029	0.00028	0.00027	0.00026	0.00025	0.00024
3.5	0.00023	0.00022	0.00022	0.00021	0.00020	0.00019	0.00019	0.00018	0.00017	0.00017
3.6	0.00016	0.00015	0.00015	0.00014	0.00014	0.00013	0.00013	0.00012	0.00012	0.00011
3.7	0.00011	0.00010	0.00010	0.00010	0.00009	0.00009	0.00008	0.00008	0.00008	0.00008
3.8	0.00007	0.00007	0.00007	0.00006	0.00006	0.00006	0.00006	0.00006	0.00005	0.00005
3.9	0.00005	0.00005	0.00004	0.00004	0.00004	0.00004	0.00004	0.00004	0.00003	0.00003
4.0	0.00003	0.00003	0.00003	0.00003	0.00003	0.00003	0.00002	0.00002	0.00002	0.00002

Statistical Table C: Student's *t*-distribution

This table gives critical values of the *t*-distribution. The two-sided or two-tailed critical value $t_{\alpha(2),\, df}$ defines the combined area α under both tails of the *t*-distribution having *df* degrees of freedom. To find a critical value in the table, select the desired value of $\alpha(2)$ along the top and the number of degrees of freedom in the far left column. For example, if *df* = 5 and $\alpha(2) = 0.05$, the critical value is 2.57; that is, the probability of a *t*-value greater than or equal to 2.57, or less than or equal to −2.57, is 0.05. The left panel of the following figure illustrates the critical value for this example.

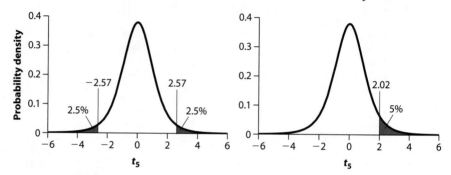

Critical values of the *t*-distribution having five degrees of freedom when $\alpha = 0.05$. The left panel shows the two-tailed case. The area in red shows the combined tail probabilities of 0.05, with 2.5% of the probability in each tail. The boundaries of the areas in red are −2.57 and 2.57. The right panel shows the one-tailed case. The area in red indicates the right tail of the distribution corresponding to 5% of the total probability. The boundary of the red section is 2.02.

The one-tailed critical value $t_{\alpha(1),\, df}$ defines the area α under the right tail of the *t*-distribution having *df* degrees of freedom. To find a critical value in the table, select the desired value of $\alpha(1)$ along the top and the number of degrees of freedom in the far left column. For example, if *df* = 5 and $\alpha(1) = 0.05$, the critical value is 2.02; that is, the probability of a value greater than or equal to 2.02 is 0.05. The right panel of the preceding figure illustrates the critical value for this example.

To calculate an exact *P*-value in R for a two-sided test when *t* = 2.57 and *df* = 5, use the command

```
2 * (1 - pt(q = abs(2.57), df = 5))
```

Replace 2.57 and 5 with your observed values. The "abs" refers to absolute value and is included in the case of a negative *t*-value.

In Excel, for the same calculation, use

```
= 2 * (1 - T.DIST(ABS(2.57), 5, TRUE))
```

	$\alpha(2)$:	0.2	0.10	0.05	0.02	0.01	0.001	0.0001
df	$\alpha(1)$:	0.1	0.05	0.025	0.01	0.005	0.0005	0.00005
1		3.08	6.31	12.71	31.82	63.66	636.62	6366.20
2		1.89	2.92	4.30	6.96	9.92	31.60	99.99
3		1.64	2.35	3.18	4.54	5.84	12.92	28.00
4		1.53	2.13	2.78	3.75	4.60	8.61	15.54
5		1.48	2.02	2.57	3.36	4.03	6.87	11.18
6		1.44	1.94	2.45	3.14	3.71	5.96	9.08
7		1.41	1.89	2.36	3.00	3.50	5.41	7.88
8		1.40	1.86	2.31	.2.90	3.36	5.04	7.12
9		1.38	1.83	2.26	2.82	3.25	4.78	6.59
10		1.37	1.81	2.23	2.76	3.17	4.59	6.21
11		1.36	1.80	2.20	2.72	3.11	4.44	5.92
12		1.36	1.78	2.18	2.68	3.05	4.32	5.69
13		1.35	1.77	2.16	2.65	3.01	4.22	5.51
14		1.35	1.76	2.14	2.62	2.98	4.14	5.36
15		1.34	1.75	2.13	2.60	2.95	4.07	5.24
16		1.34	1.75	2.12	2.58	2.92	4.01	5.13
17		1.33	1.74	2.11	2.57	2.90	3.97	5.04
18		1.33	1.73	2.10	2.55	2.88	3.92	4.97
19		1.33	1.73	2.09	2.54	2.86	3.88	4.90
20		1.33	1.72	2.09	2.53	2.85	3.85	4.84
21		1.32	1.72	2.08	2.52	2.83	3.82	4.78
22		1.32	1.72	2.07	2.51	2.82	3.79	4.74
23		1.32	1.71	2.07	2.50	2.81	3.77	4.69
24		1.32	1.71	2.06	2.49	2.80	3.75	4.65
25		1.32	1.71	2.06	2.49	2.79	3.73	4.62
26		1.31	1.71	2.06	2.48	2.78	3.71	4.59
27		1.31	1.70	2.05	2.47	2.77	3.69	4.56
28		1.31	1.70	2.05	2.47	2.76	3.67	4.53
29		1.31	1.70	2.05	2.46	2.76	3.66	4.51
30		1.31	1.70	2.04	2.46	2.75	3.65	4.48
31		1.31	1.70	2.04	2.45	2.74	3.63	4.46
32		1.31	1.69	2.04	2.45	2.74	3.62	4.44
33		1.31	1.69	2.03	2.44	2.73	3.61	4.42
34		1.31	1.69	2.03	2.44	2.73	3.60	4.41
35		1.31	1.69	2.03	2.44	2.72	3.59	4.39
36		1.31	1.69	2.03	2.43	2.72	3.58	4.37
37		1.30	1.69	2.03	2.43	2.72	3.57	4.36
38		1.30	1.69	2.02	2.43	2.71	3.57	4.35
39		1.30	1.68	2.02	2.43	2.71	3.56	4.33
40		1.30	1.68	2.02	2.42	2.70	3.55	4.32
41		1.30	1.68	2.02	2.42	2.70	3.54	4.31
42		1.30	1.68	2.02	2.42	2.70	3.54	4.30
43		1.30	1.68	2.02	2.42	2.70	3.53	4.29
44		1.30	1.68	2.02	2.41	2.69	3.53	4.28
45		1.30	1.68	2.01	2.41	2.69	3.52	4.27
46		1.30	1.68	2.01	2.41	2.69	3.51	4.26
47		1.30	1.68	2.01	2.41	2.68	3.51	4.25
48		1.30	1.68	2.01	2.41	2.68	3.51	4.24
49		1.30	1.68	2.01	2.40	2.68	3.50	4.24
50		1.30	1.68	2.01	2.40	2.68	3.50	4.23

$\alpha(2)$:	0.2	0.10	0.05	0.02	0.01	0.001	0.0001
df $\alpha(1)$:	0.1	0.05	0.025	0.01	0.005	0.0005	0.00005
51	1.30	1.68	2.01	2.40	2.68	3.49	4.22
52	1.30	1.67	2.01	2.40	2.67	3.49	4.21
53	1.30	1.67	2.01	2.40	2.67	3.48	4.21
54	1.30	1.67	2.00	2.40	2.67	3.48	4.20
55	1.30	1.67	2.00	2.40	2.67	3.48	4.20
56	1.30	1.67	2.00	2.39	2.67	3.47	4.19
57	1.30	1.67	2.00	2.39	2.66	3.47	4.18
58	1.30	1.67	2.00	2.39	2.66	3.47	4.18
59	1.30	1.67	2.00	2.39	2.66	3.46	4.17
60	1.30	1.67	2.00	2.39	2.66	3.46	4.17
61	1.30	1.67	2.00	2.39	2.66	3.46	4.16
62	1.30	1.67	2.00	2.39	2.66	3.45	4.16
63	1.30	1.67	2.00	2.39	2.66	3.45	4.15
64	1.29	1.67	2.00	2.39	2.65	3.45	4.15
65	1.29	1.67	2.00	2.39	2.65	3.45	4.15
66	1.29	1.67	2.00	2.38	2.65	3.44	4.14
67	1.29	1.67	2.00	2.38	2.65	3.44	4.14
68	1.29	1.67	2.00	2.38	2.65	3.44	4.13
69	1.29	1.67	1.99	2.38	2.65	3.44	4.13
70	1.29	1.67	1.99	2.38	2.65	3.44	4.13
71	1.29	1.67	1.99	2.38	2.65	3.43	4.12
72	1.29	1.67	1.99	2.38	2.65	3.43	4.12
73	1.29	1.67	1.99	2.38	2.64	3.43	4.12
74	1.29	1.67	1.99	2.38	2.64	3.43	4.11
75	1.29	1.67	1.99	2.38	2.64	3.43	4.11
76	1.29	1.67	1.99	2.38	2.64	3.42	4.11
77	1.29	1.66	1.99	2.38	2.64	3.42	4.10
78	1.29	1.66	1.99	2.38	2.64	3.42	4.10
79	1.29	1.66	1.99	2.37	2.64	3.42	4.10
80	1.29	1.66	1.99	2.37	2.64	3.42	4.10
81	1.29	1.66	1.99	2.37	2.64	3.41	4.09
82	1.29	1.66	1.99	2.37	2.64	3.41	4.09
83	1.29	1.66	1.99	2.37	2.64	3.41	4.09
84	1.29	1.66	1.99	2.37	2.64	3.41	4.09
85	1.29	1.66	1.99	2.37	2.63	3.41	4.08
86	1.29	1.66	1.99	2.37	2.63	3.41	4.08
87	1.29	1.66	1.99	2.37	2.63	3.41	4.08
88	1.29	1.66	1.99	2.37	2.63	3.40	4.08
89	1.29	1.66	1.99	2.37	2.63	3.40	4.07
90	1.29	1.66	1.99	2.37	2.63	3.40	4.07
100	1.29	1.66	1.98	2.36	2.63	3.39	4.05
120	1.29	1.66	1.98	2.36	2.62	3.37	4.03
140	1.29	1.66	1.98	2.35	2.61	3.36	4.01
160	1.29	1.65	1.97	2.35	2.61	3.35	3.99
180	1.29	1.65	1.97	2.35	2.60	3.35	3.98
200	1.29	1.65	1.97	2.35	2.60	3.34	3.97
400	1.28	1.65	1.97	2.34	2.59	3.32	3.93
1000	1.28	1.65	1.96	2.33	2.58	3.30	3.91

Statistical Table D: The *F*-distribution

These tables give critical values of the *F*-distribution for $\alpha(1) = 0.05$, $\alpha(1) = 0.025$, and $\alpha(1) = 0.01$. The critical value $F_{\alpha(1), df_1, df_2}$ defines the area α under the right tail of the *F*-distribution having df_1 and df_2 degrees of freedom. To find a critical value in the table, select the numerator degrees of freedom (df_1) listed across the top row and the denominator degrees of freedom (df_2) given in the first column. For example, if $df_1 = 9$, $df_2 = 12$, and $\alpha = 0.05$, then the critical value is 2.80; that is, the probability of a value greater than or equal to 2.80 is 0.05.

The critical value for the *F*-distribution having nine and 12 degrees of freedom when $\alpha = 0.05$. The area in red indicates the tail probability corresponding to 0.05, or 5%. The boundary of the red section is 2.80.

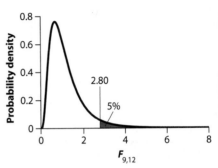

Exact probabilities under the right tail of the *F*-distribution can be calculated in the R statistical package. For example, to find the the *P*-value corresponding to $F = 2.80$ with degrees of freedom 9 (for the numerator) and 12 (for the denominator), use the command

```
1 - pf(q = 2.80, df1 = 9, df2 = 12)
```

To use this command, replace 2.80, 9, and 12 with your own values.

In Excel, enter the following in a cell for the same calculation:

```
= 1 - F.DIST(2.80, 9, 12, TRUE)
```

Critical value of F, $\alpha(1) = 0.05$, $\alpha(2) = 0.10$

Denominator df	Numerator df									
	1	**2**	**3**	**4**	**5**	**6**	**7**	**8**	**9**	**10**
1	161.45	199.50	215.71	224.58	230.16	233.99	236.77	238.88	240.54	241.88
2	18.51	19.00	19.16	19.25	19.30	19.33	19.35	19.37	19.38	19.40
3	10.13	9.55	9.28	9.12	9.01	8.94	8.89	8.85	8.81	8.79
4	7.71	6.94	6.59	6.39	6.26	6.16	6.09	6.04	6.00	5.96
5	6.61	5.79	5.41	5.19	5.05	4.95	4.88	4.82	4.77	4.74
6	5.99	5.14	4.76	4.53	4.39	4.28	4.21	4.15	4.10	4.06
7	5.59	4.74	4.35	4.12	3.97	3.87	3.79	3.73	3.68	3.64
8	5.32	4.46	4.07	3.84	3.69	3.58	3.50	3.44	3.39	3.35
9	5.12	4.26	3.86	3.63	3.48	3.37	3.29	3.23	3.18	3.14
10	4.96	4.10	3.71	3.48	3.33	3.22	3.14	3.07	3.02	2.98
11	4.84	3.98	3.59	3.36	3.20	3.09	3.01	2.95	2.90	2.85
12	4.75	3.89	3.49	3.26	3.11	3.00	2.91	2.85	2.80	2.75
13	4.67	3.81	3.41	3.18	3.03	2.92	2.83	2.77	2.71	2.67
14	4.60	3.74	3.34	3.11	2.96	2.85	2.76	2.70	2.65	2.60
15	4.54	3.68	3.29	3.06	2.90	2.79	2.71	2.64	2.59	2.54
16	4.49	3.63	3.24	3.01	2.85	2.74	2.66	2.59	2.54	2.49
17	4.45	3.59	3.20	2.96	2.81	2.70	2.61	2.55	2.49	2.45
18	4.41	3.55	3.16	2.93	2.77	2.66	2.58	2.51	2.46	2.41
19	4.38	3.52	3.13	2.90	2.74	2.63	2.54	2.48	2.42	2.38
20	4.35	3.49	3.10	2.87	2.71	2.60	2.51	2.45	2.39	2.35
21	4.32	3.47	3.07	2.84	2.68	2.57	2.49	2.42	2.37	2.32
22	4.30	3.44	3.05	2.82	2.66	2.55	2.46	2.40	2.34	2.30
23	4.28	3.42	3.03	2.80	2.64	2.53	2.44	2.37	2.32	2.27
24	4.26	3.40	3.01	2.78	2.62	2.51	2.42	2.36	2.30	2.25
25	4.24	3.39	2.99	2.76	2.60	2.49	2.40	2.34	2.28	2.24
26	4.23	3.37	2.98	2.74	2.59	2.47	2.39	2.32	2.27	2.22
27	4.21	3.35	2.96	2.73	2.57	2.46	2.37	2.31	2.25	2.20
28	4.20	3.34	2.95	2.71	2.56	2.45	2.36	2.29	2.24	2.19
29	4.18	3.33	2.93	2.70	2.55	2.43	2.35	2.28	2.22	2.18
30	4.17	3.32	2.92	2.69	2.53	2.42	2.33	2.27	2.21	2.16
40	4.08	3.23	2.84	2.61	2.45	2.34	2.25	2.18	2.12	2.08
50	4.03	3.18	2.79	2.56	2.40	2.29	2.20	2.13	2.07	2.03
60	4.00	3.15	2.76	2.53	2.37	2.25	2.17	2.10	2.04	1.99
70	3.98	3.13	2.74	2.50	2.35	2.23	2.14	2.07	2.02	1.97
80	3.96	3.11	2.72	2.49	2.33	2.21	2.13	2.06	2.00	1.95
90	3.95	3.10	2.71	2.47	2.32	2.20	2.11	2.04	1.99	1.94
100	3.94	3.09	2.70	2.46	2.31	2.19	2.10	2.03	1.97	1.93
200	3.89	3.04	2.65	2.42	2.26	2.14	2.06	1.98	1.93	1.88
400	3.86	3.02	2.63	2.39	2.24	2.12	2.03	1.96	1.90	1.85

Critical value of *F*, $\alpha(1) = 0.05$, $\alpha(2) = 0.10$ (continued)

Denominator df	Numerator *df*									
	12	15	20	30	40	60	100	200	400	1000
1	243.91	245.95	248.01	250.10	251.14	252.20	253.04	253.68	254.00	254.19
2	19.41	19.43	19.45	19.46	19.47	19.48	19.49	19.49	19.49	19.49
3	8.74	8.70	8.66	8.62	8.59	8.57	8.55	8.54	8.53	8.53
4	5.91	5.86	5.80	5.75	5.72	5.69	5.66	5.65	5.64	5.63
5	4.68	4.62	4.56	4.50	4.46	4.43	4.41	4.39	4.38	4.37
6	4.00	3.94	3.87	3.81	3.77	3.74	3.71	3.69	3.68	3.67
7	3.57	3.51	3.44	3.38	3.34	3.30	3.27	3.25	3.24	3.23
8	3.28	3.22	3.15	3.08	3.04	3.01	2.97	2.95	2.94	2.93
9	3.07	3.01	2.94	2.86	2.83	2.79	2.76	2.73	2.72	2.71
10	2.91	2.85	2.77	2.70	2.66	2.62	2.59	2.56	2.55	2.54
11	2.79	2.72	2.65	2.57	2.53	2.49	2.46	2.43	2.42	2.41
12	2.69	2.62	2.54	2.47	2.43	2.38	2.35	2.32	2.31	2.30
13	2.60	2.53	2.46	2.38	2.34	2.30	2.26	2.23	2.22	2.21
14	2.53	2.46	2.39	2.31	2.27	2.22	2.19	2.16	2.15	2.14
15	2.48	2.40	2.33	2.25	2.20	2.16	2.12	2.10	2.08	2.07
16	2.42	2.35	2.28	2.19	2.15	2.11	2.07	2.04	2.02	2.02
17	2.38	2.31	2.23	2.15	2.10	2.06	2.02	1.99	1.98	1.97
18	2.34	2.27	2.19	2.11	2.06	2.02	1.98	1.95	1.93	1.92
19	2.31	2.23	2.16	2.07	2.03	1.98	1.94	1.91	1.89	1.88
20	2.28	2.20	2.12	2.04	1.99	1.95	1.91	1.88	1.86	1.85
21	2.25	2.18	2.10	2.01	1.96	1.92	1.88	1.84	1.83	1.82
22	2.23	2.15	2.07	1.98	1.94	1.89	1.85	1.82	1.80	1.79
23	2.20	2.13	2.05	1.96	1.91	1.86	1.82	1.79	1.77	1.76
24	2.18	2.11	2.03	1.94	1.89	1.84	1.80	1.77	1.75	1.74
25	2.16	2.09	2.01	1.92	1.87	1.82	1.78	1.75	1.73	1.72
26	2.15	2.07	1.99	1.90	1.85	1.80	1.76	1.73	1.71	1.70
27	2.13	2.06	1.97	1.88	1.84	1.79	1.74	1.71	1.69	1.68
28	2.12	2.04	1.96	1.87	1.82	1.77	1.73	1.69	1.67	1.66
29	2.10	2.03	1.94	1.85	1.81	1.75	1.71	1.67	1.66	1.65
30	2.09	2.01	1.93	1.84	1.79	1.74	1.70	1.66	1.64	1.63
40	2.00	1.92	1.84	1.74	1.69	1.64	1.59	1.55	1.53	1.52
50	1.95	1.87	1.78	1.69	1.63	1.58	1.52	1.48	1.46	1.45
60	1.92	1.84	1.75	1.65	1.59	1.53	1.48	1.44	1.41	1.40
70	1.89	1.81	1.72	1.62	1.57	1.50	1.45	1.40	1.38	1.36
80	1.88	1.79	1.70	1.60	1.54	1.48	1.43	1.38	1.35	1.34
90	1.86	1.78	1.69	1.59	1.53	1.46	1.41	1.36	1.33	1.31
100	1.85	1.77	1.68	1.57	1.52	1.45	1.39	1.34	1.31	1.30
200	1.80	1.72	1.62	1.52	1.46	1.39	1.32	1.26	1.23	1.21
400	1.78	1.69	1.60	1.49	1.42	1.35	1.28	1.22	1.18	1.15

Critical value of F, $\alpha(1) = 0.025$, $\alpha(2) = 0.05$

Denominator df	Numerator df									
	1	2	3	4	5	6	7	8	9	10
1	647.79	799.5	864.16	899.58	921.85	937.11	948.22	956.66	963.28	968.63
2	38.51	39.00	39.17	39.25	39.30	39.33	39.36	39.37	39.39	39.40
3	17.44	16.04	15.44	15.10	14.88	14.73	14.62	14.54	14.47	14.42
4	12.22	10.65	9.98	9.60	9.36	9.20	9.07	8.98	8.90	8.84
5	10.01	8.43	7.76	7.39	7.15	6.98	6.85	6.76	6.68	6.62
6	8.81	7.26	6.60	6.23	5.99	5.82	5.70	5.60	5.52	5.46
7	8.07	6.54	5.89	5.52	5.29	5.12	4.99	4.90	4.82	4.76
8	7.57	6.06	5.42	5.05	4.82	4.65	4.53	4.43	4.36	4.30
9	7.21	5.71	5.08	4.72	4.48	4.32	4.20	4.10	4.03	3.96
10	6.94	5.46	4.83	4.47	4.24	4.07	3.95	3.85	3.78	3.72
11	6.72	5.26	4.63	4.28	4.04	3.88	3.76	3.66	3.59	3.53
12	6.55	5.10	4.47	4.12	3.89	3.73	3.61	3.51	3.44	3.37
13	6.41	4.97	4.35	4.00	3.77	3.60	3.48	3.39	3.31	3.25
14	6.30	4.86	4.24	3.89	3.66	3.50	3.38	3.29	3.21	3.15
15	6.20	4.77	4.15	3.80	3.58	3.41	3.29	3.20	3.12	3.06
16	6.12	4.69	4.08	3.73	3.50	3.34	3.22	3.12	3.05	2.99
17	6.04	4.62	4.01	3.66	3.44	3.28	3.16	3.06	2.98	2.92
18	5.98	4.56	3.95	3.61	3.38	3.22	3.10	3.01	2.93	2.87
19	5.92	4.51	3.90	3.56	3.33	3.17	3.05	2.96	2.88	2.82
20	5.87	4.46	3.86	3.51	3.29	3.13	3.01	2.91	2.84	2.77
21	5.83	4.42	3.82	3.48	3.25	3.09	2.97	2.87	2.80	2.73
22	5.79	4.38	3.78	3.44	3.22	3.05	2.93	2.84	2.76	2.70
23	5.75	4.35	3.75	3.41	3.18	3.02	2.90	2.81	2.73	2.67
24	5.72	4.32	3.72	3.38	3.15	2.99	2.87	2.78	2.70	2.64
25	5.69	4.29	3.69	3.35	3.13	2.97	2.85	2.75	2.68	2.61
26	5.66	4.27	3.67	3.33	3.10	2.94	2.82	2.73	2.65	2.59
27	5.63	4.24	3.65	3.31	3.08	2.92	2.80	2.71	2.63	2.57
28	5.61	4.22	3.63	3.29	3.06	2.90	2.78	2.69	2.61	2.55
29	5.59	4.20	3.61	3.27	3.04	2.88	2.76	2.67	2.59	2.53
30	5.57	4.18	3.59	3.25	3.03	2.87	2.75	2.65	2.57	2.51
40	5.42	4.05	3.46	3.13	2.90	2.74	2.62	2.53	2.45	2.39
50	5.34	3.97	3.39	3.05	2.83	2.67	2.55	2.46	2.38	2.32
60	5.29	3.93	3.34	3.01	2.79	2.63	2.51	2.41	2.33	2.27
70	5.25	3.89	3.31	2.97	2.75	2.59	2.47	2.38	2.30	2.24
80	5.22	3.86	3.28	2.95	2.73	2.57	2.45	2.35	2.28	2.21
90	5.20	3.84	3.26	2.93	2.71	2.55	2.43	2.34	2.26	2.19
100	5.18	3.83	3.25	2.92	2.70	2.54	2.42	2.32	2.24	2.18
200	5.10	3.76	3.18	2.85	2.63	2.47	2.35	2.26	2.18	2.11
400	5.06	3.72	3.15	2.82	2.60	2.44	2.32	2.22	2.15	2.08

STATISTICAL TABLES

Critical value of F, $\alpha(1) = 0.025$, $\alpha(2) = 0.05$ (continued)

Denominator df	Numerator *df*									
	12	**15**	**20**	**30**	**40**	**60**	**100**	**200**	**400**	**1000**
1	976.71	984.87	993.1	1001.41	1005.60	1009.80	1013.17	1015.71	1016.98	1017.75
2	39.41	39.43	39.45	39.46	39.47	39.48	39.49	39.49	39.5	39.5
3	14.34	14.25	14.17	14.08	14.04	13.99	13.96	13.93	13.92	13.91
4	8.75	8.66	8.56	8.46	8.41	8.36	8.32	8.29	8.27	8.26
5	6.52	6.43	6.33	6.23	6.18	6.12	6.08	6.05	6.03	6.02
6	5.37	5.27	5.17	5.07	5.01	4.96	4.92	4.88	4.87	4.86
7	4.67	4.57	4.47	4.36	4.31	4.25	4.21	4.18	4.16	4.15
8	4.20	4.10	4.00	3.89	3.84	3.78	3.74	3.70	3.69	3.68
9	3.87	3.77	3.67	3.56	3.51	3.45	3.40	3.37	3.35	3.34
10	3.62	3.52	3.42	3.31	3.26	3.20	3.15	3.12	3.10	3.09
11	3.43	3.33	3.23	3.12	3.06	3.00	2.96	2.92	2.90	2.89
12	3.28	3.18	3.07	2.96	2.91	2.85	2.80	2.76	2.74	2.73
13	3.15	3.05	2.95	2.84	2.78	2.72	2.67	2.63	2.61	2.60
14	3.05	2.95	2.84	2.73	2.67	2.61	2.56	2.53	2.51	2.50
15	2.96	2.86	2.76	2.64	2.59	2.52	2.47	2.44	2.42	2.40
16	2.89	2.79	2.68	2.57	2.51	2.45	2.40	2.36	2.34	2.32
17	2.82	2.72	2.62	2.50	2.44	2.38	2.33	2.29	2.27	2.26
18	2.77	2.67	2.56	2.44	2.38	2.32	2.27	2.23	2.21	2.20
19	2.72	2.62	2.51	2.39	2.33	2.27	2.22	2.18	2.15	2.14
20	2.68	2.57	2.46	2.35	2.29	2.22	2.17	2.13	2.11	2.09
21	2.64	2.53	2.42	2.31	2.25	2.18	2.13	2.09	2.06	2.05
22	2.60	2.50	2.39	2.27	2.21	2.14	2.09	2.05	2.03	2.01
23	2.57	2.47	2.36	2.24	2.18	2.11	2.06	2.01	1.99	1.98
24	2.54	2.44	2.33	2.21	2.15	2.08	2.02	1.98	1.96	1.94
25	2.51	2.41	2.30	2.18	2.12	2.05	2.00	1.95	1.93	1.91
26	2.49	2.39	2.28	2.16	2.09	2.03	1.97	1.92	1.90	1.89
27	2.47	2.36	2.25	2.13	2.07	2.00	1.94	1.90	1.88	1.86
28	2.45	2.34	2.23	2.11	2.05	1.98	1.92	1.88	1.85	1.84
29	2.43	2.32	2.21	2.09	2.03	1.96	1.90	1.86	1.83	1.82
30	2.41	2.31	2.20	2.07	2.01	1.94	1.88	1.84	1.81	1.80
40	2.29	2.18	2.07	1.94	1.88	1.80	1.74	1.69	1.66	1.65
50	2.22	2.11	1.99	1.87	1.80	1.72	1.66	1.60	1.57	1.56
60	2.17	2.06	1.94	1.82	1.74	1.67	1.60	1.54	1.51	1.49
70	2.14	2.03	1.91	1.78	1.71	1.63	1.56	1.50	1.47	1.45
80	2.11	2.00	1.88	1.75	1.68	1.60	1.53	1.47	1.43	1.41
90	2.09	1.98	1.86	1.73	1.66	1.58	1.50	1.44	1.41	1.39
100	2.08	1.97	1.85	1.71	1.64	1.56	1.48	1.42	1.39	1.36
200	2.01	1.90	1.78	1.64	1.56	1.47	1.39	1.32	1.28	1.25
400	1.98	1.87	1.74	1.60	1.52	1.43	1.35	1.27	1.22	1.18

Critical value of F, $\alpha(1) = 0.01$, $\alpha(2) = 0.02$

Denominator df	Numerator df									
	1	2	3	4	5	6	7	8	9	10
1	4052	4999	5403	5624	5763	5859	5928	5981	6022	6055
2	98.50	99.00	99.17	99.25	99.30	99.33	99.36	99.37	99.39	99.40
3	34.12	30.82	29.46	28.71	28.24	27.91	27.67	27.49	27.35	27.23
4	21.20	18.00	16.69	15.98	15.52	15.21	14.98	14.80	14.66	14.55
5	16.26	13.27	12.06	11.39	10.97	10.67	10.46	10.29	10.16	10.05
6	13.75	10.92	9.78	9.15	8.75	8.47	8.26	8.10	7.98	7.87
7	12.25	9.55	8.45	7.85	7.46	7.19	6.99	6.84	6.72	6.62
8	11.26	8.65	7.59	7.01	6.63	6.37	6.18	6.03	5.91	5.81
9	10.56	8.02	6.99	6.42	6.06	5.80	5.61	5.47	5.35	5.26
10	10.04	7.56	6.55	5.99	5.64	5.39	5.20	5.06	4.94	4.85
11	9.65	7.21	6.22	5.67	5.32	5.07	4.89	4.74	4.63	4.54
12	9.33	6.93	5.95	5.41	5.06	4.82	4.64	4.50	4.39	4.30
13	9.07	6.70	5.74	5.21	4.86	4.62	4.44	4.30	4.19	4.10
14	8.86	6.51	5.56	5.04	4.69	4.46	4.28	4.14	4.03	3.94
15	8.68	6.36	5.42	4.89	4.56	4.32	4.14	4.00	3.89	3.80
16	8.53	6.23	5.29	4.77	4.44	4.20	4.03	3.89	3.78	3.69
17	8.40	6.11	5.18	4.67	4.34	4.10	3.93	3.79	3.68	3.59
18	8.29	6.01	5.09	4.58	4.25	4.01	3.84	3.71	3.60	3.51
19	8.18	5.93	5.01	4.50	4.17	3.94	3.77	3.63	3.52	3.43
20	8.10	5.85	4.94	4.43	4.10	3.87	3.70	3.56	3.46	3.37
21	8.02	5.78	4.87	4.37	4.04	3.81	3.64	3.51	3.40	3.31
22	7.95	5.72	4.82	4.31	3.99	3.76	3.59	3.45	3.35	3.26
23	7.88	5.66	4.76	4.26	3.94	3.71	3.54	3.41	3.30	3.21
24	7.82	5.61	4.72	4.22	3.90	3.67	3.50	3.36	3.26	3.17
25	7.77	5.57	4.68	4.18	3.85	3.63	3.46	3.32	3.22	3.13
26	7.72	5.53	4.64	4.14	3.82	3.59	3.42	3.29	3.18	3.09
27	7.68	5.49	4.60	4.11	3.78	3.56	3.39	3.26	3.15	3.06
28	7.64	5.45	4.57	4.07	3.75	3.53	3.36	3.23	3.12	3.03
29	7.60	5.42	4.54	4.04	3.73	3.50	3.33	3.20	3.09	3.00
30	7.56	5.39	4.51	4.02	3.70	3.47	3.30	3.17	3.07	2.98
40	7.31	5.18	4.31	3.83	3.51	3.29	3.12	2.99	2.89	2.80
50	7.17	5.06	4.20	3.72	3.41	3.19	3.02	2.89	2.78	2.70
60	7.08	4.98	4.13	3.65	3.34	3.12	2.95	2.82	2.72	2.63
70	7.01	4.92	4.07	3.60	3.29	3.07	2.91	2.78	2.67	2.59
80	6.96	4.88	4.04	3.56	3.26	3.04	2.87	2.74	2.64	2.55
90	6.93	4.85	4.01	3.53	3.23	3.01	2.84	2.72	2.61	2.52
100	6.90	4.82	3.98	3.51	3.21	2.99	2.82	2.69	2.59	2.50
200	6.76	4.71	3.88	3.41	3.11	2.89	2.73	2.60	2.50	2.41
400	6.70	4.66	3.83	3.37	3.06	2.85	2.68	2.56	2.45	2.37

Critical value of *F*, $\alpha(1) = 0.01$, $\alpha(2) = 0.02$ (continued)

Denominator *df*	Numerator *df*									
	12	15	20	30	40	60	100	200	400	1000
1	6106	6157	6208	6260	6286	6313	6334	6350	6357	6362
2	99.42	99.43	99.45	99.47	99.47	99.48	99.49	99.49	99.50	99.50
3	27.05	26.87	26.69	26.50	26.41	26.32	26.24	26.18	26.15	26.14
4	14.37	14.20	14.02	13.84	13.75	13.65	13.58	13.52	13.49	13.47
5	9.89	9.72	9.55	9.38	9.29	9.20	9.13	9.08	9.05	9.03
6	7.72	7.56	7.40	7.23	7.14	7.06	6.99	6.93	6.91	6.89
7	6.47	6.31	6.16	5.99	5.91	5.82	5.75	5.70	5.68	5.66
8	5.67	5.52	5.36	5.20	5.12	5.03	4.96	4.91	4.89	4.87
9	5.11	4.96	4.81	4.65	4.57	4.48	4.41	4.36	4.34	4.32
10	4.71	4.56	4.41	4.25	4.17	4.08	4.01	3.96	3.94	3.92
11	4.40	4.25	4.10	3.94	3.86	3.78	3.71	3.66	3.63	3.61
12	4.16	4.01	3.86	3.70	3.62	3.54	3.47	3.41	3.39	3.37
13	3.96	3.82	3.66	3.51	3.43	3.34	3.27	3.22	3.19	3.18
14	3.80	3.66	3.51	3.35	3.27	3.18	3.11	3.06	3.03	3.02
15	3.67	3.52	3.37	3.21	3.13	3.05	2.98	2.92	2.90	2.88
16	3.55	3.41	3.26	3.10	3.02	2.93	2.86	2.81	2.78	2.76
17	3.46	3.31	3.16	3.00	2.92	2.83	2.76	2.71	2.68	2.66
18	3.37	3.23	3.08	2.92	2.84	2.75	2.68	2.62	2.59	2.58
19	3.30	3.15	3.00	2.84	2.76	2.67	2.60	2.55	2.52	2.50
20	3.23	3.09	2.94	2.78	2.69	2.61	2.54	2.48	2.45	2.43
21	3.17	3.03	2.88	2.72	2.64	2.55	2.48	2.42	2.39	2.37
22	3.12	2.98	2.83	2.67	2.58	2.50	2.42	2.36	2.34	2.32
23	3.07	2.93	2.78	2.62	2.54	2.45	2.37	2.32	2.29	2.27
24	3.03	2.89	2.74	2.58	2.49	2.40	2.33	2.27	2.24	2.22
25	2.99	2.85	2.70	2.54	2.45	2.36	2.29	2.23	2.20	2.18
26	2.96	2.81	2.66	2.50	2.42	2.33	2.25	2.19	2.16	2.14
27	2.93	2.78	2.63	2.47	2.38	2.29	2.22	2.16	2.13	2.11
28	2.90	2.75	2.60	2.44	2.35	2.26	2.19	2.13	2.10	2.08
29	2.87	2.73	2.57	2.41	2.33	2.23	2.16	2.10	2.07	2.05
30	2.84	2.70	2.55	2.39	2.30	2.21	2.13	2.07	2.04	2.02
40	2.66	2.52	2.37	2.20	2.11	2.02	1.94	1.87	1.84	1.82
50	2.56	2.42	2.27	2.10	2.01	1.91	1.82	1.76	1.72	1.70
60	2.50	2.35	2.20	2.03	1.94	1.84	1.75	1.68	1.64	1.62
70	2.45	2.31	2.15	1.98	1.89	1.78	1.70	1.62	1.58	1.56
80	2.42	2.27	2.12	1.94	1.85	1.75	1.65	1.58	1.54	1.51
90	2.39	2.24	2.09	1.92	1.82	1.72	1.62	1.55	1.50	1.48
100	2.37	2.22	2.07	1.89	1.80	1.69	1.60	1.52	1.47	1.45
200	2.27	2.13	1.97	1.79	1.69	1.58	1.48	1.39	1.34	1.30
400	2.23	2.08	1.92	1.75	1.64	1.53	1.42	1.32	1.26	1.2

Statistical Table E: Mann–Whitney *U*-distribution

These tables give the two-tailed critical values of the U-distribution for $\alpha = 0.05$ and $\alpha = 0.01$. U is the larger of U_1 and U_2. The critical value $U_{\alpha(2), n_1, n_2}$ defines the area α under the right tail of the U-distribution corresponding to sample sizes n_1 and n_2. To find a critical value in the table, select n_1 from the top row and n_2 from the far left column. A "—" means that it is not possible to reject a null hypothesis with that α and those sample sizes. For larger sample sizes, use the Z-approximation from Chapter 13. For example, if $n_1 = 5$, $n_2 = 7$, and $\alpha = 0.05$, then the critical value is 30; that is, the probability of a value greater than or equal to 30 is 0.05 or less (it may be less than 0.05 because the U-distribution is discrete, and no critical value may correspond exactly to 0.05). The following figure illustrates the critical value for this example.

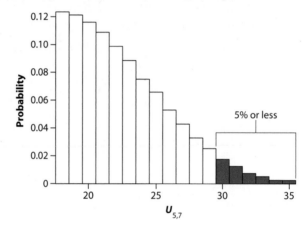

The two-tailed critical value of the Mann–Whitney *U*-distribution when $n_1 = 5$, $n_2 = 7$, and $\alpha = 0.05$. The area in red indicates the tail probability corresponding to 0.05, or 5%. The boundary of the red section is 30. Thus, the probability of a value greater than or equal to 30 is 0.05 or less. (The distribution gives the two-tailed value, because it measures the probability that either U_1 or U_2 is greater than $U_{0.05(2), n_1, n_2}$.)

$$\alpha(2) = 0.05$$

	n_1												
n_2	3	4	5	6	7	8	9	10	11	12	13	14	15
3	—	—	15	17	20	22	25	27	30	32	35	37	40
4	—	16	19	22	25	28	32	35	38	41	44	47	50
5	15	19	23	27	30	34	38	42	46	49	53	57	61
6	17	22	27	31	36	40	44	49	53	58	62	67	71
7	20	25	30	36	41	46	51	56	61	66	71	76	81
8	22	28	34	40	46	51	57	63	69	74	80	86	91
9	25	32	38	44	51	57	64	70	76	82	89	95	101
10	27	35	42	49	56	63	70	77	84	91	97	104	111
11	30	38	46	53	61	69	76	84	91	99	106	114	121
12	32	41	49	58	66	74	82	91	99	107	115	123	131
13	35	44	53	62	71	80	89	97	106	115	124	132	141
14	37	47	57	67	76	86	95	104	114	123	132	141	151
15	40	50	61	71	81	91	101	111	121	131	141	151	161

$$\alpha(2) = 0.01$$

	n_1												
n_2	3	4	5	6	7	8	9	10	11	12	13	14	15
3	—	—	—	—	—	—	27	30	33	35	38	41	43
4	—	—	—	24	28	31	35	38	42	45	49	52	55
5	—	—	25	29	34	38	42	46	50	54	58	63	67
6	—	24	29	34	39	44	49	54	59	63	68	73	78
7	—	28	34	39	45	50	56	61	67	72	78	83	89
8	—	31	38	44	50	57	63	69	75	81	87	94	100
9	27	35	42	49	56	63	70	77	83	90	97	104	111
10	30	38	46	54	61	69	77	84	92	99	106	114	121
11	33	42	50	59	67	75	83	92	100	108	116	124	132
12	35	45	54	63	72	81	90	99	108	117	125	134	143
13	38	49	58	68	78	87	97	106	116	125	135	144	153
14	41	52	63	73	83	94	104	114	124	134	144	154	164
15	43	53	67	78	89	100	111	121	132	143	153	164	174

Statistical Table F:
Tukey–Kramer q-distribution

This table gives the critical values of the Tukey–Kramer q-distribution[1] for $\alpha = 0.05$. The critical value $q_{0.05,\,k,\,N-k}$ defines the area 0.05 under the right tail of the q-distribution having k groups and $N-k$ degrees of freedom, where N is the total sample size of all groups combined. The probability of a q-value greater than or equal to $q_{0.05,\,k,\,N-k}$ is 0.05, or 5%.

---

1. Some other statistics books use a slightly different formula to calculate q (multiply our q by $\sqrt{2}$ to get the alternative version of the test statistic), which also leads to different critical values in statistical tables for the Tukey-Kramer distribution (multiply our critical values here by $\sqrt{2}$ to convert). Our formula is same as that used in the computer statistics package R.

	Number of groups (*k*)													
df_{error}	2	3	4	5	6	7	8	9	10	11	12	13	14	15
10	2.23	2.74	3.06	3.29	3.47	3.62	3.75	3.86	3.96	4.05	4.12	4.20	4.26	4.32
11	2.20	2.70	3.01	3.23	3.41	3.56	3.68	3.79	3.88	3.96	4.04	4.11	4.17	4.23
12	2.18	2.67	2.97	3.19	3.36	3.50	3.62	3.72	3.81	3.90	3.97	4.04	4.10	4.16
13	2.16	2.64	2.94	3.15	3.32	3.45	3.57	3.67	3.76	3.84	3.91	3.98	4.04	4.09
14	2.14	2.62	2.91	3.12	3.28	3.41	3.53	3.63	3.72	3.79	3.86	3.93	3.99	4.04
15	2.13	2.60	2.88	3.09	3.25	3.38	3.49	3.59	3.68	3.75	3.82	3.88	3.94	3.99
16	2.12	2.58	2.86	3.06	3.22	3.35	3.46	3.56	3.64	3.72	3.78	3.85	3.90	3.95
17	2.11	2.57	2.84	3.04	3.20	3.33	3.44	3.53	3.61	3.69	3.75	3.81	3.87	3.92
18	2.10	2.55	2.83	3.02	3.18	3.30	3.41	3.50	3.59	3.66	3.72	3.78	3.84	3.89
19	2.09	2.54	2.81	3.01	3.16	3.16	3.39	3.48	3.56	3.63	3.70	3.76	3.81	3.86
20	2.09	2.53	2.80	2.99	3.14	3.27	3.37	3.46	3.54	3.61	3.68	3.73	3.79	3.84
21	2.08	2.52	2.79	2.98	3.13	3.25	3.35	3.44	3.52	3.59	3.66	3.71	3.77	3.82
22	2.07	2.51	2.78	2.97	3.12	3.24	3.34	3.43	3.51	3.57	3.64	3.69	3.75	3.80
23	2.07	2.50	2.77	2.96	3.10	3.22	3.32	3.41	3.49	3.56	3.62	3.68	3.73	3.78
24	2.06	2.50	2.76	2.95	3.09	3.21	3.31	3.40	3.48	3.54	3.61	3.66	3.71	3.76
25	2.06	2.49	2.75	2.94	3.08	3.20	3.30	3.39	3.46	3.53	3.59	3.65	3.70	3.75
26	2.06	2.48	2.74	2.93	3.07	3.19	3.29	3.38	3.45	3.52	3.58	3.63	3.68	3.73
27	2.05	2.48	2.74	2.92	3.06	3.18	3.28	3.36	3.44	3.51	3.57	3.62	3.67	3.72
28	2.05	2.47	2.73	2.91	3.06	3.17	3.27	3.35	3.43	3.50	3.56	3.61	3.66	3.71
29	2.05	2.47	2.72	2.91	3.05	3.16	3.26	3.35	3.42	3.49	3.55	3.60	3.65	3.70
30	2.04	2.46	2.72	2.90	3.04	3.16	3.25	3.34	3.41	3.48	3.54	3.59	3.64	3.68
31	2.04	2.46	2.71	2.89	3.04	3.15	3.25	3.33	3.40	3.47	3.53	3.58	3.63	3.68
32	2.04	2.46	2.71	2.89	3.03	3.14	3.24	3.32	3.40	3.46	3.52	3.57	3.62	3.67
33	2.03	2.45	2.70	2.88	3.02	3.14	3.23	3.32	3.39	3.45	3.51	3.56	3.61	3.66
34	2.03	2.45	2.70	2.88	3.02	3.13	3.23	3.31	3.38	3.45	3.50	3.56	3.60	3.65
35	2.03	2.45	2.70	2.88	3.01	3.13	3.22	3.30	3.37	3.44	3.50	3.55	3.60	3.64
36	2.03	2.44	2.69	2.87	3.01	3.12	3.22	3.30	3.37	3.43	3.49	3.54	3.59	3.64
37	2.03	2.44	2.69	2.87	3.00	3.12	3.21	3.29	3.36	3.43	3.48	3.54	3.58	3.63
38	2.02	2.44	2.69	2.86	3.00	3.11	3.21	3.29	3.36	3.42	3.48	3.53	3.58	3.62
39	2.02	2.44	2.68	2.86	3.00	3.11	3.20	3.28	3.35	3.42	3.47	3.52	3.57	3.62
40	2.02	2.43	2.68	2.86	2.99	3.10	3.20	3.28	3.35	3.41	3.47	3.52	3.57	3.61
41	2.02	2.43	2.68	2.85	2.99	3.10	3.19	3.27	3.34	3.41	3.46	3.51	3.56	3.60
42	2.02	2.43	2.67	2.85	2.99	3.10	3.19	3.27	3.34	3.40	3.46	3.51	3.56	3.60
43	2.02	2.43	2.67	2.85	2.98	3.09	3.18	3.26	3.33	3.40	3.45	3.50	3.55	3.59
44	2.02	2.43	2.67	2.84	2.98	3.09	3.18	3.26	3.33	3.39	3.45	3.50	3.55	3.59
45	2.01	2.42	2.67	2.84	2.98	3.09	3.18	3.26	3.33	3.39	3.45	3.50	3.54	3.59
46	2.01	2.42	2.67	2.84	2.97	3.08	3.17	3.25	3.32	3.39	3.44	3.49	3.54	3.58
47	2.01	2.42	2.66	2.84	2.97	3.08	3.17	3.25	3.32	3.38	3.44	3.49	3.53	3.58
48	2.01	2.42	2.66	2.83	2.97	3.08	3.17	3.25	3.32	3.38	3.43	3.48	3.53	3.57
49	2.01	2.42	2.66	2.83	2.97	3.07	3.17	3.24	3.31	3.37	3.43	3.48	3.53	3.57
50	2.01	2.42	2.66	2.83	2.96	3.07	3.16	3.24	3.31	3.37	3.43	3.48	3.52	3.57

Statistical Table G: Critical values for the Spearman's rank correlation

This table gives two-tailed critical values of the Spearman rank correlation under the null hypothesis that the population correlation is zero. The critical value $r_{S(\alpha, n)}$ defines the combined area α under both tails of the null distribution, where n is the sample size. The probability of a value greater than or equal to $r_{S(\alpha, n)}$ or less than or equal to $-r_{S(\alpha, n)}$ is α. These critical values were obtained using the *SuppDist* package (Wheeler 2009) implemented in the statistical computer package R, according to the methods of Kendall and Smith (1939).

STATISTICAL TABLES

n	$\alpha = 0.05$	$\alpha = 0.01$	n	$\alpha = 0.05$	$\alpha = 0.01$	n	$\alpha = 0.05$	$\alpha = 0.01$
5	0.900		37	0.325	0.419	69	0.237	0.308
6	0.943	1.000	38	0.320	0.413	70	0.235	0.306
7	0.821	0.929	39	0.316	0.408	71	0.234	0.304
8	0.762	0.881	40	0.312	0.403	72	0.232	0.302
9	0.700	0.833	41	0.308	0.398	73	0.230	0.300
10	0.648	0.782	42	0.305	0.393	74	0.229	0.298
11	0.618	0.755	43	0.301	0.389	75	0.227	0.296
12	0.587	0.720	44	0.297	0.385	76	0.226	0.294
13	0.560	0.692	45	0.294	0.380	77	0.224	0.292
14	0.538	0.670	46	0.291	0.376	78	0.223	0.290
15	0.521	0.645	47	0.288	0.372	79	0.221	0.288
16	0.503	0.626	48	0.285	0.369	80	0.220	0.286
17	0.485	0.610	49	0.282	0.365	81	0.219	0.285
18	0.472	0.593	50	0.279	0.361	82	0.217	0.283
19	0.458	0.578	51	0.276	0.358	83	0.216	0.281
20	0.447	0.564	52	0.273	0.354	84	0.215	0.280
21	0.434	0.550	53	0.271	0.351	85	0.213	0.278
22	0.425	0.539	54	0.268	0.348	86	0.212	0.276
23	0.415	0.528	55	0.266	0.345	87	0.211	0.275
24	0.406	0.516	56	0.263	0.342	88	0.210	0.273
25	0.398	0.506	57	0.261	0.339	89	0.208	0.272
26	0.389	0.497	58	0.259	0.336	90	0.207	0.270
27	0.382	0.487	59	0.256	0.333	91	0.206	0.269
28	0.375	0.479	60	0.254	0.330	92	0.205	0.267
29	0.368	0.471	61	0.252	0.327	93	0.204	0.266
30	0.362	0.464	62	0.250	0.325	94	0.203	0.264
31	0.356	0.456	63	0.248	0.322	95	0.202	0.263
32	0.350	0.449	64	0.246	0.320	96	0.201	0.262
33	0.345	0.443	65	0.244	0.317	97	0.200	0.260
34	0.339	0.436	66	0.242	0.315	98	0.199	0.259
35	0.334	0.430	67	0.241	0.313	99	0.198	0.258
36	0.329	0.424	68	0.239	0.310	100	0.197	0.257

Literature cited

ABC News. 2006. Dwarfs better known than US justices: poll. http://www.abc.net.au/news/2006-08-15/dwarfs-better-known-than-us-justices-poll/1239126. Accessed February 10, 2013.

Abd-El-Al, A. M., A. M. Bayoumy, and E. A. Abou Salem. 1997. A study on *Demodex folliculorum* in rosacea. *Journal of the Egyptian Society of Parasitology* 27: 183–195.

Adolph, S. C., and J. S. Hardin. 2007. Estimating phenotypic correlations: correcting for bias due to intraindividual variability. *Functional Ecology* 21: 178–184.

Agresti, A. 2002. *Categorical Data Analysis.* Hoboken, NJ: Wiley.

Agresti, A., and B. A. Coull. 1998. Approximate is better than "exact" for interval estimation of binomial proportions. *American Statistician* 52: 119–126.

Altman, D. G., and J. M. Bland. 1998. Generalisation and extrapolation. *British Medical Journal* 317: 409–410.

Alvarez G., F. C. Ceballos, and C. Quinteiro. 2009. The role of inbreeding in the extinction of a european royal dynasty. *PLoS ONE* 4(4): e5174.

American Society of Microbiology. 2005. Women Better at Hand Hygiene Habits, Hands Down. Press release, September 25. http://www.asm.org/index.php/component/content/article/92-news-room/press-releases/1831-women-better-at-hand-hygiene-habits-hands-down. Accessed December 3, 2013.

Anderson, R. N. 2001. *Deaths: Leading Causes for 1999.* National vital statistics reports, vol. 49, no. 11. Hyattsville, MD: National Center for Health Statistics.

Andersson, M., and M. Åhlund 2012. Don't put all your eggs in one nest: spread them and cut time at risk. *American Naturalist* 180: 354–363.

Andrade, M. C. B. 1996. Sexual selection for male sacrifice in the Australian redback spider. *Science* 271: 70–72.

Anstey, M. L., S. M. Rogers, S. R. Ott, M. Burrows, and S. J. Simpson. 2009. Serotonin mediates behavioral gregarization underlying swarm formation in desert locusts. *Science* 323: 627–630.

Antiplatelet Trialists' Collaboration. 1994. Collaborative overview of randomized trials of antiplatelet therapy—I: Prevention of death, myocardial infarction, and stroke by prolonged antiplatelet therapy in various categories of patients. *British Medical Journal* 308: 81–106.

Aparicio, S., et al. 2002. Whole-genome shotgun assembly and analysis of the genome of *Fugu rubripes. Science* 297: 1301–1310.

Apiron, D., and D. Zohary. 1961. Chlorophyll lethal in natural populations of the orchard grass (*Dactylis glomerata* L.). A case of balanced polymorphism in plants. *Genetics* 46: 393–399.

Arnqvist, G., M. Edvardsson, U. Friberg, and T. Nilsson. 2000. Sexual conflict promotes speciation in insects. *Proceedings of the National Academy of Sciences (USA)* 97: 10460–10464.

Attwood, A. S., N. E. Scott-Samuel, G. Stothart, and M. R. Munafò. 2012. Glass shape influences consumption rate for alcoholic beverages. *PLoS ONE* 7: e43007.

Avon, N. 2009. Are men or women more likely to be hit by lightning? www.popsci.com/scitech/article/2009-09/are-men-or-women-more-likely-be-hit-lightning. Accessed February 24, 2013.

Balazs, A. B., et al. 2011. Antibody-based protection against HIV infection by vectored immunoprophylaxis. *Nature* 481: 81–84.

Baldwin, B. G., and M. J. Sanderson. 1998. Age and rate of diversification of the Hawaiian silversword alliance (Compositae). *Proceedings of the National Academy of Sciences (USA)* 95: 9402–9406.

Banks, T., and J. M. Dabbs Jr. 1996. Salivary testosterone and cortisol in a delinquent and violent urban subculture. *Journal of Social Psychology* 136: 49–56.

Barber, V. A., G. P. Juday, and B. P. Finney. 2000. Reduced growth of Alaskan white spruce in the twentieth century from temperature-induced drought stress. *Nature* 405: 668–673.

Barker-Plotkin, A., D. Foster, A. Lezberg, and W. Lyford. 2006. Lyford mapped tree plot. http://harvardforest.fas.harvard.edu/data/p03/

hf 032/HF032-data.html. Accessed May 26, 2006.

Barnes, A. I., S. Wigby, J. M. Boone, L. Partridge, and T. Chapman. 2008. Feeding, fecundity and lifespan in female *Drosophila melanogaster. Proceedings of the Royal Society of London, Series B: Biological Sciences* 275: 1675–1683.

Barnes, I., A. Duda, O. G. Pybus, and M. G. Thomas. 2011. Ancient urbanization predicts genetic resistance to tuberculosis. *Evolution* 65: 842–848.

Barss, P. 1984. Injuries due to falling coconuts. *Journal of Trauma* 24: 990–991.

Bass, M. S., et al. 2010. Global conservation significance of Ecuador's Yasuní National Park. *PLoS ONE* 5: e8767.

Beall, C. M., et al. 2002. An Ethiopian pattern of human adaptation to high-altitude hypoxia. *Proceedings of the National Academy of Sciences (USA)* 99: 17215–17218.

Beath, D. D. 1996. Pollination of *Amporphophallus johnsonii* (Araceae) by carrion beetles (*Phaeochrous amplus*) in a Ghanaian rain forest. *Journal of Ecological Ecology* 12: 409–418.

Beaugrand, G., K. M. Brander, J. A. Lindley, S. Souissi, and P. C. Reid. 2003. Plankton effect on cod recruitment in the North Sea. *Nature* 426: 661–664.

Beaver, J. D., et al. 2006. Individual differences in reward drive predict neural responses to images of food. *Journal of Neuroscience* 26: 5160–5166.

Beaver, L. M., and J. M. Giebultowicz. 2004. Regulation of copulation duration by period and timeless in *Drosophila melanogaster. Current Biology* 14: 1492–1497.

Bebarta, V., D. Luyten, and K. Heard. 2003. Emergency medicine animal research: does use of randomization and blinding affect the results? *Academic Emergency Medicine* 10: 684–687.

Bekelman, J. E., Y. Li, and C. P. Gross. 2003. Scope and impact of financial conflicts of interest in biomedical research: a systematic review. *Journal of the American Medical Association* 289: 454–465.

Ben-Shahar, Y., A. Robichon, M. B. Sokolowski, and G. E. Robinson. 2002. Influence of gene action across different time scales on behavior. *Science* 296: 741–744.

Bendavid, E., Y. Kaganova, J. Needleman, L. Gruenberg, and J. S. Weissman. 2007. Complica-

tion rates on weekends and weekdays in U.S. hospitals. *American Journal of Medicine* 120: 422–428.

Benjamini, Y., and Y. Hochberg. 1995. Controlling the false discovery rate: a practical and powerful approach to multiple testing. *Journal of the Royal Statistical Society, Series B* 57: 289–300.

Bereczkei, T., P. Gyuris, and G. E. Weisfeld. 2004. Sexual imprinting in human mate choice. *Proceedings of the Royal Society of London, Series B: Biological Sciences* 271: 1129–1134.

Berer, K., et al. 2011. Commensal microbiota and myelin autoantigen cooperate to trigger autoimmune demyelination. *Nature* 479: 538–541.

Berthold, P., and F. Pulido. 1994. Heritability of migratory activity in a natural bird population. *Proceedings of the Royal Society of London, Series B: Biological Sciences* 257: 311–315.

Bian, G., Y. Xu, P. Lu, Y. Xie, and Z. Xi. 2010. The endosymbiotic bacterium *Wolbachia* induces resistance to dengue virus in *Aedes aegypti. PLoS Pathogens* 6: e1000833.

Biederman, J., M. C. Monuteaux, T. Spencer, T. E. Wilens, and S. V. Faraone. 2009. Do stimulants protect against psychiatric disorders in youth with ADHD? A 10-year follow-up study. *Pediatrics* 124: 71–78.

Billet, G., et al. 2012. High morphological variation of vestibular system accompanies slow and infrequent locomotion in three-toed sloths. *Proceedings of the Royal Society of London, Series B: Biological Sciences* 279: 3932–3939.

Bisazza, A., C. Cantalupo, A. Robins, L. J. Rogers, and G. Vallortigara. 1996. Right-pawedness in toads. *Nature* 379: 408.

Blackburn, C., et al. 2003. Effect of strategies to reduce exposure of infants to environmental tobacco smoke in the home: cross sectional survey. *British Medical Journal* 327: 257–260.

Blake, E. S., E. N. Rappaport, J. D. Jarrell, and C. W. Landsea. 2005. The deadliest, costliest, and most intense United States tropical cyclones from 1851 to 2004 (and other frequently requested hurricane facts). *NOAA Technical Memorandum* NWS TPC-4.

Bland, J. M., and D. G. Altman. 1994. Matching. *British Medical Journal* 309: 1128.

Bland, M. 2000. *An Introduction to Medical Statistics*, 3rd ed. Oxford, UK: Oxford University Press.

Blas, J., G. R. Bortolotti, J. L. Tella, R. Baos, and T. A. Marchant. 2007. Stress response during development predicts fitness in a wild, long lived vertebrate. *Proceedings of the National Academy of Sciences (USA)* 104: 8880–8884.

Blaustein, A. R., J. M. Kiesecker, D. P. Chivers, and R. G. Anthony. 1997. Ambient UV-B radiation causes deformities in amphibian embryos. *Proceedings of the National Academy of Sciences (USA)* 94: 13735–13737.

Blount, J. B., N. B. Metcalfe, T. R. Birkhead, and P. F. Surai. 2003. Carotenoid modulation of immune function and sexual attractiveness in zebra finches. *Science* 300: 125–127.

Boag, P. T., and P. R. Grant. 1984. The classic case of character release: Darwin's finches (*Geozpiza*) on Isla Daphne Major, Galápagos. *Biological Journal of the Linnean Society* 22: 243–287.

Bohannon, J., R. Goldstein, and A. Herschkowitsch. 2010. Can people distinguish pâté from dog food? *Chance* 23: 43–46.

Book, A. S., K. B. Starzyk, and V. L. Quinsey. 2001. The relationship between testosterone and aggression: a meta-analysis. *Aggression and Violent Behavior* 6: 579–599.

Borenstein, M. 1997. Hypothesis testing and effect size estimation in clinical trials. *Annals of Allergy, Asthma and Immunology* 78: 5–16.

Borenstein, M., L. V. Hedges, J. P. Higgins, and H. R. Rothstein. 2009. *Introduction to Meta-analysis*. Chichester, UK: Wiley.

Bortkiewicz, L. 1898. *Das Gesetz der Kleinen Zahlen* (Teubner, Leipzig), as cited in Larsen, R. J., and M. L. Marx. 1981. *An Introduction to Mathematical Statistics and Its Applications.* Englewood Cliffs, NJ: Prentice-Hall.

Bouyer, J., M. Pruvot, Z. Bengaly, P. M. Guerin, and R. Lancelot. 2007. Learning influences host choice in tsetse. *Biology Letters* 3: 113–116.

Bowen, A., et al. 2012. Effectiveness of enhanced communication therapy in the first four months after stroke for aphasia and dysarthria: a randomised controlled trial. *British Medical Journal* 345: e4407.

Box, G. E. P., and S. L. Andersen. 1955. Permutation theory in the derivation of robust criteria and the study of departures from assumption. *Journal of the Royal Statistical Society, Series B* 17: 1–34.

Bradley, J. V. 1980. Nonrobustness in one-sample Z and t tests: A large-scale sampling study. *Bulletin of the Psychonomic Society* 15: 29–32.

Brédart, S., and R. M. French. 1999. Do babies resemble their fathers more than their mothers? A failure to replicate Christenfeld and Hill (1995). *Evolution and Human Behavior* 20: 129–135.

Brem, R. B., J. D. Storey, J. Whittle, and L. Kruglyak. 2005. Genetic interactions between polymorphisms that affect gene expression in yeast. *Nature* 436: 701–703.

Brent, D. A., et al. 1993. Firearms and adolescent suicide. A community case-control study. *American Journal of Diseases of Children* 147: 1066–1071.

Brieger, D., et al. 2004. Acute coronary syndromes without chest pain, an underdiagnosed and undertreated high-risk group—Insights from the Global Registry of Acute Coronary Events. *Chest* 126: 461–469.

Briffa, M., and J. Greenaway. 2011. High in situ repeatability of behaviour indicates animal personality in the beadlet anemone *Actinia equina* (Cnidaria). *PLoS ONE* 6: e21963.

British Columbia Ministry of Education. 2004. Your surplus, your priorities: building BC together. Government of British Columbia: http://www.gov.bc.ca/bcgov/content/docs/@2NO32_0YQtuW/middle.pdf\par.

Broadbent, N. J., L. R. Squire, and R. E. Clark. 2004. Spatial memory, recognition memory, and the hippocampus. *Proceedings of the National Academy of Sciences (USA)* 101: 14515–14520.

Brooks, R. 2000. Negative genetic correlation between male sexual attractiveness and survival. *Nature* 406: 67–70.

Brown, S. J., et al. 2002. Sequence of the *Tribolium castaneum* homeotic complex: the region corresponding to the *Drosophila melanogaster* Antennapedia complex. *Genetics* 160: 1067–1074.

Brownlee, K. A. 1955. Statistics of the 1954 polio vaccine trials. *Journal of the American Statistical Association* 50: 1005–1013.

Brutsaert, T. D., et al. 2002. Effect of menstrual cycle phase on exercise performance of high-altitude native women at 3600 m. *Journal of Experimental Biology* 205: 233–239.

Buri, P. 1956. Gene frequency in small populations of mutant *Drosophila*. *Evolution* 10: 367–402.

Cameron, E., and L. Pauling. 1976. Supplemental ascorbate in the supportive treatment of cancer: prolongation of survival times in terminal human cancer. *Proceedings of the National Academy of Sciences (USA)* 73: 3685–3689.

Campbell, S. S., and P. J. Murphy. 1998. Extraocular circadian phototransduction in humans. *Science* 279: 396–399.

Canadian Paediatric Society, Infectious Diseases and Immunization Committee. 2007. Autistic spectrum disorder: no causal relationship with vaccines. *Paediatrics and Child Health* 12: 393–395.

Candy, S., et al. 1995. A controlled double blind study of azathioprine in the management of Crohn's disease. *Gut* 37: 674–678.

Cantalupo, C., and W. D. Hopkins. 2001. Asymmetric Broca's area in great apes. *Nature* 414: 505.

Carré, J. M., and C. M. McCormick. 2008. In your face: facial metrics predict aggressive behaviour in the laboratory and in varsity and professional hockey players. *Proceedings of the Royal Society of London, Series B: Biological Sciences* 275: 2651–2656.

Catlin, J. R., and Y. Wang. 2013. Recycling gone bad: when the option to recycle increases resource consumption. *Journal of Consumer Psychology* 23: 122–127.

Cattin, M.-F., L.-F. Bersier, C. Banasek-Richter, R. Baltensperger, and J.-P. Gabriel. 2004. Phylogenetic constraints and adaptation explain food web structure. *Nature* 427: 835–839.

CBC News. November 28, 2005. Politicians facing cynical electorate as campaign opens. CBC News Online. http://www.cbc.ca/news/background/election2005/poll.html.

Chadwick Johnson, J., T. M. Ivy, and S. K. Sakaluk. 1999. Female remating propensity contingent on sexual cannibalism in sagebrush crickets, *Cyphoderis strepitans*: a mechanism of cryptic female choice. *Behavioral Ecology* 10: 227–233.

Charles, E. P. 2005. The correction for attenuation due to measurement error: clarifying concepts and creating confidence sets. *Psychological Methods* 10: 206–226.

Charlton, B. D., W. A. H. Ellis, J. Brumm, K. Nilsson, and W. T. Fitch. 2012. Female koalas prefer bellows in which lower formants indicate larger males. *Animal Behaviour* 84: 1565–1571.

Chase, J. M., and M. A. Leibold. 2002. Spatial scale dictates the productivity–biodiversity relationship. *Nature* 416: 427–430.

Chen, I-C., J. K. Hill, R. Ohlemüller, D. B. Roy, and C. D. Thomas. 2011. Rapid range shifts of species associated with high levels of climate warming. *Science* 333: 1024–1026.

Christenfeld, N., and E. Hill. 1995. Whose baby are you? *Nature* 378: 669.

Christiansen, K., and E. Winkler. 1992. Hormonal, anthropometrical, and behavioral correlates of physical aggression in !Kung San men of Namibia. *Aggressive Behavior* 18: 271–280.

Clark, D. B., and D. A. Clark. 2012. Annual tree growth, mortality, physical condition, and microsite in an old-growth tropical rain forest, 1983–2010. *Ecology* 93: 213.

Clarke, C. M., et al. 2009. Tree shrew lavatories: a novel nitrogen sequestration strategy in a tropical pitcher plant. *Biology Letters* 5: 632–635.

Clayton, N. S., and A, Dickinson. 1998. Episodic-like memory during cache recovery by scrub jays. *Nature* 395: 272–274.

Clemons, T., and M. Pagano. 1999. Are babies normal? *American Statistician* 53: 298–302.

Clubb, R., and G. Mason. 2003. Captivity effects on wide-ranging carnivores. *Nature* 425: 473–474.

Coello, C. A., et al. 2005. Adverse impact of surgical site infections in English hospitals. *Journal of Hospital Infection* 60: 93–103.

Collins S., and G. Bell. 2004. Phenotypic consequences of 1000 generations of selection at elevated CO_2 in a green alga. *Nature* 431: 566–569.

Colosimo, P. F., et al. 2004. The genetic architecture of parallel armor plate reduction in threespine sticklebacks. *PLOS Biology* 2: 635–641.

Cook, M. J., D. R. Fish, S. D. Shorvon, J. M. Stevens, and J. B. M. Kuks. 1993. Hippocampal sclerosis in epilepsy and childhood febrile seizures. *The Lancet* 342: 1391–1394.

Cook, N. R., et al. 2005. Low-dose aspirin in the primary prevention of cancer: the women's health study: a randomized controlled trial. *Journal of the American Medical Association* 294: 47–55.

Cooley, K., et al. 2009. Naturopathic care for anxiety: a randomized controlled trial ISRCTN78958974. *PLoS ONE* 4: e6628.

Cooper, H. M., and L. V. Hedges, eds. 1994. *Handbook of Research Synthesis*. New York: Sage Foundation Press.

Cooper, H., and J. C. Valentine. 2001. Using research to answer practical questions about homework. *Educational Psychologist* 36: 143–153.

Craig, J. K., and C. Foote. 2001. Countergradient variation and secondary sexual color: phenotypic convergence promotes genetic divergence in carotenoid use between sympatric anadromous and nonanadromous morphs of sockeye salmon (*Oncorhynchus nerka*). *Evolution* 55: 380–391.

Cratsley, C. K., and S. M. Lewis. 2003. Female preference for male courtship flashes in *Photinus ignitus* fireflies. *Behavioral Ecology* 14: 135–140.

Cucherousset, J., et al. 2012. "Freshwater killer whales": beaching behavior of an alien fish to hunt land birds. *PLoS ONE* 7: e50840.

Dabbs, J. M., Jr., M. F. Hargrove, and C. Huesel. 1996. Testosterone differences among college fraternities: well-behaved vs. rambunctious. *Personality and Individual Differences* 20: 157–161.

Daborn, P. J., et al. 2002. A single P450 allele associated with insecticide resistance in *Drosophila*. *Science* 297: 2253–2256.

Darnell, M. Z., and P. Munguia. 2011. Thermoregulation as an alternate function of the sexually dimorphic fiddler crab claw. *American Naturalist* 178: 419–428.

David, P., T. Bjorksten, K. Fowler, and A. Pomiankowski. 2000. Condition-dependent signaling of genetic variation in stalk-eyed flies. *Nature* 406: 186–188.

Davis, B. H., A. F. Y. Poon, and M. C. Whitlock. 2009. Compensatory mutations are repeatable and clustered within proteins. *Proceedings of the Royal Society of London, Series B: Biological Sciences* 276: 1823–1827.

Dawson, R. J. MacG. 1995. The "unusual episode" data revisited. *Journal of Statistics Education* 3(3).

Devlin, H. 2009. Want to keep your wallet? Carry a baby picture. *The Times*. http://www.thetimes.co.uk/tto/news/uk/article1967384.ece. Accessed January 9, 2014.

De Weber, K., M. Olszewski, and R. Ortolano. 2011. Knuckle cracking and hand osteoarthritis. *Journal of the American Board of Family Medicine* 24: 169–174.

Diamond, J. 1992. *The Third Chimpanzee*. New York: HarperCollins.

Diamond, J. M. 1988. Why cats have nine lives. *Nature* 332: 586–587.

Dickersin, K., Y. I. Min, and C. L. Meinert. 1992. Factors influencing publication of research results. *Journal of the American Medical Association* 267: 374–378.

Di Giusto, B., et al. 2010. Flower-scent mimicry masks a deadly trap in the carnivorous plant *Nepenthes rafflesiana*. *Journal of Ecology* 98: 845–856.

Discover Magazine 2011. http://blogs.discovermagazine.com/discoblog/2011/11/29/ncbi-rofl-taste-preference-for-brussels-sprouts-an-informal-look/. Accessed November 26, 2012.

Doll, R., R. Peto, E. Hall, K. Wheatley, and R. Gray. 1994. Mortality in relation to consumption of alcohol: 13 years' observations on male British doctors. *British Medical Journal* 309: 911–918.

Dondorp, A. M., et al. 2010. Artesunate versus quinine in the treatment of severe falciparum malaria in African children (AQUAMAT): an open-label, randomized trial. *The Lancet* 376: 1647–1657.

Duboué, E. R., A. C. Keene, and R. L. Borowsky. 2011. Evolutionary convergence on sleep loss in cavefish populations. *Current Biology* 21: 671–676.

Dunkin, R. C., W. A. McLellan, J. E. Blum, and D. A. Pabst. 2005. The ontogenetic changes in the thermal properties of blubber from Atlantic bottlenose dolphin *Tursiops truncates*. *Journal of Experimental Biology* 208: 1469–1480.

Dziekan, G., et al. 2000. Methicillin-resistant *Staphylococcus aureus* in a teaching hospital: investigation of nosocomial transmission using a matched case-control study. *Journal of Hospital Infection* 46: 263–270.

East, E. M. 1916. Studies in size inheritance in *Nicotiana*. *Genetics* 1: 164–176.

Easterbrook, P. J., J. E. Berlin, R. Gopalan, and D. R. Matthews. 1991. Publication bias in clinical research. *The Lancet* 337: 867–872.

Economist Newspaper Limited. 2005. The World in 2006. London: The Economist Newspaper Limited.

Edelaar, P., and C. W. Benkman. 2006. Replicated population divergence caused by localized coevolution? A test of three hypotheses in the red crossbill–lodgepole pine system. *Journal of Evolutionary Biology* 19: 1651–1659.

Edrey, Y. H., et al. 2012. Sustained high levels of neuregulin-1 in the longest-lived rodents; a key determinant of rodent longevity. *Aging Cell* 11: 213–222.

Edwards, A. W. F. 1992. *Likelihood*. Baltimore, MD: Johns Hopkins University Press.

Efron, B. 1979. Bootstrap methods: another look at the jackknife. *Annals of Statistics* 7: 1–26.

Efron, B., and R. J. Tibshirani. 1993. *An Introduction to the Bootstrap*. New York: Chapman & Hall.

Eggebeen, D. J., J. Dew, and C. Knoester. 2010. Fatherhood and men's lives at middle age. *Journal of Family Issues* 31: 113–130.

Eggert, L. S., J. A. Eggert, and D. S. Woodruff. 2003. Estimating population sizes for elusive animals: the forest elephants of Kakum National Park, Ghana. *Molecular Ecology* 12: 1389–1402.

Ehrenberg, A. S. C. 1977. Rudiments of numeracy. *Journal of the Royal Statistical Society, Series A* 140: 277–297.

Ehrlich, P. R., and P. H. Raven. 1964. Butterflies and plants: a study in coevolution. *Evolution* 18: 586–608.

Elgar, M. A., and D. R. Nash. 1988. Sexual cannibalism in the garden spider *Araneus diadematus*. *Animal Behaviour* 36: 1511–1517.

Elgar, M. A., and N. E. Pierce. 1988. Mating success and fecundity in an ant-tended Lycaenid butterfly. In *Reproductive Success: Studies of Selection and Adaptation in Contrasting Breeding Systems,* ed. T. H. Clutton-Broock. Chicago: University of Chicago Press, 59–75.

Elmore, J. G., K. Armstrong, C. D. Lehman, and S. W. Fletcher. 2005. Screening for breast cancer. *Journal of the American Medical Association* 293: 1245–1256.

Elstein, D. J., Ed. 1988. *Professional Judgment. A Reader in Clinical Decision Making*. Cambridge, UK: Cambridge University Press.

Emlen, D. J. 2001. Costs and the diversification of exaggerated animal structures. *Science* 291: 1534–1536.

Epel, E. S., et al. 2004. Accelerated telomere shortening in response to life stress. *Proceedings of the National Academy of Sciences (USA)* 101: 17312–17315.

Eriksson, K. 2012. The nonsense math effect. *Judgment and Decision Making* 7: 746–749.

Ernst, E., and A. R. White. 1998. Acupuncture for back pain: a meta-analysis of randomized controlled trials. *Archives of Internal Medicine* 158: 2235–2241.

Evason, K., C. Huang, I. Yamben, D. F. Covey, and K. Kornfeld. 2005. Anticonvulsant medications extend worm life-span. *Science* 307: 258–262.

Ewald, P. W. 1993. The evolution of virulence. *Scientific American* 268: 86–93.

Falconer, D. S., and T. F. C. MacKay. 1996. *Introduction to Quantitative Genetics*. Essex, UK: Longmans Green, Harlow.

Fan, Z., S. R. Shifley, M. A. Spetich, F. R. Thompson III, and D. R. Larsen. 2005. Abundance and size distribution of cavity trees in second-growth and old-growth central hardwood forests. *Northern Journal of Applied Forestry* 22: 162–169.

Fanelli, D. 2009. How many scientists fabricate and falsify research? A systematic review and meta-analysis of survey data. *PLoS ONE* 4(5): e5738.

Fang, W., et al. 2011. Development of transgenic fungi that kill human malaria parasites in mosquitoes. *Science* 331: 1074–1077.

Farrell, B. D. 1998. "Inordinate fondness" explained: why are there so many beetles? *Science* 281: 555–559.

Farrell, B. D., D. E. Dussord, and C. Mitter. 1991. Escalation of plant defense: do latex and resin canals spur plant diversification? *American Naturalist* 138: 881–900.

Farren, L., S. Shayler, and A. R. Ennos. 2004. The fracture properties and mechanical design of human fingernails. *Journal of Experimental Biology* 207: 735–741.

Farrington, D. P. 1994. Cambridge Study in Delinquent Development [Great Britain], 1961–1981. 2nd ICPSR ed. Ann Arbor, MI: Inter-university Consortium for Political and Social Research. Data distributed at http://webapp.icpsr.umich.edu/cocoon/NACJD-STUDY/08488.xml.

Farroni, T., G. Csibra, F. Simion, and M. H. Johnson. 2002. Eye contact detection in humans from birth. *Proceedings of the National Academy of Sciences (USA)* 99: 9602–9605.

Fatouros, N. E., M. E. Huigens, J. J. A. van Loon, M. Dicke, and M. Hilker. 2005. Butterfly anti-aphrodisiac lures parasitic wasps. *Nature* 433: 704.

Faurie, C., D. Pontierb, and M. Raymond. 2004. Student athletes claim to have more sexual partners than other students. *Evolution and Human Behavior* 25: 1–8.

Faurie, C., and M. Raymond. 2005. Handedness, homicide and negative frequency-dependent selection. *Proceedings of the Royal Society of London, Series B: Biological Sciences* 272: 25–28.

Felsenstein, J. 1985. Phylogenies and the comparative method. *American Naturalist* 125: 1–15.

Fernandes, C. C., J. Podos, and J. G. Lundberg. 2004. Amazonian ecology: tributaries enhance the diversity of electric fishes. *Science* 305: 1960–1962.

Flynn, J. J., M. A. Nedbal, J. W. Dragoo, and R. L. Honeycutt. 2000. Whence the red panda? *Molecular Phylogenetics and Evolution* 17: 190–199.

Fortin, N. J., K. L. Agster, and H. B. Eichenbaum. 2004. Critical role of the hippocampus in memory for sequences of events. *Nature Neuroscience* 5: 458–462.

Freakonomics. 2011. http://www.freakonomics. com/2011/07/07/another-case-of-teacher-cheating-or-is-it-just-altruism/. Accessed October 20, 2012.

Frick, R. W. 1996. The appropriate use of null hypothesis testing. *Psychological Methods* 1: 379–390.

Fricke, H. W. 1979. Mating system, resource defence and sex change in the anemonefish *Amphiprion akallopisos. Zeitschrift für Tierpsychologie* 50: 313–326.

Fritsches, K. A., R. W. Brill, and E. J. Warrant. 2005. Warm eyes provide superior vision in swordfishes. *Current Biology* 15: 55–58.

Fryer, J. D., et al. 2011. Exercise and genetic rescue of SCA1 via the transcriptional repressor Capicua. *Science* 334: 690–693.

Fuller, A., P. R. Kamerman, S. K. Maloney, G. Mitchell, and D. Mitchell. 2003. Variability in brain and arterial blood temperatures in free-ranging ostriches in their natural habitat. *Journal of Experimental Biology* 206: 1171–1181.

Fuller, R. A., K. N. Irvine, P. Devine-Wright, P. H. Warren, and K. J. Gaston. 2007. Psychological benefits of greenspace increase with biodiversity. *Biology Letters* 3: 390–394.

Fuller, R. C., L. A. Noa, and R. S. Strellner. 2010. Teasing apart the many effects of lighting environment on opsin expression and foraging pref-

erence in bluefin killifish. *American Naturalist* 176: 1–13.

Gal, R., and F. Libersat. 2010. A wasp manipulates neuronal activity in the sub-esophageal ganglion to decrease the drive for walking in its cockroach prey. *PLoS ONE* 5: e10019.

Galton, F. 1894. Note on fitting normal curves to distribution of speeds of old homing pigeons. *Homing News and Pigeon Fanciers' Journal* (April 6): 159–160. http://galton.org.

Galton. F. 1886. Regression towards mediocrity in hereditary stature. *Journal of the Anthropological Institute* 15: 246–263.

Garaci, F. G., et al. 2012. Brain hemodynamic changes associated with chronic cerebrospinal venous insufficiency are not specific to multiple sclerosis and do not increase its severity. *Radiology* 265: 233–239.

García-Planells, J., et al. 2005. Ancient origin of the CAG expansion causing Huntington disease in a Spanish population. *Human Mutation* 25: 453–459.

Gazey, W. J., and M. J. Staley. 1986. Population estimation from mark-recapture experiments using a sequential Bayes algorithm. *Ecology* 67: 941–951.

Geffeney, S., E. D. Brodie Jr., P. C. Ruben, and E. D. Brodie III. 2002. Mechanisms of adaptation in a predator-prey arms race: TTX-resistant sodium channels. *Science* 297: 1336–1339.

Gems, D., et al. 1998. Two pleiotropic classes of *daf-2* mutation affect larval arrest, adult behavior, reproduction and longevity in *Caenorhabditis elegans. Genetics* 150: 129–155.

Gettelfinger, B., and E. L. Cussler. 2004. Will humans swim faster or slower in syrup? *AIChE Journal* 50: 2646–2647.

Ghansah, A., et al. 2012. Haplotype analyses of haemoglobin C and haemoglobin S and the dynamics of the evolutionary response to malaria in Kassena-Nankana District of Ghana. *PLoS ONE* 7: e34565.

Gianoli, E. 2004. Evolution of a climbing habit promotes diversification in flowering plants. *Proceedings of the Royal Society of London, Series B: Biological Sciences* 271: 2011–2015.

Gigord, L. D. B., M. R. Macnair, and A. Smithson. 2001. Negative frequency-dependent selection maintains a dramatic flower color polymorphism in the rewardless orchid *Dactylorhiza sambucina* (L.) Soo. *Proceedings of* the *National Academy of Sciences (USA)* 98: 6253–6255.

Gilham, C., et al. 2005. Day care in infancy and risk of childhood acute lymphoblastic leukaemia: findings from UK case-control study. *British Medical Journal.* doi:10.1136/bmj.38428.521042.8F

Gilmour, J., C. Harrison, L. Asadi, M. H. Cohen, and S. Vohra. 2011. Childhood immunization: when physicians and parents disagree. *Pediatrics* 128 (Supplement 4): S167–S174.

Ginsberg, J. et al. 2009. Detecting influenza epidemics using search engine query data. *Nature* 457:1012–1014.

Glantz, M. C., et al. 2009. Gender disparity in the rate of partner abandonment in patients with serious medical illness. *Cancer* 115: 5237–5242.

Glare, P., et al. 2003. A systematic review of physicians' survival predictions in terminally ill cancer patients. *British Medical Journal* 327: 1–6.

Goldman, J. M., et al. 1988. Bone marrow transplantation for chronic myelogenous leukemia in chronic phase. Increased risk for relapse associated with T-cell depletion. *Annals of Internal Medicine* 108: 806–814.

Golenda, C. F., V. B. Solberg, R. Burge, J. M. Gambel, and R. A. Wirtz. 1999. Gender-related efficacy difference to an extended duration formulation of topical *N*, *N*-diethyl-*m*-toluamide (DEET). *American Journal of Tropical Medicine and Hygiene* 60: 654–657.

Golomb, B. A., S. Koperski, and H. L. White. 2012. Association between more frequent chocolate consumption and lower body mass index. *Archives of Internal Medicine* 172: 519–521.

Gore, S. M., I. G. Jones, and E. C. Rytter. 1977. Misuse of statistical methods: critical assessment of articles in BMJ from January to March 1976. *British Medical Journal* 1: 85–87.

Gøtzsche, P. C. 1987. Reference bias in reports of drug trials. *British Medical Journal* 295: 654–656.

Grafe, T. U., S. Döbler, and K. E. Linsenmair. 2002. Frogs flee from the sound of fire. *Proceedings of the Royal Society of London, Series B: Biological Sciences* 269: 999–1003.

Grafen, A., and R. Hails. 2002. *Modern Statistics for the Life Sciences.* Oxford, UK: Oxford University Press.

Gray, A., D. N. Jackson, and J. B. McKinlay. 1991. The relation between dominance, anger, and hormones in normally aging men: results from the Massachusetts Male Aging Study. *Psychosomatic Medicine* 53: 375–385.

Gray, E. M. 1997. Female red-winged blackbirds accrue material benefits from copulating with extra-pair males. *Animal Behaviour* 53: 625–639.

Gray, G. M., et al. 2004. Weight of the evidence evaluation of low-dose reproductive and developmental effects of bisphenol A. *Human and Ecological Risk Assessment* 10: 875–921.

Gray, R. H., et al. 2007. Limitations of rapid HIV-1 tests during screening for trials in Uganda: diagnostic test accuracy study. *British Medical Journal* 335: 188.

Gray, S. M., L. M. Dill, and J. S. McKinnon. 2007. Cuckoldry incites cannibalism: male fish turn to cannibalism when perceived certainty of paternity decreases. *American Naturalist* 169: 258–263.

Green, J. A., P. J. Butler, A. J. Woakes, I. L. Boyd, and R. L. Holder. 2001. Heart rate and rate of oxygen consumption of exercising macaroni penguins. *Journal of Experimental Biology* 204: 673–684.

Griffith, S. C., S. R. Pryke, and W. A. Buttemer. 2011. Constrained mate choice in social monogamy and the stress of having an unattractive partner. *Proceedings of the Royal of Society, London, Series B: Biological Sciences* 278: 2798–2805.

Griffith, S. C., and B. C. Sheldon. 2001. Phenotypic plasticity in the expression of sexually selected traits: neglected components of variation. *Animal Behaviour* 61: 987–993.

Gua, J.-J., et al. 2012. Wing stridulation in a Jurassic katydid (Insecta, Orthoptera) produced low-pitched musical calls to attract females. *Proceedings of the National Academy of Sciences (USA)* 109: 3868–3873.

Guelzim, N., S. Bottani, P. Bourgine, and F. Képès. 2002. Topological and causal structure of the yeast transcriptional regulatory network. *Nature Genetics* 31: 60–63.

Gundale, M. J., W. M. Jolly, and T. H. Deluca. 2005. Susceptibility of a northern hardwood forest to exotic earthworm invasion. *Conservation Biology* 19: 1075–1083.

Gunnarsson, T. G., J. A. Gill, T. Sigurbjörnsson, and W. J. Sutherland. 2004. Arrival synchrony in migratory birds. *Nature* 431: 646.

Gurevitch, J., and L. V. Hedges. 1999. Statistical issues in ecological meta-analyses. *Ecology* 80: 1142–1149.

Gurung, B., J. L. D. Smith, C. McDougal, J. B. Karkic, and A. Barlowa. 2008. Factors associated with human-killing tigers in Chitwan National Park, Nepal. *Biological Conservation* 141: 3069–3078.

Hadfield, J. D., et al. 2006. Direct versus indirect sexual selection: genetic basis of colour, size and recruitment in a wild bird. *Proceedings of the Royal Society of London, Series B: Biological Sciences* 273: 1347–1353.

Haeslery, M. P., and O. Seehausen. 2005. Inheritance of female mating preference in a sympatric sibling species pair of Lake Victoria cichlids: implications for speciation. *Proceedings of the Royal Society of London, Series B: Biological Sciences* 272: 237–245.

Hagen, M., M. Wikelski, and W. D. Kissling. 2011. Space use of bumblebees (*Bombus* spp.) revealed by radio-tracking. *PLoS ONE* 6: e19997.

Hairston, N. G., Jr., et al. 1999. Rapid evolution revealed by dormant eggs. *Nature* 401: 446.

Halpern, B. S. 2003. The impact of marine reserves: do reserves work and does reserve size matter? *Ecological Applications* 13: S117–S137.

Hama, Y., M. Uematsu, Y. Sakurai, and S. Kusano. 2001. Sex ratio in the offspring of male radiologists. *Academic Radiology* 8: 421–424.

Hamshere, M. L., et al. 2005. Genomewide linkage scan in schizoaffective disorder: significant evidence for linkage at 1q42 close to disc1, and suggestive evidence at 22q11 and 19p13. *Archives in General Psychiatry* 62: 1081–1088.

Hans, C. N. 2002. On the risk of mortality to primates exposed to anthrax spores. *Risk Analysis* 22: 189–193.

Harley, C. D. G. 2003. Abiotic stress and herbivory interact to set range limits across a two-dimensional stress gradient. *Ecology* 84: 1477–1488.

Harley, M., M. A. Mohammed, S. Hussain, J. Yates, and A. Almasri. 2005. Was Rodney Ledward a statistical outlier? Retrospective analysis using routine hospital data to identify gynaecologists' performance. *British Medical Journal* 330: 929.

Published online April 15 2005. doi:10.1136/bmj.38377.675440.8F

Harper, G. R., Jr., and D. W. Pfennig. 2008. Selection overrides gene flow to break down maladaptive mimicry. *Nature* 451: 1103–1106.

Harpole, W. S., and D. Tilman. 2007. Grassland species loss resulting from reduced niche dimension. *Nature* 446: 791–793.

Harris, J., P. Rushton, E. Hampton, and D. Jackson. 1996. Salivary testosterone and self report aggressive and pro-social personality characteristics in men and women. *Aggressive Behavior* 22: 321–331.

Hartl, D. L., and E. W. Jones. 2005. *Genetics: Analysis of Genes and Genomes*. Sudbury, MA: Jones & Bartlett.

Hasselquist, D., J. A. Marsh, P. W. Sherman, and J. C. Wingfield. 1999. Is avian humoral immunocompetence suppressed by testosterone? *Behavioral Ecology and Sociobiology* 45: 167–175.

Hayes, J. P., and C. S. O'Connor. 1999. Natural selection on thermogenic capacity of high-altitude deer mice. *Evolution* 53: 1280–1287.

Health Protection Agency, UK. 2012. http://www.hpa.org.uk/web/HPAweb&HPAwebStandard/HPAweb_C/1195733811358. Accessed March 5, 2012.

Heathcote, J. A. 1995. Why do old men have big ears? *British Medical Journal* 311: 1668.

Hedrick, P. W., and K. Ritland. 2012. Population genetics of the white-phased "Spirit" black bear of British Columbia. *Evolution* 66: 305–313.

Heffner, R. A., M. J. Butler IV, and C. K. Reilly. 1996. Pseudoreplication revisited. *Ecology* 77: 2558–2562.

Heilbuth, J. C. 2000. Lower species richness in dioecious clades. *American Naturalist* 156: 221–241.

Heiling, A. M., M. E. Herberstein, and L. Chittka. 2003. Crab spiders manipulate flower signals. *Nature* 421: 334.

Hemmingsson, T., and D. Kriebel. 2003. Smoking at age 18–20 and suicide during 26 years of follow-up—how can the association be explained? *International Journal of Epidemiology* 32: 1000–1004.

Hendricks, J. C., et al. 2001. A non-circadian role for cAMP signaling and CREB activity in *Drosophila* rest homeostasis. *Nature Neuroscience* 4: 1108–1115.

Hendry, A. P., O. K. Berg, and T. P. Quinn. 1999. Condition dependence and adaptation-by-time: breeding date, life history, and energy allocation in a population of salmon. *Oikos* 85: 499–514.

Heusner, A. A. 1991. Size and power in mammals. *Journal of Experimental Biology* 160: 25–54.

Hey, J. 1992. Using phylogenetic trees to study speciation and extinction. *Evolution* 46: 627–640.

Hidinger, L. A. 1996. Measuring the impacts of ecotourism on animal populations: a case study of Tikal National Park, Guatemala. In *The Ecotourism Equation: Measuring the Impacts*, ed. E. Malek-Zadah. New Haven, CT: Yale School of Forestry and Environmental Studies, 49–59.

Hirano, S. S., E. V. Nordheim, D. C. Arny, and C. D. Upper. 1982. Lognormal distribution of epiphytic bacterial populations on leaf surfaces. *Applied and Environmental Microbiology* 44: 695–700.

Hobbs, J.-P. A., P. L. Munday, and G. P. Jones. 2004. Social induction of maturation and sex determination in a coral reef fish. *Proceedings of the Royal Society of London, Series B: Biological Sciences* 271: 2109–2114.

Hocking, M. D., and J. D. Reynolds. 2011. Impacts of salmon on riparian plant diversity. *Science* 331: 1609–1612.

Hoogland, J. L. 1998. Why do female Gunnison's prairie dogs copulate with more than one male? *Animal Behaviour* 55: 351–359.

Horvath, G., E. Farkas, I. Boncz, M. Blaho, and G. Kriska. 2012. Cavemen were better at depicting quadruped walking than modern artists: erroneous walking illustrations in the fine arts from prehistory to today. *PLoS ONE* 7(12): e49786.

Hosken, D. J., W. U. Blanckenhorn, and T. W. J. Garner. 2002. Heteropopulation males have a fertilization advantage during sperm competition in the yellow dung fly (*Scathophaga stercoraria*). *Proceedings of the Royal Society of London, Series B: Biological Sciences* 269: 1701–1707.

Hosken, D. J., and P. I. Ward. 2001. Experimental evidence for testis size evolution via sperm competition. *Ecology Letters* 4: 10–13.

Houser, B. B. 1979. An investigation of the correlation between hormonal levels in males and mood, behavior, and physical discomfort. *Hormones and Behavior* 12: 185–197.

Hróbjartsson, A., and P. C. Gøtzsche. 2001. Is the placebo powerless? An analysis of clinical trials comparing placebo with no treatment. *Journal of the American Medical Association* 344: 1594–1602.

Hsieh, C.-H., et al. 2006. Fishing elevates variability in the abundance of exploited species. *Nature* 443: 859–862.

Hubbard, T., et al. 2005. Ensembl 2005. *Nucleic Acids Research* 33: D447–D453.

Huber, R., M. F. Ghilardi, M. Massimini, and G. Tononi. 2004. Local sleep and learning. *Nature* 430: 78–81.

Huey, R. B., and A. E. Dunham. 1987. Repeatability of locomotor performance in natural populations of the lizard *Sceloporus merriami*. *Evolution* 42: 1116–1120.

Huey, R. B., and X. Eguskitza. 2000. Supplemental oxygen and mountaineer death rates on Everest and K2. *Journal of the American Medical Association* 284: 181.

Hunt, M. 1997. *How Science Takes Stock*. New York: Sage Foundation Press.

Hurlbert, S. H. 1984. Pseudoreplication and the design of ecological field experiments. *Ecological Monographs* 54: 187–211.

Hurlbert, S. H., and M. D. White. 1993. Experiments with freshwater invertebrate zooplanctivores—quality of statistical analyses. *Bulletin of Marine Science* 53: 128–153.

Hurlburt, G. 1996. *Relative Brain Size in Recent and Fossil Amniotes: Determination and Interpretation*. Ph.D. thesis, Zoology (Vertebrate Paleontology), University of Toronto.

Hyndman, R. J., and Fan, Y. 1996. Sample quantiles in statistical packages. *American Statistician* 50: 361–365.

Inaudi, D., et al. 1995. Chaos: evidence for the butterfly effect. *Annals of Improbable Research* 1(6). Reprinted in M. Abrahams, ed. 2000. *The Best of Annals of Improbable Research*. New York: W. H. Freeman.

Ioannidis, J. P. A., J. C. Cappelleri, H. S. Sacks, and J. Lau. 1997. The relationship between study design, results, and reporting of randomized trials of HIV infection. *Controlled Clinical Trials* 18: 431–444.

Iverson, J. M., and S. Goldin-Meadow. 1998. Why people gesture when they speak. *Nature* 396: 228.

Jacobs, T. 2009. Benefits of fatherhood extend to the community. *Pacific Standard*. http://www.psmag.com/culture-society/benefits-of-

fatherhood-extend-to-the-community-5938/.
Accessed February 10, 2013.

James, R. A., P. A. Hoadley, and B. G. Sampson.
1997. Determination of postmortem interval by
sampling vitreous humour. *American Journal of
Forensic Medicine and Pathology* 18: 158–162.

Jansen, V. A. A., et al. 2003. Measles outbreaks in a
population with declining vaccine uptake.
Science 301: 804.

Jenkins, A. J. 2001. Drug contamination of US
paper currency. *Forensic Science International*
121: 189–193.

Jensen, H., et al. 2004. Lifetime reproductive success in relation to morphology in the house
sparrow *Passer domesticus. Journal of Animal
Ecology* 73: 599–611.

Jensen, K. H., T. Little, A. Skorping, and D. Ebert.
2006. Empirical support for optimal virulence
in a castrating parasite. *PLoS Biology* 4:
1265–1269.

Jerison, H. J. 2006. Paleoneurology: the study of
brain endocasts of extinct vertebrates.
http://brainmuseum.org/ Evolution/paleo/.
Accessed April 24, 2014

Jesson, L. K., and S. C. H. Barrett. 2002. The genet-
ics of mirror-image flowers. *Proceedings of the
Royal Society of London, Series B: Biological
Sciences* 269: 1835–1839.

Johnson, A. M., et al. 2001. Sexual behaviour in
Britain: partnerships, practices, and HIV risk
behaviours. *The Lancet* 358: 1835–1842.

Johnson, J. C., T. M. Ivy, and S. K. Sakaluk. 1999.
Female remating propensity contingent on sexual cannibalism in sagebrush crickets, *Cyphoderris strepitans*: a mechanism of cryptic female
choice. *Behavioral Ecology* 10: 227–233.

Johnson, J. G., et al. 2002. Television viewing and
aggressive behavior during adolescence and
adulthood. *Science* 295: 2468–2471.

Johnson, S. D., and K. E. Steiner. 1997. Long-
tongued fly pollination and evolution of floral
spur length in the *Disa draconis* complex
(Orchidaceae). *Evolution* 51: 45–53.

Jules, E. S., and B. J. Rathcke. 1999. Mechanisms of
reduced trillium recruitment along edges of old-
growth forest fragments. *Conservation Biology*
13: 784–793.

Jüni, P., D. G. Altman, and M. Egger. 2001. Assessing the quality of controlled clinical trials. *British Medical Journal* 323: 42–46.

Kacsoh, B. Z., Z. R. Lynch, N. T. Mortimer, and
T. A. Schlenke. 2013. Fruit flies medicate
offspring after seeing parasites. *Science* 339:
947–950.

Kalani, M. Y. S., et al. 2008. *Wnt*-mediated self-
renewal of neural stem/progenitor cells. *Proceedings of the National Academy of Sciences
(USA)* 105: 16970–16975.

Kanter, M. H., and J. R. Taylor. 1994. Accuracy of
statistical methods in *Transfusion*—a review of
articles from July/August 1992 through June
1993. *Transfusion* 34: 697–701.

Kanwisher, J., G. Gabrielsen, and N. Kanwisher.
1981. Free and forced diving in birds. *Science*
211: 717–719.

Keenan, J. P., et al. 2001. Self-recognition and the
right hemisphere. *Nature* 409: 305.

Kelly, C., and T. D. Price. 2005. Correcting for
regression to the mean in behavior and ecology.
American Naturalist 166: 700–707.

Kendall, M., and B. B. Smith. 1939. The problem of
m rankings. *Annals of Mathematical Statistics*
10: 275–287.

Khila, A., E. Abouheif, and L. Rowe. 2009. Evolution of a novel appendage ground plan in water
striders is driven by changes in the *Hox* gene
Ultrabithorax. PLoS Genetics 5: e1000583.

Kim, U., et al. 2003. Positional cloning of the human
quantitative trait locus underlying taste sensitivity to phenylthiocarbamide. *Science* 299:
1221–1225.

Kingsolver, J. G., et al. 2001. The strength of phenotypic selection in natural populations. *American
Naturalist* 157: 245–261.

Kirsch, I. 2010. Not all placebos are born equal. *New
Scientist* 208(2790): 30–31.

Klem, D., et al. 2004. Effects of window angling,
feeder placement, and scavengers on avian mortality at plate glass. *Wilson Bulletin* 116: 69–73.

Klinesmith, J., T. Kasser, and K. T. McAndrew.
2006. Guns, testosterone, and aggression: an
experimental test of a meditational hypothesis.
Psychological Science 17: 568–571.

Knipschild, P. 1994. Systematic reviews: some
examples. *British Medical Journal* 309:
719–721.

Kodric-Brown, A., and J. H. Brown. 1993. Highly
structured fish communities in Australian desert
springs. *Ecology* 74: 1847–1855.

Koella, J. C., F. L. Sørensen, and R. A. Anderson.
1998. The malaria parasite, *Plasmodium falci-*

parum, increases the frequency of multiple feeding of its mosquito vector, *Anopheles gambiae*. *Proceedings of the Royal Society of London, Series B: Biological Sciences* 265: 763–768.

Kong, A., et al. 2012. Rate of de novo mutations and the importance of father's age to disease risk. *Nature* 488: 471–475.

Kopczuk, W., and J. Slemrod. 2003. Dying to save taxes: evidence from estate-tax returns on the death elasticity. *Review of Economics and Statistics* 85: 256–265.

Kotiaho, J. S., L. W. Simmons, and J. L. Tomkins. 2001. Towards a resolution of the lek paradox. *Nature* 410: 684–686.

Kramer, M. S., and R. Kakuma. 2002. Optimal duration of exclusive breastfeeding. *Cochrane Database of Systematic Reviews*, Issue 1. Art. No.: CD003517. doi:10.1002/14651858.CD003517

Kramer, M. S., et al. 2002. Breastfeeding and infant growth: biology or bias? *Pediatrics* 110: 343–347.

Krebs, C. J. 1999. *Ecological Methodology*, 2nd ed. Menlo Park, CA: Benjamin Cummings.

Krochmal, A. R., G. S. Bakken, and T. J. LaDuc. 2004. Heat in evolution's kitchen: evolutionary perspectives on the functions and origin of the facial pit of pit vipers (Viperidae: Crotalinae). *Journal of Experimental Biology* 207, 4231–4238.

LaCroix, A. Z., S. G. Leveille, J. A. Hecht, L. C. Grothaus, and E. H. Wagner. 1996. Does walking decrease the risk of cardiovascular disease hospitalizations and death in older adults? *Journal of the American Geriatric Society* 44: 113–120.

Lafferty, K. D., and A. K. Morris. 1996. Altered behavior of parasitized killifish increases susceptibility to predation by bird final hosts. *Ecology* 77: 1390–1397.

Lai, K.-M., C. Bottomley, and R. McNerney. 2011. Propagation of respiratory aerosols by the vuvuzela. *PLoS ONE* 6: e20086.

LaMunyon, C. W., and S. Ward. 1998. Larger sperm outcompete smaller sperm in the nematode *Caenorhabditis elegans*. *Proceedings of the Royal Society of London, Series B: Biological Sciences* 265: 1997–2002.

Langer, E., A. Blank, and B. Chanowitz. 1978. The mindlessness of ostensibly thoughtful action: the role of "placebic" information in interpersonal interaction. *Journal of Personality and Social Psychology* 36: 635–642.

Langford, D. J., et al. 2006. Social modulation of pain as evidence for empathy in mice. *Science* 312: 1967–1970.

Lanza, F., J. Goff, C. Scowcroft, D. Jennings, and P. Greski-Rose 1994. Double-blind comparison of lansoprazole, ranitidine, and placebo in the treatment of acute duodenal ulcer. Lansoprazole study group. *American Journal of Gastroenterology* 89: 1191–1200.

Lappin, A. K., and J. F. Husak. 2005. Weapon performance, not size, determines mating success and potential reproductive output in the collared lizard (*Crotaphytus collaris*). *American Naturalist* 166: 426–436.

Laurance, W. F., et al. 1997. Biomass collapse in Amazonian forest fragments. *Science* 278: 1117–1118.

Leakey, A. D. B., J. D. Scholes, and M. C. Press. 2005. Physiological and ecological significance of sunflecks for dipterocarp seedlings. *Journal of Experimental Botany* 56: 469–482.

Leavitt, J. D., and N. J. S. Christenfeld. 2011. Story spoilers don't spoil stories. *Psychological Science* 22: 1152–1154.

Lefèvre, T. 2010. Beer consumption increases human attractiveness to malaria mosquitoes. *PLoS ONE* 5: e9546.

Lehr, R. 1992. Sixteen *s* squared over *d* squared: a relation for crude sample size estimates. *Statistics in Medicine* 11: 1099–1102.

Leopold, S. S., W. J. Warme, E. F. Braunlich, and S. Shott. 2003. Association between funding source and study outcome in orthopedic research. *Clinical Orthopaedics and Related Research* 415: 293–301.

Lesku, J. A., et al. 2012. Adaptive sleep loss in polygynous pectoral sandpipers. *Science* 337: 1654–1658.

Levey, D. J., R. S. Duncan, and C. F. Levins. 2004. Use of dung as a tool by burrowing owls. *Nature* 431: 39.

Levin, P. S., S. Achord, B. E. Fiest, and R. W. Zabel. 2002. Non-indigenous brook trout and the demise of Pacific salmon: a forgotten threat? *Proceedings of the Royal Society of London, Series B: Biological Sciences* 269: 1663–1670.

Lewis, E. B., B. D. Pfeiffer, D. R. Mathog, and S. E. Celniker. 2003. Evolution of the homeobox complex in the Diptera. *Current Biology* 13: R587–R588.

Liang, Y.-L., et al. 2001. Effects of blood pressure, smoking, and their interaction on carotid artery structure and function. *Hypertension* 37: 6–11.

Liberg, O. H., et al. 2005. Severe inbreeding depression in a wild wolf (*Canis lupus*) population. *Biology Letters* 1: 17–20.

Lim, M. M., et al. 2004. Enhanced partner preference in a promiscuous species by manipulating the expression of a single gene. *Nature* 429: 754–757.

Lindman, R., B. von der Pahlen, B. Öst, and P. Eriksson. 1992. Serum testosterone, cortisol, glucose, and ethanol in males arrested for spouse abuse. *Aggressive Behavior* 18: 393–400.

Lounibos, L. P., N. Nishimura, J. Conn, and R. Lourenco-de-Oliveira. 1995. Life history correlates of adult size in the malaria vector *Anopheles darlingi*. *Memórias do Instituto Oswaldo Cruz* 90: 769–774.

Luijckx, P., H. Fienberg, D. Duneau, and D. Ebert. 2012. Resistance to a bacterial parasite in the crustacean *Daphnia magna* shows Mendelian segregation with dominance. *Heredity* 108: 547–551.

Maddison, D., and A. Viola. 1968. The health of widows in the year following bereavement. *Journal of Psychosomatic Research* 12: 297–306.

Maddison, W. P., and D. R. Maddison. 2011. Mesquite: a modular system for evolutionary analysis. Version 2.75: http://mesquiteproject.org.

Mainland, J. D., et al. 2002. Olfactory plasticity: one nostril knows what the other learns. *Nature* 419: 802.

Malone, K. E., et al. 2006. Prevalence and predictors of *BRCA1* and *BRCA2* mutations in a population-based study of breast cancer in white and black American women ages 35 to 64 years. *Cancer Research* 16: 8297–8308.

Mantonakis, A., P. Rodero, I. Lesschaeve, and R. Hastie. 2009. Order in choice: Effects of serial position on preferences. *Psychological Science* 20: 1309–1312.

Marks, D. 2000. *The Psychology of the Psychic*. Amherst, NY: Prometheus Books.

Martin, J. A., et al. 2011. Births: Final data for 2009. *National Vital Statistics Reports* 60: 1–70.

Marzoli, D., and L. Tommasi. 2009. Side biases in humans (*Homo sapiens*): three ecological studies on hemispheric asymmetries. *Naturwissenschaften* 96: 1099–1106.

Mathevon, N., A. Koralek, M. Weldele, S. E. Glickman, and F. E. Theunissen. 2010. What the hyena's laugh tells: Sex, age, dominance and individual signature in the giggling call of *Crocuta crocuta*. *BMC Ecology* 10: 9.

Mattison, J. A., et al. 2012. Impact of caloric restriction on health and survival in rhesus monkeys from the NIA study. *Nature* 489: 318–321.

Maurer, D., T. Pathman, and C. J. Mondloch. 2006. The shape of boubas: sound–shape correspondences in toddlers and adults. *Developmental Science* 9: 316–322.

Maxwell, E. A. 1976. Analysis of contingency tables and further reasons for not using Yates' correction in 2 × 2 tables. *Canadian Journal of Statistics* 4: 277–290.

McCarthy, C. R. 1994. Historical background of clinical trials involving women and minorities. *Academic Medicine* 69: 695–698.

McDowell, M. A., C. D. Fryar, C. L. Ogden, and K. M. Flegal. 2008. Anthropometric reference data for children and adults: United States, 2003–2006. *National Health Statistics Reports* 10.

McGaha, T. L., B. Sorrentino, and J. V. Ravetch. 2005. Restoration of tolerance in lupus by targeted inhibitory receptor expression. *Science* 307: 590–593.

McGraw, K. J., and D. R. Ardia. 2003. Carotenoids, immunocompetence, and the information content of sexual colors: an experimental test. *American Naturalist* 162: 704–712.

McGuigan, S. M. 1995. The use of statistics in the *British Journal of Psychiatry*. *British Journal of Psychiatry* 167: 683–688.

McKinnon, J. S., et al. 2004. Evidence for ecology's role in speciation. *Nature* 429: 294–298.

Mead, S., et al. 2009. A novel protective prion protein variant that colocalizes with kuru exposure. *New England Journal of Medicine* 361: 2056–2065.

Mechelli, A., et al. 2004. Structural plasticity in the bilingual brain. *Nature* 431: 757.

Meeker, W. Q., and W. A. Escobar. 1995. Teaching about confidence regions based on maximum likelihood estimation. *American Statistician* 49: 48–53.

Mehl, M. R., S. Vazire, N. Ramírez-Esparza, R. B. Slatcher, and J. W. Pennebaker. 2007. Are women really more talkative than men? *Science* 317: 82.

Melander, H., J. Ahlqvist-Rastad, G. Meijer, and B. Beermann. 2003. Evidence b(i)ased medicine—elective reporting from studies sponsored by the pharmaceutical industry: review of studies in new drug applications. *British Medical Journal* 326: 1171–1175.

Mendel, G. 1866. Versuche über pflanzen-hybriden. Verhandlungen des naturforschenden vereines. *Abhand-lungen Brünn* 4: 3–47.

Messerli, F. H. 2012. Chocolate consumption, cognitive function, and Nobel laureates. *New England Journal of Medicine* 367: 1562–1564.

Michalsen, A., et al. 2003. Effectiveness of leech therapy in osteoarthritis of the knee: a randomized, controlled trial. *Annals of Internal Medicine* 139: 724–730.

Middleman, A. B., R. Anding, and C. Tung. 2010. Effect of needle length when immunizing obese adolescents with hepatitis B vaccine. *Pediatrics* 125: e508–e512.

Miller, L. M., T. Close, and A. R. Kapuscinski. 2004. Lower fitness of hatchery and hybrid rainbow trout compared to naturalized populations in Lake Superior tributaries. *Molecular Ecology* 13: 3379–3388.

Milner-Gulland, E. J., et al. 2003. Reproductive collapse in saiga antelope harems. *Nature* 422: 135.

Min, K. J., C. K. Lee, and H. N. Park. 2012. The lifespan of Korean eunuchs. *Current Biology* 22: R792–R793.

Miranda, A., O. G. Almeida1, P. C. Hubbard, E. N. Barata, and A. V. M. Canário. 2005. Olfactory discrimination of female reproductive status by male tilapia (*Oreochromis mossambicus*). *Journal of Experimental Biology* 208: 2037–2043.

Mobbs, D., C. C. Hagan, E. Azim, V. Menon, and A. L. Reiss. 2005. Personality predicts activity in reward and emotional regions associated with humor. *Proceedings of the National Academy of Sciences (USA)* 102: 16502–16506.

Modig, A. O. 1996. Effects of body size and harem size on male reproductive behaviour in the southern elephant seal. *Animal Behaviour* 51: 1295–1306.

Moertel, C. G., et al. 1985. High-dose vitamin C versus placebo in the treatment of patients with advanced cancer who have had no prior chemotherapy—a randomized double-blind comparison. *New England Journal of Medicine* 312: 137–141.

Moir, W. H., and E. P. Bachelard. 1969. Distribution of fine roots in three *Pinus radiata* plantations near Canberra, Australia. *Ecology* 50: 658–662.

Molofsky, J., and J.-B. Ferdy. 2005. Extinction dynamics in experimental metapopulations. *Proceedings of the National Academy of Sciences (USA)* 102: 3726–3731.

Mood, A. M., 1954. On the asymptotic efficiency of certain nonparametric two-sample tests. *Annals of Mathematical Statistics* 25: 514–522.

Morrow, R. L., et al. 2012. Influence of relative age on diagnosis and treatment of attention-deficit/hyperactivity disorder in children. *Canadian Medical Association Journal* 184: 755–762.

Moss-Racusin, C. A., J. F. Dovidio, V. L. Brescoll, M. J. Graham, and J. Handelsman. 2012. Science faculty's subtle gender biases favor male students. *Proceedings of the National Academy of Sciences (USA)* 109: 16474–16479.

Motulsky, H. J. 1999. Curvefit.com, the complete guide to nonlinear regression. Published by GraphPad Software Inc., San Diego, CA. Available at http://www.curvefit.com.

Moyle, L. C., M. S. Olson, and P. Tiffin. 2004. Patterns of reproductive isolation in three angiosperm genera. *Evolution* 58: 1195–1208.

Müller, M. S., et al. 2011. Maltreated nestlings exhibit correlated maltreatment as adults: evidence of a "cycle of violence" in Nazca boobies (*Sula granti*). *The Auk* 128: 615–619.

Murphy, B. F., Jr., and J. E. Heath. 1983. Temperature sensitivity in the prothoracic ganglion of the cockroach, *Periplaneta americana*, and its relationship to thermoregulation. *Journal of Experimental Biology* 105: 305–315.

Murzyn, E. 2008. Do we only dream in colour? A comparison of reported dream colour in younger and older adults with different experiences of black and white media. *Consciousness and Cognition* 17: 1228–1237.

Nagy, E. 2011. Sharing the moment: the duration of embraces in humans. *Journal of Ethology* 29: 389–393.

NASA 2004. Astronaut requirements. http://www.nasa.gov/audience/forstudents/postsecondary/features/F_Astronaut_Requirements.html. Accessed February 17, 2013.

Negre, B. S., et al. 2005. Conservation of regulatory sequences and gene expression patterns in the disintegrating *Drosophila Hox* gene complex. *Genome Research* 15: 692–700.

Negro, J. J., et al. 2002. An unusual source of essential carotenoids. *Nature* 416: 807–808.

Newberger, D. S. 2000. Down syndrome: prenatal risk assessment and diagnosis. *American Family Physician* 62: 825–832.

Newcomer, S. D., J. A. Zeh, and D. W. Zeh. 1999. Genetic benefits enhance the reproductive success of polyandrous females. *Proceedings of the National Academy of Sciences (USA)* 96: 10236–10241.

Nieuwenhuis, S., B. U. Forstmann, and E.-J. Wagenmakers. 2011. Erroneous analyses of interactions in neuroscience: a problem of significance. *Nature Neuroscience* 14: 1105–1107.

Nightingale, F. 1858. *Notes on Matters Affecting the Health, Efficiency and Hospital Administration of the British Army.* London: Harrison and Sons.

Noonan, J. P., et al. 2006. Sequencing and analysis of Neanderthal genomic DNA. *Science* 314: 1113–1118.

Norton, W. H. J., et al. 2011. Modulation of *Fgfr1a* signaling in zebrafish reveals a genetic basis for the aggression-boldness syndrome. *Journal of Neuroscience* 31: 13796–13807.

Nosil, P., and B. J. Crespi. 2006. Experimental evidence that predation promotes divergence in adaptive radiation. *Proceedings of the National Academy of Sciences (USA)* 103: 9090–9095.

Nunn, C. L., J. L. Gittleman, and J. Antonovics. 2000. Promiscuity and the primate immune system. *Science* 290: 1168–1170.

O'Reilly, M. S., et al. 1997. Endostatin: an endogenous inhibitor of angiogenesis and tumor growth. *Cell* 88: 277–285.

Olweus, D., A. Mattsson, D. Schalling, and H. Low. 1980. Testosterone, aggression, physical and personality dimensions in normal adolescent males. *Psychosomatic Medicine* 50: 261–272.

Online Mendelian Inheritance in Man. 2012. Asparagus, specific smell hypersensitivity. http://omim.org/entry/108390. Accessed December 10, 2012.

Oppliger, A., P. Christe, and H. Richner. 1996. Clutch size and malaria resistance. *Nature* 381: 565.

Orringer, J. S., et al. 2004. Treatment of acne vulgaris with a pulsed dye laser: a randomized controlled trial. *Journal of the American Medical Association* 291: 2834–2839.

Ozaki, M. A., et al. 2005. Ant nestmate and non-nestmate discrimination by a chemosensory sensillum. *Science* 309: 311–314.

Packer, C., and A. E. Pusey. 1983. Adaptations of female lions to infanticide by incoming males. *American Naturalist* 121: 716–728.

Palleroni, A., C. T. Miller, M. Hauser, and P. Marler. 2005. Prey plumage adaptation against falcon attack. *Nature* 434: 973–974.

Palmer, A. R. 1999. Detecting publication bias in meta-analysis: a case study of fluctuating asymmetry and sexual selection. *American Naturalist* 154: 220–233.

Palumbi, S. R. 1999. All males are not created equal: fertility differences depend on gamete recognition polymorphisms in sea urchins. *Proceedings of the National Academy of Sciences (USA)* 96: 12632–12637.

Panel on Scientific Responsibility and the Conduct of Research. 1992. *Responsible Science: Ensuring the Integrity of the Research Process, Vol. I.* Washington, DC: National Academy Press.

Papadakis, S., et al. 2011. A randomised controlled pilot study of standardised counselling and cost-free pharmacotherapy for smoking cessation among stroke and TIA patients. *BMJ Open* 1: e000366. Data at http://www.datadryad.org/handle/10255/dryad.35443.

Parmesan, C., et al. 1999. Poleward shifts in geographical ranges of butterfly species associated with regional warming. *Nature* 399: 579–583.

Parvanov, E. D., P. M. Petkov, and K. Paigen. 2012. *Prdm9* controls activation of mammalian recombination hotspots. *Science* 327: 835.

Patel, R. R., D. J. Murphy, and T. J. Peters. 2005. Operative delivery and postnatal depression: a cohort study. *British Medical Journal* 330: 879.

Patterson, T. B., and T. J. Givnish. 2002. Phylogeny, concerted convergence, and phylogenetic niche conservatism in the core Liliales: insights from *rbcL* and *ndhF* sequence data. *Evolution* 56: 233–252.

Pauling, L. 1986. *How to Live Longer and Feel Better.* New York: W. H. Freeman.

Pauw, A., J. Stofberg, and R. J. Waterman. 2009. Flies and flowers in Darwin's race. *Evolution* 63: 268–279.

Phillips, B. L., G. P. Brown, J. K. Webb, and R. Shine. 2006. Invasion and the evolution of speed in toads. *Nature* 439: 803.

Pinheiro, J. C., and D. M. Bates. 2000. *Mixed-Effects Models in S and S-Plus*. New York: Springer.

Pitkow, R. B. 1960. Cold death in the guppy. *Biological Bulletin* 119: 231–245.

Porter, R. H., and J. D. Moore. 1981. Human kin recognition by olfactory cues. *Physiology & Behavior* 27: 493–495.

Pounder, D. J. 1995. Postmortem changes and time of death. University of Dundee. http://www.dundee.ac.uk/forensicmedicine/llb/timedeath.htm.

Provine, R. R. 1989. Faces as releasers of contagious yawning: an approach to face detection using normal human subjects. *Bulletin of the Psychonomic Society* 27: 211–214.

Prugnolle, F., A. Manica, and F. Balloux. 2005. Geography predicts neutral genetic diversity of human populations. *Current Biology* 15: R159–R160.

Quaye, A. K., K. B. Laryea, and S. Abeney-Mickson. 2009. Soil water and nitrogen interaction effects on maize (*Zea mays* L.) grown on a vertisol. *Journal of Forestry, Horticulture, and Soil Science* 3: 1–11.

Quinn, G. P., and M. J. Keough. 2002. *Experimental Design and Data Analysis for Biologists*. Cambridge, UK: Cambridge University Press.

Ramadan, A. A., A. El-Keblawy, K. H. Shaltout, and J. Lovett-Doust. 1994. Sexual polymorphism, growth, and reproductive effort in Egyptian *Thymelaea hirsuta* (Thymelaeaceae). *American Journal of Botany* 81: 847–857.

Ramos, M., D. J. Irschick, and T. E. Christenson. 2004. Overcoming an evolutionary conflict: removal of a reproductive organ greatly increases locomotor performance. *Proceedings of the National Academy of Sciences (USA)* 101: 4883–4887.

Ramsey, P. H. 1980. Exact Type I error rates for robustness of Student's *t* test with unequal variances. *Journal of Educational Statistics* 5: 337–349.

Rannala, B., and Z. Yang. 1996. Probability distribution of molecular evolutionary trees: a new method of phylogenetic inference. *Journal of Molecular Evolution* 43: 304–311.

Raup, D. M., and J. J. Sepkoski Jr. 1982. Mass extinctions in the marine fossil record. *Science* 215: 1501–1503.

Rausher, M. D. 1984. Tradeoffs in performance in different hosts: evidence from within- and between-site variation in the beetle *Deloyala guttata*. *Evolution* 38: 582–595.

Reed, T. E., V. Grotan, S. Jenouvrier, B. Saether, and M. E. Visser. 2013. Population growth in a wild bird is buffered against phenological mismatch. *Science* 340: 488–491.

Reed T. E., V. Grotan, S. Jenouvrier, B. Saether, and M. E. Visser. 2013. Data from: Population growth in a wild bird is buffered against phenological mismatch. *Dryad Digital Repository*. doi:10.5061/dryad.8fc60

Reich, P. B., M. G. Tjoelker, J.-L. Machado, and J. Oleksyn. 2006. Universal scaling of respiratory metabolism, size and nitrogen in plants. *Nature* 439: 457–461.

Relyea, R. A. 2003. Predator cues and pesticides: a double dose of danger for amphibians. *Ecological Applications* 13: 1515–1521.

Ren, R. et al. 1991. The reconciliation behavior of golden monkeys (*Rhinopithecus roxellanae roxellanae*) in small breeding groups. *Primates* 32: 321–327.

Reusch, T. B. H., A. Ehlers, A. Hämmerli, and B. Worm. 2005. Ecosystem recovery after climatic extremes enhanced by genotypic diversity. *Proceedings of the National Academy of Sciences (USA)* 102: 2826–2831.

Ricaurte, G. A., J. Yuan, G. Hatzidimitriou, B. J. Cord and U. D. McCann. 2002. Severe dopaminergic neurotoxicity in primates after a common recreational dose regimen of MDMA ("ecstasy"). *Science* 297: 2260.

Ricaurte, G. A., J. Yuan, G. Hatzidimitriou, B. J. Cord and U. D. McCann. 2003. Retraction. *Science* 301: 1479.

Rich, S. K. 2011. A reply to B Wansink and CS Wansink. *International Journal of Obesity* 35: 462.

Richardson, D. S., J. Komdeur, and T. Burke. 2003. Altruism and infidelity among warblers. *Nature* 422: 580–581.

Richardson, T. E., and A. G. Stephenson. 1991. Effects of parentage, prior fruit set and pollen load on fruit and seed production in *Campanula americana* L. *Oecologia* 87: 80–85.

Ricklefs, R. E., and E. Bermingham. 2001. Nonequilibrium diversity dynamics of the Lesser Antillean avifauna. *Science* 294: 1522–1524.

Ridgway, P. F., et al. 2004. Perioperative diagnosis of cystosarcoma phyllodes of the breast may be enhanced by MIB-1 index. *Journal of Surgical Research* 122: 83–88.

Riskin, D. K., J. E. A. Bertram, and J. W. Hermanson. 2005. Testing the hindlimb-strength hypothesis: non-aerial locomotion by Chiroptera is not constrained by the dimensions of the femur or tibia. *Journal of Experimental Biology* 208: 1309–1319.

Rober, M. 2012. Turtles or snakes—which do cars hit more? Roadkill Experiment. http://www.youtube.com/watch?v=k-Fp7flAWMA&list =SP45865A763BAB32CA&index=5. Accessed January 10, 2012.

Rodgers, J. L., and D. Doughty. 2001. Does having boys or girls run in the family? *Chance Magazine* Fall: 8–13.

Roenneberg, T. 2012. *Internal Time: Chronotypes, Social Jet Lag, and Why You're So Tired.* Cambridge, MA: Harvard University Press. As quoted in http://www.brainpickings.org/index.php/2012/05/11/internal-time-till-roenneber/. Accessed April 24, 2014

Rogers, D. W., and R. Chase. 2001. Dart receipt promotes sperm storage in the garden snail *Helix aspersa*. *Behavioral Ecology and Sociobiology* 50: 122–127.

Rohmer, C., J. R. David, B. Moreteau, and D. Joly. 2004. Heat induced male sterility in *Drosophila melanogaster*: adaptive genetic variations among geographic populations and role of the Y chromosome. *Journal of Experimental Biology* 207: 2735–2743.

Roostalu, U., et al. 2007. Origin and expansion of haplogroup H, the dominant human mitochondrial DNA lineage in West Eurasia: the Near Eastern and Caucasian perspective. *Molecular Biology and Evolution* 24: 436–448.

Roscoe, J. T., and J. A. Byars. 1971. Sample size restraints commonly imposed on the use of the chi-square statistic. *Journal of the American Statistical Association* 66: 755–759.

Rose, G. A., and D. W. Kulka. 1999. Hyperaggregation of fish and fisheries: how catch-per-unit-effort increased as the northern cod (*Gadus morhua*) declined. *Canadian Journal of Fisheries and Aquatic Sciences* 56: 118–127.

Rosenblatt, K. A., J. R. Daling, C. Chen, K. J. Sherman, and S. M. Schwartz. 2004. Marijuana use and risk of oral squamous cell carcinoma. *Cancer Research* 64: 4049–4054.

Rossman, A. J. 1994. Televisions, physicians, and life expectancy. *Journal of Statistics Education* 2(2).

Rotton, J., and L. W. Kelly. 1985. Much ado about the full moon: a meta-analysis of lunar-lunacy research. *Psychological Bulletin* 97: 286–306.

Rousseeuw, P. J., and A. M. Leroy. 2003. *Robust Regression and Outlier Detection.* New York: Wiley.

Royal Society for the Prevention of Cruelty to Animals (RSPCA). 2005. Survey reveals the "pulling power" of pets. http://www.expertguide.com.au/news/article.aspx?ID=273. Accessed January 9, 2013.

Ruben, D. 2006. Highrise syndrome in cats: is your apartment fall-safe? http://www.petplace.com/article.aspx?id=2570. Accessed January 7, 2006

Ruff, C. B., E. Trinkaus, and T. W. Holliday. 1997. Body mass and encephalization in Pleistocene *Homo. Nature* 387: 173–176.

Runyon, J. B., M. C. Mescher, and C. M. De Moraes. 2006. Volatile chemical cues guide host location and host selection by parasitic plants. *Science* 313: 1964–1967.

Russell, A. F., and B. J. Hatchwell. 2001. Experimental evidence for kin-biased helping in a cooperatively breeding vertebrate. *Proceedings of the Royal Society of London, Series B: Biological Sciences* 268: 2169–2174.

Rutte, C., and M. Taborsky. 2007. Generalized reciprocity in rats. *PLoS Biology* 7: 1421–1425.

Rypstra, A. L. 1979. Foraging flocks of spiders: a study of aggregate behavior in *Cyrtophora citricola* Forskål (Araneae; Araneidae) in West Africa. *Behavioral Ecology and Sociobiology* 5: 291–300.

Sacktor, N. C., et al. 2000. Improvement in HIV-associated motor slowing after antiretroviral therapy including protease inhibitors. *Journal of NeuroVirology* 6: 84–88.

Sakaue, M., et al. 2001. Bisphenol-A affects spermatogenesis in the adult rat even at a low dose. *Journal of Occupational Health* 43: 185–190.

Salo, P., E. Korpimäki, P. B. Banks, M. Nordström, and C. R. Dickman. 2007. Alien predators are more dangerous than native predators to prey populations. *Proceedings of the Royal Society of London, Series B: Biological Sciences* 274: 1237–1243.

Sandell, M., J. Agrell, S. Erlinge, and J. Nelson. 1991. Adult philopatry and dispersal in the field vole *Microtus agrestis*. *Oecologia* 86: 153–158.

Sandidge, J. S. 2003. Scavenging by brown recluse spiders. *Nature* 426: 30.

Sauer, J. R., J. E. Hines, and J. Fallon. 2003. *The North American Breeding Bird Survey, Results and Analysis 1966–2002. Version 2003.1.* Laurel, MD: USGS Patuxent Wildlife Research Center.

Savage, V. M., et al. 2004. The predominance of quarter power scaling in biology. *Functional Ecology* 18: 257–282.

Scantlebury, M., J. R. Speakman, M. K. Oosthuizen, T. J. Roper, and N. C. Bennett. 2006. Energetics reveals physiologically distinct castes in a euso-cial mammal. *Nature* 440: 795–797.

Schaal, T., and G. Smith. 2000. Do baseball players regress toward the mean? *American Statistician* 54: 231–235.

Scheel, D. 1993. Profitability, encounter rates, and prey choice of African lions. *Behavioral Ecology* 4: 90–97.

Schluter, D. 1988. The evolution of finch communities on islands and continents: Kenya vs. Galápagos. *Ecological Monographs* 58: 229–249.

Schmid, P. E., M. Tokeshi, and J. M. Schmid-Araya. 2000. Relation between population density and body size in stream communities. *Science* 289: 1557–1560.

Schubert, E. L., et al. 1997. *BRCA2* in American families with four or more cases of breast or ovarian cancer: recurrent and novel mutations, variable expression, penetrance, and the possibility of families whose cancer is not attributable to *BRCA1* or *BRCA2*. *American Journal of Human Genetics* 60: 1031–1040.

Schwerzmann, M., et al. 2005. Prevalence and size of directly detected patent foramen ovale in migraine with aura. *Neurology* 65: 1415–1418.

Seaton, E. D., et al. 2003. Pulsed-dye laser treatment for inflammatory acne vulgaris: randomised controlled trial. *The Lancet* 362: 1347–1352.

Secondi J., V. Lepetz, and M. Théry. 2012. Male attractiveness is influenced by UV wavelengths in a newt species but not in its close relative. *PLoS ONE* 7: e30391.

Seltzer, C. C., R. Bosse, and A. J. Garvey. 1974. Mail response by smoking status. *American Journal of Epidemiology* 100: 453–477.

Shapiro, A. K., and E. Shapiro. 1997. The placebo: is it much ado about nothing? In *The Placebo Effect: An Interdisciplinary Exploration,* ed. A. Harrington. Cambridge, MA: Harvard University Press.

Shaw, D. J., B. T. Grenfell, and A. P. Dobson. 1998. Patterns of macroparasite aggregation in wildlife host populations. *Parasitology* 117: 597–610.

Shaw, M. R., et al. 2002. Grassland responses to global environmental changes suppressed by elevated CO_2. *Science* 298: 1987–1990.

Sheriff, M. J., et al. 2011. Phenological variation in annual timing of hibernation and breeding in nearby populations of Arctic ground squirrels. *Proceedings of the Royal Society B: Biological Sciences* 278: 2369–2375.

Sherwood, C. C., et al. 2006. Evolution of increased glia–neuron ratios in the human frontal cortex. *Proceedings of the National Academy of Sciences (USA)* 103: 13606–13611.

Sheth, S. A., et al. 2004. Linear and nonlinear relationships between neuronal activity, oxygen metabolism, and hemodynamic responses. *Neuron* 42: 347–355.

Shine, R., B. Phillips, H. Waye, M. LeMaster, and R. T. Mason. 2001. Benefits of female mimicry in snakes. *Nature* 414: 267.

Shoemaker, A. L. 1996. What's normal?—Temperature, gender, and heart rate. *Journal of Statistics Education* 4(2).

Simard, S. W., et al. 1997. Net transfer of carbon between ectomycorrhizal tree species in the field. *Nature* 388: 579–582.

Simmons, L. W., and B. Roberts. 2005. Bacterial immunity traded for sperm viability in male crickets. *Science* 309: 2031.

Simons, D. J., and C. F. Chabris. 1999. Gorillas in our midst: sustained inattentional blindness for dynamic events. *Perception* 28: 1059–1074.

Simpson, S. J., E. Despland, B. F. Hägele, and T. Dodgson. 2001. Gregarious behavior in desert locusts is evoked by touching their back legs. *Proceedings of the National Academy of Sciences (USA)* 98: 3895–3897.

Singaravelan, N., G. Nee'man, M. Inbar, and I. Izhaki. 2005. Feeding responses of free-flying honeybees to secondary compounds mimicking floral nectars. *Journal of Chemical Ecology* 31: 2791–2804.

Singh, N. P., M. T. McCoy, R. R. Tice, and E. L. Schneider. 1988. A simple technique for quantitation of low levels of DNA damage in individual cells. *Experimental Cell Research* 175: 184–191.

Sloan, R. E. G., and W. R. Keatinge. 1973. Cooling rates of young people swimming in cold water. *Journal of Applied Physiology* 35: 371–375.

Smallwood, R. H., D. J. Morgan, G. W. Mihaly, and R. A. Smallwood. 1998. Lack of linear correlation between hepatic ligand uptake rate and unbound ligand concentration does not necessarily imply receptor-mediated uptake. *Journal of Pharmacokinetics and Pharmacodynamics* 16: 397–411.

Smith, D. L., B. Lucey, L. A. Waller, J. E. Childs, and L. A. Real. 2002. Predicting the spatial dynamics of rabies epidemics on heterogeneous landscapes. *Proceedings of the National Academy of Sciences (USA)* 99: 3668–3672.

Smith, F. A., et al. 2003. Body mass of late Quaternary mammals. *Ecology* 84: 3403–3403. Data are from http://www.esapubs.org/archive/ecol/ E084/094/metadata.htm.

Smith, T. B. 1993. Disruptive selection and the genetic basis of bill size polymorphism in the African finch, *Pyrenestes*. *Nature* 363: 618–620.

Šobotník, J., et al. 2012. Explosive backpacks in old termite workers. *Science* 337: 436–436.

Socha, J. J. 2002. Gliding flight in the paradise tree snake. *Nature* 418: 603–604.

Sokal, R. R., and F. J. Rohlf. 1995. *Biometry: The Principles and Practice of Statistics in Biological Research*, 3rd ed. New York: W. H. Freeman.

Sokal, R. R., and F. J. Rohlf. 2012. *Biometry*. New York: W. H. Freeman.

Song, Y., et al. 2011. Adaptive introgression of anticoagulant rodent poison resistance by hybridization between old world mice. *Current Biology* 21: 1296–1301.

Souman, J. L., I. Frissen, M. N. Sreenivasa, and M. O. Ernst. 2009. Walking straight into circles. *Current Biology* 19: 1–5.

Spalding, K. L., B. A. Buchholz, L.-E. Bergman, H. Druid, and J. Frisén. 2005. Forensics: age written in teeth by nuclear tests. *Nature* 437: 333–334.

Spence, D. S., and E. A. Thompson. 2005. Homeopathic treatment for chronic disease: a 6-year, university-hospital outpatient observational study. *Journal of Alternative and Complementary Medicine* 11: 793–798.

Spinner-Hansen, L., S. Fernández González, S. Toft, and T. Bilde. 2008. Thanatosis as an adaptive male mating strategy in the nuptial gift–giving spider *Pisaura mirabilis*. *Behavioral Ecology* 19: 546–551.

Srivastava, D. S., and J. H. Lawton. 1998. Why more productive sites have more species: an experimental test of theory using tree-hole communities. *American Naturalist* 152: 510–529.

St. Clair, D., et al. 2005. Rates of adult schizophrenia following prenatal exposure to the Chinese famine of 1959–1961. *Journal of the American Medical Association* 294: 557–562.

Stafne, G. M., and P. R. Manger. 2004. Predominance of clockwise swimming during rest in Southern Hemisphere dolphins. *Physiology & Behavior* 82: 919–926.

Stanford Blood Center. 2012. http://bloodcenter. stanford.edu/about_blood/blood_types.html. Accessed December 4, 2012.

Steinmetz, L. M. et al. 2002. Systematic screening for human disease genes in yeast. *Nature Genetics* 31:400–404.

Stephens, R., J. Atkins, and A. Kingston. 2009. Swearing as a response to pain. *Neuroreport* 20: 1056–1060.

Stern, J. M., and R. John Simes. 1997. Publication bias: evidence of delayed publication in a cohort study of clinical research projects. *British Medical Journal* 315: 640–645.

Stigler, S. M. 1977. Do robust estimators work with real data? *Annals of Statistics* 5: 1055–1078.

Storey, J. D., and R. Tibshirani. 2003. Statistical significance for genomewide studies. *Proceedings of the National Academy of Sciences (USA)* 100: 9440–9445.

Styf, J. R., et al. 1997. Height increase, neuromuscular function, and back pain during 6 degrees head-down tilt with traction. *Aviation, Space, and Environmental Medicine* 68: 24–29.

Suárez-Rodríguez, M., I. López-Rull, and C. Macías Garcia. 2013. Incorporation of cigarette butts into nests reduces nest ectoparasite load in urban birds: new ingredients for an old recipe? *Biology Letters* 9: 20120931.

Sunda, W. G., and S. A Huntsman. 1997. Inter-related influence of iron, light and cell size on marine phytoplankton growth. *Nature* 390: 389–392.

Svanbäck, R., and D. I. Bolnick. 2007. Intraspecific competition drives increased resource use diversity within a natural population. *Proceedings of the Royal Society of London, Series B: Biological Sciences* 274: 839–844.

Svenson, N. 2006. Blink-free photos, guaranteed. *Velocity*, June 2006. http://velocity.ansto.gov.au/velocity/ans0011/article_06.asp

Sweeting, M. J., A. J. Sutton, and P. C. Lambert. 2004. What to add to nothing? Use and avoidance of continuity corrections in meta-analysis of sparse data. *Statistics in Medicine* 23: 1351–1375.

Tabershaw, I. R., and S. H. Lamm. 1977. Benzene and leukæmia. *The Lancet* 310: 867–868.

Tábora, N., A. Zelaya, J. Bakkers, W. J. G. Melchers, and A. Ferrera. 2005. *Chlamydia trachomatis* and genital human papillomavirus infections in female university students in Honduras. *American Journal of Tropical Medicine and Hygiene* 73: 50–53.

Tatem, A. J., C. A. Guerra, P. M. Atkinson, and S. I. Hay. 2004. Athletics: momentous sprint at the 2156 Olympics? *Nature* 431: 525.

Tattersall, G. J., W. K. Milsom, A. S. Abe, S. P. Brito, and D. V. Andrade. 2004. The thermo-genesis of digestion in rattlesnakes. *Journal of Experimental Biology* 207: 579–585.

Templer, D. E., D. M. Veleber, and R. K. Brooner. 1982. Geophysical variables and behavior VI. Lunar phase and accident injuries: a difference between day and night. *Perceptual and Motor Skill* 55: 280–282.

The World Bank. 2013. Health expenditure per capita (current US$). http://data.worldbank.org/indicator/SH.XPD.PCAP/. Accessed March 12, 2013.

Thomas, J. A., et al. 2004. Comparative losses of British butterflies, birds, and plants and the global extinction crisis. *Science* 303: 1879–1881.

Tilman, D., P. B. Reich, and J. M. H. Knops. 2006. Biodiversity and ecosystem stability in a decade-long grassland experiment. *Nature* 441: 629–632.

Tkachev, D., et al. 2003. Oligodendrocyte dysfunction in schizophrenia and bipolar disorder. *The Lancet* 362: 798–805.

Tomkins, J. L., and G. S. Brown. 2004. Population density drives the local evolution of a threshold dimorphism. *Nature* 431: 1099–1103.

Trichopoulou, A., et al. 2005. Modified Mediterranean diet and survival: EPIC-elderly prospective cohort study. *British Medical Journal* 330: 991–997.

Trinkaus, J., and K. Dennis. 1991. Taste preference for Brussels sprouts: an informal look. *Psychological Reports* 69: 1165–1166. As cited at *Discover Magazine*, http://blogs.discovermagazine.com/discoblog/2011/11/29/ncbi-rofl-taste-preference-for-brussels-sprouts-an-informal-look/. Accessed December 14, 2012.

Trites, A. W. 1996. Physical growth of northern fur seals (*Callorhinus ursinus*): seasonal fluctuations and migratory influences. *Journal of Zoology, London* 238: 459–482.

Tsuchida, T., et al. 2010. Symbiotic bacterium modifies aphid body color. *Science* 330: 1102–1104.

Tufte, E. R. 1983. *The Visual Display of Quantitative Information*. Cheshire, CT: Graphics Press.

Tufte, E. R. 1997. *Visual Explanations: Images and Quantities, Evidence and Narrative*. Cheshire, CT: Graphics Press.

Turner, D. C. 1975. *The Vampire Bat: A Field Study in Behavior and Ecology*. Baltimore, MD: Johns Hopkins Press.

United Nations Statistics Division. 2004. Demographic and social statistics. http://unstats.un.org/unsd/demographic/products/socind/population.htm.

U.S. Department of Transportation Traffic Safety Facts. 1999. National Center for Statistics and Analysis.

U.S. Fish and Wildlife Service. 2012. Species reports. Environmental Conservation Online System. http://ecos.fws.gov/tess_public/pub/boxScore.jsp. Accessed October 15, 2012.

Valtonen, H., K. Suominen, T. Partonen, A. Ostamo, and J. Lönnqvist. 2006. Time patterns of attempted suicide. *Journal of Affective Disorders* 90: 201–207.

Van Damme, L., et al. 2002. Effectiveness of COL-1492, a nonoxynol-9 vaginal gel, on HIV-1 transmission in female sex workers: a randomised controlled trial. *The Lancet* 360: 971–977.

van de Ven, N., L. van Rijswijk, and M. M. Roy. 2011. The return trip effect: Why the return trip often seems to take less time. *Psychonomic Bulletin & Review* 18: 827–832.

van Hylckama Vlieg, A., F. M. Helmerhorst, J. P. Vandenbroucke, C. J. M. Doggen, and F. R. Rosendaal. 2009. The venous thrombotic risk of oral contraceptives, effects of oestrogen dose and progestogen type: results of the MEGA case-control study. *British Medical Journal* 339: b2921.

Venables, W. N., and B. D. Ripley. 2002. *Modern Applied Statistics with S-PLUS*. New York: Springer Publishing.

Ventura, S. J., J. A. Martin, S. C. Curtin, F. Menacker, and B. E. Hamilton. 2001. Births: final data for 1999. *National Vital Statistics Reports* 49(1).

Villeneuve, P. J., and Y. Mao. 1994. Lifetime probability of developing lung cancer, by smoking status, Canada. *Canadian Journal of Public Health* 85: 385–388.

Visscher, P. K., and R. Dukas. 1997. Survivorship of foraging honey bees. *Insectes Sociaux* 44: 1–5.

Vnuk, D., et al. 2004. Feline high-rise syndrome: 119 cases (1998–2001). *Journal of Feline Medicine and Surgery* 6: 305–312.

Volkow, N. D., et al. 1997. Relationship between subjective effects of cocaine and dopamine transporter occupancy. *Nature* 386: 827–830.

vom Saal, F. S., and F. H. Bronson. 1980. Sexual characteristics of adult female mice are correlated with their blood testosterone levels during prenatal development. *Science* 208: 597–599.

von Arx, M., J. Goyret, G. Davidowitz, and R. A. Raguso. 2012. Floral humidity as a reliable sensory cue for profitability assessment by nectar-foraging hawkmoths. *Proceedings of the National Academy of Sciences (USA)* 109: 9471–9476.

von Beeren, C., S. Schulz, R. Hashim, and V. Witte. 2011. Acquisition of chemical recognition cues facilitates integration into ant societies. *BMC Ecology* 11: 30.

Vyas, A., S.-K. Kim, N. Giacomini, J. C. Boothroyd, and R. M. Sapolsky. 2007. Behavioral changes induced by *Toxoplasma* infection of rodents are highly specific to aversion of cat odors. *Proceedings of the National Academy of Sciences (USA)* 104: 6442–6447.

Wager, T. D., et al. 2004. Placebo-induced changes in fMRI in the anticipation and experience of pain. *Science* 303: 1162–1167.

Waide, R. B., and W. B. Reagan, eds. 1996. *The Food Web of a Tropical Rainforest*. Chicago: University of Chicago Press.

Waldron, A. 2007. Geographic range size evolution. *American Naturalist* 170: 221–231.

Walker, B. G., P. D. Boersma, and J. C. Wingfield. 2005. Physiological and behavioral differences in Magellanic penguin chicks in undisturbed and tourist-visited locations of a colony. *Conservation Biology* 19: 1571–1577.

Walters, C. and J.-J. Maguire. 1996. Lessons for stock assessment from the northern cod collapse. *Reviews in Fish Biology and Fisheries* 6: 125–137.

Wang, P. J., J. R. McCarrey, F. Yang, and D. C. Page. 2001. An abundance of X-linked genes expressed in spermatogonia. *Nature Genetics* 27: 422–426.

Wansink, B., and C. S. Wansink. 2010. The largest Last Supper: depictions of food portions and plate size increased over the millennium. *International Journal of Obesity* 34: 943–944.

Warner, S. L. 1965. Randomized response: a survey technique for eliminating evasive answer bias. *Journal of the American Statistical Association* 60: 63–69.

Waters, A. J., M. J. Jarvis, and S. R. Sutton. 1998. Nicotine withdrawal and accident rates. *Nature* 394: 137.

Waters, J. R., K. S. McKelvey, D. L. Luoam, and C. J. Zabel. 1997. Truffle production in old-growth and mature fir stands in northeastern California. *Forest Ecology and Management* 96: 155–166.

Weather Source. 2009. http://weather-warehouse.com/WeatherHistory/PastWeatherData_DeathValley_DeathValley_CA_July.html. Accessed February 19, 2013.

Weisstein, E. W. 2014. Hypothesis testing. From MathWorld—A Wolfram Web Resource: http://mathworld.wolfram.com/HypothesisTesting.html. Accessed December 21, 2012.

Wenzel, R. P. 2004. The antibiotic pipeline—challenges, costs, and values. *New England Journal of Medicine* 251: 523–526.

Werren, J. H. 1980. Sex ratio adaptations to local mate competition in a parasitic wasp. *Science* 208: 1157–1159.

West, S. A., and B. C. Sheldon. 2002. Constraints in the evolution of sex ratio adjustment. *Science* 295: 1695–1688.

Wheeler, B. 2005. The SuppDists package, version 1.0-13, April 7, 2005. Gnu Public License version 2. http://cran.r-project.org/src/contrib/SuppDists_1.1-9.1.tar.gz

Wheelwright, N. T., and B. A. Logan. 2004. Previous-year reproduction reduces photosynthetic capacity and slows lifetime growth in females of a neotropical tree. *Proceedings of the National Academy of Sciences (USA)* 101: 8051–8055.

White, A. R., H. Rampes, and J. L. Campbell. 2006. Acupuncture and related interventions for smoking cessation. *Cochrane Database of Systematic Reviews* Issue 1, Art. No.: CD000009. doi:10.1002/14651858.CD000009.pub2

White, S. A., T. Nguyen, and R. D. Fernald. 2002. Social regulation of gonadotropin-releasing hormone. *Journal of Experimental Biology* 205: 2567–2581.

Whitman, K., A. M. Starfield, H. S. Quadling, and C. Packer. 2004. Sustainable trophy hunting of African lions. *Nature* 428: 175–178.

Whitney, W. O., and C. J. Mehlhaff. 1987. High-rise syndrome in cats. *Journal of the American Veterinary Medical Association* 191: 1399–1403.

Widén, B. 1993. Demographic and genetic effects on reproduction as related to population size in a rare, perennial herb, *Senecio integrifolius* (Asteraceae). *Biological Journal of the Linnean Society* 50: 179–195.

Wikelski, M., and C. Thom. 2000. Marine iguanas shrink to survive El Niño. *Nature* 403: 37–38.

Wilkinson, G. S. 1984. Reciprocal food sharing in the vampire bat. *Nature* 308: 181–184.

Willerman, L., R. Schultz, J. N. Rutledge, and E. Bigler. 1991. In vivo brain size and intelligence. *Intelligence* 15: 223–228.

Williams, T. M., L. A. Fuiman, M. Horning, and R. W. Davis. 2004. The cost of foraging by a marine predator, the Weddell seal *Leptonychotes weddellii*: pricing by the stroke. *Journal of Experimental Biology* 207: 973–982.

Wilson, K. M., et al. 2011. Coffee consumption and prostate cancer risk and progression in the Health Professionals Follow-up Study. *Journal of the National Cancer Institute* 103: 1–9.

Winder, B., K. Ridgway, A. Nelson, and J. Baldwin. 2002. Food and drink packaging: who is complaining and who should be complaining. *Applied Ergonomics* 33: 433–438.

Wiseman, R., and P. Lamont. 1996. Unravelling the Indian rope-trick. *Nature* 383: 212–213.

Wood, E., et al. 2001. Unsafe injection practices in a cohort of injection drug users in Vancouver: could safer injecting rooms help? *Canadian Medical Association Journal* 165: 405–410.

Wright, K. P., Jr., and C. A. Czeisler 2002. Absence of circadian phase resetting in response to bright light behind the knees. *Science* 297: 571.

Yamamoto, Y., and W. R. Jeffery. 2000. Central role for the lens in cavefish eye degeneration. *Science* 289: 631–633.

Yashin, A. I., et al. 2000. Genes and longevity: lessons from studies of centenarians. *Journal of Gerontology* 55A: B319–B328.

Yates, F. 1984. Tests of significance for 2 x 2 contingency tables (with discussion). *Journal of the Royal Statistical Society, Series A* 147: 426–463.

Yeh, P. J. 2004. Rapid evolution of a sexually selected trait following population establishment in a novel habitat. *Evolution* 58: 166–174.

Yereli, K., I. C. Balcioğlu, and A. Özbilgin. 2006. Is *Toxoplasma gondii* a potential risk for traffic accidents in Turkey? *Forensic Science International* 163: 34–37.

Young, A. J., et al. 2006. Stress and the suppression of subordinate reproduction in cooperatively breeding meerkats. *Proceedings of the National Academy of Sciences (USA)* 103: 12005–12010.

Young, D. C., and E. M. Hade. 2004. Holidays, birthdays, and postponement of cancer death. *Journal of the American Medical Association* 292: 3012–3016.

Young, E. 2002. Intuitive people worse at lying. *New Scientist.* (http://www.newscientist.com/article.ns?id=dn2054). Accessed February 5, 2012.

Young, K. V., E. D. Brodie Jr., and E. D. Brodie III. 2004. How the horned lizard got its horns. *Science* 304: 65.

Zahran, S., H. W. Mielke, C. R. Gonzales, E. T. Powell, and S. Weiler. 2010. New Orleans before and after hurricanes Katrina/Rita: A quasi-experiment of the association between soil lead and children's blood lead. *Environmental Science and Technology.* 44:4433–4440.

Zhou, Z., et al. 2008. Genetic variation in human NPY expression affects stress response and emotion. *Nature* 452: 997–1001.

Zimmerman, D. W. 2003. A warning about the large-sample Wilcoxon-Mann-Whitney test. *Understanding Statistics* 2: 267–280.

Answers to practice problems

Chapter 1

1. (a) Discrete. **(b)** Continuous. **(c)** Discrete. **(d)** Continuous.

2. (a) Collectors do not usually sample randomly, but prefer rare and unusual specimens. Therefore, the moths are probably not a random sample. **(b)** A sample of convenience. **(c)** Bias (rare color types are likely to be overrepresented in the sample compared with the population because collectors spend a disproportionate amount of effort searching for them).

3. Accuracy.

4. (a) U.S. army personnel stationed in Iraq. **(b)** Yes. The number of subjects interviewed was a random sample of only 100 from the population; therefore, the responses of the particular individuals who happened to be sampled will differ from the population of interest by chance. **(c)** The advantage of random sampling is that it minimizes bias and allows the precision of estimates of stress levels to be measured. **(d)** Reduce sampling error.

5. (a) The parameter being estimated is the number of small mammal species of Kruger National Park. **(b)** No, the sample is not likely to be random. In a random sample every individual has the same chance of being selected, but some small mammals might be easier to trap than others (e.g., trapping only at night might miss individuals active only in the daytime). Also, animals caught in the same trap might not be independent if they are related or live near one another (this is harder to judge). **(c)** By chance, not all species will be represented in the sample. In this case, the number of species in the survey would likely underestimate the total number of species present. Hence the estimate of total species number would be biased.

6. (a) Neurons were not sampled randomly. Multiple neurons from the same monkey are not independent—they are likely to be more similar to one another in their measurements than neurons chosen randomly from the population. **(b)** The calculation of the precision of estimates would be erroneous.

7. (a) Discrete. **(b)** Continuous. **(c)** Continuous. **(d)** Discrete. **(e)** Continuous.

8. (a) Explanatory: reason for fetus loss, response: psychological consequences. **(b)** Observational, because the researcher did not assign treatments (reasons for fetus loss) to subjects.

9. (a) Altitude: explanatory; growth rate: response. Observational. **(b)** Drug treatment: explanatory; insulin release rate: response. Experimental. **(c)** Schizophrenia condition: explanatory; frequency of drug use: response. Observational. **(d)** Leg removal treatment: explanatory; survival: response. Experimental. **(e)** Therapy group: explanatory; communication ability: response. Experimental.

10. (a) Altitude: categorical (because there were two groups, with high and low altitude); growth rate: numerical. **(b)** Drug treatment: categorical; insulin release rate: numerical. **(c)** Schizophrenia condition: categorical; frequency of drug use: numerical. **(d)** Leg removal treatment: categorical; survival: categorical. **(e)** Therapy group: categorical; communication ability: numerical.

11. (a) No, because even though the original 500 was sampled randomly, the 80 responders are not a random sample—those who volunteered to fill out the form may differ from the general population. **(b)** Volunteer bias.

12. (a) Estimation. **(b)** Hypothesis testing. **(c)** Estimation. **(d)** Estimation. **(e)** Hypothesis testing. **(f)** Hypothesis testing.

13. (a) Subspecies (group) and wavelength of maximum sensitivity. **(b)** Group (subspecies) is the explanatory variable and wavelength is the response variable. **(c)** Observational, because the researcher has no control of which individuals belong to which subspecies.

Chapter 2

1. (a) 25%. **(b)** 33%. **(c)** 33%.

2.

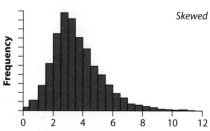

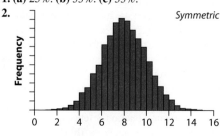

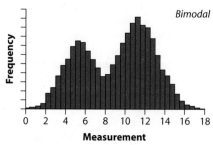

Your answers may vary.
(a) Modes: 7.5–8, 2.5–3, and 11–11.5.
(b) Positive skewed.
(c) The bimodal distribution in this example is left-skewed.

3. The lower histogram is correct. Histograms use area, not height, to represent frequency.

4. (a)

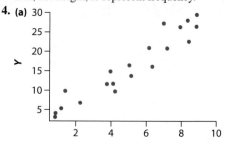

(b)

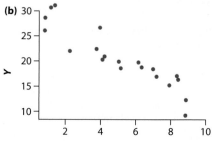

(c)

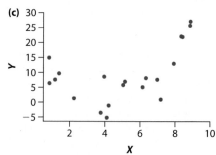

5. Contingency table.

	Source of fry		
	Hatchery	Wild	Total
Survived	27	51	78
Perished	3973	3949	7922
Total	4000	4000	8000

6. (a) The following table orders the taxa from those with the most endangered species to those with the fewest endangered species. (Other orderings could make sense, such as by phylum.)

Taxon	Number of species
Plants	782
Fish	152
Birds	93
Mammals	85
Clams	83
Insects	66
Snails	40
Reptiles	36
Amphibians	26
Crustaceans	22
Arachnids	12

(b) Frequency table.
(c) Bar graph, because data are categorical.

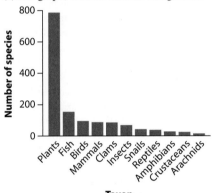

(d) The baseline should be zero so that bar height and area correspond to frequency.
(e)

Taxon	Relative frequency
Flowering Plants	0.56
Fish	0.11
Birds	0.07
Mammals	0.06
Clams	0.06
Insects	0.05
Snails	0.03
Reptiles	0.03
Amphibians	0.02
Crustaceans	0.02
Arachnids	0.01

7. (a) The variables are year and incidence of schizophrenia. **(b)** Year is numerical and discrete. Incidence of schizophrenia is categorical—nominal. **(c)** Contingency table. (*See table at bottom of page.*) **(d)** Proportions: 1956: 0.0082, 1960: 0.0140, and 1965: 0.0083. Other variations of the line graph below are valid (e.g., famine state could be the explanatory variable instead of year).

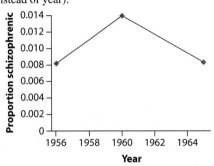

Pattern revealed: incidence of schizophrenia highest in 1960, the famine year.
8. (a) Grouped histograms. **(b)** Groups are disease types (diseases transmitted directly from one individual to another and diseases transferred by insect vectors). **(c)** The variable is the virulence of the disease. It is a continuous numerical variable. **(d)** Relative frequency, the fraction of diseases occurring in each interval of virulence. **(e)** Directly transmitted diseases tend to be less virulent than insect-transmitted diseases. Human diseases transmitted directly more frequently have low virulence, and less frequently have high virulence, than diseases transmitted by insect vector.
9. (a) Map. **(b)** Date of first appearance of rabies (measured by the number of months following March 1, 1991). **(c)** Geographic location—the township. **(d)** Roughly from west to east (from the SW of the state north, then east and south).
10. (a) Husbands of adopted women resemble a woman's adoptive father more than they resemble the women themselves or the women's adoptive mother. **(b)** First, the graph fails to "show

TABLE FOR PROBLEM 7(c)

	1956 (pre-famine)	1960 (famine)	1965 (post-famine)	Total
Schizophrenic	483	192	695	1370
Non-schizophrenic	58,605	13,556	82,841	155,002
Total	59,088	13,748	83,536	156,372

ANSWERS

the data" by using a bar graph that indicates the average score instead of a strip chart or box plot that shows the data. Second, by using an inappropriate baseline, the graph does not "represent magnitudes honestly."

11. (a) Mosaic plot. The explanatory variable is fruit set in previous years. The response variable is fruiting in the given year. Both variables are categorical. **(b)** Line graph. The explanatory variable is year. The response variable is the density of the wood. Both variables are numerical. **(c)** Strip chart. Genotype is explanatory, categorical. Gene expression is response, numerical.

12. (a) Contingency table.

(b)

Convictions

No convictions

Income level

(c) Categorical, ordered. Groups should be arranged by increasing income. **(d)** The relative frequency of conviction decreases as available income increases. **(e)** The mosaic plot, which made it easier to see the pattern. Whereas the table gives the frequencies, the graph visualizes the association between the variables.

13. (a) Multiple histograms. Explanatory variable: genotype at PTC gene. Response variable: taste sensitivity score. Genotype is categorical variable, taste sensitivity is numerical. **(b)** Scatter plot. Explanatory variable: migratory activity of parents. Response variable: migratory activity of offspring. Both variables are numerical. **(c)** Strip chart. Treatment is explanatory, categorical. Appendage size is response, numerical. **(d)** Grouped bar graph. Explanatory variable: HIV

status. Response variable: needle sharing. Both variables are categorical.

14. (a) "Show the data"—only the averages in each group are shown, not the data; "Represent magnitudes honestly"—bar area and height do not represent magnitude because of the non-zero baseline; "Draw graphical elements clearly"—horizontal axis is mislabeled and vertical axis label is missing. **(b)** Strip chart, box plot or multiple histograms. **(c)** Proposed starting salaries are higher on average for the same application if a male name is used.

15. (a) Box plot. **(b)** Median. **(c)** Upper and lower quartiles. **(d)** Whiskers, which extend to the smallest and largest non-extreme values in the data. **(e)** Yes, air source and flower visits appear to be associated (hawk moth visits to flowers differ between treatments—they visit the flowers with humidified air more frequently).

16. (a) Mosaic plot. Yes, there is a clear difference between groups in the relative frequencies of A and B. **(b)** Scatter plot. Yes, there is a negative correlation: Y tends to be smaller when X is large than when X is small. **(c)** Scatter plot. No, there is no tendency for Y to be greater or smaller when X is large than when X is small. **(d)** Strip chart. Yes, there is a clear difference between groups in Y. **(e)** Box plot. Yes, the first group is distinct from the other two. **(f)** Mosaic plot. No, the relative frequencies of A and B are identical in the two groups.

17. (a) Best method is scatter plot:

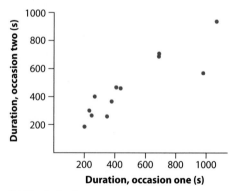

(b) Yes, indeed.

18. (a) A histogram is appropriate, though a box plot would also suffice.

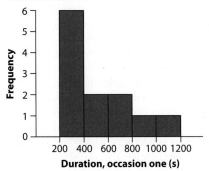

(b) Positive skew.

Chapter 3

1. (a) $n = 9$. **(b)** $\sum_{i=1}^{n} Y_i = 1179$.

(c) $\dfrac{\sum_{i=1}^{n} Y_i}{n} = 131$ mm Hg.

(d) $\sum_{i=1}^{n} (Y_i - \bar{Y})^2 = 2036$.

(e) $s^2 = \dfrac{\sum_{i=1}^{n} (Y_i - \bar{Y})^2}{n - 1} = 254.5$ mm Hg^2.

(f) $s = \sqrt{s^2} = \sqrt{254.5} = 15.95$ mm Hg.

(g) $CV = 100\% \dfrac{s}{\bar{Y}} = 12.2\%$.

2. (a) There are 101 data points, an odd number, and so the middle data point will be the 51st. In the ordered set of data, this is 123.
(b) $k = 0.75 \times 101 = 75.75$. Rounding up, we want the 76th data point. This is 131.
(c) $j = 0.25 \times 121 = 25.25$. Rounding up, we want the 26th data point. This is 115. **(d)** The interquartile range is the upper quartile minus the lower quartile, or $131 - 115 = 16$. **(e)** 1.5 times the *IQR* is 25. The upper quartile (131)

plus 24 is 155. No. Therefore there will be a dot outside the "whiskers." **(f)** $115 - 24 = 91$. No.
(g)

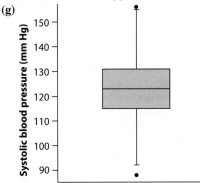

3. (a) The number of doctors studied who sterilized a given percentage of patients under age 25.
(b) The median would be more informative. The mean would be sensitive to the presence of the outlier (sterilizing more than 30% of female patients under 25), whereas the median is unaffected. **(c)** Yes, because there is a clear outlier.

4. (a) Box plot. **(b)** Median body mass of the mammals in each group. **(c)** The first and third quartiles of body mass in each group. **(d)** "Extreme values," those lying farther than 1.5 times the interquartile range from the box edge. **(e)** Whiskers. They extend to the smallest and largest values in the data, excluding extreme values (those lying farther than 1.5 times the interquartile range from the box edge). **(f)** Living mammals have the smallest median body size (with a log body mass of about 2), and mammals that went extinct in the last ice age had the largest median size (around 5.5). Mammals that went extinct recently were intermediate in size, with a median of around 3.2. **(g)** The body size distribution of living mammals is right-skewed (long tail toward larger values), whereas the size frequency distribution in mammals that went extinct in the last ice age is left-skewed. The frequency distribution of sizes is nearly symmetric in mammals that went extinct recently. **(h)** One way to answer this is to use the interquartile range as a measure of spread, in which case the mammals that went extinct in the last ice age have the lowest spread. Another way to compare spread is to calculate the standard deviation of log body size in each group, but we

ANSWERS

are unable to calculate this without the raw data. **(i)** It is likely that extinctions have reduced the median body size of mammals.

5. **(a)** Grouped histograms. **(b)** Explanatory variable: sex. Response variable: number of words per day. **(c)** Men: 8,000 − 12,000 words per day. Women: 16,000 − 20,000 words per day. **(d)** Women. **(e)** Men.

6. **(a)** The averages are 7.5 g for the crimson-rumped waxbill (CW), 15.4 g for the cutthroat finch (CF), and 37.9 g for the white-browed sparrow weaver (WS). **(b)** Standard deviations are 0.6 g for CW, 1.2 g for CF, and 3.1 g for WS. The standard deviation is greater when the mean is greater. **(c)** Coefficients of variation: 8.3% for CW, 8.0% for CF, and 8.2% for WS. The coefficients of variation are much more similar than the standard deviations. **(d)** Compare coefficients of variation to compare variation relative to the mean. Beak length: 3.8%; body mass: 8.2%. Body mass is most variable relative to the mean.

7. **(a)** cm/s.
(b)

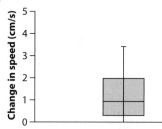

(c) Asymmetric. The range of changes in speed is much greater above the median than below the median. This is true whether we look at the third quartile or the total range. **(d)** The span of the box indicates where the middle 50% of the measurements occur (first quartile to third quartile). **(e)** Mean is 1.19 cm/s, which is greater than the median (0.94 cm/s). The distribution is right-skewed, and the large values influence the mean more than the median. **(f)** 1.15 (cm/s)². **(g)** The interval from $\bar{Y} - s$ and $\bar{Y} + s$ is 0.11 to 2.26. This includes 11 of the 16 observations, or 69%.

8. **(a)** Increase $\bar{Y}$ by 10 times (11.9 mm/s).
(b) Increase s by 10 times (10.7 mm/s).
(c) Increase median by 10 times (9.4 mm/s).
(d) Increase interquartile range by 10 times

(16.8 mm/s). **(e)** No effect. **(f)** Increase s^2 by 100 times [115.3 (mm/s)²].

9. **(a)** 5.7 hours. **(b)** 2.4 hours. **(c)** 83/114 = 0.73 (73%). **(d)** Median = 5 hours, which is less than the mean. The distribution is right-skewed, and the large values influence the mean more than the median.

10. **(a)** Cumulative frequency distribution. **(b)** The y-axis indicates the cumulative relative frequency (fraction) corresponding to quantiles of measurements on the x-axis (annual percent change in human population). The quantile of a value indicated on the x-axis is the fraction of observations less than or equal to it. **(c)** Approximately 10% of the countries had negative change in population size. **(d)** The 0.10 quantile is approximately 0 growth, the 0.50 quantile is about 1.5% growth, and the 0.90 quantile is about 3% growth. **(e)** The 60th percentile is approximately 1.75% growth.

11. Use the cumulative frequency distribution to estimate the median (0.50 quantile, approximately 1.4%), and the first and third quartiles (approximately 0.7% and 2.4%). The interquartile range is about 2.4 − 0.7 = 1.7% and 1.5 times this amount is about 2.4%, which sets the maximum length of each whisker. The smallest value is about −1.1% and the largest value is about 4.3%. Both of these values are within 1.5 times the interquartile range, so the whiskers extend from −1.1% and 4.3%. The boxplot is:

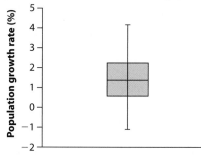

12. **(a)** When listing descriptive statistics in tables, the same descriptive statistic calculated on different groups should be put in the same column. Numbers placed side by side in a table are more difficult to compare by eye than numbers stacked vertically.

(b)

X-ray dose	Mean	Standard deviation
Control	3.70	1.10
25 rads	5.27	1.19
50 rads	12.37	4.69
100 rads	23.30	3.27
200 rads	29.80	2.99

13. **(a)** Grouped cumulative frequency distributions. **(b)** The explanatory variable is the taxonomic group (butterflies, birds, or plants). The response variable is the percent change in range size. **(c)** Group is a categorical variable. Range change is numerical.

Chapter 4

1. **(a)** The standard deviation is the square root of the variance, $\sqrt{254.5} = 15.95$. **(b)** $n = 9$. **(c)** $SE_{\bar{Y}} = s/\sqrt{n} = 15.95/\sqrt{9} = 5.32$. **(d)** The 2SE rule of thumb says that an approximate 95% confidence interval for a mean is the mean plus or minus 2 standard errors. The mean is 131.0, so the confidence interval ranges from approximately $131.0 - 2(5.32) = 120.4$ mm Hg at the lower limit to approximately $131.0 + 2(5.32) = 141.6$ mm Hg at the upper limit.

2. **(a)** 0.22 hours. **(b)** The sampling distribution of the estimates of the mean time to rigor mortis. **(c)** Random sampling from the population of corpses.

3. The median will probably be smaller than the mean. The distribution is right-skewed. Extreme values have a greater effect on the mean, pulling it upward, than on the median.

4. Sample size.

5. **(a)** False ($\hat{p}$ is just an estimate based on a sample). **(b)** True. **(c)** True. **(d)** False. This fraction is a constant (leaving aside the possibility of gene copy number variation between people). **(e)** True.

6. **(a)** The percentage of all Canadians who agree with the statement. **(b)** The sample estimate is 73%. **(c)** The sample size is 1641 people.

(d) The 95% confidence interval for the population fraction.

7. **(a)** 95.9 ms. The population mean flash duration. **(b)** No, because the calculation was based on a sample rather than on the whole population. By chance, the sample mean will differ from that of the population mean. **(c)** 1.9 ms. **(d)** The quantity estimates the spread of the sampling distribution of the sample mean. **(e)** Using the 2SE rule, 95.9 $\pm$ 2(1.9 ms) yields $92.1 < \mu < 99.7$. **(f)** The interval represents the most-plausible values for the population mean μ. In roughly 95% of random samples from the population, when we compute the 95% confidence interval, the interval will include the true population mean.

8. Part (a) is correct. The population mean is likely to be small or zero.

9. **(a)** A histogram or cumulative frequency distribution. **(b)** 8.3 genes. **(c)** 0.7 genes. **(d)** The spread of the sampling distribution of the mean number of genes regulated. **(e)** That we have a random sample of the total population of regulatory genes.

10. **(a)** Using the 2SE method, $6.9 < \mu < 9.8$ genes. **(b)** The interval between 6.9 and 9.8 represents the most-plausible values for the population mean. In roughly 95% of random samples from the population, when we compute the 95% confidence interval the interval will include the true population mean.

11. **(a)** False. **(b)** True. **(c)** False (the population proportion and the limits 0.12 and 0.28 are all constants, so no probability is involved). **(d)** False.

12. **(a)** The larger number (25.4) must be the standard deviation and the smaller (2.54) must be the standard error of the mean. **(b)** $N = 100$, because $SE_{\bar{Y}} = s/\sqrt{n}$, and here $2.54 = 25.4/\sqrt{100}$.

13. The estimates of the mean are consistently too high, meaning that the estimation of mean height was done by a process that caused bias. Perhaps the people were measured with their shoes on, or perhaps the measurement device was poorly calibrated.

14. Increase the sample size.

Chapter 5

1. (a) They are mutually exclusive because each respondent can select only one answer. Therefore, two cannot occur. **(b)** Pr[*very repulsive* or *somewhat repulsive*] = Pr[*very repulsive*] + Pr[*somewhat repulsive*] = 0.30 + 0.20 = 0.50. **(c)** Pr[*not especially delicious*] = 1 − Pr[*especially delicious*] = 1 − 0.01 = 0.99.

2. (a) 0.48. 0.52. (Probability tree is below.) **(b)** Events are mutually exclusive: 0.08 + 0.01 = 0.09. **(c)** Pr[*somewhat delicious* or *especially delicious* | *man*] = 0.09. **(d)** Pr[*somewhat delicious* or *especially delicious* | *woman*] = 0.06 + 0.01 = 0.07.

(e)

Sex		Response	Probability
	0.06	Somewhat delicious	0.0312
Woman	0.01	Especially delicious	0.0052
0.52	0.93	Other answer	0.4836
0.48	0.08	Somewhat delicious	0.0384
Man	0.01	Especially delicious	0.0048
	0.91	Other answer	0.4368

(f) Pr[*somewhat delicious* or *especially delicious*] = Pr[*woman*] Pr[*somewhat delicious* or *especially delicious* | *woman*] + Pr[*man*] Pr[*somewhat delicious* or *especially delicious* | *man*] = (0.52)(0.07) + (0.48)(0.09) = 0.0796.

3. (a) Pr[*HPV* or *Chlamydia*]. **(b)** Pr[*HPV* or *Chlamydia*] = Pr[*HPV*] + Pr[*Chlamydia*] − Pr[*HPV* and *Chlamydia*]. **(c)** Pr[*HPV*] = 0.24 + 0.04 = 0.28. Pr[*Chlamydia*] = 0.02 + 0.04 = 0.06. Pr[*HPV* or *Chlamydia*] = 0.28 + 0.06 − 0.04 = 0.30.

4. (a) Pr[*cancer* | *smoker*] = 0.172.

(b)

Smoking		Cancer	Probability
	0.172	Lung cancer	0.089
Smoker			
0.52	0.828	Not	0.431
0.48	0.013	Lung cancer	0.006
Nonsmoker			
	0.987	Not	0.474

(c) Pr[*smoker* and *cancer*] = (0.52)(0.172) = 0.089. **(d)** Pr[*smoker* and *cancer*] = Pr[*smoker*] Pr[*cancer* | *smoker*] = (0.52)(0.172) = 0.089. Yes. **(e)** Pr[*nonsmoker* and *no cancer*] = Pr[*nonsmoker*] Pr[*no cancer* | *nonsmoker*] = (0.48)(0.987) = 0.474.

5. (a) Pr[*smoker* | *cancer*] = Pr[*cancer* | *smoker*] Pr[*smoker*] / Pr[*cancer*] **(b)** Pr[*cancer*] = 0.089 + 0.006 = 0.095. **(c)** Pr[*smoker* | *cancer*] = 0.089/0.095 = 0.937.

6. (a) 5/8. **(b)** 1/4. **(c)** 7/8 (either in this case means pepperoni or anchovies or both). **(d)** No (some slices have both pepperoni and anchovies). **(e)** Yes. Olives and mushrooms are mutually exclusive. **(f)** No. Pr[*mushrooms*] = 3/8; Pr[*anchovies*] = 1/2; if independent, Pr[*mushrooms* and *anchovies*] = Pr[*mushrooms*] × Pr[*anchovies*] = 3/16. Actual probability = 1/8. Not independent. **(g)** Pr[*anchovies* | *olives*] = 1/2 (two slices have olives, and one of these two has anchovies). **(h)** Pr[*olives* | *anchovies*] = 1/4 (four slices have anchovies, and one of these has olives). **(i)** Pr[*last slice has olives*] = 1/4 (two of the eight slices have olives; you still get one slice—it doesn't matter whether your friends pick before you or after you). **(j)** Pr[*two slices with olives*] = Pr[*first slice has olives*] × Pr[*second slice has olives* | *first slice has olives*] = 2/8 × 1/7 = 1/28. **(k)** Pr[*slice without pepperoni*] = 1 − Pr[*slice with pepperoni*] = 3/8. **(l)** Each piece has either one or no topping.

7. Pr[*encounter* and *success*] = Pr[*encounter*] Pr[*capture* | *encounter*] = (0.035)(0.40) = 0.014.

8. Of 273 trees, 45 trees have cavities, so the probability of choosing a tree with a cavity is 45/273 = 0.165.

9. (a) Pr[*vowel*] = Pr[*A*] + Pr[*E*] + Pr[*I*] + Pr[*O*] + Pr[*U*] = 8.2% + 12.7% + 7.0% + 7.5% + 2.8% = 38.2%. **(b)** Pr[*five randomly chosen letters from an English text spell "STATS"*] = Pr[*S*] × Pr[*T*] × Pr[*A*] × Pr[*T*] × Pr[*S*] = 0.063 × 0.091 × 0.082 × 0.091 × 0.063 = 2.7×10^{-6}. (Each draw is independent, but all must be successful to satisfy the conditions, so we must multiply the probability of each independent event.) **(c)** Pr[*2 letters from an English text = "e"*] = 0.127 × 0.127 = 0.016.

10. (a) Pr[A_1 or A_4] = Pr[A_1] + Pr[A_4] = 0.06 + 0.03 = 0.09. **(b)** Pr[A_1 and A_1] = Pr[A_1] Pr[A_1] = 0.06 × 0.06 = 0.0036. **(c)** Pr[not (A_1 and A_1)]

$= 1 - \Pr[A_1 \text{ and } A_1] = 1 - 0.0036 = 0.9964$. **(d)** $\Pr[A_1 A_3] = (0.06)(0.84) + (0.84)(0.06) = 0.1008$.
(e) $\Pr[two\ individuals\ not\ A_1 A_1] = \Pr[not\ A_1 A_1]\ \Pr[not\ A_1 A_1] = 0.9964 \cdot 0.9964 = 0.9928$.
(f) $\Pr[at\ least\ one\ of\ two\ individuals\ is\ A_1 A_1] = 1 - \Pr[neither\ is\ A_1 A_1] = 1 - 0.9928 = 0.0072$.
(g) $\Pr[three\ individuals\ have\ no\ A_2\ or\ A_3\ alleles] = \Pr[six\ alleles\ are\ not\ A_2\ or\ A_3] = (1 - 0.84 - 0.03)^6 = (0.13)^6 = 0.0000048$.

11. (a) $\Pr[no\ dangerous\ snakes] = \Pr[not\ dangerous\ in\ the\ left\ hand] \times \Pr[not\ dangerous\ in\ the\ right\ hand] = 3/8 \times 2/7 = 6/56 = 0.107$.

(b) $\Pr[bite] = \Pr[bite\ |\ 0\ dangerous\ snakes]\ \Pr[0\ dangerous\ snakes] +$
$\qquad \Pr[bite\ |\ 1\ dangerous\ snake]\ \Pr[1\ dangerous\ snake] +$
$\qquad \Pr[bite\ |\ 2\ dangerous\ snakes]\ \Pr[2\ dangerous\ snakes]$.
$\qquad \Pr[0\ dangerous\ snakes] = 0.107$ [from part (a)].
$\qquad \Pr[1\ dangerous\ snake] = (5/8 \times 3/7) + (3/8 \times 5/7) = 0.536$.
$\qquad \Pr[2\ dangerous\ snakes] = 5/8 \times 4/7 = 0.357$.
$\qquad \Pr[bite\ |\ 0\ dangerous\ snakes] = 0$.
$\qquad \Pr[bite\ |\ 1\ dangerous\ snake] = 0.8$.
$\qquad \Pr[bite\ |\ 2\ dangerous\ snakes] = 1 - (1 - 0.8)^2 = 0.96$.
$\qquad$ Putting these all together:
$\qquad \Pr[bite] = (0 \times 0.107) + (0.8 \times 0.536) + (0.96 \times 0.357) = 0.772$.

(c) $\Pr[defanged\ |\ no\ bite] = \dfrac{\Pr[no\ bite\ |\ defanged]\ \Pr[defanged]}{\Pr[no\ bite]}$

$\Pr[no\ bite\ |\ defanged] = 1; \Pr[defanged] = 3/8;$
$\Pr[no\ bite] = \Pr[defanged]\ \Pr[no\ bite\ |\ defanged] + \Pr[dangerous]\ \Pr[no\ bite\ |\ dangerous]$
$\qquad = (3/8 \times 1) + [5/8 \times (1 - 0.8)] = 0.5$.
So, $[defanged\ |\ one\ snake\ did\ not\ bite] = (1.0 \times 3/8)/(0.5) = 0.75$.

12. (a) $\Pr[all\ five\ researchers\ calculate\ 95\%\ CI\ with\ the\ true\ value]$? Each one has a 95% chance, all samples are independent, so $\Pr = (0.95)^5 = 0.774$. **(b)** $\Pr[at\ least\ one\ does\ not\ include\ true\ parameter] = 1 - \Pr[all\ include\ true\ parameter] = 1 - 0.774 = 0.226$.

13. (a) 0.99. **(b)** $\Pr[cat\ survives\ seven\ days] = \Pr[cat\ not\ poisoned\ one\ day]^7 = (0.99)^7 = 0.932$. **(c)** $\Pr[cat\ survives\ a\ year] = \Pr[cat\ not\ poisoned\ one\ day]^{365} = (0.99)^{365} = 0.026$.
(d) $\Pr[cat\ dies\ within\ year] = 1 - \Pr[cat\ survives\ year] = 1 - 0.026 = 0.974$.

14. (a) Of the 1347 people who did not have HIV, 129 tested positive. Therefore the false-positive rate is $129/1347 = 0.096$. **(b)** Of the 170 people with HIV, 4 tested negative. The false-negative rate is $4/170 = 0.024$. **(c)** Use Bayes' theorem: $\Pr[HIV\ |\ positive\ test] = \Pr[positive\ test\ |\ HIV]\ \Pr[HIV]\ /\ \Pr[positive\ test] = (166/170)\ (170/1517)\ /\ ((129+166)/1517) = 0.56$.

15. Sampling a *Wnt*-responsive cell has probability 0.09, whereas the probability is 0.91 of a nonresponsive cell.
(a) $\Pr[WWLWWW] = 0.09^5 \times 0.91 = 5.4 \times 10^{-6}$. **(b)** $\Pr[WWWWWL] = 0.09^5 \times 0.91 = 5.4 \times 10^{-6}$.
(c) $\Pr[LWWWWW] = 0.09^5 \times 0.91 = 5.4 \times 10^{-6}$. **(d)** $\Pr[WLWLWL] = 0.09^3 \times 0.91^3 = 5.5 \times 10^{-4}$.
(e) $\Pr[WWWLLL] = 0.09^3 \times 0.91^3 = 5.5 \times 10^{-4}$. **(f)** $\Pr[WWWWWW] = 0.09^6 = 5.3 \times 10^{-7}$.
(g) $\Pr[at\ least\ one\ nonresponsive\ cell] = 1 - \Pr[WWWWWW] = 1 - 0.09^6 = 0.9999995$.

16. $\Pr[next\ person\ will\ wash\ his/her\ hands] = \Pr[wash\ |\ man] \times \Pr[man] + \Pr[wash\ |\ woman] \times \Pr[woman]$
$= 0.74 \times 0.4 + 0.83 \times 0.6 = 0.794$.

17. (a) $\Pr[one\ person\ not\ blinking] = 1 - \Pr[person\ blinks] = 1 - 0.04 = 0.96$.
(b) $\Pr[at\ least\ one\ blink\ in\ 10\ people] = 1 - \Pr[no\ one\ blinks] = 1 - (0.96)^{10} = 0.335$.

Chapter 6

1. Statement (a) is correct. If the estimate that the test is based on is biased, the estimate will on average be different from the true value. As a result, the probability of rejecting the (true) null hypothesis is increased, and therefore the Type I error rate is increased.

2. False. The Type 1 error rate is set by the experimenter, and it will be accurate provided the sample is a random sample.

3. (a) Failing to reject a false null hypothesis. (b) The probability (α) used as a criterion for rejecting the null hypothesis; if the P-value is less than or equal to α, then the null hypothesis is rejected, otherwise the null hypothesis is not rejected. (c) Setting a higher significance level, α, such as raising it to 0.05 instead of 0.01, increases the probability of failing to reject a false null hypothesis.

4. (a) True. (b) False. (c) False.

5. (a) H_0: The rate of correct guesses is 1/6. (b) H_A: The rate of correct guesses is not 1/6.

6. (a) Alternative hypothesis. (b) Alternative hypothesis. (c) Null hypothesis. (d) Alternative hypothesis. (e) Null hypothesis.

7. (a) Lowers the probability of committing a Type I error. (b) Increases the probability of committing a Type II error. (c) Lowers power of a test. (d) No effect.

8. (a) No effect. (b) Decreases the probability of committing a Type II error. (c) Increases the power of a test. (d) No effect.

9. (a) $P = 2 \times (\Pr[15] + \Pr[16] + \Pr[17] + \Pr[18]) = 0.0075$. (b) $P = 2 \times (\Pr[13] + \Pr[14] + \cdots + \Pr[18]) = 0.096$. (c) $P = 2 \times (\Pr[10] + \Pr[11] + \Pr[12] + \cdots + \Pr[18]) = 0.815$. (d) $P = 2 \times (\Pr[0] + \Pr[1] + \Pr[2] + \Pr[3] + \cdots + \Pr[7]) = 0.481$.

10. Failing to reject H_0 does not mean H_0 is correct, because the power of the test might be limited. The null hypothesis is the default and is either rejected or not rejected.

11. Begin by stating the hypotheses. H_0: Size on islands does not differ in a consistent direction from size on mainlands in Asian large mammals (i.e., $p = 0.5$); H_A: Size on islands differs in a consistent direction from size on mainlands in Asian large mammals (i.e., $p \neq 0.5$), where p is the true fraction of large mammal species that are smaller on islands than on the mainland. Note that this is a two-tailed test. The test statistic is the observed number of mammal species for which size is smaller on islands than mainland: 16. The P-value is the probability of a result as unusual as 16 out of 18 when H_0 is true: $P = 2 \times (\Pr[16] + \Pr[17] + \Pr[18]) = 0.00135$. Since $P < 0.05$, reject H_0. Conclude that size on islands is usually smaller than on mainlands in Asian large mammals.

12. (a) Not correct. The P-value does not give the size of the effect. (b) Correct. H_0 was rejected, so we conclude that there is indeed an effect. (c) Not correct. The probability of committing a Type I error is set by the significance level, 0.05, which is decided beforehand. (d) Not correct. The probability of committing a Type II error depended on the effect size, which wasn't known. (e) Correct.

13. Their test almost certainly failed to reject H_0, because the 95% confidence interval includes the value of the parameter stated in the null hypothesis (i.e., 1).

14. (a) H_0: Subjects pick the mother correctly one time in two ($p = 1/2$), H_A: Subjects pick the mother correctly more than one time in two ($p > 1/2$). (b) One-sided, because the alternative hypothesis considers parameter values on one side of the parameter value stated in the null hypothesis. This seems justified here because it is not feasible that sons would resemble their mothers less than randomly chosen women. (d) $P = 0.1214 + 0.1669 + \cdots + 0.000004 = 0.881$ (it is quicker to calculate as $1 - (0.000004 + 0.00007 + \cdots + 0.0708) = 0.881$). (e) Since $P > 0.05$, do not reject the null hypothesis. (f) Calculate a 95% confidence interval for p.

Chapter 7

1. (a) n independent trials, with each having the same probability of "success," p. Yes. (b) $p = 0.30$, $n = 7$. (c) $\Pr[5] = \binom{7}{5}(0.3)^5(0.7)^2 =$

0.0250. **(d)** $\Pr[6] = \dbinom{7}{6}(0.3)^6(0.7)^1 = 0.0036.$ $\Pr[7] = \dbinom{7}{7}(0.3)^7(0.7)^0 = 0.0002.$ **(e)** $\Pr[5 \text{ or more}] =$ $\Pr[5] + \Pr[6] + \Pr[7] = 0.0250 + 0.0036 + 0.0002 = 0.0288.$

2. (a) H_0: Probability of right turn equal to the probability of a left turn ($p = 0.5$); H_A: Probability of right turn does not equal the probability of a left turn ($p \neq 0.5$). **(b)** Number of right ears is 19. **(c)** $\Pr[19] =$ $\dbinom{25}{19}(0.5)^{19}(05)^6 = 0.00528.$ **(d)** 20, 21, 22, 23, 24, 25. **(e)** $\Pr[20] = \dbinom{25}{20}(0.5)^{20}(05)^5 = 0.00158;$ $\Pr[21]$ $= \dbinom{25}{21}(0.5)^{21}(05)^4 = 0.00038;$ $\Pr[22] = 7 \times 10^{-5};$ $\Pr[23] = 9 \times 10^{-6};$ $\Pr[24] = 7 \times 10^{-7};$ $\Pr[25] = 3 \times 10^{-8}.$ **(f)** $\Pr[19 \text{ or more}] = \Pr[19] + \Pr[20] + \cdots + \Pr[25] = 0.00528 + 0.00158 + 0.00038 + \cdots = 0.0073.$ **(g)** $P = 2(0.0073) = 0.0146.$ **(h)** P is the probability of a result as extreme or more extreme than that observed if the null hypotheses were true. **(i)** Reject H_0, because $P < 0.05$. More people use the right ear than use the left ear when listening to a stranger in the noisy nightclub.

3. (a) Estimated proportion returned $= 101/240 = 0.421.$ **(b)** $p' = \dfrac{X + 2}{n + 4} = \dfrac{101 + 2}{240 + 4} = 0.4221$

(c) The lower bound is $p' - 1.96\sqrt{\dfrac{p'(1 - p')}{n + 4}} = 0.4221 - 1.96\sqrt{\dfrac{0.4211(1 - 0.4211)}{240 + 4}} = 0.360$

(d) The upper bound is $p' + 1.96\sqrt{\dfrac{p'(1 - p')}{n + 4}} = 0.4221 + 1.96\sqrt{\dfrac{0.4211(1 - 0.4211)}{240 + 4}} = 0.484$

(e) Answers will vary. For example, $p = 0.35$ and $p = 0.45$ lie inside, whereas $p = 0.5$ and $p = 0.8$ lie outside. **(f)** It is likely that H_0: $p = 0.5$ would be rejected, because 0.5 lies outside the most-plausible range of values indicated by the 95% confidence interval.

4. (a) 91/220 had cancer, so the estimated probability of a cast or crew member developing cancer is 0.414. **(b)** The standard error of the proportion is 0.033. [The square-root of $(0.414)(1 - 0.414)/(220 - 1)$]. This quantity measures the standard deviation of the sampling distribution of the proportion. **(c)** Using the Agresti–Coull method to generate confidence intervals, we first calculate $p' = (X + 2)/(n + 4) = (91 + 2)/(220 + 4) = 0.415.$ The lower bound for the 95% confidence interval is

$0.415 - 1.96\sqrt{\dfrac{p'(1 - p')}{n + 4}} = 0.351.$ The upper bound is $0.415 + 1.96\sqrt{\dfrac{p'(1 - p')}{n + 4}} = 0.480.$ The

confidence intervals do not bracket the typical 14% cancer rate for the age group. It is unlikely that 14% is the true cancer rate for this group.

5. (a) 46/50 bills had cocaine, so the estimated proportion is 0.92. **(b)** The 95% confidence interval, calculated using the Agresti–Coull method: $p' = 48/54 = 0.89,$ $0.805 < p < 0.973.$ **(c)** We are 95% confident that the true proportion of U.S. one-dollar bills that have a measurable cocaine content lies between 0.805 and 0.973.

6. (a) No, the probability of drawing a red card changes depending on the cards that have been drawn. **(b)** Yes, the sampling is with replacement, so the probability of success remains constant. **(c)** No, the probability of drawing a red ball will change with each draw. **(d)** Yes, the number of red-eyed flies in a sample from a *large* population can be described using the binomial distribution. (Because the population sampled is large, we will assume that the sampling of each individual has a negligible effect on the probability of drawing a red fly on the next draw.) **(e)** No, the individuals within different families may have different probabilities of having red eyes due to shared genetic differences.

7. (a) This estimate pools numbers from two different groups. Women with rosacea are not a random sample of the population. **(b)** If the control group women are a random sample, $15/16 = 0.938$ have mites. **(c)** $p' = 17/20 = 0.85.$ The 95% confidence interval is 0.69 to 1.00. (Proportions cannot be larger than 1, so

the confidence interval is truncated at 1.)
(**d**) For the women with rosacea, $p' = 18/20 = 0.90$. The 95% confidence interval is $0.769 < p < 1.0$.

8. (**a**) Pr[*male is eaten*] is estimated by $\hat{p} = 21/52 = 0.404$. $p' = 23/56 = 0.411$. The 95% confidence interval is from 0.282 to 0.540. (**b**) This estimate is consistent with a 50% capture of males, but not with a 10% capture of males. (**c**) The larger sample would not change the estimate of the proportion (the fraction is exactly the same), but it would reduce width of the confidence interval. (The standard deviation of the sampling distribution for $\hat{p}$ is smaller when sample size is larger (the square root of the sample size is in the denominator in the formula for $SE_{\hat{p}}$), making the confidence interval narrower too.)

9. Estimated probability of moving north is $22/24 = 0.917$. To test the null hypothesis that north and south movements of ranges are equally probable, we need to calculate the probability of 22, 23, or 24 species moving northward if $p = 0.5$. We will make this a two-tailed test, so we'll multiply this probability by two.

$$Pr[22] = \binom{24}{22}(0.5)^{22}(1 - 0.5)^2$$
$$= 1.6 \times 10^{-5}$$

Doing the same for 23 and 24, summing all three, and multiplying by two, we find that $P = 3.6 \times 10^{-5}$. We can confidently reject the null hypothesis.

10. (**a**) The first study with the narrower confidence interval probably had the larger sample size. (**b**) The estimate with the smaller confidence interval is the more precise, in the sense that the true value is likely to be somewhere near the estimate.

11. (**a**) H$_0$: Males do not prefer one type of female over the other ($p = 0.5$). H$_A$: Males prefer one type over the other ($p \neq 0.5$). With a binomial test,

$$P = 2(Pr[19] + Pr[20] + Pr[21] + Pr[22]$$
$$+ Pr[23] + Pr[24]) = 2(0.00253 + 0.0006 +$$
$$0.0001 + \cdots) = 0.0066.$$

Therefore $P < 0.05$, and we can reject the null hypothesis that the males have no preference.

(**b**) The males may have preferred one female type over the other for a number of reasons. If the females were sisters, this would reduce the number of differences due to genes and maternal environment, and so would be more likely to test for fetal position. Ideally, this would be done with 24 sets of sisters so that each trial is independent.

12. (**a**) 30%. (**b**) The binomial distribution. (**c**) The standard deviation of the proportion of A cells is the standard error:

$$\sigma_{\hat{p}} = \sqrt{\frac{0.3(1 - 0.3)}{15}} = 0.118.$$

(**d**) 95% of the technicians should construct a confidence interval for the proportion of A cells that includes 0.3.

13. (**a**) 1/6 of the 12 dice should have "3," or 2 dice. (**b**) Pr[*no three out of 12 rolls*] = $(1 - Pr[3 \text{ on one roll}])^{12} = (5/6)^{12} = 0.112$.

(**c**) $\binom{12}{3}(0.167)^3(1 - 0.167)^9 = 0.197$.

(**d**) Each die has six faces, each with probability 1/6 of showing. The average value for the die will be the sum of each value times its probability, or $(1 \times 1/6) + (2 \times 1/6) + \cdots + (6 \times 1/6)$ or $1/6 \times 21 = 3.5$. For twelve dice, the average sum of the numbers showing will be $12 \times 3.5 = 42$. (**e**) Pr[*all dice show 1 or 6*] = $(Pr[1] + Pr[6])^{12} = 1.88 \times 10^{-6}$.

14. (**a**) 6052 out of 12,028 of the deaths occurred in the week before Christmas, or 0.503. (**b**) Using the Agresti–Coull confidence interval approximation for proportions, $p' = (X + 2)/(n + 4)$, where X is the number of successes (= "died before Christmas") and n is the total number of trials (= "died within a week of Christmas") = $6054/12,032 = 0.503$. Then the confidence interval is 0.494 to 0.512. (**c**) This interval includes 50%, which is what is expected by chance. There is no statistical support for the belief that the living hang on until that special day.

15. (**a**) We can calculate the P-value by summing the probabilities of 13, 14, 15, or 16 resin clades having more species, then multiplying by two.

$$Pr[13] = \binom{16}{13}(0.5)^{13}(1 - 0.5)^3 = 0.0085.$$

Summing the probabilities for 13 through 16, then multiplying by two, $P = 0.021$. **(b)** This is an observational study, not an experimental study. We did not randomly assign clades to have latex/resin. It is possible that clades with resin have higher diversity not due to the resin, but due to some third factor that affects both resin and the number of species.

16. (a) The best estimate of the proportion of bills with heroin is $7/50 = 0.14$.

(b) $SE_{\hat{p}} = \sqrt{\dfrac{0.14(1 - 0.14)}{50}} = 0.049$.

This measures the uncertainty of the estimate of the proportion, using the standard deviation of the sampling distribution of the proportion. **(c)** The 95% confidence interval is: $0.07 < p < 0.27$. **(d)** If estimated proportion is the true proportion, we can calculate the probability of getting exactly 7 bills with heroin out of a sample of 50.

$Pr[7] = \dbinom{50}{7}(0.14)^7 (1 - 0.14)^{43} = 0.161$.

17. (a) 1856 out of 5743 amphibian species are vulnerable, or 0.323. **(b)** This is not a sample: these are all of the known amphibian species. Because there is no sample of the population, the interpretation of the confidence interval makes no sense, so standard confidence interval calculations are not warranted.

Chapter 8

1. (a) Because the computer program encodes the assumptions of the Poisson distribution: each individual is placed independently and with equal probability anywhere on the landscape. **(b)** H_0: The number of individuals per block has Poisson distribution; H_A: The number of individuals per block does not have a Poisson distribution.

(c)

Number of individuals	Observed number of blocks	Expected number of blocks
0	118	120.0
1	64	61.2
2	16	15.6
3	2	2.6
>3	0	0.6
Total	200	200

(d) $\bar{X} = \dfrac{118(0) + 64(1) + 16(2) + 2(3)}{200} =$ 0.51. **(e)** Substitute the sample mean in Poisson formula:

$Pr[0] = \dfrac{e^{-0.51}(0.51)^0}{0!} = 0.600$,

$Pr[1] = \dfrac{e^{-0.51}(0.51)^1}{1!} = 0.306$,

$Pr[2] = \dfrac{e^{-0.51}(0.51)^2}{2!} = 0.078$,

$Pr[3] = \dfrac{e^{-0.51}(0.51)^3}{3!} = 0.013$.

(f) Results added to table: $200 \times 0.600 = 120.0$, $200 \times 0.306 = 61.2$, $200 \times 0.078 = 15.6$, $200 \times 0.013 = 2.6$. **(g)** We are missing the expected frequency of blocks with more than 3 individuals. Obtain by subtraction: $200 - 120 - 61.2 - 15.6 - 2.6 = 0.6$. Now we see that one of the expected values is less than 1 and more than 20% are less than 5. **(h)** To solve both problems combine the last three categories as follows:

Number of individuals	Observed number of blocks	Expected number of blocks
0	118	120.0
1	64	61.2
>1	18	18.8
Total	200	200

(i) We lose an extra degree of freedom because we estimated mean from data: $3 - 1 - 1 = 1$.

(j) $\chi^2 = \sum \dfrac{(O - E)^2}{E} = \dfrac{(118 - 120.0)^2}{120.0} +$ $\dfrac{(64 - 61.2)^2}{61.2} + \dfrac{(18 - 18.8)^2}{18.8} = 0.195$.

(k) $\chi^2_{0.05,1} = 3.84$. Since $\chi^2 < 3.84$, $P > 0.05$ (on the computer, $P = 0.66$). Do not reject H_0. **(l)** The data are compatible with the Poisson distribution and the hypothesis that the program is placing individuals randomly in space.

2. (a) Histogram:

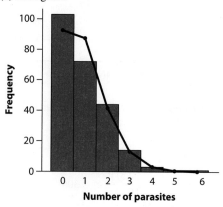

Number of parasites

(b) First we need to estimate the mean number of nematodes per fish: $\bar{X} = 103(0) + 72(1) + 44(2) + 14(3) + 3(4) + 1(5) + 1(6) = 0.945$. Use $\bar{X}$ in formula for Poisson to get expected probabilities and expected frequencies of fish with 0 to 6 nematodes: 92.58, 87.35, 41.41, 13.09, 3.09, 0.48, 0.00. **(c)** (See graph). The observed frequency of fish with 0 nematodes is slightly more than expected, whereas the frequency with 1 nematode is somewhat less than expected. Observed and expected frequencies are similar in fish with 2 to 6 nematodes.
(d) H_0: The number of nematodes per fish has a Poisson distribution. H_A: The number of nematodes per fish does not have a Poisson distribution. We need to combine the frequencies corresponding to 4, 5, and 6 parasites to meet the χ^2 criteria, yielding

Nematodes	Frequency	Expected
0	103	92.58
1	72	87.35
2	44	41.41
3	14	13.09
≥ 4	5	3.57

$\chi^2 = 4.67$, $df = 5 - 1 - 1 = 3$. $\chi^2_{0.05,3} = 7.81$. Since χ^2 not greater than or equal to the critical value, $P > 0.05$ and we do not reject H_0

(on the computer, $P = 0.20$). There is no evidence that nematodes do not worm their way into fish at random.
3. (a) $\hat{p} = 57/71 = 0.80$. Using the Agresti–Coull method, the 95% confidence interval for p is $0.69 < p < 0.88$. Yes, the plausible range includes the value 0.75, the predicted proportion under the simple genetic model. **(b)** Answers will vary; e.g., the proportions $p = 0.70$ and $p = 0.80$ are also consistent with the data.
(c) H_0: Resistant and susceptible offspring occur in a 3:1 ratio ($p = 0.75$). H_A: Resistant and susceptible offspring do not occur in a 3:1 ratio ($p \neq 0.75$). The observed frequencies are 57, 14. The expected frequencies are 53.25, 17.75. $\chi^2 = 1.06$, $df = 1$. The critical value $\chi^2_{0.05,1} = 3.84$. Since χ^2 is not greater than or equal to 3.84, do not reject H_0. ($P = 0.30$.) This is compatible with the findings in (a) that the plausible range of values for p includes 0.75.
(d) No, the results are consistent with the predictions of the genetic model, but they do not confirm it because failure to reject H_0 does not mean it is correct. The 95% confidence interval in (a) indicates the values for p that remain consistent with the data.
4. (a) Use a histogram.

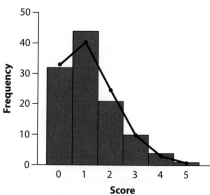

Score

(b) $\bar{X} = 1.223$. **(c)** Expected frequencies of games with 0–5 goals is 33.040, 40.320, 24.640, 9.968, 3.024, 0.784. **(d)** See graph in part (a) for one method to compare observed and expected. A grouped bar graph will also work. The expected frequencies from the Poisson distribution are close to those observed (the observed goals per game have slightly more 1's and slightly fewer 2's than predicted). **(e)** Greater than. $s^2 = 1.256$, is very close to the mean! The ratio $s^2/\overline{X} = 1.03$, which is very close to 1.

5. (a) 5. There would be six categories, one for each outcome, and so five degrees of freedom, since no parameters are estimated. **(b)** 10. There are 11 categories (from 0 heads to 10 heads), and 10 degrees of freedom, since no parameters are estimated. **(c)** 9. There are now nine degrees of freedom, since p must be estimated from the data. **(d)** 3. There would be five categories (0 to 4 insect heads per sample), and three degrees of freedom, since the mean number of heads per sample would need to be estimated from the data.

6. (a) H_0: The number of heads has a binomial distribution with $p = 0.5$. H_A: The number of heads does not have a binomial distribution with $p = 0.5$. Use the binomial distribution to calculate the following expected frequencies.

Number of heads	Observed	Binomial expectation	Expected
0	6	0.0039	3.91
1	32	0.0313	31.25
2	105	0.1094	109.38
3	186	0.2188	218.75
4	236	0.2734	273.44
5	201	0.2188	218.75
6	98	0.1094	109.38
7	33	0.0313	31.25
8	103	0.0039	3.91
Total	1000	1.0	1000

The expected values for 0 and 8 are less than 5, so we combined (0 and 1) and (7 and 8) to produce:

Number of heads	Observed	Expected	$\dfrac{(Observed - Expected)^2}{Expected}$
0 or 1	38	35.16	0.23
2	105	109.38	0.18
3	186	218.75	4.90
4	236	273.44	5.13
5	201	218.75	1.44
6	98	109.38	1.18
7 or 8	136	35.16	289.2
Total	1000	1000	$\chi^2 = 302.27$

A χ^2 goodness-of-fit test gives $\chi^2 = 302.3$. The critical value for χ^2 with $df = 6$ and $\alpha = 0.05$ is 12.59; the critical value for $\alpha = 0.001$ is 22.46. Therefore, we reject the null hypothesis with $P < 0.001$ (on the computer, $P < 10^{-10}$). **(b)** If you compare the observed and expected frequencies with a histogram, you can see very easily that there is an excess of coins yielding eight heads. **(c)** With $p = 0.5$ we would expect the observed number of coins yielding eight heads to be roughly equal to the number of coins yielding zero heads, or around four. This means that there is an excess of two-headed coins of roughly $103 - 4 = 99$.

7. H_0: The probability of death by cancer was $p = 0.14$. H_A: The probability of death by cancer was not $p = 0.14$.

	Observed	14% expected	$\dfrac{(Observed - Expected)^2}{Expected}$
Cancer death	91	30.8	117.67
Noncancer death	129	189.2	19.15
Total	220	220	$\chi^2 = 136.82$

$df = 1$; $\chi^2 = 136.82$ is much greater than 10.83, the critical value for χ^2 at $\alpha = 0.001$, so we can reject the hypothesis that the cast and crew suffered the population-wide cancer mortality rate, $P < 0.001$. ($P < 10^{-10}$.)

8. The average is 20 admissions per night. What is the probability of five or fewer admissions? Assume that each admission is independent (no riots, please). Then, the number of admissions per night should have a Poisson distribution. To find the probability of a quiet night, we sum the probabilities of 0 to 5 admissions.

Admissions	Probability
0	2.06×10^{-9}
1	4.12×10^{-8}
2	4.12×10^{-7}
3	2.75×10^{-6}
4	1.37×10^{-5}
5	5.50×10^{-5}

Total probability $= 7.2 \times 10^{-5}$.

9. The probabilities that bound the test statistic are given below (along with the precise values calculated using a computer program).

df	χ^2	P from statistical Table A	P from computer
1	4.12	$P < 0.05$	0.042
4	1.02	$P > 0.05$	0.907
2	9.50	$P < 0.01$	0.009
10	12.40	$P > 0.05$	0.259
1	2.48	$P > 0.05$	0.115

10. We will use a goodness-of-fit test to test the hypotheses. H_0: The death rate is the same before Christmas as after Christmas. H_A: The death rate before Christmas is different from that after Christmas.

	Observed	Expected	$\dfrac{(Observed - Expected)^2}{Expected}$
Die before Christmas	6052	6014	0.24
Die after Christmas	5976	6014	0.24
Total	12,028	12,028	0.48

$\chi^2 = 0.48$, and $df = 1$. $0.48 < 3.84$, the critical value for $\alpha = 0.05$, so we do *not* reject the null hypothesis of equal rates of cancer deaths around Christmas ($P = 0.49$).

Chapter 9

1. **(a)** We follow the convention recommended in Chapter 2, that that the explanatory variable (here, coffee intake) be in the columns and the response variable (here, cancer) be in the rows.

	High coffee	No coffee	Total
Cancer	19	122	141
No cancer	2473	7768	10241
Total	2492	7890	10382

It is difficult to tell from the contingency table alone whether there is a strong pattern in these data.

ANSWERS

(b) $19/2492 = 0.0076$. **(c)** $122/7890 = 0.0155$. **(d)** $RR = $ Risk with coffee/risk without coffee $= 0.0076/0.0155 = 0.49$. **(e)** $19/2473 = 0.00768$. **(f)** $122/7768 = 0.0157$. **(g)** $0.00768/0.0157 = 0.489$. **(h)** -0.715. **(i)** $SE[\ln(\widehat{OR})] = \sqrt{\frac{1}{19} + \frac{1}{122} + \frac{1}{2473} + \frac{1}{7768}} = 0.2477$ **(j)** $-0.715 \pm 1.96(0.2477)$; $-1.20 < \ln[OR] < -0.23$. **(k)** $e^{-1.20} < OR < e^{-0.23}$; $0.30 < OR < 0.79$. **(l)** The 95% confidence interval for the OR does not include 1, so cancer rates in high coffee drinkers and non-coffee-drinkers are unlikely to be identical. The date are consistent with a broad range of magnitudes of reduction in the risk of prostate cancer with coffee drinking.

2. (a)

	Man patient	Woman patient	Total
Divorce	7	53	60
No divorce	254	201	455
Total	261	254	515

The frequency of divorce appears to be greater when the woman is the patient than when the man is the patient. **(b)** H_0: Divorces happen with equal probability when the woman is sick as when the man is sick. H_A: The proportion of divorces differs depending on which sex is ill. **(c)** $261/515 = 0.507$. **(d)** $60/515 = 0.117$. **(e)** The expected proportion of couples where the man was the diagnosed patient that resulted in divorce is $0.507 \times 0.117 = 0.059$. The other combinations can be calculated in a similar way:

	Man patient	Woman patient	Total
Divorce	0.059	0.057	0.117
No divorce	0.448	0.436	0.883
Total	0.507	0.493	1

(f) 515. **(g)** The expected number of couples where the man was the diagnosed patient that resulted in divorce is $0.059 \times 515 = 30.4$. Expected frequencies:

	Man patient	Woman patient	Total
Divorce	30.4	29.6	60
No divorce	230.6	224.4	455
Total	261	254	515

(h) Yes, because all expected values are greater than 5. **(i)** $\chi^2 = \frac{(7 - 30.4)^2}{30.4} + \frac{(53 - 29.6)^2}{29.6} + \frac{(254 - 230.6)^2}{230.6} + \frac{(201 - 224.4)^2}{224.4} = 41.3$ **(j)** $(r - 1)(c - 1) = 1$ degree of freedom. **(k)** The critical value of $\chi^2_{0.05,1} = 3.84$. **(l)** $P < 0.001$, based on Statistical Table A, because 41.3 is further in the tail than the largest critical value in the table at $\alpha = 0.001$. (On the computer, P is calculated as 1.3×10^{-10}). **(m)** Reject the null hypothesis. They are not independent. Female patients are more likely to be divorced by their spouses after a serious diagnosis than are male patients.

3. (a) 2. **(b)** 0.5. **(c)** 1.

4. (a) A much higher proportion of the captured pigeons are blues.

	White	Blue	Total
Captured	9	92	101
Not captured	92	10	102
Total	101	102	203

(b) H_0: The two types of pigeons have the same capture probability. H_A: The two types of pigeons do not have the same capture probability. Expected values:

	White	Blue	Total
Captured	50.25	50.75	101
Not captured	50.75	51.25	102
Total	101	102	203

$\chi^2 = 134.13$, $df = 1$. The critical value $\chi^2_{0.05,1} = 3.84$. Since $\chi^2 > 3.84$, so $P < 0.05$ (on the computer, $P < 10^{-10}$. Reject H_0. Blue pigeons are more vulnerable. **(c)** $\widehat{OR} = ad/bc = 9 \times 10/92 \times 92 = 0.011$. For the 95% confidence interval, we use the log odds ratio, $\ln(0.011) = -4.54$.

$$SE[\ln(\widehat{OR})] = \sqrt{\frac{1}{9} + \frac{1}{92} + \frac{1}{92} + \frac{1}{10}} = 0.48$$

For the 95% confidence interval, we use $Z = 1.96$.

$$-4.54 - 1.96(0.48) < \ln(OR) < -4.54 + 1.96(0.48)$$
$$-5.49 < \ln(OR) < -3.60$$
$$0.004 < OR < 0.027$$

5. (a) Answers may vary (grouped bar plot also acceptable, with number of persons bitten as the response variable).

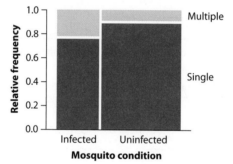

(b) H_0: There is no association between mosquito condition (infected, uninfected) and number of people bitten. H_A: There is an association between mosquito condition and number of people bitten. Observed frequencies:

	Infected	Uninfected	Total
Multiple	20	16	36
Single	69	157	226
Total	89	173	262

Expected frequencies:

	Infected	Uninfected	Total
Multiple	12.23	23.77	36
Single	76.77	149.23	226
Total	89.00	173.00	262

$\chi^2 = 8.67$, $df = 1$. Critical value $\chi^2_{0.05,1} = 3.84$. Since $8.67 > 3.84$, $P < 0.05$. Reject H_0. ($P = 0.003$.)

ANSWERS

6. (a) The odds of the second suitor being accepted if the first was eaten are 0.5, while the odds of the second suitor being accepted if the first escaped are 22.0. There is a much higher probability of the second suitor being rejected if the first was eaten. **(b)** There are only four cells, so if any cell has an expected frequency of less than five, the assumptions of the χ^2 contingency test will be violated. In this case, the expected value is much less than five, so it would be necessary to use Fisher's exact test in this case.

7. (a) Frequency table is below. Frogs appear to leave most frequently when noise of fire is played.

												$\dfrac{(Observed - Expected)^2}{Expected}$		
	Observed					**Expected**						**Expected**		
		Reverse					**Reverse**						**Reverse**		
	Fire	fire	White	Total		Fire	fire	White	Total			Fire	fire	White	Total
Leave	18	6	0	24	Leave	8	8	8	24		Leave	12.5	0.5	8	21
Stay	2	14	20	36	Stay	12	12	12	36		Stay	8.33	0.333	5.33	14
Total	20	20	20	60	Total	20	20	20	60		Total	20.8	0.833	13.3	35

(b) H_0: The frequency of frogs that leave and stay is the same when different noises are played. H_A: The frequency of frogs that stay or leave changes with the noise played. $\chi^2 = 35.0$ for 2 df [there are two rows and three columns, so $(2 - 1)(3 - 1) = 2\ df$]. $35.0 > 5.99$, the critical value for $P = 0.05$ for 2 df, so $P < 0.05$. Reject H_0 ($P = 2.5 \times 10^{-8}$.) Yes, reed frogs change their behavior in response to the sound of fire.

8.

	Observed				**Expected**		
	Adult male	Adult female	Total		Adult male	Adult female	Total
Juv male	1	6	7	Juv male	3.82	3.18	7
Juv female	11	4	15	Juv female	8.18	6.82	15
Total	12	10	22	Total	12	10	22

(a) Two of the expected values are less than 5, so we cannot use a χ^2 contingency analysis. Fisher's exact test is a good alternative. **(b)** The effects are in the expected direction.

9. (a) Proportions of kids in each TV viewing class having violent records eight years later, and confidence intervals (using Agresti–Coull calculations from Chapter 7):

	Record	Total	Proportion	p'	Low CI	High CI
1 hr	5	88	0.057	0.0761	0.022	0.130
1–3 hr	87	386	0.225	0.2282	0.187	0.270
3+ hr	67	233	0.288	0.2911	0.233	0.349

(b) Contingency test of independence of TV watching and later violent record. H_0: Amount of TV watching and violence are not associated. H_A: Amount of TV and later violence are associated.

											$\dfrac{(Observed - Expected)^2}{Expected}$		
	Observed					**Expected**					**Expected**		
Class	No record	Record	Total	Class	No record	Record	Total		Class	No record	Record	Total
1 hr	83	5	88	1 hr	68.21	19.79	88		1 hr	3.21	11.05	
1–3 hr	299	87	386	1–3 hr	299.19	86.81	386		1–3 hr	0.0001	0.0004	
3+ hr	166	67	233	3+ hr	180.60	52.40	233		3+ hr	1.18	4.07	
Total	548	159	707	Total	548	159	707		Total			19.5

TABLE FOR PROBLEM 10(a)

| | Observed | | | | Expected | | | | $\frac{(Observed - Expected)^2}{Expected}$ | | |
	Arrest	Healthy	Total		Arrest	Healthy	Total		Arrest	Healthy	Total
Abstainers	12	197	209	Abstainers	10.7	198.3	209	Abstainers	0.16	0.01	0.17
Drinkers	9	192	201	Drinkers	10.3	190.7	201	Drinkers	0.16	0.01	0.17
Total	21	389	410	Total	21	389	203	Total	0.32	0.02	0.34

$\chi^2 = 19.5$ for 2 df [there are three rows and two columns, so $(3-1)(2-1) = 2\ df$]. $35.0 > 5.99$, the critical value for $P = 0.05$ for 2 df, so $P < 0.05$. Reject H_0. ($P = 0.00006$.) There is a relationship between watching TV and future violence: those that watched less than one hour of TV were less likely to have a record, while those that watched three or more hours were more likely to have a record. **(c)** No, this does not prove that TV watching causes aggression in kids. This was an observational study, not an experimental one, so we do not know if kids watching more TV have other factors in common (e.g., lower parental supervision, etc.).

10. (a) H_0: Alcohol and heart disease are not associated. H_A: Alcohol and heart disease are not associated.(*See the table at the top of the page.*) $\chi^2 = 0.34$ for 1 df $<<$ 3.84 (critical value for $P = 0.05$), so we cannot reject the null hypothesis that there is no difference in heart attack risk between drinkers and non-drinkers ($P = 0.56$). **(b)** No, this does not imply that drinking has no effect on cardiac arrest, only that this study found no evidence for a connection. This study was observational: non-drinkers at the age of 40 may have been heavy drinkers at the age of 20, or may have never consumed alcohol. Drinking may have an effect on heart attacks, but with too small an effect to be detected by this sample size.

11. (a) Answers may vary [grouped bar plot, mosaic plot] but depression should be the response variable. Occurrence of depression is slightly higher in the C-section group.

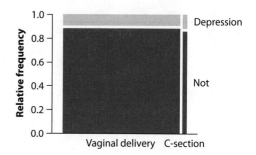

(b) $\widehat{OR} = 0.76$. **(c)** $0.56 < OR < 1.04$. **(d)** No, it would likely not be rejected, because the 95% confidence interval for the odds ratio includes 1. **(e)** $\widehat{RR} = 0.79$. This is similar to the estimate of OR (0.76), which is expected if the probability of postpartum depression is low.

12. (a) Observational study (and a case-control study). **(b)** We recommend the modified mosaic plot in which the cases (migraines) and controls (no migraines) are grouped by categories of the explanatory variable (cardiac shunt vs no shunt), as follows.

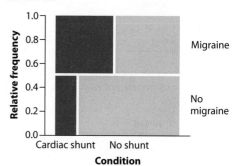

(c) The contingency table:

	Shunt	No shunt
Migraine	44	49
No migraine	16	77

The odds ratio of migraine is $\widehat{OR} = ad/bc = (44)(77)/(16)(49) = 4.32$. **(d)** To find the confidence interval, we first take the log of the odds ratio, $\ln(4.32) = 1.464$ and calculate the SE of the log-odds ratio, $\sqrt{1/44 + 1/49 + 1/16 + 1/77} = 0.344$. The confidence interval for the log-odds ratio is $1.464 - 1.96(0.344)$ to $1.464 + 1.96(0.344)$, or 0.79 to 2.14. Taking the exponential of these to find the confidence interval of the odds ratio, we get $\exp(0.79) = 2.2$, and $\exp(2.14) = 8.5$. $2.2 < OR < 8.5$.

13. The contingency table presented in the question assumes that each climber has an independent probability of mortality. However, many factors affect mortality, and on high mountains, weather can have a very serious effect and would hit all members of a climbing party at once. Similarly, an entire team might have the same acclimation regime, which might affect mortality. Therefore, this table is guilty of pseudo-replication: the proper sample size would be 159 (the number of teams—although this too might be pseudo-replicated if multiple teams were on the mountain at the same time.)

14. **(a)** The proportion of smokers and former smokers declines with increasing Mediterranean diet.

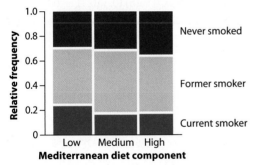

Mediterranean diet component

(b) H_0: There is no association between diet and smoking. H_A: These is an association between diet and smoking.

Observed values

	Low	Med	High	Total
Never	2516	2920	2417	7853
Former	3657	4653	3449	11759
Current	2012	1627	1294	4933
Total	8185	9200	7160	24545

Expected values

	Low	Med	High	Total
Never	2618.7	2943.5	2290.8	7853
Former	3921.3	4407.5	3430.2	11759
Current	1645.0	1849.0	1439.0	4933
Total	8185	9200	7160	24545

$$\frac{(Observed - Expected)^4}{Expected}$$

	Low	Med	High	Total
Never	4.03	0.19	6.95	
Former	17.81	13.67	0.10	
Current	81.88	26.65	14.60	
Total				166

$\chi^2 = 166$. There are $(3 - 1)(3 - 1) = 4$ degrees of freedom in this test. Our test statistic is much greater than the critical value for $P = 0.05$ ($\chi^2_{0.05,4} = 9.49$), so we reject H_0. We conclude that there is a non-random association between diet and smoking: current smokers appear more likely to have a "low" component of the Mediterranean diet in their daily meals. ($P < 10^{-10}$.)

Review 1

1. **(a)** The probability that a single fly survives one day is $1 - 0.03 = 0.97$. **(b)** The probability that a single fly survives 80 days is $(0.97)^{80} = 0.0874$. **(c)** The probability that a fly has died by 80 days is $1 - 0.0874 = 0.9126$, so the probability that all 50 flies have died by 80 days is $(0.9126)^{50} = 0.0103$. There is only about a 1% chance that her experiment will have ended by the times she goes on vacation.

2. **(a)** To be an experimental study, the researchers must assign the treatments (smoking and nonsmoking) to subjects, whereas in this study the researchers had no control over who smoked and two did not. It is a case-control study: subjects were sampled according to whether they were cases (finger defects) or controls (no defects), and then were assessed for expo-

sure to the potential causal factor (smoking).
(b) Answers may vary. We used the mosaic
plot modified for case-control studies (grouped
bar plot is OK too, with finger condition as the
response variable).

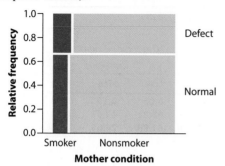

(c) Use the Agresti–Coull method.
$\hat{p} = 0.000953$, $p' = 0.000953$,
$SE_{\hat{p}} = 0.0000118$. $0.000931 < p < 0.000977$.
(d) Observed frequencies:

	Smoker	Nonsmoker
Defect	805	4366
Normal	1280	9062

$\widehat{OR} = (805)(9062)/(4366)(1280) = 1.305$.
95% CI: $1.19 < OR < 1.44$. Smoking mothers
have the higher odds of a baby with a finger
defect. **(e)** It is reasonable to consider the $\widehat{OR}$
of 1.305 also to be an estimate of relative risk,

because the focal condition (finger defect) is so
rare.
3. (a) If pine seedlings are distributed randomly,
the frequencies should fit the Poisson distribution. **(b)** H_0: Number of seedlings per quadrat
has a Poisson distribution. H_A: Number of seedlings does not have a Poisson distribution. The
estimated mean number of seedlings per quadrat
is $\overline{X} = 106/80 = 1.325$.

Number of seedlings	Observed	Expected
0	47	21.26
1	6	28.18
2	5	18.67
3	8	8.24
>4	14	3.65

The expected counts for 5–7 were below 1, so
4 through 7 were combined. $\chi^2 = 88.0$, with
$df = 3$ (five categories, minus one, minus one
parameter estimated). $\chi^2_{0.001, 3} = 16.27$, and
χ^2 is much greater than this critical value, so
$P < 0.001$. Reject H_0 ($P < 10^{-10}$). **(c)** The
variance, 3.49, is greater than the mean (1.325),
indicating that the probability of having seedlings is higher for some quadrats than others:
seedlings are clumped.
4. (a) $1 - 0.7 = 0.3$. **(b)** $(0.3)(0.3) = 0.09$.
(c) $\Pr[2 \text{ on suitable}] = 0.063 + 0.063 + 0.063$
$= 0.189$. (*See the figure at the bottom of the
page.*)

FIGURE FOR PROBLEM 4(c)

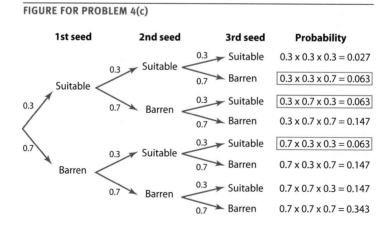

5. Their test almost certainly rejected H_0, because the 95% confidence interval does not include the value of the parameter stated in the null hypothesis (i.e., 1).

6. (a) The estimate is likely to be biased for at least two reasons: students might be less likely to respond honestly to questions about drug use with their parents present, and students with high drug use might be less likely to volunteer for the study than students with lower drug use. **(b)** An estimate is expected to have greater sampling error with a small sample size than with a larger sample size.

7. (a) 0.5. **(b)** Standard deviation $= \sqrt{p(1-p)/n} = \sqrt{0.5(1-0.5)/32} = 0.088$.

(c) $1 - 0.6 = 0.4$. **(d)** $\Pr[16 \text{ } red\text{-}eyed \text{ } alleles] = \binom{32}{16}(0.5)^{16}(1-0.5)^{32-16} = 0.13995$.

(e) Using the binomial distribution, $\Pr[30 \text{ } red] + \Pr[31 \text{ } red] + \Pr[32 \text{ } red] =$

$\binom{32}{30}(0.5)^{30}(1-0.5)^2 + \binom{32}{31}(0.5)^{31}(1-0.5)^1 + \binom{32}{32}(0.5)^{32}(1-0.5)^0 = 1.23 \times 10^{-7}$.

8. (a) First new group $\hat{p} = \frac{9}{32} = 0.281$, second new group $\hat{p} = \frac{26}{32} = 0.8125$. Confidence intervals: first new group, $0.155 < p < 0.456$; second new group, $0.642 < p < 0.914$. **(b)** These do not overlap, either with each other or with the known true proportion of 0.5. By chance, one out of 20 such 95% confidence intervals will on average not include the true parameter being estimated.

9. (a) H_0: Chest pain and death are not associated. H_A: Chest pain and death are associated.

Observed frequencies				Expected frequencies			
	Chest pain	No pain	Total		Chest pain	No pain	Total
Died	822	229	1051	Died	962.3	88.7	1051
Survived	18296	1534	19830	Survived	18155.7	1674.3	19830
Total	19118	1763	20881	Total	19118.0	1763.0	20881

The test statistic $\chi^2 = 254.99$ is much great than 3.84, the critical value for $\alpha = 0.05$ for 1 *df*, so $P < 0.05$. Reject H_0. $(P < 10^{-10}.)$ We conclude that chest pain is associated with a lower probability of death from heart attack. **(b)** This is not a case-control study (researchers measured mortality in people with and without chest pain), so either relative risk or odds ratio is appropriate. $\widehat{OR} = 0.30$. $0.26 < OR < 0.35$. $\widehat{RR} = 0.33$. $0.29 < RR < 0.38$.

10. (a) 5.5. **(b)** 0.255. **(c)** $39/39 = 1.0$ (100%).

11. H_0: The frequency of people preferring the 4 glasses is the same. H_A: The frequency of people preferring the 4 glasses is not the same.

Preferred glass	Observed frequency of participants	Expected frequency
First	15	8.25
Second	5	8.25
Third	2	8.25
Fourth	11	8.25
Total	33	33.00

$\chi^2 = 12.45$, $df = 3$. Critical value $\chi^2_{0.05, 3} = 7.81$. $P < 0.05$ ($P = 0.006$). Reject H_0. Position in the sequence has an effect on preferences (there is an advantage to being first).

12. (a) $\widehat{RR} = 0.74$. **(b)** Fisher's exact test is required (χ^2 contingency test cannot be used because too many of the expected frequencies are less than 5). **(c)** The confidence interval covers a very broad range of values for the relative risk, for which the scientific interpretations vary greatly. The confidence interval needs to be narrower to be useful. Hence, the biggest limitation to the study is that its sample size is so small.

ANSWERS

ANSWERS

13. (a) Histogram shows a sharply right-skewed frequency distribution of ages, with the mode at a young age. There might be a second, low peak at intermediate ages.

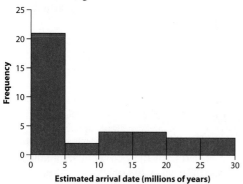

(b) Median appears to be between 0 and 5 million years ago (mya), whereas mean is between 5 and 10 mya. The mean is greater than the median because the distribution is right-skewed: the large values influence the mean more than the median. **(c)** Mean (8.66 mya) is indeed greater than the median (3.51 mya). **(d)** First quartile: 1.25 mya; third quartile: 17.30 mya; interquartile range: 16.05 mya. **(e)** Box plot:

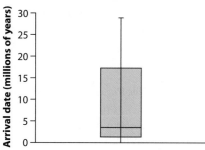

14. False. The given hypothesis is either right or wrong. Hypotheses aren't variables subject to chance. The P-value is a measure of how unusual the data are if the null hypothesis is true.

15. (a) 3D effects violate "Make patterns in the data easy to see." Small fonts violate "Draw graphical elements clearly." **(b)** Bar graph.

16. (a) Histogram of whale numbers, χ^2 goodness-of-fit test to the Poisson distribution. **(b)** Mosaic plot or grouped bar graph, χ^2 contingency test or Fisher's exact test. **(c)** Mosaic plot or grouped bar graph, Fisher's exact test. **(d)** Histogram, χ^2 goodness-of-fit test to the Poisson distribution. **(e)** Mosaic plot or grouped bar graph, χ^2 contingency test or Fisher's exact test. **(f)** Bar graph, binomial test (or χ^2 goodness-of-fit test if the sample size is large enough). **(g)** Bar graph, binomial test (or χ^2 goodness-of-fit test if the sample size is large enough).

Chapter 10

1. (a)

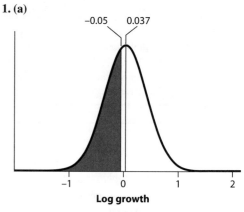

(b) $Z = (Y - \text{mean})/(\text{standard deviation}) = (-0.05 - 0.037)/(0.385) = -0.226$.
(c) $\Pr[Z < -0.226] = \Pr[Z > 0.226]$, because the standard normal distribution is symmetrical around zero. **(d)** From Statistical Table B, 0.397. **(e)** Same as (d): 0.397. **(f)**. Same as (e): 0.397.
2. (a) H_0: The proportion of people who are struck by lightning who are men is 0.5. **(b)** The mean should be $np = 648(0.5) = 324$.
(c) The standard deviation should be $\sqrt{np(1 - p)} = \sqrt{648(0.5)(0.5)} = 12.7$.
(d) 531 is above the mean value in the null hypothesis (324), so we subtract ½ to get 530.5 with the continuity correction.
(e) $Z = (530.5 - 324)/12.7 = 16.2$.
(f) $\Pr[\textit{number of men} \geq 531] = \Pr[Z \geq 16.2]$. Using Statistical Table A, this is off the chart, so we know that this probability is < 0.00002.
(g) The two-tailed value is twice the probability found in the previous step, so $P < 0.00004$.
(h) We conclude that the proportion of men struck by lightning is greater than 0.5.

3. From Statistical Table B. **(a)** 0.090. **(b)** 0.090.
(c) $\Pr[Z > -2.15] = 1 - \Pr[Z < -2.15] = 1 - \Pr[Z > 2.15] = 1 - 0.016 = 0.984$.
(d) $\Pr[Z < 1.2] = 1 - \Pr[Z > 1.2] = 1 - 0.11507 = 0.885$.
(e) $\Pr[0.52 < Z < 2.34] = \Pr[0.52 < Z] - \Pr[2.34 < Z]$, because the first value is the area under the curve from 0.52 to infinity, and the second value is the area under the curve from 2.34 to infinity. The difference will be the area under the curve from 0.52 to 2.34. $0.302 - 0.010 = 0.292$.
(f) $\Pr[-2.34 < Z < -0.52] = 0.292$: the normal distribution is symmetrical on either side of 0.
(g) $\Pr[Z < -0.93] = \Pr[Z > 0.93] = 0.176$.
(h) $\Pr[-1.57 < Z < 0.32] = (1 - \Pr[Z > 1.57]) - \Pr[Z > 0.32] = (1 - 0.058) - 0.37 = 0.567$.

4. (a) To determine the proportion of men excluded, we convert the height limit into standard normal deviates. $(180.3 - 177.0)/7.1 = 0.46$. $\Pr[Z > 0.46] = 0.323$, so roughly one-third of British men are excluded from applying. **(b)** $(172.7 - 163.3)/6.4 = 1.47$. $\Pr[Z > 1.47] = 0.071$, which is the proportion of British women excluded. $1 - 0.071 = 0.929$ is the proportion of British women acceptable to MI5. **(c)** $(183.4 - 180.3)/7.1 = 0.44$ standard deviation units above the height limit.

5. (a) *ii* is most like the normal distribution. *i* is bimodal, while *iii* is skewed. **(b)** All three would generate approximately normal distributions of sample means due to the central limit theorem.

6. (a) $\Pr[weight > 0.5 \text{ kg}]$: Transform to standard normal deviate: $(5 - 3.296)/0.560 = 3.04$. $\Pr[Z > 3.04] = 0.00118$.
(b) $\Pr[3 < birth\ weight < 4]$: Transform both to standard normal deviates, and subtract probabilities of the Z-values. $(3 - 3.296)/0.560 = -0.53$. $(4 - 3.296)/0.560 = 1.28$. $\Pr[Z < -0.53] = 0.29806$, $\Pr[Z > 1.28] = 0.10027$. $1 - 0.29806 - 0.0027 = 0.602$. **(c)** 0.06681 babies are more than 1.5 standard deviations above, with the same fraction below, so 0.13362 of babies are more than 1.5 standard deviations in either direction. **(d)** First, transform 1.5 kg into normal standard deviates: $1.5/0.560 = 2.68$ standard deviations. $\Pr[Z > 2.68] = 0.00368$. Since the distribution is symmetric, we

multiply this by two to reflect the probability of being 2.68 standard deviations above or below the mean: $(0.00368)(2) = 0.00736$. **(e)** The standard error is the same as the standard deviation of the mean. It is equal to the standard deviation divided by the square root of *n*, or $0.560/\sqrt{10} = 0.18$ kg. To find the probability that the mean of a sample of 10 babies is greater than 3.5 kg, we transform this mean into a Z-score. $(3.5 - 3.296)/0.18 = 1.13$. $\Pr[Z > 1.13] = 0.129$.

7. (a) The lower graph has the higher mean (ca. 20 vs. 10), whereas the upper graph has the higher standard deviation. **(b)** The lower graph has the higher mean (ca. 15 vs. 10) and the higher standard deviation (ca. 5 vs. 2.5).

8. The standard deviation is approximately 10, as the region within one standard deviation from the mean will contain roughly 2/3 of the data points.

9. (a) In a normal distribution, the modal value occurs at the mean, so the mode is 35 mm. **(b)** A normal distribution is symmetric, so the middle data point is the mean, 35 mm. **(c)** Twenty percent of the distribution is less than 20 mm in size. (Why? Normal distributions are symmetric, so if 20% of the distribution is 15 mm or larger than the mean, 20% must be 15 mm or smaller than the mean.)

10. (a) The lower distribution (*ii*) would have sample means that had a more normal distribution, because the initial distribution is closer to normal. Both distributions would converge to a normal distribution if the sample size were sufficiently large. **(b)** The distribution of the sums of samples from a distribution will be normally distributed, given a sufficiently large number of samples.

11. (a) $\Pr[Y \geq 180] = \Pr[Z > (179.5 - np)/\sqrt{np(1 - p)}] = \Pr[Z > (179.5 - (400)(0.4))/\sqrt{(400)(0.4)(1 - 0.4)}] = \Pr[Z > 1.99] = 0.023$. **(b)** $\Pr[Y \geq 130] = \Pr[Z > (129.5 - (400)(0.4))/\sqrt{(400)(0.4)(1 - 0.4)}] = \Pr[Z > -3.11] = 1 - \Pr[Z > 3.11] = 0.999$. **(c)** $\Pr[155 \leq Y \leq 170] = \Pr[Y \geq 155] - \Pr[Y \geq 171]$. $\Pr[Y \geq 155] = \Pr[Z > (154.5 - (400)(0.4))/\sqrt{(400)(0.4)(1 - 0.4)}] = \Pr[Z > -0.56] = 1 - \Pr[Z > 0.56] = 0.712$

ANSWERS

(0.713 if calculated with a computer). $\Pr[Y \geq 171] =$
$\Pr[Z > (170.5 - (400)(0.4))/\sqrt{(400)(0.4)(1 - 0.4)}] =$
$\Pr[Z > 1.07] = 0.142.$ $\Pr[155 \leq Y \leq 170] = 0.69497 - 0.13136 = 0.570.$

12. Average expected number of cancer victims $= 0.14 \times 220 = 30.8.$ Standard deviation $= \sqrt{n\,p(1 - p)}$
$= \sqrt{220(0.14)(0.86)} = 5.15.$ Convert actual number of victims, 91, into standard normal deviate:
$(91 - \frac{1}{2} - 30.8)/5.15 = 11.60.$ This is off the chart: $P < 0.00002.$

13. $SE = s/\sqrt{n}.$ $Z = (Y - \text{mean})/SE.$

Mean	SD	Value	SE_{10}	Z_{10}	$\Pr(\bar{Y} > Value)$	SE_{30}	Z_{30}	$\Pr(\bar{Y} > Value)$
14	5	15	1.58	0.63	0.264	0.91	1.10	0.136
15	3	15.5	0.95	0.53	0.298	0.54	0.91	0.181
−23	4	−22	1.26	0.79	0.215	0.73	1.37	0.085
72	50	45	15.81	−1.71	0.956	9.18	−2.96	0.998

Chapter 11

1. (a) $n = 31.$ (b) $\bar{Y} = 39.329$ m.
(c) $s = 30.663$ m.
(d) $SE_{\bar{Y}} = s/\sqrt{n} = 30.633/5.567 = 5.507.$
(e) $df = n - 1 = 30.$ (f) $\alpha = 0.05.$
(g) $t_{0.05(2),30} = 204.$ (h) Variable has a normal
distribution in the population, and sample is random. (i) $\bar{Y} \pm t_{0.05(2),\,30}\,s,$ $28.09 < \mu < 50.56.$
(j) H_0: The mean elevation change is 0 ($\mu = 0$).
H_A: The mean elevation change is not 0
($\mu \neq 0$). (k) $t = (\bar{Y} - 0)/SE_{\bar{Y}} = 7.141.$
(l) Variable has a normal distribution in
the population, and sample is random. (m)
$t_{0.0001(2),\,30} = 4.48.$ Since $t > 4.48,$ $P < 0.0001$
(on the computer, $P = 6.06 \times 10^{-8}$). (n) Yes,
the average elevational range shifted upward.

2. (a) The population variance. (b) Variable has a
normal distribution in the population and sample
is random. (c) $s^2 = 940.23.$ (d) $df = 30.$
(e) $\alpha = 0.05.$ (f) $\chi^2_{0.025,\,30} = 46.98.$
(g) $\chi^2_{0.975,\,30} = 16.79.$
(h) $df\,s^2/\chi^2_{0.025,\,30} < \sigma^2 < df\,s^2/\chi^2_{0.975,\,30}.$
$600.4 < \sigma^2 < 1680.0.$

3. (a) $t_{0.05(2),\,11} = 2.20.$ (b) $t_{0.05(2),\,31} = 2.04.$
(c) $t_{0.01(2),\,100} = 2.63$ (is the same as $t_{0.01(2),\,120}$).
(d) $t_{0.05(2),\,22} = 2.07.$ (e) $t_{0.01(2),\,7} = 3.50.$

4. (a) $\chi^2_{0.025,\,11} = 21.92.$ $\chi^2_{0.975,\,11} = 3.82.$
(b) $\chi^2_{0.025,\,31} = 48.23.$ $\chi^2_{0.975,\,31} = 17.54.$

(c) $\chi^2_{0.005,\,100} = 140.17.$ $\chi^2_{0.995,\,100} = 67.33.$
(d) $\chi^2_{0.025,\,22} = 36.78.$ $\chi^2_{0.975,\,22} = 10.98.$
(e) $\chi^2_{0.005,\,7} = 20.28.$ $\chi^2_{0.995,\,7} = 0.99.$

5. The 99% confidence interval must be *larger* than
the 95% confidence interval in order to have a
higher probability of capturing the true mean.

6. (a) Answers will vary.

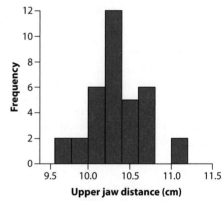

(b) The sample mean is 10.32 cm and the standard
error of the mean is 0.056 cm. The standard error
estimates the standard deviation of the sampling
distribution of sample means. (c) The 95% confidence interval for the mean is $10.32 \pm (0.056$
$t_{0.05(2),34}).$ $t = 2.03,$ so the 95% confidence interval
is 10.21 cm $< \mu < 10.44$ cm. (d) The variance is
0.11005. There are 35 individuals, so 34 $df.$ The
χ^2 critical values are looked up in Statistical Table

A for $\alpha = 0.01/2$ and $\alpha = 1 - (0.01/2)$; they are 58.96 and 16.50, respectively. We now calculate the lower bound of the confidence interval: $34(0.11005)/58.96 = 0.063$. The upper bound is $34(0.11005)/16.5 = 0.227$. $0.063 < \sigma^2 < 0.227$. **(e)** The 99% confidence interval of the standard deviation of the sample is found by taking the square root of the variance confidence interval: $0.252 < \sigma < 0.476$, around the estimated standard deviation of 0.332.

7. No, using $n = 70$ would assume that the right distance was independent of the left distance on the same animal, which is not true.

8. (a) Answers will vary. We've drawn a strip chart; grouped histograms and box plots would also work (although there are rather few data points). Polygamy tends to be associated with larger testes size than monogamy.

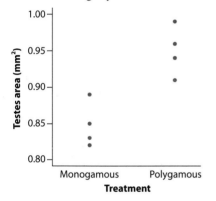

(b) The mean testes area for monogamous lines is 0.848, while the mean area for polyandrous lines is 0.950. The standard deviation was 0.031 and 0.034, respectively, for the two sets. **(c)** The standard error of the mean is 0.015 for the monogamous lines and 0.017 for the polyandrous lines. **(d)** The 95% confidence interval for the mean in the polyandrous line is $0.95 \pm 0.017 t_{0.05(2),3} = 0.95 \pm 3.18 (0.017)$, or $0.90 < \mu < 1.00$. **(e)** The 99% CI for the standard deviation of testes area among monogamous lines requires the sample variance (0.001), the degrees of freedom (3), and the critical values of the χ^2 distribution for the appropriate α and df. The χ^2

critical values are looked up in Statistical Table A for $\alpha = 0.01/2$ and $1 - (0.01/2)$; they are 0.07 and 12.84, respectively. Then, the 99% CI for the variance is $3(0.001)/0.07$ and $3(0.001)/12.84$, or 0.0002 to 0.04. To get the 99% CI for the standard deviation, we take the square root of each of these, or $0.015 < \sigma < 0.20$.

9. H_0: Mean continuity score is zero ($\mu = 0$). H_A: Mean continuity score is not zero ($\mu \neq 0$). $\overline{Y} = 0.183$, $s = 0.135$, $SE_{\overline{Y}} = 0.051$, $t = 3.59$, $df = 6$, $t_{0.05(2),6} = 2.45$. Since $t > 2.45$, reject H_0 ($P = 0.011$). Mean discontinuity score in natural food webs is greater than zero.

10. H_0: Rats do not choose one odor over another ($\mu = 50\%$). H_A: Rats choose one odor over another ($\mu \neq 50\%$). In this case, we want to see if rats do better than the chance performance, so we are comparing their scores with $\mu_0 = 50\%$. The mean is 68.4% and the standard deviation is 7.1%. There were seven rats, so $SE = 7.1/\sqrt{7} = 2.7$. $t = (68.4 - 50)/2.7 = 6.82$. The one-tailed critical value $t_{0.05(1),6} = 1.94$; $t > 1.94$, so we reject the null hypothesis that rats were doing as expected by chance ($P = 0.00024$). Assume that the seven rats were random, independent samples and that performance scores were normally distributed. Rats are able to remember the odor previously presented.

11. The confidence interval for the mean is symmetric, so the lower bound must be 0.3 kg below the mean, or at 2.9 kg.

12. The confidence interval of the variance is the square of the bounds for the standard deviation.

Standard deviation	Variance
$2.22 < \sigma < 4.78$	$4.93 < \sigma^2 < 22.85$
$20.6 < \sigma < 26.1$	$425.4 < \sigma^2 < 678.8$
$36.4 < \sigma < 59.6$	$1325.0 < \sigma^2 < 3552.2$
$13.63 < \sigma < 16.70$	$185.8 < \sigma^2 < 279.0$

13. (a) Answers may vary. A histogram is probably the best, but box plot and strip chart can also be used. Histogram is shown here. Graphing the data is a good idea because it allows us to easily see if there are outliers or the data depart drastically in other ways from the assumption of a normal population. The graph also allows us to see the trend in the data (values greater than 1 predominate, but the trend is not strong) and the variability present (most values are very close to 1).

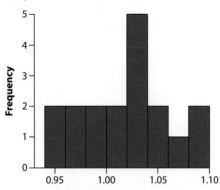

Relative swim speed in syrup

(b) H_0: Mean relative swim speed is 1 ($\mu = 1$). H_A: Mean relative swim speed is not 1 ($\mu \neq 1$). In this case, we wish to test whether the relative swimming speed in the syrup pool is different from $\mu_0 = 1$. We calculate t from the mean (1.0117) and the standard error (0.00997): $t = (1.0117 - 1)/0.00997 = 1.17$. The critical value for $\alpha(2) = 0.05$ and 17 df is 2.11, so we do *not* reject the null hypothesis that swimming speed remained constant in the two environments ($P = 0.26$). **(c)** The 99% confidence interval for the relative swimming speed in syrup is $1.01 \pm 0.01 \, t_{0.01(2),17} = 1.01 \pm 0.01$ (2.9), or $0.98 < \mu < 1.04$.

Chapter 12

1. (a) H_0: Death rate is the same under higher and lower estate tax rates ($\mu_d = 0$). H_A: Death rate differs under higher and lower estate tax rates ($\mu_d \neq 0$). **(b)** The study design is paired: each measurement of death rate from lower and higher tax rate periods is from the same year. **(c)** 2.72, 1.14, 1.72, -1, 2.43, 0.64, 3.71, -1.35, -0.07, -1.29, 6.29. **(d)** $\bar{d} = 1.358$ deaths/day. **(e)** $s_d = 2.356$. **(f)** $n = 11$. **(g)** $SE_{\bar{d}} = s_d/\sqrt{n} = 2.356/3.3166 = 0.7103$. **(h)** $t = (\bar{d} - 0)/SE_{\bar{d}} = 1.912$. **(i)** $df = 10$. **(j)** $t_{0.05(2), 10} = 2.23$. **(k)** Since $|t| < 2.23$, $P > 0.05$. **(l)** Do not reject H_0 ($P = 0.085$). We

are unable to reject the possibility that death rate changes when tax regime changes.

2. (a) H_0: The mean percentage of dreams in color is the same for younger and older people. H_A: The mean percentage of dreams in color differs between younger and older people. **(b)** Assumptions are (1) both samples are random samples, (2) both populations have a normal distribution for this variable, and (3) the variance in both populations for this variable is the same. The data are normally distributed, and the standard deviations are approximately equal in the samples (31.8 vs. 36.9). **(c)** $31.8^2 = 1011.24$ and $36.9^2 = 1361.61$. **(d)** $30 - 1 = 29$ for the young group and $30 - 1 = 29$ for the older group. **(e)** $s_p^2 = \dfrac{df_1 \, s_1^2 + df_2 \, s_2^2}{df_1 + df_2} =$

$\dfrac{29(1011.24) + 29(1361.61)}{29 + 29} = 1186.425$.

(f) $SE_{\bar{Y}_1 - \bar{Y}_2} = \sqrt{s_p^2 \left(\dfrac{1}{n_1} + \dfrac{1}{n_2} \right)} =$

$\sqrt{1186.425 \left(\dfrac{1}{30} + \dfrac{1}{30} \right)} = 8.89$.

(g) $t = \dfrac{\bar{Y}_1 - \bar{Y}_2}{SE_{\bar{Y}_1 - \bar{Y}_2}} = \dfrac{68.4 - 33.9}{8.89} = 3.88$.

(h) $df = n_1 + n_2 - 2 = 30 + 30 - 2 = 58$. **(i)** $t_{0.05(2), 58} = 2.00$. **(j)** $t = 3.88$ is greater than $t_{0.001(2), 58} = 3.47$, so we know that $P < 0.001$. **(k)** We reject the null hypothesis ($P = 0.0003$). These data provide evidence that the older group of people dream in color less often than the younger group. This is consistent with the hypothesis suggested by the researchers about exposure to black and white or color TVs.

3. (a) 8.89. **(b)** $df = n_1 + n_2 - 2 = 30 + 30 - 2 = 58$ **(c)** $1 - 0.95 = 0.05$. **(d)** 2.00 **(e)** $68.4 - 33.9 = 34.5$. **(f)** $\bar{Y}_1 - \bar{Y}_2 - t_{0.05(2), df} \, SE_{\bar{Y}_1 - \bar{Y}_2} < \mu_1 - \mu_2 < \bar{Y}_1 - \bar{Y}_2 + t_{0.05(2), df} \, SE_{\bar{Y}_1 - \bar{Y}_2}$ $34.5 - 2.00(8.89) < \mu_1 - \mu_2 < 34.5 + 2.00(8.89)$. $16.7 < \mu_1 - \mu_2 < 52.3$.

4. (a) Paired t-test. **(b)** Paired t-test. **(c)** Two-sample t-test. **(d)** Paired t-test. **(e)** Two-sample t-test. **(f)** Paired t-test. **(g)** Two-sample t-test. **(h)** Paired t-test.

5. The two-sample t-test cannot be used to test for differences in the means, since the standard deviations are more than threefold different between the two groups. Instead, a Welch's approximate t-test is appropriate.

6. The monogamous flies had a mean testes size of 0.8475 mm², and the polyandrous of 0.95 mm², for a difference of 0.1025 mm². $s_1^2 = 0.0010$ and $s_2^2 = 0.0011$. $SE_{\bar{Y}_1 - \bar{Y}_2} = 0.023$, $df = 6$. The confidence interval for the difference is the standard error times $t_{0.05(2),6} = 2.45$, so $0.1025 \pm (2.45 \times 0.023)$, or 0.046 to 0.159. **(b)** H_0: There is no difference between the monogamous and polyandrous treatments in testes size $(\mu_1 - \mu_2 = 0)$. H_A: Mean testes size differs between treatments $(\mu_1 - \mu_2 \neq 0)$. $t =$ the difference in means = 0.1025, over the SE of the difference, 0.023. $t = 4.48$, for 6 df. $t_{0.05(2),6} = 2.45$. Since $t > 2.45$, $P < 0.05$. Reject H_0 $(P = 0.042)$. The mean testes sizes are significantly different.

7. (a) On average, 33% more of the male bodies were covered if they emitted pheromones. $s_p = 26.5\%$; $SE_{\bar{Y}_1 - \bar{Y}_2} = 6.02\%$. $df_1 = 48$ and $df_2 = 31$, so $df = 79$. $t_{0.05(2), 79\ df} = 1.99$, so the confidence interval is $21\% < \mu_1 - \mu_2 < 45\%$. **(b)** Using a two-sample t-test, we will assume that the percent coverage is normally distributed, that each snake is independent, and that the standard deviations are not different (they are not more than threefold different). H_0: The means do not differ between the males emitting pheromones and those not, $\mu_1 - \mu_2 = 0$. H_A: The means differ between the two types of males $(\mu_1 - \mu_2 \neq 0)$. $t = 0.33/0.0602 = 5.48 > 1.99$, the critical value for $\alpha(2) = 0.05$ for 79 df, so we reject the null hypothesis, with $P < 0.05$ $(P = 0.0000005)$.

8. As described, this test assumes that the eight open-water samples were independent of the eight nearshore samples, because it uses a two-sample t-test (and so would have 14 df). Differences in growth rate could be due to differences between lakes, so the two samples within each lake are not independent. The paired t-test would better reflect this (and would have 7 df).

9. (a) Answers will vary. Multiple histograms is one of three effective methods (the others being strip chart and box plots). The two distributions have similar mean preferences, but the variance appears highest in the F_2 hybrids.

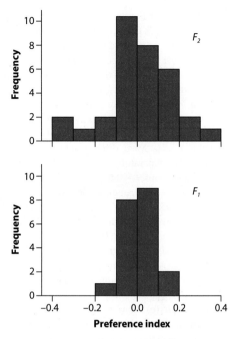

(b) $s_1^2 = 0.00411$, $s_2^2 = 0.02509$. The F_2 hybrids indeed have a higher sample variance in female preference index. **(c)** H_0: Variance of preference is the same between the two crosses $(\sigma_1^2 = \sigma_2^2)$. H_A: Variance of preference is different between the two crosses $(\sigma_1^2 \neq \sigma_2^2)$. $F = s_2^2/s_1^2 = 0.02509/0.00411 = 6.09$ (6.10 with rounding error). $df = 32, 19$. $F_{0.05(2), 32, 19} = 2.38$. Since $6.09 > 2.38$, $P < 0.05$. Reject H_0 $(P = 0.00012)$. Conclude that the variance is higher in the F_2 cross (suggesting a relatively small number of genes underlying the difference in preference between the two species).

10. H_0: Mean difference in number of promises is 0. H_A: Mean difference in the number of promises is not zero. $\bar{d} = 22.25$, $s_d = 13.98$, $SE_{\bar{d}} = 4.94$, $df = 7$, $t = 4.50$, $t_{0.05(2),7} = 2.36$. Since $t > 2.36$, $P < 0.05$. Reject H_0 $(P = 0.0028)$. On average, the winners make more promises.

11. (a) We would use Welch's approximate t-test for this comparison since the standard deviations differ between *Pteronotus* and the vampires by more than threefold. **(b)** The average strength was *higher* for *Pteronotus*, which is in the opposite direction as predicted by the model.

12. **(a)** We can use a paired *t*-test in this case, because the body temperature measurements are taken on the same individuals as the brain temperature measurements. To do this, we calculate the difference in temperature between brains and bodies for each ostrich, find the mean difference (0.648°C) and the standard error of the difference (0.116°C). H_0: Mean difference between brain and body temperature is zero. H_A: Mean difference is not zero. $t = 0.648/0.116 = 5.6$ with 5 *df*, which is greater than the critical value for $\alpha(2) = 0.05, 2.57$, so $P < 0.05$. We reject the null hypothesis of no difference between brain and body temperature ($P = 0.0025$). **(b)** While our test is significant, the deviation is the opposite of that predicted from observations of mammals in similar environments: brains are hotter than bodies in ostriches, not cooler.

13. **(a)** *ii*. **(b)** *ii*. **(c)** From *ii*, we can still be fairly confident that the groups are different, but we need to mentally double the size of the error bars to make this determination. **(d)** With sample sizes of 100, the standard errors will be one-tenth as great as the standard deviations. Graphs *i*, *ii*, and *iii* will be significantly different.

14. **(a)** Use a 95% confidence interval for the difference: $-10.68 < \mu_d < -0.94$ mm. This assumes that the distribution of change in length in the population is normal and that the iguanas are a random sample. **(b)** All values within the most-plausible range are negative and include, e.g., -1 mm, -10 mm, etc. The data are not consistent with no change or an increase in the mean. **(c)** $16.6 < \sigma_d < 23.6$ mm. **(d)** H_0: The mean change in length is zero ($\mu_d = 0$). H_A: The mean change in length is not zero. $t = -2.38, df = 63, t_{0.05(2), 63} = 2.00$. Reject H_0 ($P = 0.020$). Iguanas really did shrink.

15. H_0: Mean male aggressiveness is the same in the two groups. H_A: Mean aggressiveness is not equal. Two-sample *t*-test: $s_p^2 = 0.498$, $SE_{\bar{Y}_1 - \bar{Y}_2} = 0.252, t = 9.85, df = 44$, $t_{0.05(2), 44} = 2.02$. Since $t > 2.02, P < 0.05$. Reject H_0 ($P < 10^{-10}$). Males are more aggressive when mated with a neighboring female.

16. **(a)** Strip chart, box plot, and multiple histograms. The strip chart is shown below. There is a trend for the activation to be greater in the beer

group. The beer group also appears to be more variable.

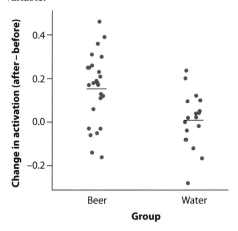

(b) The standard deviations of the two groups are not hugely different (0.162 vs. 0.126) so ordinary two-sample *t*-test OK. H_0: Mean changes for beer and water are not different ($\mu_1 - \mu_2 = 0$). H_A: Mean changes for beer and water are different ($\mu_1 - \mu_2 \neq 0$). $\bar{Y}_1 = 0.1544, \bar{Y}_2 = 0.0081$, $\bar{Y}_1 - \bar{Y}_2 = 0.1463, n_1 = 25, n_2 = 18$, $s_1^2 = 0.02633, s_2^2 = 0.01591, s_p^2 = 0.02201$, $SE_{\bar{Y}_1 - \bar{Y}_2} = 0.04586, t = 3.191, df = 41$, $t_{0.05(2), 41} = 2.02$. Since $3.191 > 2.02$, $P < 0.05$. Reject H_0 ($P = 0.0027$).

17. **(a)**

(b) Antibody mice: $\bar{Y}_1 = 83.125, s_1 = 13.58$, $SE_{\bar{Y}_1} = 4.80, df_1 = 7, t_{0.05(2), 7} = 2.36$, $71.8 < \mu_1 < 94.5$. Control mice: $\bar{Y}_2 = 9.333$, $s_2 = 7.50, SE_{\bar{Y}_2} = 3.06, df_2 = 5, t_{0.05(2), 5} = 2.57, 1.46 < \mu_2 < 17.20$. **(c)** Added to the plot in part (a). **(d)** The null hypothesis would be rejected, because the 95% confidence intervals do not overlap.

Chapter 13

1. (a) Distribution is heavily skewed to the right rather than symmetric.

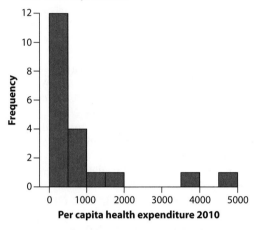

Per capita health expenditure 2010

(b) Distribution has positive skew (to the right) rather than negative skew, and all values are greater than zero. **(c)** Rounded to 2 decimals: 6.76, 5.77, 5.48, 6.78, 6.16, 4.28, 4.09, 6.41, 8.29, 8.45, 3.85, 5.19, 5.51, 4.69, 7.44, 5.00, 5.89, 4.34, 7.31, 6.52. **(d)** $n = 20$.
(e) $\bar{Y} = 5.910$. **(f)** $s = 1.346$.
(g) $SE_{\bar{Y}} = 0.301$. **(h)** Need $t_{0.05(2), 19} = 2.09$ to get $5.28 < \mu < 6.54$. **(i)** $196.6 < \mu < 691.7$.

2. (a) A histogram is probably the best choice. It shows that the data aren't skewed, but neither is the distribution bell-shaped. A high fraction of the values are either very close to the middle or in the tails of the distribution.

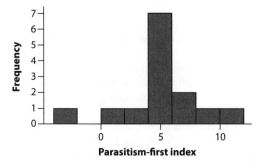

Parasitism-first index

(b) It is suitable as a substitute for the one-sample t-test, which compares a sample mean to a null hypothesized constant, when the data are not normal. **(c)** H_0: The median parasitism-first index is 0. H_A: The median parasitism-first index is not 0. **(d)** There are 12 positive values for the index, 1 negative value, and one zero.
(e) $n = 13$. **(f)** Pr[12 *or more positives*] =

$$\text{Pr}[12] + \text{Pr}[13] = \binom{13}{12}(0.5)^{12}(0.5) + \binom{13}{13}(0.5)^{13}(0.5)^0 = 0.0017.$$

Therefore $P = 2(0.0017) = 0.0034$. Reject the null hypothesis. **(g)** The goldeneye females tend to lay parasitic eggs first before raising their own offspring.

3. (a) The distributions are similar insofar as they are both have positive skew. However, the group with the recycle bin has a couple of outliers that greatly elevate the spread of the frequency distribution.

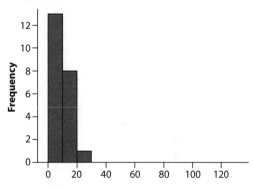

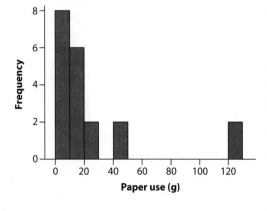

Paper use (g)

(b) A transformation might work (e.g., the log transformation) to render the distributions more symmetric and more similar in spread. If not, we should use a Mann-Whitney U-test or a permutation test. **(c)** H_0: Paper use in the two groups has the same distribution. H_A: Paper use in the two groups does not have the same distribution. **(d)** No bin (group 1): 4.5, 4.5, 4.5, 4.5, 4.5, 4.5, 4.5, 9.5, 12.5, 18, 18, 18, 18, 23, 23, 25, 28.5, 28.5, 28.5, 28.5, 32.5, 36.5. With bin (group 2): 4.5, 9.5, 12.5, 12.5, 12.5, 18, 18, 18, 23, 28.5, 28.5, 32.5, 34, 35, 36.5, 38, 39, 40, 41, 42. **(e)** $n_1 = 22$, $R_1 = 379.5$, $n_2 = 20$, $R_2 = 523.5$. **(f)** $U_1 = 313.5$. **(g)** $U_2 = 126.5$. **(h)** $U = 313.5$. **(i)** Use normal approximation because n_1 or $n_2 > 15$. $Z = 2.34$. $P = 2 \times 0.00964 = 0.01928$. Since $P < 0.05$, reject H_0 (exact $P = 0.0182$, obtained on a computer). People use more paper when there's a recycle bin.

4. **(a)** Not normal. The points show strong curvature. **(b)** Not normal. The curve bends steeply at high and low values. **(c)** Approximately normal. **(d)** Not normal. The points follow an S-shape, not a straight line.

5. **(a)** *i*. No, this is a uniform distribution, not a normal distribution.
ii. No, this plot is bimodal, not normal.
iii. No, this plot is right-skewed. It looks log-normal rather than normal.
iv. Yes, this is a normal distribution.
(b) *i*. The sign test would be best for these data, as they are unlikely to transform into anything resembling normal.
ii. The sign test would probably be best for this distribution as well. Bimodal distributions are tough to transform into anything else.
iii. These data could probably be tested by a one-sample t-test after transformation (probably a log transformation, as it is right-skewed).
iv. These data could be tested by a one-sample t-test, because they are a normal distribution.
(c) *i*: b. *ii*: d. *iii*: a. *iv*: c.

6. **(a)** Mean: 2.75; 95% CI: $-3.28 < \mu_{\log[X]} < 8.78$. **(b)** Mean: 1.86; 95% CI: $-0.02 < \mu_{\log[X]} < 3.74$. **(c)** Not possible: cannot use ln transformation on negative values. **(d)** Mean: 4.23; 95% CI: $-2.04 < \mu_{\log[X]} < 10.5$. **(e)** Here, we must use $Y' =$

$\ln[Y+1]$. Mean: 0.98; 95% CI: $-0.23 < \mu_{\log[X]} < 2.2$.

7. **(a)** Divide percentages by 100 to obtain fractions before transforming. Then take square root before applying the arcsine function. The transformed data are 0.577, 0.727, 0.869, 1.146, 1.085, 1.013 (displaying only the first 3 decimals). $\overline{Y}' = 0.903$, $s = 0.220$. **(b)** $df = 5$, $t_{0.05(2), 5} = 2.57$, $0.67 < \mu_{Y'} < 1.13$. **(c)** Convert by applying sine function, taking square, and then multiplying by 100 to get percent: $38.8 < \mu_Y < 82.1$.

8. H_0: Delay to reproduction is the same when cubs die from infanticide and accidentally. H_A: Delay to reproduction is not the same when cubs die from infanticide and accidentally. Mann–Whitney U-test: $U_1 = 39$. $U_2 = 6$, so $U = 39$. The critical value for $n_1 = 5$ and $n_2 = 9$ is 38. U is larger, so $P < 0.05$. Reject H_0 ($P = 0.029$). The delay is longer when cubs die by infanticide. If you used a permutation test on the difference between means, you would obtain a P-value of approximately 0.01.

9. **(a)** Benton: $\overline{Y}_1 = 0.67$ and $s_1 = 0.274$; Warrenton: $\overline{Y}_2 = 0.244$ and $s_2 = 0.0814$. We used a strip chart, but histograms would also work.

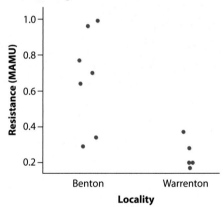

The standard deviations are more than three-fold different, so a two-sample t-test is inappropriate. There is also some skew, but it is not large. **(b)** There are at least four options:
1. Log transformation, followed by a two-sample t-test. 2. Welch's t-test on the non-transformed data, which does not require equal variances. 3. Mann–Whitney U-test. 4. Permutation test.
(c) A log transformation might work because

the population with the larger mean has the larger variance, and all the measurements are greater than zero. H_0: The means are the same ($\mu'_1 - \mu'_2 = 0$). H_A: The means are different ($\mu'_1 - \mu'_2 \neq 0$). $\overline{Y}'_1 = -0.490$, $\overline{Y}'_2 = -1.451$, $\overline{Y}'_1 - \overline{Y}'_2 = 0.961$. $SE_{\overline{Y}'_1-\overline{Y}'_2} = 0.249$, $df = 10$, $t = 0.961/0.249 = 3.87$, $t_{0.05(2), 11} = 2.23$. Since $t > 2.23$, reject H_0 ($P = 0.003$). (d) $0.407 < \mu'_1 - \mu'_2 < 1.515$. (e) The back-transformed confidence interval is $1.50 < \mu_1 - \mu_2 < 4.55$. This interval represents the most-plausible range of values for the difference between means. If we took many samples of the same size from the same population, 95% of the confidence intervals from these samples would contain the true mean.

10. Two-sample t-tests and confidence intervals are robust to violations of equal standard deviations so long as the sample sizes of the two groups are roughly equal and the standard deviations are within three times of one another.

11. (a) For example: Singleton: 3.5, 3.5, 2.6, 4.4; Twin: 3.4, 4.2, 3.4.; or Singleton: 3.5, 4.2, 4.4, 3.4; Twin: 2.7, 2.6, 1.7. (b) No.

12. (a) Two-sample t-test. (b) These distributions are skewed right, so a log-transformation could be tried. (c) This one is tough: neither distribution is normal, but they deviate in different ways, so a transformation that would improve the fit for one would not help with the other. Also, the standard deviations will be very different. The differences in shape make the rank-sum test inappropriate. A permutation test is best. (d) Two-sample t-test. (e) Two-sample t-test.

13. (a) The null hypothesis was that the spermatids of males and hermaphrodites did not differ in size; the alternative hypothesis is that they do differ in size. (b) $U = 35910$, $n_1 = 211$, $n_2 = 700$. Use the normal approximation:

$$Z = \frac{2U - n_1 n_2}{\sqrt{n_1 n_2 (n_1 + n_2 + 1)/3}} = -11.32.$$

$P < 0.00002$, which is the smallest value available in Statistical Table B. (c) Because the distributions are not the same shape. Equal shape is an assumption of the U-test when used to compare means or medians.

14. The differences are not normally distributed, but skewed right. We need a sign test. H_0: The median difference in the number of species between groups is zero. H_A: The median difference is not zero. There are 6 "−" and 22 "+," where a "+" means more species in the monoecious group.

$$P = 2\left[\sum_{X=22}^{28} \binom{28}{X}(0.5)^{28-X}(0.5)^X \right] = 0.0037$$

Since $P < 0.05$, we reject the null hypothesis.

15. First we transform the data by taking the natural log of each weight. The mean ln-transformed weights are -1.37 for females and -1.76 for males. Since the transformed weights are normally distributed, we can use the two-sample t-test. H_0: Male and female means are equal. H_A: The means are not equal. $t = 3.51$ with $df = 18$. The critical value for $\alpha(2) = 0.01$ for 18 df is 2.88, so $P < 0.01$. Since $P < 0.05$, we reject the null hypothesis ($P = 0.0025$). Female mosquitoes weigh more than males.

16. (a) No, the differences are probably not normally distributed: two values less than 500 and three values above 24,000 would not be expected from a normal distribution. (b) We can use a sign test: we have five pairs where there are more species feeding on angiosperms than on gymnosperms. The probability of five out of five in the binomial distribution, assuming equal probabilities of either outcome, is $(0.5)^5$, or 0.031. However, for a two-tailed test, we must double this, so $P = 0.062$. We are unable to reject the null hypothesis with the usual significance level. (c) The number of species that feed on angiosperms was often two orders of magnitude greater, and all pairs had a higher number of species in the angiosperm group. It is impossible to get a P-value under 5% with only five data points in a sign test, even with all the data in a consistent direction.

17. Graph (ii) has the real data. In (i), there is little or no relationship between the two variables, whereas (ii) shows a strong relationship. Since permutation tends to break up associations, it is more likely that (ii) is the data and (i) is a permutation.

ANSWERS

Review 2

1. (a) $0.5/100 = 0.005$. (b) $(1 - 0.005)^{10} = 0.951$. (c) $1 - (1 - 0.005)^{20} = 0.095$.
(d) $(.005)(0.80) = 0.004$. (e) $(.005)(0.80) + (0.995)(0.12) = 0.123$. (f) Apply Bayes' Theorem: $\Pr[carrier \mid cancer] = \Pr[cancer \mid carrier] \Pr[carrier]/\Pr[cancer] = (0.80)(0.005)/((.005)(0.80) + (0.995)(0.12)) = 0.032$.

2. H_0: The mean of measurements is 10.0. H_A: The mean is not 10.0. $\bar{Y} = 10.01$ µg, $s = 0.2$ µg and since $n = 30$, $SE_{\bar{Y}} = 0.2/\sqrt{30} = 0.0365$ µg. We test whether the mean sampled weight, 10.01 µg, differs from the expected weight of 10 µg. $t = (10.01 - 10.0)/0.0365 = 0.274$, $df = 29$. Since t not greater than $t_{0.05(2), 29} = 2.05$, do not reject the null hypothesis ($P = 0.786$).

3. (a) Histogram of differences in species number for climbing and non-climbing clades.

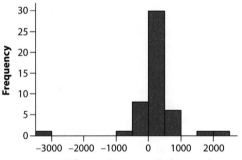

Difference between climbing and non-climbing species number

(b) The sign test is appropriate. H_0: The median difference in the number of species is 0. H_A: The median difference is not 0. There are 48 genera. 10 have more species in the non-climbing clade, 38 have more in the climbing clade.

$$P = 2\left(\sum_{x=38}^{48} \binom{48}{x}(0.5)^{48-x}(0.5)^{x}\right) = 0.00006.$$

Since $P < 0.05$, reject H_0. Genera with climb-

ing vines have more species than expected by chance.

4. (a) Leaving out rows where there are zero F_1 plants, sample mean is $\bar{Y}_1 = [55(4) + 58(10) + 61(41) + \cdots + 70(3)]/173 = 63.53$. Sample variance is $s_1^2 = [(55 - \bar{Y}_1)^2(4) + (58 - \bar{Y}_1)^2(10) + \cdots + (70 - \bar{Y}_1)^2(3)]/(173 - 1) = 8.62$. (b) H_0: The variances of the F_1 and F_2 populations are the same ($\sigma_1^2 = \sigma_2^2$). H_A: The variances of the F_1 and F_2 populations are not the same ($\sigma_1^2 \neq \sigma_2^2$). Assume that data are from normally distributed populations and use the F-test. $F = 42.37/8.62 = 4.92$ (4.91 if no rounding used). $df = 443, 172$. The closest we can get to the correct degrees of freedom in Statistical Table D is $F_{0.05(2), 400, 100} = 1.39$ (round down the df to be conservative; the correct critical value for $df = 443, 172$ is 1.24). Since $4.92 > 1.39$, $P < 0.05$. Reject H_0 ($P < 10^{-10}$). Conclude that the variance of the F_2 plants is greater than the variance of the F_1 plants.

5. (a) Bar graph is best:

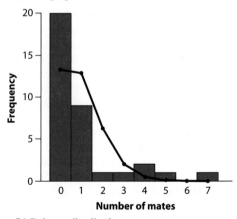

Number of mates

(b) Poisson distribution.
(c) $\bar{X} = 0.971$. Applying the Poisson formula yields the expected frequencies shown in the table at the bottom of the page.
(d) Compared with the Poisson distribution, the data include too few males with an intermediate

TABLE FOR PROBLEM 5(c)								
Number of mates	0	1	2	3	4	5	6	7
Observed frequency	20	9	1	1	2	1	0	1
Expected frequency	13.25	12.87	6.25	2.02	0.49	0.10	0.02	0.00

number of mates, too many males having 0 mates, and too many males having a large number of mates. **(e)** H_0: The number of mates follows a Poisson distribution. H_A: The number of mates does not fit a Poisson distribution. Categories need to be grouped so that no more than 20% of expected frequencies are less than 5. Group data into three categories: 0, 1, and ≥ 2. The observed and expected values for the category $\geq$ 2 are 6 and 8.88. $\chi^2 = 5.54$, $df = 1$, $\chi^2_{0.05, 1} = 3.84$. Since $\chi^2 > 3.84$, $P < 0.05$. Reject H_0 ($P = 0.019$). Number of mates does not follow a Poisson distribution.

6. (a) A log-transformation could make the data more normally distributed, allowing use of a 2-sample t-test. Alternatively, a Mann-Whitney U-test could be used if the data are still highly skewed after transformation. **(b)** After log-transformation the standard deviations are similar (0.397 and 0.540) and the distributions are not too skewed, so a two-sample t-test on the log-transformed data is reasonable. H_0: Mean GnRH mRNA levels are the same in territorial and non-territorial fish ($\mu_1 - \mu_2 = 0$). H_A: Mean GnRH mRNA levels differs between territorial and non-territorial fish ($\mu_1 - \mu_2 \neq 0$). For non-territorial fish $\overline{Y}_1 = -0.359$, $s_1 = 0.397$. For territorial fish, $\overline{Y}_2 = 0.460$, $s_2 = 0.540$. $s_p^2 = 0.217$, $SE_{\overline{Y}_1 - \overline{Y}_2} = 0.282$. $t = -2.90$, $df = 9$. $t_{0.05(2), 9} = 2.26$. Since $|t| > 2.26$, $P < 0.05$. Reject H_0 ($P = 0.018$). Mean hormone levels are higher in territorial males. **(c)** $-1.46 < \mu_1 - \mu_2 < -0.181$.

7. (a) Calculate $d =$ injuries on No Smoking Day minus injuries before. $\overline{d} = 25.0$. **(b)** Calculate as $\overline{d} \pm t_{0.01(2), 9} SE_{\overline{d}}$. $t_{0.01(2), 9} = 3.25$. $SE_{\overline{d}} = 10.22$, yielding $-8.2 < \mu_d < 58.2$. **(c)** If 100 random samples are collected from the same population and a 99% confidence interval is computed each time, on average we expect that 99 of the confidence intervals would include the true mean difference μ_d. **(d)** Use the paired t-test at significance level $\alpha = 0.05$. H_0: The mean difference μ_d is zero. H_A: The mean difference μ_d is not zero. $t = (25 - 0)/10.22 = 2.45$, $df = 9$, $t_{0.05(2), 9} = 2.26$. Since $t > 2.26$, $P < 0.05$. Reject H_0 ($P = 0.037$). We conclude that the accident rate is higher on No Smoking Day.

8. (a)

The probability that the match lasts two sets is $1/4 + 1/4 = 1/2$. The probability that it lasts three games is $1/8 + 1/8 + 1/8 + 1/8 = 1/2$.

(b)

The probability of the weaker player winning is $(0.45^2 \times 0.55) + (0.45^2 \times 0.55) + (0.45)^2 = 0.425$.

9. Use a binomial test. H_0: The ants have no preference ($p = 0.50$). H_A: Ants prefer one arm more than the other arm ($p \neq 0.50$). $P = 2 \times \Pr[X \geq 14] = 2(\Pr[14] + \Pr[15] + \cdots + \Pr[20])$. Use the binomial distribution to calculate

$$\Pr[14] = \binom{19}{14}(0.50)^{14}(0.50)^5 = 0.0222, \text{ etc.}$$

$P = 2 \times 0.0318 = 0.0636$. Since $P < 0.05$, do not reject H_0. These data do not indicate a preference for one arm over the other.

10. (a) H_0: Stickleback numbers per square meter fit a Poisson distribution. χ^2 goodness-of fit test. **(b)** H_0: The proportion of infected trees is equal for Douglas-fir and Western hemlock. χ^2 contingency analysis. **(c)** H_0: The proportion of infected trees is equal for Douglas-fir and Western hemlock. Fisher's exact test. **(d)** H_0: Change in patients' body mass is on average zero. Paired t-test. **(e)** H_0: Amount of rainfall per day in a rainforest has a normal distribution. Shapiro-Wilk test. **(f)** H_0: Male and female bald eagles weigh the same on average. Welch's t-test. **(g)** H_0: Male and female sperm whales travel the same distance per day. Mann–Whitney U-test. **(h)** H_0: Cats have equal strength in their two front paws. Paired t-test. **(i)** H_0: Chirps per minute are the same at 15°C and at 25°C. Paired t-test. **(j)** H_0: Mean number of bacteria per ml is 130 individuals per 100 l. One-sample t-test.

11. Precision. Non-independence tends to lead to standard errors and confidence intervals that are too small.

12. (a) $\Pr[Y < 110] = \Pr[Z < (110 - 124.6)/6.5] = \Pr[Z < -2.25] = \Pr[Z > 2.25] = 0.01222$. **(b)** $\Pr[130 < Y < 140] = \Pr[(130 - 124.6)/6.5 < Z < (140 - 124.6)/6.5] = \Pr[0.83 < Z < 2.37] = \Pr[Z > 0.83] - \Pr[Z > 2.37] = 0.20327 - 0.00889 = 0.19438$. **(c)** $\Pr[120 < Y < 125] = 1 - \Pr[Y < 120] - \Pr[Y > 125]$. $\Pr[Y < 120] = \Pr[Z < (120 - 124.6)/6.5] = \Pr[Z < -0.71] = \Pr[Z > 0.71] = 0.23885$. $\Pr[Y > 125] = \Pr[Z > (125 - 124.6)/6.5] = \Pr[Z > 0.06] = 0.47608$. $1 - 0.23885 - 0.47608 = 0.28507$.

13. H_0: The number of gestures is the same in sighted and blind people. H_A: The number of gestures is different between sighted and blind people. The Mann–Whitney U-test is appropriate, but there will be many ties, so the test will be less powerful than it might be. We assign a rank of "3" to the zeros (the average of 1–5), a rank of 11.5 to the 1's (the average of 6–17), a rank of 19.5 to the 2's (the average of 18–21), and a rank of 23 to the 3's (the average of 22–24). $R_1 = 140.5$, $R_2 = 159.5$, $U_1 = 81.5$, $U_2 = 62.5$, $U = 81.5$. $U_{0.05(2), 12, 12} = 107$. U is not ≥ 107, so we do not reject the null hypothesis (approximate $P = 0.58$). (A permutation test yields approximately $P = 0.66$).

14. (a) Mosaic plot or grouped bar graph. **(b)** The frequency table, with row and column sums is

	Knuckle-crackers	Not	Sum
Osteoarthritis	24	111	135
No osteoarthritis	19	61	80
Sum	43	172	215

$\widehat{OR} = 24 \times 61/(111 \times 19) = 0.694$.
(c) $0.35 < OR < 1.37$. **(d)** H_0: Knuckle-cracking and osteoarthritis are independent. H_A: Knuckle-cracking and osteoarthritis are not independent. Expected frequencies are

	Knuckle-crackers	Not	Sum
Osteoarthritis	27	108	135
No osteoarthritis	16	64	80
Sum	43	172	215

$\chi^2 = 1.12$, $df = 1$, $\chi^2_{0.05, 1} = 3.84$. Since χ^2 not greater than 3.84, do not reject H_0 ($P = 0.29$).

15. (a) -0.55. **(b)** $-1.07 < \mu < -0.03$. **(c)** $s^2 = 4.66$; $3.42 < \sigma^2 < 6.73$.

16. (a) *Spd* mutant: $\bar{Y}_1 = 142.1$, Wild type: $\bar{Y}_2 = 74.0$. Difference: $\bar{Y}_1 - \bar{Y}_2 = 68.1$, $s_1 = 37.74$, $s_2 = 52.51$, $s_p^2 = 2126.0$, $df = 19$, $SE_{\bar{Y}_1 - \bar{Y}_2} = 20.1$, $t_{0.05(2), 19} = 2.09$. $26.0 < \mu_1 - \mu_2 < 110.2$ ($25.9 < \mu_1 - \mu_2 < 110.3$ if calculated in a computer package, avoiding rounding error). **(b)** H_0: The two genotypes have the same mean time spent in aggressive activity ($\mu_1 - \mu_2 = 0$). H_A: The two genotypes do not have the same mean time spent in aggressive activity

$(\mu_1 \neq \mu_2 = 0)$. $t = 3.38$. Since $t > t_{0.05(2), 19}$, $P < 0.05$. Reject H_0. The weight of evidence against H_0 is indicated by the P-value, which on a computer is calculated to be $P = 0.0031$. Using the $t_{\alpha(2), 19}$ values in Statistical Table C, P is found to be between 0.01 and 0.001.

Chapter 14

1. **(a)** Limit sampling error. **(b)** Reduce bias. **(c)** Reduce sampling error. **(d)** Reduce bias. **(e)** Reduce bias.
2. **(a)** [Answers will vary.] T, H, H, H, H, T, T, H. **(b)** [Answers will vary.] No. **(c)** $1 - \Pr[\text{exactly 4 heads in 8 tosses}]$

$$= 1 - \binom{8}{4}(0.5)^8 = 0.727.$$

(d) [Answers will vary.] Assign a random number between zero and one to each unit. Assign the first treatment to the units with the four smallest random numbers and the second treatment to the remaining units.
3. Use a randomized block design, where each block is a position along the moisture gradient. Place three plots in each block, one for each of the three fertilizer treatments (call the fertilizers A, B, and C). Within each block, randomly assign the three fertilizers plots. The figure below illustrates the design for six blocks.

Block

Moisture gradient ⟶

4. The researchers planned their sample size assuming a significance level of 0.05 and an 80% probability of rejecting a false null hypothesis (for a specified difference between treatment means).

5. Observational study. The treatments, presence and absence of brook trout, were not randomly assigned to the units (streams)—the trout were already present in the streams prior to the study. Potential confounding factors (water temperature, stream depth, food supply) might differ between streams with and without brook trout, and randomization was not used to break their association with the treatment variable.
6. **(a)** Decrease bias (reduces effects of confounding variables); decrease sampling error (by grouping similar units into pairs). **(b)** No effect on bias; reduce sampling error. **(c)** Decrease bias (corrects for effects of age, a possible confounding variable); no effect on sampling error. **(d)** No effect on bias; reduce sampling error.
7. **(a)** No. In an experimental study, the experimenter assigns two or more treatments to subjects. Here, there was only one treatment. **(b)** Have two treatments: the salt infusion and a placebo control (e.g., distilled-water infusions). Assign treatments randomly to a sample of severe pneumonia patients. Ensure equal numbers of patients in each treatment. Keep patients unaware of which treatment they are receiving. A clinician unaware of which patient received which treatment should record their subsequent condition.
8. **(a)** Observational study: cancer treatments were not assigned randomly to subjects. **(b)** Yes, it compares marijuana use of cancer patients and non-cancer (control) patients. **(c)** Reducing bias. Age and sex might be confounding variables, affecting both marijuana use and cancer incidence. Using only subjects similar in age and sex reduces the effect of these confounding variables on the association between marijuana use and cancer. **(d)** First, it is not wise to "accept" the null hypothesis, because the study might not have had sufficient power to detect an effect. The 95% confidence interval for the odds ratio ranged from 0.6 to 1.3, which still includes the possibility of a moderate effect. Second, observational studies such as this one cannot decide causation because of confounding variables. Perhaps marijuana use does increase cancer risk, but marijuana users may have lifestyle differences that diminish cancer risk, offsetting any marijuana effect.

ANSWERS

9. (a) The stings were applied to two volunteers, which means that the reactions to the 40 stings were probably not independent. Treating the sample size as 40 is a case of pseudoreplication. (b) More conservatively and appropriately, this study should be treated as a paired design with two samples. The mean difference in swelling would be found for each subject, and then the average of the two subjects would be tested to see if it was significantly different from zero. (c) Since each subject adds just one data point, there is no need to inflict 20 stings on each subject. Fewer stings per subject would be less cruel, and more subjects would allow more replicates and thus more power.

10. Extreme doses increase power and so enhance the probability of detecting an effect. However, the effects of a large dose might be very different from effects of a smaller, more realistic dose. If an effect is detected, then studies of the effects of more realistic doses would be the next step.

11. (a) 13 plots per treatment. The "margin of error" is $0.4/2 = 0.2$, so $n = 8(0.25/0.2)^2 = 12.5$; round up to 13. (b) 50 plots per treatment. The "margin of error" is $0.2/2 = 0.1$, so $n = 8(0.25/0.1)^2 = 50$. (c) A greater total sample size would be needed if the design were not balanced. For a given total sample size, the expected width of a confidence interval for the difference between two means increases as the design is more imbalanced (because the precision of the treatment mean having the lower sample size is greatly reduced). Achieving the same confidence interval width as a balanced design will therefore require a greater total sample size. (d) Because environmental differences between the normal-corn plot and the Bt-corn plot would be confounding variables. In effect, such a design would lack replication because plants in the same plot are not independent.

12. 16 plots per treatment. $n = 16(0.25/0.25)^2 = 16$.

13. The study should have a control group receiving a placebo treatment. Without it we cannot estimate the effect of the treatment.

14. Replication; balance (same numbers of treated and untreated eyes); blocking (treated and untreated eyes were paired); control (untreated eyes; sham surgery or transplant from a blind cave fish would have provided a more complete control); randomization (eye to be treated was chosen randomly on each fish). Ironically, blinding was not used.

15. (a) Blocking. (b) Reduce sampling error (by eliminating the effect of date on the response variable).

16. (a) The second study used randomization and double-blinding, whereas the first study did not, It also used an appropriate control group, in contrast to the Cameron and Pauling study, which compared living patients given the vitamin C treatment to a sample of dead people. (b) The second study should be more reliable. Randomization removes effects of confounding variables, and blinding avoids unconscious bias. The poor choice of a control group by Cameron and Pauling increased the possibility of confounding by unmeasured variables.[1]

Chapter 15

1. (a) H_0: Mean amount of nectar doesn't differ among caffeine concentrations. H_A: Mean amount of nectar differs among caffeine concentrations.

(b)

Caffeine (ppm)	$\bar{Y}_i$	s_i	n_i
50	0.008	0.289	5
100	−0.172	0.169	5
150	0.376	0.309	5
200	0.378	0.393	5

(c)–(i) $\bar{Y} = 0.1475$.

Source	Sum of Squares	df	Mean Squares	F
Groups	1.1344	3	0.3781	4.18
Error	1.4482	16	0.0905	
Total	2.5826	19		

(j) $F_{0.05(1), 3, 16} = 3.24$. Since $F > 3.24$, $P < 0.05$. Reject H_0 ($P = 0.023$).

1. As Paul Rosenbaum (2002, p.4) says, "*in the control group, one can say with total confidence, without reservation or caveat, that the prognosis of a patient already dead is not good.*"

ANSWERS

2. (a) H_0: The pair of population means is different, $\mu_i = \mu_j$ for all $j > i$.

(b)–(f)

Groups	100	50	150	200
$\bar{Y}_i$	−0.172	0.008	0.376	0.378
	a	*a, b*	*b*	*b*

(g) The means for treatment "100" and the treatments "150" and "200" fall into distinct groups, whereas placement of the mean for "50" is ambiguous, as it cannot be distinguished statistically from either group.

3. (a) $s_A^2 = \dfrac{\text{MS}_{\text{groups}} - \text{MS}_{\text{error}}}{n} =$ $(163.71 - 119.56)/2 = 22.07$. This estimates the variance among individuals.
(b) Repeatability $= s_A^2/(s_A^2 + \text{MS}_{\text{error}}) =$ $22.07/(22.07 + 119.56) = 0.156$. **(c)** About 16% of the variation in the data is due to real differences between individuals, and about 84% of the variation is caused by difference between tests for the same student. **(d)** Assume that the distribution of test scores within students is normal, that the students are random sample of the population of students, and that mean exam grades of different students follow a normal distribution.

4. (a) Explanatory: population isolation. Response: generations persisted. **(b)** Experimental study: treatments were assigned to plants by the experimenters. **(c)** We assume that the variable has a normal distribution. (We also assume random samples.)

Treatment	Mean	95% CI
Isolated	9.25	$5.3 < \mu < 13.2$
Medium	14.50	$11.5 < \mu < 17.5$
Long	10.75	$8.0 < \mu < 13.5$
Continuous	12.75	$8.2 < \mu < 17.3$

(d) [Answers may vary.] In the following figure, open circles are the data (offset where needed to minimize overlap). Means are filled circles. Vertical lines indicate 95% confidence intervals.

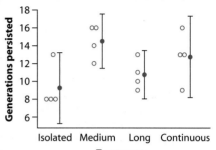

5. (a) H_0: The mean persistence times of the four isolation treatments are equal ($\mu_1 = \mu_2 = \mu_3 = \mu_4$). H_A: At least one of the means μ_i is different. **(b)** See the table at the bottom of the page. **(c)** F-distribution with 3 and 12 degrees of freedom. **(d)** The probability of obtaining an F-ratio statistic as large as or larger than the value observed when the null hypothesis is true. **(e)** The sums of squares measure the variation among all individuals ($\text{SS}_{\text{groups}}$), which is separated into the variation among individuals within groups and variation among groups. **(f)** Use R^2, the ratio of the group sum of squares and the total sum of squares. **(g)** $R^2 = 63.188/126.438 = 0.50$.

6. (a) H_0: The mean of group i equals the mean of group j, for all pairs of means $j > i$. H_A: The mean of group i does not equal the mean of group j, for all pairs of means. **(b)** These are unplanned comparisons, because they are intended to search for differences among all pairs of means. Planned comparisons must be few and identified as crucial in advance of gathering and analyzing the data. **(c)** Failure to reject a null hypothesis that the difference between a given pair of means is zero does not imply that the means are equal, because power is not necessarily high, especially when the differ-

TABLE FOR PROBLEM 5(b)

Source of variation	Sum of squares	df	Mean squares	F-ratio	P
Groups	63.188	3	21.063	3.996	$P < 0.05$ ($P = 0.035$)
Error	63.250	12	5.271		
Total	126.438	15			

ences are small. If the means of the "medium" and "isolated" treatments differ from one another, then one or both of them must differ from the means from the other two groups, but we don't know which.

(d)

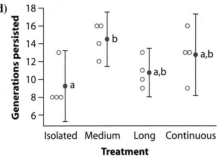

(e) The critical value for a t-test has a Type I error rate of 0.05 when comparing two means. The critical value for the Tukey comparison of all pairs of means is larger so that the probability of making at least one Type I error in *all* of the comparisons is only 0.05. (Note that degrees of freedom are 12 in either case because we are using the error mean square to calculate the standard error of the difference between means.)

7. (a) Transformations or a nonparametric Kruskal–Wallis test. **(b)** The transformation should be attempted first, because this would yield the more powerful test.

8. (a) $\bar{Y}_1 - \bar{Y}_2 = -0.004 - (-0.195) = 0.191$. $MS_{error} = 0.0345$, $df = 42$, $SE = 0.0679$, $t_{0.05(2),\,42} = 2.02$, $0.054 < \mu_1 - \mu_2 < 0.328$. **(b)** Yes, because the study was designed mainly to compare *PLP1* gene expression in persons with schizophrenia to that of control individuals. It was a single focused comparison, not a broad search for differences between groups. **(c)** The expression measurements are normally distributed in the populations with equal variances. (We also assume random samples.)

9. (a) H_0: Mean *PLP1* gene expression is equal in the three groups ($\mu_1 = \mu_2 = \mu_3$).
H_A: At least one of the group means μ_i is different.

Source of variation	Sum of squares	df	Mean squares	F-ratio	P
Groups	0.5403	2	0.2701	7.82	0.0013
Error	1.4502	42	0.0345		
Total	1.9905	44			

The critical value is $F_{0.05(1),\,2,\,42} = 3.22$. Since $F > 3.22$, $P < 0.05$, we reject H_0. Conclude that mean *PLP1* expression differs among groups. **(b)** The expression measurements are normally distributed in the populations with equal standard deviations. **(c)** Fixed-effects ANOVA: we are comparing predetermined and repeatable treatment groups, not a random selection of groups in a population. **(d)** Use R^2 to describe the fraction of the variance explained by group differences: $R^2 = 0.27$. **(e)** Use the Tukey–Kramer method.

10. (a) H_0: Mean carotenoid plasma concentration in vultures is the same among the four sites.
H_A: Mean carotenoid plasma concentration in vultures differs among the four sites.

Source of variation	Sum of squares	df	Mean squares	F-ratio	P
Groups (sites)	712.25	3	237.42	30.44	0.0001
Error	1388.26	178	7.80		
Total	2100.51	181			

The critical value is $F_{0.05(1),\,3,\,178} = 2.66$. Since $F > 2.66$, $P < 0.05$, we reject H_0. Conclude that sites vary in the mean carotenoid plasma concentration. **(b)** We assume that carotenoid plasma concentration has a normal distribution in every population with equal variances. (We also assume random samples.)

11. (a) Random-effects ANOVA: the males were chosen at random from the population. A given male is not a specific, repeatable treatment. The goal is to generalize to the population of males (the hypothesis state-

ANSWERS

ments should reflect this). **(b)** See the table at the bottom of the page. $s_A^2 = (0.9036 - 0.1951)/3 = 0.236$. Repeatability $= 0.236/(0.236 + 0.195) = 0.548$.

12. (a) ANOVA. **(b)** Try transforming the data to better meet the assumptions of normality and equal variances. If this fails, use the Kruskal–Wallis test if the distributions have equal shape. **(c)** ANOVA is appropriate if sample size is large enough (appealing to the Central Limit Theorem). **(d)** Tukey–Kramer test of all pairs of means.

13. (a) Histogram is shown. Boxplot or strip chart is also effective. Body color is more green in the infected group compared to the original and uninfected groups.

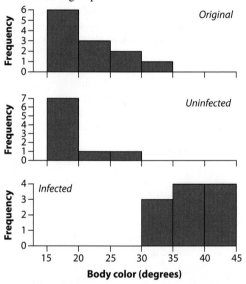

Body color (degrees)

(b) The data in the original and uninfected groups have positive skew (skewed to the right). The data in the infected group have a uniform (flat) distribution rather than bell-shaped, but sample size is small. **(c)** H_0: The distribution of body color is the same among groups. H_A: The

distribution of body color is not the same. **(d)** $df = 2$, $\chi^2_{0.05, 2} = 5.99$. Since $H > 5.99$, $P < 0.05$. Reject H_0 ($P = 0.000026$). Conclude that the groups differ in color. **(e)** The ranks (in particular, the rank sums of each group). **(f)** That the distributions have the same shape. It doesn't appear that the distributions here have the same shape (although sample size is not large enough to be absolutely sure).

14. (a) The figure below shows the proportion of flies that took a second blood meal from cows in the two groups.

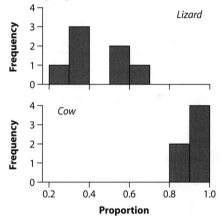

Proportion

(b) ANOVA assumes that the measurements in the two populations are normally distributed with equal variance. These assumptions do not appear to be met in the present data. In flies given a first blood meal from a cow, the measurements do not appear to be normally distributed and the variance is low compared with flies given a first blood meal from a lizard. **(c)** The data are proportions, so the arcsine transformation is the logical first choice for a transformation. This does indeed largely fix the problem: the data appear more normal and the variances are similar in the two groups. **(d)** H_0: The means of the two treatment groups are the same ($\mu_1 = \mu_2$).

TABLE FOR PROBLEM 11(b)

Source of variation	Sum of squares	df	Mean squares
Groups (males)	9.9401	11	0.9036
Error	4.682	24	0.1951
Total	14.622	35	

ANSWERS

H_A: The means of the two treatment groups are different ($\mu_1 \neq \mu_2$). You can use either a two-sample t-test or ANOVA. The ANOVA results are:

Source of variation	Sum of squares	df	Mean squares	F-ratio	P
Groups	1.258154	1	1.258154	56.34	0.00001
Error	0.245619	11	0.022329		
Total	1.503773	12			

The critical value $F_{0.05(1), 1, 11} = 4.84$. Since $F > 4.84$, $P < 0.05$, reject H_0. First blood meal affects the mean proportion of flies taking their second blood meal from cows.

15. (a) H_0: The mean number of shoots is equal in the three treatments ($\mu_1 = \mu_2 = \mu_3$). H_A: At least one of the treatment means μ_i is different.

(b)

Source of variation	Sum of squares	df	Mean squares	F-ratio	P
Groups (treatments)	2952.808	2	1476.40	5.32	0.011
Error	8049.067	29	277.55		
Total	11,001.875	3			

The critical value is $F_{0.05(1), 2, 29} = 3.33$. Since $F > 3.33$, $P < 0.05$, we reject H_0. Conclude that there are differences between treatments in mean shoot number. **(c)** The variable is normally distributed with equal variance in the three treatment populations. **(d)** Fixed-effects ANOVA. The levels chosen were set by the researcher and are repeatable—they were not randomly sampled from a population of treatments.

Chapter 16

1. (a) Scatter plot. Negative. **(b)** $\sum(X - \bar{X})^2 = 380.4375$. **(c)** $\sum(Y - \bar{Y})^2 = 174175$.
(d) $\sum(X - \bar{X})(Y - \bar{Y}) = -4898.75$. **(e)** $r = -0.602$. **(f)** $z = -0.696$. **(g)** 0.277. **(h)** 1.96.
(i) $-1.240 < \zeta < -0.152$. **(j)** $-0.845 < \rho < -0.151$.

2. (a) H_0: The population correlation ρ is zero. H_A: The population correlation ρ is not zero.
(b) $SE_r = 0.213$. **(c)** $t = -2.819$. **(d)** $n = 16$. $df = 14$. **(e)** $t_{0.05(2), 14} = 2.14$. **(f)** Since $t < -2.14$, $P < 0.05$. Reject H_0, conclude there is a negative correlation in the population ($P = 0.036$).
3. (a) 17, 16, 15, 14, 13, 12, 11, 10, 9, 8, 7, 6, 5, 4, 3, 2, 1. **(b)** 15, 16.5, 12, 9, 11, 7, 16.5, 3, 13, 14, 8, 10, 5, 6, 1, 2, 4. **(c)** SS for duration: 408.0, SS for allele frequency: 407.5, sum of products: 291.5. **(d)** $r_S = 0.715$.
(e) $n = 17$. **(f)** H_0: There is no correlation between allele frequency and duration of settlement ($\rho_S = 0$). H_A: Allele frequency and duration of settlement are correlated ($\rho_S \neq 0$). **(g)** $r_{S(0.05, 17)} = 0.485$. **(h)** Since $r_S > 0.485$, $P < 0.05$. Reject H_0 ($P = 0.0012$).
4. (a) $r = 0$. **(b)** $r = -0.8$. **(c)** $r = 0.5$. **(d)** $r = 0$.
5. (a) Scatter plot:

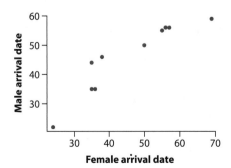

(b) The relationship is linear, positive, and strong. **(c)** $r = 0.93$. $SE_r = 0.13$. **(d)** The standard error is the standard deviation of the sampling distribution of r. **(e)** $0.72 < \rho < 0.98$.

6. **(a)** No change to the correlation coefficient ($r = 0.93$). Adding a constant to one of the variables does not alter the correlation coefficient. **(b)** No change to the correlation coefficient ($r = 0.93$). Dividing one of the variables by a constant does not alter the correlation coefficient.

7. We can use a paired t-test (Chapter 12). H_0: Mean arrival date of male and female partners is the same ($\mu_d = 0$). H_A: Mean arrival date of male and female partners is different ($\mu_d \neq 0$). $\bar{d} = 0.3$ days (males are slightly earlier on average), $s_d = 1.667$, $t = 0.18$, $df = 9$, and $P = 0.86$. $t_{0.05(2), 9} = 2.26$; since observed t is less than $t_{0.05(2), 9}$, $P > 0.05$, we do not reject H_0 ($P = 0.86$). Conclude that we cannot reject null hypothesis of equal mean arrival times of males and females. Assume a normal distribution of differences between arrival rates of males and females.

8. When there is measurement error in one or both of the variables X and Y.

9. A narrower range of values for inbreeding coefficient should lower the correlation with the number of surviving pups compared with a wider range of inbreeding coefficients.

10. **(a)** H_0: There is no correlation between earwig density and proportion of males with forceps ($\rho_S = 0$). H_A: There is a correlation between earwig density and proportion of males with forceps ($\rho_S \neq 0$). Spearman's $r_S = 0.66$, $n = 22$, $r_{S(0.05, 22)} = 0.425$. Since $r_S > 0.425$, $P < 0.05$, reject the null hypothesis that the two variables are not associated ($P = 0.0008$). **(b)** Spearman's rank correlation assumes that the data points are a random sample and that the relationship between the ranks of the two variables is linear.

11. Sampling error in the estimates of earwig density and the proportion of males with forceps means that true density and proportion on an island are measured with error. Measurement error will tend to decrease the estimated correlation. Therefore, the actual correlation is expected to be higher on average than the estimated correlation.

12. **(a)** The assumption of bivariate normality is violated: there is an outlier. **(b)** Using a rank correlation would be appropriate. **(c)** H_0: The population rank correlation is zero ($\rho_S = 0$). H_A: The population rank correlation is not zero ($\rho_S \neq 0$). $r_S = 0.30$. $r_{S(0.05, 41)} = 0.308$. Since r_S is not greater than or equal to $r_{S(0.05, 41)}$, $P > 0.05$, we do *not* reject H_0 ($P = 0.053$). Conclude that we cannot reject the null hypothesis of zero correlation. **(d)** Assume a random sample and a linear relationship between the ranks of the two variables.

13. **(a)**

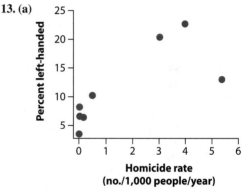

(b) The assumption of bivariate normally is violated. The frequency distribution of each variable is positively skewed, and there is higher variance in Y for large X than for small X. **(c)** Transform one or both variables. The log transformation is a good one to try when variables have positive skew and values are greater than zero. The arcsine transformation is also an obvious one to try on the variable "percent left-handed" (don't forget to divide by 100 and take the square root). **(d)** A log transformation of both variables yielded a satisfactory outcome:

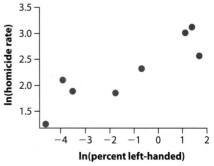

A log transformation of homicide rate and an arcsine transformation of percent left-handed (after dividing by 100) also gave a satisfactory outcome (though the log appeared slightly better). **(e)** H_0: There is no correlation between homicide rate and percent left-handed individuals ($r = 0$). H_A: There is a correlation between homicide rate and percent left-handed individuals ($r \neq 0$). Results using the log-log transformation: $r = 0.88$, $SE_r = 0.19$, $t = 4.59$, $df = 6$, $t_{0.05(2), 6} = 2.45$. Since $t > t_{0.05(2), 6}$, $P < 0.05$. Reject H_0 ($P = 0.0034$). Conclude that homicide rate and percent left-handed individuals in societies are correlated. Results using log transformation of homicide and arcsine transformation of percent left-handed: $r = 0.87$, $SE_r = 0.20$, $t = 4.29$, $df = 6$, $t_{0.05(2), 6} = 2.45$. Since t is greater than $t_{0.05(2), 6}$, $P < 0.05$. Reject H_0 ($P = 0.0048$). Conclude that homicide rate and percent left-handed individuals in societies are correlated.

14. **(a)** There is a negative linear relationship between telomere length and chronicity, but it is not strong. **(b)** -0.43. **(c)** $-0.66 < \rho < -0.13$. **(d)** It is the range of most-plausible values for the parameter ρ. If you were to repeatedly and randomly sample individuals from the same population and compute the 95% confidence interval each time, 19 out of 20 of the intervals are expected to include the population correlation ρ. **(e)** Assume random sampling, and that the two variables have a bivariate normal distribution in the population. The scatter plot suggests that the relationship between telomere length and chronicity might be mildly nonlinear, which would violate the assumption of bivariate normality.

Chapter 17

1. **(a)**

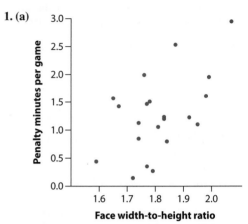

(b) The data seem to fit the assumptions. The residuals are symmetric and don't show any obvious non-normality. The variance of the residuals does not appear to change greatly for different values of X. **(c)** Mean face ratio: 1.81. Mean penalty minutes: 1.28. $\sum X = 38.07$, $\sum Y = 26.81$. **(d)** $\sum X^2 = 69.2997$, $\sum Y^2 = 44.2563$, $\sum XY = 49.5091$. **(e)** Sum of products: $\sum (X - \bar{X})(Y - \bar{Y}) =$

$$\sum (XY) - \frac{\left(\sum X\right)\left(\sum Y\right)}{n} =$$

$$49.5091 - \frac{(38.07)(26.81)}{21} = 0.9064 .$$

This can also be calculated by adding $(X_i - \bar{X})(Y_i - \bar{Y})$ for all data points i. For example, for the first data point this would be $(1.59 - 1.81)(0.44 - 1.28) = 0.1848$. Sum of squares for face ratio:

$$\sum (X - \bar{X})^2 = \sum (X^2) - \frac{\left(\sum X\right)^2}{n} =$$

$$69.2997 - \frac{(38.07)^2}{21} = 0.2842.$$

(f) $b = \dfrac{\sum (X_i - \bar{X})(Y_i - \bar{Y})}{\sum (X_i - \bar{X})^2} = \dfrac{0.9064}{0.2842} =$

3.19. The slope is estimated to be positive, consistent with the pattern in the graph.
(g) $a = \bar{Y} - b\bar{X} = 1.28 - 3.19(1.81) = -4.50$. **(h)** $Y = -4.50 + 3.19X$.

2. (a) $\sum(Y - \bar{Y})^2 = \sum(Y^2) - \dfrac{\left(\sum Y\right)^2}{21} =$

$44.2563 - \dfrac{(26.81)^2}{21} = 10.0289$.

(b) $MS_{residual} =$

$\dfrac{\sum(Y - \bar{Y})^2 - b\sum(X - \bar{X})(Y - \bar{Y})}{n - 2} =$

$\dfrac{10.0289 - 3.19(0.9064)}{21 - 2} = 0.3757$.

(c) $SE_b = \sqrt{\dfrac{MS_{residual}}{\sum(X - \bar{X})}} = \sqrt{\dfrac{0.3757}{0.2845}} =$

1.15. **(d)** $df = n - 2 = 21 - 2 = 19$.
(e) $t_{\alpha(2), df} = t_{0.05(2), 19} = 2.09$.
(f) $b - t_{\alpha(2), 19}SE_b < \beta < b + t_{\alpha(2), 19}SE_b$;
$3.19 - 2.09(1.15) < \beta < 3.19 + 2.09(1.15)$;
$0.79 < \beta < 5.59$.

3. (a) H_0: The slope of the line predicting penalty minutes from face ratio is 0. H_A: The slope of the line predicting penalty minutes from face ratio is not 0. **(b)** $\beta_0 = 0$.

(c) $t = \dfrac{b - \beta_0}{SE_b} = \dfrac{3.19 - 0}{1.15} = 2.77$.

(d) $t_{\alpha(2), df} = t_{0.05(2), 19} = 2.09$. **(e)** Yes, the observed t exceeds the critical value. Therefore, we can conclude that $P < 0.05$ and reject the null hypothesis that the slope is zero. Penalty minutes can be at least partially predicted by the face width-to-height ratio for hockey players.
(f) $P = 0.012$. From Statistical Table C, we can conclude that $P < 0.02$, because the observed t exceeds the critical value for $\alpha = 0.02$, but not for $\alpha = 0.01$. **(g)** $R^2 = SS_{regression}/SS_{total} = 2.8932/10.0326 = 0.288$. About 29% of the variation in penalty minutes per game is accounted for by face ratio.

4. No, because the slope of the relationship is negative, meaning that people who eat more chocolate on average have lower BMI. The P-value is small ($P = 0.008$), so this pattern is statistically significant.

5. (a) $Y = 2/3X + 1$. **(b)** $Y = X - 1$.
(c) $Y = -0.5X + 2$. **(d)** $Y = X - 5$.

6. (a)

The percent infant mortality increases approximately linearly with the log of the home-range size. **(b)** Mortality $= 16.37 + 10.26$(log home range). **(c)** H_0: Home-range size does not predict infant mortality ($\beta = 0$). H_A: Home-range size predicts infant mortality ($\beta \neq 0$). $b = 10.26$, $SE_b = 2.69$, $t = 3.81$, $df = 18$, and $t_{0.05(2), 18} = 2.10$. Since $t > 2.10$, $P < 0.05$. Reject H_0 ($P = 0.0013$). Conclude that home-range size predicts infant mortality. **(d)** Mortality $= 17.51 + 6.60$ (log home range). The slope is much lower with the polar bear removed (6.60 instead of 10.26, a reduction of more than a third).

7. (a) The "least-squares" regression line is the one the minimizes the sum of squared differences between the predicted Y-values on the regression line for each X and the observed Y-values.
(b) Residuals are the differences between predicted Y-values on the estimated regression line and the observed Y-values. **(c)** The most conspicuous problem is that the variance of the residuals increases with increasing progesterone concentration, violating the assumption that the variance of Y is the same at all values of X. There are no conspicuous departures from the other two assumptions, linearity and a normal distribution of Y-values at every X.

8. Caution is warranted because the prediction is based on extrapolation, which is risky because the relationship between winning time and year might not be linear beyond the range of the existing data.

ANSWERS

9. (d).

10. (a) The arcsine transformation is a good bet for data that are proportions. (b) $Y = 0.416 + 7.10X$, where X is genetic distance and Y is arcsine transformed proportion sterile.

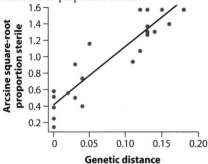

(c) $5.67 < \beta < 8.52$.

11. (a) Such a complicated curve is unwarranted by the data. It would do a poor job of predicting new observations because it does not describe a general trend. A curve fit should be as simple as possible. (b) First try to transform the data to make the relationship linear. If that fails, consider nonlinear regression.

12. (a) $b = 91.56$ and $SE_b = 42.55$. (b) $-46.70 < \beta < 229.83$. (c) The range of most-plausible values for the parameter. In 99% of random samples, the confidence interval will bracket the population value for the slope. (d) Measurement error in the X-variable (bite force) will tend to lead to an underestimation of the population slope. (e) Measurement error in the Y-variable (territory area) will not bias the estimate of slope (though it will increase its uncertainty).

13. (a) See the table at the bottom of the page. (b) H_0: The slope of the regression of territory area on bite force is zero ($\beta = 0$). H_A: The slope of the regression of territory area on bite force is not zero ($\beta \neq 0$). $F = 4.63$, $df = 1,9$. $F_{0.05(1), 1, 9} = 5.12$. Since $F < 5.12$, $P > 0.05$. Do not reject H_0 ($P = 0.060$). We are unable to

conclude that the slope differs from zero. (c) That the relationship between X and Y is linear; for each X there is a normal distribution of Y-values in the population, of which we have a random sample; the variance of Y is the same at all values of X. (d) The variance of the residuals. (e) $R^2 = 0.340$. It measures the fraction of the variation in Y that is explained by X.

14. (a) First, check the data to ensure this individual was not entered incorrectly. Perform the analysis with and without the outlier included in the data set to determine whether it has an influence on the outcome. If it has a big influence, then it is probably wise to leave it out and limit predictions to the range of X-values between 0 and about 200 (and urge them to obtain more data at the higher X-values). (b) Confidence bands give the confidence interval for the predicted *mean* time since death for a given hypoxanthine concentration. (c) Confidence bands. (d) The prediction interval, because it measures uncertainty when predicting the time of death of a single individual.

15. (a)

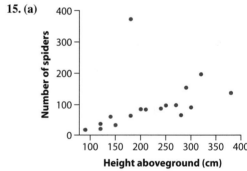

(b) The assumptions of equal variance of residuals, and of a normal distribution of Y-values at each X, are not met because of the presence of an outlier. (c) A transformation of the data might improve matters, but success is doubtful. Robust regression methods is another option (not covered here).

TABLE FOR PROBLEM 13(a)

Source of variation	Sum of squares	df	Mean squares	F-ratio
Regression	3758539	1	3758539	4.63
Residual	7303662	9	811518	
Total	11062201	10		

16. (a) The residuals are not normally distributed, and the variance in Y is not the same for all X, because of the outlier. **(b)** The relationship between X and Y is not linear. **(c)** The residuals are not normally distributed. **(d)** The variance in Y is not equal for all X, but increases with increasing X.

17. (a)

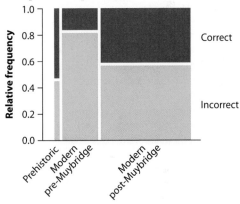

Number of earthworm species

(b) $Y = 0.152 + 0.028X$. **(c)** % nitrogen per earthworm species. **(d)** $\hat{Y} = 0.294$. **(e)** $SE_b = 0.0096$. **(f)** $0.009 < \beta < 0.048$.

18. (a) The power law can be converted to a linear relationship by taking logs of both sides, yielding $\ln(Y) = \ln(\alpha) + \beta \ln(X)$. The slope of the line is the power law exponent. Take the log of both variables and use this log-transformed data to calculate $b = 0.857$. $SE_b = 0.157$.

(b)

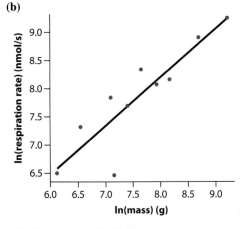

ln(mass) (g)

(c) $0.50 < \beta < 1.22$. This most-plausible range for β includes the value ¾. **(d)** H_0: $\beta = ¾$. H_A: $\beta \neq ¾$. $t = 0.684$, $df = 8$, $t_{0.05(2), 8} = 2.31$. Since $t < 2.31$, $P > 0.05$. Do not reject H_0 ($P = 0.51$). We are unable to reject the null hypothesis that the exponent is ¾. **(e)** The

exponent (slope) is likely to be underestimated if there is measurement error in the explanatory variable, tree mass (and ln mass). The exponent is not biased by measurement error in the response variable, respiration rate (and ln respiration rate), but the uncertainty of the estimate is expected to be increased by measurement error.

Review 3

1. (a) Answers will vary (grouped bar graph also acceptable). The graph reveals that while the proportion of paintings with correct posture increased after Muybridge, it remains below a half and has hardly attained the level of accuracy of prehistoric paintings.

Date

(b) H_0: The frequency of modern images with correct postures is no different pre- and post-Muybridge. H_A: The frequency of modern images with correct postures is different between these two periods. $\chi^2 = 56.1$, $df = 1$, $\chi^2_{0.05, 1} = 3.84$. Since $\chi^2 > 3.84$, $P < 0.05$. Reject H_0 ($P < 10^{-10}$). **(c)** That we have a random sample of images from both periods and that depiction of posture in images is independent (unlikely to be strictly true, because a given painter might have multiple images in the sample and because later painters tend to copy prior images). **(d)** $0.39 < p < 0.68$.

ANSWERS

2. Because the hairs appear with equal probability and independently over his scalp, their numbers per square cm should follow a Poisson distribution. Therefore $\Pr[4] = \dfrac{e^{-2.3} 2.3^4}{4!} = 0.1169$.

3. (a) $\Pr[0.3 < X < 0.5] = 1 - \Pr[X < 0.3] - \Pr[X > 0.5] = 1 - \Pr[Z < 0] - \Pr[Z > 2.5] = 1 - 0.5 - 0.0062 = 0.4938$. (b) $(1 - 0.4938)^5 = 0.0332$.

4. First, the graph fails to "show the data." Use a strip chart, box plot, or multiple histograms of injury scores instead. Second, the bar graph violates "represent magnitudes honestly" by using a nonzero baseline. Bar graphs require a zero baseline so that bar area represents magnitudes.

5. (a) The log transformation is appropriate and reduces the skew. (b) Scatter plot. The graph suggests a negative relationship between the two variables.

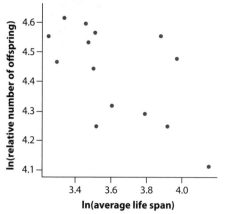

(c) H_0: The two variables are uncorrelated ($\rho = 0$). The two variables are correlated ($\rho \neq 0$). $r = -0.610$, $SE_r = 0.229$, $df = 12$, $t = -2.67$, $t_{0.05(2),\ 12} = 2.18$. Since $|t| > 2.18$, $P < 0.05$. Reject H_0 ($P = 0.021$). Conclude that the variables are negatively correlated.

6. H_0: Offspring occur in the predicted frequencies (9:3:3:1). H_A: Offspring do not occur in the predicted frequencies. Using the χ^2 goodness-of-fit-test: $\chi^2 = 0.47$, $df = 3$, $\chi^2_{0.05,\ 3} = 7.81$. Since $\chi^2 < 7.81$, $P > 0.05$. Do not reject H_0 ($P = 0.93$). The null hypothesis given by Mendel's prediction is not rejected, so the data are consistent with the prediction.

7. (a) Case-control study. (b) [Answers may vary.] Shown below is a mosaic plot (in the style recommended for case-control studies). The women who do not take oral contraceptives have the higher rate of thrombosis.

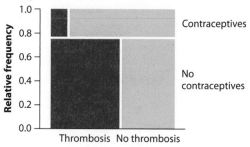

(c) H_0: There is no association between oral contraceptive use and thrombosis. H_A: There is an association between oral contraceptive use and thrombosis.

Observed frequencies:

	Thrombosis	No thrombosis	Row sums
Contraceptives	103	658	761
No contraceptives	1421	1102	2523
Column sums	1524	1760	3284

Expected frequencies:

	Thrombosis	No thrombosis	Row sums
Contraceptives	353.1559	407.8441	761
No contraceptives	170.8441	1352.1559	2523
Column sums	1524.0000	1760.0000	3284

$\chi^2 = 430.36$, $df = 1$, $\chi^2_{0.05,\ 1} = 3.84$. Since $\chi^2 > 3.84$, $P < 0.05$. Reject H_0 ($P < 10^{-10}$). (d) $\widehat{OR} = 0.121$, 95% CI: $0.097 < OR < 0.152$. (e) If the proportion of women with thrombosis in the population is very low.

8. (a) $MS_{error} = 0.1524$. $s_A^2 = 0.1167$. (b) Repeatability $= 0.4337$. Only about 43% of the variation is among individuals, the rest is within-mouse measurement variability. (c) Mice are randomly sampled and mouse mean measurements have a normal distribution in the population.

9. Association between treatment (a categorical variable with two groups) is measured by the

difference between the means of the two groups rather than with a correlation coefficient. The difference is tested with a two-sample t-test (or ANOVA). H_0: Mean growth rate is the same in the two CO_2 groups ($\mu_1 = \mu_2$). H_A: Mean growth rate differs between the two CO_2 groups ($\mu_1 = \mu_2$). $\overline{Y}_1 = 1.66$ (Normal), $\overline{Y}_2 = 1.53$ (High), $SE_{\overline{Y}_1 - \overline{Y}_2} = 0.237$, $t = 0.54$, $df = 12$. $t_{0.05(2),\ 12} = 2.18$. Since t is not greater than or equal to $t_{0.05(2),\ 12}$, $P > 0.05$. Do not reject H_0 ($P = 0.60$). Conclude that the null hypothesis of no difference between the means of the two groups is not rejected by these data.

10. (a)

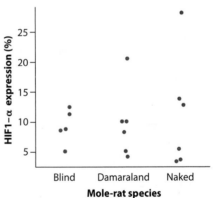

(b) H_0: Rate of heat loss does not change with body leanness ($\beta = 0$). H_A: Rate of heat loss changes with body leanness ($\beta \neq 0$). $b = 0.0190$, $SE_b = 0.0023$, $t = 8.29$, $df = 10$, $t_{0.05(2),\ 18} = 2.29$. Since $t > 2.29$, $P < 0.05$. Reject H_0 ($P = 0.000009$). Conclude that rate of heat loss increases with body leanness. **(c)** $0.0139 < \beta < 0.0241$. **(d)** For every X there is a normal distribution of Y-values, of which we have a random sample; the relationship between leanness and heat loss is linear; the variance of Y is the same for all values of X. **(e)** $MS_{residual} = 0.0002154$, $MS_{regression} = 0.014806$, $R^2 = 0.873$.

11. (a) Odds ratio **(b)** Difference between means of number of bacteria between people who wash and people who don't wash. **(c)** Difference between means of mitochondria number per muscle cell. **(d)** Mean of the difference of mitochondria number between dominant and subdominant arms. **(e)** Proportion of males. **(f)** Odds ratio. **(g)** Variance or standard deviation of weight. **(h)** Correlation coefficient.

12. (a) χ^2 contingency test. **(b)** Analysis of variance. **(c)** Linear regression, t-test or F-test of slope. **(d)** Two-sample t-test ($=$ANOVA with two groups). A paired design is also possible, but tricky to implement. **(e)** Two-sample t-test ($=$ANOVA with two groups). **(f)** Two-sample t-test ($=$ANOVA with two groups). **(g)** Two-sample t-test (or ANOVA with two groups). **(h)** Mann–Whitney U-test or a permutation test. **(i)** Paired t-test. **(j)** Binomial test (or goodness-of-fit test with two categories if sample size is large). **(k)** Contingency test. **(l)** Paired t-test. **(m)** χ^2 goodness-of-fit test to Poisson distribution. **(n)** Shapiro–Wilk test. **(o)** F-test or Levene's test. **(p)** Mann–Whitney U-test or a permutation test. **(q)** Linear regression, t-test or F-test of slope. **(r)** Logistic regression, χ^2 analysis of deviance.

13. (a) Strip chart or multiple histograms would work; data are too few for a box plot. Variance is heterogeneous among groups, and distributions in the "naked" and "Damaraland" groups are skewed.

(b) H_0: The species do not differ in mean amount of HIF1-α. H_A: Mean amount of HIF1-α is not the same for all species.

Source	df	SS	MS	F	P
Species	2	0.0084	0.0042	0.011	0.99
Error	14	5.4903	0.3922		
Total	16	5.4987			

$F_{0.05(1),\ 2,\ 14} = 3.74$, $df = 2, 14$. Since $F < 3.74$, do not reject H_0 ($P = 0.99$).

14. (a) Let the short-needle group be group 1. $\overline{Y}_1 = 209.22$, $\overline{Y}_2 = 340.76$, $\overline{Y}_1 - \overline{Y}_2 =$

-131.54 mIU/ml, $s_1 = 114.50$, $s_2 = 157.99$, $s_p^2 = 20111.79$, $SE_{\bar{Y}_1 - \bar{Y}_2} = 58.72$, $df = 22$, $t_{0.05(2), 22} = 2.07$. $-253.082 < \mu_1 - \mu_2 < -9.99$ mIU/ml ($-253.31 < \mu_1 - \mu_2 < -9.76$ mIU/ml without rounding). **(b)** H_0: Mean antibody titer is the same in the two needle groups ($\mu_1 = \mu_2$). H_A: Mean titer is not the same in the two groups ($\mu_1 \neq \mu_2$). $t = -2.24$, $t_{0.05(2), 22} = 2.07$. Since $|t| > 2.07$, $P < 0.05$. Reject H_0 ($P = 0.036$). Longer needles lead to higher mean antibody titers (and a more effective vaccination). **(c)** $249.54 < \mu < 431.98$.

15. The graph fails to "show the data." The data points should be included when possible to provide visual information on the shapes of the distributions, the spread of data points, and the strength of the association. A strip chart would be a good choice here.

16. **(a)** The distribution of coefficients has negative skew (skewed to the left), so the one-sample t-test is not appropriate. We also cannot use the Wilcoxon signed-rank test, which assumes that the distribution is symmetric. The sign test is most appropriate instead. **(b)** H_0: The median coefficient is 0. H_A: The median coefficient is not 0. Thirteen of the 15 samples have positive coefficients, so $P = 2(\text{Pr}[13] + \text{Pr}[14] + \text{Pr}[15]) = 0.0074$. Since $P < 0.05$, reject the null hypothesis. The median coefficient is greater than zero.

17. **(a)** The experimental study, because randomly assigning smoked or unsmoked cigarette butts to nests breaks any association between ectoparasite numbers and smoked cigarette butts caused by confounding variables. **(b)** Blinding and blocking were not used. Blinding can be incorporated if the person retrieving the traps and counting the parasites is unaware of the treatment, avoiding unconscious bias when counting. Blocking could be incorporated if nests are grouped by shared features, such as nesting materials, or spatial location, that might affect ectoparasite numbers. By blocking, the effects of these other sources of variation are accounted for, yielding a more precise estimate of treatment effect and a more powerful test of difference.

18. **(a)** $p' = 0.833$, $0.684 < p < 0.982$. **(b)** H_0: The proportion of people calling the angular shape kiki is $p = 0.50$. H_A: The proportion p is not 0.50. $X = 18$, $n = 20$. Binomial test, $P = 0.0004$. Since $P < 0.05$, reject H_0.

19. **(a)** The difference in white blood cell count is 1.87 cells/nanoliter, with the more promiscuous species having the higher count. **(b)** $s_d = 1.57$, $SE_{\bar{d}} = 0.525$, $df = 8$, $t_{0.01(2), 8} = 3.36$, $0.10 < \mu_d < 3.63$. **(c)** H_0: There is no difference in WBC count between more and less promiscuous species. H_A: There is a difference in mean WBC count. $t = 1.87/0.525 = 3.56$. Since $t > t_{0.05(2), 8} = 2.31$, $P < 0.05$. Reject the null hypothesis ($P = 0.0074$): promiscuous primates have higher mean white blood cell counts. **(d)** Primate pairs are randomly sampled and differences in WBC counts of pairs are normally distributed.

Chapter 18

1. **(a)** A main effect of A and main effect of B are present. There is no interaction. **(b)** A main effect of A is present but there is no main effect of B or interaction. **(c)** A main effect of B is present and an interaction is present. There is no main effect of A evident.

2. **(a)** HOURS = CONSTANT + MUTANT. HOURS is hours of resting. CONSTANT is the grand mean. MUTANT is the effect of each line of mutant flies. **(b)** HOURS = CONSTANT. **(c)** Long horizontal line indicates the grand mean, which is the predicted value for the mean of each group under the null hypothesis. The short horizontal lines give the group means, which are the predicted values under the "full" general linear model including the MUTANT term.

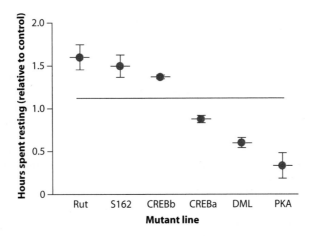

(d) *F*.

3. (a) Plasma corticosterone concentration. (b) Chick age group and disturbance regime. (c) The two explanatory variables interact to affect corticosterone concentrations. (d) Observational. The penguins were not assigned by the researcher to age groups or disturbance regimes. (e) Yes. Every combination of the two factors (age group and disturbance regime) is included in the design.

4. (a) CORTICOSTERONE = CONSTANT + AGE + DISTURBANCE + AGE*DISTURBANCE. CORTICOSTERONE is the corticosterone concentration, CONSTANT is the grand mean corticosterone concentration, AGE is the age group of the penguin, DISTURBANCE indicates whether the penguin lived in the undisturbed or tourist-disturbed area, and AGE*DISTURBANCE is the interaction between age and disturbance. (b) H_0: Penguin age group has no effect on mean corticosterone concentration. H_0: Disturbance regime has no effect on mean corticosterone concentration. H_0: There is no interaction between penguin age group and disturbance regime. (c) Penguins were randomly sampled from each area and each age class. Corticosterone concentrations are normally distributed within age class and disturbance regime. The variance of the corticosterone concentration is the same for each combination of age and disturbance.

5. Three of the following: a. To investigate the effect of more than one variable with the same experimental design; b. to include the effects of blocking in an experimental design; c. to investigate interactions between the effects of different factors; and d. to control for the effects of a confounding variable by including it as a covariate.

6. (a) MEMORY = CONSTANT + LESION. MEMORY is the spatial memory score of a rat, CONSTANT is a constant indicating the value of MEMORY when LESION is zero (*Y*-intercept), and LESION is the extent of the lesion. (b) MEMORY = CONSTANT. (c) The line having negative slope in the following graph represents the predicted values from the "full" general linear model. The flat (horizontal) line represents the predicted values under the null model.

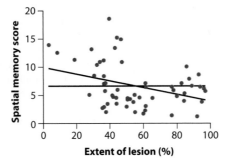

ANSWERS

7. (a) A factorial design (with two factors).
(b) Response variable: life span. Explanatory variables: male treatment and female fertility.
(c) LIFESPAN = TREATMENT + FERTILITY + TREATMENT * FERTILITY.
(d) Females in the low-cost treatment seem to live longer than those in the high-cost treatment (difference is about 7 days). There seems to be a small effect of fertility on life span (difference looks like it is about 2 days). Hence the main effect of male treatment appears to be larger than the effect of female fertility. There does not seem to be an interaction between TREATMENT and FERTILITY (lines are roughly parallel).
(e) H_0: Mean female life span is the same in each male treatment. H_0: Mean female life span is the same in each female fertility group. H_0: Mean female life span is not changed by an interaction between male treatment and fertility group (the effect of male treatment does not depend on female fertility). **(f)** Male treatment and female fertility both have statistically significant effects on female life span. There is no significant interaction between treatment and fertility. These findings agree with the visual assessment.
(g) No, the ANOVA table merely shows that there is some effect of female fertility. **(h)** If the hypothesis is correct that SP reduces female life span by causing females to produce more eggs, we would expect male treatment to affect life span of fertile females but not of sterile females. In other words, there should be an interaction between male treatment and female fertility. There is no evidence of an interaction term, and male treatment seems to affect the sterile and fertile females similarly. This does not support the hypothesis.

8. (a)

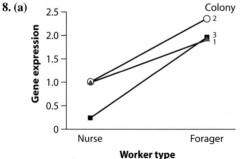

(b) EXPRESSION = CONSTANT + WORKERTYPE + BLOCK.
EXPRESSION is the *for* gene expression, CONSTANT is the grand mean of *for* gene expression, WORKERTYPE is the worker type, and BLOCK identifies which colony the bee comes from. **(c)** EXPRESSION = CONSTANT + BLOCK. **(d)** Fixed effect, because the types are repeatable and of direct interest. The worker types in the analysis are not randomly sampled from a population of worker types. **(e)** Blocking variables allow extraneous variation caused by the block variable to be accounted for in the analysis and eliminated. When a block variable is included, the error variance is smaller, making it easier to detect real effects of the factor.
9. (a) It measures whether the full linear model that include the term WORKERTYPE is a significantly better fit to the data than a model lacking the term (but including all other terms).
(b) It measures whether the full linear model that includes the BLOCK term is a significantly better fit to the data than a model lacking the term (but including all other terms). **(c)** It should be retained because it was part of the design, and because it might still improve the ability to detect an effect of the factor of interest.
(d) The residuals are the differences between the observed values and the values predicted by the model. **(e)** The residuals are plotted along the Y-axis. The predicted values from the model are plotted along the X axis.

Chapter 19

1. (a) For example: Singleton: 3.5, 3.5, 2.6, 4.4; Twin: 3.4, 4.2, 3.4. **(b)** No.
2. (a) Yes—by chance, bootstrap samples might contain the same observations as the data.
(b) No. The bootstrap sample for a group can contain only data from that group. **(c)** No. Sample sizes of groups must be the same in each bootstrap replicate as in the data. **(d)** Yes. **(e)** No, 3.8 is not in the data. **(f)** Yes.
3. $10.20 < \mu < 10.40$.
4. Approximately 10 units. About 2/3 of the bootstraps lie within ± 10 units of the mean.

ANSWERS

5. (a) The null distribution for the ratio.
(b) H_0: Mean ratio of range size equals that expected of a randomly broken stick. H_A: Mean ratio of range size differs from that expected of a randomly broken stick. P is approximately $2 \times (42/10{,}000) = 0.0084$. Since $P < 0.05$, reject H_0. Conclude that mean ratio in birds exceeds that expected from a randomly broken stick.

6. (a) 98. **(b)** 100. **(c)** "Giant pandas are most closely related to bears."

7. (a) For example:

Group A			Group B		
7.8	4.5	2.1	12.4	8.9	8.9
7.8	7.8	4.5	12.4	12.4	12.4
4.5	2.1	7.8	8.9	8.9	12.4
7.8	2.1	4.5	10.8	12.4	10.8
7.8	2.1	4.5	12.4	8.9	10.8
7.8	4.5	7.8	8.9	10.8	10.8
4.5	4.5	7.8	10.8	12.4	8.9
2.1	7.8	4.5	8.9	8.9	10.8
4.5	7.8	4.5	10.8	10.8	12.4
2.1	7.8	7.8	8.9	12.4	12.4

(b) Difference in median (B minus A) for these 10 bootstrap replicates:
3.0, 4.4, 4.4, 4.4, 4.6, 4.6, 6.3, 6.3, 6.3, 6.3.
4.4 < difference between population medians < 6.3.

8. Yes.

9. 10.4 seconds.

10. Welch's approximate t-test.

11. (a) No, this is not a bootstrap sample. All the individuals' measurements of shell volume are included. All the measurements of sperm stored are also present but they have been permuted.
(b) Yes. Each row of the data set is a different

individual sampled from the original sample of individuals. Sampling is with replacement because the same individual sometimes occurs more than once. **(c)** The linear correlation coefficient, r.

Chapter 20

1. (a) Binomial distribution with $n = 47$. **(b)**
$$L[p \mid 12 \text{ heterozygotes}] = \binom{47}{12} p^{12}(1 - p)^{35}.$$
It measures the probability of getting 12 heterozygotes (the data) given that the true value of the proportion of heterozygotes is p.
(c) $\ln L[p] = \ln\left[\binom{47}{12}\right] + 12 \ln[p] + 35 \ln[1 - p]$. **(d)** -7.90.

2. (a) 0.26. **(b)** $0.14 < p < 0.40$.

3. (a) The log of the probability of the DNA data given different possible times since the split between Neanderthals and modern humans.
(b) 710 thousand years. **(c)** $470 <$ time since split < 1000 (in thousands of years). **(d)** This interval is a 95% confidence interval. It describes the most-plausible values of the time since the split between humans and Neanderthals.

4. (a) -4400. **(b)** -4396.5. **(c)** $G = 2[-4396.5 - (-4400)] = 7$. **(d)** χ^2 with $df = 1$.
(e) $\chi^2_{0.05,\,1} = 3.84$. Since G is greater than 3.84, $P < 0.05$ ($P = 0.008$). Reject H_0. Conclude that heterogeneity is not zero.

5. (a) $\Pr[\text{Yes}] = 1/2 + s/2 = (1 + s)/2$. (See figure below.)

FIGURE FOR PROBLEM 5(a)

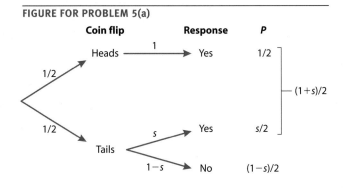

ANSWERS

(b) $L[s \mid 113 \text{ yeses}] = \binom{185}{113}\left(\dfrac{1+s}{2}\right)^{113}\left(\dfrac{1-s}{2}\right)^{72}$.

(c) $\ln L[s \mid 113 \text{ yeses}] = \ln\left[\binom{185}{113}\right] + 113\ln[(1+s)/2] + 72\ln[(1-s)/2]$.

(d) -7.39.

6. (a) The maximum likelihood estimate is $\hat{s} = 0.22$. **(b)** $0.07 < s < 0.36$. **(c)** H_0: The fraction of thieves s is zero. H_A: The fraction of thieves s is not zero. $G = 2[-2.81 - (-7.39)] = 9.16$. $\chi^2_{0.05,\,1} = 3.84$. Since $G > 3.84$, $P < 0.05$ ($P = 0.0025$). Reject H_0. Conclude that there are thieves among us.

7. (a)

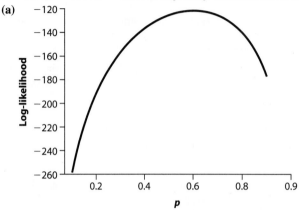

(b) 0.60. **(c)** -121.653. **(d)** The data seem to fit a geometric distribution very well.

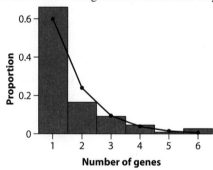

(e) χ^2 goodness-of-fit test.

8. (a)

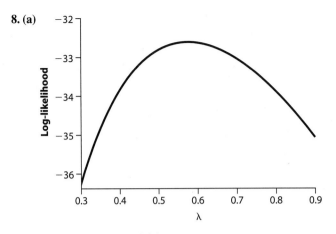

(b) 0.58. **(c)** $0.36 < \lambda < 0.86$.

Chapter 21

1. Results are seen as more interesting or exciting; results are more likely to be accepted for publication; results are more likely to be accepted in a better journal.

2. **(a)** Yes, average heritability declines with increasing sample size. **(b)** Small-sample studies yielding low heritability estimates are unlikely to be submitted for publication (researchers) or accepted for publication (editors).

3. The value of including this type of study is low because we would need to assume that success in tennis is a measure of aggression. Including it might be an act of desperation, resulting from a shortage of human studies that directly measure aggression.

4.

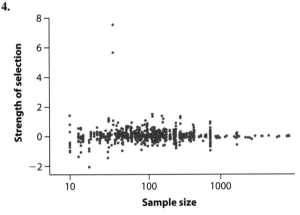

5. [Answers will vary.]

6. The smaller effect size with larger studies suggests that there is a publication bias affecting the estimates: small studies yielding low effect sizes are less likely to be appear in the published literature.

7. Even if H_0 is true, some studies might reject it by chance. If these are the only studies available for review (because of publication bias), then meta-analysis would conclude that H_0 is false.

LIBRARY, UNIVERSITY OF CHESTER

Photo credits

Page ii (left to right): Snake with fangs © Mark Laita, courtesy Fahey/Klein Gallery; Honeybee on flower © Ian A. Kirk (CC-BY 2.0) / flickr; Human tuberculosis © Alfred Pasieka / Science Source; Golden monkeys © Cyril Ruoso / Minden Pictures

Page iii (left to right): Drosophila melanogaster with red eyes © Eye of Science / Science Source; *Rubella* (German measles) virus © PASIEKA/SCIENCE PHOTO LIBRARY; Sea star © Colin Bates / www.coastalimageworks.com; DNA single crystal © Molecular Expressions

Chapter 1
Page 1: Leafcutter ant © Bence Mate / naturepl.com. Page 3: Fat cat capsizing © photos courtesy of Richard Watherwax / watherwax.com. Page 14: Large-beaked ground finch © Dolph Schluter. Page 17: Peppered moth (*Biston betularia*) © John Mason / ardea.com. Page 20: Vermillion flycatcher © 2012 Jeff Whitlock / theonlinezoo.com. Page 23: Sir Ronald Fisher © Antony Barrington Brown, reproduced with permission of the Fisher Memorial Trust.

Chapter 2
Page 31: Bengal tiger © iStockphoto / kevdog818. Page 34: White-winged dove © 2011 Jeff Whitlock / theonlinezoo.com. Page 39: Great tit (*Parus major*) © Karel Gallas / Shutterstock.com. Page 42: Wild guppy © Kim Hughes. Page 46: *Rubella* (German measles) virus © PASIEKA / SCIENCE PHOTO LIBRARY. Page 57: Hawk moth (*Hyles lineata*) © Malcolm Schuyl / flpa-images.co.uk. Page 59: Black-bellied seed crackers © photo courtesy of Thomas B. Smith.

Chapter 3
Page 65: Saiga antelope male © Paul Johnson / naturepl.com. Page 66: Flying snake © Jake Socha. Page 74: *Tidarren* spider, Ramos M., Irschick D.J., Christenson T.E., Overcoming an evolutionary conflict: Removal of a reproductive organ greatly increases locomotor performance. *PNAS* 101(14):4883–4887. Copyright 2004 National Academy of Sciences, U.S.A. Page 78: Threespine stickleback fish, Threespine sticklebacks reproduced with permission from K. B. Marchinko and D. Schluter (2007) [*Evolution* 61:1084–90, Wiley-Blackwell Publishing Ltd.] Page 87: Cutthroat finch © John Kormendy. Page 93: One tree reef. sea urchin, Dr. Dwayne Meadows, NOAA / NMFS / OPR.

Chapter 4
Page 95: DNA single crystal © Molecular Expressions. Page 113: *top* Mountain tree shrew (*Tupaia montana*) © Ch'ien Lee / Minden Pictures; *bottom* Bumble bee of genus *Bombus* © tr3gin / Shutterstock.com

Chapter 5
Page 117: White-breasted nuthatch © 2012 Jeff Whitlock / theonlinezoo.com. Page 131: Jewel wasp © photo courtesy of John H. Werren. Page 142: Snake with fangs © Mark Laita, courtesy Fahey / Klein Gallery. Page 146: Black jack hand © iStockphoto / davidjmorgan. Page 147: Sickle-cell anemia © David Mack / Science Source.

Chapter 6
Page 149: *Cyanella alba* flower © photo courtesy of Lawrence Harder. Page 153: European toad (*Bufo bufo*) © Hintau Aliaksei / Shutterstock.com. Page 161: *Heteranthera alba* © photo courtesy of Spencer C. H. Barrett. Page 169: Land mammoth and island pygmy mammoth, Artwork by Carl Buell from Giants and Pygmies: Mammoths of Santa Rosa Island, California (USA) Larry D. Agenbroad, *Quaternary International* Volume 255, 26 March 2012, Pages 2–8. Page 171: Ocellated turkey © Gerry Ellis / Minden Pictures. Page 174: *Triturus montandoni Lissotriton montandoni* (newt) © Petr Mückstein / bio-foto.com.

Chapter 7
Page 179: Follicle mites (*Demodex folliculorum*) © Andrew Syred / Science Source. Page 186: Mouse chromosomes © Look At Sciences / Science Photo Library. Page 198: Gorilla with students © photo courtesy of Daniel Simons [Simons and Chabris (1999)] Figure provided by Daniel Simons, www.theinvisiblegorilla.com. Page 199: 11 tits with parent feeding © Gary Shilton / RSPB.

Chapter 8
Page 203: Marine diatom fossil © Stephen S. Nagy, M.D. Page 213: Human X chromosome © Andrew Syred / Science Source. Page 219: Marine diatom fossil © Stephen S. Nagy, M.D. Page 228: Mother bear with black cub © Wayne McCrory / Valhalla Wilderness Society. Page 229: Birds being killed by windows © Kenneth Herdy / FLAP Canada. Page 232: Dodder (*Cuscuta salina*) © Colin Purrington.

Chapter 9
Page 235: Goby © photo courtesty of Philip L. Munday. Page 243: © Jennifer Gordon and Wandy

Beatty. Page 246: *top* California killifish © photo courtesy of Drew Talley; *bottom* Great blue heron © Jeff Whitlock / theonlinezoo.com. Page 252: Vampire bat © Ron Austing / flpa-images.co.uk. Page 258: Caffeine crystals © DAVID PARKER / SCIENCE PHOTO LIBRARY. Page 266: Golden monkeys © Cyril Ruoso / Minden Pictures. Page 267: *Disa draconis* © Peter Swart.

Review 01
Page 270: *Drosophila melanogaster* with red eyes © Eye of Science / Science Source.

Chapter 10
Page 273: Crab spider © Ed Nieuwenhuys. Page 287: Camp Funston 1918 Flu epidemic Photo by courtesy National Museum of Health and Medicine, AFIP (Washington, D.C.). Page 290: Brown recluse spider © S. Camazine / K. Visscher / Science Source. Page 294: Lichtenberg figure on arm from lightning © Winston Kemp. Page 297: Rewardless orchid, *Dactylorhiza sambucina* © Kristian Lindqvist (CC-BY 2.0) / flickr. Page 299: European earwig © B. Borrell Casals / flpa-images.co.uk.

Chapter 11
Page 303: Stalk-eyed flies © photos courtesy of Sam Cotton. Page 307: Stalk-eyed flies © photos courtesy of Sam Cotton. Page 310: Human body temperature © Dr. Ray Clark FRPS & Mervyn de Calcina-Goff FRPS / Science Source. Page 323: Western diamondback rattlesnake © Ralph Arveson. Page 324: *left* Blind tetra (*Astyanax mexicanus*) Kazakov Maksim / Shutterstock.com; *right* Walk in Forest This graph is excerpted from Fig 1 in Souman, J. L., I. Frissen, M. N. Sreenivasa, and M. O. Ernst. 2009. Walking straight into circles. *Current Biology* 19: 1–5. © 2007 Google, Image © 2008 GeoContent, © 2008 European Space Agency, © Tele Atlas, Dr. Jan L. Souman, Max Planck Institute for Biological Cybernetics, Tübingen. Page 325: Three-toed sloth (*Bradypus variegatus*) © Nacho Such / Shutterstock.com.

Chapter 12
Page 327: Galápagos marine iguana © David Hosking / flpa-images.co.uk. Page 330: Redwing blackbird © Paul Bannick / www.PaulBannick.com. Page 335: *left* Lizard skull © Butch Brodie; *center* Horned lizard © Butch Brodie; *right* Loggerhead shrike © iStockphoto / Paul Wolf. Page 342: Brook Trout image prepared by Ellen Edmondson as part of the 1927–1940 New York Biological Survey. Permission for use in this book is granted by the New York State Department of Environmental Conservation. Page 335: Burrowing owl © Ronald G. Wolff. Page 356: Cichlid from Lake Victoria © Ole Seehausen and Oliver Selz. Page 357: Ostrich head © iStockphoto / ra-photos. Page 361: Tilapia © Ammit Jack / Shutterstock.com. Page 362: Weddell seals © Paul Ward.

Chapter 13
Page 369: Sea star © Colin Bates / www.coastalimage works.com. Page 373: Lightfoot crab © Sally Otto / theonlinezoo.com. Page 384: Dragonflies © Gary Cox. Page 388: Sagebrush crickets © David H. Funk. Page 391: Praying mantis cartoon © Danny Shanahan / The Cartoon Bank. Page 401: Common goldeneye female duck © David Dohnal / Shutterstock. com. Page 404: Lion committing infanticide © Tim Caro. Page 408: Zebra finch © Jeff Whitlock / theonlinezoo.com. Page 409: Gouldian finches © Sarah Pryke. Page 412: Pseudoscorpion © Jeanne Zeh. Page 414: Silverfish *Malayatelura ponerophila* in colonies of the Southeast Asia army ant, *Leptogenys distinguenda* © Cristoph von Beeren.

Review 02
Page 418: Australian butterfly © dhobern (CC-BY 2.0) / flickr. Page 420: Pitcher plant © Biopix: JK Overgaard.

Chapter 14
Page 423: Frog deformities © Stanley K. Sessions. Page 426: HIV Virus Los Alamos Labs / Public Domain Unless otherwise indicated, this information has been authored by an employee or employees of the University of California, operator of the Los Alamos National Laboratory under Contract No. W-7405-ENG-36 with the U.S. Department of Energy. The U.S. Government has rights to use, reproduce, and distribute this information. The public may copy and use this information without charge, provided that this Notice and any statement of authorship are reproduced on all copies. Neither the Government nor the University makes any warranty, express or implied, or assumes any liability or responsibility for the use of this information. Page 436: Experimental tree-holes © Diane Srivastava. Page 437: Rat © iStockphoto / Sergey Goruppa. Page 439: Bullfrog tadpole © iStockphoto / Ron Brancato. Page 451: Monarch caterpillar © Mark Plonsky. Page 455: Meercat © iStockphoto / Gordana Sermek. Page 456: Butterfly © iStockphoto / cotesebastien.

Chapter 15
Page 459: Egyptian vulture © Diana Hromish / LunarEye Photography. Page 460: Clocks © Alan Crosthwaite / Absolute Stock Photo. Page 474: Mycorrhizae Mycorrhizal fungi, Reprinted by permission from Macmillan Publishers Ltd: Mycorrhizal fungi The ties that bind (1997), David Read, *Nature* (vol. 388), issue 6642, Figure 1, Nature Publishing Group. Page 478: Walking stick insect © Patrick Nosil. Page 490: Dung beetle © Frank Kohler. Page 491: 6 Aphids; Kacsoh, Balint. Z., Lynch, Zachary R., Mortimer, Nathan T., Schlenke, Todd A. 2013. Fruit Flies Medicate Offspring After Seeing Parasites. *Science* Vol. 339 (6122), pp. 947–950. Reprinted with permission from AAAS. Page 492: Tsetse fly © Ray Wilson. Page 495: Lodgepole pine © S. Rae (CC-BY 2.0) / flickr.

Chapter 16
Page 503: Western trillium © Dolph Schluter.
Page 506: Nazca boobies Lindsey Kramer / U.S. Fish
and Wildlife Service. Page 510: Wolf © iStockphoto /
Karel Broz. Page 516: Rope Trick Poster Image cour-
tesy of The Nielsen Magic Poster Collection. Page 525:
Human tuberculosis © Alfred Pasieka / Science
Source. Page 527: Black-tailed godwits © Steve Gant-
lett. Page 530: Salmonberry © jfh686 (CC-BY 2.0) /
flickr.

Chapter 17
Page 539: *Silene dioica* © haraldmuc / Shutterstock.
com. Page 541: African lion © Deborah Kolb / Shutter-
stock.com. Page 551: *top* Baby with big ears © Mike
Whitlock / theonlinezoo.com; *bottom* Biodiversity plots
© Cedar Creek Ecosystem Science Reserve. Page 558:
Dark-eyed junco © Janine Russell. Page 566: Northern
fur seal © Andrew Trites. Page 580: Lizard, A. Kristo-
pher Lappin and Jerry F. Husak, "Weapon Performance,
Not Size, Determines Mating Success and Potential
Reproductive Output in the Collared Lizard (*Crotaphytus collaris*)" The American Naturalist 2005: 166(3):
426–36, The University of Chicago Press. Page 583:
Perodicticus potto © Mark Dumont (CC-BY 2.0) /
flickr. Page 585: Macaroni penguins © iStock
photo / Geno Sajko. Page 588: Bomb test mushroom
cloud Photo courtesy of AJ Software & Multimedia.
Page 589: *top* Last supper from early painting (by Ugo-
liano di Nerio) Image copyright © The Metropolitan
Museum of Art. Image source: Art Resource, NY Ugolino da Siena (d.c.1339). *The Last Supper*. Tempera and
gold on wood, Overall, with engaged (modern) frame,
15 x 22 1/4 in. (38.1 x 56.5 cm); painted surface 13 1/2
x 20 3/4 in. (34.3 x 52.7 cm). Robert Lehman Collection, 1975 (1975.1.7); *bottom left* Scarlet king snake
© David W. Pfennig; *bottom right* Coral snake © Matt
Jeppson / Shutterstock.com. Page 593: *left* Blue bead
lily © Harvey Kirsch Rivernen; *right* Turk's-cap lily
© iStockphoto / Michael Hare.

Review 03
Page 597: Lascaux cave painting of horse, A Horse,
c.15,000–10,000BC (pigments on stone), Prehistoric /
Caves of Lascaux, Dordogne, France / The Bridgeman
Art Library. Page 601: Naked mole rat © Ron Aust-
ing / flpa-images.co.uk. Page 603: From The Shape of
Boubas: Sound-Shape Correspondences in Toddlers

and Adults, Maurer, D., Pathman, T., and Mondloch,
Catherine J. *Developmental Science* 9:3 (2006), Fig. 1,
p 318. © 2006 The Authors. Journal compilation
© 2006 Blackwell Publishing Ltd.

Chapter 18
Page 605: Zooplankton © Wim van Egmond / Visuals
Unlimited. Page 615: Study plot of red alga © Chris
Harley. Page 620: Damaraland mole rat © Wendy
Dennis / flpa-images.co.uk. Page 626: Magellinic pen-
guin © iStockphoto / Alexander Hafemann. Page 628:
Honeybee on flower © Ian A. Kirk (CC-BY 2.0) /
flickr. Page 632: *top* Tortoise beetle © Edward Tram-
mel; *bottom* Yellow dung fly © Dr. Alan J. Silverside,
University of the West of Scotland. Page 633: Bluefin
killifish © Brian Gratwicke (CC-BY 2.0) / flickr.

Chapter 19
Page 635: Red Panda © Jeff Whitlock / theonlinezoo.
com. Page 641: Chimp communicating © Frans
Lanting / Frans Lanting Stock. Page 647: Prairie vole
© Michael R. Jeffords. Page 650: Snail, Ronald Chase
and Katrina C. Blanchard, The snail's love-dart deliv-
ers mucus to increase paternity, Figure 1, (2006) 273,
1471–1475, *Proceedings of the Royal Society B*, by
permission of the Royal Society.

Chapter 20
Page 655: Chimpanzee © iStockphoto / Kitch Bain.
Page 659: Wasp on cabbage white butterfly, Reprinted
by permission from Macmillan Publishers Ltd: Chemical communication: Butterfly anti-aphrodisiac lures
parasitic wasps (2005), Nina E. Fatouros, Martinus E.
Huigens, Joop J. A. van Loon, Marcel Dicke and Mon-
ika Hilker, *Nature* vol. 433, issue 7027, p. 704, Nature
Publishing Group. Page 665: Elephant © john michael
evan potter / Shutterstock.com. Page 678: Eye-gazing
in humans Copyright 2002 National Academy of Sciences, U.S.A. *PNAS* 2002, vol. 99, no. 14, 9602–9605,
Eye contact detection in humans from birth by Teresa
Farroni, Gergely Csibra, Francesca Simion and Mark
H. Johnson.

Chapter 21
Page 681: Blood Red Moon © Fred Espenak,
www.MrEclipse.com. Page 688: Testosterone
© Molecular Expressions

Index